Die Grundlehren der mathematischen Wissenschaften

in Einzeldarstellungen
mit besonderer Berücksichtigung
der Anwendungsgebiete

Band 202

Leopold Schmetterer

Introduction to Mathematical Statistics

Translated from the German
by Kenneth Wickwire

Springer-Verlag
Berlin Heidelberg New York 1974

Leopold Schmetterer
Professor of Statistics and Mathematics
at the University of Vienna

Translator:

Kenneth Wickwire
Department of Mathematics, University of Manchester

Title of the German Original Edition:
Einführung in die mathematische Statistik,
2. verbesserte und wesentlich erweiterte Auflage
Springer-Verlag Wien New York 1966

With 11 figures

AMS Subject Classification (1970): 62-01, 62 A xx, 62 B xx, 62 C xx, 62 D 03, 62 E xx, 62 F xx, 62 G xx, 62 H xx

ISBN-13: 978-3-642-65544-9 e-ISBN-13: 978-3-642-65542-5
DOI: 10.1007/978-3-642-65542-5

Softcover reprint of the hardcover 1st edition 1974
Bookbinding: Konrad Triltsch, Würzburg.

Preface

I have used the opportunity of the second edition of the German version being translated into English to alter and improve some details. Of course I tried to correct misprints and errata of the original version. Moreover some proofs have been slightly changed and I hope thereby improved. In many instances more recent results have been inserted in the text, for example the treatment of the infinite dimensional linear programms in the third chapter and the investigation of superefficiency in the 5th Chapter. On the whole I enlarged the references by pointing to more recent papers. In doing so I have considered the literature up to 1971. The 7th chapter is left unchanged although there has been particular progress in developing the theory of non-parametric statistical inference: Analysis and Probability theory as treated in the Introduction and in the first chapter would have been an insufficient background for this study. At least the concept of the contiguity of measures would have had to be introduced. Moreover the excellent book by J. Hájek and J. Šidak, "Theory of Rank Tests" has in the meantime been published.

I am greatly indebted to Mr. Wickwire who translated almost the entire original text into English.—Mr. Wickwire has succeeded in rendering an almost literal translation of often linguistically difficult German structures. Let me then heartly thank my colleague Professor Pfanzagl, University of Cologne, who read the entire English manuscript and gave me several important hints. Dr. Sendler, University of Dortmund, accepted the cumbersome task of reading the galley proofs. Many improvements are due to his efforts. Mess. Grossmann and Pflug, University of Vienna assisted in the compilation of the voluminuous references. I should also like to express my gratitude to Professor

Eckmann and Professor van der Waerden for their decision to incorporate this translated version in the "Grundlehren". In 1973 I spent several months at Bowling Green State University during which time I was able to read and correct a large portion of the page proofs. I should like to express my appreciation to Bowling Green University and in particular to Professor E. Lukacs.

Last but not least I owe many thanks to Springer-Verlag. Not only did they overcome all difficulties which arose in connection with the translation but they also always respected all my wishes.

I hope very much that the English version will be accepted as kindly as the German was.

Vienna, in Summer 1974

L. Schmetterer

Table of Contents

Notation and Preliminary Remarks

1. The Euclidean Space R_n. The collection of ordered n-tuples of real numbers $(x_1,\ldots,x_n)$ is denoted by R_n. We sometimes call R_n n-dimensional space; in doing so, we will refer to the n-tuples as the coordinates of a point. If $n=1$, R_1 is the set of real numbers.

2. Distance in R_n. In the following we will often use vector notation and denote an ordered n-tuple of real numbers $(x_1,\ldots,x_n)$ by x. We will then refer to $x_1,\ldots,x_n$ as the components of the vector x. Equality and addition of vectors along with multiplication of vectors by real numbers are defined in the usual way: The vectors a and b are equal, $a=b$, if and only if (iff) $a_i=b_i$, $i=1,\ldots,n$. By $a+b$ we will understand the n-tuple $(a_1+b_1,\ldots,a_n+b_n)$. If γ is a real number, then γa is the n-tuple $(\gamma a_1,\ldots,\gamma a_n)$. Further, to each vector a we associate an absolute value (a norm), denoted by $|a|$, and defined by setting $|a|=\sqrt{a_1^2+\cdots+a_n^2}$. Then the important triangular inequality

$$|a+b|\leqslant|a|+|b|$$

is valid. Moreover, we have $|a|\leqslant|a_1|+\cdots+|a_n|\leqslant\sqrt{n}\,|a|$. If $x=(x_1,\ldots,x_n)$ is an n-tuple and $y=(y_1,\ldots,y_m)$ an m-tuple, then (x,y) will denote the $n+m$-tuple $(x_1,\ldots,x_n,y_1,\ldots,y_m)$. We introduce an ordering on the set of all n-tuples by writing $a<b$ if and only if $a_1<b_1,\ldots,a_n<b_n$. Then $a<b$ and $b<c$ imply that $a<c$; however, for arbitrary $a,b\in R_n$, it is clear that neither $a<b$, nor $b<a$, nor $a=b$ need hold. We also write $a<+\infty$ if $a_1<+\infty,\ldots,a_n<+\infty$ with a similar agreement for $-\infty<a$.

3. Operations on sets. Let M be an arbitrary set. If m is an element of this set, we write $m\in M$. We write $M_1=M_2$ if the sets M_1 and M_2 are identical, i.e., if they contain exactly the same elements. If M_1 is a subset of M_2 (i.e., if all the elements of M_1 are also elements of M_2), then we write $M_1\subseteq M_2$.[1] If $M_1\subseteq M_2$ and $M_2\subseteq M_1$, then $M_1=M_2$.

[1] If $M_1\subseteq M_2$, but $M_1\neq M_2$, then we write $M_1\subset M_2$.

It is useful to introduce a set which contains no elements. We call such a set the empty set and denote it by $\emptyset$.

The *intersection* of M_1 and M_2 is denoted by $M_1 \cap M_2$. We write $\bigcap_{i=1}^{n} M_i$ for the intersection of finitely many sets, using a similar notation for infinitely many sets.

Finitely or infinitely many sets M_i are said to be *pairwise disjoint* if $M_i \cap M_j = \emptyset$, whenever $i \neq j$.

The *union* of M_1 and M_2 is denoted by $M_1 \cup M_2$. The extension to a countable number of sets is self-evident.

The operations of intersection and union on sets are commutative, associative and distributive. For example, we recall the latter rule: $A \cap \bigcup_{i=1}^{\infty} B_i = \bigcup_{i=1}^{\infty} (A \cap B_i)$ and $A \cup \bigcap_{i=1}^{\infty} B_i = \bigcap_{i=1}^{\infty} (A \cup B_i)$.

The collection of elements that belong to M_1, but not to M_2, is called the *difference* of M_1 and M_2 and is denoted by $M_1 - M_2$. If M and N are arbitrary sets, then the collection of all ordered pairs (m, n), where $m \in M$ and $n \in N$, is called the *Cartesian product* of M and N; this set is denoted by $M \times N$. If one of these sets is the empty set, then the Cartesian product is also understood to be empty. The extension of these concepts to finitely or infinitely many sets is straightforward. We will write $\prod_{i=1}^{n} M_i$ for the Cartesian product of $M_1, \ldots, M_n$.

Let M be an arbitrary set and let E characterize some property such that for each $m \in M$, either m has the property E or not. Then the well-defined set of all m with property E is denoted by $\{m : E(m)\}$.

Important examples of sets are the subsets of R_n. R_n is obviously the n-fold cartesian product of R_1 with itself.

Let a be a real number or $a = -\infty$ and b a real number or $b = \infty$ with $a < b$. The set of points x of R_1 which satisfy the inequality $a < x < b$ is called an open interval and is denoted by (a, b). The set $\{x : a \leqslant x \leqslant b\}$ is called a closed interval and is written $[a, b]$. The meaning of $(a, b]$ or $[a, b)$ should be clear[2].

Each Cartesian product of intervals of R_1 is referred to as an interval in R_n. If, in particular, $a_i < b_i$, $i = 1, \ldots, n$, then the set $(a_1, b_1) \times (a_2, b_2) \times \cdots \times (a_n, b_n)$ is called an open interval in R_n. This is just the set of n-tuples $(x_1, \ldots, x_n) \in R_n$ which satisfy the inequality $a_i < x_i < b_i$, $i = 1, \ldots, n$. The Cartesian product $[a_1, b_1] \times \cdots \times [a_n, b_n]$ is called a closed interval in R_n. For $n = 2$ we refer to intervals as rectangles; for $n \geqslant 3$, we also use the term cubes in lieu of intervals.

[2] Whenever the equality sign occurs in the definition of these intervals it is supposed that $-\infty < a$ respectively $b < \infty$.

A subset M of R_n is called *open* if for each $x_0 \in M$ there exists an open sphere $K_{x_0} = \{x: |x - x_0| < r_0\}$, $r_0 > 0$, such that $K_{x_0} \subseteq M$. Each open set containing x is called a *neighborhood* of x. M is said to be *closed* if $R_n - M$ is open; M is called *compact* if M is closed and bounded. For compact sets M we have the

Borel Covering Theorem. Each collection of open sets which covers M contains a finite subcollection which also covers M.[3]

M is said to be *convex*, if the line-segment $tx + (1-t)y$, $0 \leqslant t \leqslant 1$, belongs to M whenever $x, y \in M$.

4. Rings and Algebras of sets. Let R be a non-empty set and H a set of subsets of R. We note the

Definition. H is said to be a *semiring*, if the intersection of two sets from H belongs to H and the following "chain-condition" is satisfied: if $M_1, M_2 \in H$ and $M_1 \subseteq M_2$, then there exists a natural number k and sets $N_0, N_1, \ldots, N_k$ in H with $M_1 = N_0 \subset N_1 \cdots \subset N_k = M_2$ such that $N_i - N_{i-1} \in H$ for $i = 1, 2, \ldots, n$.

We also give the definition of a ring:

Definition. A set $\mathfrak{R}$ of subsets of R is called a *ring* if the union of finitely many sets from $\mathfrak{R}$ and the difference of any pair of sets from $\mathfrak{R}$ belong to $\mathfrak{R}$. $\mathfrak{R}$ is called a *σ-ring* if $\mathfrak{R}$ is a ring and the union of any denumerable collection of sets from $\mathfrak{R}$ is a member of $\mathfrak{R}$. A semiring (ring, σ-ring) is said to be a *semialgebra (algebra, σ-algebra)*, if it has a set which contains all other members as subsets. In this case we shall always assume that this largest set is R.

Each semiring (ring, σ-ring) contains the empty set. Each ring contains the intersection of finitely many of its members while each σ-algebra contains the intersection of any denumerable subcollection.

Each σ-ring is a ring, each ring a semiring. Corresponding statements hold for algebras.

The intersection of rings (σ-rings) of subsets of R is again a ring (σ-ring) and the set of all subsets of a set R is a σ-ring. Hence, if $\mathfrak{M}$ is a nonempty family of subsets of R one can easily show that there exists a smallest ring, as well as a smallest σ-ring, of subsets of R which contains $\mathfrak{M}$. This ring (σ-ring) is called the ring (σ-ring) generated by $\mathfrak{M}$.

An important example of a semiring is the set of all intervals $(a, b]$ of R_1. If we add R_1 to this semiring we obtain a semialgebra $\mathfrak{H}_1$.

---

[3] This property characterizes the compact sets and is frequently used as definition of compactness which, in this form, can easily be carried over to more general spaces.

Similarly, the collection of all cubes of the form $(a_1,b_1]\times\cdots\times(a_n,b_n]$ from R_n is a semiring. We will denote by $\mathfrak{H}_n$ the semialgebra obtained by adding R_n to this collection.

The σ-algebra generated by $\mathfrak{H}_1$ is called the σ-algebra of Borel sets, $\mathfrak{B}_1$, of R_1. $\mathfrak{B}_1$ is generated by the set of all open intervals as well as by the set of all closed intervals. In particular, the Borel sets contain all open and all closed sets of R_1.

The Borel sets of R_n are defined in a completely analogous manner. We denote the collection of Borel sets of R_n by $\mathfrak{B}_n$.

5. Mappings and functions. Let M and N be arbitrary non-empty sets. We say that ϕ is a mapping of M into N if ϕ associates with each $m\in M$ exactly one element $n\in N$, $\phi(m)=n$. We then say that ϕ is defined on M. Sometimes it is convenient to denote such a mapping by writing $m\to\phi(m)$. Important examples of mappings are obtained when $N=R_1$; such mappings are called real-valued functions or, more briefly, functions. For such functions, one defines addition and multiplication, and division when the denominator does not vanish, in the usual way. If ϕ and ψ are two functions defined on M, we write $\phi+\psi$ for the sum, $\phi\psi$ for the product and ϕ/ψ for the quotient of ϕ and ψ. For example, $\phi+\psi$ is the mapping which associates with each $m\in M$ the value $\phi(m)+\psi(m)$. The arrow notation above turns out to be especially useful, if for example ϕ is a mapping from R_n into R_1 and we want to consider it as a function on R_1 by fixing all but the ith component:

$$x_i\to\phi(x_1,\ldots,x_i,\ldots,x_n).$$

For each subset $P\subseteq M$, the set $\{n:\phi(m)=n, m\in P\}$ is denoted by $\phi(P)$. $\phi(M)$ is called the image of M in N under the mapping ϕ. If ϕ maps M into N and ψ is defined on $\phi(M)$, mapping it into a set Q, then we denote the composite mapping by $\psi\circ\phi$. Hence, this mapping associates with each $m\in M$ the point $\psi(\phi(m))$ in Q. The set $\{m:\phi(m)\in N_1\}$, where N_1 is a subset of N, is called the inverse image of N_1 under the mapping ϕ and is denoted by $\phi^{-1}(N_1)$. Inverse image mappings satisfy the important

Theorem I. $\phi^{-1}(N_1-N_2)=\phi^{-1}(N_1)-\phi^{-1}(N_2)$,

$$\phi^{-1}\left(\bigcup_{i=1}^{\infty}N_i\right)=\bigcup_{i=1}^{\infty}\phi^{-1}(N_i)\quad\textit{and}$$

$$\phi^{-1}\left(\bigcap_{i=1}^{\infty}N_i\right)=\bigcap_{i=1}^{\infty}\phi^{-1}(N_i).$$

An important example of a function is the *indicator function*. Let R be an arbitrary non-emply set and $M \subseteq R$. The function c_M on R is called the indicator function[4] of M if

$$c_M(x)=1, \quad \text{if } x \in M,$$
$$c_M(x)=0, \quad \text{if } x \in R-M.$$

6. Measures.[5] Let R be an arbitrary set and $\mathcal{S}$ a σ-algebra of subsets of R. The pair $(R,\mathcal{S})$ is called a *measurable space*. Each mapping μ of $\mathcal{S}$ into the non-negative real numbers is called measure over the measurable space $(R,\mathcal{S})$, if it is *completely additive or σ-additive*; that is, for any pairwise disjoint sets $A_i \in \mathcal{S}$, $i=1,2,\dots, \mu\left(\bigcup_{i=1}^{\infty} A_i\right)=\sum_{i=1}^{\infty} \mu(A_i)$. If the sets $A_i \in \mathcal{S}$ are not pairwise disjoint, then it follows that $\mu\left(\bigcup_{i=1}^{\infty} A_i\right) \leqslant \sum_{i=1}^{\infty} \mu(A_i)$. $\mu(A)=+\infty$ is allowed for some but not all $A \in \mathcal{S}$.

If $A,B \in \mathcal{S}$ and $A \subseteq B$, then $\mu(A) \leqslant \mu(B)$.

One can easily show that $\mu(\emptyset)=0$ for any measure.

Two measures μ_1, μ_2 on $\mathcal{S}$ are said to be equal if $\mu_1(B)=\mu_2(B)$ for all $B \in \mathcal{S}$.

If $\mu(R)$ is finite, then the measure μ is said to be bounded.

A measure for which $\mu(R)=1$ is called a *probability measure*.

The important continuity property of measures is a consequence of complete additivity: $A_1 \subseteq A_2 \subseteq \cdots$ implies $\lim_{i \to +\infty} \mu(A_i)=\mu\left(\bigcup_{i=1}^{\infty} A_i\right)$. If $A_1 \supseteq A_2 \supseteq \cdots$ and $\mu(A_j)<+\infty$ for some j, then $\lim_{i \to +\infty} \mu(A_i)=\mu\left(\bigcap_{i=1}^{\infty} A_i\right)$.

The measure μ is called *σ-finite*, if for each $A \in \mathcal{S}$ with $\mu(A)=+\infty$, there exists a sequence $\{A_i\}$, $A_i \in \mathcal{S}$, such that $\bigcup_{i=1}^{\infty} A_i \supseteq A$ and $\mu(A_i)$ is finite for each i, $i=1,2,\dots$.

Important examples of measurable spaces are furnished by $(R_1, \mathfrak{B}_1)$ and $(R_n, \mathfrak{B}_n)$. Lebesgue measure is the most important example of a measure over $(R_1, \mathfrak{B}_1)$; this measure is σ-finite. Each interval of length one has Lebesgue measure one. n-dimensional Lebesgue measure, defined over $(R_n, \mathfrak{B}_n)$, has analogous properties. Each n-dimensional unit-cube is assigned n-dimensional Lebesgue measure one.

---

[4] c_M is also called the characteristic function of M.

[5] For more detailed information, see P. Halmos, Measure Theory, D. Van Nostrand, New York 1950, and H. Richter, Wahrscheinlichkeitstheorie, Second Edition, Bd. 66 Springer-Verlag, Berlin-New York 1966.

Let μ be a measure over $(R,\mathscr{S})$. Each set $A\in\mathscr{S}$ for which $\mu(A)=0$ is called a μ-null set. If μ is Lebesgue measure, then A is referred to simply as a null set. The union of at most a denumerable number of μ-null sets is again a μ-null set.

If E is a property which holds for all $x\in R$ with the exception of a μ-null set, then we say that E holds almost everywhere (μ), (μ-a.e.), or almost everywhere (a.e.) if μ is Lebesgue measure.

A set $B\in\mathscr{S}$ is called an *atom* of the measure μ if $0<\mu(B)$ and whenever $A\subseteq B$, $A\in\mathscr{S}$, it follows that either $\mu(A)=0$ or $\mu(A)=\mu(B)$. If atoms exist for the measure μ, then μ is said to be atomic; if μ has no atoms it is said to be non-atomic. The following important theorem is due to Ljapunov.

Theorem II.[6] *Let* $\mu_1,\ldots,\mu_k$, $k\geqslant 1$, *be finitely many non-atomic, bounded measures over a measurable space* $(R,\mathscr{S})$. *Then the set of k-tuples* $(\mu_1(B),\ldots,\mu_k(B))$, $B\in\mathscr{S}$, *is a compact, convex subset of* R_k.

A very important theorem in the construction of probability theory is the so-called extension theorem for measures.

Theorem III.[7] *Let* $\mathfrak{H}$ *be a semialgebra of subsets of R and* μ *a mapping of* $\mathfrak{H}$ *into the non-negative real numbers which satisfies the complete additivity condition for pairwise disjoint sets* $A_i\in\mathfrak{H}$, $i=1,2,\ldots$, *with* $\bigcup_{i=1}^{\infty} A_i\in\mathfrak{H}$. *Moreover, suppose that there exists a sequence* $\{M_i\}$ *with* $M_i\in\mathfrak{H}$, $\bigcup_{i=1}^{\infty} M_i=R$ *and* $\mu(M_i)<+\infty$, *for each i,* $i=1,2,\ldots$. *Then* μ *can be extended to a measure* μ^* *on the* σ*-algebra generated by* $\mathfrak{H}$ *and this extension is unique. Therefore, for all* $A\in\mathfrak{H}$, $\mu(A)=\mu^*(A)$.

Hence, in particular, each such mapping μ *defined on* $\mathfrak{H}_1$ *may be extended to a measure on* $\mathfrak{B}_1$; *a similar statement holds for such mappings on* $\mathfrak{H}_n$.

We now consider the construction of product measures: Let

$$(R^{(i)},\mathscr{S}^{(i)}),\qquad i=1,\ldots,n,\qquad n\geqslant 2$$

be measurable spaces and μ_i measures over $(R^{(i)},\mathscr{S}^{(i)})$. We denote by $\mathfrak{M}$ the collection of all sets $\prod_{i=1}^{n} M_i$, with $M_i\in\mathscr{S}^{(i)}$. The σ-algebra generated by $\mathfrak{M}$ is called the product σ-algebra generated by the $\mathscr{S}^{(i)}$ and is denoted

---

[6] Cf. A. Ljapunov, Bull. Acad. Sci. URSS, Ser. Math. 4, 465–478 (1940). See also J. Lindenstrauss, J. Math. Mech. 15, 971–972 (1966).

[7] This theorem may be found, for example, in the book by K. Krickeberg, Probability Theory, Addison-Wesley, Reading-London, 1965.

by $\bigotimes_{i=1}^{n} \mathscr{S}^{(i)}$. A measure μ may be constructed over the measurable space $\left(\prod_{i=1}^{n} R^{(i)}, \bigotimes_{i=1}^{n} \mathscr{S}^{(i)}\right)$ in the following manner: For each $M = \prod_{i=1}^{n} M_i$, $\mu(M)$ is defined as $\mu_1(M_1)\mu_2(M_2)\ldots\mu_n(M_n)$ and, by an application of the extension theorem, μ is defined uniquely over $\bigotimes_{i=1}^{n} \mathscr{S}^{(i)}$. μ is called the product measure of $\mu_1,\ldots,\mu_n$ and is denoted by $\mu=\mu_1\times\mu_2\times\cdots\times\mu_n$.

Further, we mention that if $\mathfrak{H}_i$, $1\leqslant i\leqslant n$, are semialgebras which generate the σ-algebras $\mathscr{S}^{(i)}$, then $\prod_{i=1}^{n} \mathfrak{H}_i$, i.e., the set of all $\prod_{i=1}^{n} M_i$ with $M_i\in\mathfrak{H}_i$, is likewise a semialgebra which generates $\bigotimes_{i=1}^{n} \mathscr{S}^{(i)}$.

This construction can be extended to infinitely many measurable spaces $(R^{(i)}, \mathscr{S}^{(i)})$, $i\in I$, if the corresponding measures μ_i are probability measures. Then one considers the smallest σ-algebra over $\prod_{i\in 1} R^{(i)}$ which is generated by the family $\mathfrak{M} = \left\{\prod_{i\in I} M^{(i)}: M^{(i)}\in\mathscr{S}^{(i)};\ M^{(i)}\neq R^{(i)}\text{ for only finitely many values of } i\right\}$. This σ-algebra is denoted by $\bigotimes_{i\in I} \mathscr{S}^{(i)}$. If $\prod_{i\in I} M^{(i)}\in\mathfrak{M}$ and $M^{(i)}\neq R^{(i)}$ iff $i\in\{i_1,\ldots,i_r\}\subset I$, then $\mu=\prod_{i\in I}\mu_i$ is defined by setting $\mu\left(\prod_{i\in I} M^{(i)}\right) = \prod_{j=1}^{r} \mu_{i_j}(M^{(i_j)})$ and once more applying the extension theorem.

n-dimensional Lebesgue measure is the product of n one-dimensional Lebesgue measures.

7. Measurable functions. Let $(R,\mathscr{S})$ be a measurable space and ϕ a function defined over R. ϕ is said to be *$\mathscr{S}$-measurable* if $\phi^{-1}((-\infty,\alpha))$ belongs to $\mathscr{S}$ for all real α. According to Theorem I, we could just as well give the definition in terms of bounded, open, closed or half-open intervals. Since it follows from Theorem I that the collection of sets whose inverse images under ϕ belong to $\mathscr{S}$ forms a σ-algebra, the inverse images of all sets from $\mathfrak{B}_1$ belong to $\mathscr{S}$. If $R=R_n$, $n\geqslant 1$, and $\mathscr{S}=\mathfrak{B}_n$, then $\mathfrak{B}_n$-measurable functions are called Borel-measurable. Sums, products and quotients (if defined) of two $\mathscr{S}$-measurable functions are again $\mathscr{S}$-measurable. If ϕ is an $\mathscr{S}$-measurable mapping of R into R_1 and ψ is a Borel-measurable function of R_1 into R_1, then $\psi\circ\phi$ is also $\mathscr{S}$-measurable. (Cf. p. 42ff.)

The supremum and infimum of a sequence f_i, $i=1,2,\ldots$, of $\mathscr{S}$-measurable functions are $\mathscr{S}$-measurable. Thus, if $\lim_{i\to+\infty} f_i(x)=f(x)$, for each $x\in R$, then f is also $\mathscr{S}$-measurable

If ϕ is a mapping from R into R_n, $n \geqslant 2$, then ϕ is said to be measurable if the inverse images under ϕ of all intervals—or equivalently, the inverse images of all intervals $(-\infty, \alpha_1) \times \cdots \times (-\infty, \alpha_n)$, α_i real, $1 \leqslant i \leqslant n$—belong to $\mathscr{S}$. The inverse images of all sets from $\mathfrak{B}_n$ then belong to $\mathscr{S}$ as well.

The following concept is more general: Let $(R, \mathscr{S})$ and $(Q, \mathfrak{Q})$ be measurable spaces. Let T be a mapping from R into Q. T is called $(\mathscr{S}, \mathfrak{Q})$-measurable if $T^{-1}(M) \in \mathscr{S}$ for every $M \in \mathfrak{Q}$. Using this definition, an $\mathscr{S}$-measurable function should be referred to as an $(\mathscr{S}, \mathfrak{B}_1)$-measurable function.

Let us note the following fact: If Q is an arbitrary set and T an arbitrary mapping from R into Q it is always possible to construct a σ-algebra $\mathfrak{Q}$ of subsets of Q such that T is $(\mathscr{S}, \mathfrak{Q})$-measurable. For this it is enough to consider the class $\mathfrak{Q}$ of all subsets $M \subseteq Q$ such that $T^{-1}(M) \in \mathscr{S}$. Then $\mathfrak{Q}$ as well as $\{T^{-1}(M) : M \in \mathfrak{Q}\}$ is a σ-algebra and T is of course $(\mathscr{S}, \mathfrak{Q})$-measurable.

8. The integral with respect to a measure. Let μ be an arbitrary measure over $(R, \mathscr{S})$. The integral $\int_R f \, d\mu$, which we also write as $\int_R f(x) \, d\mu(x)$, is defined for a subfamily L_μ of $\mathscr{S}$-measurable functions. L_μ is called the set of μ-integrable functions, or more precisely, functions μ-integrable over R.

For all indicator functions c_E with $\mu(E) < +\infty$, we define $\int_R c_E \, d\mu = \mu(E)$. For all mappings ϕ which have the form

$$\sum_{i=1}^{n} \alpha_i c_{E_i}, \tag{I}$$

where $n \geqslant 1$, α_i arbitrary real numbers, $E_i \cap E_j = \emptyset$, $i \neq j$ and $\mu(E_i) < \infty$, $1 \leqslant i \leqslant n$ (the so-called integrable step-functions), we define

$$\int_R \phi \, d\mu = \sum_{i=1}^{n} \alpha_i \mu(E_i).$$

It is easy to show that this definition is independent of the representation of ϕ in (I). Finally, let $\{\phi_i\}$ be a sequence of mappings of the form (I) with $0 \leqslant \phi_i \leqslant \phi_{i+1}$, $i = 1, 2, \ldots$. Then $\lim_{i \to +\infty} \phi_i(x) = f(x)$ exists for each $x \in R$ as a finite or infinite limit. If $\lim_{i \to +\infty} \int_R \phi_i \, d\mu$ (which always exists) is finite, we define

$$\int_R f \, d\mu = \lim_{i \to +\infty} \int_R \phi_i \, d\mu.$$

Once again, the integral can be shown to be uniquely defined. The set L_μ is then the set of all f with $f=f_1-f_2$, $f_i\geqslant 0$, $i=1,2$, for which $\int_R f_i\,d\mu$ exists.

It is easily shown that for arbitrary real numbers α_1, α_2 and $f_1, f_2 \in L_\mu$, $\alpha_1 f_1+\alpha_2 f_2 \in L_\mu$ and $\int_R (\alpha_1 f_1+\alpha_2 f_2)\,d\mu=\alpha_1 \int_R f_1\,d\mu+\alpha_2 \int_R f_2\,d\mu$.

$|f|\in L_\mu$, whenever $f\in L_\mu$, and the following important inequality holds: $\left|\int_R f\,d\mu\right| \leqslant \int_R |f|\,d\mu$.

If g is a bounded and $\mathscr{S}$-measurable mapping from R into R_1, then $gf\in L_\mu$ whenever $f\in L_\mu$. For each $f\in L_\mu$ and $E\in\mathscr{S}$, $\int_E f\,d\mu$ is defined by $\int_R f c_E\,d\mu$. Since, c_E is bounded, this definition makes sense. Somewhat more generally, one defines $\int_E f\,d\mu=\int_R f c_E\,d\mu$ for each $\mathscr{S}$-measurable f with $f c_E\in L_\mu$. f is then said to be μ-integrable over E. If $f\in L_\mu$ and $f(x)=g(x)$ for all $x\in R$ except for a μ-null set, then $g\in L_\mu$ also and $\int_R f\,d\mu=\int_R g\,d\mu$, as well as $\int_E f\,d\mu=\int_E g\,d\mu$ for all $E\in\mathscr{S}$. Therefore, the definition of a function f on a μ-null set does not affect its membership in L_μ or the value of its integral. If $\mu(N)=0$ and f is μ-integrable, then $\int_N f\,d\mu=0$. Each $f\in L_\mu$ is finite, with the exception of at most a μ-null set. If $f\geqslant 0$ and $\int_R f\,d\mu=0$, then it follows that $f(x)=0$ for all $x\in R$ μ-a.e.

We add the following *remarks*: Let h be an $\mathscr{S}$-measurable function defined over R. If for each sequence of μ-integrable step functions g_i (see p. 8), which converge to h on R, we have $\lim_{i\to+\infty}\int_R g_i\,d\mu=\infty$, respectively, $=-\infty$, then it is customary to say that $\int_R h\,d\mu=\infty$, respectively, $=-\infty$. However, in what follows, when we say that $\int_R h\,d\mu$ exists, we will always mean that this value is finite. If $\int_R h\,d\mu=\infty$ and $A_M=\{x\colon -\infty<h(x)\leqslant M\}$, where M is an arbitrary real number, then $\int_{A_M} h\,d\mu$ always exists when $\mu(A_M)<+\infty$.

We will often make use of

Theorem IV. *If f is $\mathscr{S}$-measurable and $|f|\leqslant g$, with g integrable with respect to μ, then f is also integrable with respect to μ.*

Therefore, if f and g are $\mathscr{S}$-measurable and f^2 and g^2 are μ-integrable, then fg is also μ-integrable.

Further, Schwarz's *inequality* holds:[8]

$$\left(\int_R |fg|\,d\mu\right)^2 \leqslant \int_R f^2\,d\mu \int_R g^2\,d\mu .$$

The equality relation holds iff there exists a real number λ such that $f=\lambda g$ μ-a.e.[9]

Let $A_1, A_2, \ldots$ be a finite or denumerably infinite number of pairwise disjoint members of $\mathscr{S}$. Then if $f \in L_\mu$

$$\int_{\bigcup_i A_i} f\,d\mu = \sum_i \int_{A_i} f\,d\mu . \tag{II}$$

A particularly important example of an integral is obtained when μ is Lebesgue measure on $R_n, n \geqslant 1$. The resulting integral is called the Lebesgue integral. For brevity, a function from L_μ is then said to be integrable. If f is a function defined on R_n, then the Lebesgue integral of f over R_n may be written as $\int_{R_n} f\,dx$, or $\int_{R_n} f(x)\,dx$, or

$$\int_{-\infty}^{+\infty} \cdots \int_{-\infty}^{+\infty} f(x_1, \ldots, x_n)\,dx_1 \ldots dx_n .$$

In the same way, we abbreviate

$$\int_{y_1}^{z_1} \cdots \int_{y_n}^{z_n} f(x_1, \ldots, x_n)\,dx_1 \ldots dx_n$$

to $\int_y^z f(x)\,dx$ and understand analogous expressions in a similar way.

The Lebesgue integral can be regarded as an extension of the Riemann integral. Each Riemann integrable function is Lebesgue integrable, but not conversely. The Lebesgue integral itself can be generalized to the Radon-Stieltjes integral. This integral in turn represents only a special case of the more general method of integration mentioned above. In the simplest case, let G be a bounded nondecreasing function defined on R_1. Set $G(-\infty) = \lim_{x \to -\infty} G(x)$. For each real α, the non-negative number $G(\alpha+0) - G(-\infty)$ is the "measure" assigned to the interval $(-\infty, \alpha]$. It is easy to see that this correspondence defines a completely

[8] This inequality sometimes goes under the name of Bunjakowski.

[9] More generally, we have the Hölder inequality:

$$\int_R |fg|\,d\mu \leqslant \left(\int_R |f|^p\,d\mu\right)^{1/p} \left(\int_R |g|^q\,d\mu\right)^{1/q}$$

with $p,q>1$ and $1/p+1/q=1$.

additive measure over the semi-algebra $\mathfrak{H}_1$, (see p. 3), which, as a consequence of Theorem III, can be extended to a completely additive measure μ_G on $\mathfrak{B}_1$. If $f \in L_{\mu_G}$, then we also write $\int_{R_1} f(x)dG(x)$ in place of $\int_{R_1} f d\mu_G$. This notation is suggested by the definition of the Riemann-Stieltjes integral.

If G' exists and is Lebesgue integrable and $G(x) = \int_{-\infty}^{x} G'(y)dy$ for all $x \in R_1$, then

$$\int_{R_1} f(x)dG(x) = \int_{R_1} f(x)G'(x)dx, \tag{III}$$

where the integral on the right is a Lebesgue integral.

These considerations can be extended to R_n, $n \geqslant 2$.[10]

The Radon-Stieltjes integral is (at least for continuous G) an extension of the Riemann-Stieltjes integral, the well-known definition of which goes as follows:

Let f be continuous and G monotone (e.g., increasing) on the interval $[a,b]$; G need not be continuous. We form the generalized Riemann sums

$$\sum_{i=1}^{n} f(\xi_i)(G(x_i) - G(x_{i-1})), \tag{IV}$$

$x_0 = a < x_1 < \cdots < x_n = b$, $x_{i-1} \leqslant \xi_i \leqslant x_i$. It can be shown that the sum in (IV) approaches a limit for each partition of $[a,b]$ and every choice of the points ξ_i as $n \to +\infty$, provided only that $\max_i |x_i - x_{i-1}| \to 0$ as $n \to +\infty$. This limit is unique and is called the Riemann-Stieltjes integral (denoted by $\int_a^b f(x)dG(x)$ or $\int_a^b f dG$). This integral possesses the usual properties. In particular, the formula for partial integration holds.

If G has finitely or infinitely many discontinuity points $x_1, x_2, \ldots$ and is constant between any two successive points, then $\int_a^b f dG = \sum_i f(x_i)c_i$, where c_i is the magnitude of the jump at the discontinuity point x_i.

Further, it can be shown that in lieu of (III) the following holds: If G has a Riemann integrable derivative G', then

$$\int_a^b f(x)dG(x) = \int_a^b f(x)G'(x)dx. \tag{V}$$

[10] This was first carried out by J. Radon: J. Radon, Österreich. Akad. Wiss., math.-naturw. Kl., S.-Ber. 122, Abt. II a, 1295–1438 (1913).

The extension of the Riemann-Stieltjes integral to infinite intervals is accomplished in the usual manner.

If G is the difference of two nondecreasing (or nonincreasing) bounded functions, and so is a function of bounded variation, then for continuous f the integral $\int_a^b f\,dG$ is defined as the difference of two Riemann-Stieltjes integrals. The properties of the Riemann-Stieltjes integral easily carry over to this integral.

We note the important Lemma of Fatou

Theorem V. *Let $(R,\mathscr{S})$ be a measurable space and μ a measure on $\mathscr{S}$. Let $\{f_i\}$ be a sequence of non-negative $\mathscr{S}$-measurable functions which are defined on R. If there exists a function $g\in L$ such that $f_n\leqslant g$ μ-a.e. for all $n\geqslant 1$, then $\int_R \varliminf_{n\to\infty} f_n\,d\mu \leqslant \varliminf_{n\to\infty}\int_R f_n\,d\mu \leqslant \varlimsup_{n\to\infty}\int_R f_n\,d\mu \leqslant \int_R \varlimsup_{n\to\infty} f_n\,d\mu$. These relations hold also if the value ∞ for the integrals is permitted.* (Cf. p. 9).

Closely related is the Lebesgue theorem:

Theorem VI. *Let $\{f_i\}$ be a sequence of functions from L_μ. Suppose that $f_i(x)$ converges μ-a.e. to a limit $f(x)$ and there exists a function $g\in L_\mu$ such that $|f_i(x)|\leqslant g(x)$ for $i=1,2,\ldots$ and all $x\in R$, with the exception of at most a μ-null set. Then $f\in L_\mu$ and $\lim_{i\to+\infty}\int_R f_i\,d\mu=\int_R f\,d\mu$.*

We now point out two significant consequences of Lebesgue's theorem.

Theorem VII. *Let $(R,\mathscr{S})$ be a measurable space, $A\subseteq R_n$ an open set and $(t,x)\to g(t,x)$ a mapping from $A\times R$ into R_1, the mapping $t\to g(t,x)$ being continuous. Suppose that $x\to g(t,x)$ is a function from L_μ.*

Suppose that f is a μ-integrable function over R such that $|g(t,x)|\leqslant f(x)$ for all $x\in R$ (or μ-a.e.) and $t\in A$. Then the mapping $t\to\int_R g(t,x)\,d\mu(x)$ is continuous.

Theorem VIII. *Let $A\subseteq R_n$ be open and g be defined as in Theorem VII over $A\times R$. Suppose that $x\to g(t,x)$ is a function from L_μ. Let $\frac{\partial}{\partial t_1}g(t,x)$ exist for each $t\in A$ and $x\in R$. Further, suppose that $\left|\frac{\partial}{\partial t_1}g(t,x)\right|\leqslant f(x)$ for each $t\in A$ and all $x\in R$ (or μ-a.e.), where f is a μ-integrable function over R. Then the mapping $t\to\int_R g(t,x)\,d\mu(x)$ is differentiable at t, and $\frac{\partial}{\partial t_1}\int_R g(t,x)\,d\mu(x)=\int_R \frac{\partial}{\partial t_1}g(t,x)\,d\mu(x)$ for all $x\in A$.*

Obviously, an entirely analogous theorem holds for each other component t_i of t, $2 \leqslant i \leqslant n$.

Further, we mention a theorem concerning the transformation of Lebesgue integral.

Theorem IX. *Let $g_1, \ldots, g_n$ be mappings of the open set $A_n \subseteq R_n$ into R_1, with each g_i having continuous partial derivatives of the first order in each of the variables. Moreover, suppose that the Jacobian $\frac{\partial(g_1, \ldots, g_n)}{\partial(y_1, \ldots, y_n)}$ is different from zero on A_n and the mapping $g = (g_1, \ldots, g_n)$ of A_n onto $g(A_n) = B_n$ determines a $1-1$ correspondence between these two sets. Then B_n is also open. Let the function f be integrable over B_n. Then, setting $k = f \circ g$, $k \left| \frac{\partial(g_1, \ldots, g_n)}{\partial(y_1, \ldots, y_n)} \right|$ is integrable and*

$$\int_{B_n} f(x_1, \ldots, x_n)\, dx_1 \ldots dx_n = \int_{A_n} k(y_1, \ldots, y_n) \left| \frac{\partial(g_1, \ldots, g_n)}{\partial(y_1, \ldots, y_n)} \right| dy_1 \ldots dy_n .$$

For an integral corresponding to a product measure we have the Fubini theorem:

Theorem X. *Let $\mu = \mu_1 \times \mu_2$ be a product measure defined on $\left(R^{(1)} \times R^{(2)}, \bigotimes_{i=1}^{2} \mathscr{S}^{(i)}\right)$ and f an $\bigotimes_{i=1}^{2} \mathscr{S}^{(i)}$-measurable function defined over $R^{(1)} \times R^{(2)}$ which is integrable with respect to μ. Moreover, suppose that μ_1 and μ_2 are σ-finite. Then the mapping defined on $R^{(1)}$ by $x^{(1)} \to f(x^{(1)}, x^{(2)})$ is μ_1-integrable for almost all (relative to μ_2) $x^{(2)} \in R^{(2)}$ and the mapping $x^{(2)} \to \int_{R^{(1)}} f(x^{(1)}, x^{(2)})\, d\mu_1(x^{(1)})$ (which is defined for $x^{(2)} \in R^{(2)}$* a.e.-μ_2) *is integrable relative to μ_2 and*

$$\int_{R^{(1)} \times R^{(2)}} f\, d\mu = \int_{R^{(2)}} \int_{R^{(1)}} f(x^{(1)}, x^{(2)})\, d\mu_1(x^{(1)})\, d\mu_2(x^{(2)}) . \qquad \text{(VI)}$$

Conversely, if f is $\bigotimes_{i=1}^{2} \mathscr{S}^{(i)}$-measurable and the integral

$$\int_{R^{(2)}} \int_{R^{(1)}} |f(x^{(1)}, x^{(2)})|\, d\mu_1(x^{(1)})\, d\mu_2(x^{(2)})$$

exists, then f is also integrable with respect to μ and equation (VI) *holds.*

Obviously the theorem holds if the roles of μ_1 and μ_2 are interchanged. An integral of the form $\int_{R^{(2)}} \int_{R^{(1)}} f(x^{(1)}, x^{(2)})\, d\mu_1(x^{(1)})\, d\mu_2(x^{(2)})$ is called an iterated integral. If we apply this theorem to n-dimensional

Lebesgue integrals of the form $\int\limits_{-\infty}^{+\infty} \dots \int\limits_{-\infty}^{+\infty} f(x_1,\dots,x_n)\,dx_1 \dots dx_n$, then Fubini's theorem states that this integral can also be understood as an iterated integral which does not depend on the order of integration. This justifies the notation used for n-dimensional Lebesgue integrals.

Occasionally we will require

Theorem XI.[11] *Let μ be a σ-finite measure over $(R,\mathscr{S})$, L_μ the set of μ-integrable functions and Φ a non-empty set of uniformly bounded $\mathscr{S}$-measurable functions ϕ over R with $\sup_{x\in R_n}|\phi(x)|\leqslant M$ for all $\phi\in\Phi$. Then each sequence $\phi_n\in\Phi$ contains a subsequence ϕ_{n_i} to which there corresponds an $\mathscr{S}$-measurable function ψ such that $\int\limits_R \phi_{n_i} f\,d\mu \to \int\limits_R \psi f\,d\mu$ for all $f\in L_\mu$ and $\sup_{x\in R}|\psi(x)|\leqslant M$.*

9. Absolutely continuous set functions. Let μ be a measure over $(R,\mathscr{S})$ and ν a completely additive set function over $\mathscr{S}$, i.e., a mapping $B\to\nu(B)$ of $\mathscr{S}$ into R_1 which satisfies $\nu\left(\bigcup\limits_{i=1}^{\infty} A_i\right)=\sum\limits_{i=1}^{\infty}\nu(A_i)$ for all $A_i\in\mathscr{S}$ and $A_i\cap A_j=\emptyset$ for $i\neq j$. ν is said to be *absolutely continuous* with respect to (w.r.t.) μ, if $\mu(A)=0$ implies $\nu(A)=0$. If ν is bounded, so that $|\nu(B)|<+\infty$ for each $B\in\mathscr{S}$, then this is equivalent to the following: to each $\varepsilon>0$ there corresponds a $\delta(\varepsilon)>0$ such that $|\nu(A)|<\varepsilon$, whenever $\mu(A)<\delta(\varepsilon)$, $A\in\mathscr{S}$.

Theorem XII. *Let μ be a measure over $(R,\mathscr{S})$ and f be μ-integrable. Then the set function ν, defined for each $B\in\mathscr{S}$ by setting $\nu(B)=\int\limits_B f\,d\mu$, is absolutely continuous* w.r.t. *to μ.*

The reader will be able to find a converse to this theorem in I, Theorem 6.1.

If ν is absolutely continuous w.r.t. a measure μ, we write $\nu\ll\mu$. We now give the following

Definition. Let V be a non-empty set of measures ν. Then a *measure μ is said to dominate the set V if $\nu\ll\mu$ for all $\nu\in V$.*

It is easy to show

Theorem XIII. *Let a finite or countably infinite number of probability measure ν_i be given. Then there exists a probability measure μ which dominates the set $\{\nu_i\}$.*

---

[11] For example, see E.L. Lehmann, Testing Statistical Hypotheses, John Wiley & Sons, New York 1959, 354. See also G. Nölle und D. Plachky, Z. Wahrscheinlichkeitstheorie und Verw. Gebiete 8, 182–184 (1967).

For example, in the case of infinitely many ν_i it is sufficient to choose $\mu = \sum_{\nu=1}^{\infty} p_i \nu_i$ with $p_i > 0$, $1 \leqslant i$ and

$$\sum_{i=1}^{\infty} p_i = 1. \tag{VII}$$

$\sum_{i=1}^{\infty} p_i \nu_i$ is understood as the measure given by $\lim_{n \to +\infty} \sum_{i=1}^{n} p_i \nu_i(A)$ for each $A \in \mathscr{S}$. The absolute convergence of $\sum_{i=1}^{\infty} p_i \nu_i(A)$ for each $A \in \mathscr{S}$ guarantees that μ is a measure; it is a probability measure because of (VII).

10. Complex-valued functions. A mapping of a set R into the complex plane is called a complex-valued function. Such a mapping k is sometimes represented in the form

$$k = k_1 + i k_2, \tag{VIII}$$

where k_1 and k_2 are real functions. $k_1 = \Re(k)$ is the real part of k and $k_2 = \Im(k)$ is the imaginary part. The representation (VIII) makes it possible to carry over, in a natural manner, certain properties of real functions and arithmetical operations with real functions to complex-valued functions.

A complex valued function k defined on R_n is, for example, continuous iff $\Re(k)$ and $\Im(k)$ are continuous; differentiable iff $k_1 = \Re(k)$ and $k_2 = \Im(k)$ are differentiable and by definition e.g.

$$\frac{\partial k(x_1, \ldots, x_n)}{\partial x_1} = \frac{\partial k_1(x_1, \ldots, x_n)}{\partial x_1} + i \frac{\partial k_2(x_1, \ldots, x_n)}{\partial x_1}.$$

Similarly, k is said to be integrable iff k_1 and k_2 are integrable and one defines

$$\underbrace{\int \ldots \int}_{n} k \, dx_1 \ldots dx_n = \underbrace{\int \ldots \int}_{n} k_1 \, dx_1 \ldots dx_n + i \underbrace{\int \ldots \int}_{n} k_2 \, dx_1 \ldots dx_n.$$

The absolute value or modulus of k is understood to be $|k| = \sqrt{k_1^2 + k_2^2}$. On the basis of these definitions we note that calculations on complex valued functions can be carried out as with real functions, bearing in mind that $i^2 = -1$. The proof that the inequality $\left| \int_M k(x) \, dx \right| \leqslant \int_M |k(x)| \, dx$

is also valid for complex valued functions, $M\in\mathfrak{B}_n$, presents a slight difficulty.[12]

11. Matrices. Matrices will usually be denoted by capital Latin letters. Let A be a matrix with n rows and m columns, briefly an $n\times m$ matrix of the form

$$A=\begin{pmatrix} a_{11} \dots a_{1m} \\ \cdot\ \cdot\ \cdot\ \cdot\ \cdot\ \cdot \\ a_{n1} \dots a_{nm} \end{pmatrix}.$$

Then we also use the symbol $(a_{ij})_{1n}^{1m}$ for A. We denote the transpose of A by A':

$$A'=\begin{pmatrix} a_{11} \dots a_{n1} \\ \cdot\ \cdot\ \cdot\ \cdot\ \cdot\ \cdot \\ a_{1m} \dots a_{nm} \end{pmatrix}.$$

The inverse of A, if it exists, is denoted by A^{-1}.

If $x=(x_1,\dots,x_n)$ is an n-tuple of real numbers, then it will often be convenient to regard x as an $n\times 1$ matrix without explicit reference to this fact. Thus, if $A=(a_{ij})_{1m}^{1n}$, then Ax is understood to be the matrix

$$\begin{pmatrix} a_{11}x_1+\dots+a_{1n}x_n \\ \cdot\ \cdot\ \cdot\ \cdot\ \cdot\ \cdot\ \cdot\ \cdot\ \cdot\ \cdot\ \cdot \\ a_{m1}x_1+\dots+a_{mn}x_n \end{pmatrix}.$$

In particular, if $x=(x_1,\dots,x_n)$ and $y=(y_1,\dots,y_n)$, then on the basis of this agreement $x'y$ represents $\sum_{i=1}^{n} x_i y_i$.

Let $A=(a_{ij})_{1n}^{1n}$ be an $n\times n$ matrix, then we denote the determinant of A by $|A|$ or $|a_{ij}|_{1n}^{1n}$.

Finally, we mention the following

Theorem XIV. *Let $0<r<n$ and $(c_{ij})_{1r}^{1n}$ be an $r\times n$ matrix with*

$$\sum_{j=1}^{n} c_{ij}c_{kj}=\begin{cases}1, & i=k\\ 0, & i\neq k\end{cases},\qquad 1\leqslant i,\quad k\leqslant r.$$

Then there exists an $(n-r)\times n$ matrix $(c_{ij})_{(r+1)n}^{1n}$, such that $(c_{ij})_{1n}^{1n}$ is an orthogonal matrix; this matrix is not in general unique.

---

[12] The proof takes the following form: First we notice immediately that $c\int_M k(x)\,dx = \int_M ck(x)\,dx$, for any complex constant c. Then, setting $\phi=\arctan\left(\mathfrak{I}\left(\int_M k(x)\,dx\right)/\mathfrak{R}\left(\int_M k(x)\,dx\right)\right)$, we have $\left|\int_M k(x)\,dx\right|=\mathfrak{R}\left(e^{-i\phi}\int_M k(x)\,dx\right)=\int_M \mathfrak{R}(e^{-i\phi}k(x))\,dx\leqslant\int_M |k(x)|\,dx$.

12. Definition of the Gamma function for positive arguments. We consider $\int_0^\infty e^{-t} t^{x-1}\, dt$. This integral exists for all $x>0$ and defines the *Gamma function* $\Gamma(x)$. Hence, for $x>0$

$$\int_0^\infty e^{-t} t^{x-1} dt = \Gamma(x).$$

By integration by parts, one obtains

$$\Gamma(x+1) = x\Gamma(x), \quad \text{for } x>0.$$

For positive integers n, it follows that

$$\Gamma(n+1) = n!$$

In I, p. 75 it is shown that $\int_0^\infty e^{-t^2/2} dt = \sqrt{\pi/2}$. It follows easily that

$$\Gamma(\tfrac{1}{2}) = \sqrt{\pi}.$$

Finally, we note the following important formula: For $a>0$ and $b>0$,

$$\int_0^1 (1-x)^{a-1} x^{b-1} dx = \frac{\Gamma(a)\Gamma(b)}{\Gamma(a+b)}. \qquad \text{(IX)}$$

13. The Landau-symbol. Let $\{a_n\}$ and $\{b_n\}$ be sequences of real numbers. One writes $a_n = O(b_n)$, if $|a_n| < K|b_n|$, for some $K>0$ and all sufficiently large n, while $a_n = o(b_n)$, if $a_n/b_n \to 0$ as $n \to \infty$.

This notation also applies to functions: e.g. $f(x) = O(g(x))$ as $x \to \infty$, if $|f(x)| < K|g(x)|$ for all sufficiently large x.

Introduction

A short description of the content of statistics is not easy to give. An examination of the historical development of the notion statistics[1] indicates that for a long time it was taken to mean the description of "national peculiarities" (such as population size, land conditions and economic data collection). Only in recent times have statistical concepts penetrated into the natural sciences (Boltzmann, Gibbs, Maxwell). Resting on the fundament of probability theory, which has rapidly developed since the turn of the century, mathematical statistics has also grown exceedingly rapidly in the last forty years and its methods have been enriched by a store of ideas that threatens to defy cataloging. Statistical considerations appear today in the most varied fields of astronomy, biology, medicine, psychology, physics and sociology.

Although, as said, it is not easy to characterize the general concept of statistics in few words, it would not be wrong to say that it treats the study of phenomena which either concern a large number of individuals or in some way summarize numerous individual phenomena. One can thus consider statistics as the study of mass phenomena. Experience shows that uniformities can be demonstrated for such mass phenomena which have no counterpart in the individual occurences. The study of these empirical laws which can be observed in the mass phenomena is the task of formal statistics.[2] The task of mathematical statistics, which shall concern us exclusively here, could be called the construction of a calculus whose theorems agree with the statements of formal statistics when given an appropriate interpretation.

An example will show how rough versions of the empirical laws of statistics have become almost obvious everyday ideas. The fact that the sex ratio of new-born children is approximately 1:1 is of great importance for all areas of human life and a significant disturbance of this relation

[1] See W. Winkler, Grundriß der Statistik I, 2nd ed.: Manzsche Verlagsbuchhandlung, Vienna 1947, 1 ff. which also contains a number of historical remarks on statistics.

[2] loc. cit.[1], 96—97.

would have serious consequences. However, it would occur to no one to become anxious in this regard because he has just learned of a family all of whose children (5, say) are boys. We know indeed from experience that when only one case is observed, large deviations from the relations holding for the totality can occur, which cannot be explained in detail. In this connection, one speaks of random deviations from the relations pertaining to the total population and means thereby that there is no reason to assume that these deviations are caused by a definite common source. For example, one cannot speak of random deviations if one observes the sex ratio of the pedestrians on some street and finds that it is considerably altered in the vicinity of a boys' school shortly before 9 a.m. The avoidance of such systematic deviations, as well as the judgement of whether deviations are random or systematic, are two of the most important tasks of the statistician.

In the exaggerated example cited, one would not hesitate to call the sex ratio near the boys' school a systematic deviation from that in the total human population. In less extreme cases, one cannot determine the validity of conclusions drawn from observations without more subtle statistical methods. If one assumes for example that the average height of an adult male Austrian is 170 cm, then one cannot in general decide at first glance whether a number of corresponding observations supports this hypothesis or not. We will later develope statistical methods which will answer such questions in a definite sense. Viewed empirically, these rest on a fact which we will initially explain by means of the example just mentioned. Assume that we have a large number of male subjects chosen "at random". This means, say, that obvious preferences have been avoided in their choice. We denote the height of the Austrian men, measured in cm (fractions of a centimeter rounded up or down) as the characteristic under observation. If we have chosen a man whose height is for example 168 cm, we will say that the event $h = 168$ cm has been realized. Since the realization of this event is not definitely determined but is rather guided by chance, we must speak a bit more precisely of a random event. At first we are interested in the (absolute) frequency of occurence of the various events of this type within the series of observations of Austrian men on hand. We then relate these frequencies to the total number of observations and thereby obtain the relative frequency. This allows a comparison of two series of observations of the same characteristic: If, for example, 100 men of height 170 cm appear in a first series of observations and 400 of the same height in a second, and if we know that the first series comprises 1500 and the second 6200 observations, then only after the given absolute frequencies are related to the total number of observations in each case will one be able to state that both series of observations are essentially the same with regard to the frequency of occurance of the event $h=170$ cm. Indeed, 100/1500 and 400/6200, the relative frequencies of the height 170 in the first and second series of observations, resp., are not essentially different. Experience now allows one to state that the relative frequency of an event within a series of observations varies about a fixed value when the observations are continued or repeated under the same conditions. These variations are within bounds which become smaller when the number of observations within a series becomes larger. The number of observations is also called the size of the observation series and we call the above constant value the empirical probability of the (random) event in question. In practical work, one can identify without misgivings the empirical probability of an event with the relative frequency of its occurance in a series of observations of sufficiently large size. These relations become especially clear in the case of games of chance. Consider the example of roulette. The observed characteristics here are the numbers 1—36. Observations yielding 0 (zéro) will not be considered. We are interested in the relative frequency of the event: "an even integer occurs" or "an odd number appears". In the table on page 21 we present the results of roulette table No. 1 at the Salzburg casino (1—6 December 1952)[3].

On the basis of these results, one would tend to give the empirical probability of the event "even", as well as that of the event "odd", as 0.5000... .

The decrease in the deviations of the relative frequencies about the value 1/2 becomes especially clear in Fig. 1.

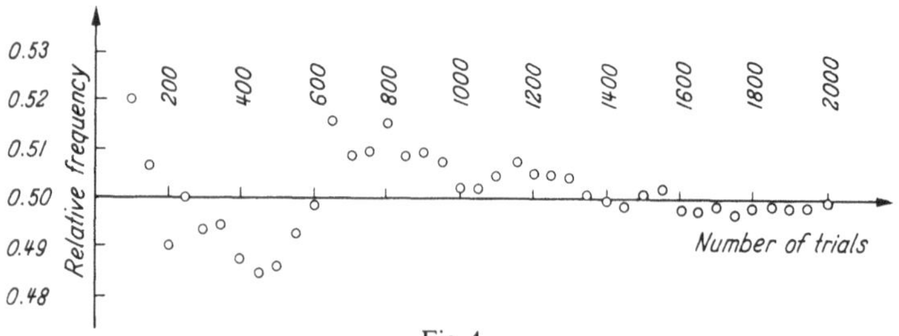

Fig. 1

We point out that in these statements and considerations we need not at all restrict ourselves to the case where the characteristics are always given by a single measurement or datum, as is the case in the examples treated up to now. Hence, the results of a series of throws of two dice are most conveniently characterized by giving both numbers of points cast in each case. The observed characteristics are thus de-

[3] From "Permanenzen der Österreichischen Casino AG." December 1952.

scribed here by giving two numbers. If we denote the event of some such throw by w_1, w_2, then we are interested for example in the realization of the event $\{w_1=2, w_2=5\}$.

number of observations	Absolute frequency even	odd	Relative frequency even	odd
50	29	21	0.5800	0.4200
100	52	48	5200	4800
150	76	74	5067	4933
200	98	102	4900	5100
250	125	125	5000	5000
300	148	152	4933	5067
350	173	177	4943	5057
400	195	205	4875	5125
450	218	232	4844	5156
500	243	257	4860	5140
550	271	279	4927	5073
600	299	301	4983	5017
650	335	315	5154	4846
700	356	344	5086	4914
750	382	368	5093	4907
800	412	388	5150	4850
850	432	418	5082	4918
900	458	442	5089	4911
950	482	468	5074	4926
1000	502	498	5020	4980
1050	527	523	5019	4981
1100	555	545	5045	4955
1150	583	567	5070	4930
1200	606	594	5050	4950
1250	631	619	5048	4952
1300	655	645	5038	4962
1350	676	674	5007	4993
1400	699	701	4993	5007
1450	723	727	4986	5014
1500	751	749	5007	4993
1550	778	772	5019	4981
1600	796	804	4975	5025
1650	820	830	4970	5030
1700	847	853	4982	5018
1750	869	881	4966	5034
1800	897	903	4983	5017
1850	922	928	4984	5016
1900	946	954	4979	5021
1950	972	978	4985	5015
2000	998	1002	4990	5010

With logical or mathematical operations we can derive new events from those given. Thus, in roulette, a realization of the event "even" is characterized by the fact that either 2 or 4 or 6 etc., or 36 appears. In the game with two dice it can thus be of interest to observe the sum of cast points in each case. Then 2→12 are the possible values of the sum. We ask, say, for the relative frequency of events typified by $\{w_1+w_2=6\}$ etc.

The basic ideas of the viewpoint sketched here are due to R. v. Mises[4] who for the first time clearly emphasized the importance of the frequency interpretation of probability for all applied questions in the theory. v. Mises defined more precisely as we have done here what is to be understood by a series of observations (a collective) which is in principle arbitrarily continuable. An essential role is played here by the so-called irregularity axiom which says that the empirical probability of an event does not change when one arbitrarily neglects observations from the potentially infinite series of observations on hand. v. Mises then defines the mathematical probability as the limit of the relative frequency of an event when the number of observations grows without bound. With this theory, v. Mises gave impetous to the modern development of probability theory, although he rejected the idea of a mathematical probability. Note, however, that in our presentation the notion of relative frequency was only called upon in order to describe an empirical situation. As we will see in Chapter 1, the behavior of the relative frequency of an event will not be directly used in the definition of probability. The reason for this is that the definition of mathematical probability according to v. Mises leads to difficulties of a logical nature.[5] It may at first be surprising that it is not immediately possible to base a mathematical definition of probability on the description of empirical data. However, consideration of the deductive-abstract character of mathematics makes it clear that one cannot proceed at once from the facts of experience to the corresponding mathematical idea. We mention in this connection that even the use of the limit notion in

---

[4] R. v. Mises, Math. Z. 4, 1–97 (1919) or Wahrscheinlichkeitsrechnung und ihre Anwendungen in der Statistik und theoretischen Physik. (Vorlesungen aus dem Gebiet der angewandten Mathematik, Vol. I), Deuticke, Leipzig-Wien 1931.

[5] For investigations in this direction see P. Cantelli, W. Feller, M. Fréchet, R. v. Mises, J. F. Steffensen and A. Wald, Les Fondements du Calcul des Probabilités (Actualités scientifiques et industrielles 735). Hermann & Cie, Paris 1938.
Recently these investigations have been reconsidered and essentially enriched by adding new ideas. Cf. C. P. Schnorr: Zufälligkeit und Wahrscheinlichkeit. Eine algorithmische Begründung der Wahrscheinlichkeitstheorie (Lecture Notes in Mathematics) Springer-Verlag, Berlin-Heidelberg-New York 1971.

v. Mises' definition is a mathematical idealization of experience arising from mathematical requirements.

In what follows, we will give a definition of probability due essentially to Kolmogorov.[6] Kolmogorov followed a path which, on the one hand, appears to conform to the modern demands for rigor in mathematics, but which is still modelled after the empirical structure of the frequency interpretation. This happens in such a way that the basic assumptions or axioms, upon which the probability theory is built, become abstractions of properties of the relative frequency of events in series of trials. We therefore want to briefly study certain properties of relative frequencies of events in such a series. First, we remark that $A \cap B$ denotes the event which occurs iff A as well as B occur, while $A \cup B$ denotes that event which is realized iff at least one of the events A or B occurs. Different experimental results can arise from the same events. If we denote by $\mathfrak{A}$ the set of all events which imply the event A and if $\mathfrak{B}$ has a similar meaning, where $\mathfrak{A}$ and $\mathfrak{B}$ are subsets of a set $\mathfrak{R}$ of possible results, then the set $\mathfrak{A} \cup \mathfrak{B}$ corresponds to the event $A \cup B$ and $\mathfrak{A} \cap \mathfrak{B}$ to the event $A \cap B$. Hence, in this sense one can identify events with subsets of a set. This makes the concept of mathematical probability theory intuitive, where one speaks of the probability of sets. These considerations can be sharpened considerably by calling on the notion of a Boolean algebra of events.[7]

Let A and B be two disjoint events in a series of n trials. Let the absolute frequency of A be n_1. The relative frequency of the appearance of A is thus n_1/n. Since n_1 is always $\leqslant n$, for the relative frequency of an event A we always have

1. $0 \leqslant n_1/n \leqslant 1$. If A always occurs, then the relative frequency is $n/n = 1$.

Let the absolute frequency of B be n_2 so that its relative frequency is n_2/n.

Since A and B are supposed to be disjoint, the frequency of $A \cup B$ is given by $n_1 + n_2$; hence, the relative frequency by $\frac{n_1 + n_2}{n}$. But $\frac{n_1 + n_2}{n} = \frac{n_1}{n} + \frac{n_2}{n}$.

This simple circumstance can be expressed as follows:

[6] A.N. Kolmogorov, Grundbegriffe der Wahrscheinlichkeitsrechnung (Ergebnisse der Mathematik und ihrer Grenzgebiete), Springer-Verlag, Berlin 1933.

[7] See A. Rényi, Wahrscheinlichkeitsrechnung. Mit einem Anhang über Informationstheorie (Hochschulbücher für Mathematik, Vol. 54), VEB Deutscher Verlag der Wissenschaften, Berlin 1962. 1 ff.

2. The relative frequency of the event $A \cup B$ is equal to the sum of the relative frequencies of the disjoint events A and B.

Since for sufficiently large n we can identify the relative frequency of an event with its empirical probability, we view properties 1. and 2. as also satisfied for the empirical probabilities.

Chapter I

Introduction to Probability Theory

1. The axioms of probability theory. We have already mentioned in the introduction that the axioms of mathematical probability[1] are to be so chosen that they reflect empirical situations when given an appropriate interpretation. We have seen that a characterization of mass phenomena can be given in a certain sense by the empirical probabilities of the events occuring. It is thus desirable to choose the notion of mathema-

---

[1] We give here a summary of the most important texts on probability theory by means of which the reader can fill in any gaps left here and deepen his knowledge. Bauer, Heinz: Wahrscheinlichkeitstheorie und Grundzüge der Maßtheorie. Walter de Gruyter & Co., Berlin, 1968. Breiman, Leo: Probability, Addison-Wesley Publ. Comp., 1968. Chung, Kai Lai: A course in probability theory. Harcourt, Brace & World, Inc., New-York, 1968. Feller, William: An introduction to probability theory and its applications. Vol. I. 3rd ed. and Vol. II. John Wiley & Sons, Inc., New York-London-Sydney, 1966. Fisz, Marek: Probability Theory and Mathematical Statistics. 3rd ed., John Wiley & Sons, New York 1963. Gnedenko, B.V.: The theory of probability. Chelsea Publ., Co., New York, 1967. Hennequin, P. L. et Tortrat A.: Théorie des probabilités et quelques applications. Màsson et Cie., Éditeurs, Paris, 1965. Métivier M.: Notions fondamentales de la théorie des probabilités. Maitrises de mathématiques. Dunod, Paris, 1968. Neveu, J.: Mathematical foundations of the Calculus of Probability. Holden Day, Inc., San Francisco-London-Amsterdam, 1965. Parzen, Emanuel: Modern Probability theory and its applications. A. Wiley Publication in Mathematical Statistics. John Wiley & Sons, Inc., New York-London, 1960. L. Schmetterer and R. Stender: Grundzüge der Mathematik, Vol. 3 (Analysis), Vandenhoeck & Ruprecht, Göttingen 1962. Kappos, Demetrios A.: Strukturtheorie der Wahrscheinlichkeitsfelder und -Räume. Ergebnisse der Mathematik und ihrer Grenzgebiete. Neue Folge, Heft 24. Springer-Verlag, Berlin-Göttingen-Heidelberg. 1960. A. Blanc-Lapierre et R. Fortet: Théorie des Fonctions aléatoires. Masson & Cie., Éditeurs. Paris. 1953. Chung, Kai Lai: Markov chains with stationary transition probabilities. Springer-Verlag, Berlin-Göttingen-Heidelberg, 1960. Doob, J.L.: Stochastic processes. John Wiley & Sons, Inc., New York; Chapman & Hall, Ltd., London, 1953. Dynkin, E. B.: Markov Processes. Vol. I. Springer-Verlag, Berlin-Göttingen-Heidelberg, 1965. Dynkin, E.B.: Die Grundlagen der Theorie der Markoffschen Prozesse. Die Grundlehren der mathematischen Wissenschaften. Bd. 108, Springer-Verlag, Berlin-Göttingen-Heidelberg, 1961. Lévy, Paul: Processus stochastiques et mouvement brownien. Gauthier-Villars & Cie., Paris, 1965. Meyer, Paul-André: Probabilités et potentiel. Actualités Scientifiques et Industrielles, No. 1318. Hermann, Paris. 1966. M. Loéve: Probability Theory. 3rd Ed., D. van Nostrand, Comp. Inc., Princeton-Toronto-New York-London. 1963.

tical probability in such a way that the theorems of the mathematical theory yield empirically verifiable facts if the mathematical probability is replaced by the empirical. We then speak briefly of the frequency interpretation of the mathematical theory. The simplest calculation rules of empirical probability are expressed by 1. and 2. (p. 23). These serve as model for the axioms of mathematical probability. In this chapter, we will discuss the most important facts of probability theory. However, we should point out at once that our program is not a complete construction of the theory. Since the main emphasis in this book is on the application of probability in mathematical statistics, many of the important theorems in this chapter will be given without proof.

For the mathematical foundations of probability theory we begin with a class $\mathcal{S}$ of sets and assume that this class is a σ-algebra of subsets of a set R.

The sets in $\mathcal{S}$ will often be interpreted in the sequel as "events".

Axiom I. Each set $A \in \mathcal{S}$ is associated with a uniquely determined real number $P(A)$ satisfying $0 \leqslant P(A) \leqslant 1$. $P(A)$ is called the *probability* of A. For the probability of R we assume $P(R)=1$.

The all-inclusive set R hence, intuitively, the all-inclusive event corresponds to the sure event with which we associate the mathematical probability 1 in agreement with 1. (p. 23).

Axiom II. Whenever $A_1, A_2, \ldots$ are countably many pairwise disjoint sets from $\mathcal{S}$,

$$P\left(\bigcup_{i=1}^{\infty} A_i\right) = \sum_{i=1}^{\infty} P(A_i)\,. \tag{1.1}$$

P thus defines a probability measure over the measurable space $(R, \mathcal{S})$. The triple $(R, \mathcal{S}, P)$ is called a *probability space.*

As an important example of a probability space where R is a finite set, we consider Laplace's definition of probability. As is well known, this definition is as follows: The probability of an event E is given by

$$P(E) = \frac{\text{number of cases favorable for the occurance of } E}{\text{number of possible cases}}\,.$$

We arrive at this definition when we begin in the sense of our interpretation with a set R with $n \geqslant 1$ elements and take $\mathcal{S}$ as the set of all subsets of R. Associate the same probability with each element $x_i \in R$ $(i=1,\ldots,n)$. Let E be a subset of R with $k \leqslant n$ elements. By Theorem 1.1 (p. 27)

$$\sum_{i=1}^{n} P(\{x_i\}) = 1\,, \quad \text{so that } P(\{x_i\}) = 1/n\,, \quad 1 \leqslant i \leqslant n\,.$$

Hence, again by Theorem 1.1, $P(E)=k/n$. Thus, behind the Laplace definition which now results as a theorem, there stands the assumption that the same probability is associated with each set $\{x_i\}$. Similar remarks hold for the so-called urn schemes of the Laplacian probability calculus which are always based on a finite probability space, so that the same probability is associated with each element.

It follows immediately from Axioms I and II, as already mentioned on p. 5, that $P(\emptyset)=0$. The empty set corresponds to an impossible event. Hence, in agreement with the frequency interpretation, one must associate the probability 0 with impossible events. We mention that the converse does not hold, i. e., $P(A)=0$ does not necessarily imply that A is the empty set, that is, an impossible event. (See, for example, Theorem 5.1.)

In this connection, we briefly treat the meaning (within the frequency interpretation) of probabilities which are "very small" or which lie "very close" to one: Events whose relative frequency within an extended series of trials is very small appear extremely rarely relative to the number of trials. Very small probabilities can thus be interpreted by considering the corresponding event as practically non-occurring within an extended series of observations. This is essentially the content of the principle of d'Alembert-Borel. Correspondingly, probabilities close to 1 are to be viewed in the sense that the associated event appears almost every time within such a series of trials. The methods of mathematical statistics which will be described later owe their practical applicability to this circumstance.

If one chooses $A_{n+1}=A_{n+2}=\cdots=\emptyset$ in Axiom II, then one gets because of $P(\emptyset)=0$ the following

Theorem 1.1. *Let* $A_1,\ldots,A_n$ *be pairwise disjoint sets from* $\mathscr{S}$. *Then* $P(A_1\cup A_2\cup\cdots\cup A_n)=\sum_{i=1}^{n} P(A_i)$.

This theorem is called the *addition theorem* of the probability calculus. It is evident that an immediate translation of property 2 of the relative frequency (p. 23) to mathematical Probability leads to Theorem 1.1 and not to Axiom II. Axiom II requires the complete additivity of probabilities, hence, the validity of the addition theorem for countably many sets. But it is precisely this extended version of the addition theorem—to which there corresponds no immediate empirical fact—which puts the theory on a wide enough basis. However, for many elementary questions of probability theory it is sufficient to replace Axiom II by Theorem 1.1. Trivially, when R is a finite set. Theorem 1.1 will be sufficient.

2. Independent sets and the definition of conditional probability. It frequently occurs in practice that two events A and B are independent of one another in the following sense: The knowledge of the outcome of A has no bearing on a prediction of the result of B. This occurs for example when A and B are the results of two throws of a die. If the first throw has yielded a 6, the chance of a particular event occuring in the second throw is not at all changed.

If one considers a series of trials of size n within which E_1 has occured n_1 times, and a series of size m within which the event E_2 has occured n_2 times, one can then combine the two series in such a way that each result of the first series is combined into a pair with each result of the second series. One obtains in this way nm pairs. If E_1 and E_2 are independent in the above sense, one can view these nm pairs as a series of trials for the occurence of paired events. If one is now interested in the event $E_1 \cap E_2$, then one must seek pairs of the form (E_1, E_2). For their relative frequency one obviously gets $n_1 n_2/nm$. This expression is also equal to the product of the relative frequencies of $E_1(E_2)$ in the first (second) trial series.

These remarks serve as guideline for the following

Definition. Two sets A and $B \in \mathscr{S}$ are said to be *independent* if $P(A \cap B) = P(A) \cdot P(B)$.

For $n \geqslant 2$ one has the

Definition. $A_1, \ldots, A_n$ [2] are *mutually independent* if for *each nonempty set of indices* $i_1, \ldots, i_l$, $i_j \neq i_k$, whenever $j \neq k$, $1 \leqslant i_j \leqslant n$ the relation

$$P(A_{i_1} \cap \cdots \cap A_{i_l}) = \prod_{j=1}^{l} P(A_{i_j}) \tag{2.1}$$

holds.

This definition can be immediately extended to arbitrarily many sets. One need only require that (2.1) hold for all possible finite subclasses of sets.

We introduce the notion of *conditional probability*. Let A and B be any sets with $P(A) \neq 0$. By the probability of B under the condition A, i. e., by the conditional probability of B for given A, we understand the quotient $P(A \cap B)/P(A)$, which will be denoted by $P(B|A)$. Hence,

$$P(B|A) = P(A \cap B)/P(A). \tag{2.2}$$

If $P(B)$ is also $\neq 0$, one can define $P(A|B)$ as well, and since $P(A \cap B) = P(A|B)P(A)$ follows from (2.2), one gets

$$P(A|B) = \frac{P(B|A)P(A)}{P(B)}. \tag{2.3}$$

(2.3) can be viewed as the simplest case of Bayes' Theorem[3]. This can

[2] All sets here belong to $\mathscr{S}$ even when this is not explicitly stated.

[3] Th. Bayes, Philos. Trans. Roy. Soc. 53, 376–398 (1763) and 54, 298–310 (1764).

be extended to infinitely many pairwise disjoint A_i with

$$\bigcup_{i=1}^{\infty} A_i = R \tag{2.4}$$

and $P(A_i) \neq 0$ for all $i=1,2,\ldots$. Indeed, if B is some set, then (2.4) implies that $B = B \cap R = B \cap \bigcup_{i=1}^{\infty} A_i = \bigcup_{i=1}^{\infty} (B \cap A_i)$, so that $P(B) = P\left(\bigcup_{i=1}^{\infty} (B \cap A_i)\right) = \sum_{i=1}^{\infty} P(B \cap A_i)$, since the sets $B \cap A_i$ are also pairwise disjoint. Hence, for $A = A_i$ we have from (2.2) that $P(B) = \sum_{i=1}^{\infty} P(B|A_i)P(A_i)$. If $P(B) \neq 0$, then we get

Bayes' Theorem. For $i=1,2,\ldots$, we have under the given conditions:

$$P(A_i|B) = \frac{P(B|A_i)P(A_i)}{\sum_{j=1}^{\infty} P(B|A_j)P(A_j)}. \tag{2.5}$$

It's easy to show that $P(B|A)$ satisfies the requirements of Axioms I and II for fixed A as a function of B, so that it actually represents a probability. Indeed, it follows at once from (2.2) that $P(R|A)=1$. Since $A \cap B \subseteq A$, we have $P(A) \geq P(A \cap B)$ [4] and since $P(A) \neq 0$ by assumption, $0 \leq P(A \cap B)/P(A) \leq 1$.

It follows just as simply from Axiom II that $P\left(\bigcup_{i=1}^{\infty} B_i | A\right) = \sum_{i=1}^{\infty} P(B_i|A)$ if the B_i are pairwise disjoint.

If A and B are independent (and e.g. $P(A) \neq 0$) then $P(B|A) = P(A \cap B)/P(A)$, and by the definition of independence, this equals $P(A)P(B)/P(A) = P(B)$. Hence, in this case, the probability of B does not depend on the condition A. In this connection, see what was said at the beginning of **2**.

Definition (2.2) has an intuitive interpretation within the frequency framework. Indeed, assume we have a series of n trials. Assume the event A occurs n_1 times in this series and the event B n_2 times and let the absolute frequency of $A \cap B$ be m. Naturally, $m \leq n_1$ and $m \leq n_2$. The relative frequency of the event A is n_1/n, of B n_2/n and of $A \cap B$ m/n. Replacing the quotient on the right side of (2.2) by the expression in terms of relative frequencies, one obtains $(m/n):(n_1/n) = m/n_1$. m/n_1 can be viewed as the relative frequency of the event $A \cap B$, provided one considers only those n_1 trials which have yielded the event A. One can thus speak of m/n_1 as the "relative frequency of the event B under the hypothesis A". This agrees with the definition of the probability of B under the hypothesis A. The relative frequency m/n_2 can be interpreted similarly.

[4] See p. 5.

3. The notion of a random variable (r.v.). In practical applications, the results of experiments are often described by numerical data which reflect the random character of the experiments. These data often depend functionally on other (not immediately observable) random events. For example, one measures the more or less randomly conditioned defense mechanism against human infection, namely fever, by means of a scale of real numbers on a thermometer. Such randomly conditioned numbers are called random variables, whose exact mathematical definition we now give.

We consider maps ξ from R into the real numbers, i. e., functions which associate a number $\xi(\omega) \in R_1$ with each element $\omega \in R$. Such a map ξ is called $\mathcal{S}$-measurable (see p. 7 ff.), if for each real number α, the set $\{\omega: \xi(\omega) \leqslant \alpha\}$ belongs to $\mathcal{S}$. That is, the inverse image of $(-\infty, \alpha]$ under ξ belongs to $\mathcal{S}$. We now give the

Definition. Let $(R, \mathcal{S}, P)$ be a probability space. Then, *each $\mathcal{S}$-measurable map* of R into R_1 is called a *(one-dimensional) random variable.* We will also occasionally say that ξ defines a r.v. over the probability space $(R, \mathcal{S}, P)$.

With each r.v. is connected its probability distribution: Let ξ be a r.v. We consider the class K of all sets of R_1 whose inverse images under ξ lie in $\mathcal{S}$. K is then a σ-algebra of sets of R_1 (see Theorem I), which, according to the above definition of a r.v., contains all Borel sets of R_1.

Hence, if $B \in \mathfrak{B}_1$, then $P(\xi^{-1}(B))$ is always defined. This will be used to define a probability measure over $\mathfrak{B}_1$—the *probability distribution* of ξ: For each $B \in \mathfrak{B}_1$, set

$$P_\xi(B) = P(\xi^{-1}(B)). \tag{3.1}$$

P_ξ is in fact a probability measure which is defined over the Borel set $\mathfrak{B}_1$ of R_1 since $0 \leqslant P_\xi(B) \leqslant 1$ for each $B \in \mathfrak{B}_1$ and $P_\xi(R_1) = 1$ which follows at once from (3.1). Likewise, $P_\xi\left(\bigcup_{i=1}^{\infty} B_i\right) = \sum_{i=1}^{\infty} P_\xi(B_i)$ for pairwise disjoint Borel sets $B_i \in \mathfrak{B}_1$ $(i = 1, 2, \ldots)$. (See Theorem I.)

It is easy to interpret P_ξ intuitively. One considers the r.v. ξ as a outcome of some experiment. $P_\xi(B)$ with $B \in \mathfrak{B}_1$ is then the probability that the outcome ξ in an experiment lies in B. In this sense, the notation $P(\xi \in B)$ will often be used instead of $P_\xi(B)$. In particular, if B is an interval (a, b), then one could say that $P_\xi(a, b)$ gives the probability that the outcome of an experiment lies between a and b. We also write $P(a < \xi < b)$ for this probability. Analogous terminology in the sequel is to be understood in the same spirit.

4. The distribution function of a random variable. Let ξ be a r.v. and P_ξ be its probability distribution. If x is an arbitrary real number, then $P(-\infty<\xi\leqslant x)$ obviously depends only on x. This observation leads to the following

Definition. The function F, defined for all real x by setting

$$F(x)=P(-\infty<\xi\leqslant x) \tag{4.1}$$

is called the *distribution function* of the r.v. ξ.[5]

The distribution function (also written d.f.) is one of the most important concepts in the elementary calculus of probability. If we again interpret the random variable ξ as a outcome of some experiment, then we can view the d.f. as the probability that the observed outcome ξ does not exceed the real number x. We immediately obtain

Theorem 4.1. *If F is the d.f. of a random variable, then*

$$P(a<\xi\leqslant b)=F(b)-F(a),$$

for $a<b$.

Proof. The interval $(-\infty,b]$ is the union of the disjoint intervals $(-\infty,a]$ and $(a,b]$. Therefore, it follows from Theorem 1.1 that $P(-\infty<\xi\leqslant b)=P(-\infty<\xi\leqslant a)+P(a<\xi\leqslant b)$. The assertion is then a result of Eq. (4.1).

5. The most important properties of the d.f. of a random variable. Let F be the d.f. of a random variable. Then

1. F is nondecreasing, i. e. if

$$x_1<x_2, \tag{5.1}$$

then

$$F(x_1)\leqslant F(x_2). \tag{5.2}$$

2. F is right continuous. Thus, for each real x

$$F(x+h)-F(x)\to 0 \tag{5.3}$$

as $h\to 0$ and $h>0$.

3. F satisfies

$$\lim_{x\to-\infty} F(x)=0, \tag{5.4}$$

and

4. F satisfies

$$\lim_{x\to+\infty} F(x)=1. \tag{5.5}$$

---

[5] Frequently, $P(-\infty<\xi<x)$ is defined to be the distribution function of ξ, e.g. by Kolmogorov, l.c. Intro.[6]

Proof of 1. From Theorem 4.1, $F(x_2)-F(x_1)=P(x_1<\xi\leqslant x_2)$. However, this is $\geqslant 0$ by Axiom I and so (5.2) follows.

Proof of 2. By Theorem 4.1, it is sufficient to prove that

$$\lim_{h\to 0, h>0} P(x<\xi\leqslant x+h)=0. \tag{5.6}$$

First, let $\{h_n\}$ be a monotone sequence of positive numbers converging to zero; without loss of generality we may assume that $0<h_{n+1}<h_n$, $n=1,2,\dots$. Denote the interval $(x+h_{i+1}, x+h_i]$, $i=1,2,\dots$ by E_i. Clearly,

$$\bigcup_{i=1}^{\infty} E_i=(x, x+h_1].$$

Since the E_i are pairwise disjoint, because of the monotonicity of the sequence $\{h_n\}$, we have from Axioms I and II that

$$1\geqslant P(x<\xi\leqslant x+h_1)=P_\xi\left(\bigcup_{i=1}^{\infty} E_i\right)=\sum_{i=1}^{\infty} P_\xi(E_i).$$

It follows that

$$\sum_{i=n}^{\infty} P_\xi(E_i)\to 0 \tag{5.7}$$

as $n\to+\infty$. By Axiom II, (5.7) is equivalent to $P_\xi\left(\bigcup_{i=n}^{\infty} E_i\right)\to 0$. Therefore, from the definition of E_i, $P(x<\xi\leqslant x+h_n)\to 0$ as $n\to+\infty$ and this is what was to be shown. Now, if $h_n>0$ is an arbitrary null-sequence, then set $l_n=\sup_{k\geqslant n} h_k>0$. l_n is obviously a monotone null-sequence and so

$$0\leqslant P(x<\xi\leqslant x+h_n)\leqslant P(x<\xi\leqslant x+l_n)\to 0, \quad \text{as } n\to+\infty.$$

This proof is based on the continuity property of probability measures (cf. p. 5), which we have adapted for this special case.

The proof shows that Axiom II makes rigorous the following heuristic argument:

As $h_n\to 0$, $h_n>0$, the event $x<\xi\leqslant x+h_n$ approaches the impossible event $x<\xi\leqslant x$ and, therefore, $P(x<\xi\leqslant x+h_n)\to 0$.

The *proof* of 3. and 4. can be patterned along the lines of the proof of 2. Also, using reasoning similar to that in the proof of 2., it can be shown that

$$\lim_{h\to 0, h>0} (F(x)-F(x-h))=P(\xi=x). \tag{5.8}$$

According to the usual notation for right and left-hand limits, Property 2. and relation (5.8) for the d.f. of a r.v. can be formulated as follows:

$$F(x+0)=F(x); \qquad F(x-0)=F(x)-P(\xi=x).$$

From this we easily obtain

Theorem 5.1. F *is continuous at* x *if, and only if,* $P(\xi=x)=0$.

Proof. If $P(\xi=x)=0$, then at the point x $F(x+0)=F(x-0)=F(x)$ and F is continuous at x. However, if F is continuous at x, then in particular $F(x-0)=F(x)$, so that $P(\xi=x)=0$.

It follows from 1. that F is continuous except for an at most countable number of points. The discontinuity points are jump points.

The significance of Properties 1. to 4. of a d.f. is brought to light in the following important theorem:

Theorem 5.2. *Each real-valued function* F *on* R_1 *which satisfies conditions* 1. *to* 4. *defines a probability measure* P_F *over* $(R_1, \mathfrak{B}_1)$; *hence,* F *can be considered to be the d.f. of some random variable.*

We will sketch the proof of this theorem. To begin with, the probability measure $P_F\{(a,b]\}$, associated with each interval of the form $(a,b]$ is taken to be the real number $F(b)-F(a)$; this is always non-negative because of 1. With this agreement, it is easy to show that the probability measure defined over all intervals of the specified form is completely additive. Therefore, by the general extension theorem for measures (cf. Theorem III), P_F can be extended to all Borel sets of R_1 in one and only one way. The identity mapping which assigns to each $x\in R_1$ the value x defines a random variable ξ over the probability field $(R_1, \mathfrak{B}_1, P_F)$. For this random variable, $P(\xi\leqslant y)=P_F(\xi^{-1}(-\infty, y])$ $=P_F((-\infty, y])$ for each real y; but then $P(\xi\leqslant y)=F(y)$, so that F is the d.f. of the random variable ξ.

The essential point to be gleaned from this is the following: If the d.f. of a random variable ξ is given, then it can always be assumed that the probability $P(\xi\in B)$ is defined for all B from $\mathfrak{B}_1$.

6. The distribution density of a one-dimensional random variable. We consider a d.f. F which is differentiable at the point x. Let $f(x)=F'(x)$. We shall refer to $f(x)$ as the distribution density or, briefly, the density of the corresponding one-dimensional random variable at the point x.

In many important cases $f(x)$ exists for each real x, and moreover,

$$F(x)=\int_{-\infty}^{x} f(t)\,dt. \tag{6.1}$$

for all $x \in R_1$. If ξ is a r.v. with d.f. F and F has the representation (6.1), then it follows that

$$P(\xi \in B) = \int_B f(x)\,dx \tag{6.2}$$

for each $B \in \mathfrak{B}_1$.

More generally, we will speak of a *distribution density* if F has a derivative f almost everywhere and the representation (6.1) holds for all x.[6] Then the distribution density f possesses the following properties:

$$1.\quad \int_{-\infty}^{+\infty} f(x)\,dx = 1 \tag{6.3}$$

$$2.\quad f \geqslant 0 \tag{6.4}$$

(6.3) follows immediately from (6.1) and (5.5). (6.4) is a result of the monotonicity of F. Conversely, if a function f is integrable over R_1, satisfies (6.3) and obeys (6.4), at least up to a set of Lebesgue measure zero, then the function F defined by (6.1) satisfies Properties 1. to 4. Therefore, F can be viewed as the d.f. of a r.v. ξ. Further, each d.f. which is representable in the form (6.1) is continuous, so that $P(\xi = x) = 0$ for all $x \in R_1$, from Theorem 5.1.

This concept of a density is only a special case of a more general idea. If μ is a measure defined on the measurable space $(R, \mathscr{S})$ and f is a μ-integrable real-valued function on R, then setting $\nu(A) = \int_A f\,d\mu$, $A \in \mathscr{S}$, defines a set function which is absolutely continuous w.r.t. μ. (Cf. **9**, p. 14.) If $f \geqslant 0$, then ν is nonnegative and is therefore a measure. f is then called the density of ν w.r.t. μ. Especially important is the fact that this result has a far-reaching converse in the following

Theorem 6.1. *If ν and μ are σ-finite measures and ν is absolutely continuous with respect to μ, then there exists a real-valued, $\mathscr{S}$-measurable function $f \geqslant 0$ defined on R, which is uniquely determined up to μ-null sets, and satisfies*

$$\nu(A) = \int_A f\,d\mu \tag{6.5}$$

for all $A \in \mathscr{S}$.

This is the Radon-Nikodym Theorem. f is sometimes denoted by $d\nu/d\mu$. This notation is justified by the following

[6] This is precisely the case when F is *absolutely continuous*, i. e., to each $\varepsilon > 0$ there corresponds a $\delta > 0$, such that for each finite or countably infinite set of pairwise disjoint intervals (x_i, y_i), $\sum |F(y_i) - F(x_i)| < \varepsilon$ if $\sum |y_i - x_i| < \delta$.

Theorem 6.2. *Let ρ, ν and μ be σ-finite measures with ν absolutely continuous with respect to μ and ρ absolutely continuous w.r.t. ν. Then ρ is also absolutely continuous with respect to μ. If f is the*[7] *density of ν relative to μ and g the density of ρ relative to ν, then the density h of ρ relative to μ satisfies $h=fg$ or $d\rho/d\mu=(d\rho/d\nu)(d\nu/d\mu)$, μ-a.e.*

In this connection the distribution density defined on p. 34 can be referred to as the Radon-Nikodym density[8] relative to Lebesgue measure. When we refer simply to a *density*, we will always mean the density relative to Lebesgue measure.

We will now clarify the probabilistic significance of the density $f(x)$ of a random variable at the point x. By definition, the derivative of the corresponding d.f. F is given by

$$\frac{F(x+h)-F(x)}{h}=f(x)+\varepsilon(h), \tag{6.6}$$

where $\varepsilon(h)\to 0$ as $h\to 0$. Now, it follows for $h>0$, say, from (6.6) and Theorem 4.1 that

$$\frac{1}{h}P(x<\xi\leqslant x+h)=f(x)+\varepsilon(h). \tag{6.7}$$

The left-hand side of (6.7) can be thought of as the "average probability" that ξ lies in the interval $(x, x+h]$. This average probability is approximated better by $f(x)$ as h becomes smaller. Similar remarks can be made, of course, for $h<0$. Heuristically, one says that the probability that "ξ is x or lies in an interval of length dx around the point x" is given by $f(x)|dx|$. The precise meaning of this statement is given by the relation (6.7).

7. Discrete and continuous type d.f.'s. For most applications to questions in mathematical statistics, it is sufficient to consider two types of d.f.'s.

I. The discrete type. Let a finite[9] or denumerable number of points in R_1, say $\ldots, x_{-n}, \ldots, x_0, \ldots, x_n, \ldots$ be given. Further, let p_i, $i=\ldots, -n, \ldots, 0, \ldots, n, \ldots$ be positive real numbers with

$$\sum_{i=-\infty}^{+\infty} p_i=1. \tag{7.1}$$

[7] Properly, it should be referred to as "a" density; however, when no misunderstanding is likely—here and in similar cases—we apply the definite article.

[8] Briefly, we usually write R.-N.-density.

[9] In this case, trivial changes in notation have to be introduced.

Then the d.f. is defined by $F=\sum_{i=-\infty}^{+\infty} p_i c_{[x_i,\infty)}$, where $c_{[x_i,\infty)}$ is the indicator function of the interval $[x_i,\infty)$ (cf. p. 5). F is obviously nondecreasing: if $x'<x''$, then $F(x')=\sum_{x_l\leqslant x'} p_l c_{[x_l,\infty)}$ and $F(x'')=\sum_{x_l\leqslant x''} p_l c_{[x_l,\infty)}$ so that $F(x'')-F(x')=\sum_{x'<x_l\leqslant x''} p_l c_{[x_l,\infty)}\geqslant 0$. F is right continuous, since obviously $F(x_i)=F(x_i+0)$ for $i=0,\pm 1,\ldots$. Further $\lim_{x\to\infty} F(x)$ $=\lim_{x\to\infty}\sum_{i=-\infty}^{+\infty} p_i c_{[x_i,\infty)}(x)=\lim_{l\to\infty}\sum_{i=-l}^{l} p_i$ and similarly $\lim_{x\to-\infty} F(x)=0$.

In order to illustrate the notion of probability space and random variable, we shall now obtain this d.f. in another manner: Let the x_i be as previously defined and denote by R the set of all x_i. The set of all subsets of R is denoted by $\mathscr{S}$ (cf. p. 5). The p_i are taken to have the same meaning as above and we define a probability measure P over the σ-algebra $\mathscr{S}$ by setting $P(\{x_i\})=p_i$. For sets in $\mathscr{S}$ with more than one element P is defined in accordance with Axiom II. We now consider the mapping ξ of R into R_1 which associates the value $x_i\in R_1$ with each $x_i\in R$. Clearly, ξ is $\mathscr{S}$-measurable since each subset of R belongs to $\mathscr{S}$. The d.f. of ξ is then, by definition, given by $P(\xi\leqslant x)=P(\{x_i\colon x_i\leqslant x\})$ for each $x\in R_1$. However, $P(\{x_i\colon x_i\leqslant x\})$ is given by $\sum_{-\infty}^{+\infty} p_i c_{[x_i,\infty)}(x)$ and so we arrive at the d.f. defined earlier. Clearly, $P(\xi=x_i)=p_i$ for all i and $P(\xi\in B)=0$ for each $B\in\mathfrak{B}_1$ which contains none of the points x_i. We will say that such a random variable and its corresponding d.f. are of the discrete type or that ξ possesses a discrete distribution.

We shall illustrate this definition by an especially simple example involving exactly n points, so that $P(\xi=x_i)=p_i\neq 0$ and $\sum_{i=1}^{n} p_i=1$. The x_i are ordered according to their magnitudes: $x_1<x_2<\cdots<x_n$. The function $F=\sum_{i=1}^{n} p_i c_{[x_i,\infty)}$ is then a step-function with jumps at x_i of magnitude p_i, $1\leqslant i\leqslant n$, and otherwise is constant. Therefore,

$$\begin{array}{ll}
F(x)=0 & \text{for } x<x_1\\
F(x)=p_1 & \text{for } x_1\leqslant x<x_2\\
F(x)=p_1+p_2 & \text{for } x_2\leqslant x<x_3\\
\cdots\cdots\cdots & \cdots\cdots\cdots\\
F(x)=p_1+\cdots+p_{n-1} & \text{for } x_{n-1}\leqslant x<x_n\\
F(x)=p_1+\cdots+p_n=1 & \text{for } x_n\leqslant x\,.
\end{array}$$

See Fig. 2.

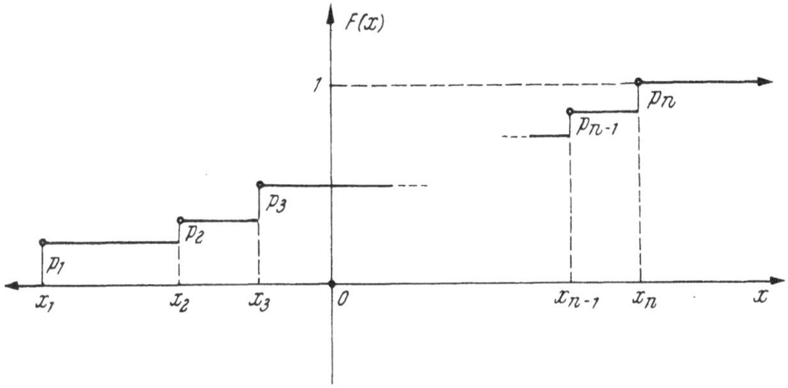

Fig. 2

In particular, if $p_i = \cdots = p_n = 1/n$, then one obtains the so-called discrete *uniform distribution*. In this special case, therefore, the Laplace definition can be applied (cf. p. 26). For $n=1$, we obtain $P(\xi = x_1) = 1$. Here, the "entire probability distribution is concentrated at x_1". This distribution is called the *degenerate distribution* or *Dirac-distribution* at x_1.

Important examples of discrete probability distributions are also treated in **32**, **33** and **34**.

II. The continuous type. The distribution of a r.v. is said to be of the continuous type if the d.f. F has a distribution density f. Then F has a representation of the form (6.1).

An especially simple but important example is represented by the (continuous) *uniform distribution*: Let a and b be real numbers with $a<b$ and

$$f(x) = \begin{cases} 1/(b-a), & a<x<b \\ 0, & \text{otherwise}. \end{cases}$$

It is immediate that f is a density and defines a distribution of the continuous type.

The statement that F is of the continuous type is of course different from the assumption that F is continuous (see [6]). Indeed, each distribution of the continuous type possesses a continuous d.f., but not conversely. Many important distribution densities f of distributions of the continuous type occurring in practice are differentiable at all except an at most finite number of points. We refer to **25, 28, 29** and **30.**

8. Multidimensional random variables. The concept of a one-dimensional r.v. defined in **3** has a natural generalization: If $(R, \mathcal{S}, P)$ is a probability space, then each $\mathcal{S}$-measurable mapping $\xi = (\xi_1, \ldots, \xi_n)$

of R into R_n is called an n-dimensional r.v. Therefore, it follows from the definition of $\mathcal{S}$-measurability that the sets $\{\omega: \xi_1(\omega)\leqslant\alpha_1, \ldots, \xi_n(\omega) \leqslant\alpha_n\}$ belong to $\mathcal{S}$ for all n-tuples $\alpha_1, \ldots, \alpha_n$ of real numbers. Hence, the inverse image under ξ of any n-dimensional interval as well as that of any Borel set $B\in\mathfrak{B}_n$ are $\mathcal{S}$-measurable. Each n-tuple of one-dimensional random variables can therefore be considered to be an n-dimensional r.v. The probability distribution P_ξ of an n-dimensional r.v. ξ can be defined over $\mathfrak{B}_n$ in complete analogy to (3.1). That is, by setting

$$P_\xi(B)=P(\xi^{-1}(B)) \tag{8.1}$$

for all $B\in\mathfrak{B}_n$.

Actually, it is easy to show that P_ξ is a probability measure over $(R_n, \mathfrak{B}_n)$ and satisfies Axioms I and II. P_ξ is also referred to as the *joint distribution* of $\xi_1, \ldots, \xi_n$. Once more we write $P(\xi\in B)$ in place of $P_\xi(B)$ and $P(-\infty<\xi_1\leqslant x_1, \ldots, -\infty<\xi_n\leqslant x_n)$ in place of $P_\xi((-\infty, x_1]\times\cdots \times(-\infty, x_n])$ and so forth.

9. The d.f. of a multidimensional random variable. Let $\xi=(\xi_1,\ldots,\xi_n)$ be an n-dimensional r.v. and $x_1, \ldots, x_n$ be arbitrary real numbers. The d.f. F of the r.v. ξ is then given by

$$F(x_1, x_2, \ldots, x_n)=P(-\infty<\xi_1\leqslant x_1, -\infty<\xi_2\leqslant x_2, \ldots, -\infty<\xi_n\leqslant x_n). \tag{9.1}$$

The value assigned to the function $F(x_1, x_2, \ldots, x_n)$, therefore represents the probability that not only ξ_1 is less than x_1 but also ξ_2 is less than $x_2, \ldots, \xi_n$ is less than x_n. The d.f. of a multidimensional random variable is characterized by a series of properties. These generalize properties 1. to 4. of one-dimensional d.f.'s (p. 31).

For any function G defined over R_n let

$$\Delta_i G(x_1, \ldots, x_i, \ldots, x_n)=G(x_1, \ldots, x_i+h_i, \ldots, x_n)-G(x_1, \ldots, x_i, \ldots, x_n), \tag{9.2}$$

where h_i is an arbitrary real number, $1\leqslant i\leqslant n$.

We shall also describe this definition in terms of an operator Δ: if the Δ-operator is applied to the ith argument of G, then the result is the right side of equation (9.2).

Let $\{i_1, \ldots, i_k\}, k\geqslant 1$, be a subset of $\{1, \ldots, n\}$. Then

$$1\,\text{m}.\quad \Delta_{i_1}\cdots\Delta_{i_k} F(x_1, \ldots, x_n)\geqslant 0 \tag{9.3}$$

if the real numbers $h_{i_1},\ldots,h_{i_k}$ occurring in (9.3) according to definition (9.2) are all assumed to be positive.

2 m. F is right continuous in each of its arguments with the others held fixed. That is, for each i, $1\leqslant i\leqslant n$, the mapping $x_i\to F(x_1,\ldots,x_i,\ldots,x_n)$ from R_1 into R_1 is right continuous.

3m. If at least one of the variables approaches $-\infty$, then $F(x_1,\ldots x_n)$ approaches zero; moreover, the convergence is uniform in the other variables.

4m. $\lim\limits_{x_1\to\infty,\ldots,x_n\to\infty} F(x_1,\ldots,x_n)=1$.

Proof of 1m. In order to demonstrate (9.3) we first show that applying the Δ-operator first to the ith argument and then to the kth argument of F produces the same result as first applying it to the kth and then to the ith argument. By means of (9.2), we immediately obtain

$$\begin{aligned}
&\Delta_k\Delta_i F(x_1,\ldots,x_i,\ldots,x_k,\ldots,x_n)\\
&=F(x_1,\ldots,x_i+h_i,\ldots,x_k+h_k,\ldots,x_n)-F(x_1,\ldots,x_i+h_i,\ldots,x_k,\ldots,x_n)\\
&\quad -F(x_1,\ldots,x_i,\ldots,x_k+h_k,\ldots,x_n)+F(x_1,\ldots,x_i,\ldots,x_k,\ldots,x_n)\\
&=\Delta_i\Delta_k F(x_1,\ldots,x_i,\ldots,x_k,\ldots,x_n).
\end{aligned}$$

Therefore, the Δ-operations are interchangeable with one another; this can easily be shown by induction for any arbitrary finite number of Δ-operators. Here, however, we are only interested in the case where the Δ-operator is applied to each of the variables x_i at most once. (9.3) will be proved if it is shown that

$$\begin{aligned}
\Delta_{i_1}\ldots\Delta_{i_k}F(x_1,\ldots,x_n)=P(-\infty<\xi_1\leqslant x_1,\ldots,x_{i_1}<\xi_{j_1}\leqslant x_{i_1}+h_{i_1},\ldots,x_{i_k}\\
<\xi_{i_k}\leqslant x_{i_k}+h_{i_k},\ldots,-\infty<\xi_n\leqslant x_n),\\
h_{i_j}>0,\quad j=1,\ldots,k.
\end{aligned}\tag{9.4}$$

Indeed, the right side of (9.4) is $\geqslant 0$ by Axiom I. The identity (9.4) follows immediately by induction. For $k=1$.

$$\begin{aligned}
&\Delta_{i_1}F(x_1,\ldots,x_{i_1},\ldots,x_n)\\
&=F(x_1,\ldots,x_{i_1}+h_{i_1},\ldots,x_n)-F(x_1,\ldots,x_{i_1},\ldots,x_n).
\end{aligned}$$

From the definition (9.1), it follows that the right side of this equation is equal to

$$\begin{aligned}
&P(\xi_1\leqslant x_1,\ldots,\xi_{i_1}\leqslant x_{i_1}+h_{i_1},\ldots,\xi_n\leqslant x_n)\\
&-P(\xi_1\leqslant x_1,\ldots,\xi_{i_1}\leqslant x_{i_1},\ldots,\xi_n\leqslant x_n).
\end{aligned}\tag{9.5}$$

The interval $(-\infty,x_1]\times\cdots\times(-\infty,x_{i_1}+h_{i_1}]\times\cdots\times(-\infty,x_n]$ is denoted by I_1. Obviously,

$$I_1=I_2\cup I_3,$$

where

$$I_2=(-\infty,x_1]\times\cdots\times(-\infty,x_{i_1}]\times\cdots\times(-\infty,x_n]$$

and

$$I_3=(-\infty,x_1]\times\cdots\times(x_{i_1},x_{i_1}+h_{i_1}]\times\cdots\times(-\infty,x_n].$$

The intervals I_2 and I_3 are disjoint so that the expression in (9.5)$=P(I_3)$ by Theorem 1.1. Hence (9.4) is shown for $k=1$.

Suppose that (9.4) holds for $k-1$ operators. Then

$$\begin{aligned}&\Delta_{i_1}\dots\Delta_{i_{k-1}}F(x_1,\dots,x_n)\\&=P(x_{i_1}<\xi_{i_1}\leqslant x_{i_1}+h_{i_1},\dots,x_{i_{k-1}}<\xi_{i_{k-1}}\leqslant x_{i_{k-1}}+h_{i_{k-1}}),\end{aligned}\tag{9.6}$$

where we have neglected to write the random variables ξ_i satisfying an inequality of the form $\xi_i\leqslant x_i$. We will also follow this practice in the following chain of equations. Applying Δ_{i_k} to (9.6), we obtain

$$\begin{aligned}&\Delta_{i_k}\Delta_{i_1}\dots\Delta_{i_{k-1}}F(x_1,\dots,x_n)\\&=\Delta_{i_k}P(x_{i_1}<\xi_{i_1}\leqslant x_{i_1}+h_{i_1},\dots,x_{i_{k-1}}<\xi_{i_{k-1}}\leqslant x_{i_{k-1}}+h_{i_{k-1}})\\&=P(x_{i_1}<\xi_{i_1}\leqslant x_{i_1}+h_{i_1},\dots,x_{i_{k-1}}<\xi_{i_{k-1}}\\&\leqslant x_{i_{k-1}}+h_{i_{k-1}},\dots,-\infty<\xi_{i_k}\leqslant x_{i_k}+h_{i_k})\\&\quad-P(x_{i_1}<\xi_{i_1}\leqslant x_{i_1}+h_{i_1},\dots,x_{i_{k-1}}<\xi_{i_{k-1}}\leqslant x_{i_{k-1}}+h_{i_{k-1}})\\&=P(x_{i_1}<\xi_{i_1}\leqslant x_{i_1}+h_{i_1},\dots,x_{i_{k-1}}<\xi_{i_{k-1}}\\&\leqslant x_{i_{k-1}}+h_{i_{k-1}},\dots,x_{i_k}<\xi_{i_k}\leqslant x_{i_k}+h_{i_k}).\end{aligned}$$

However, since the Δ-operators are commutative, the proof of (9.4) is complete.

It follows from 1m. that the mapping $x_i\to F(x_1,\dots,x_i,\dots,x_n)$ is non-decreasing for all i.

The *proof of* 2m. can be carried out in exactly the same manner as the proof of 2., p. 32.

The *proof of* 3m. is easily accomplished by showing that $P(\xi_1\in R_1,\dots,\xi_i\leqslant x_i^{(k)},\dots,\xi_n\in R_1)\to 0$ as $x_i^{(k)}\to-\infty$ for $k\to+\infty$ and observing that

$$P(\xi_1\leqslant x_1,\dots,\xi_i\leqslant x_i,\dots,\xi_n\leqslant x_n)\leqslant P(\xi_1\in R_1,\dots,\xi_i\leqslant x_i,\dots,\xi_n\in R_1)$$

for all $(x_1,\dots,x_i,\dots,x_n)\in R_n$.

The *proof of* 4m. is straightforward, essentially a consequence of Axiom II, and is not reproduced here.

As in the one-dimensional case, if F is continuous, then $P(\xi=x)=0$ for all $x\in R_n$. However, if $n\geqslant 2$, the converse is not correct. The exact analogue to Theorem 5.2 holds:

Theorem 9.1. *Each real-valued function F defined over R_n which satisfies conditions* 1m. *to* 4m. *is the d.f. of an n-dimensional r.v.*

If an n-dimensional d.f. F possesses partial derivatives $\dfrac{\partial^n F}{\partial x_1\dots\partial x_n}$ at the point x, then we again refer to this derivative as the distribution density at x. If this derivative, which we denote by f, exists everywhere or a. e. in R_n and if f is integrable and the relation

$$\int_{-\infty}^{x_1} \cdots \int_{-\infty}^{x_n} f(t_1,\ldots,t_n)\,dt_1\ldots dt_n = F(x_1,\ldots,x_n) \quad (9.7)$$

[10]

holds for all $(x_1,\ldots,x_n)$, then, in particular, it follows that

$$\int_{R_n} f(x)\,dx = 1 \quad (9.8)$$

and

$$f(x) \geqslant 0 \quad (9.9)$$

for all $x \in R_n$ (up to a set of measure zero). Once more, it is only in this case that we will refer simply to a density.

10. Discrete and continuous type multidimensional d.f.'s. The two types of d.f.'s are also of particular importance in the multidimensional case.

I. The discrete type. The distribution of an n-dimensional r.v. ξ is said to be of the discrete type, if there exists a finite or countably infinite number of n-tuples $x^{(1)}, x^{(2)},\ldots,$ such that $P(\xi = x^{(i)}) = p_i > 0$ with $\sum_{i=1}^{\infty} p_i = 1$; therefore, for any Borel set $B \in \mathfrak{B}_n$ which contains none of the n-tuples $x^{(1)}, x^{(2)},\ldots, P(\xi \in B) = 0$. If $x^{(i)} = (x_1^{(i)},\ldots,x_n^{(i)})$ and I_i denotes the interval $[x_1^{(i)},\infty) \times \cdots \times [x_n^{(i)},\infty)$, then the corresponding d.f. is given by $F = \sum_{i=1}^{\infty} c_{I_i} p_i$. An important example of a discrete multidimensional distribution can be found in **37**.

II. The continuous type. We assume that the d.f. F of an n-dimensional random variable ξ possesses a density $\dfrac{\partial^n F}{\partial x_1 \ldots \partial x_n} = f$ throughout R_n. It follows that equation (9.7) holds for all $x_1,\ldots,x_n$. An important example is furnished by the *general uniform distribution.* Let $B \in \mathfrak{B}_n$ and suppose that its measure $L(B) \neq 0$.[11] Then the density of the general uniform distribution is defined by setting

$$f(x) = \begin{cases} 1/L(B), & x \in B \\ 0, & x \in R_n - B. \end{cases}$$

[10] Again this is precisely the case when F is absolutely continuous. f is the Radon-Nikodym density relative to n-dimensional Lebesgue measure.

[11] Naturally, $L(B)$ denotes the Lebesgue measure of B.

Further important examples of multidimensional distributions of the continuous type are given in **35** and VI.

11. Functions of random variables. Let g be a real-valued Borel-measurable function defined on R_n and $\omega \to \xi(\omega) = (\xi_1(\omega), \ldots, \xi_n(\omega))$ be an n-dimensional r.v. as defined in **8**.

We assert that the mapping $g \circ \xi$,[12] i.e. the mapping which associates with each $\omega \in R$ the real number $g(\xi_1(\omega), \ldots, \xi_n(\omega))$ is also a r.v. It is sufficient to show that $g \circ \xi$ is $\mathscr{S}$-measurable. Now, for arbitrary $B \in \mathfrak{B}_1$

$$\{\omega : g(\xi_1(\omega), \ldots, \xi_n(\omega)) \in B\} = \{\omega : (\xi_1(\omega), \ldots, \xi_n(\omega)) \in g^{-1}(B)\}.$$

However, since g is Borel measurable, $g^{-1}(B)$ is a Borel set of R_n. Therefore, the set $\{\omega : (\xi_1(\omega), \ldots, \xi_n(\omega)) \in g^{-1}(B)\}$ lies in $\mathscr{S}$, since ξ is a r.v. From (4.1), the d.f. G of $\eta = g \circ \xi$ is given by $G(x) = P(\eta \leqslant x)$ for each real x. If we denote $g^{-1}((-\infty, y])$ by M_y, then obviously we also have

$$G(y) = P(\xi \in M_y). \tag{11.1}$$

This makes it possible to calculate the d.f. G and also, by Theorem 5.2, the probability distribution of η, if the probability distribution of ξ is given. Equation (11.1) can also be written as: $G(y) = \int_{M_y} dP$.

An example of the construction of such a set M_y when ξ is a one-dimensional r.v. is shown below in Fig. 3. In this case M_y consists of four intervals which are indicated by heavy lines in the figure.

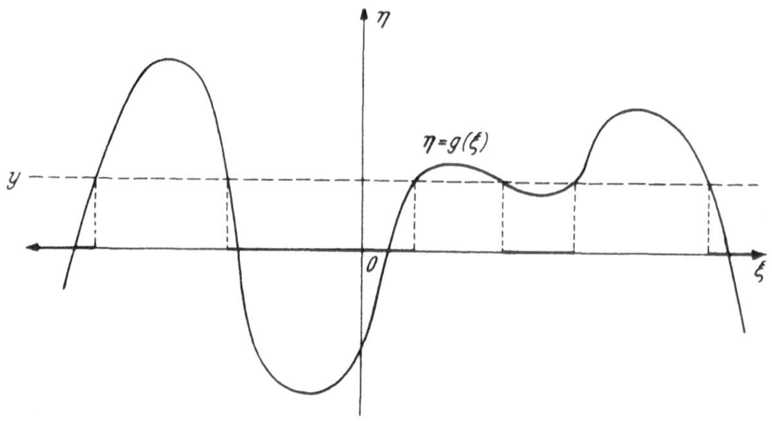

Fig. 3

[12] It is convenient to agree also to write $g(\xi)$ or $g(\xi_1, \ldots, \xi_n)$.

Continuous functions or functions which are continuous except at a finite number of points are important examples of Borel measurable functions. Likewise, all functions satisfying (9.3), therefore all monotone functions in the one-dimensional case, are important examples of Borel-measurable functions.

We will now illustrate the above comments with several simple, but important, examples. Let g be a function defined, continuous and strictly increasing on R_1. In particular, then, $x'<x''$ implies $g(x')<g(x'')$. Moreover, set

$$\inf_{-\infty<x<\infty} g(x)=m, \qquad \sup_{-\infty<x<\infty} g(x)=M. \tag{11.2}$$

Let ξ be a one-dimensional r.v. with d.f. F. We defined the r.v. $\eta=g\circ\xi$. As a consequence of (11.1), in order to determine the d.f. G of η, we consider the set M_y consisting of those real numbers x satisfying the inequality $g(x)\leqslant y$. Now, if $y\leqslant m$, then M_y is the empty set; that is, there are no points x satisfying the inequality $g(x)\leqslant y$. However, if $y\geqslant M$, which is only possible if $M<+\infty$, then M_y is identical with R_1. Finally, if $m<y<M$, then to begin with we consider the equation $g(x)=y$. Because of the continuity of g, each such equation has a solution and it is unique because of the assumed strict monotonicity. We denote it by x_y. The monotonicity implies that the sets $M_y=\{x\colon g(x)\leqslant y\}$ and $\{x\colon x\leqslant x_y\}$ are equal. Therefore, if for example $m>-\infty$ and $M<+\infty$, then we obtain

$$G(y)=\begin{cases}0, & y\leqslant m\\ F(x_y), & m<y<M.\\ 1, & y\geqslant M\end{cases}$$

This can also be expressed in the following manner: from the assumptions the inverse mapping g^{-1} of g exists, is continuous, strictly increasing and maps (m,M) onto $(-\infty,+\infty)$. Therefore, $G(y)=F\circ g^{-1}(y)$. for $m<y<M$.

An analogous result is obtained if g is assumed to be strictly decreasing.

Example: Let $a>0$ and $g(x)=ax+b$ for each x. We consider the r.v. $\eta=a\xi+b$. The solution of the equation $y=ax+g$ is $x_y=(y-b)/a$ for all $y\in R_1$. The d.f. G is therefore given by $G(y)=F(x_y)=F((y-b)/a)$ for all $y\in R_1$.

However, if $a<0$, then the sets $\{x\colon ax+b\leqslant y\}$ and $\{x\colon(y-b)/a\leqslant x\}$ are equal for each $y\in R_1$. Therefore, it follows from Theorem 1.1 that $P(a\xi+b\leqslant y)=P((y-b)/a\leqslant\xi)=P(\xi\in R_1)-P(\xi\leqslant(y-b)/a)+P(\xi=(y-b)/a)$. Hence, it follows from (4.1) and (5.8) that $P(\xi\in M_y)=1-F((y-b)/a-0)$, so that $G(y)=1-F((y-b)/a-0)$ for each $y\in R_1$.

To illustrate further, we choose $g(x)=x^2$ for each $x\in R_1$. Clearly g is not monotone. Once more, let F be the d.f. of ξ and denote the d.f. of $\eta=\xi^2$ by G. If $y<0$, then $\{x: x^2\leqslant y\}$ is the empty set so that $G(y)=0$ for $y<0$. For $y\geqslant 0$, the sets $\{x: x^2\leqslant y\}$ and $\{x: -\sqrt{y}\leqslant x\leqslant\sqrt{y}\}$ are equal. Therefore, $G(y)=P(-\sqrt{y}\leqslant\xi\leqslant\sqrt{y})=F(\sqrt{y})-F(-\sqrt{y}-0)$, for $y\geqslant 0$. In particular, if F is continuous, then

$$G(y)=\begin{cases}0, & y\leqslant 0\\ F(\sqrt{y})-F(-\sqrt{y}), & y>0.\end{cases} \tag{11.3}$$

As an additional example we set $g=F$, where F is a continuous, strictly increasing d.f. Let ξ be a r.v. with the d.f. F. Then F^{-1} exists and maps $(0,1)$ into $(-\infty,+\infty)$. Therefore, the d.f. of $\eta=F\circ\xi$ is given by $G(y)=F(F^{-1}(y))=y$ for $0<y<1$. Hence, η has a uniform distribution. If F is continuous and strictly monotone in a finite interval $[a,b]$ and $F(a)=0$ as well as $F(b)=1$, then an analogous argument leads to the same result. Therefore,

Theorem 11.1. *If ξ is a r.v. whose d.f. F satisfies the above assumptions, then $\eta=F\circ\xi$ is distributed according to a uniform distribution.*

Until now we have only considered mappings from R_n into R_1. Frequently, however, more general situations must be considered. Let g_i, $i=1,\dots,m$, be a Borel-measurable mappings from R_n into R_1. As we have shown, if $\xi=(\xi_1,\dots,\xi_n)$ is an n-dimensional r.v., then $\eta_i=g_i\circ\xi$ is a r.v. for each i. We form the m-dimensional r.v. $\eta=(\eta_1,\dots,\eta_m)$ and are interested in its joint distribution. This can be obtained formally in the following manner. If the sets $B_1,\dots,B_m$ belongs to $\mathfrak{B}_1$, then $P(\eta_1\in B_1,\dots,\eta_m\in B_m)$ is given by $P(g_1\circ\xi\in B_1,\dots,g_m\circ\xi\in B_m)$ and this latter expression is nothing more than $P(\xi\in g_1^{-1}(B_1),\dots,\xi\in g_m^{-1}(B_m))$, so that $P(\eta_1\in B_1,\dots,\eta_m\in B_m)$ is given by $P(\xi\in g_1^{-1}(B_1)\cap\cdots\cap g_m^{-1}(B_m))$. Important special cases of these general considerations will be treated in the following paragraphs.

12. The distribution of differentiable functions of random variables of the continuous type. Let ξ be a r.v. with d.f. F. Suppose that F is of the continuous type, so that it has a density f everywhere. Let the function g be defined and continuously differentiable on R_1; in addition, assume that g is one to one. Moreover, let $\inf\limits_{-\infty<x<\infty} g(x)=m$ and $\sup\limits_{-\infty<x<\infty} g(x)=M$, where $m=-\infty$ or $M=+\infty$ are permitted. The inverse function g^{-1} of g, defined over (m,M), is denoted by h. By means of g we define the r.v. $\eta=g\circ\xi$.

Theorem 12.1. *The random variable η possesses a probability density a.e. given by*

$$f(h(y))\,|h'(y)| \tag{12.1}$$

on $m<y<M$. The density of η vanishes for $y<m$ or $y>M$.

Proof. According to (11.1), the d.f. G of η is given by $P(\xi\in M_y)$ for all real y, where $M_y=g^{-1}((-\infty,y])$. From (6.2), this probability equals

$$\int_{M_y} f(x)\,dx. \tag{12.2}$$

Suppose that $m<y<M$. If we make the transformation $x=h(t)$ in this integral, then the set M_y is mapped onto the interval $(m,y]$. Therefore, we obtain $\int_m^y f(h(t))|h'(t)|\,dt$ in place of (12.2). This proves the non-trivial part of the Theorem; the statement about the density of η for $y<m$ or $y>M$ is obvious.

Theorem 12.1 can be generalized to multidimensional random variables in the following manner:

Theorem 12.2. *Suppose that $\xi=(\xi_1,\dots,\xi_n)$ is an n-dimensional r.v. possessing a d.f. of the continuous type. The distribution density, denoted by f, is assumed to vanish outside an open set $M_1\subseteq R_n$. Suppose that n functions g_i are defined and possess continuous partial derivatives with respect to each variable on M_1. Moreover, assume that their Jacobian is different from 0 on M_1 and that $g=(g_1,\dots,g_n)$ is a $1-1$ mapping from M_1 onto a set M_2 of R_n, which necessarily is open. The inverse mapping is denoted by $(h_1,\dots,h_n)$. If n random variables are defined by setting $\eta_i=g_i\circ\xi$, then the density of the n-dimensional r.v. $\eta=(\eta_1,\dots,\eta_n)$ is given by*

$$f(h_1(y),\dots,h_n(y))\left|\frac{\partial(h_1,\dots,h_n)}{\partial(y_1,\dots,y_n)}\right|, \tag{12.3}$$

for all $y=(y_1,\dots,y_n)\in M_2$, while the density is identically zero outside of M_2.

The *proof* is a result of the transformation theory of multiple integrals (cf. Theorem IX) and an argument analogous to that of Theorem 12.1.

We will apply this theorem many times; hence, it will be convenient to have a brief statement of its meaning. In order to determine the density of $(\eta_1,\dots,\eta_n)$ from the density $(x_1,\dots,x_n)\to f(x_1,\dots,x_n)$ of $(\xi_1,\dots,\xi_n)$, the transformation $y_i=g_i(x_1,\dots,x_n)$ is applied to f. The result (12.3) of this transformation yields the density of $(\eta_1,\dots,\eta_n)$, provided it does not vanish.

13. Independent random variables. Let $\xi^{(1)},\ldots,\xi^{(k)}$ be arbitrary, not necessarily one-dimensional random variables. Their d.f.'s are denoted by $F^{(1)},\ldots,F^{(k)}$, respectively. Further, we consider the r.v. $\xi=(\xi^{(1)},\ldots,\xi^{(k)})$ and denote its d.f. by F.

We give the following

Definition. $\xi^{(1)},\ldots,\xi^{(k)}$ are said to be *stochastically independent* or, more briefly, *independent* if

$$F(x)=\prod_{i=1}^{k} F^{(i)}(x^{(i)}) \tag{13.1}$$

holds, for all $x=(x^{(1)},\ldots,x^{(k)})$. [13]

Therefore, when $\xi^{(1)},\ldots,\xi^{(k)}$ are independent,

$$P(\xi\leqslant x)=P(\xi^{(1)}\leqslant x^{(1)})\cdots P(\xi^{(k)}\leqslant x^{(k)}) \tag{13.2}$$

for all x. It is immediately verified that the function F defined by the right side of equation (13.1) is a d.f. The product of finitely many d.f.'s satisfies conditions 1 m.—4 m., p. 38—39, and so represents a d.f. As a consequence of (13.2) and the extension theorem for measures (Theorem III), we have the equation

$$P(\xi\in B)=P(\xi^{(1)}\in B^{(1)})\ldots P(\xi^{(k)}\in B^{(k)}) \tag{13.3}$$

for all Borel sets of the form $B=B^{(1)}\times\cdots\times B^{(k)}$, where $B^{(i)}$ is a Borel set of an R_i having the same dimension as the r.v. $\xi^{(i)}$, $1\leqslant i\leqslant k$.

Conversely, it is clear that (13.2) and therefore (13.1) follow from (13.3). If the d.f.'s $F^{(1)},\ldots,F^{(k)}$ are of the continuous type, then by differentiation of (13.1) we obtain, using easily understood notation, the distribution density:

$$f(x)=f^{(1)}(x^{(1)})\,f^{(2)}(x^{(2)})\ldots f^{(k)}(x^{(k)}), \tag{13.4}$$

a.e. On the other hand, if the densities of d.f.'s of the continuous type satisfy the relation in (13.4), then by integration and an application of Fubini's theorem we obtain

$$\int_{-\infty}^{x} f(y)\,dy=\int_{-\infty}^{x^{(1)}} f^{(1)}(y^{(1)})\,dy^{(1)}\int_{-\infty}^{x^{(2)}} f^{(2)}(y^{(2)})\,dy^{(2)}\ldots\int_{-\infty}^{x^{(k)}} f^{(k)}(y^{(k)})\,dy^{(k)},$$

so that the corresponding d.f.'s satisfy (13.1). Therefore, if they are of the continuous type, the random variables $\xi^{(1)},\ldots,\xi^{(k)}$ are independent iff (13.4) holds. In what follows we will very frequently consider independent random variables. The d.f. of their joint distribution is then always given by a product of the form (13.1).

---

[13] This definition can easily be extended to infinitely many random variables. Cf. the remark following (2.1).

It is clear from (13.3) that there is a very close connection between independence and the concept of a product measure. We will pursue this briefly. Let $F_1,\dots,F_n$ be any one-dimensional d.f.'s. According to Theorem 5.2, F_i, $1 \leqslant i \leqslant n$, can be considered to be the d.f. of a r.v. This raises the question as to whether a probability field can be constructed over which random variables $\xi_1,\dots,\xi_n$ can be defined in such a way that they are independent in the sense of Definition (13.2) and the d.f. of ξ_i is F_i, $1 \leqslant i \leqslant n$. This is indeed the case. Let P_i, for $i=1,\dots,n$, be the probability measure corresponding to F_i. We consider the probability field $(R_n, \mathfrak{B}_n, P_1 \times \cdots \times P_n)$. Further, for each i, $1 \leqslant i \leqslant n$, we define the mapping ξ_i of R_n into R_1 by $(x_1,\dots,x_i,\dots,x_n) \to x_i$. This mapping is Borel-measurable, i.e., a r.v. The d.f. of ξ_i is determined by $\{(x_1,\dots,x_n)\colon \xi_i(x_1,\dots,x_n) \leqslant y\}$ for each $y \in R_1$. Therefore, since $\xi_i^{-1}((-\infty, y])$ is of the form $R_1 \times \cdots \times (-\infty, y] \times \cdots \times R_n$, the $P_1 \times \cdots \times P_n$-measure of this set is just $F_i(y)$. It follows that the d.f. of the r.v. $\xi = (\xi_1,\dots,\xi_n)$ is given by $\prod_{i=1}^{n} F_i$, so that, according to (13.3), $\xi_1,\dots,\xi_n$ are independent relative to the probability field

$$(R_n, \mathfrak{B}_n, P_1 \times \cdots \times P_n).$$

This construction can be extended to infinitely many random variables. If an arbitrary sequence $\xi_1, \xi_2, \dots$ of independent random variables is given, then one obtains the probability field consisting of R_∞, i.e. the set of all sequences $x_1, x_2, \dots$ of real numbers, the σ-algebra $\mathfrak{B}_\infty = \bigotimes_{i=1}^{\infty} \mathfrak{B}^{(i)}$, with $\mathfrak{B}^{(i)} = \mathfrak{B}_1$ for $i = 1, 2, \dots$, and the probability measure $\prod_{i=1}^{\infty} P_{\xi_i} = P_{\xi_1} \times P_{\xi_2} \times \cdots$ over this σ-algebra.

The following is a simple generalization of (13.4): If P_{ξ_i}, $1 \leqslant i \leqslant n$, possesses a R.-N.-density f_i relative to a measure μ_i, then a R.-N.-density of $P_{\xi_1} \times \cdots \times P_{\xi_n}$ relative to $\mu_1 \times \cdots \times \mu_n$ is given by $\prod_{i=1}^{n} f_i(x_i)$ for all $(x_1,\dots,x_n) \in R_n$.

Theorem 13.1. *Suppose that $\xi^{(i)}$, $i=1,\dots,k$, are independent random variables of dimension n_i and that g_i, $i=1,\dots,k$, are functions from R_{n_i} into R_1. Then the k random variables $g_i \circ \xi^{(i)}$ are also independent.*

According to (13.3), for the *proof* it is sufficient to show the following: If $B_i \in \mathfrak{B}_1$, $1 \leqslant i \leqslant k$, then

$$P(g_1(\xi^{(1)}) \in B_1, \dots, g_k(\xi^{(k)}) \in B_k) = \prod_{i=1}^{k} P(g_i(\xi^{(i)}) \in B_i)$$

is equivalent to

$$P(\xi^{(1)}\in g_1^{-1}(B_1),\dots,\xi^{(k)}\in g_k^{-1}(B_k)) = \prod_{i=1}^{k} P(\xi^{(i)}\in g_i^{-1}(B_i)).$$

However, since the $\xi^{(i)}$ are independent and the inverse images $g_i^{-1}(B_i)$ belong to $\mathfrak{B}_{n_i}$, this last equation is correct because of (13.3).

14. The concept of a marginal distribution. Let $\xi=(\xi_1,\dots,\xi_n)$ be an n-dimensional r.v. over the probability field $(R,\mathcal{S},P)$ and F be the d.f. of ξ. We now investigate the joint distribution of an arbitrary subset of $(\xi_1,\dots,\xi_n)$, e.g. the distribution of $(\xi_1,\dots,\xi_k)$, $1\leqslant k\leqslant n$. The sets

$$M_k=\{\omega\colon (\xi_1(\omega),\dots,\xi_k(\omega))\in B\}$$

and

$$N_{k,n}=\{\omega\colon (\xi_1(\omega),\dots,\xi_k(\omega))\in B, (\xi_{k+1}(\omega),\dots,\xi_n(\omega))\in R_{n-k}\}$$

for $B\in\mathfrak{B}_k$ are obviously identical and so their probabilities are equal. However, as a consequence of (8.1), $P_{(\xi_1,\dots,\xi_k)}(B)=P(M_k)$ and $P_\xi(B\times R_{n-k})=P(N_{k,n})$; therefore

$$P_{(\xi_1,\dots,\xi_k)}(B)=P_\xi(B\times R_{n-k}), \tag{14.1}$$

for all $B\in\mathfrak{B}_k$. If to each $B\in\mathfrak{B}_k$ we associate the real number $P_\xi(B\times R_{n-k})$, then we again obtain a probability distribution which we refer to as the projection of P_ξ onto R_k or as the marginal distribution of $(\xi_1,\dots,\xi_k)$[14]. In particular, if B is of the form $(-\infty,x_1]\times\cdots\times(-\infty,x_k]=I_{x_1,\dots,x_k}$ and G is the d.f. of $(\xi_1,\dots,\xi_k)$, then as a result of (14.1)

$$G(x_1,\dots,x_k)=P_\xi(I_{x_1,\dots,x_k}\times R_{n-k}) \tag{14.2}$$

for all $(x_1,\dots,x_k)\in R_k$.

Thus, we have

Theorem 14.1. *For each* $(x_1,\dots,x_k)\in R_k$,

$$G(x_1,\dots,x_k)=\lim_{x_{k+1}\to\infty,\dots,x_n\to\infty} F(x_1,\dots,x_k,x_{k+1},\dots,x_n). \tag{14.3}$$

In particular, if F possesses a density f, then G has a density g a.e. in R_k and for all such $(x_1,\dots,x_k)$

$$g(x_1,\dots,x_k)=\int_{-\infty}^{+\infty}\dots\int_{-\infty}^{+\infty} f(x_1,\dots,x_k,t_{k+1},\dots,t_n)\,dt_{k+1}\dots dt_n. \tag{14.4}$$

For the *proof*, it is sufficient to show that the right side of (14.2) is the same as the right side of (14.3). This follows in exactly the same way as property 4m. of a d.f. We omit the proof.

---

[14] A better terminology would be marginal distribution of $(\xi_1,\dots,\xi_n)$ relative to $(\xi_1,\dots,\xi_k)$, but the expression employed here has established itself in the literature.

We write $F(x_1, \ldots, x_k, \infty, \ldots, \infty)$ for the right side of (14.3). In this context we shall refer to the left side of (14.3) as the marginal d.f. and the left side of (14.4) as the marginal distribution density. If $G_1, \ldots, G_k$ are d.f.'s of $\xi_1, \ldots, \xi_k$, then as is easily seen, $G_1 \ldots G_k$ is also a d.f., but in general $G \neq G_1 \cdots G_k$. However, if $\xi_1, \ldots, \xi_n$ are independent random variables with d.f.'s $F_1, \ldots, F_n$, so that the joint d.f. is given by $F = F_1 \cdots F_n$, then it follows that the marginal d.f. G of $(\xi_1, \ldots, \xi_k)$ is given by

$$G(x_1, \ldots, x_k) = \prod_{i=1}^{k} F(x_i) = F(x_1, \ldots, x_k, \infty, \ldots, \infty),$$

for all $(x_1, \ldots, x_k) \in R_k$.

The situation is analogous for the marginal distribution density.

15. The interpretation of a probability distribution as a mass distribution. Many definitions and theorems in probability become clear when a probability distribution on R_n is viewed as a distribution of mass over R_n. This is accomplished by associating the mass $P(B)$ with each $B \in \mathfrak{B}_n$. The total mass in R_n is then 1. If a distribution density f exists, it corresponds to a mass density. A discrete mass distribution corresponds to a discrete probability distribution. If the d.f.'s G and F, defined in **14** and related by (14.3), are considered, then in terms of the mass interpretation we say that $G(x_1, \ldots, x_k)$, for each $(x_1, \ldots, x_k) \in R_k$, is obtained by projecting the mass $F(x_1, \ldots, x_k, \infty, \ldots, \infty)$ onto the k-dimensional region $I_{x_1 \ldots x_k}$. This justifies the use of the terminology marginal distribution. For an illustration when $n=3$, $k=2$, see Fig. 4.

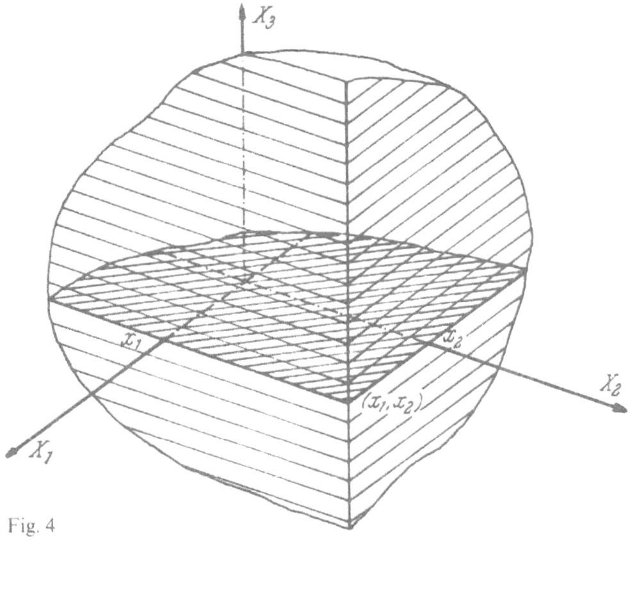

Fig. 4

16. Expectation. Let ξ be a r.v. over the probability space $(R, \mathcal{S}, P)$. We now assume that ξ is integrable with respect to P and give the following

Definition. $\int_R \xi(\omega)\,dP(\omega)$ is called *the expectation of the r.v.* ξ. The expectation shall be denoted by $E(\xi)$, or, when the probability measure must be in evidence, by $E(\xi; P)$.

Before we present several important properties of expectations, we mention a useful theorem which shows that the expectation of ξ can be obtained by means of the probability distribution P_ξ as indicated by equation (16.2) below.

Theorem 16.1. *If g is a measurable function defined over R_n and ξ is an n-dimensional r.v., then*

$$\int_R (g\circ\xi)\,dP = \int_{R_n} g\,dP_\xi, \tag{16.1}$$

where the existence of either integral implies the existence of the other.

In particular, if ξ is a one-dimensional r.v. and g is the identity mapping, then

$$\int_R \xi\,dP = \int_{-\infty}^{+\infty} x\,dP_\xi(x). \tag{16.2}$$

This theorem follows easily from the definition of the integral by first proving it for indicator functions, then for linear combinations of such functions and finally for general functions g. Thus, Theorem 16.1 states: To calculate the expectation of a function g of an n-dimensional random variable ξ, it is sufficient to consider $\int_{-\infty}^{+\infty} g\,dP_\xi$.

If F is the d.f. of ξ, then we also write $\int_{R_n} g\,dF$ in place of $\int_{R_n} g\,dP$ (cf. p. 11).

Theorem 16.2. *If the probability distribution of a r.v. ξ possesses a density f, then*

$$E(\xi) = \int_{-\infty}^{+\infty} x f(x)\,dx. \tag{16.3}$$

If ξ is an n-dimensional r.v. and g is a measurable function from R_n into R_1 and ξ has a density f, then the expectation of $g(\xi)$ is given by $\int_{R_n} g(x) f(x)\,dx$ (provided it exists).

It follows immediately from the definition of expectation that

$$E(1) = 1, \tag{16.4}$$

i.e., the expectation of the random variable ξ, which is equal to the constant 1, always exists and has the value one.

From the properties of an integral we have

Theorem 16.3. *Let g and h be functions over R_n and ξ be an n-dimensional random variable, $n \geqslant 1$. Suppose that the expectations of $g \circ \xi$ and $h \circ \xi$ exist. Then for arbitrary real numbers γ_1, γ_2*

$$E(\gamma_1 g \circ \xi + \gamma_2 h \circ \xi) = \gamma_1 E(g \circ \xi) + \gamma_2 E(h \circ \xi).$$

The following important theorem is based on the independence of random variables:

Theorem 16.4. *Let ξ_1, ξ_2 be independent random variables of arbitrary dimension and $g \circ \xi_1$ and $h \circ \xi_2$ functions of these random variables. Then*

$$E((g \circ \xi_1) \cdot (h \circ \xi_2)) = E(g \circ \xi_1) E(h \circ \xi_2), \tag{16.5}$$

where the existence of the left side is implied by that of the right side.

For the *proof* one has only to rely on the definition of independence and the properties of product measures. If P_i is the probability measure of ξ_i, $i=1,2$, then according to **13** the probability measure of (ξ_1, ξ_2) is given by $P_1 \times P_2$. The existence of $\int_{-\infty}^{+\infty} |g(x_1)| dP_1(x_1) \int_{-\infty}^{+\infty} |h(x_2)| dP_2(x_2)$ is a consequence of the existence of $E(g \circ \xi_1) E(h \circ \xi_2) = \int_{-\infty}^{+\infty} g(x_1) dP_1(x_1) \cdot \int_{-\infty}^{+\infty} h(x_2) dP_2(x_2)$. Therefore, we can apply Fubini's Theorem and obtain $E(g \circ \xi_1) E(h \circ \xi_2) = \int_{-\infty}^{+\infty} g(x_1) h(x_2) dP_1 \times P_2(x)$, where $x = (x_1, x_2)$. But this is just the left side of (16.5).

17. Mean value, variance, moments. We consider an n-dimensional random variable $\xi = (\xi_1, \ldots, \xi_n)$ and assume that all expectations occurring in the following definitions exist.

Definition. $E(\xi_i) = a_i$ is called the *mean value* of ξ_i, $i = 1, \ldots, n$. In general, $E(\xi_i^k)$, $k = 1, 2, \ldots$, is called the kth *moment* of ξ_i; more precisely the kth *moment relative to zero.*

The kth moment can also be defined for $k=0$; then $E(\xi_i^0) = E(1) = 1$ from (16.4). For $k=1$, the kth moment agrees with the mean value. If γ is an arbitrary real number, then $E[(\xi_i - \gamma)^k)]$ is the kth moment of ξ_i relative to γ.

The second moment of ξ_i relative to a_i is of special significance.

Definition. $E[(\xi_i - a_i)^2] = \sigma_i^2$ is called the *variance* of ξ_i.

The variance has an important *minimal property*:

Theorem 17.1. $E[(\xi_i-\gamma)^2] \geqslant E[(\xi_i-a_i)^2]$ *for each real number* γ *with equality holding only for* $\gamma = a_i$.

For the *proof* we write $E[(\xi_i-\gamma)^2] = E[(\xi_i - a_i - \gamma + a_i)^2]$ and apply Theorem 16.3. Hence, taking into account $E(\xi_i - a_i) = 0$, it follows that

$$E[(\xi_i-\gamma)^2] = E[(\xi_i-a_i)^2] + (\gamma - a_i)^2. \tag{17.1}$$

The assertion is a result of this identity.

The connection between σ_i^2 and the second moment is given in the following theorem.[15]

Theorem 17.2. $\sigma_i^2 = E(\xi_i^2) - a_i^2$.

The *proof* consists in setting $\gamma = 0$ in (17.1).

Definition. By the *covariance* of random variables $\xi_i, \xi_j, i \neq j$, we mean

$$\sigma_{ij} = E[(\xi_i - a_i)(\xi_j - a_j)].$$

Obviously, $\sigma_{ij} = \sigma_{ji}$. Sometimes it is convenient to apply this notation to the variance of ξ_i and denote it by σ_{ii}.

Theorem 17.3. *The existence of variances* σ_{ii} *and* $\sigma_{jj}, i \neq j$, *implies the existence of the covariance of* ξ_i *and* ξ_j.

The *proof* utilizes Schwarz's inequality. (Cf. p. 10.)

We now give the definition of the *correlation coefficient* ρ_{ij}: Let $\sigma_{ii}, \sigma_{jj} > 0$, then $\rho_{ij} = \sigma_{ij}/\sqrt{\sigma_{ii}\sigma_{jj}}, i \neq j$, is called the correlation coefficient of ξ_i, ξ_j. Obviously, $\rho_{ij} = \rho_{ji}$. ξ_i and ξ_j are said to be uncorrelated if $\rho_{ij} = 0$.

Theorem 17.4. *If* ξ_1 *and* ξ_2 *are independent random variables whose variance exist, then they are uncorrelated.*

The *proof* is an easy application of (16.5), that is,

$$E[(\xi_1 - a_1)\cdot(\xi_2 - a_2)] = E(\xi_1 - a_1)\cdot E(\xi_2 - a_2) = 0.$$

It is not difficult to construct examples which show that the converse of this theorem does not hold.

It is easy to show that the matrix of all covariances and variances of a random variable $\xi = (\xi_1, \ldots, \xi_n)$, $n \geqslant 2$, which we call the *covariance matrix* is positive semidefinite. More precisely, we have

Theorem 17.5. *Let* $\xi = (\xi_1, \ldots, \xi_n)$ *be a random variable; its covariance matrix is defined by* $(\sigma_{ij})_{1n}^{1n}$. *This matrix is positive semi-definite.*

---

[15] See also Theorem 17.7.

Proof. For arbitrary real $u_1, \dots, u_n$

$$\sum_{i,j=i}^{n} u_i u_j \sigma_{ij} = E[[u_1(\xi_1 - E(\xi_1)) + \cdots + u_n(\xi_n - E(\xi_n))]^2] \geqslant 0 .$$

We now prove

Theorem 17.6. *Suppose that* $\xi = (\xi_1, \dots, \xi_n)$ *is a random variable with a distribution for which all mean values and variances exist. If the determinant of the covariance matrix vanishes, then with probability one there is a linear relation between the variables, i.e. there exists an n-tuple* $u_1^{(0)}, \dots, u_n^{(0)}$ *of real numbers, such that*

$$P(u_1^{(0)} \xi_1 + \cdots + u_n^{(0)} \xi_n = \text{constant}) = 1 . \tag{17.2}$$

Proof. Let $a = (a_1, \dots, a_n)$ be the mean value vector of ξ. Since by Theorem 17.5 the covariance matrix is positive semidefinite, it follows from the vanishing of $|\sigma_{ij}|_{1n}^{1n}$ that there exists at least one vector $u^{(0)\prime} = (u_1^{(0)}, \dots, u_n^{(0)})$ with $u^{(0)\prime} \cdot u^{(0)} \neq 0$, such that $\sum_{i,j=1}^{n} \sigma_{ij} u_i^{(0)} u_j^{(0)} = 0$. Then $E\left[\left(\sum_{i=1}^{n} u_i^{(0)}(\xi_i - a_i)\right)^2\right] = 0$. From this it follows that $\sum_{i=1}^{n} u_i^{(0)}(\xi_i - a_i) = 0$ P-a.e. (cf. p. 9) and this is an equivalent form of the assertion.

The covariance satisfies a relation analogous to the Theorem 17.2:

$$\sigma_{ij} = E(\xi_i \xi_j) - a_i a_j , \tag{17.3}$$

as is easily checked.

If the covariance matrix $(\sigma_{ij})_{1n}^{1n}$ of an n-dimensional random variable ξ has a non-vanishing determinant,[16] then the inverse matrix $(\tau_{ij})_{1n}^{1n}$ exists. The set of all real n-tuples $(\lambda_1, \dots, \lambda_n)$ with $\sum_{i,j=1}^{n} \lambda_i \lambda_j \tau_{ij} = n+2$ is called the concentration ellipsoid of ξ induced by the probability distribution P_ξ. The volumn of this ellipsoid is

$$(\Gamma(n/2+1))^{-1}(n+2)^{n/2}\, \pi^{n/2} \sqrt{|\sigma_{ij}|_{1n}^{1n}} .$$

The covariance matrix of the uniform distribution over this ellipsoid is the same as the covariance matrix of P_ξ.

$E(\xi_i \xi_j)$ is called the second mixed moment of ξ_i and ξ_j. An obvious generalization of this concept to several variables and higher order moments is obtained from $E(\xi_{i_1}^{r_1} \dots \xi_{i_l}^{r_l})$, where $r_1, \dots, r_l$ are positive integers and $\xi_{i_1}, \dots, \xi_{i_l}$ represent variables chosen from $\xi_1, \dots, \xi_n$. We remark again that $E(|\xi_{i_1}|^{r_1} \dots |\xi_{i_l}|^{r_l})$ exists along with $E(\xi_{i_1}^{r_1} \dots \xi_{i_l}^{r_l})$ and refer to this as an absolute moment. In the definition of absolute moments

[16] We will also refer to $(\sigma_{ij})_{1n}^{1n}$ as the covariance matrix of P_ξ.

it is no longer necessary that the $r_1, \ldots, r_l$ be integers, indeed, they can be arbitrary positive real numbers.

We now make a few remarks about the moments of one-dimensional random variables. First we show

Theorem 17.7. *If the* k^{th} *moment, k a positive integer, of a r.v.* ξ *exists, then all* j^{th} *moments,* $0 \leqslant j \leqslant k$, *exist.*

Proof. $\int_{-\infty}^{+\infty} x^k \, dP_\xi(x)$ exists. Now, if $|x| \leqslant 1$, then $|x^j| \leqslant 1$ and if $|x| > 1$, then $|x^j| < |x^k|$. This demonstrates that the function $x \to x^j$, $0 \leqslant j \leqslant k-1$ is P_ξ-integrable (cf. Theorem IV).

It follows from Theorem 17.7 that $E[(\xi+\gamma)^k]$ exists for any real number γ, if the kth moment of ξ exists, k a positive integer. This is obvious from

$$E[(\xi+\gamma)^k] = \sum_{l=0}^{k} \binom{k}{l} E(\xi^l)\gamma^{k-l}. \tag{17.4}$$

We now define two quantities connected with the 3rd, resp., 4th, moments; however, these do not have the same significance in mathematical statistics today as they had formerly.

Definition. Let a be the mean value and σ^2 be the variance of a r.v. ξ. The quantity

$$v = E[(\xi - a)^3]/\sigma^3, \qquad \sigma > 0$$

is called the *skewness of the distribution of* ξ.

It is easy to show that v is invariant with respect to translations of the origin and changes in scale of the r.v. ξ. That is, if we consider the r.v. $\eta = (\xi - b)/c$, where b, c are real with $c > 0$, it can be shown that the skewness of η agrees with v. This follows from a repeated application of Theorem 16.3.

We give further

Definition. *The distribution of a r.v.* ξ with d.f. F is said to be *symmetric* with respect to the real number γ if $F(\gamma - x) = 1 - F(\gamma + x - 0)$ for each real x.

This means that a pair of intervals located symmetrically with respect to γ have the same P_ξ-measure. Hence, it follows more generally that if $B \in \mathfrak{B}_1$ and $B_s = \{y: y = 2\gamma - x, x \in B\}$, then $P_\xi(B) = P_\xi(B_s)$.

If F has a density f, then for a symmetric distribution we have

$$f(\gamma - x) = f(\gamma + x) \tag{17.5}$$

for all $x \in R_1$ (or more exactly for almost all x in R_1).

If F is an arbitrary d.f., then we refer to each solution of the equation $F(x)=1/2$ as *median*, each solution of $F(x)=p$, $0<p<1$, as a *p-quantile*. Clearly, these need not exist. If F is symmetric with respect to γ and continuous at γ, then γ is a median of F.

It follows from the definition of skewness, that it vanishes for each distribution which is symmetric with respect to its mean value. To see this, we have only to note that

$$\begin{aligned} E[(\xi-a)^3] &= \int_{-\infty}^{+\infty} (x-a)^3\,dF(x) \\ &= \int_{\{x:x<a\}} (x-a)^3\,dF + \int_{\{x:x>a\}} (x-a)^3\,dF + \int_{\{x:x=a\}} 0\,dF = 0\,.\end{aligned}$$ [17]

We now explain what is meant by the excess of the distribution of a r.v.

Definition. The *excess* of a distribution with mean value a and variance σ^2 is given by

$$\varepsilon = (1/\sigma^4)E[(\xi-a)^4]-3\,. \tag{17.6}$$

Again, it can easily be shown that ε is invariant under translations and changes in scale of the random variable ξ. (Cf. p. 76.)

18. Conditional Probability Distributions. We have introduced conditional probabilities according to (2.2) where it was essential that $P(A)\neq 0$. In many practical problems (see **21** and **VI**) one considers hypotheses A which are of the form $\{\omega: \xi(\omega)=x\}$, where x is a real number and ξ is a r.v. Intuitively speaking, we have observed a value x of some outcome and would like to make a probabilistic statement based on this observation. If the set $\{\omega: \xi(\omega)=x\}$ has non-zero probability, there are no difficulties. In general, however, this probability will be zero (see Theorem 5.1). In spite of this, most people would still tend to assign a meaning to the conditional probability in this case. We want to consider briefly the extent to which this can be justified within the framework of mathematical probability theory.

Let $A\in\mathcal{S}$ and $B\in\mathcal{S}$ with $P(B)\neq 0$ and $P(R-B)\neq 0$ and consider the σ-algebra $\mathfrak{S}$ generated by B which consists of the sets $\emptyset$, B, $R-B$ and R. We disregard the sets $\emptyset$ and R and define

$$P(A\,|\,\mathfrak{S})=P(A\,|\,B)c_B+P(A\,|\,R-B)c_{R-B}\,, \tag{18.1}$$

$P(A\,|\,\mathfrak{S})$ is called the *conditional probability of* A *w.r.t. the* σ*-algebra* $\mathfrak{S}$.

[17] Obviously, each moment of odd order $E[(\xi-a)^{2n+1}]$, $n\geqslant 0$, of a distribution which is symmetric with respect to a vanishes whenever it exists.

$P(A|\mathfrak{S})$ is thus defined as a function over R (and not, as one might think, as a function over $\mathfrak{S}$) and is trivially $\mathfrak{S}$-measurable, hence $\mathscr{S}$-measurable.

Since $\int_B c_B dP = P(B)$ and $\int_B c_{R-B} dP = 0$, we have from (18.1): $\int_B P(A|\mathfrak{S})dP = P(A|B)P(B)$. Hence also

$$P(A\cap B) = \int_B P(A|\mathfrak{S})dP \tag{18.2}$$

and

$$P(A\cap(R-B)) = \int_{R-B} P(A|\mathfrak{S})dP. \tag{18.3}$$

These arguments can be immediately extended to countably many disjoint sets $B_i \in \mathscr{S}$, $i=1,2,\ldots$, with $\bigcup_{i=1}^{\infty} B_i = R$, $P(B_i)\neq 0$ for all i. Denoting the σ-algebra generated by the B_i again by $\mathfrak{S}$, we define for a given $A\in\mathscr{S}$ a mapping $P(A|\mathfrak{S})$ from R into the real numbers by means of

$$P(A|\mathfrak{S}) = \sum_{i=1}^{\infty} P(A|B_i)c_{B_i}. \tag{18.4}$$

For each set $B\in\mathfrak{S}$ (B can in fact always be written as the union of suitable B_i's) we again have

$$P(A\cap B) = \int_B P(A|\mathfrak{S})dP \tag{18.5}$$

which is a generalization of (18.2) and (18.3). Naturally, $P(A|\mathfrak{S})$ is again $\mathfrak{S}$-measurable.

Now a decisive step follows. If we disregard the condition $P(B_i)\neq 0$ for all i, then certain $P(A|B_i)$ become meaningless in definition (18.4). However, the union N of all B_i with $P(B_i)=0$ is itself a P-null set since it is the union of at most countably many P-null sets. But this means that (18.4) is still usable as definition of $P(A|\mathfrak{S})$ provided that one dispenses with the definition of $P(A|\mathfrak{S})$ on a P-null set. Relation (18.5) still holds without change since the value of the integral does not depend on the definition of $P(A|\mathfrak{S})$ on a null set. Likewise, $P(A|\mathfrak{S})$ is $\mathfrak{S}$-measurable if we define $P(A|\mathfrak{S})$ appropriately on the exceptional null set N.

On the other hand, for an arbitrary sub-σ-algebra $\mathfrak{S}$ of $\mathscr{S}$, we can no longer use (18.4) to define $P(A|\mathfrak{S})$ over R for each $A\in\mathscr{S}$. However, by means of the Radon-Nikodym theorem we can demonstrate the existence of a $\mathfrak{S}$-measurable function $P(A|\mathfrak{S})$ which fulfills (18.5) for all $B\in\mathfrak{S}$. We need only that $P(B)=0$ always implies $P(A\cap B)=0$, so that the set function $B\to P(A\cap B)$ is absolutely continuous over $\mathfrak{S}$

w.r.t. P. However, $P(A|\mathfrak{S})$ is uniquely determined only up to P-null sets from $\mathfrak{S}$.[18] These null sets depend in general on the choice of the $A\in\mathscr{S}$. This circumstance causes certain difficulties,[19] in particular, when we want to investigate the question whether or not the mapping $A\to P(A|\mathfrak{S})(\omega)$ is a probability over $\mathscr{S}$ for each $\omega\in R$. In general, it is not (on the other hand, see p. 29, where we showed that $A\to P(A|B)$ with $P(B)\neq 0$ is always a probability measure over $\mathscr{S}$). However, we can prove by means of (18.5) that $0\leqslant P(A|\mathfrak{S})\leqslant 1$ and $P(R|\mathfrak{S})=1$ hold for each $A\in\mathscr{S}$ P-a.e. Furthermore, for countably many pairwise disjoint sets A_i, we have $P\left(\bigcup_{i=1}^{\infty} A_i|\mathfrak{S}\right)=\sum_{i=1}^{\infty} P(A_i|\mathfrak{S})$ up to P-null sets. Indeed, only countably many exceptional P-null sets play a role in this relation and their union is again a P-null set.

To return from these general considerations to our original problem, we assume that ξ is a one-dimensional r.v. By definition, each $B\in\mathfrak{B}_1$ possesses a measurable inverse image under ξ and the totality of the measurable inverse images is again a σ-algebra, which we will denote by $\xi^{-1}(\mathfrak{B}_1)$. $\xi^{-1}(\mathfrak{B}_1)$ is naturally a sub-σ-algebra of $\mathscr{S}$ so that a conditional probability $P(A|\xi^{-1}(B_1))$ is defined for each $A\in\mathscr{S}$. One also writes $P(A|\xi)$ for this mapping which is also called the probability of A under the condition that ξ is given. The notation as well as the name can be justified. Indeed, for all ω with $\xi(\omega)=x$, this mapping has the same value, which we write as $P(A|\xi=x)$. Furthermore, if P denotes the probability distribution of ξ, then one sees immediately that the set function $B_1\to P(A\cap\xi^{-1}(B_1))$ defined over $\mathfrak{B}_1$ is absolutely continuous w.r.t. P_ξ since $P(A\cap\xi^{-1}(B_1))\leqslant P(\xi^{-1}(B_1))=P_\xi(B_1)$. But then we can apply the Radon-Nikodym theorem: there exists a uniquely defined (up to P_ξ-null sets), Borel-measurable mapping $P_{R_1}(A|\xi)$ from R_1 into R_1 which fulfills

$$P(A\cap\xi^{-1}(B_1))=\int_{B_1} P_{R_1}(A|\xi)\,dP_\xi \tag{18.6}$$

for all $B_1\in\mathfrak{B}_1$. Using (16.1) we can immediately see the connection between $P(A|\xi)$ and $P_{R_1}(A|\xi)$: the mappings $P(A|\xi)$ and $P_{R_1}(A|\xi)\circ\xi$ are identical (up to null sets). Thus, the values of $P(A|\xi)$ actually depend only on the real values assumed by ξ.

---

[18] We also say: all versions of $P(A|\mathfrak{S})$ differ from each other only on P-null sets.

[19] See for this and related problems D.H. Blackwell, Proceedings of the Third Berkeley Symposium on Mathematical Statistics and Probability 1954—1955 Vol. II, pp. 1—6, University of California Press, Berkeley and Los Angeles (1956) and D.H. Blackwell and C. Ryll-Nardzewski, Ann. Math. Statist. 34, 223—225 (1963).

Let η be a r.v. and $A_y = \eta^{-1}((-\infty, y])$ for each real y. Then $P(A_y|B)$ is defined according to (2.2) for each $B \in \mathscr{S}$ with $P(B) \neq 0$. The function F_B defined for each real y by means of

$$F_B(y) = P(A_y|B) \tag{18.7}$$

is called the conditional distribution function of η under the condition B. It is easy to see that F_B satisfies Conditions 1.—4., p. 31 for each fixed B. If ξ is another r.v., then the previous arguments show that, for each fixed $y \in R_1$, one can define the mapping $P_{R_1}(A_y|\xi)$ which satisfies (18.6) for $A = A_y$. This suggests introducing "a conditional distribution of η under the condition $\xi = z$". The idea here is to define a function on R_2 which maps (y, z) into, say, $F(y|z)$ in such a way that $y \to F(y|z)$ is a distribution function and $z \to F(y|z)$ is a Borel-measurable function satisfying $P(A_y \cap \xi^{-1}(B_1)) = \int_{B_1} F(y|z)\, dP_\xi(z)$ for all $y \in R_1$ and all $B_1 \in \mathfrak{B}_1$. A natural choice for this is the definition $F(y|z) = P_{R_1}(A_y|\xi)(z)$ for all $(y, z) \in R_2$.[20]

However, we still have to deal with the previously mentioned difficulties (p. 57) connected with the fact that the mapping $P_{R_1}(A_y|\xi)$ is, for each fixed y, only uniquely defined up to P_ξ-null sets, and it is by no means clear that one can choose such versions of $P_{R_1}(A_y|\xi)$ that a distribution function is defined for each z (possibly up to a fixed null set) by $y \to P_{R_1}(A_y|\xi)(z)$. However, it turns out that this is actually possible. To prove this, one uses the fact that for the set of all rational numbers y_i in R_1, for example, which is a countable set, the definition is possible up to a fixed null set N. Indeed, to each y_i there corresponds an exceptional set of elements z and the union of all these null sets is again a null set N. For irrational y, one uses the fact that the rationals are dense in R_1, i.e., each irrational y is the limit of a sequence of rational $y_i > y$, and then defines $F(y|z) = \inf_{y_i > y} P_{R_1}(A_{y_i}|\xi)(z)$ for all $z \in R_1 - N$. If $z \in N$ then one may define $F(y|z) = P_{R_1}(A_y|\xi)(z) = P(A_y)$ for $y \in R_1$. It is easy to see that the map $(y, z) \to F(y|z)$ possesses the desired properties. $z \to F(y|z)$ is then a suitable version of $P_{R_1}(A_y|\xi)$.

We have thus essentially proved

Theorem 18.1. *It is possible to choose suitable versions $P_{R_1}(A|\xi)$ such that for all $z \in R_1$ the mapping $A \to P_{R_1}(A|\xi = z)$ is a probability measure over $(R, \mathscr{S})$.*[21]

[20] In place of $P_{R_1}(A_y|\xi)(z)$ we also write $P_{R_1}(A_y|\xi = z)$. See p. 57 and p. 60.

[21] For more general investigations see M. Jiřina, Czechosl. Math. J. 4, (79) 372—380 (1954) and Czechosl. Math. J. 9, (84) 445—451 (1959).

For applications it is, of course, especially important to know how to calculate $F(y|z)$. We have the following important

Theorem 18.2. *If the probability distribution of the r.v. (η,ξ) possesses a density f, then for all real y and all z*

$$F(y|z)=\int_{-\infty}^{y} f(v,z)\,dv\bigg/\int_{-\infty}^{+\infty} f(v,z)\,dv \tag{18.8}$$

as long as the denominator does not vanish.

The function

$$f(y|z)=\frac{f(y,z)}{\int_{-\infty}^{+\infty} f(v,z)\,dv}, \tag{18.9}$$

defined for all y and all z for which the denominator does not vanish is called the *conditional density* of η under the hypothesis $\xi=z$. Naturally, because of (18.8) we have $F(y|z)=\int_{-\infty}^{y} f(v|z)\,dv$ for all $y\in R_1$ and all $z\in R_1$ up to a P_ξ-null set.

Theorems 18.1 and 18.2 can be extended without change to multidimensional r.v.'s ξ and η. Examples of the calculation of conditional distribution functions and conditional densities can be found in **19.** and on p. 99.

We will now consider a notion occupying a position between the concept of conditional probability "given a σ-algebra" and conditional probability "given a r.v.". Let $(R,\mathscr{S},P)$ be a probability space and $(Q,\mathfrak{Q})$ a measurable space. If T is a $(\mathscr{S},\mathfrak{Q})$-measurable mapping from R into Q, then a probability measure P_T on $(Q,\mathfrak{Q})$ induced by T can be defined exactly as in (3.1): For each $M\in\mathfrak{Q}$ let

$$P_T(M)=P(T^{-1}(M)). \tag{18.10}$$

One sees without difficulty that P_T satisfies Axioms I and II. We can also generalize the important formula (16.1):

If g is a $\mathfrak{Q}$-measurable function defined over Q, then

$$\int_R g\circ T\,dP=\int_Q g\,dP_T \tag{18.11}$$

in the sense that the existence of one of the integrals implies that of the other.

Consider the set $\mathscr{S}_0$ of all $T^{-1}(M)$ with $M\in\mathfrak{Q}$. $\mathscr{S}_0$ is a subalgebra of $\mathscr{S}$ (see Theorem I) and forms the natural starting point for the following

Definition. Let T be a $(\mathscr{S}, \mathfrak{Q})$-measurable mapping from R into Q. Let $A \in \mathscr{S}$ and assume that $\mathscr{S}_0$ is defined as above. Then the *conditional probability* $P(A|T)$ *of* A *under the condition* T is understood as *the conditional probability* $P(A|\mathscr{S}_0)$. [22]

$P(A|T)$ is thus defined (up to P-null sets) as an $\mathscr{S}_0$-measurable mapping from R into R_1. It is often useful—just as in the case of a r.v.—to consider the conditional probability of A under the condition T as a $\mathfrak{Q}$-measurable function over Q. In this case we write $P_T(A|T)$ for this conditional probability. The connection between the two definitions is provided by (18.11): $P_T(A|T) \circ T = P(A|T)$ up to P-null sets. Usually, it is clear from the context whether one wants to view the conditional probability under the condition T as a function over R or over Q. Hence, we will usually suppress the index T in $P_T(A|T)$. Closely related to (18.11) is the following

Theorem 18.3. *Let T be a $(\mathscr{S}, \mathfrak{Q})$-measurable mapping from R onto Q, i.e. $T(R)=Q$,*[23] *and assume (in obvious notation) that $\mathscr{S}_0 = T^{-1}(\mathfrak{Q})$. Further, let f be a function defined on R. Then there exists a $\mathfrak{Q}$-measurable function g defined over Q for which $f = g \circ T$ iff f is $\mathscr{S}_0$-measurable.*

Proof. Let A be an arbitrary set in R_1. Then, $\{x: g(T(x)) \in A\} = T^{-1}\{y: g(y) \in A\}$. Indeed, if $x_0 \in \{x: g(T(x)) \in A\}$, then $g(T(x_0)) \in A$ and $T(x_0) \in \{y: g(y) \in A\}$ or finally $x_0 \in T^{-1}\{y: g(y) \in A\}$. The converse is clear. Hence, $g \circ T$ is $\mathscr{S}_0$-measurable for $\mathfrak{Q}$-measurable g. Assume now that f is $\mathscr{S}_0$-measurable. If $f = g \circ T$, then $\{x: g(T(x)) \in B\} \in \mathscr{S}_0$ for $B \in \mathfrak{B}_1$, hence also $T^{-1}\{y: g(y) \in B\} \in \mathscr{S}_0$. Because of $T(R)=Q$ we have $T(T^{-1}(N)) = N$ for every $N \subseteq Q$. Hence, according to the definition of $\mathscr{S}_0$ we have also $\{y: g(y) \in B\} \in Q$, i.e. g is $\mathfrak{Q}$-measurable. We now show that under the assumption of the $\mathscr{S}_0$-measurability of f, there actually exists a g defined over Q for which $f = g \circ T$. Let $y_0 \in Q$ and $M_{y_0} = T^{-1}(\{y_0\})$. Since $T(R)=Q$, $M_{y_0} \neq \emptyset$. Let $x_0 \in M_{y_0}$ and $B_{x_0} = \{x: f(x) = f(x_0)\}$. We naturally have $B_{x_0} \in \mathscr{S}_0$ so that there exists an $M \in \mathfrak{Q}$ with $B_{x_0} = T^{-1}(M)$. Since $x_0 \in B_{x_0} \cap M_{y_0}$, we have $y_0 \in M$ and hence $M_{y_0} \subseteq B_{x_0}$. Hence, f is constant on M_{y_0}. If we now define $g(y_0) = f(x_0)$, then g is well defined and $g(T(x)) = f(x)$ for each $x \in R$.

19. An Example.[24] We illustrate the notions of marginal and conditional distribution by means of an example of a discrete dis-

[22] This concept is equivalent to that of the conditional probability given a σ-algebra as it is easy to see. T is called a statistic, if $\mathfrak{Q} = \{A \subseteq Q: T^{-1}(A) \in \mathscr{S}\}$. The obvious question as to whether the concept of the conditional probability "given a statistic" is equivalent to that of the conditional probability "given a σ-algebra" must be answered in the negative. See R. R. Bahadur and E. L. Lehmann, Ann. Math. Statist. 26, 139—142 (1955).

[23] This condition can be dispensed with. See E. L. Lehmann, l.c. Not.[11], 37—38.

[24] See II. **12**.

tribution. Let $x_1,\dots,x_N$, $N\geqslant 2$, be real numbers. We assume that $x_1<x_2<\dots<x_N$ and define an n-dimensional r.v. $\xi=(\xi_1,\dots,\xi_n)$, $n\leqslant N$, with discrete distribution. Let the mass points be of the form $(x_{i_1},\dots,x_{i_n})$, where $\{i_1,\dots,i_n\}$ is a subset of the set $\{1,\dots,N\}$. Suppose the same probability is assigned to each point so that we have a (discrete) multi-dimensional uniform distribution defined on these mass points. Since one immediately sees that there are $N!/(N-n)!$ different mass points, one gets

$$P(\xi_1=x_{i_1},\xi_2=x_{i_2},\dots,\xi_n=x_{i_n})=[N(N-1)\dots(N-n+1)]^{-1}. \quad (19.1)$$

Hence, with the notation $I_{(i_1,\dots,i_n)}=[x_{i_1},\infty)\times\dots\times[x_{i_n},\infty)$, we obtain for the distribution function of $\xi=(\xi_1,\dots,\xi_n)$

$$F=\frac{1}{N(N-1)\dots(N-n+1)}\sum_{(i_1,\dots,i_n)} c_{I_{(i_1,\dots,i_n)}}.$$

We are interested in the marginal distribution of ξ_i, $1\leqslant i\leqslant n$. Consider for a real x the subset M_x of R_n given by

$$\{y\colon -\infty<y_1<\infty,\dots,-\infty<y_{i-1}<\infty, y_i\leqslant x,$$
$$-\infty<y_{i+1}<\infty,\dots,-\infty<y_n<\infty\}.$$

Applying (14.2), we obtain for the distribution function of ξ_i: $G_i(x)=P_\xi(M_x)$. However, for fixed x_j, the set $R_1\times\dots\times R_1\times\{x_j\}\times R_1\times\dots\times R_1$ contains exactly $(N-1)(N-2)\cdots(N-n+1)$ mass points of ξ. Hence, taking (19.1) into consideration, we obtain $G_i=\left(\sum_{j=1}^{N} c_{I_j}\right)\Big/N$ with $I_j=[x_j,\infty)$. The right side is independent of i and each ξ_i has for $1\leqslant i\leqslant n$ the same marginal distribution with

$$P(\xi_i=x_j)=1/N, \quad j=1,\dots,N. \quad (19.2)$$

Let us also determine the marginal distribution of the two-dimensional r.v.'s (ξ_i,ξ_j), $i\neq j$. One can proceed exactly as before taking into consideration that the sets $R_1\times\dots\times\{x_{i_1}\}\times\dots\times\{x_{i_2}\}\times\dots\times R_1$ contain $(N-2)\cdots(N-n+1)$ mass points of the probability distribution of ξ for fixed $i_1\neq i_2$.

Thus, (ξ_i,ξ_j) has a discrete marginal distribution independent of i,j when $i\neq j$, which is given by

$$P(\xi_i=x_k,\xi_j=x_l)=[N(N-1)]^{-1}, \quad k,l=1,\dots,N, \quad k\neq l. \quad (19.3)$$

We are now interested in the conditional distribution of ξ_i under the condition ξ_j. Thus, for arbitrary real x and y, we consider $P(\xi_i \leqslant x | \xi_j = y) = F(x|y)$. From (19.2) it follows that $F(x|y)$ is only defined if $y = x_j$, $1 \leqslant j \leqslant N$. Let $x_i \leqslant x < x_{i-1}$ and $y = x_l$. One then concludes from (19.3) and (19.2) that

$$F(x|x_l) = \sum_{\substack{1 \leqslant k \leqslant i \\ k \neq l}} \frac{1}{N(N-1)} \Big/ \frac{1}{N},$$

so that

$$F(x|x_l) = \begin{cases} (i-1)/(N-1) & l \leqslant i \\ i/(N-1) & l > i. \end{cases}$$

20. Conditional Expectation. Let ξ be a r.v. whose expectation exists. Let $B \in \mathscr{S}$ and $P(B) \neq 0$. A natural way to define the conditional expectation of ξ under the hypothesis B is by setting

$$E(\xi|B) = \frac{1}{P(B)} \int_B \xi \, dP. \tag{20.1}$$

Obviously, (20.1) implies

$$\int_B E(\xi|B) \, dP = \int_B \xi \, dP. \tag{20.2}$$

We now proceed as in the introduction of conditional probability in **18.** It will turn out later that the concept of conditional probability is only a special case of conditional expectation. We consider a countably infinite number of pairwise disjoint sets $B_i \in \mathscr{S}$ with $\bigcup_{i=1}^{\infty} B_i = R$. Further, let $\mathfrak{S}$ have the same meaning as in **18**. We define $E(\xi|\mathfrak{S})$ as a real-valued function over R (where again the definition is only determined up to a P-null set) by setting

$$E(\xi|\mathfrak{S}) = \sum_{i=1}^{\infty} c_{B_i} E(\xi|B_i). \tag{20.3}$$

$E(\xi|\mathfrak{S})$ is called the *conditional expectation of* ξ *under the condition* $\mathfrak{S}$. $E(\xi|\mathfrak{S})$ is clearly an $\mathfrak{S}$-measurable function and therefore also an $\mathscr{S}$-measurable function over R (if we neglect P-null sets). For each set $B \in \mathfrak{S}$, it follows from (20.3) that

$$\int_B E(\xi|\mathfrak{S}) \, dP = \int_B \xi \, dP. \tag{20.4}$$

(20.4) is the analogue of (20.2).

Equation (20.4) can be utilized as the starting point for the definition of $E(\xi|\mathfrak{S})$ when $\mathfrak{S}$ is an arbitrary sub-σ-algebra of $\mathscr{S}$. Applying the Radon-Nikodym Theorem[25], we obtain a unique, up to P-null sets, $\mathfrak{S}$-measurable function $E(\xi|\mathfrak{S})$ which satisfies (20.4) for all $B\in\mathfrak{S}$. In particular, if we choose $\xi=c_A$, $A\in\mathscr{S}$, then we obtain $E(c_A|\mathfrak{S})=P(A|\mathfrak{S})$, where strictly speaking the equality only holds up to P-null sets.

The important relation

$$E(E(\xi|\mathfrak{S}))=E(\xi), \tag{20.5}$$

follows from (20.4) by setting $B=R$.

Just as conditional probability is "essentially" a probability for each $\omega\in R$, so also is conditional expectation "essentially" an expectation. The following theorem is a consequence of (20.4).

Theorem 20.1. *Let ξ and η be two random variables whose expectations exist. Then $E(\xi+\eta|\mathfrak{S})=E(\xi|\mathfrak{S})+E(\eta|\mathfrak{S})$ P-a.e. For each real number γ $E(\gamma\xi|\mathfrak{S})=\gamma E(\xi|\mathfrak{S})$ P-a.e. From $\xi\leqslant\eta$ P-a.e., it follows that $E(\xi|\mathfrak{S})\leqslant E(\eta|\mathfrak{S})$ P-a.e. and therefore also $|E(\xi|\mathfrak{S})|\leqslant E(|\xi|\,|\mathfrak{S})$ P-a.e.*

Suppose that $\xi_1\leqslant\xi_2\leqslant\cdots$ is a sequence of random variables which converges to a r.v. ξ such that $E(\xi_i)$, $1\leqslant i$ and $E(\xi)$ exist. Then $E(\xi_i|\mathfrak{S})\to E(\xi|\mathfrak{S})$ P-a.e.

Another important property is stated in

Theorem 20.2. *Let ξ be as in Theorem 20.1 and η be a bounded $\mathfrak{S}$-measurable function. Then $E(\xi\cdot\eta|\mathfrak{S})=\eta E(\xi|\mathfrak{S})$ P-a.e.*

The same result holds without boundedness of η if $E(\xi^2)$ and $E(\eta^2)$ exist.

If $\mathfrak{S}$ is generated by the inverse images of a r.v. ξ (cf. Theorem I) then we write $E(\eta|\xi)$. This notation can be justified by changing over to a Borel-measurable mapping of R_1 into R_1, denoted by $E_{R_1}(\eta|\xi)$, and defined by means of the Radon-Nikodym Theorem by $\int_{\xi^{-1}(B_1)}\eta\,dP=\int_{B_1}E_{R_1}(\eta|\xi)\,dP_\xi$ for all $B_1\in\mathfrak{B}_1$ (up to P_ξ-null sets). Then $E(\eta|\xi)=E_{R_1}(\eta|\xi)\circ\xi$, P_ξ-a.e.

In this case, it follows easily from Theorem 18.1, that the conditional expectation may be calculated directly in terms of the conditional probability. Moreover the following result holds

---

[25] Of course, one cannot manage with the form of Theorem 6.1 given here, but requires a generalization of this result to set functions which are not necessarily non-negative. However, this generalization can easily be obtained from Theorem 6.1.

Theorem 20.3. *Let ξ and η be random variables of dimension n and m, respectively, and f be their joint density. If h is a function from R_n into R_1 and $E(h\circ\xi)$ exists, then*

$$E(h\circ\xi|\eta=y)=\int_{R_n} h(x)f(x|y)dx$$

for all y with the exception of a P_η-null set, where $f(x|y)$ denotes the conditional density of ξ at the point x under the condition $\eta=y$.

If T is an $(\mathscr{S},\mathfrak{Q})$-measurable mapping of R into a set Q, then the conditional expectation $E(\eta|T)$ is defined for each η whose expectation exists. A comparison with the corresponding concept of conditional probability "under the condition T" and the explanation given previously in this paragraph make it completely obvious how the definition of $E(\eta|T)$ is to be formulated. $E(\eta|T)$ is, for example, a function of T and all properties expressed in Theorems 20.1 and 20.2 carry over in an obvious manner.

21. Regression Surfaces. Let F be the d.f. of an n-dimensional r.v. $(\xi_1,\ldots,\xi_n)$. Suppose that $E(\xi_1)$ exists. Then, except for a null set, $E(\xi_1|\xi_2=x_2,\ldots,\xi_n=x_n)$ also exists, and we will abbreviate this expression by $E(\xi_1|x_2,\ldots,x_n)$. If we neglect exceptional sets, which have $P_{(\xi_2,\ldots,\xi_n)}$-measure 0, then a function ρ on R_{n-1} is defined by $(x_2,\ldots,x_n)\to E(\xi_1|x_2,\ldots,x_n)$. ρ is referred to as the *regression function* or the *regression surface* (in R_n) of ξ_1 relative to $\xi_2,\ldots,\xi_n$. This regression function has a minimum property analogous to that of Theorem 17.1, as the next theorem shows.

In connection with this, we note that the existence of $E(\xi_1^2)$ implies the existence of $E(\rho^2(\xi_2,\ldots,\xi_n))$. This is because it follows from Theorem 20.1 that Schwarz's inequality also holds[26] for conditional expectations; hence,

$$E(\rho^2(\xi_2,\ldots,\xi_n))=E\left((E(\xi_1|\xi_2,\ldots,\xi_n))^2\right)\leqslant E(E(\xi_1^2|\xi_2,\ldots,\xi_n)).$$

Therefore, it follows from (20.5) that $E(\rho^2(\xi_2,\ldots,\xi_n))\leqslant E(\xi_1^2)$, that is, $E(\rho^2(\xi_2,\ldots,\xi_n))$ exists.

Theorem 21.1. *Let F be the distribution function of an n-dimensional r.v. $(\xi_1,\ldots,\xi_n)$ and ρ be the regression function defined above. Moreover, suppose that $E(\xi_1^2)$ exists. Then, for all functions $(x_2,\ldots,x_n)\to h(x_2,\ldots,x_n)$ for which $E(h^2(\xi_2,\ldots,\xi_n))$ exists,*

---

[26] More precisely: If ξ and η are r.v.'s and $E(\xi^2)$ and $E(\eta^2)$ exist, then (in the notation of **20**) $(E(|\xi\eta||\mathfrak{S}))^2\leqslant E(\xi^2|\mathfrak{S})E(\eta^2|\mathfrak{S})$ P-a.e.

$$E((\xi_1-h(\xi_2,\dots,\xi_n))^2)\geqslant E((\xi_1-\rho(\xi_2,\dots,\xi_n))^2). \tag{21.1}$$

The equality holds iff $h(x_2,\dots,x_n)=\rho(x_2,\dots,x_n)$ $P_{(\xi_2,\dots,\xi_n)}$-*a.e.*[27]

Proof. On the basis of the assumptions, and due to the preceding remarks and Theorem 17.3 applied to ξ_1 and $h\circ(\xi_2,\dots,\xi_n)$, the expectations listed in (21.1) make sense.

Further,

$$\begin{aligned}
&E((\xi_1-h(\xi_2,\dots,\xi_n))^2|\xi_2,\dots,\xi_n)\\
&=E((\xi_1-\rho(\xi_2,\dots,\xi_n)+\rho(\xi_2,\dots,\xi_n)\\
&\quad-h(\xi_2,\dots,\xi_n))^2|\xi_2,\dots,\xi_n)\\
&=E((\xi_1-\rho(\xi_2,\dots,\xi_n))^2|\xi_2,\dots,\xi_n)\\
&\quad+(\rho(\xi_2,\dots,\xi_n)-h(\xi_2,\dots,\xi_n))^2
\end{aligned}$$

where we have made use of Theorems 20.1 and 20.2 and (20.5). Applying (20.5) once more, it follows that

$$\begin{aligned}
&E(\xi_1-h(\xi_2,\dots,\xi_n))^2=E(E(\xi_1-h(\xi_2,\dots,\xi_n))^2|\xi_2,\dots,\xi_n)\\
&=E(\xi_1-\rho(\xi_2,\dots,\xi_n))^2+E(\rho(\xi_2,\dots,\xi_n)-h(\xi_2,\dots,\xi_n))^2,
\end{aligned}$$

whence all the assertions of the Theorem are proved.

If the distribution function F has a density f, then as a consequence of Theorem 20.3, the regression function can be calculated as follows:

$$\rho(x_2,\dots,x_n)=\int_{-\infty}^{+\infty} x_1\, f(x_1|x_2,\dots,x_n)\,dx_1, \tag{21.2}$$

where $f(x_1|x_2,\dots,x_n)$ is, in accordance with (18.9), the conditional density of ξ_1 under the condition $\xi_2=x_2,\dots,\xi_n=x_n$. Moreover, in this case (21.1) can be written in the form

$$\int_{R_n}(x_1-h(x_2,\dots,x_n))^2 f(x)\,dx\geqslant\int_{R_n}(x_1-\rho(x_2,\dots,x_n))^2 f(x)\,dx. \tag{21.3}$$

Regression theory for multidimensional normal distributions is developed further in VI.

[27] Thus, in somewhat more general formulation, Theorem 21.1 states that if ξ is a r.v. and $E(\xi^2)$ exists, then $E(\xi\,|\,\mathfrak{S})$ is the orthogonal projection of ξ onto the set of $\mathfrak{S}$-measurable functions.

22. Čebyšev's Inequality. We first prove the following theorem:

Theorem 22.1. *Let* ξ *be a r.v. and* h *a non-negative function. Set* $b(c)=\inf_{|y|\geq c} h(y)$ *for* $c\geq 0$. *If* $E(h\circ\xi)$ *exists and* $b(c)>0$, *then*

$$P(|\xi|\geq c)\leq E(h\circ\xi)/b(c). \tag{22.1}$$

Proof. $E(h\circ\xi)=\int_{-\infty}^{+\infty} h(x)\,dP_\xi(x)\geq \int_{\{x:|x|\geq c\}} h(x)\,dP_\xi(x)\geq b(c)P(|\xi|\geq c)$.

In particular, if $h(y)=y^2$ for $y\in R_1$, then Čebyšev's inequality follows: If a r.v. η has finite variance $\sigma^2\neq 0$ and its mean value is denoted by a, then for each real $t>0$

$$P(|\eta-a|\geq t\sigma)\leq 1/t^2. \tag{22.2}$$

Because of Axiom I (p. 26), the inequality is trivial if $0<t<1$. For $t>1$, set $c=t\sigma$, so that $b(c)=t^2\sigma^2$, and apply (22.1) to the r.v. $\xi=\eta-a$.

If $\sigma=0$, then (22.2) obviously does not hold. The particular significance of (22.2) lies in the fact that the inequality holds for all distributions whose second moment exists and whose variance does not vanish. From the proof of (22.1) it can be seen immediately that there are distribution functions for which (22.2) cannot be sharpened. Therefore, the equality sign on the right side of (22.2) cannot in general be omitted.

23. The Characteristic Function. Let ξ be an n-dimensional random variable, which we shall think of as an $n\times 1$ matrix.

We give the following

Definition. The expected value $E(e^{it'\xi})=\phi(t)$ (cf. p. 15) defines a function for all $t\in R_n$ called *the characteristic function* of the r.v. ξ.

This definition utilizes the (as yet unproved) fact that $\phi(t)$ is defined for all $t\in R_n$. We establish this in

Theorem 23.1. *The characteristic function* $\phi(t)=E(e^{it'\xi})$ *of a r.v.* ξ *exists for all* $t\in R_n$ *and is, moreover, a uniformly continuous function on* R_n.

Proof. We have

$$|e^{it'x}|=1 \tag{23.1}$$

for all $t,x\in R_n$. Therefore, the existence of the characteristic function is proved. The continuity follows from (23.1), the continuity of $t\to e^{it'x}$ and Theorem VII. The uniform continuity is a consequence of $|e^{it_1'x}-e^{it_2'x}|\leq|1-e^{i(t_1-t_2)'x}|$ for all $t_1,t_2\in R_n$ and $x\in R_n$.

Obviously,

$$|\phi(t)|\leq 1, \tag{23.2}$$

for all $t\in R_n$.

That the characteristic function is of fundamental significance in probability theory is revealed in part in the following theorems. We begin with

Theorem 23.2. *Let* $\xi=(\xi_1,\dots,\xi_n)$ *be an n-dimensional r.v. and* ϕ *denote the characteristic function of* ξ. *Suppose that* $E(\xi_1^{r_1}\dots\xi_n^{r_n})$ *exists, where the* r_i *are non-negative integers and* $\sum_{i=1}^{n} r_i=r$. *Then the derivative* $\frac{\partial^r\phi(t)}{\partial t_1^{r_1}\dots\partial t_n^{r_n}}=\psi(t)$ *exists for all* $t=(t_1,\dots,t_n)$ *and* ψ *is a continuous function of t. If* $r_i=0$, *then the derivative does not involve* t_i, *so that* $\phi=\psi$ *for* $r=0$. *Further, for each t*

$$\psi(t)=i^r\int_{R_n} x_1^{r_1}\dots x_n^{r_n}e^{it'x}\,dP_\xi(x) \tag{23.3}$$

and, in particular,

$$i^r E(\xi_1^{r_1}\dots\xi_n^{r_n})=\left.\frac{\partial^r\phi(t)}{\partial t_1^{r_1}\dots\partial t_n^{r_n}}\right|_{t=(0,\dots,0)}. \tag{23.4}$$

Proof. Except for the uniformity statement, if $r=0$, the assertion is equivalent to Theorem 23.1.

Therefore, suppose that $r>0$. Since $|x_1^{r_1}\dots x_n^{r_n}e^{it'x}|=|x_1^{r_1}\dots x_n^{r_n}|$ is P_ξ-integrable by assumption, Theorem VIII may be applied to obtain the existence of ψ and formula (23.3). The continuity of ψ follows in the same way as the continuity of ϕ in Theorem 23.1. The relation (23.4) is an immediate consequence of (23.3).

We now state a theorem the basic part of which establishes that a probability distribution of a r.v. is uniquely determined by its characteristic function.

Theorem 23.3[28] *(Uniqueness Theorem). Let* ϕ *be the characteristic function of a r.v.* ξ *with d.f. F. Then F is uniquely determined by* ϕ *and conversely.*

We restrict ourselves to an outline of the *proof* for one-dimensional random variables. Thus, let $\phi(t)=\int_{-\infty}^{+\infty} e^{itx}\,dF(x)$, for $t\in R_1$. Suppose that $a<b$ $(a,b\in R_1)$ and that a and b are continuity points of F. Then

$$F(b)-F(a)=\lim_{N\to+\infty}\frac{1}{2\pi}\int_{-N}^{N}\frac{e^{-ita}-e^{-itb}}{it}\phi(t)\,dt. \tag{23.5}$$

[28] This Theorem is due to P. Lévy: P. Lévy, Calcul des Probabilités, Gauthier-Villars et Cie., Paris, 1925, 166 ff.

For a modern and thourough treatment of characteristic functions see: E. Lukacs, Characteristic functions, Second Edition, Griffin, London, 1970.

Note that from Fubini's Theorem

$$\int_{-N}^{N} \frac{e^{-ita}-e^{-itb}}{it}\phi(t)dt = 2\int_{R_1}\int_{N(y-b)}^{N(y-a)} \frac{\sin t}{t}\,dt\,dF(y).$$

Now $\int_{N(y-b)}^{N(y-a)} \frac{\sin t}{t}dt$ converges to a limit as $N\to+\infty$; it converges to π if $a<y<b$, to $\pi/2$ if either $y=a$ or $y=b$ and to 0 if either $y<a$ or $y>b$. Moreover, $\left|\int_{y'}^{y''} \frac{\sin t}{t}dt\right|$ is uniformly bounded for $y', y''\in R_1$ and so, from Lebesgue's Theorem, we can pass to the limit under the first integral sign. This, together with the assumption that a and b are continuity points of F, yields the assertion in (23.5).

Therefore, according to Property 2 (p. 31), F is uniquely determined for all $x\in R_1$. Hence, knowledge of the characteristic function is equivalent to knowledge of the probability distribution P_ξ on $\mathfrak{B}_1$. Another very important property of characteristic functions is expressed in the Continuity Theorem. We first give the following

Definition. *A sequence of distribution functions* $\{F_n\}$ is said to *converge weakly to* a distribution function F, if $F_n(x)$ converges to $F(x)$ at each *continuity point of* F.

Since $0\leqslant F_n\leqslant 1$, one might suppose that each convergent sequence of distribution functions converges to a distribution function. However, this conclusion is false, as is shown by the following simple example:

For each $k, k=1,2,\dots,$ let F_k be the distribution function of a distribution degenerate at the point k. It follows immediately that the sequence F_k converges to 0 for each x. More precisely, if x is an arbitrary point in R_1, then $F_k(x)=0$ for all $k>k(x)$ if e.g. $k(x)=x$. Nevertheless, under certain assumptions we can conclude that the (necessarily non-decreasing) limiting function determined by a convergent sequence of distribution functions at the continuity points of the limiting function is again a distribution function. For this purpose we state, without proof, the following

Lemma 23.1. *If a sequence of distribution functions* $\{F_n\}$ *converges weakly to a distribution function* F, *then* $\int_{R_1} f\,dF_n \to \int_{R_1} f\,dF$ *for any bounded, continuous function* f *defined on* R_1. *The converse of this statement holds: If* $\int_{R_1} f\,dF_n \to \int_{R_1} f\,dF$ *for any bounded, continuous function* f *on* R_1, *then* F *is a distribution function and* $\{F_n\}$ *converges weakly to* F.

Now we obtain

Theorem 23.4.[29] *If a sequence of k-dimensional distribution functions $\{F_n\}$ converges weakly to a distribution function F, then the corresponding sequence of characteristic functions $\{\phi_n\}$ converges at each t to the characteristic function ϕ of F. Conversely, if a sequence of characteristic functions $\{\phi_n\}$ converges for each t to a function ϕ continuous at 0, then the corresponding sequence of distribution functions $\{F_n\}$ converges weakly to a distribution function F and ϕ is the characteristic function of F.*

We shall be content once again with an outline of the *proof* in the one-dimensional case. An application of Lemma 23.1 to $x \to e^{itx}$ which is a bounded continuous function for every t, furnishes the proof of the first part of the Theorem. For the converse, we use the *Theorem of Helly*:

Lemma 23.2. *Each sequence of distribution functions contains a subsequence which converges to a nondecreasing function at each of its continuity points.*[30]

This will also not be proved here.[31]

Let F_n be the d.f. corresponding to the characteristic function ϕ_n for each n. We can choose a subsequence $\{F_{n_i}\}$ from the sequence $\{F_n\}$ which converges to a limit function F at each of its continuity points. Naturally, F satisfies $0 \leqslant F \leqslant 1$. Suppose that $\lim_{x \to +\infty} F(x) - \lim_{x \to -\infty} F(x) = d < 1$. We choose an ε with $0 < 4\varepsilon < 1 - d$. Because ϕ is continuous and $\phi(0) = 1$, there exists a $u > 0$ such that

$$\left| \int_0^u \phi(t)\, dt \right| \Big/ u > d + 4\varepsilon. \tag{23.6}$$

We now choose $y_1 \geqslant 1/u\varepsilon$ in such a way that F is continuous at y_1 and $-y_1$. Since $|F(y_1) - F(-y_1)| \leqslant d$, we also have

$$|F_{n_k}(y_1) - F_{n_k}(-y_1)| \leqslant d + \varepsilon \tag{23.7}$$

for all $n_k \geqslant N(\varepsilon)$. Now, for $y \geqslant y_1$ and all $u \in R_1$

$$\left| \int_0^u e^{ity}\, dt \right| \leqslant 2/y_1 \tag{23.8}$$

and for all real y and $u \in R_1$ obviously

$$\left| \int_0^u e^{ity}\, dt \right| \leqslant u. \tag{23.9}$$

[29] P. Lévy, l.c.[28] 195ff.

[30] Moreover, it can always be assumed that this function is right-continuous.

[31] From Lemma 23.2 and Lemma 23.1 one can easily infer Theorem 23.4.

As a consequence of Fubini's Theorem we obtain

$$\left|\int_0^u \phi_{n_k}(t)\,dt\right| \Big/ u = \left|\int_{R_1}\int_0^u e^{ity}\,dt\,dF_{n_k}(y)\right| \Big/ u .$$

This expression is

$$\leqslant \left|\int_{|y|\leqslant y_1}\int_0^u e^{ity}\,dt\,dF_{n_k}(y)\right| \Big/ u + \left|\int_{|y|>y_1}\int_0^u e^{ity}\,dt\,dF_{n_k}(y)\right| \Big/ u .$$

From (23.7) and (23.9) we obtain the upper bound $d+\varepsilon$ for the first summand and, for the second summand, the upper bound 2ε from (23.8). Thus, $\left|\int_0^u \phi_{n_k}(t)\,dt\right| \big/ u \leqslant d+3\varepsilon$ for $n_k \geqslant N(\varepsilon)$. However, since ϕ_{n_k} converges to ϕ and $|\phi_{n_k}| \leqslant 1$, from (23.2), it follows from Lebesgue's Theorem that $\int_0^u \phi_{n_k}(t)\,dt \to \int_0^u \phi(t)\,dt$. Therefore, $\left|\int_0^u \phi(t)\,dt\right| \big/ u \leqslant d+3\varepsilon$ in contradiction to (23.6). Hence, we must have $\lim_{x\to+\infty} F(x) - \lim_{x\to-\infty} F(x) = 1$ and then the first part of the continuity theorem shows that ϕ is the characteristic function of F. Consequently, if the sequence $\{F_n\}$ itself did not converge weakly to F, there would exist a subsequence $\{F_{m_j}\}$ which converges weakly to a distribution function different from F, but having the same characteristic function. This would be a contradiction of Theorem 23.3.

From Theorem 23.4 we immediately obtain

Theorem 23.5. *Let* $(\xi_1^{(n)},\ldots,\xi_k^{(n)})$ *be a sequence of k-dimensional random variables and suppose that the corresponding sequence of distribution functions* $\{F_n\}$ *converges weakly to a distribution function F. Let* $r_1,\ldots,r_l$, $l \leqslant k$, *be a subset of* $\{1,\ldots,k\}$ *and* G_n *be the marginal distribution function of* $(\xi_{r_1}^{(n)},\ldots,\xi_{r_l}^{(n)})$ *for each* $n \geqslant 1$. *Then* $\{G_n\}$ *converges weakly to the corresponding marginal distribution function G, obtained from F by means of* (14.3).

Proof. For the sake of simplicity, let $r_j = j$, $j=1,\ldots,l$. The characteristic functions ϕ_n of F_n and ϕ of F satisfy $\phi_n(t)\to\phi(t)$ for all $t\in R_k$. If we choose $t_{l+1}=\cdots=t_k=0$, then as $n\to\infty$

$$E\left(e^{i(t_1\xi_1^{(n)}+\cdots+t_l\xi_l^{(n)})}\right) \to \int_{-\infty}^{+\infty}\!\!\cdots\int_{-\infty}^{+\infty} e^{i(t_1y_1+\cdots+t_ly_l)}\,dG(y_1,\ldots,y_l)$$

for all $t_i\in R_1$, $1\leqslant i\leqslant l$.

Therefore, as follows from Theorem 23.4, $G_n\to G$ at all continuity points of G.

Let ξ be a r.v. with characteristic function ϕ. If we define the r.v. $\eta=(\xi-b)/c$, where b and c are real numbers with $c\neq 0$, then

$$E(e^{i\eta t})=E(e^{it(\xi-b)/c})=E(e^{(it\xi)/c}e^{(-itb)/c})=e^{(-itb)/c}E(e^{(it\xi)/c}).$$

Therefore, the characteristic function of η is given by

$$\phi(t/c)e^{(-itb)/c} \tag{23.10}$$

for each $t\in R_1$.

Examples for the calculation of various characteristic functions can be found on pp. 77, 82, 89, 92, and 98.

24. Sums of random variables. If $\xi_1,\ldots,\xi_n$ are random variables, then $\zeta_n=\xi_1+\cdots+\xi_n$ is also, from **11** (p. 42), a r.v.

We have

Theorem 24.1. *Suppose the mean values a_i and the variances σ_{ii} of ξ_i exist. Then the mean value and variance of ζ_n exist and $E(\zeta_n)=\sum_{i=1}^{n} a_i$ and $E\left[\left(\zeta_n-\sum_{i=1}^{n} a_i\right)^2\right]=\sum_{i,j=1}^{n}\sigma_{ij}$, where σ_{ij} is the covariance*[32] *of ξ_i and ξ_j.*

Proof. The assertion concerning the mean of ζ_n follows immediately from Theorem 16.3, and that concerning the variance is a consequence of

$$E\left[\left(\zeta_n-\sum_{i=1}^{n} a_i\right)^2\right]=E\left[\sum_{i,j=1}^{n}(\xi_i-a_i)(\xi_j-a_j)\right]$$
$$=\sum_{i,j=1}^{n}E[(\xi_i-a_i)(\xi_j-a_j)]=\sum_{i,j=1}^{n}\sigma_{ij}.$$

If the joint distribution of $(\xi_1,\ldots,\xi_n)$ is known, then the probability distribution of ζ_n can be determined from **11** (p. 42). Of particular importance is the case where $\xi_1,\ldots,\xi_n$ are independent of one another. Then the following theorem, which we state for $n=2$, holds;

Theorem 24.2. *Let ξ_1 and ξ_2 be independent random variables and let the distribution function of ξ_i be F_i, $i=1,2$. Set $\zeta=\xi_1+\xi_2$. Then the d.f. F of ζ is given by*

$$F(x)=\int_{-\infty}^{+\infty}F_1(x-y)\,dF_2(y)=\int_{-\infty}^{+\infty}F_2(x-y)\,dF_1(y)$$

[32] This exists for all i,j from Theorem 17.3.

for each $x \in R_1$. *In particular, if* ξ_i *has a density* f_i, $i=1,2$, *then* ζ *also has a density* f *and*

$$f(x) = \int_{-\infty}^{+\infty} f_1(x-y) f_2(y) dy. \tag{24.1}$$

We shall restrict our *proof* to the case where ξ_1 and ξ_2 are of the continuous type. Then (24.1) can be proved by means of Theorem 12.2. For this purpose we consider the two-dimensional r.v. (ξ_1, ξ_2) whose density is given by $f_1(x_1) f_2(x_2)$ for all $(x_1, x_2) \in R_2$ (cf. (13.4)). Then we define, for all $(x_1, x_2) \in R_2$, the mappings $g_1(x_1, x_2) = x_1 + x_2$ and $g_2(x_1, x_2) = x_2$. The mapping (g_1, g_2) of R_2 onto R_2 has a unique inverse. The inverse mapping (h_1, h_2) is given by $h_1(y_1, y_2) = y_1 - y_2$ and $h_2(y_1, y_2) = y_2$ for all $(y_1, y_2) \in R_2$ and $\left|\frac{\partial(h_1, h_2)}{\partial(y_1, y_2)}\right| = 1$. Therefore, setting $\eta_1 = g_1(\xi_1, \xi_2)$ and $\eta_2 = \xi_2$ the density of (η_1, η_2) is $f_1(y_1 - y_2) f_2(y_2)$ for all $(y_1, y_2) \in R_2$. In order to obtain the density of η_1 we must go to the appropriate marginal distribution; thus we obtain (24.1) by an application of (14.4).

The function f defined by (24.1) is called the *convolution* of f_1 and f_2. Consequently,

Theorem 24.3. *The distribution of a sum of independent random variables, each of which possesses a probability density, is obtained by successive convolution of these densities. It does not depend on the order in which these convolutions are taken.*

The last statement of this theorem follows from Theorem 24.2.

It is evident that the definition of sums of random variables is not restricted to the assumption that the variables are one-dimensional. Let ξ_1 and ξ_2 be two independent, n-dimensional random variables with densities f_1 and f_2, respectively. We are interested in the characteristic function ψ of the sum $\zeta = \xi_1 + \xi_2$. For each $t \in R_n$

$$\psi(t) = E(e^{it'\zeta}) = \int_{R_{2n}} e^{it'(x_1 + x_2)} f_1(x_1) f_2(x_2) dx_1 \, dx_2 .$$

To obtain this result we have used the fact that the density of (ξ_1, ξ_2) is given by $f_1(x_1) f_2(x_2)$ for all $(x_1, x_2) \in R_2$ and have employed (16.1). From Fubini's Theorem it follows that

$$\psi(t) = \prod_{j=1}^{2} \int_{R_n} e^{it'x_j} f_j(x_j) dx_j .$$

Hence, the characteristic function ψ is equal to the product of the characteristic functions of ξ_1 and ξ_2. This result holds more generally; we state it as a theorem:

Theorem 24.4. *The characteristic function of a finite sum of independent random variables is equal to the product of the characteristic functions of the summands.*

To illustrate this result, we consider an example[33] concerning the distribution of a sum of discrete independent random variables. Let r be a non-negative integer and let n independent random variables ξ_i be given, each having the same uniform distribution with discrete mass points $0, 1, \dots, r$. Let $\zeta_n = \xi_1 + \dots + \xi_n$. The characteristic function of ξ_i is except for the factor $1/(1+r)$, given by $1 + e^{it} + \dots + e^{irt}$, for each real t. Therefore, using Theorem 24.4, the characteristic function of ζ_n is $(1+r)^{-n}(1 + e^{it} + \dots + e^{irt})^n$ for each real t. This is equal to $(1+r)^{-n} \sum_{j=0}^{nr} q_j e^{itj}$. It follows from Theorem 23.3 that ζ_n is a discrete random variable with mass points $0, 1, \dots, nr$, where $P(\zeta_n = j) = q_j/(1+r)^n$, $1 \leqslant j \leqslant nr$. In order to facilitate finding the quantities q_j, we replace e^{it} by x, i.e., we use the method of moment generating functions. For $|x| < 1$, we have the identity

$$\begin{aligned}(1 + x + \dots + x^r)^n &= (1 - x^{r+1})^n/(1-x)^n \\ &= \sum_{k=0}^{n} (-1)^k \binom{n}{k} x^{(r+1)k} \sum_{l=0}^{\infty} \binom{-n}{l} (-1)^l x^l.\end{aligned}$$

Now, $(-1)^l \binom{-n}{l} = \binom{n+l-1}{l}$.

Therefore, we obtain the following expression for the coefficient q_j; Let $j = (r+1)m + p$; m, p integers, $0 \leqslant m \leqslant (n-1)$, $0 \leqslant p < r+1$, $j \leqslant rn$. Then

$$q_j = \sum_{k=0}^{m} (-1)^k \binom{n}{k} \binom{n + (m-k)(r+1) + p - 1}{(m-k)(r+1) + p}.$$

25. The normal distribution. One of the most important distributions is the so-called *normal* or *Gaussian distribution*. It is of the continuous type and its density is given by

$$\frac{1}{\sqrt{2\pi}\,\sigma} e^{-\frac{(x-a)^2}{2\sigma^2}}, \tag{25.1}$$

$x \in R_1$, where a is an arbitrary, and σ a positive real number. a and σ^2 are called the parameters of the normal distribution. Their significance will be made clear soon. First we will show that (25.1) satisfies Conditions 1 and 2 of p. 34. Clearly, the expression in (25.1) is >0 for all $x \in R_1$. We show that (6.3) is fulfilled. Consider

$$\frac{1}{\sqrt{2\pi}\,\sigma} \int_{-\infty}^{+\infty} e^{-\frac{(x-a)^2}{2\sigma^2}} dx. \tag{25.2}$$

[33] Another example is given e.g. on p. 81.

As a result of the substitution $u=(x-a)/\sigma$ we arrive at the integral

$$J=\frac{1}{\sqrt{2\pi}}\int_{-\infty}^{+\infty} e^{-u^2/2}\,du. \tag{25.3}$$

(25.3) shall be evaluated according to a method which originated with Stieltjes[34] and, in addition to the value of J, also gives all the moments of the normal distribution. First, we note that the Wallis product

$$W_k=\frac{(2\cdot 4\cdot\ldots\cdot 2k)^2}{(1\cdot 3\cdot\ldots\cdot(2k-1))^2(2k+1)}$$

satisfies

$$\lim_{k\to+\infty} W_k=\pi/2. \tag{25.4}$$

For each natural number $n\geqslant 0$, set $I_n=\int_0^\infty u^n e^{-u^2/2}\,du$. This integral obviously converges and, as a result of an integration by parts, we obtain

$$I_n=\frac{1}{n+1}I_{n+2}. \tag{25.5}$$

Employing this relation for $n=0,2,\ldots,2k-2$, $k\geqslant 1$, it follows that

$$I_{2k}=1\cdot 3\cdot\ldots\cdot(2k-1)\,I_0. \tag{25.6}$$

Similarly, for odd n

$$I_{2k+1}=2\cdot 4\cdot\ldots\cdot 2k\,I_1, \quad k\geqslant 1. \tag{25.7}$$

However,

$$I_1=\int_0^\infty u e^{-u^2/2}\,du=\int_0^\infty e^{-u^2/2}\,d(u^2/2)=1$$

and from (25.7)

$$I_{2k+1}=2\cdot 4\cdot\ldots\cdot 2k. \tag{25.8}$$

For all real z,

$$I_{n+1}+2zI_n+z^2I_{n-1}=\int_0^\infty u^{n-1}(u+z)^2e^{-u^2/2}\,du>0.$$

Hence, the discriminant of the quadratic in z is <0, i.e.,

$$I_n^2<I_{n-1}I_{n+1}. \tag{25.9}$$

[34] Stieltjes, T.J. Nouv. Ann. Math., ser. 3, 9, 479—480 (1890).

From (25.9) and (25.5) (with $n-1$ in place of n), we obtain

$$I_n^2 < n I_{n-1}^2. \tag{25.10}$$

Now we combine (25.9) for $n=2k$ and (25.10) with $n=2k+1$ to obtain

$$I_{2k+1}^2 \cdot (1/(2k+1)) < I_{2k}^2 < I_{2k-1} I_{2k+1}, \qquad k \geqslant 1.$$

From (25.8) and (25.6)

$$(2\cdot 4 \cdot \ldots \cdot 2k)^2 (1/(2k+1)) < (1\cdot 3 \cdot \ldots \cdot 2k-1)^2 \cdot I_0^2$$
$$< (2\cdot 4 \cdot \ldots \cdot (2k-2)) \cdot 2 \cdot 4 \cdot \ldots \cdot 2k,$$

or

$$\frac{(2\cdot 4 \cdot \ldots \cdot 2k)^2}{(1\cdot 3 \cdot \ldots \cdot (2k-1))^2 (2k+1)} < I_0^2 < \frac{(2\cdot 4 \cdot \ldots \cdot 2k)^2}{(1\cdot 3 \cdot \ldots \cdot (2k-1))^2 (2k+1)} \cdot \frac{2k+1}{2k},$$

i.e., we have $W_k < I_0^2 < W_k(2k+1)/2k$, so that from (25.4), we obtain, letting $k \to +\infty$, $I_0^2 = \pi/2$. Further, since $I_0 > 0$, $I_0 = \sqrt{\pi/2}$ and we find that $J=1$. Hence this is also the case for the integral in (25.2).

Now, if ξ is a r.v. whose density is given by $\dfrac{1}{\sqrt{2\pi}} e^{-x^2/2}$, $x \in R_1$, then all moments exist and all odd-order moments have the value

$$E(\xi^{2k+1}) = 0, \qquad k \geqslant 1, \tag{25.11}$$

since the density of the normal distribution is symmetric with respect to the origin. In particular $E(\xi)=0$. To calculate the even-order moments, we use (25.6) and, taking into account the value of $I_0 = \sqrt{\pi/2}$, obtain

$$E(\xi^{2k}) = 1 \cdot 3 \cdot \ldots \cdot (2k-1), \qquad k \geqslant 1. \tag{25.12}$$

From (25.12), it follows for $k=1$ that

$$E(\xi^2) = 1, \tag{25.13}$$

and therefore, since $E(\xi)=0$, we have found the value of the variance of the r.v. ξ.

More generally we have

Theorem 25.1. *If the r.v. ξ has a distribution with density given by* (25.1) *for each* $x \in R_1$, *then*

$$E(\xi) = a \tag{25.14}$$

and

$$E[(\xi - a)^2] = \sigma^2. \tag{25.15}$$

For the *proof* we observe that

$$E(\xi)=\frac{1}{\sqrt{2\pi}\,\sigma}\int_{-\infty}^{+\infty} x e^{-(x-a)^2/2\sigma^2}\,dx$$

$$=\frac{1}{\sqrt{2\pi}\,\sigma}\int_{-\infty}^{+\infty}(x-a)e^{-(x-a)^2/2\sigma^2}\,dx+a\frac{1}{\sqrt{2\pi}\,\sigma}\int_{-\infty}^{+\infty}e^{-(x-a)^2/2\sigma^2}\,dx$$

$$=\frac{\sigma}{\sqrt{2\pi}}\cdot\int_{-\infty}^{+\infty}u e^{-u^2/2}\,du+a$$

and this equals a since the last integral vanishes. Further,

$$E[(\xi-a)^2]=\frac{1}{\sqrt{2\pi}\,\sigma}\int_{-\infty}^{+\infty}(x-a)^2 e^{-(x-a)^2/2\sigma^2}\,dx$$

$$=\frac{\sigma^2}{\sqrt{2\pi}}\int_{-\infty}^{+\infty}u^2 e^{-u^2/2}\,du=\sigma^2$$

because of (25.13).

This explains the significance of the parameters in (25.1): a is the *mean value* and σ^2 the *variance* of the normal distribution. A normal distribution is uniquely determined by specifying both of these parameters. From now on we will denote a normal distribution with mean value a and variance σ^2 by the symbol $N(a,\sigma^2)$. Finally, we give the moments for $N(a,\sigma^2)$ relative to the mean value a. They are, reading off from (25.11) and (25.12),

$$E[(\xi-a)^{2k+1}]=0, \quad k\geqslant 0, \tag{25.16}$$

and

$$E[(\xi-a)^{2k}]=1\cdot 3\cdot\ldots\cdot(2k-1)\sigma^{2k}, \quad k\geqslant 1. \tag{25.17}$$

The mean value and the variance of a normal distribution are of importance for the geometrical description of the density (25.1). The mean value a is the point where the density assumes its unique maximum and $a\pm\sigma$ are the inflection points of the density (cf. Fig. 5).

The skewness of a normal distribution is zero since the distribution is symmetric about its mean value; the excess of the normal distribution is $3(\sigma^4/\sigma^4)-3=0$.

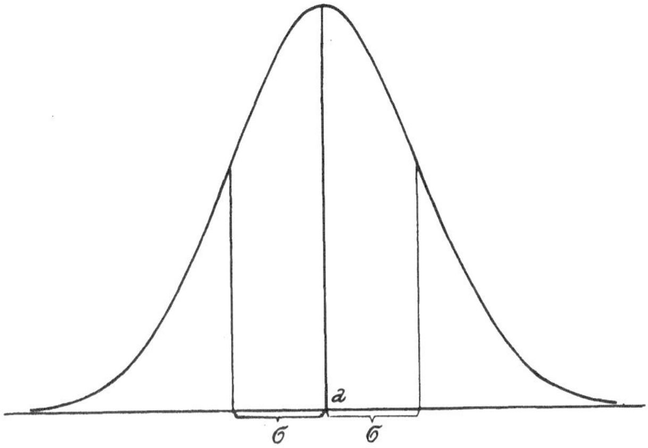

Fig. 5

If we consider an arbitrary probability density which has the same mean value a and the same variance as $N(a, \sigma^2)$, then, roughly speaking, a positive excess indicates that the graph of this density in a neighborhood of a is higher than the graph of the density of $N(a,\sigma^2)$ and a negative excess indicates a corresponding flattening.

26. The characteristic function of the normal distribution. We show

Theorem 26.1. *The characteristic function of an $N(0,1)$-distribution is given by $e^{-t^2/2}$ for all real t.*

Proof. We must compute

$$E(e^{i\xi t}) = \frac{1}{\sqrt{2\pi}} \int_{-\infty}^{+\infty} e^{-x^2/2} e^{itx} dx .$$

It is well known that e^{itx} has the representation $\sum_{k=0}^{\infty} \frac{(itx)^k}{k!}$ for each real t and all $x \in R_1$. Substituting this expression into the integral and assuming for the moment that the summation and integration operations can be interchanged, and taking (25.11) and (25.12) into account, it follows that

$$\begin{aligned} E(e^{i\xi t}) &= \sum_{k=0}^{\infty} (-1)^k \frac{1\cdot 3\cdot \ldots \cdot (2k-1)}{(2k)!} t^{2k} \\ &= \sum_{k=0}^{\infty} (-1)^k t^{2k}/(k!\, 2^k) \\ &= e^{-t^2/2} . \end{aligned}$$

Therefore, it only remains to show that the stated interchange of operations is indeed allowable. To accomplish this, it is sufficient to observe that

$$\sum_{k=0}^{n}\left|\frac{(i\,t\,x)^k}{k!}\right| e^{-x^2/2} \leqslant \sum_{k=0}^{\infty}\frac{|t\,x|^k}{k!}\,e^{-x^2/2} \leqslant e^{|tx|-x^2/2},$$

for each natural number $n \geqslant 1$, each $t \in R_1$ and all $x \in R_1$. The function $x \to e^{|tx|-x^2/2}$ is obviously integrable and, therefore, the theorem of Lebesgue can be applied.

Combining Theorem 26.1 with an appropriate application of (23.10), we obtain the characteristic function of an $N(a, \sigma^2)$ distribution:

$$\phi(t) = e^{iat - t^2\sigma^2/2} \tag{26.1}$$

for each real t.

If we use Theorem 23.2, then proceeding from (26.1) we can once more find the moments of an $N(a, \sigma^2)$-distributed r.v. ξ. For example, for all $t \in R_1$

$$\phi'(t) = (i\,a - t\,\sigma^2)\,e^{iat - t^2\sigma^2/2},$$

so that $\phi'(0) = i\,a$ and therefore, $E(\xi) = a$. Further, since

$$\phi''(t) = [-\sigma^2 + (i\,a - t\,\sigma^2)^2]\,e^{iat - t^2\sigma^2/2}$$

for each $t \in R_1$ and so $\phi''(0) = -\sigma^2 - a^2$, it follows that $E(\xi^2) = \sigma^2 + a^2$ and, applying Theorem 17.2, $E[(\xi - a)^2] = \sigma^2$.

27. The reproductive property of the normal distribution. Let ξ_1, ξ_2 be independent random variables, distributed according to $N(a_i, \sigma_i^2)$, $i = 1, 2$. We consider the sum $\zeta_2 = \xi_1 + \xi_2$ and assert the following reproductive property of the normal distribution:

Theorem 27.1. *The sum of two (and therefore finitely many) independent normally distributed random variables is again normally distributed.*

Proof. We apply Theorem 24.2 to obtain the density g of ζ_2:

$$g(z) = \frac{1}{2\pi \cdot \sigma_1 \cdot \sigma_2} \int_{-\infty}^{+\infty} e^{-(z-y-a_1)^2/2\sigma_1^2}\, e^{-(y-a_2)^2/2\sigma_2^2}\, d\,y$$

for each $z \in R_1$.

After an easy calculation we find that

$$g(z) = \frac{1}{\sqrt{2\pi(\sigma_1^2 + \sigma_2^2)}}\, e^{-(z-a_1-a_2)^2/2(\sigma_1^2+\sigma_2^2)}.$$

Therefore, g is the density of a r.v. distributed according to $N(a_1+a_2, \sigma_1^2+\sigma_2^2)$. This illustrates the statement in Theorem 24.1 concerning the mean value and variance of ζ_2 in the case of independent random variables. Further, we remark that Theorem 27.1 can be proved using the properties of characteristic functions. Namely, if m independent, $N(a_i, \sigma_i^2)$-distributed random variables ξ_i are given, then it follows from Theorem 24.4, using (26.1), that the characteristic function of $\sum_{i=1}^{m} \xi_i$ is just $\prod_{k=1}^{m} e^{ia_k t - t^2\sigma_k^2/2}$, for each $t \in R_1$. According to Theorem 23.3 the corresponding probability distribution is uniquely determined; therefore it is $N\left(\sum_{i=1}^{m} a_i, \sum_{i=1}^{m} \sigma_i^2\right)$.

Now we will prove the following Theorem:

Theorem 27.2. *Let $\xi_1, \ldots, \xi_n$ be n independent, $N(0,1)$-distributed random variables and $(o_{ij})_{1n}^{1n}$ be an orthogonal matrix. We define the random variables $\eta_i = \sum_{l=1}^{n} o_{il}\xi_l$, $i=1,\ldots,n$ and assert that the random variables $\eta_1, \ldots, \eta_n$ are also independent and distributed according to $N(0,1)$.*

Proof. By assumption, the density of the joint distribution of $(\xi_1, \ldots, \xi_n)$ is

$$\frac{1}{(\sqrt{2\pi})^n} e^{-\sum_{i=1}^{n} x_i^2/2}, \tag{27.1}$$

for each $(x_1, \ldots, x_n)$. We now apply Theorem 12.2. If $g_i(x_1, \ldots, x_n) = o_{i1}x_1 + \cdots + o_{in}x_n$ for each $(x_1, \ldots, x_n) \in R_n$, then $g=(g_1, \ldots, g_n)$ establishes a one-to-one correspondence of R_n onto itself. The inverse mapping is also linear and is determined by the inverse of $(o_{ij})_{1n}^{1n}$. The absolute value of the determinant of this inverse is 1; this is then also the absolute value of the Jacobian of the transformation g. Thus, the density of the joint distribution of $\eta_1, \ldots, \eta_n$ is given by $\frac{1}{(\sqrt{2\pi})^n} e^{-\sum_{i=1}^{n} y_i^2/2}$ for each $(y_1, \ldots, y_n) \in R_n$. Whence, the assertion follows (cf. p. 46).

Let c be a real number $\neq 0$. Then it is immediate that $c\xi$ is normally distributed whenever ξ is. Therefore, it follows directly from Theorem 27.1 that each linear-combination[35] of independent normally-distributed random variables is again normally-distributed (cf. also Theorem 36.2).

A series of important distributions connected with the normal distribution will be considered in some detail in the following paragraphs.

---

[35] Not all the coefficients in these linear combinations should be zero.

28. The chi-square distribution of Helmert-Pearson.[36] Let $\xi_1, \ldots, \xi_n$ be n independent $N(0, 1)$-distributed random variables, $n \geqslant 1$. We inquire into the distribution of the r.v. η_n, defined by

$$\eta_n = \xi_1^2 + \cdots + \xi_n^2 . \tag{28.1}$$

We claim that η_n has the following density:

$$g_n(y) = \begin{cases} \frac{1}{2^{n/2}\,\Gamma(n/2)}\, y^{n/2-1}\, e^{-y/2}, & y > 0 \\ 0, & y \leqslant 0 \end{cases} . \tag{28.2}$$

Cf. Fig. 6.

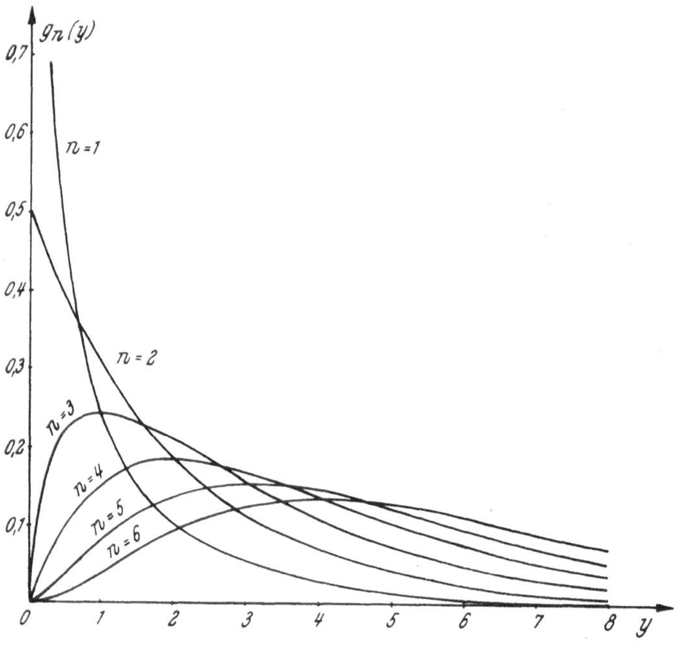

Fig. 6 Density of the χ^2-distribution $g_n(y)$ for $y \geqslant 0$ and $n = 1, \ldots, 6$

The symbol χ_n^2 is frequently used for this distribution in order to emphasize that the r.v. η_n is positive with probability one. By an appropriate application of (11.3), it follows immediately that $g_n(y) = 0$, if $y \leqslant 0$. In

[36] F. R. Helmert, Zeitschrift für Math. und Physik 21, 192—219 (1876). K. Pearson, Philos. Mag. 50. Ser. 5, 157—175 (1900).

order to prove (28.2) for $y>0$, we make use of the induction principle. For $n=1$, we have from (11.3)

$$\int_0^y g_1(x)\,dx=\frac{1}{\sqrt{2\pi}}\left[\int_{-\infty}^{\sqrt{y}} e^{-x^2/2}\,dx-\int_{-\infty}^{-\sqrt{y}} e^{-x^2/2}\,dx\right]=\sqrt{\frac{2}{\pi}}\int_0^{\sqrt{y}} e^{-x^2/2}\,dx.$$

Therefore, using the fact that $\sqrt{\pi}=\Gamma(1/2)$, we obtain (28.2) for $y>0$ and $n=1$ by differentiation. As the induction hypothesis we assume that the statement is correct for $n-1$. We must show that under this assumption for $y>0$ the density of the random variable η_n defined by (28.1) is given by (28.2). However,

$$\eta_n=\xi_1^2+\cdots+\xi_{n-1}^2+\xi_n^2=\eta_{n-1}+\xi_n^2.$$

Since the random variables $\xi_i, 1\leqslant i\leqslant n$, are independent, it follows from Theorem 13.1 that η_{n-1} and ξ_n^2 are also independent. The density can therefore be obtained by an application of (24.1). We find

$$\begin{aligned}g_n(y)&=\int_0^y g_{n-1}(z)g_1(y-z)\,dz\\&=2^{-n/2}(\Gamma((n-1)/2)\Gamma(1/2))^{-1}e^{-y/2}\int_0^y z^{(n-3)/2}(y-z)^{-1/2}\,dz,\end{aligned}$$

for $y>0$. Now, if we make the substitution $z=yx$ and apply (IX) for $a=1/2$ and $b=(n-1)/2$, we obtain the result.

The distribution defined by (28.2) is customarily called the *Helmert-Pearson χ^2-distribution (chi-square-distribution) with n degrees of freedom.*

It is seen immediately that g_n is not continuous at the point $y=0$ for $n=1,2$; however, for $n\geqslant 3$ it is continuous and for $n\geqslant 5$ differentiable on R_1. For the mean value of η_n,

$$E(\eta_n)=\int_0^\infty y\,g_n(y)\,dy=2[\Gamma(n/2)]^{-1}\Gamma((n/2)+1)=2(n/2),$$

so that

$$E(\eta_n)=n,\qquad n\geqslant 1. \tag{28.3}$$

For the second moment, an equally simple calculation gives $E(\eta_n^2)=(2+n)/n$ and therefore for the variance

$$E[(\eta_n-n)^2]=2n. \tag{28.4}$$

In general, from the definition of the Γ-function, we obtain the kth moment,

$$E(\eta_n^k)=2^k\,\Gamma(k+n/2)[\Gamma(n/2)]^{-1}. \tag{28.5}$$

In addition, we will compute the characteristic function of the χ^2-distribution with n degrees of freedom. In analogy to the proof of Theorem 26.1 we obtain first for $|t|<1/2$.

$$\begin{aligned}
E(e^{i\eta_n t}) &= (2^{n/2}\Gamma(n/2))^{-1}\int_0^\infty e^{iyt}e^{-y/2}y^{n/2-1}\,dy \\
&= (2^{n/2}\Gamma(n/2))^{-1}\sum_{k=0}^\infty \int_0^\infty \frac{(iyt)^k}{k!}y^{n/2-1}e^{-y/2}\,dy \\
&= (\Gamma(n/2))^{-1}\sum_{k=0}^\infty \frac{(it)^k}{k!}2^k\Gamma(k+n/2) \\
&= (1-2it)^{-n/2}.
\end{aligned}$$

Considering the integral $\int_0^\infty e^{iyt}e^{-y/2}y^{n/2-1}\,dy$ for complex values of t, we see that it represents an analytic function of t for all t with $|\Im(t)|<1/2$. However, $(1-2it)^{-n/2}$ is also an analytic function for these values of t. Since the set $\{t: |\Im(t)|<1/2\}$ contains the set of real numbers, it follows from the uniqueness theorem for analytic functions that

$$E(e^{i\eta_n t}) = (1-2it)^{-n/2} \tag{28.6}$$

for all real t.

From the definition of the Helmert-Pearson distribution or from (28.6) combined with Theorem 23.3 it follows that the *reproductive property* holds for the chi-square-distribution:

Theorem 28.1. *If a chi-square distributed r.v. with n degrees of freedom is independent of, and added to, a r.v. having a chi-square distribution with m degrees of freedom, then the sum has a chi-square distribution with $m+n$ degrees of freedom.*

We note further that the chi-square distribution is a special case of a more general class of distributions, the so-called *Gamma distributions.* Let $0<\beta<+\infty$ and $\gamma>0$. Then the distribution with density defined by

$$\begin{cases} \dfrac{\gamma^\beta x^{\beta-1}e^{-\gamma x}}{\Gamma(\beta)}, & x>0 \\ 0, & x\leqslant 0 \end{cases} \tag{28.7}$$

is called the Gamma distribution with parameters β and γ. For $\gamma=1/2$ and $\beta=n/2$ we obtain a chi-square distribution with n degrees of freedom.

29. The *t* or Student distribution[37]. Let ξ be an $N(0,1)$-distributed r.v. independent of η_n, a chi-square distributed r.v. with $n \geqslant 1$ degrees of freedom.

We define the random variable

$$t = \xi / \sqrt{\eta_n / n}. \tag{29.1}$$

It is only defined for $\eta_n > 0$, but according to (28.2) $P(\eta_n \leqslant 0) = 0$ so that this fact will be of no consequence in the sequel. Because of the independence of ξ and η_n, the density of the joint distribution of (ξ, η_n) is

$$\begin{matrix} (2^{n/2}\,\Gamma(n/2))^{-1}\, e^{-x_2/2}\, x_2^{n/2-1}\, \dfrac{e^{-x_1^2/2}}{\sqrt{2\pi}}, & -\infty < x_1 < \infty, & x_2 > 0 \\ 0, & -\infty < x_1 < \infty, & x_2 \leqslant 0. \end{matrix} \tag{29.2}$$

If we make the following one-to-one transformation

$$\begin{matrix} y_1 = x_1 / \sqrt{x_2/n}, \\ y_2 = x_2, \end{matrix} \qquad -\infty < x_1 < \infty, \quad x_2 > 0$$

then for the absolute value of the Jacobian we find $\sqrt{y_2/n}$. Whence we obtain a new two-dimensional density

$$\begin{cases} [2^{n/2}\,\Gamma(n/2)]^{-1}\,(2\pi n)^{-1/2}\, e^{-1/2\,(y_1^2 y_2/n + y_2)}\, y_2^{n/2-1} \sqrt{y_2}, & -\infty < y_1 < \infty,\ y_2 > 0 \\ 0, & -\infty < y_1 < \infty,\ y_2 \leqslant 0. \end{cases}$$

In order to find the density h_n of t, we turn to the marginal distribution and obtain

$$h_n(t) = \frac{1}{\sqrt{\pi n}\,\Gamma(n/2)} \int_0^\infty e^{-(u/2)(1+t^2/n)}\,(u/2)^{(n-1)/2}\, du/2$$

for each real t and from this by a simple calculation

$$h_n(t) = \frac{\Gamma((n+1)/2)\,(1+t^2/2n)^{-(n+1)/2}}{\Gamma(1/2)\,\Gamma(n/2)\sqrt{n}}, \qquad -\infty < t < \infty, \qquad n \geqslant 1. \tag{29.3}$$

The expression in (29.3) is referred to as the density of a *t-distribution with n degrees of freedom* or as the density of the *Student distribution.* This distribution is also sometimes named for Fisher. The number of degrees of freedom is that of the chi-square distribution occurring in the definition of the t-distribution, It is immediately seen that the t-distribution is symmetric with respect to the origin. (Cf. Fig. 7.)

---

[37] "Student", Biometrika 6, 1—25 (1908), (Student is a pseudonym for W.S. Gosset). R.A. Fisher, Biometrika 10, 507—521 (1915).

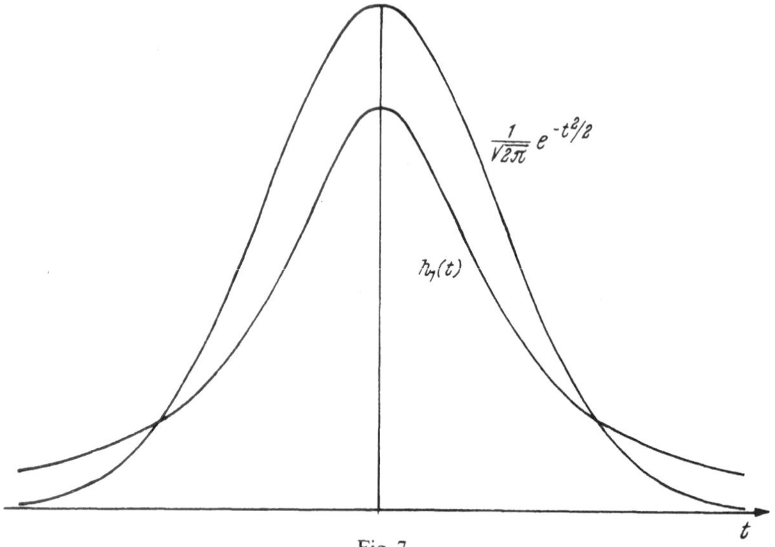

Fig. 7

We consider briefly the case $n=1$. Then $h_1(t)=(1/\pi)(1/(1+t^2))$ for all real t. This density is a special case of a more general class of densities whose distributions are referred to as *Cauchy-distributions*. The general form of their density is for each real t given by $(a/\pi)(1/[(t-b)^2+a^2])$, $a>0$, $-\infty<b<\infty$. The mean value, and therefore also any higher order moments, does not exist for these distributions; hence, in particular, they do not exist for the t-distribution with one degree of freedom.

Moreover the distributions defined by (29.3) have moments up to, but not including, order n. That is,

$$\int_0^\infty t^k(1+t^2/n)^{-(n+1)/2}\,dt$$

converges for (all real) k with $-1<k<n$, but not for $k=n$.

Since the t-distribution is symmetric, the odd order moments vanish so far as they exist. For the purpose of calculating the moments of even order, we consider the integral

$$\int_0^\infty t^{2m}(1+t^2/n)^{-k}\,dt, \qquad 0\leqslant m<(k-1/2), \qquad k>1/2, \qquad n\geqslant 1.$$

Taking (IX) into consideration, we find that the substitution $1+t^2/n=1/z$ yields

$$\frac{n^{m+1/2}}{2}\,\frac{\Gamma(m+1/2)\,\Gamma(k-m-1/2)}{\Gamma(k)}$$

for this integral. If we choose $k=(n+1)/2$, then it follows from this result that

$$E(t^{2m})=\frac{(2m-1)\cdot\ldots\cdot 3\cdot 1\cdot n^m}{(n-2)\cdot\ldots\cdot(n-2m)} \tag{29.4}$$

for $2m<n$.

30. The *F*-distribution.[38] Let η_n be a chi-square distributed r.v. with $n, n\geqslant 1$, degrees of freedom and η_m be chi-square distributed with m degrees of freedom, independently of η_n. We define a new r.v. $\boldsymbol{F}$ by setting

$$\boldsymbol{F}=\frac{\eta_n}{n}\Big/\frac{\eta_m}{m}. \tag{30.1}$$

As in the case of the t-distribution, the vanishing of the denominator presents no difficulty in the definition. In order to derive the density of F we again proceed from the joint density of the random variable (η_n, η_m) which, because of the independence assumption, is given by

$$\begin{matrix} 2^{-(n+m)/2}(\Gamma(n/2)\Gamma(m/2))^{-1}x_1^{n/2-1}e^{-x_1/2}x_2^{m/2-1}e^{-x_2/2}, & x_1>0, \quad x_2>0, \\ 0, & \text{otherwise.}\end{matrix}$$

If we now make the transformation $(x_1/n)(x_2/m)^{-1}=y_1$, $x_2=y_2$, $x_1>0$, $x_2>0$, it follows that the absolute value of the Jacobian of this transformation is $(n/m)y_2$. Hence, we obtain a new two-dimensional density and, turning to the marginal distribution, we obtain, after an easy calculation, the density of $\boldsymbol{F}$:

$$k_{n,m}(F)=\begin{cases}\dfrac{\Gamma((n+m)/2)\,(n/m)^{n/2}}{\Gamma(n/2)\Gamma(m/2)}F^{(n/2)-1}(1+F(n/m))^{-(m+n)/2}, & F>0,\\ 0, & F\leqslant 0.\end{cases} \tag{30.2}$$

The function defined in (30.2) is called the density of the *F-distribution with (n, m) degrees of freedom.* From the definition of the F-distribution it follows immediately that

Theorem 30.1. *If* $\boldsymbol{F}$ *has an F-distribution with (n, m) degrees of freedom, then the r.v.* $1/\boldsymbol{F}$ *has an F-distribution with (m, n) degrees of freedom.*

Various simple transformations of (30.2) lead to other useful distributions. We consider first the random variable[39]

$$\zeta=(\log \boldsymbol{F})/2,$$

[38] This distribution is also named for Snedecor.

[39] R.A. Fisher, Metron 1, 1—32 (1921).

where F has the density (30.2). It is easily shown that the distribution of ζ has the following density:

$$\frac{2\Gamma((n+m)/2)\,(n/m)^{n/2}}{\Gamma(n/2)\,\Gamma(m/2)}\,e^{nz}(1+(n/m)\,e^{2z})^{-(m+n)/2}, \qquad -\infty<z<+\infty.$$

This distribution is called the *z-distribution of Fisher.*

If we consider the random variable $\xi=\dfrac{(n/m)\boldsymbol{F}}{1+(n/m)\boldsymbol{F}}$, then we obtain a so-called Beta distribution whose density is given by

$$\begin{cases} \dfrac{\Gamma((n+m)/2)}{\Gamma(n/2)\,\Gamma(m/2)}\,x^{n/2-1}(1-x)^{m/2-1}, & 0<x<1 \\ 0, & \text{otherwise.} \end{cases} \tag{30.3}$$

We denote this distribution briefly by $B(n/2,m/2)$. More generally, the *Beta distribution* $B(\alpha,\beta)$, $\alpha>0$, $\beta>0$ is determined by the density

$$\begin{cases} \dfrac{\Gamma(\alpha+\beta)}{\Gamma(\alpha)\,\Gamma(\beta)}\,x^{\alpha-1}(1-x)^{\beta-1}, & 0<x<1 \\ 0, & \text{otherwise.} \end{cases} \tag{30.4}$$

Notice that by setting $n=1$ in (30.2) we obtain a distribution which is closely related to the t-distribution with m degrees of freedom. The r.v. $\eta_1/(\eta_m/m)$ is obviously the square of a r.v. with density h_m. This can also be demonstrated by an appropriate application of (11.3). Consider for each $y>0$ the expression

$$m^{-1/2}\,\Gamma((m+1)/2)\,[\Gamma(1/2)\,\Gamma(m/2)]^{-1}$$
$$\times\left\{\int_{-\infty}^{\sqrt{y}}(1+t^2/m)^{-(m+1)/2}\,dt-\int_{-\infty}^{-\sqrt{y}}(1+t^2/m)^{-(m+1)/2}\,dt\right\};$$

by differentiation w.r.t y, we obtain

$$m^{-1/2}\,\Gamma((m+1)/2)\,[\Gamma(1/2)\,\Gamma(m/2)]^{-1}(1+y/m)^{-(m+1)/2}\,y^{-1/2}.$$

However, this is just $k_{1,m}(y)$ for $y>0$.

31. The Pearson type distributions. All the special distributions considered up to now in **28**, **29** and **30** and a number of other practically important distributions can be obtained formally from a more general point of view. This is motivated by the following fact: the density of an $N(a,\sigma^2)$ distribution satisfies the following first order, linear and homogeneous differential equation

$$y'=-\frac{(x-a)}{\sigma^2}\,y, \qquad -\infty<x<\infty, \qquad -\infty<y<\infty. \tag{31.1}$$

We now generalize (31.1) somewhat and consider the differential equation

$$y' = \frac{(x+a_1)y}{b_1+b_2x+b_3x^2}, \tag{31.2}$$

where the right side is defined for all real x and y which do not permit the denominator to vanish and a_1, b_1, b_2 and b_3 are four arbitrary real numbers. By varying these four numbers an abundance of typical solutions of (31.2) can be obtained. By suitable normalization and restriction to appropriate intervals we get the so-called Pearson-type[40] distributions. They owe their origin chiefly to the desire to approximate empirically acquired frequency distributions by distributions of the continuous type. Through proper choice of the four parameters a_1, b_1, b_2, b_3 it is, neglecting exceptional cases, always possible to produce a distribution density whose first four moments are given in advance. We will not go into this and will for brevity give only a classification according to seven types of densities defined by means of (31.2). The Pearson type curves are characterized by the behavior of the denominators $b_1+b_2x+b_3x^2$ on the right side of (31.2).

Type I. If we suppose that $b_3 \neq 0$, then we can write

$$b_1+b_2x+b_3x^2 = b_3(c_1+c_2x+x^2).$$

Now, we assume that $c_1+c_2x+x^2=0$ has real roots α_1, α_2 and also that $\alpha_1<\alpha_2$. In place of (31.2) we obtain, for $y \neq 0$,

$$\frac{d\log y}{dx} = \frac{x+a_1}{b_1+b_2x+b_3x^2}. \tag{31.3}$$

By a partial fraction decomposition we find that a particular solution y of (31.2) has the form

$$\log y(x) = (A/b_3)\log|x-\alpha_1| + (B/b_3)\log|x-\alpha_2| \quad \text{for } x \neq \alpha_1, \alpha_2$$

with appropriate constants A and B. The general solution is therefore of the form

$$y(x) = C|x-\alpha_1|^{\beta_1}|x-\alpha_2|^{\beta_2}, \qquad x \neq \alpha_1, \alpha_2,$$

where C is an arbitrary real number. This leads to distribution densities of the following kind:

$$\begin{cases} 0, & x \leqslant \alpha_1, \\ C(x-\alpha_1)^{\beta_1}(\alpha_2-x)^{\beta_2}, & \alpha_1<x<\alpha_2, \quad \beta_1>-1, \quad \beta_2>-1 \\ 0, & x \geqslant \alpha_2. \end{cases} \tag{31.4}$$

Naturally, C is to be chosen so that $C\int_{\alpha_1}^{\alpha_2}(x-\alpha_1)^{\beta_1}(\alpha_2-x)^{\beta_2}dx=1$. For $\alpha_1=0$, $\alpha_2=1$, (31.4) yields the Beta distribution (30.4).

Type II. These distribution densities represent only a special case of those of having the form (31.4). As such, they are not suitable for characterizing a type; yet the case $\alpha_1=-\alpha_2=\alpha$, and thus, densities of the form

$$C(\alpha^2-x^2)^{\beta}, \qquad -\alpha<x<\alpha, \qquad \beta>-1/2 \tag{31.5}$$

have traditionally been used to denote type II.[41]

---

[40] K. Pearson, Philos. Trans. Roy. Soc. London, Ser. A 185, 71—110 (1894).

[41] We will no longer state the intervals over which the densities vanish. The constant C is always to be chosen in such a way that (6.3) holds in each case.

Type III. Let $b_3=0$, $b_2 \neq 0$. Then we obtain distribution densities representing type III in the form

$$Ce^{-\gamma x}(b_1+b_2 x)^\beta, \quad x>-b_1/b_2, \quad \beta>-1, \quad \gamma>0.$$

The Gamma distribution (28.7) is obtained as a special case when $b_1=0$, $b_2=1$.

Type IV. Again, let $b_3 \neq 0$. However, now suppose, in contrast to type I, that the polynomial $c_1+c_2 x+x^2$ is always positive on $-\infty<x<\infty$, so that $c_2^2-4c_1<0$. Then $c_1>0$. A simple integration of (31.2) yields the distribution densities of type IV:

$$C(x^2+c_2 x+c_1)^\beta e^{\gamma \arctan[(x+c_2/2)\sqrt{c_1-c_2^2/4/2}]},$$

$-\infty<x<\infty$, $\beta<-1/2$, $-\infty<\gamma<\infty$.

Type V. Let $b_3 \neq 0$. Suppose that the polynomial $c_1+c_2 x+x^2$ has a (real) double zero α: $c_1+c_2 x+x^2=(x-\alpha)^2$. The densities then have the form

$$C(x-\alpha)^\beta e^{-\gamma/(x-\alpha)}, \quad x>\alpha, \quad \beta>-1, \quad \gamma>0.$$

Type VI. This type is closely related to type I. Suppose that all the assumptions made when we considered type I are satisfied. Using the notation introduced there, we also obtain densities of the form

$$C(x-\alpha_1)^{\beta_1}(x-\alpha_2)^{\beta_2}, \quad x>\alpha_2, \quad \beta_2>-1, \quad \beta_1+\beta_2<-1.$$

Type VII. This type is a special case of type IV. It is now assumed that the polynomial $c_1+c_2 x+x^2$ has two purely imaginary roots; hence $c_2=0$.

We write down only the special case $b_2=0$, $b_1=1$, $b_3=1/n$, $\beta=-(n+1)/2$, $\gamma=0$ which yields the density h_n of the t-distribution.

We mention that the normal distribution is often viewed as limiting case of type VII (see also p. 115). Actually, by appropriate choice of the constants and then by a limiting process, one can obtain the normal distribution from each type. For the densities of type I or II for example, we choose the notation $C(1-x^2/\alpha^2)^\beta$, set $\alpha^2=2n$, $\beta=n$ and let $n\to\infty$.

For type IV or VII, write the densities in the form $C(1+x^2/\alpha^2)^\beta$, choose $\alpha^2=2n$, $\beta=-n$ and let $n\to\infty$ again. Similar results hold for the rest.

32. The binomial distribution. Up to now we have only especially important examples of continuous distributions. We turn now to some important discrete distributions. Perhaps the best known is the Bernoulli or binomial, which one can obtain as follows:

Let $\xi_1,\dots,\xi_n$ be n, $n\geqslant 1$, independent r.v.'s with the same discrete distribution given by

$$P(\xi_i=0)=q, \quad P(\xi_i=1)=p, \quad p,q>0, \quad p+q=1. \tag{32.1}$$

Consider the sum

$$\zeta_n=\xi_1+\dots+\xi_n, \quad n\geqslant 1. \tag{32.2}$$

We claim that ζ_n possesses the following discrete distribution:

For each integer k, $0\leqslant k\leqslant n$

$$P(\zeta_n=k)=\binom{n}{k}p^k q^{n-k}. \tag{32.3}$$

This distribution is called the *binomial distribution*, which we abbreviate by $B_n(p)$. To show (32.3), we use induction, noting immediately that it holds for $n=1$. Let ζ_{n-1} be $B_{n-1}(p)$-distributed. Since the ξ_i are independent, ξ_n and ζ_{n-1} are also independent. Thus, for each integer $k, 0 \leqslant k \leqslant n$

$$P(\zeta_n = k) = P(\zeta_{n-1} = k)P(\xi_n = 0) + P(\zeta_{n-1} = k-1)P(\xi_n = 1)$$
$$= \binom{n-1}{k} p^k q^{n-k-1} q + \binom{n-1}{k-1} p^{k-1} q^{n-k} p = \binom{n}{k} p^k q^{n-k}$$

which is the right side of (32.3). This conclusion holds for $1 \leqslant k \leqslant n-1$, since $\binom{n-1}{n} = \binom{n-1}{-1} = 0$ it is also formally correct for $k=0$ and n.

This can also be shown in the following more intuitive fashion: The probability that $\zeta_n = k$ is the same as the probability that any k of the ξ_i assume the value 1 and the rest of the ξ_i's the value 0. This can happen in $\binom{n}{k}$ ways. But the probability that k given r.v.'s assume the value 1 and the rest 0, is given by $p^k q^{n-k}$ by reason of the assumed independence. Theorem 1.1 then leads to (32.3).

An application of Theorem 24.1 and 17.4 allows us to give the mean and variance of $B_n(p)$ at once. For the r.v. ξ_i we have by (32.1)

$$E(\xi_i) = 0 \cdot q + 1 \cdot p = p \quad \text{and } E(\xi_i^2) = p, \quad \text{thus } E[(\xi_i - p)^2] = pq,$$

so that

$$E(\zeta_n) = np \tag{32.4}$$

and

$$E[(\zeta_n - np)^2] = npq. \tag{32.5}$$

For the characteristic function of $B_n(p)$ one gets for each real t

$$\phi(t) = \sum_{k=0}^{n} e^{ikt} \binom{n}{k} p^k q^{n-k} = \sum_{k=0}^{n} (e^{it} p)^k q^{n-k} \binom{n}{k},$$

thus,

$$\phi(t) = (p e^{it} + q)^n, \qquad -\infty < t < \infty. \tag{32.6}$$

From either the definition, or using the characteristic function we immediately obtain the reproductive property of the binomial distribution:

Theorem 32.1. *The sum of two independent $B_n(p)$- and $B_m(p)$-distributed r.v.'s is $B_{n+m}(p)$-distributed.*

Applying Čebyšev's inequality (22.2) to $B_n(p)$, one gets the so-called *Theorem of Bernoulli:*

Theorem 32.2. *Let ζ_n be a $B_n(p)$-distributed r.v. The probability that the r.v. ζ_n/n deviates arbitrarily little from p can be made arbitrarily close to 1 by chosing n large enough.*

Proof. From (32.4) and (32.5) we get, applying Čebyšev's inequality:

$$P(|\zeta_n - np| \leqslant t\sqrt{npq}) \geqslant 1 - 1/t^2, \qquad t>0$$

or

$$P(|\zeta_n/n - p| \leqslant t\sqrt{pq/n}) \geqslant 1 - 1/t^2.$$

Now let $\varepsilon>0$ be given and set $t\sqrt{pq/n}=\varepsilon$, thus $t^2=\varepsilon^2 n/pq$. Then, for arbitrary given $\delta>0$

$$P(|\zeta_n/n - p| \leqslant \varepsilon) \geqslant 1-\delta$$

provided one chooses n large enough that

$$pq/\varepsilon^2 n \leqslant \delta.$$

The derivation of the binomial distribution given on p. 89 can obviously be interpreted as follows: ζ_n is the absolute frequency of the event "1" in a series of n independent Bernoulli trials. Correspondingly, ζ_n/n represents the relative frequency of this event. The statement of Theorem 32.2 can thus be viewed in the sense of this interpretation as "justification" of the frequency interpretation of the probability calculus. In the sense of the interpretation on p. 27 of probabilities which come arbitrarily close to 1, the statement of Theorem 32.2 can be formulated as follows: The relative frequency of the occurrence of the event "1" agrees "almost always" with p provided the series of trials is long enough (see also p. 20 and 103).

We prove a theorem which we will need later (see II, p. 146, IV, p. 260) and which compares the values of the distribution functions of two binomial distributions.

Theorem 32.3. *Let $0\leqslant k<n$ and $0<p<p_1<1, q=1-p, q_1=1-p_1$. Then*

$$\sum_{r=0}^{k}\binom{n}{r}p^r q^{n-r} > \sum_{r=0}^{k}\binom{n}{r}p_1^r q_1^{n-r}. \tag{32.7}$$

Proof: For $0\leqslant p\leqslant 1$, $0\leqslant k<n$, let $S_k(p)=\sum_{r=0}^{k}\binom{n}{r}p^r q^{n-r}$. Then, for $0<p<1$

$$S_k'(p)=\sum_{r=1}^{k} r\binom{n}{r}p^{r-1}q^{n-r} - \sum_{r=0}^{k}(n-r)\binom{n}{r}p^r q^{n-r-1}.$$

Further

$$S_k'(p)=n\sum_{r=1}^{k}\binom{n-1}{r-1}p^{r-1}q^{n-r} - n\sum_{r=0}^{k}\binom{n-1}{n-r-1}p^r q^{n-r-1}$$

or

$$S_k'(p) = -n\binom{n-1}{k}p^k q^{n-k-1}. \tag{32.8}$$

But then $S_k'(p)<0$ in $0<p<1$, so that (32.7) is proved.[42]

Now $S_k(0)=1$ and thus from (32.8), one also has

$$S_k(p) = 1 - n\binom{n-1}{k}\int_0^p x^k(1-x)^{n-k-1}\,dx \tag{32.9}$$

or

$$S_k(p) = 1 - \int_0^p x^k(1-x)^{n-k-1}\,dx \Big/ \int_0^1 x^k(1-x)^{n-k-1}\,dx.$$ [43]

Form this one can, again, immediately see the correctness of Theorem 32.3.

33. The Poisson distribution. Let $a>0$. We define a r.v. ξ by the following formula:

$$P(\xi=r) = e^{-a}a^r/r!, \qquad r=0,1,2,\ldots. \tag{33.1}$$

One sees immediately that $\sum_{r=0}^{\infty} e^{-a}a^r/r! = 1$ so that (33.1) actually defines a distribution, called the *Poisson distribution.* The development of probability theory has reflected the crucial role played by this distribution which becomes especially clear in limit theorems[44]. We will not go into this here but merely show how one can obtain the Poisson distribution from the Bernoulli distribution by a limiting process. For $a>0$ and

---

42 We can also show this without any calculations: Let $p_1>p$. Then it is possible to define r.v.'s ξ_i, resp. $\xi_i^{(1)}$ for $i=1,\ldots,n$ over the same probability space in such a way that they are distributed by $B_1(p)$, resp., $B_1(p_1)$, and so that $\{\omega:\xi_i(\omega)=1\}\subset\{\omega:\xi_i^{(1)}(\omega)=1\}$. Then with $\zeta_n=\sum_{i=1}^{n}\xi_i$ and $\zeta_n^{(1)}=\sum_{i=1}^{n}\xi_i^{(1)}$ we naturally have $P(\zeta_n^{(1)}>k)>P(\zeta_n>k)$ if $0\leqslant k<n$, whence (32.7) follows.

This result is related to the fact that the set of all $B_n(p)$, $0<p<1$, is a class with monotone likelihood ratios. (See III, p. 186.) If a d.f. F_2 has a monotone likelihood ratio w.r.t. a d.f. F_1 then $F_2(x)\leqslant F_1(x)$ for all $x\in R_1$.

43 This formula already appears in A. Meyer, Vorlesungen uber Wahrscheinlichkeitsrechnung, B.G. Teubner, Leipzig 1879. The integral in the numerator is called the incomplete Beta-function and has been widely tabulated, for example, K. Pearson, Tables of the Incomplete *B*-function, Cambridge University Press, London 1934.

44 For a thorough treatement of limit theorems see B. V. Gnedenko and A. Kolmogoroff, Limit Distributions for Sums of Independent Random Variables, Cambridge, Mass., 1954.

$n=1,2,\ldots,$ consider $B_n(a/n)$. Then, for each fixed, nonnegative integer r, we get for large enough n

$$\binom{n}{r}(a/n)^r(1-a/n)^{n-r} = \frac{a^r}{r!}(1-a/n)^n(1-a/n)^{-r}\frac{n(n-1)\cdots(n-r+1)}{n^r}$$

and this tends to $e^{-a}a^r/r!$ for $n\to\infty$ since $\dfrac{n-i}{n}\to 1$ for $i=0,1,\ldots,r-1$.

For the mean of a Poisson variable we get

$$E(\xi)=\sum_{r=0}^{\infty} r e^{-a}\frac{a^r}{r!} = a e^{-a}\sum_{r=1}^{\infty}\frac{a^{r-1}}{(r-1)!}$$

so that

$$E(\xi)=a. \tag{33.2}$$

For the variance one easily finds

$$E[(\xi-a)^2]=a. \tag{33.3}$$

The characteristic function is also easy to obtain: for each $t\in R_1$,

$$E(e^{i\xi t})=\sum_{r=0}^{\infty} e^{irt}e^{-a}\frac{a^r}{r!} = \sum_{r=0}^{\infty} e^{-a}\frac{(a e^{it})^r}{r!},$$

so that

$$E(e^{i\xi t})=e^{-a}e^{ae^{it}}. \tag{33.4}$$

The Poisson distribution has the *reproductive property:*

Theorem 33.1. *The sum of two independent Poisson variables is also a Poisson variable.*

Proof. This is shown either by (33.4) and Theorem 24.4 or directly. We will do the latter. Let ξ_i, $i=1,2$ be independent Poisson r.v.'s with means a_i. Let $\zeta_2=\xi_1+\xi_2$. For $k\geqslant 0$

$$\begin{aligned} P(\zeta_2=k) &= \sum_{r=0}^{k} e^{-a_1}\frac{a_1^{k-r}}{(k-r)!}e^{-a_2}a_2^r/r! \\ &= e^{-(a_1+a_2)}/k!\sum_{r=0}^{k}\frac{a_1^{k-r}a_2^r}{r!(k-r)!}k! \\ &= e^{-(a_1+a_2)}(a_1+a_2)^k/k!\,. \end{aligned}$$

There exists an analogue to Theorem 32.3:

Theorem 33.2. *For* $0<a<\infty$, $0\leqslant k<\infty$ *let*

$$S_k(a)=\sum_{r=0}^{k} e^{-a}a^r/r!\,.$$

Then for each k *the map* $a\to S_k(a)$ *is strictly monotone decreasing.*

Proof. Let $0<a<\infty$. Then, $S_k'(a)=-e^{-a}a^k/k!$, i.e., always <0, from which the claim follows. We get, in addition, the formula

$$S_k(a)=1-\int_0^a e^{-x}(x^k/k!)\,dx.$$

34. The hypergeometric distribution. We recall the example considered in **19.** and use the terminology introduced there. Let M be an integer with $1\leqslant M<N$. Define r.v.'s η_i, $1\leqslant i\leqslant n$ as follows:

$$\eta_i=\begin{cases}0, & \text{if } \xi_i>x_M\\ 1, & \text{if } \xi_i\leqslant x_M\end{cases}.$$

We write $n_1=\max(0,n+M-N)$ and $n_2=\min(n,M)$. From (19.1) and the definition of the η_i it is easy to see that if

$$n_1\leqslant m\leqslant n_2 \tag{34.1}$$

and $\{i_1,\ldots,i_m\}$ is a subset of $\{1,\ldots,n\}$, then the probability that the r.v.'s $\eta_{i_1},\ldots,\eta_{i_m}$ all assume the value 1 and the $n-m$ remaining the value 0 is given by

$$\frac{M(M-1)\cdots(M-m+1)(N-M)(N-M-1)\cdots(N-M-(n-m)+1)}{N(N-1)\cdots(N-n+1)}$$
$$=\frac{M!(N-M)!(N-n)!}{(M-m)!(N-M-(n-m))!N!}.$$

Considering now the r.v. $\eta=\eta_1+\cdots+\eta_n$, we have for all m satisfying (34.1)

$$P(\eta=m)=\binom{n}{m}\frac{M!(N-M)!(N-n)!}{(M-m)!(N-M-(n-m))!N!},$$

thus,

$$P(\eta=m)=\binom{M}{m}\binom{N-M}{n-m}\bigg/\binom{N}{n}. \tag{34.2}$$

We can easily check that (34.2) defines a distribution. Indeed, comparing the coefficient of x^n in the binomial expansion of $(1+x)^N$ with that in the expansion of $(1+x)^M(1+x)^{N-M}$, we get

$$\binom{N}{n}=\sum_{n_1\leqslant m\leqslant n_2}\binom{M}{m}\binom{N-M}{n-m}. \tag{34.3}$$

The distribution defined by (34.2) is called the *hypergeometric distribution.*

For its mean we have

$$E(\eta)=\sum_{n_1\leqslant m\leqslant n_2} m\binom{M}{m}\binom{N-M}{n-m}\bigg/\binom{N}{n}=M\sum\frac{\binom{M-1}{m-1}\binom{(N-1)-(M-1)}{(n-1)-(m-1)}}{\binom{N}{n}},$$

where the summation in the last sum is over

$$\max(0,(n-1)+(M-N))\leqslant m-1\leqslant \min(n-1,M-1).$$

Applying (34.3) then gives $E(\eta)=Mn/N$ or, letting $M/N=p$,

$$E(\eta)=np. \tag{34.4}$$

A simple calculation yields, with $1-p=q$:

$$E[(\eta-np)^2]=\frac{N-n}{N-1}npq. \tag{34.5}$$

The derivation of the hypergeometric distribution and the formulas (34.2), (34.4) and (34.5) illustrate the analogy to the binomial distribution.

In fact, the considerations leading to the hypergeometric distribution can also be interpreted as follows: Assume we have a finite set of N elements, M elements of which belong to one class, and the rest $(N-M)$, to another. One draws a sample of size n from this population *(without replacement)* and asks for the probability that exactly $m\leqslant n$ elements in this sample are from the first class and $n-m$ from the second. This probability is given by the right side of (34.2). The relation between this question and the interpretation on p. 89 of the binomial distribution is obvious. In short: One gets the *hypergeometric distribution* if one draws a sample from a finite population *without replacement*, and the *binomial* if one *replaces each element* drawn before taking the next one.

The binomial distribution is easily seen to be a limiting case of the hypergeometric distribution. We have

Theorem 34.1. *Letting* $M,N\to\infty$ *in* (34.2) *in such a way that*

$$M/N\to p, \quad 0<p<1 \tag{34.6}$$

and m *and* n *remain fixed, one obtains the probability of a* $B_n(p)$, *where* p *is given by* (34.6).

For the *proof*, note that it follows from (34.6) that $N-M$ also tends to ∞. Now apply Stirling's formula (p. 111) to (34.2). After easy manipulations we get

$$\binom{n}{m}A_{M,N}\frac{M^M(N-M)^{N-M}(N-n)^{N-n}}{(M-m)^{M-m}(N-M-n+m)^{N-M-n+m}N^N}.$$

$A_{M,N}$ consists of a finite set of factors which are easily seen to tend to 1. Gathering terms appropriately, we find

$$\binom{n}{m} A_{M,N} \left(\frac{N-n}{N}\right)^N \left(\frac{M}{M-m}\right)^{M-m} \left(\frac{N-M}{N-M-n+m}\right)^{N-M-n+m}$$
$$\times \left(\frac{M}{N}\right)^m \left(1-\frac{M}{N}\right)^{n-m} \left(1-\frac{n}{N}\right)^{-n}.$$

These factors tend in order for $M, N \to \infty$ and $M/N \to p$ to

$$\binom{n}{m},\ 1,\ e^{-n},\ e^m,\ e^{n-m},\ p^m,\ q^{n-m} \text{ and } 1,$$

where $q = 1-p$. Hence,

$$\frac{\binom{M}{N}\binom{N-M}{n-m}}{\binom{N}{n}} \to \binom{n}{m} p^m q^{n-m}$$

when we pass to the limit as explained above.

We shall now give an analogue to Theorem 32.3 which will be used later on.

Theorem 34.2. *Let k, n, M, M' and N be nonnegative integers satisfying $0 \leqslant k \leqslant \min(n, M)$, $\max(n, M) \leqslant N$, $M' > M$ and $n \leqslant N - M'$. Then*

$$\sum_{0 \leqslant r \leqslant k} \binom{M'}{r}\binom{N-M'}{n-r} \leqslant \sum_{0 \leqslant r \leqslant k} \binom{M}{r}\binom{N-M}{n-r} \tag{34.7}$$

and equality holds, if at all, only for $k = \min(n, M)$.

For the *proof* note that an application of (34.3) leads to

$$\binom{M'}{r} = \sum_{0 \leqslant l \leqslant r} \binom{M}{l}\binom{M'-M}{r-l}^{45} \tag{34.8}$$

and

$$\binom{N-M}{n-r} = \sum_{0 \leqslant t \leqslant n-r} \binom{N-M'}{n-r-t}\binom{M'-M}{t} \tag{34.9}$$

for $0 \leqslant r \leqslant n$.

---

[45] If $r-l > M'-M$, then $\binom{M'-M}{r-l} = 0$.

Hence, by (34.8)

$$\sum_{0\leqslant r\leqslant k}\binom{M'}{r}\binom{N-M'}{n-r}=\sum_{0\leqslant r\leqslant k}\sum_{0\leqslant l\leqslant r}\binom{M}{l}\binom{M'-M}{r-l}\binom{N-M'}{n-r}$$
$$=\sum_{0\leqslant l\leqslant k}\binom{M}{l}\sum_{0\leqslant r\leqslant k}\binom{M'-M}{r-l}\binom{N-M'}{n-r}$$

by interchanging the order of summation. Replacing $r-l$ by t, we get

$$\sum_{0\leqslant l\leqslant k}\binom{M}{l}\sum_{0\leqslant t\leqslant k-l}\binom{M'-M}{t}\binom{N-M'}{n-l-t}.$$

On the other hand, by (34.9),

$$\sum_{0\leqslant r\leqslant k}\binom{M}{r}\binom{N-M}{n-r}=\sum_{0\leqslant l\leqslant k}\binom{M}{l}\sum_{0\leqslant t\leqslant n-l}\binom{M'-M}{t}\binom{N-M'}{n-l-t}.$$

But since $\binom{M}{l}$ as well as $\sum_{0\leqslant t\leqslant n-l}\binom{N-M'}{n-l-t}\binom{M'-M}{t}$ are always >0 by assumption, (34.7) holds with equality at most for $k=\min(n,M)$.

35. The multidimensional normal distribution. Let $\sum_{i,j=1}^{n} a_{ij}x_i x_j$, $n\geqslant 1$ be a positive definite quadratic form which we will also write as $x'Ax$ with $A=(a_{ij})_{1n}^{1n}$. By assumption, $|A|>0$. We now claim that with $a\in R_n$ and $x\in R_n$

$$x\to|A|^{1/2}(2\pi)^{-n/2}\,e^{-(x-a)'A(x-a)/2} \tag{35.1}$$

defines the density of a continuous distribution called the *n-dimensional normal distribution.* The nonnegativity of (35.1) is trivial. We still need to verify that

$$|A|^{1/2}(2\pi)^{-n/2}\int_{R_n} e^{-(x-a)'A(x-a)/2}\,dx=1. \tag{35.2}$$

For this make the transformation $y=x-a$ whose Jacobian has absolute value 1. Then it suffices to show that

$$\int_{R_n} e^{-y'Ay/2}\,dy=(2\pi)^{n/2}A^{-1/2}. \tag{35.3}$$

For this purpose apply an orthogonal transformation with matrix $\mathfrak{D}$ which takes $y'Ay$ into $z'\Lambda z$, where

$$\mathfrak{D}A\mathfrak{D}=\Lambda=\begin{pmatrix}\lambda_1 & 0 & \dots & 0\\ 0 & \lambda_2 & \dots & 0\\ \cdot & \cdot & \cdot & \cdot\\ 0 & \dots & \dots & \lambda_n\end{pmatrix}$$

is a diagonal matrix with diagonal elements $\lambda_i>0$, $i=1,\dots,n$. Such an orthogonal transformation exists because A is symmetric and $|A|>0$. Since the absolute value of the Jacobian of $z=\mathfrak{D}^{-1}y$ is 1, we get

$$\int_{R_n} e^{-y'Ay/2}\,dy=\int_{R_n} e^{-z'\Lambda z/2}\,dz=\prod_{j=1}^{n}\int_{R_1} e^{-\lambda_j z_j^2/2}\,dz_j=\prod_{j=1}^{n}(2\pi/\lambda_j)^{1/2}.$$

For the last equality see p. 75. But $\prod_{j=1}^{n}\lambda_j=|A|$, whence (35.3) and hence also (35.2).

Let $\xi=(\xi_1,\dots,\xi_n)$ be a r.v. distributed by the density (35.1). Then

$$E(\xi_i)=a_i,\quad i=1,\dots,n. \tag{35.4}$$

Indeed, from (35.2)

$$\int_{R_n} e^{-(x-a)'A(x-a)/2}\,dx=(2\pi)^{n/2}|A|^{-1/2}. \tag{35.5}$$

(35.5) is an identity which holds for all $a\in R_n$. Differentiating both sides w.r.t. a_j, $j=1,\dots,n$, one gets

$$\int_{R_n} e^{-(x-a)'A(x-a)/2}\sum_{k=1}^{n}a_{jk}(x_k-a_k)\,dx=0,\quad j=1,\dots,n. \tag{35.6}$$

The differentiation under the integral is allowed in every compact cube $\{a:|a_i|\leqslant M,\ M>0\}$ since the integral on the left in (35.6) converges uniformly in this cube. After easy calculation, one gets from (35.6)

$$\sum_{k=1}^{n}a_{jk}E(\xi_k-a_k)=0,\quad j=1,\dots,n\,.$$

These are n homogeneous linear equations in the variables $E(\xi_k-a_k)$. Since $|A|>0$, they have the unique solution $E(\xi_k-a_k)=0$, $k=1,\dots,n$, which proves (35.4). To see the meaning of the a_{ik} we proceed quite similarly. We view (35.5) for fixed a as an identity in the a_{ik}, $i,k=1,\dots,n$. Since $a_{ik}=a_{ki}$ for $i,k=1,\dots,n$, each A can be viewed as an element of an $R_{n(n+1)/2}$ which fulfills $|a_{ij}|_{1n}^{1n}>0$. Since a determinant is a continuous function of its elements, one sees immediately that $\{a_{jk}: a_{jk}=a_{kj},\ k,j=1,\dots,n,\ |a_{jk}|_{1n}^{1n}>0\}$ is an open set of $R_{n(n+1)/2}$. The identity (35.5) can thus be differentiated w.r.t. a_{ij} and we get for $i,j=1,\dots,n$

$$\int_{R_n} e^{-(x-a)'A(x-a)/2}(x_i-a_i)(x_j-a_j)\,dx=(2\pi)^{n/2}|A|^{3/2}|A^{ij}|. \tag{35.7}$$

Here, $|A^{ij}|$ is the determinant resulting from A with the ith row and j^{th} column deleted and the resulting determinant multiplied by $(-1)^{i+j}$. $|A^{ij}|$ is thus the algebraic complement of a_{ij}. Since A is symmetric,

$|A^{ij}|=|A^{ji}|$. The differentiation under the integral sign can again be easily justified. (35.7) means the same as

$$E[(\xi_i - a_i)(\xi_j - a_j)] = |A^{ij}|/|A|. \tag{35.8}$$

The matrix $(|A^{ij}|/|A|)_{1n}^{1n}$ is the inverse of A and is thus the covariance matrix of the n-dimensional normal distribution. A^{-1} is positive definite, which illustrates Theorem 17.5.

Let us now find the characteristic function of a normal distribution with the density (35.1). For each $t \in R_n$, this function is given by

$$\phi(t) = |A|^{1/2}(2\pi)^{-n/2} \int_{R_n} e^{it'x-(x-a)'A(x-a)/2}\,dx. \tag{35.9}$$

Substituting $x-a=y$ and then making the transformation $y=\mathfrak{O}z$ given on p. 96, we get with $\mathfrak{O}'t=u$:

$$\phi(t) = |A|^{1/2}(2\pi)^{-n/2}\, e^{it'a} \int_{R_n} e^{iu'z-z'\Lambda z/2}\,dz. \tag{35.10}$$

Now

$$\int_{R_n} e^{iu'z-z'\Lambda z/2}\,dz = \prod_{j=1}^{n} \int_{R_1} e^{iu_j z_j - \lambda_j z_j^2/2}\,dz_j = (2\pi)^{n/2} \prod_{j=1}^{n} \lambda_j^{-1/2}\, e^{-u_j^2/2\lambda_j}$$

which is easy to see. Further

$$e^{-(1/2)\sum_{j=1}^{n} u_j^2/\lambda_j} = e^{-u'\Lambda^{-1}u/2}.$$

Hence, using $\mathfrak{O}A^{-1}\mathfrak{O} = \Lambda^{-1}$ and (35.10), we get from (35.9)

$$\phi(t) = e^{it'a - t'A^{-1}t/2}. \tag{35.11}$$

The n-dimensional normal distribution for $n \geqslant 1$ is easily recognized as a generalization of the one-dimensional normal distribution, whose density is given by (25.1) for each $x \in R_1$. When $n=1$, the vector a of means is to be identified with the mean a and the matrix A with $1/\sigma^2$. Correspondingly, (35.11) coincides with (26.1) when one identifies A^{-1} with σ^2.

Let $p \geqslant 1$ and the density of the joint distribution of the r.v.'s $(\xi_1, \dots, \xi_{p+1})$ be given for $x \in R_{p+1}$ by $|D|^{1/2}(2\pi)^{-(p+1)/2}\,e^{-(x-a)'D(a-x)/2}$. Here, let $a \in R_{p+1}$ and $D=(d_{ij})_{1\,p+1}^{1\,p+1}$ be positive definite.

As indicated on p. 59 for each real x_{p+1} and $(x_1, \dots, x_p) \in R_p$, the conditional density $f(x_{p+1}|x_1, \dots, x_p)$ of ξ_{p+1}, given $(\xi_1, \dots, \xi_p)$, is given by

$$e^{-(x-a)'D(x-a)/2} \left(\int_{-\infty}^{+\infty} e^{-1/2 \sum_{i,j=1}^{p+1} d_{ij}(x_i-a_i)(x_j-a_j)}\,dx_{p+1} \right)^{-1}.$$

For the integral in the denominator we get after easy manipulation

$$(2\pi)^{1/2}(d_{p+1\,p+1})^{-1/2}\exp\left(\frac{1}{2d_{p+1\,p+1}}\left(\sum_{i=1}^{p} d_{i\,p+1}(x_i-a_i)\right)^2\right)$$
$$\times\exp\left(-\left(\sum_{i,j=1}^{p} d_{ij}(x_i-a_i)(x_j-a_j)\right)\Big/2\right).$$

Hence,

$$f(x_{p+1}|x_1,\dots,x_p)=\sqrt{d_{p+1\,p+1}}\,(2\pi)^{-1/2}$$
$$\times\exp\{-\tfrac{1}{2}d_{p+1\,p+1}[(x_{p+1}-a_{p+1})+(x_1-a_1)d_{1\,p+1}/d_{p+1\,p+1} \tag{35.12}$$
$$+\cdots+(x_p-a_p)d_{pp+1}/d_{p+1\,p+1}]^2\}.$$

36. Linear combinations of normally distributed r. v.'s. Using (35.11), Theorem 23.3 and Theorem 24.4, one can easily prove the *reproductive property of the multi-dimensional normal distribution:*

Theorem 36.1. *Let ξ_1 and ξ_2 be two independent r.v.'s of the same dimension which are both normally distributed. Then their sum $\xi_1+\xi_2$ is also normally distributed.*

A still more general theorem holds:

Theorem 36.2. *Let $\xi=(\xi_1,\dots,\xi_n)$ be a r.v. with density (35.1). Let $C=(c_{ij})_{1m}^{1n}$ be a matrix of real numbers and $|CA^{-1}C'|\neq 0$. By means of the transformation*

$$\eta_1=c_{11}\xi_1+\cdots+c_{1n}\xi_n$$
$$\dots\dots\dots\dots\dots$$
$$\eta_m=c_{m1}\xi_1+\cdots+c_{mn}\xi_n$$

we define m r.v.'s $\eta_1,\dots,\eta_m$. Then, the distribution of $\eta=(\eta_1,\dots,\eta_m)$ is again normal with mean Ca and covariance matrix $CA^{-1}C'$.

Proof. We have $\eta=C\xi$. Let $u\in R_m$. Then by (35.11)

$$E(e^{iu'\eta})=E(e^{iu'C\xi})=e^{iu'Ca-u'CA^{-1}C'u/2}$$

so that η is normally distributed as indicated above.

If $CA^{-1}C'=0$, then one gets formally the same result. But then we no longer have a normal distribution in the usual sense: η is distributed by a so-called *degenerate normal distribution.* Indeed, we know from Theorem 17.6 that in this case the r. v.'s $\eta_1,\dots,\eta_m$ must satisfy at least one linear relation with probability 1. If the rank of $CA^{-1}C'$ is greater than 1, then there exists at least one r. v. η_i which is not a linear function of the remaining r.v.'s η_i, $i\neq j$ and is then normally distributed.

37. The multinomial distribution. A k-dimensional r.v. $\eta=(\eta_1,\dots,\eta_k)$, $k\geqslant 2$, is said to be multinomially distributed if it possesses a discrete distribution of the following type:

Let n and r_i for $i=1,\dots,k$ be integers with

$$0\leqslant r_i\leqslant n, \qquad r_1+\dots+r_k=n. \tag{37.1}$$

Let $p_i>0$, $i=1,\dots,k$, with $\sum_{i=1}^{k} p_i=1$ and set

$$P(\eta_1=r_1,\eta_2=r_2,\dots,\eta_k=r_k)=\frac{n!}{r_1!\,r_2!\dots r_k!}\,p_1^{r_1}p_2^{r_2}\dots p_k^{r_k}. \tag{37.2}$$

One sees immediately that for $k=2$, this is the binomial distribution in somewhat different notation. The multinomial distribution can also be derived in a way similar to that for the binomial distribution. To do this, we start with n independent k-dimensional r.v.'s ξ_l with the same discrete distribution given by

$$\begin{aligned}
&P(\xi_l=(1,0,\dots,0))=p_1\\
&P(\xi_l=(0,1,\dots,0))=p_2,\\
&\dots\dots\dots\dots\dots\\
&P(\xi_l=(0,0,\dots,1))=p_k.
\end{aligned}$$

Now consider the r.v. $\eta=\xi_1+\dots+\xi_n$. One easily shows that the distribution of the k-dimensional r.v. η is given by (37.2) because of the independence of the ξ_l.

As in the case of the Bernoulli distribution, the multinomial can also be interpreted as follows: Perform n independent trials which are all determined by the same random mechanism. The result of each trial consists of k alternatives symbolized by the numbers $1,\dots,k$, with the probability of each given by p_i, $i=1,\dots,k$. The left side of (37.2) then gives the probability that in the course of these n trials the event "1" occurs exactly r_1 times, "2" exactly r_2 times, ..., and "k" r_k times.

Let us calculate the mean and covariance matrix of a multinomially distributed r.v. η: We have

$$E(\eta_j)=n\,p_j, \qquad 1\leqslant j\leqslant k \tag{37.3}$$

which follows at once from

$$\sum_{\substack{0\leqslant r_i\leqslant n\\ r_1+\dots+r_k=n}} r_j\frac{n!}{r_1!\dots r_j!\dots r_k!}\,p_1^{r_1}\dots p_j^{r_j}\dots p_k^{r_k}$$

$$=p_j\sum\frac{n!}{r_1!\dots(r_j-1)!\dots r_k!}\,p_1^{r_1}\dots p_j^{r_j-1}\dots p_k^{r_k}$$

where the summation extends over $0\leqslant r_i\leqslant n$, $i\neq j$, $1\leqslant r_j\leqslant n$, $\sum_{i=1}^{k} r_i=n$.

Writing r_j again in place of r_j-1, we get

$$n p_j \sum_{\substack{0 \leqslant r_i \leqslant n-1 \\ r_1+\cdots+r_k=n-1}} \frac{(n-1)!}{r_1!\ldots r_j!\ldots r_k!} p_1^{r_1}\ldots p_j^{r_j}\ldots p_k^{r_k}=n p_j.$$

Analogously, with $q_j=1-p_j$, $1 \leqslant j \leqslant k$, we find

$$E[(\eta_j-n p_j)^2]=n p_j q_j \tag{37.4}$$

and

$$E[(\eta_i-n p_i)(\eta_j-n p_j)]=n p_i p_j, \quad i \neq j, \quad i,j=1,\ldots,k. \tag{37.5}$$

We now evaluate the determinant of the covariance matrix:

$$\begin{vmatrix} n p_1 q_1 & -n p_1 p_2 & \cdots & -n p_1 p_k \\ -n p_2 p_1 & n p_2 q_2 & \cdots & -n p_2 p_k \\ \cdots & \cdots & \cdots & \cdots \\ -n p_k p_1 & -n p_k p_2 & \cdots & n p_k q_k \end{vmatrix}. \tag{37.6}$$

Add the 2nd through k^{th} row to the first and get for this first modified row:

$$n p_1-n p_1 \sum_{i=1}^{k} p_i \quad n p_2-n p_2 \sum_{i=1}^{k} p_i \ldots n p_k-n p_k \sum_{i=1}^{k} p_i.$$

Since $\sum_{i=1}^{k} p_i=1$, all of these elements vanish, so that (37.6) = 0. According to Theorem 17.6, there thus exists at least one linear relation among the r.v.'s η_i, $1 \leqslant i \leqslant k$. This is, however, trivial in this case since $\eta_1+\cdots+\eta_k=n$ follows from (37.1) with probability 1.

38. Convergence in probability and convergence with probability 1. Let $\xi_1, \xi_2, \ldots$ be any r.v.'s over a probability space $(R, \mathscr{S}, P)$. Below we give the first of several ways of defining the convergence of a sequence of r.v.'s. We will soon see that this notion is closely connected with the weak convergence of a sequence of distributions functions (see p. 102).

Definition. The *sequence* $\{\xi_n\}$ is said to *converge stochastically, or in probability* to a r.v. η if for each $\varepsilon>0$ and $\delta>0$ there exists a positive number $n(\varepsilon, \delta)$ such that

$$P(|\xi_n-\eta|>\varepsilon)<\delta \tag{38.1}$$

for $n \geqslant n(\varepsilon, \delta)$.

This can also be expressed as follows: For every $\varepsilon>0$

$$\lim_{n \to \infty} P(|\xi_n-\eta|>\varepsilon)=0. \tag{38.2}$$

We also write: $\xi_n \to \eta$ in probability.

If $\{\xi_i\}$ converges stochastically to η, then the r.v. η is uniquely determined w.p. 1. Thus, if $\{\xi_i\}$ also converges in probability to a r.v. η^* which is different from η, then

$$P(\eta-\eta^*=0)=1. \tag{38.3}$$

Indeed, for $\varepsilon>0$ and $\delta>0$ let (38.1) hold and

$$P(|\xi_n-\eta^*|>\varepsilon)<\delta \tag{38.4}$$

hold for $n\geqslant n'(\varepsilon,\delta)$. Then for $n\geqslant \max(n(\varepsilon,\delta), n'(\varepsilon,\delta))$, with

$$\begin{aligned} A_n &= \{\omega\in R: |\xi_n(\omega)-\eta(\omega)|\leqslant\varepsilon\} \\ A_n^* &= \{\omega\in R: |\xi_n(\omega)-\eta^*(\omega)|\leqslant\varepsilon\}, \end{aligned}$$

we have from (38.1) and (38.4) because of $A_n\cap A_n^*=R-((R-A_n)\cup(R-A_n^*))$, that $P(A_n\cap A_n^*)\geqslant 1-2\delta$. For $\omega\in A_n\cap A_n^*$, however, $|\eta(\omega)-\eta^*(\omega)|\leqslant 2\varepsilon$, so that $P(|\eta-\eta^*|\leqslant 2\varepsilon)\geqslant 1-2\delta$ and since this is correct for each ε and $\delta>0$, (38.3) follows.

If for a real c, $\eta(\omega)=c$ holds for all $\omega\in R$, then one says that the sequence converges in probability to c. We prove

Theorem 38.1. *Let F_i be the d.f. of ξ_i, $i=1,2,\ldots$. The sequence of the ξ_i converges in probability to c iff the sequence of the F_i converges weakly to the d.f. of the Dirac distribution at c.*

Proof. Denote the d.f. of the Dirac distribution at c by G_c. Let y be a real number satisfying $y<c$. Then for each sufficiently small $\varepsilon>0$

$$y+\varepsilon<c. \tag{38.5}$$

By assumption, $\lim\limits_{n\to\infty} P(|\xi_n-c|>\varepsilon)=0$ holds according to (38.2). If $\varepsilon>0$ is chosen according to (38.5), then from $\xi_n\leqslant y$ it always follows that $\xi_n<c-\varepsilon$, and so also $|\xi_n-c|<\varepsilon$. Hence, $0\leqslant P(\xi_n\leqslant y)\leqslant P(|\xi_n-c|>\varepsilon)\to 0$. For each real $y<c$, we thus have $F_n(y)\to G_c(y)$. If $y>c$ then analogously, $F_n(y)\to 1$, i.e., $F_n(y)\to G_c(y)$.

Conversely, assume $F_n(y)\to G_c(y)$ for each real $y\neq c$. Then, in particular, for $\varepsilon>0$, $F_n(c+\varepsilon)-F_n(c-\varepsilon)\to 1$, thus, $P(c-\varepsilon<\xi_n\leqslant c+\varepsilon)\to 1$ and hence also $P(c-\varepsilon\leqslant\xi_n\leqslant c+\varepsilon)\to 1$, so that $P(|\xi_n-c|>\varepsilon)\to 0$.

The theorem of Bernoulli yields an example of a sequence of r.v.'s which converges stochastically to a real number. Theorem 32.2 says precisely that ζ_n/n converges stochastically to p. Bernoulli's theorem is, however, only a special case of a more general statement referred to as the *weak law of large numbers*.

Theorem 38.2.[46] *Let $\{\xi_i\}$ be a sequence of independent r.v.'s. Let each ξ_i, $i=1,2,\dots$ have the same probability distribution and assume that $E(\xi_i)=a$ exists. Then $(\xi_1+\dots+\xi_n)/n$ converges stochastically to a.*

Proof. Considering $\xi_i - a$, one sees that it suffices to assume $a=0$. By Theorem 38.1, we must show that the sequence of distribution functions of the $(\xi_1+\dots+\xi_n)/n$, $n=1,2,\dots$, converges weakly to the distribution function of the degenerate distribution at 0. If ϕ is the characteristic function of ξ_i, then $\xi_1+\dots+\xi_n$ possesses, according to Theorem 24.4, the characteristic function ϕ^n. The characteristic function ψ of $(\xi_1+\dots+\xi_n)/n$ is then given for each real t by $\psi(t)=\phi^n(t/n)$. The characteristic function of the degenerate distribution at 0 is given by $\chi(t)=1$, t real. Thus, it is enough to show that $\phi^n(t/n)\to 1$ for each real t because of Theorem 23.4. By Theorem 23.2, ϕ' exists, is continuous, and $\phi'(0)=0$. By Taylor's theorem: $\phi(t/n)=1+\varepsilon(t/n)\cdot(t/n)$, where $\varepsilon(t/n)\to 0$ when $n\to\infty$ follows for each real t. But this yields $\phi^n(t/n) =(1+\varepsilon(t/n)\cdot(t/n))^n\to 1$ for each real t, which was to be shown.

If we also assume that $E[(\xi_i-a)^2]$ exists, then Theorem 38.2 can also easily be proved be means of Čebyšev's inequality, as we illustrated in the proof of Bernoulli's theorem.

We can also prove the following

Supplement to Theorem 38.2. *Let $\{\xi_i\}$ be a sequence of independent r.v.'s. Let each ξ_i, $i=1,2,\dots$ have the same probability distribution and let $E(\xi_i)=\infty$ in the sense of the definition on p. 9. Then for arbitrarily large $G>0$*

$$\lim_{n\to\infty} P((\xi_1+\dots+\xi_n)/n>G)=1. \tag{38.6}$$

The *proof* can be carried out by means of the important "truncation" method. Define for each $k=1,2,\dots$ a sequence of r.v.'s $\{\eta_i^{(k)}\}$ by

$$\eta_i^{(k)}=\begin{cases} k & \xi_i>k \\ \xi_i & -\infty<\xi_i\leqslant k \end{cases} \qquad i=1,2,\dots.$$

For each k, the r.v.'s $\eta_i^{(k)}$, $i=1,2,\dots$ are independent and have the same distribution. Moreover, $E(\eta_i^{(k)})$ exist and

$$\lim_{k\to\infty} E(\eta_i^{(k)})=\infty. \tag{38.7}$$

Applying Theorem 38.2 to the sequence $\{\eta_i^{(k)}\}$, we find for each $\varepsilon>0$ and $k=1,2,\dots$, that

$$\lim_{n\to\infty} P((\eta_1^{(k)}+\dots+\eta_n^{(k)})/n>E(\eta_i^{(k)})-\varepsilon)=1.$$

[46] A.J. Hinčin, C.R. Acad. Sci., Paris 189, 477—479 (1929).

But for each k and $n=1,2,\dots$,

$$(\xi_1+\cdots+\xi_n)/n \geqslant (\eta_1^{(k)}+\cdots+\eta_n^{(k)})/n$$

so that also

$$\lim_{n\to\infty} P((\xi_1+\cdots+\xi_n)/n > E(\eta_i^{(k)})-\varepsilon)=1.$$

Because of (38.7), we have thus proved (38.6).

A similar result holds for $E(\xi_i)=-\infty$.

The usual notion of convergence allows statements such as: The sum or product of two convergent sequences converges, etc. We will see that analogous results also hold for stochastic convergence.[47] First we want to note a trivial generalization of the notion of stochastic convergence: If $\xi_1, \xi_2, \dots$ is a sequence of k-dimensional ($k \geqslant 2$) r.v.'s, then the definition of stochastic convergence according to (38.1) or (38.2) remains unchanged, provided the absolute value is understood as the distance in R_k.

We now prove

Theorem 38.3. *Let* $\{\xi_i\}$ *be a sequence of k-dimensional r.v.'s,* $a \in R_k$ *and* $\gamma \in R_1$. *From* $\xi_i \to a$ *in probability there follows* $\gamma\xi_i \to \gamma a$ *in probability. If* $\{\eta_i\}$ *is another sequence of k-dimensional r.v.'s and* $b \in R_k$, *then,* $\xi_i \to a$ *and* $\eta_i \to b$ *in probability imply* $\xi_i+\eta_i \to a+b$ *and, provided that* $k=1$, $\xi_i\eta_i \to ab$ *in probability.*

Proof. The first claim can be proved by means of Theorem 38.1, which, as is easy to see, also holds for multidimensional r.v.'s. Indeed, by assumption, for each $t \in R_k$, $\lim_{j\to\infty} E(e^{it'\xi_j})=e^{it'a}$, thus also $\lim_{j\to\infty} E(e^{i\gamma t'\xi_j})=e^{i\gamma t'a}$.

Now for the second claim. For each ε and $\delta>0$, $P(|\xi_n-a|\leqslant\varepsilon)\geqslant 1-\delta$ and $P(|\eta_n-b|\leqslant\varepsilon)\geqslant 1-\delta$ for sufficiently large n. On the other hand, $|\xi_n+\eta_n-(a+b)| \leqslant |\xi_n-a|+|\eta_n-b|$ always holds, and thus, with the same argument as on p. 102, $P(|\xi_n+\eta_n-(a+b)|\leqslant 2\varepsilon)\geqslant 1-2\delta$ for all large enough n. Further,

$$\xi_n\eta_n-ab=(\xi_n-a)\eta_n-a(b-\eta_n).$$

But for each $\varepsilon>0$

$$\{\omega: |\eta_n(\omega)-b|\leqslant\varepsilon\} \subseteq \{\omega: |\eta_n(\omega)|\leqslant |b|+\varepsilon\}$$

[47] A systematic treatment of the properties of this notion and of related concepts can be found in E. Lukacs, Stochastic Convergence, Math. Monographs, D. C. Heath, Lexington, Mass. 1968.

and thus, for all ω from the set

$$\{\omega:|\xi_n(\omega)-a|\leqslant\varepsilon\}\cap\{\omega:|\eta_n(\omega)-b|\leqslant\varepsilon\},$$

$$|\xi_n\eta_n-ab|\leqslant\varepsilon(|b|+\varepsilon)+|a|\varepsilon \quad \text{holds, i.e.,}$$

$$\lim_{n\to\infty} P(|\xi_n\eta_n-ab|\leqslant\varepsilon)=1 \quad \text{for each } \varepsilon>0.$$

We prove the following very general theorem for one-dimensional r.v.'s only, although it can be easily extended to the multi-dimensional case.

Theorem 38.4. *Let $\xi_n\to a$ in probability and let f be a function defined over R_1 which is continuous at a. Then $f\circ\xi_n\to f(a)$ in probability.*

Proof.[48] We have to show that for each ε, $\eta>0$ there exists an $n(\varepsilon,\eta)$ such that for all $n\geqslant n(\varepsilon,\eta)$,

$$P(|f(\xi_n)-f(a)|<\varepsilon)>1-\eta. \tag{38.8}$$

Since f is continuous at a, there exists a $\delta_\varepsilon>0$ such that for all ω with $|\xi_n(\omega)-a|<\delta_\varepsilon$, $|f(\xi_n(\omega))-f(a)|<\varepsilon$ also holds. Hence,

$$P(|f(\xi_n)-f(a)|<\varepsilon)\geqslant P(|\xi_n-a|<\delta_\varepsilon). \tag{38.9}$$

Since $\xi_n\to a$ in probability, there exists an $n'(\delta_\varepsilon,\eta)$ such that for all $n\geqslant n'(\delta_\varepsilon,\eta)$,

$$P(|\xi_n-a|<\delta_\varepsilon)>1-\eta. \tag{38.10}$$

If we set $n(\varepsilon,\eta)=n'(\delta_\varepsilon,\eta)$, the sought-for inequality follows from (38.9) and (38.10).

One can thus largely operate with sequences of r.v.'s converging in probability to a real number in the same way as with the usual notion of convergence.

We now introduce a further convergence concept.

Definition. A sequence of r.v.'s $\xi_1,\xi_2,\ldots$ over a probability space $(R,\mathcal{S},P)$ is said to *converge to a r.v. η with probability one* (w.p. 1), or to be *convergent P-a.e.* if $\lim_{n\to\infty}\xi_n(\omega)=\eta(\omega)$ for all $\omega\in R$ with the exception of at most a P-null set.

One easily sees that convergence of a sequence $\{\xi_i\}$ to η w.p. 1 implies its convergence to η in probability.

With this convergence idea we can now formulate the so-called *Strong Law of Large Numbers*, which we will give here without proof. With the notation of Theorem 38.2 we have

---

[48] This short proof is due to Borges. Theorem 38.3 as well as Theorem 38.4 remain correct if the stochastic convergence to a real number (or to a k-tuple of real numbers) in the assumption and claim is replaced by stochastic convergence to a r.v. See K. Krickeberg, l.c. Not.[7]. We will make no use of this fact here.

Theorem 38.5. *Let the assuptions of Theorem 38.2 be fulfilled. Then the sequence* $(\xi_1+\cdots+\xi_n)/n$ *converges w.p. 1 to a.*

There again also holds the following

Supplement to Theorem 38.5. *If* $E(\xi_i)=\pm\infty$, *then* $(\xi_1+\cdots+\xi_n)/n \to \pm\infty$ *w.p. 1.*

The complete analogy between Theorems 38.2 and 38.5 makes it possible to state many theorems of mathematical statistics in either "strong" or "weak" versions. (See especially V.)

39. Some limit theorems of probability theory[49]**.** Experience shows that many randomly generated phenomena are approximately normally distributed. This is especially the case when the random mechanism arises from a large number of more or less independent random events which are additively superposed. Limit theorems in probability describe this situation; in particular, the so-called *Central Limit Theorem.* We first prove

Theorem 39.1[50]**.** *Let* $\{\xi_i\}$ *be a sequence of independent r.v.'s all with the same probability distribution. Let* $E(\xi_i)=a$ *and* $E[(\xi_i-a)^2]=\sigma^2$, $\sigma>0$, *exist. For* $n\geqslant 1$ *let* $\zeta_n=\xi_1+\cdots+\xi_n$ *and* $\eta_n=(\zeta_n-na)/\sigma\sqrt{n}$, *the so-called "standardized" r.v. Let* F_n *be the distribution function of* η_n. *Then for each* $x\in R_1$

$$F_n(x)\to\frac{1}{\sqrt{2\pi}}\int_{-\infty}^{x}e^{-y^2/2}\,dy$$

for $n\to\infty$.

Proof. This goes like that of Theorem 38.2. However, we have here the 2nd moment available and can thus use Taylor's formula up to terms of 2nd order. Let ϕ be the characteristic function of ξ_i-a. Since $\phi(0)=1$, $\phi'(0)=0$, $\phi''(0)=-\sigma^2$, and since ϕ'' is continuous (Theorem 23.2), we have for each $t\in R_1$

$$\phi(t)=1-\sigma^2t^2/2+\varepsilon(t)t^2 \tag{39.1}$$

with $\varepsilon(t)\to 0$ for $t\to 0$. But now for each $t\in R_1$, $E(e^{i\eta_n t})=(\phi(t/\sigma\sqrt{n}))^n$. From (39.1)

$$(\phi(t/\sigma\sqrt{n}))^n=[1-t^2/2n+\varepsilon(t/\sigma\sqrt{n})t^2/\sigma^2n]^n.$$

---

[49] See loc. cit.[44] and P. Lévy, Théorie de l'addition des variables aléatoires. Gauthier-Villars 2nd Ed., Paris 1954.

[50] P. Lévy, loc. cit.[28], 233ff. It is not hard to show that the convergence of the sequence $\{F_n\}$ is even uniform in $-\infty<x<\infty$.

For each fixed $t \in R_1$, $t/\sigma\sqrt{n} \to 0$, hence also $\varepsilon(t/\sigma\sqrt{n}) \to 0$ for $n \to \infty$. Thus, for each $t \in R_1$

$$\lim_{n\to\infty} E(e^{i\eta_n t}) = \lim_{n\to\infty} [1 - t^2/2n + \varepsilon(t/\sigma\sqrt{n})t^2/\sigma^2 n]^n = e^{-t^2/2},$$

which was to be proved.

We can also express the content of Theorem 39.1 as follows: The r.v. ζ_n is *asymptotically* $N(na, n\sigma^2)$*-distributed.* We will use the same terminology in similar cases.

Remark: By Theorem 39.1 we obviously have $\lim_{n\to\infty} P(\zeta_n \leqslant na + x\sigma\sqrt{n}) = (1/\sqrt{2\pi}) \int_{-\infty}^{x} e^{-y^2/2} dy$ for each $x \in R_1$. However, Theorem 39.1 makes only trivial statements about the behavior of the probability of "large" deviations of the r.v. ζ_n from the mean na. We will prove soon a simple theorem of this type (see Theorem 39.5).

Theorem 39.1 admits a generalization which can be proved in almost the same way:

Theorem 39.2. *Let*

$$\begin{matrix} \xi_{11} & & & \\ \xi_{21} & \xi_{22} & & \\ \cdots & \cdots & \cdots & \cdots \\ \xi_{n1} & \xi_{n2} & \cdots & \xi_{nn} \\ \cdots & \cdots & \cdots & \cdots \end{matrix}$$

be an infinite matrix of r.v.'s. Let the r.v.'s ξ_{ji}, $1 \leqslant i \leqslant j$, $j = 1, 2, \ldots$ *in each row be independent and identically distributed. Suppose that all r.v.'s have finite second moment. Let* $E(\xi_{ji}) = 0$ *and* $E(\xi_{ji}^2) = \sigma_j^2$, $1 \leqslant i \leqslant j$, $j = 1, 2, \ldots$. *Assume further that there exists a positive real number* a *such that* $a \leqslant \sigma_j^2$ *as* $j = 1, 2, \ldots$. *Let* ϕ_j *be the characteristic function of* ξ_{ji}, $1 \leqslant i \leqslant j$ *and suppose that the* ϕ_j''*'s,* $j = 1, 2, \ldots$ *are equicontinuous at* 0.[51] *Then* $\sum_{i=1}^{n} \xi_{ni}/(\sigma_n\sqrt{n})$ *is asymptotically* $N(0,1)$*-distributed.*

Theorem 39.1 is based on the assumption that the r.v.'s ξ_i have the same distribution for $i = 1, 2, \ldots$. If we drop this assumption we obtain the so-called *Central Limit Theorem of Ljapunov* which we will formulate in slightly more general form for future application.

[51] That is: For every $\varepsilon > 0$ there exists a $\delta > 0$ such that $|\phi_j''(0) - \phi_j''(t)| < \varepsilon$ if $|t| < \delta$ uniformly for $j = 1, 2, \ldots$.

Theorem 39.3. *Let*

$$\begin{matrix} \xi_{11} & & & \\ \xi_{21} & \xi_{22} & & \\ \cdots & \cdots & \cdots & \cdots \\ \xi_{n1} & \xi_{n2} & \cdots & \xi_{nn} \\ \cdots & \cdots & \cdots & \cdots \end{matrix}$$

be an infinite matrix of r.v.'s. Let the r.v.'s ξ_{ji}, $1 \leqslant i \leqslant j$, $j = 1, 2, \ldots$ *in each row be independent and let all of them have finite first through third moments. Write* $E(\xi_{ji}) = a_{ji}$, $E[(\xi_{ji} - a_{ji})^2] = \sigma_{ji}^2$ *and* $E[|\xi_{ji} - a_{ji}|^3] = v_{ji}$. *Set*

$$a_j = \sum_{i=1}^{j} a_{ji}, \quad \sigma_j^2 = \sum_{i=1}^{j} \sigma_{ji}^2, \quad v_j = \sum_{i=1}^{j} v_{ji}, \quad 1 \leqslant i \leqslant j, \quad j = 1, 2, \ldots .$$

Then, under the assumption $\lim_{n\to\infty} v_n \sigma_n^{-3} = 0$, *the sum* $\zeta_n = \xi_{n1} + \cdots + \xi_{nn}$, $n \geqslant 1$, *is asymptotically* $N(a_n, \sigma_n^2)$*-distributed.*

The *proof* can again be carried out according to the pattern of the proof of Theorem 39.1. However, one must now consider products of, in general, different characteristic functions which are best treated by going over to logarithms. On the other hand, one now has the Taylor expansion available up to terms of 3rd order.

Theorem 39.1 has a number of important applications in mathematical statistics. For example, it follows immediately from the definition of the χ^2-distribution that the r.v. η_n defined by (28.1) is asymptotically $N(n, 2n)$-distributed. For the Bernoulli distribution we have according to its definition in (32.2) that a $B_n(p)$ r.v. is asymptotically $N(np, npq)$-distributed. This theorem, called the *Theorem of Laplace*, will be proved later in another way. As a further example consider the Poisson distribution defined in **33**. Let $b > 0$ and ζ_n be a Poisson-distributed r.v. with mean nb, $n \geqslant 1$. Then by Theorem 33.1, ζ_n can be viewed as the sum of n independent r.v.'s with mean b. An application of Theorem 39.1 thus yields, in somewhat intuitive language: For $a \to \infty$, a Poisson-distributed r.v. with mean a is asymptotically $N(a, a)$-distributed.

Theorem 39.1 can be extended almost verbatim to multi-dimensional r.v.'s. One has

Theorem 39.4. *Let* $\xi_1, \xi_2, \ldots$ *be* k*-dimensional* $(k \geqslant 2)$ *independent r.v.'s, all with the same distribution. Let all means and variances exist. Let the vector of means be denoted by* a, *and the covariance matrix by* A^{-1} *with* A *positive definite. Then the sequence of distribution functions of* $\zeta_n = (\xi_1 + \cdots + \xi_n - na)/\sqrt{n}$ *converges for* $n \to \infty$ *and each* $x \in R_k$ *to the distribution function of a normal distribution with mean* 0 *and covariance matrix* A^{-1}.

The *proof* follows again from Theorem 24.4 and 23.4.

We also have the following

Supplement to Theorem 39.4. *Let the assumptions of Theorem 39.4 be fulfilled with $a=0$ for the sake of simplicity. Let m be an arbitrary natural number and B an $m \times k$ matrix with $|BA^{-1}B'| \neq 0$. Then the sequence of distribution functions of $B\zeta_n$ tends for each $x \in R_k$ to the normal distribution function with mean vector 0 and covariance matrix $BA^{-1}B'$.*

The *proof* is almost obvious: If ϕ_n is the characteristic function of ζ_n, then by Theorem 39.4 for each $t \in R_k$, $\phi_n(t) = e^{-t'A^{-1}t/2}$. Hence, also

$$E(e^{it'B\zeta_n}) = \phi_n(B't) \to e^{-t'BA^{-1}B't/2} \quad \text{for } n \to \infty .$$

Error estimates are quite important for the practical application of these limit theorems. Such estimates have also been given for the more difficult multi-dimensional case[52]. We will give later some idea of such an error estimate for the special case of $B_n(p)$ (see Theorem 39.7). We first turn, however, to the treatment of "*large deviations*". We give only the following theorem which we will need later.

Theorem 39.5[53]. *Let ξ be a r.v. with d.f. F, which is assumed not to be the d.f. of the Dirac distribution at c (see below). Let there exist an open interval I such that $I \cap [0, \infty] \neq \emptyset$ and for all $t \in I$ let the moment generating function*

$$\psi(t) = E(e^{t\xi})$$

exist. Let c be a nonnegative number and k_n a sequence of positive numbers with

$$\lim_{n\to\infty} k_n/n = c. \tag{39.2}$$

Further let there exist a $u \in I$, $u > 0$, such that

$$\psi'(u)/\psi(u) = c. \tag{39.3}$$

Let $\xi_1, \xi_2, \ldots$ be a sequence of independent, identically distributed r.v.'s with d.f. F and set $\zeta_n = \xi_1 + \cdots + \xi_n$, $n \geqslant 1$. For each $t \in I$ let

$$\chi(t) = e^{-tc}\psi(t). \tag{39.4}$$

---

52 We refer to the fundamental paper of C. G. Esseen, Acta Math. 77, 1—125 (1944) and to generalizations by E. Hlawka, Monatsh. Math. 55, 105—137 (1951). An important reference work is Ibragimov, I. A. and Linnik, Ju. V. Independent and stationary sequences of random variables. Ed. by F. C. Kingman, Groningen, Wolters-Noordhoff 1971.

53 A good outline of this problem area is in Ju. V. Linnik, Proc. Fourth Berkeley Sympos. Math. Statist. and Prob., Vol. II, pp. 289—306. Univ. California Press, Berkeley, Calif, (1960). See also W. Richter, Wiss. Z. Techn. Hochsch. Dresden 10, 7—14 (1961). For the proof here, see R.R. Bahadur and R. Ranga Rao, Ann. Math. Statist. 1015—1027 (1960).

We write $m=\chi(u)$. *Then*

$$m=\min_{t\in I}\chi(t) \tag{39.5}$$

and

$$\lim_{n\to\infty} n^{-1}\log P(\zeta_n \geqslant k_n)=\log m. \tag{39.6}$$

We sketch the *proof*: ψ is infinitely often differentiable in I. Since

$$\chi''(t)=\int_{-\infty}^{+\infty}(x-c)^2 e^{t(x-c)}dF(x) \quad \text{for all } t\in I, \quad \text{one has } \chi''(t)>0.$$

Hence, (39.5) follows from (39.3). (39.6) makes sense since one easily sees that $m>0$. Now for each $y\in R_1$ let

$$G(y)=(1/m)\int_{-\infty}^{y} e^{ux}dF(x+c). \tag{39.7}$$

Then G is the d.f. of a r.v. η. From (39.3)

$$E(\eta)=0 \tag{39.8}$$

and, moreover,

$$E(\eta^2)=\sigma^2 \tag{39.9}$$

exists. In addition, G is not the d.f. of a Dirac distribution. Now let $\eta_1, \eta_2, \ldots$ be a sequence of independent, identically distributed r.v.'s with the same distribution as η. For $n=1,2,\ldots$ let $\chi_n=\sum_{i=1}^{n}\eta_i/\sigma\sqrt{n}$. Denote by H_n the d.f. of χ_n. At first let $k_n=nc$, $n\geqslant 1$. Then

$$P(\zeta_n\geqslant nc)=\int_0^\infty\ldots\int_0^\infty dF(x_1+c)\ldots dF(x_n+c)$$

$$=m^n \int\limits_{y_1+\cdots+y_n\geqslant 0}\ldots\int e^{-u(y_1+\cdots+y_n)}dG(y_1)\ldots dG(y_n)=m^n\int_0^\infty e^{-u\sqrt{n}\sigma x}dH_n(x).$$

It follows that

$$\log P(\zeta_n\geqslant nc)=n\log m+\log\int_0^\infty e^{-u\sqrt{n}\sigma x}dH_n(x).$$

It thus suffices to prove that

$$\log\int_0^\infty e^{-u\sqrt{n}\sigma x}dH_n(x)=o(n) \tag{39.10}$$

for $n\to\infty$.

Trivially, $\left|\int\limits_0^\infty e^{-u\sqrt{n}\sigma x} dH_n(x)\right| \leqslant 1$. On the other hand, by partial integration,

$$\int\limits_0^\infty e^{-u\sqrt{n}\sigma x} dH_n(x) = u\sqrt{n}\sigma \int\limits_0^\infty e^{-u\sqrt{n}\sigma x}(H_n(x)-H_n(0))\,dx,$$

which for fixed $a>0$ is not smaller than $(H_n(a)-H_n(0))e^{-u\sqrt{n}\sigma a}$. But from Theorem 39.1, H_n converges for each $x \in R_1$ to the d.f. of $N(0,1)$. Hence, $\lim\limits_{n\to\infty}(H_n(a)-H_n(0))>0$, whence (39.10). The transition to the more general condition (39.2) is simple and will not be given here.

Up to now we have investigated only the convergence of a sequence of d.f.'s to the d.f. of a normal distribution. Whether or not similar results hold for densities is also of practical and theoretical interest. This leads to so-called *local limit theorems*. We illustrate with several important examples.

We first prove *Stirling's formula:*

Theorem 39.6. *For* $n=1,2,\ldots$ *and* $0 \leqslant \Theta_n \leqslant 1$

$$n! = n^n e^{-n}\sqrt{2\pi n}\, e^{\Theta_n/12n}. \tag{39.11}$$

Proof. We first show that if $\lim\limits_{n\to\infty} n!/(n^n e^{-n}\sqrt{n}) = c$ with finite c, then

$$c = \sqrt{2\pi}. \tag{39.12}$$

Indeed, let $n! = n^{n+1/2} e^{-n} a_n$ with

$$\lim_{n\to\infty} a_n = c. \tag{39.13}$$

Since for $n \geqslant 1$ $a_n \neq 0$ naturally holds, we have

$$\log(2n)! = (2n+1/2)\log 2n - 2n + \log a_{2n} \tag{39.14}$$

and likewise

$$\log n! = (n+1/2)\log n - n + \log a_n. \tag{39.15}$$

From (39.15) we easily get

$$\log(2\cdot 4\cdots 2n) = (n+1/2)\log n - n + n\log 2 + \log a_n. \tag{39.16}$$

Hence, from (39.14)

$$\log(1\cdot 3\cdots(2n-1)) = n\log n - n + (n+1/2)\log 2 + \log a_{2n} - \log a_n. \tag{39.17}$$

From (39.16) and (39.17)

$$\log\frac{2\cdot 4\cdots 2n}{1\cdot 3\cdots(2n-1)\sqrt{2n+1}} + \frac{1}{2}\log\frac{2n+1}{n} + \frac{1}{2}\log 2 = 2\log a_n - \log a_{2n}.$$

Then, from (25.4) and (39.13)

$$\log(\pi/2)^{1/2} + \log 2 = \log c$$

which proves (39.12).

We now show in an elementary way[54] that a_n tends to a limit and will also justify the remainder term $e^{\Theta_n/12n}$. We show that

$$a_{n+1}/a_n < 1, \quad n \geqslant 1. \tag{39.18}$$

This will prove the convergence claim.

It is known that for $|x| < 1$

$$\log[(1+x)/(1-x)] = 2\sum_{k=1}^{\infty} x^{2k-1}/(2k-1),$$

whence, for $x > 0$

$$2x < \log[(1+x)/(1-x)] < 2\left(x + \frac{x^3}{3}\sum_{k=0}^{\infty} x^{2k}\right) = 2\left(x + \frac{x^3}{3}\frac{1}{1-x^2}\right).$$

Hence, for $x = 1/(2n+1)$ one gets

$$\begin{aligned} 2/(2n+1) &< \log(1+1/n) \\ &< 2(1/(2n+1) + 1/[3(2n+1)((2n+1)^2-1)]). \end{aligned} \tag{39.19}$$

But now

$$\log\frac{a_n}{a_{n+1}} = (n+1/2)\log(1+1/n) - 1, \tag{39.20}$$

so that from (39.19) and (39.20)

$$0 < \log\frac{a_n}{a_{n+1}} < \frac{1}{12}\left(\frac{1}{n} - \frac{1}{n+1}\right). \tag{39.21}$$

The inequality on the left in (39.21) is the same as (39.18).

From (39.21) we have for natural m and $k = 1, 2, \ldots$

$$0 < \log\frac{a_m}{a_{m+k}} < \frac{1}{12}\left(\frac{1}{m} - \frac{1}{m+k}\right). \tag{39.22}$$

For $k \to \infty$, we have from (39.22) because of (39.12) and (39.13)

$$0 \leqslant \log\frac{a_m}{\sqrt{2\pi}} \leqslant 1/12m,$$

so that (39.11) holds.

---

[54] See B. L. van der Waerden, Nieuw Arch. Wisk. 18, 40—45 (1936).

We now prove

Theorem 39.7[55]. *Let $p, q>0$ and $p+q=1$. Let x be an arbitrary positive real number and n, r nonnegative integers with $n \geqslant 1$, $r \leqslant n$ and*

$$|r-np|/\sqrt{n} \leqslant x. \tag{39.23}$$

Then

$$\binom{n}{r} p^r q^{n-r} = (2\pi npq)^{-1/2} e^{-\frac{(r-np)^2}{2npq}} + O(1/n). \tag{39.24}$$

Proof. From (39.11)

$$\frac{n!}{r!(n-r)!} = \frac{n^{n+1/2} e^{\Theta_n/12n}}{\sqrt{2\pi} r^{r+1/2} e^{\Theta_r/12r} (n-r)^{(n-r+1/2)} e^{\Theta_{(n-r)}/[12(n-r)]}}. \tag{39.25}$$

Let $A_n = (np/r)^r (nq/(n-r))^{n-r}$. For $|x|<1$

$$\log(1+x) = x - x^2/2 + x^3/[3(1+\Theta x)^3] \quad \text{with } |\Theta|<1. \tag{39.26}$$

Writing

$$r - np = s, \tag{39.27}$$

we get from (39.23)

$$s/n = O(1/\sqrt{n}). \tag{39.28}$$

Hence, taking n sufficiently large and applying (39.26), we get

$$\begin{aligned} \log A_n = & -(s+np)(s/np - s^2/2n^2p^2 + O(s^3/n^3)) \\ & -(nq-s)(-s/nq - s^2/2n^2q^2 + O(s^3/n^3)) \\ = & -s^2/2npq + O(s^3/n^3). \end{aligned}$$

Thus,

$$A_n = e^{-s^2/2npq} e^{O(s^3/n^2)}. \tag{39.29}$$

Further

$$n^{1/2} r^{-1/2} (n-r)^{-1/2} = (npq)^{-1/2} (1+s/np)^{-1/2} (1-s/nq)^{-1/2}.$$

Hence, for large enough n we get from an application of Taylor's theorem and using (39.28):

$$n^{1/2} r^{-1/2} (n-r)^{-1/2} = (npq)^{-1/2} + O(1/n). \tag{39.30}$$

[55] In a quite analogous way one shows that the multinomial distribution (see **37**) can be approximated by a $(k-1)$-dimensional normal distribution. We mention for the sake of completeness that one can arrive at a multi-dimensional Poisson distribution by means of another passage to the limit. (See p. 92.) More precise and general results of this type can be found in M. Fisz, Studia Math. 14, 272—275 (1954).

From (39.23) and (39.28) we also get easily

$$1/r = O(1/n) \tag{39.31}$$

and

$$1/(n-r) = O(1/n). \tag{39.32}$$

From (39.31)

$$e^{\Theta_r/12r} = 1 + O(1/n), \tag{39.33}$$

and from (39.32)

$$e^{\Theta_{(n-r)}/12(n-r)} = 1 + O(1/n), \tag{39.34}$$

and naturally also

$$e^{\Theta_n/12n} = 1 + O(1/n). \tag{39.35}$$

Because of (39.23), $s^3/n^2 = O(1/\sqrt{n})$ and hence also

$$e^{O(s^3/n^2)} = 1 + O(1/\sqrt{n}). \tag{39.36}$$

One can thus write according to (39.29) and (39.36)

$$A_n = e^{-s^2/2npq}(1 + O(1/\sqrt{n})). \tag{39.37}$$

Now combine (39.30), (39.33), (39.34), (39.35) and (39.37) under consideration of (39.25) and obtain

$$\binom{n}{r} p^r q^{n-r} = (2\pi pqn)^{-1/2} e^{-(r-np)^2/2pqn}(1 + O(1/\sqrt{n}))$$

which means the same as (39.24).

We can now also prove the Theorem of Laplace which we mentioned on p. 108. Indeed, with the notation introduced in (39.27):

$$(2\pi)^{-1/2} \int_{\frac{s-1/2}{\sqrt{npq}}}^{\frac{s+1/2}{\sqrt{npq}}} e^{-x^2/2} dx = (2\pi npq)^{-1/2} e^{-s^2/2npq} + O(1/n). \tag{39.38}$$

For, from the Mean Value Theorem for integrals we get at once

$$(2\pi)^{-1/2} \int_{\frac{s-1/2}{\sqrt{npq}}}^{\frac{s+1/2}{\sqrt{npq}}} e^{-x^2/2} dx = (2\pi npq)^{-1/2} e^{-(s+\theta_1 r)^2/2npq}, \qquad |\theta_1| < 1/2$$

and hence also (39.38) with an application of (39.28). Hence, in place of (39.24) we can also write

$$\binom{n}{r} p^r q^{n-r} = (2\pi)^{-1/2} \int_{\frac{s-1/2}{\sqrt{npq}}}^{\frac{s+1/2}{\sqrt{npq}}} e^{-x^2/2} dx + O(1/n). \tag{39.39}$$

Now let r_1, r_2 be integers with say $0<r_1 \leqslant r_2<n$, both fulfilling (39.23). Then the latter also holds for all integers r with $r_1 \leqslant r \leqslant r_2$ and we get from (39.39)

$$\sum_{k=r_1}^{r_2} \binom{n}{k} p^k q^{n-k} = (2\pi)^{-1/2} \int\limits_{\frac{r_1-np-1/2}{\sqrt{npq}}}^{\frac{r_2-np+1/2}{\sqrt{npq}}} e^{-x^2/2}\,dx + (r_2-r_1)O(1/n).$$

Since r_1 and r_2 fulfill (39.23), $r_2-r_1=O(\sqrt{n})$ and we obtain

$$\sum_{k=r_1}^{r_2} \binom{n}{k} p^k q^{n-k} = (2\pi)^{-1/2} \int\limits_{\frac{r_1-np-1/2}{\sqrt{npq}}}^{\frac{r_2-np+1/2}{\sqrt{npq}}} e^{-x^2/2}\,dx + O(1/\sqrt{n}). \quad (39.40)$$

An application of the theorem mentioned on p. 108 would yield as an estimate for the left side of (39.40) only $(1/\sqrt{2\pi}) \int\limits_{\frac{r_1-np}{\sqrt{npq}}}^{\frac{r_2-np}{\sqrt{npq}}} e^{-x^2/2}\,dx$. However, it's easy to convince oneself that the right side of (39.40), with error omitted, yields a considerably better approximation, even for quite small values of n.[56] (See Fig. 8.)

Finally, we give another simple example of a local limit theorem: By means of Stirling's formula for the Γ-function[57] one easily sees that for densities of the t-distribution defined in (29.3) and each $t \in R_1$

$$\lim_{n\to\infty} h_n(t) = (1/\sqrt{2\pi})e^{-t^2/2}.$$

It is easy to complete this statement by one on the corresponding distribution function. Indeed, since for $n \geqslant 1$ and $-\infty<t<\infty$ the inequality $(1+t^2/2)^{-1} \geqslant (1+t^2/n)^{-(n+1)/2}$ holds, one can use Lebesgue's theorem and finds that the distribution function of a t-distribution converges to $N(0,1)$ for $n\to\infty$.

40. Some theorems of Cramér. We give here several important theorems useful in the determination of the asymptotic distribution of sequences of r.v.'s.

---

[56] There are many subtle investigations of this question. We mention only W. Feller, Ann. Math. Statist. 16, 319—329 (1945) and Ann. Math. Statist. 21, 301 (1950). See also Ibragimov, I. A. and Linnik, Ju. V. l. c.[52].

[57] This is an extension of Formula (39.11). See, for example, Lösch-Schoblik, Die Fakultät und verwandte Funktionen, Teubner, Leipzig 1951, 30.

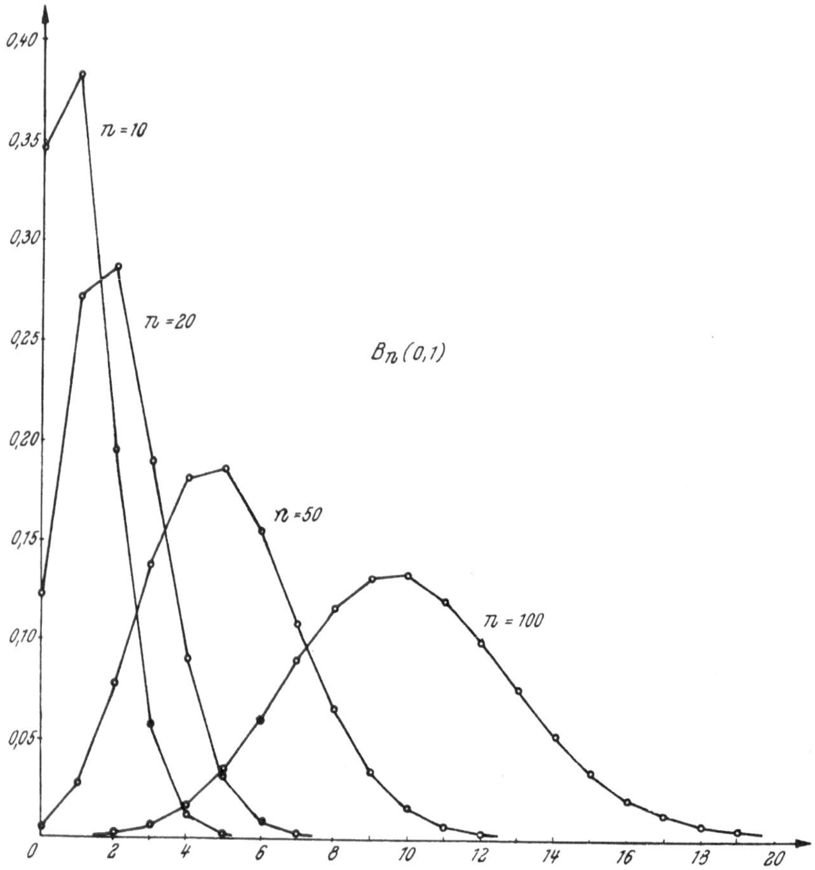

Fig. 8 Approach of a binomial distribution to the normal for increasing n

Theorem 40.1[58]. *Let $\xi_1, \xi_2, \ldots$ be a sequence of r.v.'s of the same dimension $k \geqslant 1$ and let ξ_n be asymptotically distributed with d.f. F. Let $\{\eta_i\}$ also be a sequence of r.v.'s and assume $\{\xi_i - \eta_i\}$ converges stochastically to 0 (i.e., to the k-dimensional null vector). Then, η_n is asymptotically distributed with d.f. F.*

Proof. For the sake of simplicity, take $k=1$; we need to show that for each $x \in R_1$ at which F is continuous

$$\lim_{n \to \infty} P(\eta_n \leqslant x) = F(x). \tag{40.1}$$

---

[58] For this and the following theorems, see H. Cramér, Mathematical Methods of Statistics, Princeton Univ. Press, Princeton 1946. See also E. Lukacs loc. cit.[47].

Let $\varepsilon>0$, $\delta>0$. Then, for large enough n,

$$P(|\xi_n-\eta_n|\leqslant\varepsilon)\geqslant 1-\delta. \tag{40.2}$$

Let $(R,\mathscr{S},P)$ be the given probability space. Let

$$\{\omega: |\xi_n(\omega)-\eta_n(\omega)|\leqslant\varepsilon\} = A_n.$$

We have

$$P(\eta_n\leqslant x)=P(\{\omega: \eta_n(\omega)\leqslant x\}\cap A_n)+P(\{\omega: \eta_n(\omega)\leqslant x\}\cap(R-A_n)).$$

Since $P(\{\omega: \eta_n(\omega)\leqslant x\}\cap(R-A_n))\leqslant P(R-A_n)$, by (40.2)

$$P(\{\omega: \eta_n(\omega)\leqslant x\}\cap(R-A_n))\leqslant\delta. \tag{40.3}$$

On the other hand, $P(\{\omega: \eta_n(\omega)\leqslant x\}\cap A_n)\leqslant P(\xi_n\leqslant x+\varepsilon)$ so that with (40.3), one gets

$$-\delta+P(\eta_n\leqslant x)\leqslant P(\xi_n\leqslant x+\varepsilon). \tag{40.4}$$

Replacing x by $x-\varepsilon$ and interchanging the roles of η_n and ξ_n we get in exactly the same way

$$-\delta+P(\xi_n\leqslant x-\varepsilon)\leqslant P(\eta_n\leqslant x). \tag{40.5}$$

But from (40.4) and (40.5)

$$-2\delta+P(\xi_n\leqslant x-\varepsilon)\leqslant -\delta+P(\eta_n\leqslant x)\leqslant P(\xi_n\leqslant x+\varepsilon). \tag{40.6}$$

Taking ε arbitrarily small, but such that x, $x-\varepsilon$ and $x+\varepsilon$ are points of continuity of F,[59] we find that (40.6) yields

$$-2\delta+F(x-\varepsilon)\leqslant -\delta+\varliminf_{n\to\infty}P(\eta_n\leqslant x)\leqslant -\delta+\varlimsup_{n\to\infty}P(\eta_n\leqslant x)\leqslant F(x+\varepsilon).$$

The claim follows by letting ε and $\delta\to 0$.

We also prove

Theorem 40.2. *Let $\{\xi_i\}$ be a sequence of r.v.'s. For $n=1,2,\ldots$, denote the d.f. of ξ_n by F_n and F be a d.f. to which F_n weakly converges. Moreover, let $\{\eta_i\}$ be a sequence of r.v.'s with $\eta_n\to 0$ in probability. Then $\{\xi_i\eta_i\}$ also converges stochastically to* 0.

Proof. We have to show that for each $\varepsilon>0$ and $\delta>0$ there exists an $n(\varepsilon,\delta)$ such that

$$P(|\xi_n\eta_n|\leqslant\varepsilon)\geqslant 1-\delta \tag{40.7}$$

for $n\geqslant n(\varepsilon,\delta)$.

---

[59] F possesse only countably many discontinuities (see p. 33), whence follows the possibility of such a choice for infinitely many $\varepsilon>0$ with $\varepsilon\to 0$.

Now, for $\varepsilon_1 = \delta/12$ there exists an x_{ε_1} such that F is continuous at x_{ε_1} and $-x_{\varepsilon_1}$ and such that $0 \leqslant 1 - F(x_{\varepsilon_1}) + F(-x_{\varepsilon_1}) < \varepsilon_1$. Since $F_n(-x_{\varepsilon_1}) \to F(-x_{\varepsilon_1})$ it follows that for all large enough n

$$0 \leqslant F_n(-x_{\varepsilon_1}) < 2\varepsilon_1 \tag{40.8}$$

and likewise

$$0 \leqslant 1 - F_n(x_{\varepsilon_1}) < 2\varepsilon_1. \tag{40.9}$$

Now hold x_{ε_1} fixed and choose $\varepsilon_2 = \varepsilon/x_{\varepsilon_1}$ and $\delta_2 = \delta/3$. By assumption, for large enough n

$$P(|\eta_n| \leqslant \varepsilon_2) \geqslant 1 - \delta_2. \tag{40.10}$$

Now with $A_n = \{\omega\colon |\eta_n(\omega)| \leqslant \varepsilon_2\}$ and $B_n = \{\omega\colon |\xi_n(\omega)| > x_{\varepsilon_1}\}$ we have

$$1 = P(A_n \cap (R - B_n)) + P((R - A_n) \cap (R - B_n)) + P(A_n \cap B_n) + P((R - A_n) \cap B_n).$$

From this, (40.8), (40.9) and (40.10)

$$1 \leqslant P(|\xi_n \eta_n| \leqslant \varepsilon_2 x_{\varepsilon_1}) + \delta_2 + 4\varepsilon_1 + \delta_2$$

for all large enough n, or

$$P(|\xi_n \eta_n| \leqslant \varepsilon) \geqslant 1 - 2\delta_2 - 4\varepsilon_1.$$

This is essentially the claim.

The *proof* of the following theorem is quite similar.

Theorem 40.3. *Let the sequence* $\{\xi_i\}$ *satisfy the assumptions of Theorem* 40.2 *and assume that* $\{\eta_i\}$ *converges stochastically to* $c > 0$. *Then at every point of continuity of* F

$$\lim_{n \to \infty} P(\xi_n \eta_n \leqslant x) = F(x/c).$$

Theorems 40.2 and 40.3 thus say the following: Let ξ be a r.v. with d.f. F and let the assumptions of Theorem 40.3 hold with $c \geqslant 0$. Then, the sequence of distribution functions of $\xi_n \eta_n$ converges weakly to the distribution function of ξc. [60]

[60] K. Krickeberg, Metrika 10, 179—181 (1966), has pointed out that one can prove the following theorem, which contains the cited results of Cramér: Let $\xi_1, \xi_2, \ldots$ and ξ be k-dimensional r.v.'s. Let the d.f. of ξ_i be F_i, $i = 1, 2, \ldots$ and F that of ξ. Assume that $\{F_i\}$ converges weakly to F and assume given a sequence of l-dimensional r.v.'s η_i which converges stochastically to the constant vector c. Let ϕ be a continuous map from R_{k+l} into R_m. Then the sequence of distribution functions of $\phi(\xi_n, \eta_n)$ converges weakly to the distribution function of $\phi(\xi, c)$. The proof is based on Lemma 23.1.

41. The moment problem. We will give only a cursory treatment and restrict ourselves to those results which we will need later[61]. In short, we have the *following problem*: Let $c_0=1$, $c_1, c_2, \dots$ be infinitely many real numbers. Does there exist a d.f. whose n^{th} moment is c_n for $n=0,1,\dots$?

One can easily find *necessary conditions* that a sequence c_i be a moment sequence. Let ξ be a r.v. all of whose moments m_i, $i=1,2,\dots$ exist. Let k and l be nonnegative integers and $u_0,\dots,u_k$ arbitrary real numbers. Then we always have

$$E[(u_0\xi^l+u_1\xi^{l+1}+\cdots+u_k\xi^{l+k})^2]\geqslant 0. \tag{41.1}$$

It is easy to see that the left side of (41.1) is also equal to $\sum_{i,j=0}^{k} u_i u_j m_{i+j+2l}$. This quadratic form in the variables $u_0,\dots,u_k$ is positive semi-definite, whence follows that all principal minors of the determinant $|m_{i+j+2l}|_{0k}^{0k}$ are nonnegative. One thus gets

$$\left.\begin{aligned}
&\begin{vmatrix} m_{2l} & m_{2l+1} & \cdots & m_{2l+k} \\ m_{2l+1} & m_{2l+2} & \cdots & m_{2l+k+1} \\ \cdot & \cdot & \cdot & \cdot \\ m_{2l+k} & m_{2l+k+1} & \cdots & m_{2l+2k} \end{vmatrix} \geqslant 0 \\
&\quad \cdot\ \cdot\ \cdot\ \cdot\ \cdot\ \cdot\ \cdot\ \cdot\ \cdot\ \cdot \\
&\begin{vmatrix} m_{2l} & m_{2l+1} \\ m_{2l+1} & m_{2l+2} \end{vmatrix} \geqslant 0 \\
&|m_{2l}| \geqslant 0
\end{aligned}\right\} \tag{41.2}$$

which hold for all k, $l\geqslant 0$. On the other hand, one can show that for each sequence $\{c_i\}$ satisfying the inequalities analogous to (41.2), there exists a d.f. whose n^{th} moment coincides with c_n. This is, in somewhat loose form, the main result of a series of investigations by Hamburger[62].

We especially want to point out the so-called *Moment Problem of Hausdorff* which is a special case of the *Hamburger Moment Problem.* Let $\mathfrak{F}$ be the set of all distribution functions F satisfying $F(x)=0$, $x<0$, $F(x)=1$, $x>1$. The Hausdorff Moment problem[63] concerns finding

---

[61] A detailed presentation is J.A. Shohat and J.D. Tamarkin, The Problem of Moments (Mathematical Surveys, Vol. I), Amer. Math. Soc., New York: 1943 and 1950.

[62] H. Hamburger, Math. Z. 4, 186—222 (1919), Math. Ann. 81, 31—45, 235—319 (1920); 82, 120—164, 168—187 (1921).

[63] F. Hausdorff, Math. Z. 9, 74—109 (1921). Also see S. Karlin and L.S. Shaple, Geometry of Moment Spaces, Mem. Amer. Math. Soc. No. 12, Providence 1953.

conditions under which distribution functions from $\mathfrak{F}$ exist for a given moment sequence $\{c_i\}$. This problem is therefore also called the *restricted moment problem.* If $F \in \mathfrak{F}$ is a d.f. with

$$c_i = \int_0^1 x^i \, dF(x), \qquad i = 0, 1, 2, \ldots,$$

then one immediately sees by considering $\int_0^1 (1-x)^k x^l \, dF(x)$ for integers $k, l > 0$, that the following conditions must necessarily hold for all integers k, l:

$$\sum_{j=0}^{k} \binom{k}{j} c_{k+l-j}(-1)^{k-j} \geqslant 0. \tag{41.3}$$

These are, however, also sufficient and determine exactly one distribution function.

This moment problem (and naturally, also the general Hamburger problem) can be modified by giving only the first n moments $c_1, \ldots, c_n$, $n \geqslant 1$. The conditions (41.3) in which only $c_0 = 1$, $c_1, \ldots, c_n$ appear, are again necessary but no longer sufficient for the existence of a distribution function whose i^{th} moment coincides with c_i for $i = 1, \ldots, n$. A condition which is also sufficient is obtained as follows: All determinants (41.2) with $m_{2l+i} = c_i$ for $l = 0$, $0 \leqslant k \leqslant [n/2]$ must be nonnegative along with the determinants $|a_{rs}|_{1j}^{1j}$ with $a_{rs} = c_{r+s-1}$ and $a_{rs} = c_{r+s-2} - c_{r+s-1}$ for $1 \leqslant j \leqslant [(n+1)/2]$ and $a_{rs} = c_{r+s-1} - c_{r+s}$ with $1 \leqslant j \leqslant [n/2]$.

However, the distribution function is no longer uniquely determined in this case.

42. Cumulants or Semi-invariants. Let ψ be the characteristic function of a r.v. ξ whose first k moments exist. By Theorem 23.2,

$$E(\xi^j)\, i^j = \left. \frac{d^j \psi}{dt^j} \right|_{t=0}, \qquad j = 1, \ldots, k,$$

so that from the continuity of $\dfrac{d^k \psi}{dt^k}$ and Taylor's theorem

$$\psi(t) = 1 + \sum_{j=1}^{k} (it)^j \frac{E(\xi^j)}{j!} + o(t^k) \tag{42.1}$$

for $t \to 0$.

Now consider the r.v. $\eta = \xi + c$ in place of ξ, where $c \neq 0$ is a real number. The characteristic function of η is $E(e^{it\eta}) = e^{itc} \psi(t)$ for each $t \in R_1$. Using (42.1), we get

$$\left(1 + \sum_{j=1}^{k} (it)^j c^j / j! + o(t^k)\right)\left(1 + \sum_{j=1}^{k} (it)^j E(\xi^j)/j! + o(t^k)\right) = e^{ict} \psi(t). \tag{42.2}$$

The left side of (42.2) can be written as

$$1+\sum_{j=1}^{k}(t\,i)^j\sum_{l=0}^{j}\frac{c^l}{l!}\frac{E(\xi^{j-l})}{(j-l)!}+\varepsilon_3(t)\,t^k. \tag{42.3}$$

Here, $\varepsilon_3(t)=o(1)\psi(t)+o(1)e^{ict}+b_1t+\cdots+b_kt^k$, and the coefficients b_i can be easily determined by comparison with (42.2). In any case, it follows that $\varepsilon_3(t)=o(1)$.

A comparison of (42.1) with (42.3) shows that the coefficients of the expansion (42.1) do not in general remain invariant w.r.t. zero point translations of the r.v.'s.

Considering instead the expansion of $\log\psi$ in a neighborhood of zero, one easily finds that the coefficients of powers of t, except for that of the linear term, are invariant w.r.t. translations of the r.v.'s.

$\log\psi$ is defined in a neighborhood of zero since $\psi(t)\neq 0$ because of $\psi(0)=1$ and the continuity of ψ in a suitable neighborhood of $t=0$. Fix the value of $\log\psi(0)$ as 0. Since by assumption $\dfrac{d^j\psi}{dt^j}$ exists for $j=1,\dots,k$, the existence of $\dfrac{d^j\log\psi}{dt^j}$, $j=1,\dots,k$, in a neighborhood of zero is also assured. In particular,

$$\frac{d\log\psi(t)}{dt}=\frac{\psi'(t)}{\psi(t)}.$$

Denoting the coefficients of the expansion of $\log\psi$ at the point $t=0$ by K_j, we see that $K_1=E(\xi)$. For each t in a neighborhood of $t=0$

$$\log\psi(t)=\sum_{j=1}^{k}K_j(t\,i)^j+o(t^k) \tag{42.4}$$

for $t\to 0$. Denote the characteristic function of $\eta=\xi+c$ by ϕ. Then

$$\log\phi(t)=c\,i\,t+\log\psi(t)$$

so that by (42.4)

$$\log\phi(t)=(c+K_1)\,i\,t+\sum_{j=2}^{k}K_j(t\,i)^j+o(t^k)$$

which shows the announced invariance principle. If the r.v. $b\xi$, with $b\neq 0$ a real number, is considered in place of ξ, then we get for the characteristic function, for each $t\in R_1$,

$$E(e^{it\xi b})=\psi(b\,t)$$

and hence for its logarithm in a suitable neighborhood of zero:

$$\log\psi(b\,t)=\sum_{j=1}^{k}K_j(i\,b\,t)^j+o(t^k)$$

for $t\to 0$.

Thus, the coefficients of the expansion of the logarithm of the characteristic function of a r.v. ξ are multiplied by b^j, $j=1,\dots,k$, when the r.v. ξ is replaced by $b\xi$, $b\neq 0$.

This property is also enjoyed by the characteristic function itself, which is immediately clear from $E[(b\,\xi)^j]=b^jE(\xi^j)$.

Summarizing, we can thus say that the coefficients of (42.4) are transformed by the transition to the r.v. $b\xi+c$ in such a way that except for the coefficient of the linear term, the jth coefficient remains unchanged up to the multiplication by b^j $(j=2,\dots,k)$. This fact has given rise to the term *semi-invariants* for the K_j's. It is preferable to call them *cumulants*.

Chapter II

Elementary Sampling Theory[1]

1. Introduction. In connection with the notion of probability we described the following situation: Assume we have a population of observations which are related to certain measurable outcomes. This population is taken to be infinite in the sense that the observations are always reproducible according to a fixed prescription, for example, an infinite series of throws of a die. From this population one now chooses a series of observations "at random". If there are enough observations, then the relative frequencies of events related to the outcome under observation deviate in general only slightly from a constant value, which we have called the empirical probability (see p. 20). It is not easy to give empirical criteria for deciding when a sample from a population can be viewed as random. One often satisfies oneself with the somewhat vague formulation that a random sample has been realized when there is no reason to believe that the choice of any particular sample is more probable than the rest. In this connection, one often calls on an "urn model". The urn, or better, its contents (for example, equal balls) represents the population and balls are then drawn from it, making sure

---

[1] A number of excellent works treat applied sampling theory: A. Linder, Statistische Methoden für Naturwissenschaftler, Mediziner und Ingenieure, 3rd ed., Birkhäuser, Basel. J. Neyman, First Cours in Probability and Statistics, John Wiley & Sons, New York-London 1962. J. Pfanzagl, Allgemeine Methodenlehre der Statistik, I, 5th ed., II, 3rd ed., Sammlung Göschen, Walter de Gruyter & Co., Berlin 1972, 1968. D. Morgenstern, Einführung in die Wahrscheinlichkeitsrechnung und Mathematische Statistik, Springer-Verlag, Berlin-Göttingen-Heidelberg, 1964. The book: B. L. van der Waerden, Mathematical Statistics, Springer-Verlag, New York-Heidelberg-Berlin 1969, is also of theoretical interest. Further treatements with emphasis on mathematical methods are: H. Cramér, Loc. cit. I[58]. D. Dugué, Traité de statistique théorique et appliquée, Masson & Cie, Paris 1958. A. M. Mood, Introduction to the Theory of Statistics, McGraw-Hill, New York 1950. S. S. Wilks, Mathematical Statistics, John Wiley & Sons, New York-London 1962. H. Witting, Mathematische Statistik, B. G. Teubner, Stuttgart 1966. H. Witting und G. Nölle, Angewandte Mathematische Statistik, B. G. Teubner, Stuttgart 1970. A compendium with extensive bibliography is: M. Kendall and A. Stuart, The Advanced Theory of Statistics, I: 1969, II: 1967, III: 1968, Griffin, London.

that they are always "well-mixed" before each draw. The drawn ball is viewed as a random choice from the urn. We recall what has already been said about the urn scheme; in particular, to what these ideas correspond in the calculus of probability (see p. 27).

The goal of taking such random samples is the formulation of statements about the structure of the population, in particular, about the magnitude of the empirical probabilities that are associated with it. We are guided here by the notion that one "could exactly determine" the empirical probabilities if one performed infinitely many experiments. Practically speaking, however, we encounter difficulties of the most varied sorts when we try to carry out an arbitrary long series of experiments or even a large number of observations. For example, a large number of observations might be so limited by technical or financial considerations so that one has to be satisfied with a smaller number. The nearly ideal experimental conditions obtained in most games of chance are usually not realized in practical cases.

The following terminology has established itself: The (hypothetical) infinite set of possible observations is called the *population* and the (randomly) drawn observations from this set a *sample* from this population. The number of observations contained in the sample is called the *sample size*. The notion of an infinite population represents an idealization of the actual situation, even when one imagines a "potentially infinite" population in the sense of unlimited reproducibility of the random experiments. In practical work, one views every population which is "very large" in relation to the size of the drawn sample as infinite.

2. Introduction to the terminology. We now want to give an adequate description of the ideas treated above empirically, in terms of the calculus of probability, and thereby lay the cornerstone (in the sense of what was said at the beginning of the introduction (p. 18)) for all of mathematical statistics.

We emphasize that we will pay no attention to the actual gathering of the samples. The practically very important problem of how one obtains samples which are as informative as possible for the smallest possible financial or labor outlay as well as the problem of how one practically realizes a "random sample" will only be touched on to the extent that they involve mathematically relevant questions.

If $\xi_1,\ldots,\xi_n$ are (not necessarily one-dimensional) r.v.'s (over the same probability space) and the events $(\xi_1=x_1),\ldots,(\xi_n=x_n)$ have been observed, we say that $(x_1,\ldots,x_n)$ is a *realization* of $(\xi_1,\ldots,\xi_n)$. Let $\xi_1,\ldots,\xi_n$ be n r.v.'s which have the same probability distribution and let $(x_1,\ldots,x_n)$ be a realization of $(\xi_1,\ldots,\xi_n)$. Assume that the probability distribution of the ξ_i is unknown. We ask what can be learned about this distribution on the basis of a knowledge of the $(x_1,\dots,x_n)$.

In order to see that this question is connected with the contents of **1**, one must identify the probability distribution of the ξ_i with the empirical distribution which results from the consideration of the population from which the $(x_1,\dots,x_n)$ has been drawn as sample.

We will often call the r.v.'s $\xi_1,\dots,\xi_n$ *sample variables* and the observed realizations $x_1,\dots,x_n$ which are not to be confused with the former, *sample values*. Their totality is the sample. If $(x_1,\dots,x_n)$ is a realization of independent r.v.'s with the same distribution P_ξ, then, with regard to what we have just said, we will also use the following more intuitive notation: "$x_1,\dots,x_n$ is a sample from a population distributed by P_ξ".

In later chapters, however, we will use the notion of a sample in a more general sense which we want to mention here along with the idea of a sample space. We want to interpret a sample (e. g., of one dimensional sample values) of size n as a point in R_n, which we denote as the *sample space*. Each probability distribution defined over the Borel sets $\mathfrak{B}_n$ can be viewed as the probability distribution corresponding to the sample space. This viewpoint is, however, not bound to the assumptions already made. If $(R,\mathcal{S})$ is an arbitrary measurable space, then each $x\in R$ can, under certain circumstances, represent the result of a sample. Then we denote R again as sample space and a probability measure P defined over $(R,\mathcal{S})$ as the probability distribution corresponding to the sample space.

3. Testing a hypothesis on the mean of a normal distribution with known variance. Here we want to illustrate the still somewhat vague ideas formulated in **2** with the concrete example of the normal distribution. We have theoretical reasons for first treating a normal distribution, whose density for all $x\in R_1$ is given by I. (25.1): many questions can be especially simply and completely treated for the normal distribution. On the other hand, it turns out that in many practical situations, one can, at least approximately, assume that the population is normally distributed. See I, p. 106 and **11**. We now present an example which we have already encountered (p. 19). The distribution of the heights of men has been found to be approximately normal.

W. Winkler[2] has shown this for an example of 906 recruits enlisted in 1913. He found the height distribution given in the table on p. 125.

We note first in connection with our claim of approximate normality of the heights, that these can *never* be strictly normally distributed.

[2] W. Winkler, loc. cit. Intro.[1], 35.

First, the event that a height is <0 is impossible (has probability 0), while the density I. (25.1) is always positive in the interval $-\infty<x<\infty$. The claim is thus to be understood in the frequency interpretation sense. Since we have agreed (p. 19) to round off values for instance

cm	frequency	cm	frequency	cm	frequency
147	1	159	22	171	48
148	0	160	30	172	36
149	0	161	35	173	31
150	2	162	43	174	33
151	4	163	48	175	21
152	3	164	47	176	24
153	4	165	60	177	13
154	7	166	63	178	9
155	6	167	74	179	9
156	12	168	60	180	3
157	14	169	64	181	3
158	25	170	47	182	4
				183	1

between 170.5 and 171.5 cm to 171 cm, we see from the above data that the relative frequency of the event that the observated heights in 906 experiments lie in the interval (170.5, 171.5), is 48/906.

Grouping the observations into sets of width 3 cm and associating the relative frequencies corresponding to the groups with the center of each group, we obtain Fig. 9.

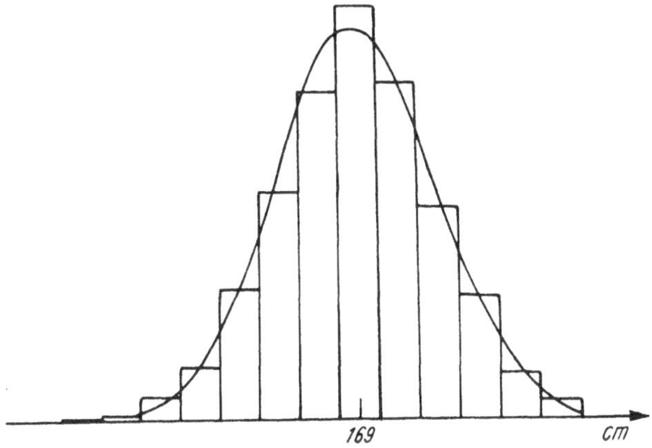

Fig. 9

The graphed density of a normal distribution fits the frequency distribution very well and this can serve as empirical basis for the hypothesis that the heights can, to good approximation, be taken as normally distributed.

Now assume that the population is normally distributed with variance 36. We want to investigate whether the observed results allow the assumption that the 906 height measurements represent a sample from such a population with mean 166.

We now formulate the problem more generally, within the framework of mathematical statistics.

Assume we have a sample $x_1,\ldots,x_n$ of size n. Let the sample variables ξ_i be independently $N(a,\sigma^2)$-distributed, with σ^2 a given, positive number.

Suppose we hypothesized that $a=a_0$. We are interested in to what extent one is justified in viewing the hypothesis $a=a_0$ as false or correct on the basis of the sample $x_1,\ldots,x_n$. In short: how does one test $a=a_0$ on the basis of a sample? We will develope a test procedure that must appear rather arbitrary from the mathematical standpoint. This applies to all of the procedures treated in this chapter. This means initially to study elementary or naive sample theory, although in the following chapters we will classify these test procedures from various standpoints. The considerations below, however, already show the practical value of the method.

Anticipating the sequel, we remark that (in the notation of III) the problem can be posed as follows: Let $\xi_1,\ldots,\xi_n$ be n independent r.v.'s with the same distribution function, which is assumed to belong to the set of normal distributions with given variance σ^2. How can one test the assumption $a=a_0$ on the basis of a sample $x_1,\ldots,x_n$, i.e., the assumption that the distribution function is $N(a_0,\sigma^2)$?

We first prove

Theorem 3.1. *Let* $\xi_1,\ldots,\xi_n$ *be* n *independent r.v.'s with the same probability distribution. Let* $E(\xi_i)=a$ *and* $E[(\xi_i-a)^2]=\sigma^2$. *Then for the r.v.* $\bar{\xi}=(\xi_1+\cdots+\xi_n)/n$ *we have*

$$E(\bar{\xi})=a \tag{3.1}$$

and

$$E[(\bar{\xi}-a)^2]=\sigma^2/n. \tag{3.2}$$

The *proof* follows immediately from I, Theorem 16.3 and I, Theorem 17.4.

Applying this to the normal distribution, we get with the help of the remark on p. 79 on linear combinations of independent normally distributed r.v.'s the following

Theorem 3.2. *Let* $\xi_1,\ldots,\xi_n$ *be independent* $N(a,\sigma^2)$*-distributed r.v.'s. Then, the r.v.* $\bar{\xi}=(\xi_1+\cdots+\xi_n)/n$ *is* $N(a,\sigma^2/n)$*-distributed.*

This theorem forms the basis of our test procedure. Indeed, if the hypothesis $a=a_0$ holds, then the standardized r.v. $\eta=(\bar{\xi}-a_0)\sqrt{n}/\sigma$ is $N(0,1)$-distributed, and from Čebyšev's inequality I, (22.2) we have, quite generally, that "large" deviations of η from the mean 0 occur with only small probability. Considering what we have already said about the interpretation of small probabilities (p. 27), we thus arrive at the following test procedure: let a real number α with $0<\alpha<1$, the so-called *level of significance*, be arbitrarily given. To each number α there corresponds exactly one number κ_α [3] for which $\alpha=1-\left(\int_{-\kappa_\alpha}^{\kappa_\alpha} e^{-x^2/2}\,dx\right)\Big/\sqrt{2\pi}$ holds. Hence, $P(|\eta|\geqslant\kappa_\alpha)=\alpha$. On the basis of the given sample values $x_1,\dots,x_n$ we now form the quantity $\bar{x}=(x_1+\cdots+x_n)/n$ and accept the hypothesis $a=a_0$ when $|(\bar{x}-a_0)\sqrt{n}/\sigma|<\kappa_\alpha$ holds, and reject it when $|(\bar{x}-a_0)\sqrt{n}/\sigma|\geqslant\kappa_\alpha$. Taking what was said above into consideration, it is obvious that it is almost always reasonable to choose α small. Typical values are $\alpha=0.05, 0.01, 0.001$. We will also call κ_α the *significance limit* for the level of significance α.

In the example of the recruits we thus proceed as follows: Choose $\alpha=0.01$. This corresponds with sufficient precision to $\kappa_\alpha=2.576$. We have $n=906$, $\bar{x}=166.77$ and with $a_0=166$ and $\sigma=6$, $|(\bar{x}-a_0)\sqrt{n}/\sigma|=3.86\ldots$. If we follow the previous prescription, we must reject the hypothesis $a_0=166$. On the other hand, on the basis of this test, we can accept for example the hypothesis $a=167$.

Summary. In order to test a hypothesis $a=a_0$ on the mean of a normal distribution with known variance σ^2 on the basis of a sample of size n for given level of significance α, one can proceed as follows: Form $|(\bar{x}-a_0)\sqrt{n}/\sigma|$ and determine κ_α from the expression

$$\alpha=1-\left(\int_{-\kappa_\alpha}^{\kappa_\alpha} e^{-x^2/2}\,dx\right)\Big/\sqrt{2\pi}. \tag{3.3}$$

If $|(\bar{x}-a_0)\sqrt{n}/\sigma|<\kappa_\alpha$, accept $a=a_0$, if $|(\bar{x}-a_0)\sqrt{n}/\sigma|\geqslant\kappa_\alpha$, reject $a=a_0$.

We remark that this argument can be interpreted as follows: If a_0 is the correct value of a, i.e., if the given sample actually arises from a normally distributed population with mean a_0 and if in spite of this, $|(\bar{x}-a_0)\sqrt{n}/\sigma|\geqslant\kappa_\alpha$, then note that $|(\xi-a_0)\sqrt{n}/\sigma|\geqslant\kappa_\alpha$ can occur, on the basis of the determination of κ_α by (3.3), with probability α; one can thus say, in the sense of the frequency interpretation, that if one uses the test procedure sufficiently

[3] The arbitrarity of this test procedure becomes clear when one notes that for given α, one can choose arbitrarily many pairs of real numbers $(\kappa'_\alpha, \kappa''_\alpha)$ for which $1-\left(\int_{\kappa'_\alpha}^{\kappa''_\alpha} e^{-x^2/2}\,dx\right)\Big/\sqrt{2\pi}=\alpha$. We have chosen $\kappa''_\alpha=-\kappa'_\alpha=\kappa_\alpha$ for no other reason (for the moment), than the fact that the symmetry thus obtained is convenient. See, however, III, p. 203.

often, one runs the danger—when $a=a_0$ is the true parameter value—of rejecting the hypothesis $a=a_0$ $100\,\alpha$ times in 100 decisions of this type, although it is correct. An analogous interpretation applies to all of the remaining test procedures in this chapter. Nevertheless, even in the case $a \neq a_0$, the probability of acceptance of the hypothesis $a=a_0$ can be "large".

We point out that one can also use Theorem 3.1 to test a hypothesis on the mean of an arbitrary distribution function F of the population. One uses Čebyšev's inequality I. (22.2), where it is assumed that one has some idea of the size of the (assumed existent) variance σ^2, say, $\sigma^2 \leqslant \sigma_1^2$, with known σ_1^2. One can then construct a (rough, but previously widely used) test procedure with a level of significance $\leqslant \alpha$. Indeed, with $t^{-2}=\alpha$ we have, under the assumption $a=a_0$,

$$P(|\bar{\xi}-a_0| \geqslant \sigma_1/\sqrt{\alpha n}) \leqslant P(|\bar{\xi}-a_0| \geqslant \sigma/\sqrt{\alpha n}) \leqslant \alpha .$$

4. Testing a hypothesis on the mean of a normal distribution with unknown variance. The recruit example in **3** contained the artificial assumption that the variance σ^2 was known. In few cases does one actually know the value of the variance σ^2 in the (assumed normally distributed) population. We will show that one can give a test procedure for the mean a of a $N(a,\sigma^2)$-distributed population which does not assume a priori knowledge of the variance σ^2. Assume we have a sample $x_1,\dots,x_n$ from an $N(a,\sigma^2)$-distributed population. We are interested in a test for the hypothesis $a=a_0$ when σ^2 is unknown. Before we can find such a test, we must turn to some deeper mathematical considerations. To this end, we prove a general theorem due to Cochran[4].

Theorem 4.1. *Let $\xi_1,\dots,\xi_n$ be independent $N(0,1)$-distributed r.v.'s and $q_1,\dots,q_k$, $k \geqslant 1$, quadratic forms over R_n, $n \geqslant 1$. Let q_j have rank n_j, $n_j \geqslant 1$ $(j=1,\dots,k)$ and assume*

$$\sum_{j=1}^{k} q_j(x_1,\dots,x_n) = \sum_{i=1}^{n} x_i^2 . \tag{4.1}$$

Let $\boldsymbol{q}_j$ denote the r.v. $q_j(\xi_1,\dots,\xi_n)$. Then:

1. *From*

$$\sum_{j=1}^{k} n_j = n \tag{4.2}$$

it follows that each $\boldsymbol{q}_j$ is χ^2-distributed with n_j degrees of freedom and the $\boldsymbol{q}_j$ are independent, $1 \leqslant j \leqslant k$.

2. *If each $\boldsymbol{q}_j$ is χ^2-distributed, then necessarily with n_j degrees of freedom; the $\boldsymbol{q}_j$ are then also independent for $j=1,\dots,k$ and, furthermore,* (4.2) *holds.*

---

[4] W.G. Cochran, Proc. Cambridge Philos. Soc. 30, 178—191 (1933—1934).

3. *If the* $\boldsymbol{q}_j$ *are independent, then each* $\boldsymbol{q}_j$ *is* χ^2*-distributed with* n_j *degrees of freedom,* $1 \leqslant j \leqslant k$, *and* (4.2) *holds.*

We will essentially follow the proof of James[5].

1. *Proof by induction* on the number of quadratic forms. For $k=1$, the claim for arbitrary n has already been proved (see I.**28**). Assume 1. has been demonstrated for $k-1$ forms q_j and arbitrary n. We show that it is then also correct for k forms.

In fact, since q_1 has rank n_1, there exists an orthogonal transformation

$$x = A\,y \tag{4.3}$$

such that

$$q_1 = c_1\,y_1^2 + \cdots + c_{n_1}\,y_{n_1}^2 \qquad (c_i \neq 0,\ i=1,\dots,n_1). \tag{4.4}$$

By means of (4.3), $\sum_{i=1}^{n} x_i^2$ goes into $\sum_{i=1}^{n} y_i^2$. Hence, from (4.1)

$$\sum_{j=2}^{k} q_j(y_1,\dots,y_n) = (1-c_1)\,y_1^2 + \cdots + (1-c_{n_1})\,y_{n_1}^2 + \sum_{j=n_1+1}^{n} y_j^2 . \tag{4.5}$$

Since each q_j has rank n_j, the rank r of $\sum_{j=2}^{k} q_j$ is $\leqslant n - n_1$ because of (4.2). But this is not contradicted by (4.5) iff $c_i = 1$ for $i=1,\dots,n_1$ and

$$r = n - n_1 . \tag{4.6}$$

Thus,

$$q_1(y_1,\dots,y_n) = \sum_{j=1}^{n_1} y_j^2 \tag{4.7}$$

and

$$\sum_{j=2}^{n} q_j(y_1,\dots,y_n) = \sum_{j=n_1+1}^{n} y_j^2 . \tag{4.8}$$

By I, Theorem 27.2, it follows that the components η_i $(i=1,\dots,n)$ of the n-dimensional r.v.'s $\eta = A^{-1}\xi$ are independently $N(0,1)$-distributed. By the induction assumption, we have from (4.6) and (4.8) that each $\boldsymbol{q}_j$ $(j=2,\dots,k)$ is χ^2-distributed with n_j degrees of freedom and the $\boldsymbol{q}_j$ are independently distributed. But from (4.7), $\boldsymbol{q}_1$ is χ^2-distributed with n_1 degrees of freedom, and because of (4.7) and (4.8), $\boldsymbol{q}_1$ is also independent of all $\boldsymbol{q}_j$ with $2 \leqslant j \leqslant k$.

2. In the same notation one can express $\boldsymbol{q}_1$ by means of (4.3) again in the form (4.4). We proceed again by induction:

For $k=1$ everything is clear.

---

[5] G. S. James, Proc. Cambridge Philos. Soc. 48, 443—446 (1952).

If 2. has already been proved for $k-1$ forms q_j, then the proof for k forms goes as follows:

We have

$$E(e^{itq_1})=E(e^{it(c_1\eta_1^2+\cdots+c_{n_1}\eta_{n_1}^2)})=\prod_{m=1}^{n_1} E(e^{itc_m\eta_m^2}),$$

where the last equality comes from the independence of the η_i. But $E(e^{itc_m\eta_m^2})=(1-2c_m it)^{-1/2}$ for all real t. To show this, one need only apply I. (28.6) for $n=1$ with t replaced by $c_m t$. It follows that

$$E(e^{itq_1})=\prod_{m=1}^{n_1}(1-2c_m it)^{-1/2}.$$

On the other hand, we have assumed that q_1 is χ^2-distributed with, say, f_1 degrees of freedom, so that $E(e^{itq_1})=(1-2it)^{-f_1/2}$. Comparing the two expressions for $E(e^{itq_1})$ we find that

$$\prod_{m=1}^{n_1}(1-2c_m it)^{-1/2}=(1-2it)^{-f_1/2}$$

for all real t.

Since the two polynomials $\prod_{m=1}^{n_1}(1-2c_m it)$ and $(1-2it)^{f_1}$ coincide, we have, since $c_m\neq 0$, that

$$f_1=n_1 \quad \text{and} \quad c_m=1, \; m=1,\ldots,n_1.$$

Because of (4.4) and (4.5) this yields (4.7) and (4.8). Thus, by the induction hypothesis, each q_j is χ^2-distributed with n_j degrees of freedom for $2\leqslant j\leqslant k$ and $\sum_{j=2}^{k} n_j=n-n_1$. Hence, (4.2) holds and, by 1., we are finished.

3. As at 1., one can assume that (4.4) and (4.5) hold. Let $Q=\sum_{l=2}^{k} q_l$ and consider the two-dimensional r.v. (q_1, Q). Its characteristic function is

$$E(e^{i(q_1t_1+Qt_2)}), \quad -\infty<t_i<+\infty, \quad i=1,2.$$

Since the q_j, $1\leqslant j\leqslant k$, are assumed to be independent, q_1 and Q are independent. Hence (see I. (16.5))

$$E(e^{i(q_1t_1+Qt_2)})=E(e^{iq_1t_1})E(e^{iQt_2}) \tag{4.9}$$

for all $(t_1, t_2)\in R_2$.

Now

$$E\left(\exp\left(it_1\sum_{l=1}^{n_1} c_l\eta_l^2+it_2\left[\sum_{m=1}^{n_1}(1-c_m)\eta_m^2+\sum_{m=n_1+1}^{n}\eta_m^2\right]\right)\right)$$
$$=\left[(1-2it_2)^{n-n_1}\prod_{m=1}^{n_1}\left(1-2ic_mt_1-2i(1-c_m)t_2\right)\right]^{-1/2}.$$

This follows from the independence of the η_i and an application of I. (28.6). Calculating the right side of (4.9) in a similar way, one gets in its place

$$\prod_{j=1}^{n_1}(1-2ic_jt_1-2i(1-c_j)t_2)^{-1/2}[(1-2it_2)^{n-n_1}]^{-1/2}$$
$$=\prod_{j=1}^{n_1}(1-2ic_jt_1)^{-1/2}\prod_{m=1}^{n_1}(1-2i(1-c_m)t_2)^{-1/2}[(1-2it_2)]^{-(n-n_1)/2}. \tag{4.10}$$

Since $(1-2it_2)^{-(n-n_1)/2}\neq 0$ for $t_2\in R_2$, we obtain

$$\prod_{j=1}^{n_1}(1-2ic_jt_1-2i(1-c_j)t_2)=\prod_{j=1}^{n_1}(1-2ic_jt_1)\prod_{m=1}^{n_1}(1-2i(1-c_m)t_2)$$

for all $(t_1, t_2)\in R_2$.

The two polynomials in the equation above are identical. Comparing coefficients of the highest powers in t_1 we have, identically in t_2

$$\prod_{j=1}^{n_1}(-2ic_j)=\prod_{m=1}^{n_1}(1-2i(1-c_m)t_2)\prod_{j=1}^{n_1}(-2ic_j).$$

But since $c_j\neq 0$, we get $c_m=1, m=1,\ldots,n_1$.

This means that $\boldsymbol{q}_1$ is χ^2-distributed with n_1 degrees of freedom and hence, by 2. and 1., 3. is proved.

The question posed on p. 128 requires further preparation. We first agree on the following *notational convention* which we have already used in part: If $x_1,\ldots,x_n$ are real numbers, we write $\bar{x}$ or $\bar{x}_n$ instead of $(x_1+\cdots+x_n)/n$. If $\xi_1,\ldots,\xi_n$ are r.v.'s, we write analogously $\bar{\xi}$ in place of $(\xi_1+\cdots+\xi_n)/n$. We also write s_x^2 (or even s^2) in place of $\sum_{i=1}^{n}(x_i-\bar{x})^2/(n-1)$ and $\boldsymbol{s}^2$ instead of $\sum_{i=1}^{n}(\xi_i-\bar{\xi})^2/(n-1)$. It always will be clear from the context to which r.v.'s we are referring.

We now prove

Theorem 4.2. *Let* $\xi_1,\ldots,\xi_n$, $n\geqslant 2$ *be independent, identically distributed r.v.'s. Assume their first four moments exist and that* $E(\xi_i)=a$, $E[(\xi_i-a)^2]=\sigma^2$ *and* $E[(\xi_i-a)^4]=m_4$.

Then, $E(s^2)=\sigma^2$ *and*

$$E[(s^2-\sigma^2)^2]=\frac{m_4}{n}-\frac{n-3}{n(n-1)}\sigma^4. \tag{4.11}$$

For the *proof* note that

$$(n-1)s^2=\sum_{i=1}^{n}\xi_i^2-2\bar{\xi}\sum_{i=1}^{n}\xi_i+n\bar{\xi}^2=\sum_{i=1}^{n}\xi_i^2-n\bar{\xi}^2. \tag{4.12}$$

Taking I, Theorem 17.2 into account along with (3.1) and (3.2), we find

$$(n-1)E(s^2)=n(\sigma^2+a^2)-n(\sigma^2/n+a^2),$$

so that

$$E(s^2)=\sigma^2. \tag{4.13}$$

For the rest, note that s^2 does not change when one replaces the variables ξ_i by ξ_i-a. Writing $\xi_i-a=\eta_i$ for the moment, we have

$$E(\eta_i)=0,\quad E[\eta_i^2]=\sigma^2,\quad E[\eta_i^4]=m_4,$$

and

$$(n-1)^2s^4=\left(\sum_{i=1}^{n}\eta_i^2-n\bar{\eta}^2\right)^2$$

$$=\sum_{i,k=1}^{n}\eta_i^2\eta_k^2-\left(2n\sum_{i,j,k=1}^{n}\eta_i\eta_j\eta_k^2\right)\Big/n^2+\left(n^2\sum_{i,j,k,l=1}^{n}\eta_i\eta_j\eta_k\eta_l\right)\Big/n^4$$

because of

$$\bar{\eta}^2=\left(\sum_{i,j=1}^{n}\eta_i\eta_j\right)\Big/n^2.$$

To calculate the expectation of s^4, note that the expected value of terms of the form $\eta_i\eta_j\eta_k^2$ and $\eta_i\eta_j\eta_k\eta_l$ vanishes, provided at least one index of a linear factor is different from all of the rest, since $E(\eta_i)=0$ and the η_i are independent.

Therefore,

$$(n-1)^2E(s^4)$$

$$=E\left(\sum_{i,j=1}^{n}\eta_i^2\eta_j^2\right)-2\left[E\left(\sum_{i,j,k=1}^{n}\eta_i\eta_j\eta_k^2\right)\right]\Big/n+\left[E\left(\sum_{i,j,k,l=1}^{n}\eta_i\eta_j\eta_k\eta_j\right)\right]\Big/n^2$$

$$=\sum_{i\neq j}E(\eta_i^2\eta_j^2)+\sum_{i=1}^{n}E(\eta_i^4)-2\left[\sum_{i\neq j}E(\eta_i^2\eta_j^2)+\sum_{j=1}^{n}E(\eta_i^4)\right]\Big/n$$

$$+\left[3\sum_{i\neq j}E(\eta_i^2\eta_j^2)+\sum_{i=1}^{n}E(\eta_i^4)\right]\Big/n^2$$

$$=n(n-1)\sigma^4+nm_4-2[n(n-1)\sigma^4+nm_4]/n+[3n(n-1)\sigma^4+nm_4]/n^2.$$

This gives

$$E(s^4)=\frac{m_4}{n}+\frac{n^2-2n+3}{n(n-1)}\sigma^4.$$

I, Theorem 17.2 now yields (4.11) by means of (4.13).

It immediately follows from (4.11) that $E[(s^2-\sigma^2)^2]$ tends to 0 for $n\to\infty$.

We now consider as a special case n independent, $N(0,1)$-distributed r.v.'s $\xi_1,\ldots,\xi_n$, form the r.v. s^2 and prove

Theorem 4.3. *Let $\xi_1,\ldots,\xi_n$, $n\geqslant 2$, be n independent $N(0,1)$-distributed r.v.'s. Then $(n-1)s^2$ is χ^2-distributed with $n-1$ degrees of freedom. The r.v. s^2 is distributed independently of $\bar{\xi}$.*

Proof. It follows from (4.12) that $(n-1)s^2+n\bar{x}^2=\sum_{i=1}^{n}x_i^2$. Thus, with $k=2$, $q_1=(n-1)s^2$ and $q_2=n\bar{x}^2$ the assumptions of Theorem 4.1 are satisfied. Since q_2 has rank 1 and q_1, as is easily seen, rank $n-1$, assumption (4.2) is fulfilled. Moreover, the proof of 1. (p. 129ff.) shows that there exists an orthogonal transformation of the form (4.3) such that $(n-1)s^2=\sum_{i=1}^{n-1}\eta_i^2$ and $\bar{\xi}=\eta_n$, where the r.v.'s η_i, $1\leqslant i\leqslant n$, are mutually independent. The proof of Theorem 4.3 is thus complete.

Theorem 4.3 can be immediately extended to a sample from an $N(a,\sigma^2)$-distributed population: Let $\xi_1,\ldots,\xi_n$ be independent and $N(a,\sigma^2)$-distributed. Form the r.v.'s $\bar{\xi}$ and s^2. To apply Theorem 4.3, consider in place of the ξ_i $(i=1,\ldots,n)$ the standardized variables

$$\eta_i=\frac{\xi_i-a}{\sigma},\qquad i=1,\ldots,n.$$

Theorem 4.3 can now be immediately applied to the r.v.'s

$$\bar{\eta}=\frac{\bar{\xi}-a}{\sigma}\quad\text{and}\quad s'^2=\frac{s^2}{\sigma^2}.$$

We then obtain the

Corollary. *Let $\xi_1,\ldots,\xi_n$, $n\geqslant 2$, be independent and $N(a,\sigma^2)$-distributed. Then s^2 and $\bar{\xi}$ are independently distributed and $(n-1)s^2/\sigma^2$ possesses a χ^2-distribution with $n-1$ degrees of freedom.*

We remark that one can prove the following converse:

Let F be a d.f. We do not assume the existence of any moments.

Let $\xi_1,\ldots,\xi_n$ be $n \geq 2$ independent identically distributed r.v.'s with d.f. F. Let the r.v.'s $\bar{\xi}$ and s^2 be independent. Then the ξ_i are normally distributed[6].

We now proceed from the assumptions of the corollary. By Theorem 3.2, $\frac{\bar{\xi}-a}{\sigma}\sqrt{n}$ is $N(0,1)$-distributed and, as we just saw, $(n-1)s^2/\sigma^2$ is independently χ^2-distributed with $n-1$ degrees of freedom. This allows us to call on the t-distribution. Indeed,

$$\frac{\bar{\xi}-a}{\sigma}\sqrt{n}\Bigg/\sqrt{\frac{(n-1)s^2}{(n-1)\sigma^2}} = \frac{\bar{\xi}-a}{s}\sqrt{n} \tag{4.14}$$

is, by I.**29** t-distributed with $n-1$ degrees of freedom. We then get the important

Theorem 4.4. *Let $\xi_1,\ldots,\xi_n$ be independent, $N(a,\sigma^2)$-distributed r.v.'s. Then, the quotient $\frac{\bar{\xi}-a}{s}\sqrt{n}$ is t-distributed with $n-1$ degrees of freedom.*

Note that in (4.14) the expression on the right no longer depends on σ and this is a very important fact.

We now want to apply Theorem 4.4 to the problem posed at the beginning of this section: Assume we have a sample $x_1,\ldots,x_n$ from an $N(a,\sigma^2)$-distributed population. We seek a test for the hypothesis $a=a_0$ when the variance σ^2 is unknown. We proceed as before (p. 127): A level of significance α is given and t_α is uniquely determined by

$$1-\int_{-t_\alpha}^{t_\alpha} h_{n-1}(t)\,dt=\alpha \tag{4.15}$$

where h_n is defined by I. (29.3). From the given sample values we form the quantities $\bar{x}$ and s^2 and accept, using Theorem 4.4, the hypothesis $a=a_0$ if

$$|(\bar{x}-a_0)\sqrt{n}/s| < t_\alpha .$$

We reject it if

$$|(\bar{x}-a_0)\sqrt{n}/s| \geq t_\alpha . \tag{4.16}$$

---

[6] In this generality the theorem is due to T. Kawata and H. Sakamoto, J. Math. Soc. Japan 1, 111—115 (1949). For the case where the variance of F exists, the theorem was first proved by E. Lukacs, Ann. Math. Statist. 13, 91—93 (1942). See also R. C. Geary, J. Roy. Statist. Soc. Supp. 3, 178 (1936).

Let us consider a numerical example: The diameter of an axle should be 22 mm. Ten measurements yield

$$\begin{array}{ll} x_1=22.04 & x_6=21.99 \\ x_2=22.08 & x_7=22.02 \\ x_3=22.01 & x_8=22.03 \\ x_4=21.97 & x_9=22.00 \\ x_5=22.02 & x_{10}=22.01 \end{array}$$

whence $\bar{x}=22.017$, $s^2=0.008010/9=0.00089$.

Thus with $a=22$, $(\bar{x}-a)\sqrt{n}/s=(0.017)(3.1623)/(0.02983)$ and this is, to sufficient precision, 1.802. We choose $\alpha=5/100$. For 9 degrees of freedom $t_{5/100}=2.26\ldots$. The hypothesis that the desired value of the diameter of the axles is 22 mm can thus be accepted.

We summarize: In order to test a hypothesis on the mean of a normal distribution $a=a_0$ with unknown variance σ^2 on the basis of a sample of size n for given level of significance α, one can proceed as follows: Form the expression $(\bar{x}-a_0)\sqrt{n}/s$ and determine t_α by (4.15). If $|(\bar{x}-a_0)\sqrt{n}/s|<t_\alpha$, accept the hypothesis. If $|(\bar{x}-a_0)\sqrt{n}/s|\geq t_\alpha$, reject it.

5. Testing the difference between the means of two independent normal distributions with known variances. The methods previously developed also allow a treatement of the following problem: Let $x_1,\ldots,x_n$ be a sample taken from an $N(a_1,\sigma_1^2)$-distributed population. Let $y_1,\ldots,y_m$ be a sample independent of the first one taken from an $N(a_2,\sigma_2^2)$-distributed population. More precisely: Assume we have n realizations of independent r.v.'s $\xi_1,\ldots,\xi_n$, all of which are $N(a_1,\sigma_1^2)$-distributed. Further assume given m realizations of independent $N(a_2,\sigma_2^2)$-distributed r.v.'s $\eta_1,\ldots,\eta_m$ which are independent of all the ξ_i $(i=1,\ldots,n)$.[7]

We turn to the question of how one tests the hypothesis $a_1=a_2$, i. e., the hypothesis that the two samples come from populations with the same mean values.

We first treat the case in which σ_1^2 as well as σ_2^2 are known. By Theorem 3.2, $\bar{\xi}$ is $N(a_1,\sigma_1^2/n)$-, and $\bar{\eta}$ is $N(a_2,\sigma_2^2/m)$-distributed, so that $\bar{\xi}-\bar{\eta}$ is $N(a_1-a_2,\sigma_1^2/n+\sigma_2^2/m)$-distributed.

If $a_1=a_2$ holds, then

$$\zeta_1=\frac{(\bar{\xi}-\bar{\eta})\sqrt{mn}}{\sqrt{m\sigma_1^2+n\sigma_2^2}}$$

is $N(0,1)$-distributed.

Following the procedure described on p. 127, we immediately obtain a test for the hypothesis $a_1=a_2$ with given level of significance α. The only difference between the two is that the r.v. ζ_1 replaces the r.v. η defined on p. 127.

---

[7] This elucidates the significance of the heading of **5** and, of course, also illuminates those of **6**, **9** and **10**.

6. Testing the difference between the means of two independent normal distributions with the same unknown variance. We now drop the assumption that σ_1^2 and σ_2^2 are known and replace it by the assumption that $\sigma_1^2=\sigma_2^2=\sigma^2$, where the common value σ^2 is unknown. Thus, let $x_1,\dots,x_n$, $n\geqslant 2$, be a sample from an $N(a_1,\sigma^2)$ population and $y_1,\dots,y_m$, $m\geqslant 2$, one from an $N(a_2,\sigma^2)$ population drawn independently of the first sample. In order to develope a test procedure for the hypothesis $a_1=a_2$, we make use of the corollary to Theorem 4.3. Let

$$s_\xi^2=\frac{\sum_{i=1}^{n}(\xi_i-\bar{\xi})^2}{n-1}, \qquad s_\eta^2=\frac{\sum_{i=1}^{m}(\eta_i-\bar{\eta})^2}{m-1}.$$

Then $s_\xi^2(n-1)/\sigma^2$, resp., $s_\eta^2(m-1)/\sigma^2$ has a Helmert-Pearson distribution with $n-1$, resp., $m-1$ degrees of freedom. By assumption, the r.v.'s $s_\xi^2(n-1)/\sigma^2$ and $s_\eta^2(m-1)/\sigma^2$ are independent.

Applying the reproductive property of the χ^2-distribution (I, Theorem 28.1), we find that

$$s_\zeta^2=\left[s_\xi^2(n-1)+s_\eta^2(m-1)\right]/\sigma^2 \tag{6.1}$$

is Helmert-Pearson distributed with $m+n-2$ degrees of freedom.

If the hypothesis $a_1=a_2$ is correct, then using $\sigma_1^2=\sigma_2^2=\sigma^2$, we find, as previously seen, that

$$\zeta=\frac{(\bar{\xi}-\bar{\eta})\sqrt{mn}}{\sigma\sqrt{m+n}} \tag{6.2}$$

is $N(0,1)$-distributed. By assumption and from the corollary to Theorem 4.3, ζ and s_ζ^2 are distributed independently of each other. Hence,

$$\zeta\left(\sqrt{s_\zeta^2/(m+n-2)}\right)^{-1}$$

is t-distributed with $m+n-2$ degrees of freedom. Thus:

Under the assumption $a_1=a_2$, the r.v.

$$Q=\frac{(\bar{\xi}-\bar{\eta})\sqrt{mn}\sqrt{m+n-2}}{\sqrt{\left[s_\xi^2(n-1)+s_\eta^2(m-1)\right]}\sqrt{m+n}} \tag{6.3}$$

is t-distributed with $m+n-2$ degrees of freedom. This leads to the following procedure[8]:

Let $x_1,\dots,x_n$ be a sample from an $N(a_1,\sigma^2)$-distributed population, $y_1,\dots,y_m$ a sample from an $N(a_2,\sigma^2)$-distributed population independ-

[8] This and the other applications of the t-distribution are clearly presented in R. A. Fisher, Metron 5, 90—104 (1925).

ent from the first one. Assume σ^2 is unknown. We want to test $a_1 = a_2$. We give a level of significance α and determine t_α according to

$$1 - \int_{-t_\alpha}^{t_\alpha} h_{m+n-2}(t)\,dt = \alpha. \tag{6.4}$$

Now calculate the value Q of $\boldsymbol{Q}$ for the given sample and accept the hypothesis if

$$|Q| < t_\alpha.$$

We reject it if

$$|Q| \geqslant t_\alpha.$$

The problem treated here is a special case of the so-called two-sample problem which can be formulated in practical language as follows: Let $x_1,\ldots,x_n$ and $y_1,\ldots,y_m$ be two samples, possibly taken from different populations. By means of these samples we want to test the assumption that the two populations are identical (see VII.**4**. as well as III, p. 225).

7. Testing a hypothesis on the variance of a normal distribution with known mean. Let n $N(a,\sigma^2)$-distributed r.v.'s $\xi_1,\ldots,\xi_n$ be given. The variables $(\xi_i - a)/\sigma$ are independent and $N(0,1)$-distributed and $\left(\sum_{i=1}^{n}(\xi_i - a)^2\right)\Big/\sigma^2$ is thus χ^2-distributed with n degrees of freedom, which results from I.**28**. We will use this to develope a test for the following problem: Let $x_1,\ldots,x_n$ be a sample from an $N(a_0,\sigma^2)$-distributed population. Let a_0 be known. We want to test $\sigma^2 = \sigma_0^2$ with $\sigma_0^2 > 0$.

Determine $\lambda_{\alpha/2}$ for a given level of significance α according to I. (28.2) from

$$\frac{\alpha}{2} = \frac{1}{2^{n/2}\Gamma(n/2)} \int_{\lambda_{\alpha/2}}^{\infty} e^{-x/2} x^{n/2-1}\,dx \tag{7.1}$$

and $\bar{\lambda}_{\alpha/2}$ from

$$\frac{\alpha}{2} = \frac{1}{2^{n/2}\Gamma(n/2)} \int_{0}^{\bar{\lambda}_{\alpha/2}} e^{-x/2} x^{n/2-1}\,dx. \tag{7.2}$$

Now note that for each sample $x_1,\ldots,x_n$ it always follows from $\sigma^2 \gtrless \sigma_0^2$ that

$$\frac{1}{\sigma^2}\sum_{i=1}^{n}(x_i - a_0)^2 \lessgtr \frac{1}{\sigma_0^2}\sum_{i=1}^{n}(x_i - a_0)^2$$

and conversely, provided $\sum_{i=1}^{n}(x_i - a_0)^2 \neq 0$. Thus, values of $\sum_{i=1}^{n}(x_i - a_0)^2/\sigma_0^2$

that are "too small" support a presumption that the "true" variance is $<\sigma_0^2$, values "too large", that the "true" variance is $>\sigma_0^2$. This suggests constructing the test as follows:

The hypothesis $\sigma^2=\sigma_0^2$ is accepted if for the values determined by (7.1) and (7.2),

$$\bar{\lambda}_{\alpha/2} < \frac{1}{\sigma_0^2} \sum_{i=1}^{n} (x_i - a_0)^2 < \lambda_{\alpha/2}$$

holds, and if not, rejected.

8. Testing a hypothesis on the variance of a normal distribution with unknown mean. The procedure given in 7 was based essentially on the assumption that the mean a_0 of the underlying normal distribution was known. We now drop this assumption. Let $x_1, \ldots, x_n$ be a sample from an $N(a,\sigma^2)$-distributed population. Let a be unknown and suppose we want to test the hypothesis $\sigma^2=\sigma_0^2$. The basis of such a test will be the corollary to Theorem 4.3 and, in particular, the fact that $s^2(n-1)/\sigma^2$ is χ^2-distributed with $n-1$ degrees of freedom. Taking the previous case as pattern, determine $\lambda_{\alpha/2}$ by

$$\alpha/2 = 2^{-(n-1)/2} \left(\Gamma((n-1)/2)\right)^{-1} \int_{\lambda_{\alpha/2}}^{\infty} e^{-x/2} x^{(n-3)/2} dx \tag{8.1}$$

and $\bar{\lambda}_{\alpha/2}$ by

$$\alpha/2 = 2^{-(n-1)/2} \left(\Gamma((n-1)/2)\right)^{-1} \int_{0}^{\bar{\lambda}_{\alpha/2}} e^{-x/2} x^{(n-3)/2} dx. \tag{8.2}$$

The hypothesis $\sigma^2=\sigma_0^2$ is accepted if

$$\bar{\lambda}_{\alpha/2} < s^2(n-1)/\sigma_0^2 < \lambda_{\alpha/2}$$

and rejected if this is not the case.

9. Testing the difference between the variances of two independent normal distributions with known means. Suppose we have a sample $x_1, \ldots, x_n$ from an $N(a_1,\sigma_1^2)$-distributed population and $y_1, \ldots, y_m$ a sample from an $N(a_2,\sigma_2^2)$-population independent of the first one. We want to test the hypothesis $\sigma_1^2=\sigma_2^2$.

Assume that a_1 and a_2 are known. In any case, the r.v.'s

$$\sum_{i=1}^{n} (\xi_i - a_1)^2/\sigma_1^2, \quad \text{resp.,} \quad \sum_{i=1}^{m} (\eta_i - a_2)^2/\sigma_2^2$$

are χ^2-distributed with n, resp., m degrees of freedom. They are (corresponding to the assumption) independent of each other. On the basis of the definition of the F-distribution, the r.v.

$$\frac{\sum_{i=1}^{n}(\xi_i-a_1)^2}{n\sigma_1^2}\;\frac{m\sigma_2^2}{\sum_{i=1}^{m}(\eta_i-a_2)^2} \tag{9.1}$$

is F-distributed with (n,m) degrees of freedom, i. e., (9.1) has the density $k_{n,m}$ defined by I. (30.2). In particular, if $\sigma_1^2=\sigma_2^2$, then (9.1) goes into

$$\frac{\sum_{i=1}^{n}(\xi_i-a_1)^2}{n}\;\frac{m}{\sum_{i=1}^{m}(\eta_i-a_2)^2}.$$

If $\sigma_1^2 \gtrless \sigma_2^2$, then for fixed x_i $(i=1,\ldots,n)$ and y_k $(k=1,\ldots,m)$

$$\frac{m}{n}\frac{\sum_{i=1}^{n}(x_i-a_1)^2}{\sum_{i=1}^{m}(y_i-a_2)^2}\frac{\sigma_2^2}{\sigma_1^2} \lessgtr \frac{m}{n}\frac{\sum_{i=1}^{n}(x_i-a_1)^2}{\sum_{i=1}^{n}(y_i-a_2)^2}. \tag{9.2}$$

Hence, we test $\sigma_1^2=\sigma_2^2$ under the given assumption as follows. For given α, determine the positive real number $\kappa_{\alpha/2}(n,m)$ from

$$\int_{\kappa_{\alpha/2}(n,m)}^{\infty} k_{n,m}(z)\,dz=\alpha/2$$

and $\bar{\kappa}_{\alpha/2}(n,m)$ from

$$\int_{0}^{\bar{\kappa}_{\alpha/2}(n,m)} k_{n,m}(z)\,dz=\alpha/2.$$

One accepts the hypothesis if

$$\bar{\kappa}_{\alpha/2}(n,m)<\frac{\sum_{i=1}^{n}(x_i-a_1)^2}{n}\;\frac{m}{\sum_{i=1}^{m}(y_i-a_2)^2}<\kappa_{\alpha/2}(n,m)$$

and it otherwise rejects.

10. Testing the difference between the variances of two independent normal distributions with unknown means. Let $\xi_1,\ldots,\xi_n$, $n\geqslant 2$ be independent $N(a_1,\sigma^2)$-distributed r.v.'s $\eta_1,\ldots,\eta_m$, $m\geqslant 2$ independent $N(a_2,\sigma^2)$-

distributed r.v.'s which are independent of all the ξ_i, $1 \leqslant i \leqslant n$. From the corollary to Theorem 4.3, the r.v.'s $s_\xi^2(n-1)/\sigma_1^2$ and $s_\eta^2(m-1)/\sigma_2^2$ are independently χ^2-distributed with $n-1$ and $m-1$ degrees of freedom respectively. Hence, $(s_\xi^2/\sigma_1^2)(s_\eta^2/\sigma_2^2)^{-1}$ is F-distributed with $(n-1, m-1)$ degrees of freedom. In particular if $\sigma_1^2 = \sigma_2^2$, one finds that s_ξ^2/s_η^2 is F-distributed with $(n-1, m-1)$ degrees of freedom.

Determining the practical procedure for carrying out a test of the above hypothesis should now cause no difficulty.

We remark that one can easily determine the value of $\kappa_{\alpha/2}(n,m)$ from a knowledge of $\bar{\kappa}_{\alpha/2}(n,m)$ for $n, m \geqslant 1$. Indeed, from the definition of the F-distribution, $\kappa_{\alpha/2}(n,m) = 1/\bar{\kappa}_{\alpha/2}(m,n)$.

11. The role of the central limit theorem. For the test procedures above one requires tables which allow, for given level of significance α, the determination of κ_α (p. 127); t_α (p. 134) for n degrees of freedom, $n \geqslant 1$; $\lambda_{\alpha/2}$ and $\bar{\lambda}_{\alpha/2}$ (p. 138) for $n \geqslant 1$ degrees of freedom; and $\kappa_{\alpha/2}(n, m)$ (p. 139) for $n, m \geqslant 1$ degrees of freedom. Such tables are contained in most of the works mentioned in the footnote at the beginning of the chapter.

However, for "large sample sizes" one is usually satisfied by a replacement of the t- or χ^2-distribution by the normal distribution. For the t-distribution this is justified to a large extent by the result mentioned on p. 115. Thus, for $n \geqslant 30$, one views the r.v.'s $(\bar{\xi} - a)\sqrt{n}/s$ mentioned on p. 134 as $N(0,1)$-distributed. For the χ^2-distribution one uses the result on p. 108. Hence, for large enough n, one need not use tables for the t- and χ^2-distributions. However, one should always convince oneself that the sample size is "large enough" to allow such approximations. On the other hand, progress in the mechanization of function tabulation is rapidly reducing the importance of these questions. It is much more important to convince oneself that the assumption of a normal distribution, on which assumption almost all of the previous tests are based, is fulfilled. The question to what extent deviations from this assumption are admissable is connected with the so-called robustness of tests.[9] Roughly speaking, I. Theorem 39.1, 39.2 and 39.3 say that the mentioned procedures can still be applied even without the assumption of a normal distribution for "sufficiently large" sample sizes. We will illustrate this with the practically important example of the binomial distribution. Let $\xi_1, \ldots, \xi_n$, $n \geqslant 1$, be independent r.v.'s, whose distribution is given by I. (32.1). Then, as follows from the remark on p. 108, the r.v. $(\bar{\xi} - p)\sqrt{n}/\sqrt{pq}$ is asymptotically $N(0,1)$-distributed. Thus, for "suf-

[9] See P. J. Huber, Théorie de l'inférence statistique robuste. (Séminaire de mathématiques supérieures 31.) Montréal, Canada: Les Presses de l'Université de Montreal 1969.

ficiently large" n we have for the hypothesis $p=p_0$ the following test procedure: Let α be a given level of significance. Determine κ_α as on p. 127. Let $x_1,\dots,x_n$ be a sample from a population distributed by I. (32.1). Accept $p=p_0$ or not according as $|((\bar{x}-p_0)\sqrt{n})/\sqrt{p_0(1-p_0)}|<\kappa_\alpha$ or not.

A test procedure applicable for each $n\geqslant 1$ for the given hypothesis can be modelled according to IV, p. 260.

12. The sampling theory of finite populations. We begin with a r.v. ξ which has a discrete uniform distribution with mass points $x_1,\dots,x_N$, $N\geqslant 1$. Hence

$$P(\xi=x_i)=1/N, \quad 1\leqslant i\leqslant N. \tag{12.1}$$

Denote the mean $E(\xi)$ by a and the variance $E[(\xi-a)^2]$ by σ^2. Consider the discrete n-dimensional $(n\leqslant N)$ r.v. $(\xi_1,\dots,\xi_n)$ which we defined in I.**19**. Its distribution is given by I. (19.1). For the r.v. $\bar{\xi}=(\xi_1+\dots+\xi_n)/n$ one immediately calculates with the help of I, Theorem 16.3, that

$$E(\bar{\xi})=a. \tag{12.2}$$

For the variance of $\bar{\xi}$ we get

$$\begin{aligned} E[(\bar{\xi}-a)^2] &= E\left[\left(\sum_{i=1}^{n}(\xi_i-a)\right)^2\right]\Big/n^2 \\ &= E\left[\sum_{i,j=1}^{n}(\xi_i-a)(\xi_j-a)\right]\Big/n^2 \\ &= \sum_{i,j=1}^{n} E[(\xi_i-a)(\xi_j-a)]/n^2. \end{aligned}$$

If $i=j$ in this last sum, then one must calculate $E[(\xi_i-a)^2]$. By I. (19.2), this is equal to $\sum_{k=1}^{N}(x_k-a)^2/N$, so that

$$E[(\xi_i-a)^2]=\sigma^2, \quad 1\leqslant i\leqslant n. \tag{12.3}$$

For $i\neq j$ one must calculate $E[(\xi_i-a)(\xi_j-a)]$. For this purpose we use I. (19.3) and obtain

$$\begin{aligned} E[(\xi_i-a)(\xi_j-a)] &= \sum_{\substack{k,l=1\\ k\neq l}}^{N}\frac{1}{N(N-1)}(x_k-a)(x_l-a) \\ &= \frac{1}{N(N-1)}\sum_{k,l=1}^{N}(x_k-a)(x_l-a)-\frac{1}{N(N-1)}\sum_{l=1}^{N}(x_l-a)^2 \\ &= 0-\sigma^2/(N-1). \end{aligned}$$

Hence,

$$E[(\xi_i - a)(\xi_j - a)] = -\sigma^2/(N-1), \qquad 1 \leqslant i < j \leqslant n. \tag{12.4}$$

From (12.3) and (12.4) we find that

$$E[(\bar{\xi} - a)^2] = \frac{\sigma^2}{n} \frac{N-n}{N-1}. \tag{12.5}$$

For fixed n, the right side of (12.5) tends to σ^2/n for $N \to \infty$. Compare with the result (3.2) of Theorem 3.1.

The previous results obtained in this section can be interpreted as follows: Assume we have a finite population containing N different elements $x_1, \ldots, x_N$. A sample of size n is taken "without replacement" from this population, i.e., each removed element x_i no longer appears among the remaining elements of the population. We then also say that we have a sample from a finite population. This sample is then a realization of the r.v.'s $(\xi_1, \ldots, \xi_n)$ defined in I.**19**. Each possible choice of this sample is viewed as equally probable. $\bar{\xi}$ can be viewed as the mean of the sample variables; its expectation coincides with that of the population. The remark following (12.5) on the variance can now be expressed as follows: For $N \to \infty$, "the finite population goes into an infinite one". The variance of the sample mean then goes into the variance of the mean for an infinite population given in Theorem 3.1.

Also note that for $n = N$, $E[(\bar{\xi} - a)^2] = 0$ follows from (12.5) and $\bar{\xi} = a$ (with probability 1). In this case, "the sample exhausts the finite population".

From (12.2) and (12.5) we have, following the pattern of the remark on p. 128, a test procedure for an hypothesis on the mean a.

Now, we will consider the r.v.

$$S^2 = \frac{\sum_{i=1}^{n} (\xi_i - \bar{\xi})^2}{n-1}. \tag{12.6}$$

We have deviated from our notational convention here (p. 131) in order not to cause confusion with the r.v.'s considered in Theorem 4.2. We remark again, that $\xi_1, \ldots, \xi_n$ are not independent. We have

$$\begin{aligned} E(S^2) &= \frac{1}{n-1} E\left[\sum_{i=1}^{n} (\xi_i - \bar{\xi})^2\right] = \frac{1}{n-1} E\left[\left(\sum_{i=1}^{n} \xi_i^2 - n\bar{\xi}^2\right)\right] \\ &= \frac{1}{n-1}\left[\sum_{i=1}^{n} E(\xi_i^2) - nE(\bar{\xi}^2)\right]. \end{aligned}$$

Now, $E(\xi_i^2) = \sigma^2 + a^2$ and likewise by (12.2) and (12.5),

$$E(\bar{\xi}^2) = \frac{\sigma^2}{n} \frac{N-n}{N-1} + a^2,$$

so that

$$E(S^2) = \frac{1}{n-1}\left[n\sigma^2 - \sigma^2 \frac{N-n}{N-1}\right] = \frac{N\sigma^2}{N-1} \tag{12.7}$$

which differs from (4.13). From (12.7)

$$E\left(\frac{N-1}{N} S^2\right) = \sigma^2 .$$

An elementary, but long calculation delivers, with $m_4 = \frac{1}{N}\sum_{i=1}^{n}(x_i - a)^4$:

$$E\left[\left(\frac{N-1}{N}S^2 - \sigma^2\right)^2\right] = \frac{m_4}{n}\,\frac{N-n}{N}\left(1 + \frac{3nN - 5(N+n)+7}{(n-1)(N-2)(N-3)}\right)$$
$$- \frac{N-n}{N}\,\frac{\sigma^4}{n}\left(\frac{n-3}{n-1} + \frac{5Nn - 9N - 9n + 15}{(n-1)(N-2)(N-3)}\right).$$

13. The single sampling procedure. The hypergeometric distribution derived in I.**34** and its interpretation turn out to be important in practical work, for example, in the wide area of quality control of manufactured products. Assume that the products are delivered in lots of N pieces (for example, boxes each containing N light bulbs). Each lot contains a certain percentage of defective pieces (for example, bulbs with unsatisfactory burning life), which naturally can also be 0. The number of defectives in each lot will be denoted by $M = pN$. p will vary in general from lot to lot but should not (according to terms of delivery) exceed a prescribed fraction p_0. In order to determine if this has happened, one must either examine all N articles, which can often be expensive or—in destructive testing—impossible (for example the burning life of light bulbs), or satisfies oneself with the selection of a sample. The question now arises how one can check whether or not $p \leqslant p_0$ from the results of the sample.

We will now use the more general terminology introduced on p. 94 and assume that a finite population of N elements is on hand. Let $M = Np$ be defined as there. A sample of size n will be taken "without replacement" from the population. The distribution of the sample variables is given by I. (34.2). For given p_0, $0 \leqslant p_0 \leqslant 1$, we want to give a test procedure for the hypothesis $p \leqslant p_0$. Here, only those p's resp. p_0's make sense for which Np resp. Np_0 is an integer. For given level of significance α, $0 < \alpha < 1$, we choose the smallest integer $k_\alpha \geqslant 0$ such that

$$\sum_{r=k_\alpha+1}^{n} \binom{Np_0}{r}\binom{N-Np_0}{n-r}\Big/\binom{N}{n} \leqslant \alpha . \tag{13.1}$$

Note that equality in (13.1) is not always attainable since one has only finitely many values $k_\alpha < n$ available and α can be chosen in infinitely many ways. The hypothesis $p \leqslant p_0$ is accepted if the number r of elements in the first class (cf. p. 94) does not exceed k_α and otherwise rejected.

Thus, in the quality control example, the number of defectives should not exceed k_α. This procedure becomes clear when one considers that the probability that the number of defectives is $>k_\alpha$ is given for some p with $0<p<1$, and such that Np is an integer, by

$$\sum_{r=k_\alpha+1}^{n} \binom{Np}{r}\binom{N-Np}{n-r}\Big/\binom{N}{n}. \tag{13.2}$$

By I, Theorem 34.2, this expression is a non-decreasing function of p. Thus, the smaller p, the smaller the probability that the hypothesis will be rejected if it is true.

These considerations form the basis of the single sampling procedure developed by Dodge and Romig for quality control. Choose p_0 as above, and for given α and n, the number k_α according to (13.1). p_0 is called the *acceptable quality level.* If the number r of defectives in the sample remains $\leqslant k_\alpha$, then the lot is assumed to correspond to the delivery condition $p \leqslant p_0$ after the r defectives found have been replaced by non-defectives. However, if $r>k_\alpha$, then a complete check of all N articles in the lot is carried out, with all defectives being replaced by non-defectives. (This assumes, however, that the check does not destroy the pieces, so that this procedure is not applicable to light bulbs.)

We now determine how many defectives the consumer can expect on the average when each lot contains Np defectives. More precisely: Denoting the discretely distributed r.v. which reflects the number of defectives remaining after the described procedure by η^*, one then has

$$P(\eta^*=Np-r)=\binom{Np}{r}\binom{N-Np}{n-r}\Big/\binom{N}{n}, \quad \text{for } r \leqslant k_\alpha$$

and

$$P(\eta^*=0)=\sum_{r=k_\alpha+1}^{n} \binom{Np}{r}\binom{N-Np}{n-r}\Big/\binom{N}{n}.$$

Thus, for the mean we get

$$E(\eta^*)=\sum_{r\leqslant k_\alpha} (Np-r)\binom{Np}{r}\binom{N-Np}{n-r}\Big/\binom{N}{n}.$$

$E(\eta^*)$ thus depends on N, n, k_α and p. In order to eliminate the annoying dependence on the proportion of defectives p, one considers $\max\limits_{0\leqslant p\leqslant 1} E(\eta^*)$. p assumes only finitely many values M/N with $0 \leqslant M \leqslant N$, so that there exists at least one value $\tilde{M}=\tilde{M}(N,n,k_\alpha)$ which maximises $E(\eta^*)$ as function of $p=M/N$. Let $\max\limits_{0\leqslant p\leqslant 1} E(\eta^*)=E^*$. E^*/N is called the *average outgoing quality limit*, which is justified by the fact that $E(\eta^*)/N$ can be viewed as a scale for the average quality of lots of N pieces actually received by the customer.

If we hold N fixed and give α and p_0, then we can still choose the sample size n. It is obviously desirable for economic reasons to set up the control procedure in such a way that the number of pieces to be inspected is as small as possible.

More precisely: If n is fixed, then so is k_α by (13.1). We investigate the mean I of the number of pieces per lot to be inspected on the basis of the control procedure for fixed defective proportion p.

In any case, the n pieces in the sample are checked. If the number of defectives is then $\leqslant k_\alpha$—this event has probability $\sum_{r\leqslant k_\alpha}\binom{Np}{r}\binom{N-Np}{n-r}\Big/\binom{N}{n}$—then no further pieces will be inspected; if, however, $r\geqslant k_\alpha+1$, then the remaining $N-n$ pieces are checked. We obtain for the desired mean:

$$\left.\begin{aligned} I &= n+0\cdot\sum_{r\leqslant k_\alpha}\frac{\binom{Np}{r}\binom{N-Np}{n-r}}{\binom{N}{n}}+(N-n)\sum_{r\geqslant k_\alpha+1}\frac{\binom{Np}{r}\binom{N-Np}{n-r}}{\binom{N}{n}} \\ &= n+(N-n)\sum_{r\geqslant k_\alpha+1}\frac{\binom{Np}{r}\binom{N-Np}{n-r}}{\binom{N}{n}}. \end{aligned}\right\}\quad(13.3)$$

One is interested in a choice of n and k_α which satisfies (13.1) on the one hand and which makes I as small as possible.

Dodge and Romig have tabulated pairs of values n, k_α which minimize I for $\alpha=0.10$, given quality p, acceptable quality p_0 and lot size N.

For some purposes it is important to give E^* instead of α and p_0. For given p and N, Dodge and Romig have also tabulated pairs of values n, k_α which minimize I and which correspond to given $E^*=E^*(N,n,k_\alpha)$.[10]

We have already mentioned on p. 144 that the control procedure assumes non-destructive testing. If this is not the case. then one must in any case be satisfied with sample-control. The arguments on p. 144 are still valid, but the r.v. η^* has a slightly different meaning. It gives the number of accepted defectives in a lot of size $N-n$. The quantity defined by (13.3) now no longer makes sense. The number of checked pieces among N articles is always n. Compare the contents of this section with III, p. 175 and V, p. 271 ff.

---

[10] H. F. Dodge and H. G. Romig, Sampling Inspection Tables (Single and Double Sampling), 2nd ed., John Wiley & Sons-Chapman & Hall, Ltd. London 1959.

These results can also be applied to the binomial distribution. By I, Theorem 34.1, one will use the binomial distribution when the lot size is "infinitely large" compared with the sample size.

14. Stratified sampling. Let (ξ, η) be a two-dimensional r.v. and assume the marginal distribution of η is discrete:

$$P(\eta=i)=p_i, \quad p_i \neq 0, \quad 1 \leqslant i \leqslant k, \quad \sum_{i=1}^{k} p_i = 1, \quad k \geqslant 2. \quad (14.1)$$

For each real x denote the conditional probability of $\{\xi \leqslant x \mid \eta=i\}$ by $F_i(x)$, $1 \leqslant i \leqslant k$. F_i is the conditional d.f. of ξ under the hypothesis $\eta=i$. From I, p. 29 we have for the distribution function of the marginal distribution of ξ

$$F = \sum_{i=1}^{k} p_i F_i. \quad (14.2)$$

Let $E(\xi)$ exist and set

$$E(\xi) = a \quad (14.3)$$

and

$$E(\xi \mid \eta=i) = a_i, \quad 1 \leqslant i \leqslant k. \quad (14.4)$$

From (14.2), (14.3) and (14.4) we have

$$a = \sum_{i=1}^{k} p_i a_i. \quad (14.5)$$

Similarly, for the second moments (if they exist), we have in obvious notation:

$$m_2 = \sum_{i=1}^{k} p_i m_2^{(i)}.$$

Then from I, Theorem 17.2 we get for the variance

$$\sigma^2 = \sum_{i=1}^{k} p_i(\sigma_i^2 + a_i^2) - a^2. \quad (14.6)$$

The previously developed relations, especially (14.2), can be interpreted as follows: Assume we have a population with distribution function F. Let the former be partitioned into k disjoint sub-populations. The probability of "drawing an element from the i-th sub-population" is taken to be p_i and let the distribution of the elements in the i-th sub-population be given by F_i. In this framework one arrives at (14.2): The probability of "drawing an element from the population" whose characteristic ξ is $\leqslant x$ is given by F. On the other hand, this element necessarily belongs to exactly one sub-population—e.g., to the l-th with probability p_l $(l=1,\ldots,k)$. If it lies for example in the i-th sub-population then the probability that the characteristic ξ is $\leqslant x$, is given by F_i. The probability of both of these events occurring is given by $p_i F_i$. (14.2) thus results. The sub-populations are called the *strata* of the population.

We turn to the sample theory of stratified populations. Because of its intuitiveness, we retain the viewpoint just introduced.

Let $\xi_1,\ldots,\xi_n$ be sample variables from a stratified population with d.f. F. Thus, a representation of the form (14.2) holds for F and conditions (14.5) and (14.6) also hold. In any case, by Theorem 3.1

$$E(\bar{\xi}) = a \tag{14.7}$$

and

$$E[(\bar{\xi}-a)^2] = \sigma^2/n. \tag{14.8}$$

However, this result does not take into account the knowledge that the population is stratified, i. e., that (14.2) holds. We thus carry out another sampling procedure which is called *stratified sampling*. From the i-th sub-population U_i one draws exactly $n_i \geqslant 1$ elements with

$$\sum_{i=1}^{k} n_i = n. \tag{14.9}$$

More precisely, we have the following situation: Let $\xi_1^{(i)},\ldots,\xi_{n_i}^{(i)}$ be the sample variables of the choice from U_i, i. e., $\xi_1^{(i)},\ldots,\xi_{n_i}^{(i)}$ are, for $i=1,\ldots,k$, mutually independent r.v.'s with d.f. F_i such that the set of all sample variables is given by $\xi_1^{(1)},\ldots,\xi_{n_1}^{(1)},\ldots,\xi_1^{(k)},\ldots,\xi_{n_k}^{(k)}$ and all these are mutually independent. Let

$$\bar{\xi}^{(i)} = \sum_{j=1}^{n_i} \xi_j^{(i)} \Big/ n_i, \qquad 1 \leqslant i \leqslant k. \tag{14.10}$$

Consider the r.v. $\bar{\xi}_s = \sum_{i=1}^{k} c_i \bar{\xi}^{(i)}$, where the c_i are initially arbitrary real numbers. We have $E(\bar{\xi}_s) = \sum_{i=1}^{k} c_i E(\bar{\xi}^{(i)})$, so that by (14.4)

$$E(\bar{\xi}_s) = \sum_{i=1}^{k} c_i a_i. \tag{14.11}$$

We now want to determine the $c_i, 1 \leqslant i \leqslant k$, in such a way that $E(\bar{\xi}_s) = a$ identically in the a_i. Thus, by (14.5) for $-\infty < a_i < \infty$, $1 \leqslant i \leqslant k$, $\sum_{i=1}^{k} c_i a_i = \sum_{i=1}^{k} p_i a_i$ should hold[11]. This implies $c_i = p_i, 1 \leqslant i \leqslant k$. We write $\bar{\xi}_r = \sum_{i=1}^{k} p_i \bar{\xi}^{(i)}$ and have

$$E(\bar{\xi}_r) = a. \tag{14.12}$$

[11] See V.1.

For the variance, we have from (14.5)

$$E[(\bar{\xi}_r-a)^2]=E\left[\left(\sum_{i=1}^{k} p_i(\bar{\xi}^{(i)}-a_i)\right)^2\right],$$

thus,

$$E[(\bar{\xi}_r-a)^2]=\sum_{i,j=1}^{k} p_i p_j E[(\bar{\xi}^{(i)}-a_i)(\bar{\xi}^{(j)}-a_j)].$$

Because of the independence of the $\xi^{(i)}$, $1\leqslant i\leqslant k$, we get

$$E[(\bar{\xi}_r-a)^2]=\sum_{i=1}^{k} p_i^2\,\frac{\sigma_i^2}{n_i}. \tag{14.13}$$

15. Proportional sampling. In many practical cases, one can arbitrarily choose the $n_i\geqslant 1$ subject to the restriction (14.9). This suggests a choice which makes (14.13) as small as possible.

In this connection we first discuss so-called *proportional sampling*: For the sake of simplicity, assume that the quantities np_i, $i=1,\dots,k$ are integers and choose

$$n_i=np_i,\qquad i=1,\dots,k. \tag{15.1}$$

The term "proportional sample" is based on the frequency interpretation according to which the p_i represent the ratio of the "number of elements in the U_i" to the "number of elements in the population"; the number of sample values taken from the U_i stands in the same ratio to the total sample size. If we specialize the r.v. $\bar{\xi}_r$ by choosing the n_i according to (15.1), we get a r.v. denoted by $\bar{\xi}_p$. Hence,

$$\bar{\xi}_p=\sum_{i=1}^{k} p_i\bar{\xi}^{(i)},$$

where the $\xi^{(i)}$ are defined according to (14.10) and the n_i according to (15.1) for $1\leqslant i\leqslant k$. By (14.12) one has $E(\bar{\xi}_p)=a$, and according to (14.13),

$$E[(\bar{\xi}_p-a)^2]=\sum_{i=1}^{k}\frac{p_i\sigma_i^2}{n} \tag{15.2}$$

where we have used (15.1). We compare (15.2) with the variance (14.8) of $\bar{\xi}$ and claim that

$$E[(\bar{\xi}-a)^2]\geqslant E[(\bar{\xi}_p-a)^2]. \tag{15.3}$$

Indeed, from Schwarz' inequality and (14.1)

$$\left(\sum_{i=1}^{k} a_i p_i\right)^2\leqslant\sum_{i=1}^{k} p_i a_i^2\sum_{i=1}^{k} p_i=\sum_{i=1}^{k} p_i a_i^2. \tag{15.4}$$

Equality holds in (15.4) iff for a real λ

$$a_i p_i^{1/2} = \lambda p_i^{1/2}, \qquad 1 \leqslant i \leqslant k.$$

Hence, in this case $a_1 = a_2 = \cdots = a_k$. But from (15.4)

$$\sum_{i=1}^{k} p_i(\sigma_i^2 + a_i^2) - \left(\sum_{i=1}^{k} a_i p_i\right)^2 \geqslant \sum_{i=1}^{k} p_i \sigma_i^2$$

which is precisely (15.3) because of (14.6) and (14.5). Equality holds iff all of the a_i are equal.

16. Optimal sampling.[12] The choice of the n_i according to (15.1) assumes knowledge of the p_i but not of the σ_i^2.

We want to show that n_i can be chosen as a function of p_i and σ_i^2 in such a way that they fulfill (14.9) for given n, and that the r.v. $\bar{\xi}_r$ formed by this choice, which we will denote by $\bar{\xi}_0$, fulfills

$$E[(\bar{\xi}_p - a)^2] \geqslant E[(\bar{\xi}_0 - a)^2]. \tag{16.1}$$

Since $\bar{\xi}_0$ is of the form $\sum_{i=1}^{k} p_i \bar{\xi}^{(i)}$, (14.12) is fulfilled, i. e., $E(\bar{\xi}_0) = a$.

We first have

Lemma 16.1. *Let*

$$\alpha_i \geqslant 0, \qquad \beta_i > 0, \qquad \sum_{i=1}^{k} \alpha_i = \sum_{i=1}^{k} \beta_i = 1. \tag{16.2}$$

Then,

$$\sum_{i=1}^{k} \frac{\alpha_i^2}{\beta_i} \geqslant \sum_{i=1}^{k} \alpha_i = 1 \tag{16.3}$$

and equality holds iff $\beta_i = \alpha_i$, $i = 1, \ldots, k$.

The *proof* is easy and rests on Schwarz' inequality. We have

$$1 = \sum_{i=1}^{k} \alpha_i = \left(\sum_{i=1}^{k} \alpha_i\right)^2 = \left(\sum_{i=1}^{k} \frac{\alpha_i}{\beta_i^{1/2}} \beta_i^{1/2}\right)^2 \leqslant \sum_{i=1}^{k} \frac{\alpha_i^2}{\beta_i} \cdot \sum_{i=1}^{k} \beta_i = \sum_{i=1}^{k} \frac{\alpha_i^2}{\beta_i}.$$

Equality holds iff for all i with $1 \leqslant i \leqslant k$ and a real number λ, the equations $\frac{\alpha_i}{\beta_i^{1/2}} = \lambda \beta_i^{1/2}$, resp., $\alpha_i = \lambda \beta_i$ hold, and from (16.2), $\lambda = 1$.

We define with

$$\sum_{i=1}^{k} p_i \sigma_i = B \tag{16.4}$$

---

[12] J. Neyman, J. Roy. Statist. Soc. 97, 558—606 (1934).

the numbers

$$n_i' = \frac{np_i\sigma_i}{B}, \qquad i=1,\dots,k \tag{16.5}$$

and recall that (14.9) and

$$0 < n_i < n \tag{16.6}$$

hold.

With $(p_i\sigma_i)/B=\alpha_i$, $n_i/n=\beta_i$, $1 \leqslant i \leqslant k$, Lemma 16.1 is applicable since its assumptions (16.2) are fulfilled according to (16.4), (14.9) and (16.6). Hence

$$\sum_{i=1}^{k}\left(\frac{p_i\sigma_i}{B}\right)^2\frac{n}{n_i} \geqslant 1$$

or

$$\sum_{i=1}^{k}\frac{(p_i\sigma_i)^2}{n_i} \geqslant \frac{B^2}{n} = \frac{\left(\sum_{i=1}^{k}p_i\sigma_i\right)^2}{n}.$$

Thus, according to (14.13), $\left(\sum_{i=1}^{k}p_i\sigma_i\right)^2\Big/n$ is the minimum value of the variance of $\bar{\xi}_r$.

Lemma 16.1 says the lower bound in this inequality is assumed for $n_i=n_i'$, if the n_i' are chosen according to (16.5). Assuming again that the n_i' are integers one obtains the r.v. $\bar{\xi}_0$ mentioned above from $\bar{\xi}_r$ by means of the choice $n_i=n_i'$. Our argument then says that the variance of $\bar{\xi}_0$ is given by $\left(\sum_{i=1}^{k}p_i\sigma_i\right)^2\Big/n$. The correctness of (16.1) is now immediate.

By Schwarz' inequality we have at once:

$$\sum_{i=1}^{k}p_i\sigma_i^2 \geqslant \left(\sum_{i=1}^{k}p_i\sigma_i\right)^2.$$

Equality holds only for $\sigma_1=\sigma_2=\dots=\sigma_k$. This shows (16.1) because of (15.2). The validity of (16.1) has brought about the designation "optimal sample" for the sample chosen by means of the n_i'.

17. Stratified sampling in the finite population case. The investigations carried out in **14**—**16** are quite similar for finite populations. We want to elaborate on this here. Assume that the population G contains N elements, i.e., we consider a uniform distribution with discrete mass points $x_1,\dots,x_N$ (see **12**). The sub-population U_i $(i=1,\dots,k)$, $k \geqslant 2$, is assumed to consist of $N_i \geqslant 2$ elements with

$$\sum_{i=1}^{k}N_i = N. \tag{17.1}$$

Because of the uniform distribution,

$$p_i = \frac{N_i}{N} \tag{17.2}$$

is the probability of "choosing an element from U_i". For the mean of G we have $a = \left(\sum_{i=1}^{N} x_i\right) \Big/ N$. Denoting the discrete mass points of U_i by $x_1^{(i)}, \ldots, x_{N_i}^{(i)}$ we then have for the mean a_i in U_i:

$$a_i = \frac{x_1^{(i)} + \cdots + x_{N_i}^{(i)}}{N_i}, \qquad i = 1, \ldots, k \tag{17.3}$$

thus, one again has the relation (14.5) and, as is easy to see by appropriate transfer of notation, also (14.6) for the variance.

Let us draw a sample of size n from G. By (12.2) we have for the mean $\bar{\xi}$ of the sample variables $\xi_1, \ldots, \xi_n$

$$E(\bar{\xi}) = a \tag{17.4}$$

and by (12.5)

$$E[(\bar{\xi} - a)^2] = \frac{\sigma^2}{n} \frac{N-n}{N-1}. \tag{17.5}$$

We again take a stratified sample. From the sub-population U_i we draw a sample of size n_i $(0 < n_i \leqslant N_i,\ i = 1, \ldots, k)$.

From the corresponding sample variables $\xi_1^{(i)}, \ldots, \xi_{n_i}^{(i)}$ we form the means $\bar{\xi}^{(i)}$ according to (14.10) and from them, the r.v.

$$\bar{\xi}_r = \sum_{i=1}^{k} p_i \bar{\xi}^{(i)}.$$

Its mean satisfies (14.12), where now a, the a_i and the p_i have the meaning given in this section. Since the $\bar{\xi}^{(i)}$ are independent r.v.'s for $1 \leqslant i \leqslant k$, we get from an application of (12.5) to them:

$$E[(\bar{\xi}_r - a)^2] = \sum_{i=1}^{k} p_i^2 \frac{\sigma_i^2}{n_i} \frac{N_i - n_i}{N_i - 1}. \tag{17.6}$$

The right side of (17.6) is now written in the form

$$\sum_{i=1}^{k} p_i^2 \frac{\sigma_i^2}{n_i} \frac{N_i}{N_i - 1} - \sum_{i=1}^{k} p_i^2 \frac{\sigma_i^2}{N_i - 1}. \tag{17.7}$$

We assume knowledge of the quantities p_i defined by (17.2) and the variances σ_i^2 of the U_i and will generalize (14.9) somewhat. (This generalization can naturally also be carried out analogously in **16**.) Let

$l_1, \ldots, l_k$ as well as L be given positive real numbers. We consider only those n_i which satisfy

$$n_1 l_1 + \cdots + n_k l_k = L. \tag{17.8}$$

(17.8) can be interpreted as follows: Let the l_i for $i=1,\ldots,k$ be the costs of choosing an element from the sub-population U_i. Then L represents the total cost of the sample. We consider the problem of minimizing (17.7) as function of the n_i when (17.8) is fulfilled. Note that the second summand in (17.7) is independent of n_i. Abbreviate

$$\sum_{i=1}^{k} p_i \sigma_i \sqrt{l_i \frac{N_i}{N_i-1}}$$

by P_1 and set

$$\frac{p_i \sigma_i \sqrt{\frac{l_i N_i}{N_i-1}}}{P_1} = \alpha_i, \qquad \frac{n_i l_i}{L} = \beta_i.$$

We see immediately that the hypotheses of Lemma 16.1 hold and so

$$\frac{L}{P_1^2} \sum_{i=1}^{k} \frac{p_i^2 \sigma_i^2}{n_i} \frac{N_i}{N_i-1} = \sum_{i=1}^{k} \left(\frac{p_i \sigma_i \sqrt{l_i \frac{N_i}{N_i-1}}}{P_1} \right)^2 \cdot \left(\frac{L}{n_i l_i} \right) \geqslant 1.$$

The lower bound can only be attained for

$$n_i' = L \frac{p_i \sigma_i}{P_1} \sqrt{\frac{N_i}{N_i-1}} \frac{1}{\sqrt{l_i}}.$$ [13]

If we assume that the numbers n_i' so defined are integers and $\leqslant N_i$, then we can define a r.v. $\bar{\xi}_0$ following the pattern of **16** which minimizes (17.6) under the condition that the sample costs are equal to the available resources L. Considering, for example, the special case $l_1 = \cdots = l_k = 1$ and $L=n$, we see that the minimal variance obtained from (17.6) can be larger then (17.5). The "random" choice can thus be superior to the "optimal" one in this sense, which was not the case in **16**.[14]

18. Multistage sampling. Consider a 2-dimensional r.v. (ξ, η) with discrete distribution. Let the marginal distribution of ξ be a uniform distribution:

$$P(\xi = i) = 1/M \qquad (i=1,\ldots,M). \tag{18.1}$$

[13] This result can also be obtained from the minimum value of the variance of $\bar{\xi}_r$ by setting $\sigma_i \sqrt{\frac{l_i N_i}{N_i-1}}$ for σ_i, $n_i l_i$ for n_i (p. 150) and L for n.

[14] Details are in P. Armitage, Biometrika 34, 273—280 (1947).

For the conditional distribution of η given $\xi=i$, assume that

$$P(\eta=x_{ij}|\xi=i)=1/N_i, \quad j=1,\dots,N_i, \quad i=1,\dots,M. \tag{18.2}$$

Here, $x_{ij}=x_{kl}$ is taken to hold iff $i=k$ and $j=l$. From (18.1) and (18.2)

$$P(\xi=i,\eta=x_{ij})=1/MN_i, \quad j=1,\dots,N_i, \quad i=1,\dots,M.$$

Thus, the marginal distribution of η is likewise given by

$$P(\eta=x_{ij})=1/MN_i, \quad j=1,\dots,N_i, \quad i=1,\dots,M. \tag{18.3}$$

Starting from (18.1) we consider a r.v.

$$(\xi_1,\dots,\xi_m), \quad m\leqslant M \tag{18.4}$$

of the type treated in **12**.

I. (19.1) now takes the form

$$P(\xi_1=i_1,\xi_2=i_2,\dots,\xi_m=i_m)=\frac{1}{M(M-1)\dots(M-m+1)}. \tag{18.5}$$

Here, $\{i_1,\dots,i_m\}$ is a subset of the set $\{1,\dots,M\}$.

For $l=1,\dots,m$ we define a r.v. $(\eta_{1,l},\dots,\eta_{n,l})$ which is related to the r.v. (18.4) in a way similar to the connection between the r.v.'s η and ξ distributed by (18.1). Let $n=\sum_{i=1}^{M} n_i$ where n_i is an integer with $0<n_i\leqslant N_i$, $1\leqslant i\leqslant M$. Moreover, let $n_0=0$ and set

$$\begin{aligned}P(\eta_{n_1+\dots+n_{i-1}+1,l}=x_{ij_1},\dots,\eta_{n_1+\dots+n_{i-1}+n_i,l}=x_{ij_{n_i}}|\xi_l=i)&\\ =1/[N_i(N_i-1)\dots(N_i-n_i+1)].&\end{aligned} \tag{18.6}$$

Here, $j_k\neq j_r$ for $k\neq r$ and $j_1,\dots,j_{n_i}$ are any n_i indices from the set $\{1,\dots,N_i\}$.

Before we continue, we want to interpret the previous definitions of this section in analogy to p. 146.

Let the distribution (18.3) of η be associated with a finite population of elements x_{ij}, $1\leqslant j\leqslant N_i$, $1\leqslant i\leqslant M$. The population is stratified. As (18.1) shows, there exist M sub-populations U_i and an element drawn from the population belongs to a U_i with the same probability $1/M$ for each i. The distribution within a U_i is also uniform by (18.2). Draw a sample from the population as follows: First draw randomly m sub-populations U_i ($m\leqslant M$). (18.4) corresponds as sample variable to this procedure. From the chosen sub-populations U_i we now draw samples of sizes $n_i\leqslant N_i$. Denoting the randomly chosen sub-populations with the indices $i_1,\dots,i_m$, we then choose from U_{i_1}, say, the elements $x_{i_1 1},\dots,x_{i_1 n_{i_1}}$, from U_{i_2} the elements $x_{i_2 1},\dots,x_{i_2 n_{i_2}},\dots$, and from U_{i_m} $x_{i_m 1},\dots,x_{i_m n_{i_m}}$. Each choice within U_{i_j} occurs with the same probability. This is the meaning of (18.6). The complete sample of size n is obtained in the form $x_{i_1 1},\dots,x_{i_1 n_{i_1}},\dots,x_{i_m 1},\dots,x_{i_m n_{i_m}}$. The justification for calling this procedure a *two-stage* choice is obvious. Continuing in this way, one can define a 3-*stage* and, in general a *multistage* procedure.

Let

$$a_i = \frac{1}{N_i} \sum_{j=1}^{N_i} x_{ij} \tag{18.7}$$

be the mean in the population U_i. With σ_i^2 we denote the variance $\left(\sum_{j=1}^{N_i} (x_{ij}-a_i)^2\right) \Big/ N_i$, $1 \leqslant i \leqslant M$. For the mean a of the entire population one has from (18.3)

$$a = \frac{1}{M} \sum_{i=1}^{M} \sum_{j=1}^{N_i} x_{ij}/N_i = \frac{1}{M} \sum_{i=1}^{M} a_i. \tag{18.8}$$

For $l=1,\ldots,m$ define the r.v.'s

$$\bar{\eta}_{\xi_l} = \frac{\eta_{n_1+\cdots+n_{\xi_{l-1}}+1} + \cdots + \eta_{n_1+\cdots+n_{\xi_l}}}{n_{\xi_l}}. \tag{18.9}$$

Their distribution can be calculated from the known distribution of ξ_l and (18.6). Hence, the r.v.

$$\bar{\eta} = \frac{1}{m} \sum_{l=1}^{m} \bar{\eta}_{\xi_l} \tag{18.10}$$

is also defined.

As realization of (18.9) one gets the sample mean

$$\bar{x}_i = \frac{x_{ij_1} + \cdots + x_{ij_{n_i}}}{n_i}, \qquad i=1,\ldots,M.$$

We show that

$$E(\bar{\eta}) = a. \tag{18.11}$$

Note that

$$E(\bar{\eta}) = \frac{1}{m} \sum_{l=1}^{m} E(\bar{\eta}_{\xi_l}) = \frac{1}{m} \sum_{l=1}^{m} E(E(\bar{\eta}_{\xi_l} \mid \xi_l))$$

by I. (20.5)

Because of (18.6), an application of (12.2) yields $E(\bar{\eta}_{\xi_l} \mid \xi_l) = a_{\xi_l}$,[15] so that

$$E(\bar{\eta}) = \frac{1}{m} \sum_{l=1}^{m} E(a_{\xi_l}).$$

Applying I. (19.2) gives

$$P(\xi_i = j) = 1/M, \qquad i=1,\ldots,m, \qquad j=1,\ldots,M. \tag{18.12}$$

[15] More precisely, this means that $E(\bar{\eta}_{\xi_l} \mid \xi_l = i) = a_i$ for $1 \leqslant i \leqslant M$.

Hence, by (18.8), $E(\bar{\eta})=\frac{1}{m}\sum_{l=1}^{m}\frac{1}{M}\sum_{i=1}^{M}a_i=a$, which is (18.11). For the variance $E[(\bar{\eta}-a)^2]$ we have

$$E[(\bar{\eta}-a)^2]=\frac{M-m}{mM(M-1)}\sum_{i=1}^{M}(a_i-a)^2+\frac{1}{mM}\sum_{i=1}^{M}\frac{\sigma_i^2}{n_i}\,\frac{N_i-n_i}{N_i-1}. \quad (18.13)$$

To check (18.13), we consider $E(\bar{\eta}^2)$ and call on the Theorem 17.2. First, for $l\neq k$,

$$E(\bar{\eta}_{\xi_l}\bar{\eta}_{\xi_k})=E[E(\bar{\eta}_{\xi_l}\bar{\eta}_{\xi_k}\,|\,\xi_l,\xi_k)]=E(\alpha_{\xi_l}a_{\xi_k}).$$

The marginal distribution of (ξ_l,ξ_k) has already been studied (in somewhat different notation) on p. 61. We use this and get:

$$E(a_{\xi_l}a_{\xi_k})=\frac{1}{M(M-1)}\sum_{\substack{i,j=1\\ i\neq j}}^{M}a_i a_j.$$

Further

$$E(\bar{\eta}_{\xi_i}^2)=E[E(\bar{\eta}_{\xi_i}^2\,|\,\xi_i)]=E\left[\frac{\sigma_{\xi_i}^2}{n_{\xi_i}}\,\frac{N_{\xi_i}-n_{\xi_i}}{N_{\xi_i}-1}+a_{\xi_i}^2\right]$$

by (12.2), resp., (12.5) and Theorem 17.2.

We also find

$$E(\bar{\eta}_{\xi_i}^2)=\frac{1}{M}\sum_{l=1}^{M}\left[\frac{\sigma_l^2}{n_l}\,\frac{N_l-n_l}{N_l-1}+a_l^2\right].$$

Now

$$E(\bar{\eta}^2)=\frac{1}{m^2}\sum_{i,j=1}^{m}E(\bar{\eta}_{\xi_i}\bar{\eta}_{\xi_j})=\frac{1}{m^2}\sum_{\substack{i,j=1\\ i\neq j}}^{m}E[E(\bar{\eta}_{\xi_i}\bar{\eta}_{\xi_j}\,|\,\xi_i,\xi_j)]$$
$$+\frac{1}{m^2}\sum_{i=1}^{m}E[E(\eta_{\xi_i}^2\,|\,\xi_i)].$$

Using the results just obtained, we get

$$E(\bar{\eta}^2)=\frac{1}{m^2}\left[\frac{m(m-1)}{M(M-1)}\sum_{\substack{k,l=1\\ k\neq l}}^{M}a_k a_l+m\frac{1}{M}\sum_{l=1}^{M}\left(\left(\frac{\sigma_l^2}{n_l}\,\frac{N_l-n_l}{N_l-1}\right)+a_l^2\right)\right].$$

which we re-write as

$$\frac{1}{m^2}\left[\frac{m(m-1)}{M(M-1)}\left(\sum_{k=1}^{M}a_k\right)^2-\frac{m(m-1)}{M(M-1)}\sum_{k=1}^{M}a_k^2\right.$$
$$\left.+\frac{m}{M}\sum_{l=1}^{M}\frac{\sigma_l^2}{n_l}\,\frac{N_l-n_l}{N_l-1}+\frac{m}{M}\sum_{k=1}^{M}a_k^2\right].$$

Taking (18.8) into consideration we obtain

$$E((\bar{\eta}-a)^2) = \frac{1}{m^2}\left[m\,\frac{m-M}{M-1}\,a^2 + \frac{m}{M}\,\frac{M-m}{M-1}\sum_{k=1}^{M} a_k^2 + \frac{m}{M}\sum_{l=1}^{M}\frac{\sigma_l^2}{n_l}\,\frac{N_l-n_l}{N_l-1}\right].$$

An easy manipulation leads to

$$\frac{1}{m^2}\left[\frac{m(M-m)}{(M-1)M}\left(\sum_{k=1}^{M} a_k^2 - M a^2\right) + \frac{m}{M}\sum_{l=1}^{M}\frac{\sigma_l^2}{n_l}\,\frac{N_l-n_l}{N_l-1}\right]$$

and hence to (18.13).

For practical purposes, these considerations are modified in many different ways. We give a typical example.

Let

$$A = \sum_{i=1}^{N}\sum_{j=1}^{N_i} x_{ij}. \tag{18.14}$$

Also

$$\sum_{j=1}^{N_i} x_{ij} = A_i, \qquad \frac{1}{M}\sum_{i=1}^{M} A_i = \bar{A}.$$

By means of (18.9) we define the r.v.

$$\bar{X} = \frac{M}{m}\sum_{i=1}^{m} N_{\xi_i}\bar{\eta}_{\xi_i}. \tag{18.15}$$

Then

$$E(\bar{X}) = A$$

and

$$E[(\bar{X}-A)^2] = \left(\frac{M}{m}\right)^2\left(\frac{M-m}{M-1}\,\frac{m}{M}\sum_{i=1}^{M}(A_i-\bar{A})^2 + \frac{m}{M}\sum_{i=1}^{M}\left(\frac{N^i}{n_i}\right)^2\sigma_i^2\,n_i\,\frac{N^i-n_i}{N_i-1}\right)$$

which follows from the given pattern or directly from (18.13).

19. Multistage sampling with proportional choice. Hurwitz and Hansen[16] have suggested replacing the uniform distribution in the choice of the m sub-populations by the assumption that sub-population U_i is chosen with probability N_i/N $(i=1,\dots,M)$, $\sum_{i=1}^{M} N_i = N$ (see p. 153). The probability of a unit being chosen is thus proportional to the number of elements it contains. In most cases, this is a more realistic reflection of the idea of a "weighted" choice of the sub-populations than the uniform distribution. We turn briefly to a treatment of these two-stage procedures in the more general version due to Midzuno[17] and will use the notation of **18**, p. 153ff. and the interpretation given there.

---

[16] M.H. Hansen and W.N. Hurwitz, Ann. Math. Statist. 14, 332—362 (1943).

[17] H. Midzuno, Ann. Inst. Statist. Math. 3, 99—107 (1951/52).

Choose m sub-populations from the set of M available. Let the probability of the choice of the sub-populations with indices $i_1,\ldots,i_m$ be

$$\sum_{k=1}^{m} N_{i_k} \Big/ \binom{M-1}{m-1} N .$$

Then the probability of choosing $U_{i_1},\ldots,U_{i_m}$ is proportional to the sum of the number of their elements $N_{i_1}+\cdots+N_{i_m}$. Hence, in place of the variable (18.4) with distribution (18.5) we have an m-dimensional r.v. $(\xi_1,\ldots,\xi_m)$ with the distribution

$$P(\xi_1=i_1,\xi_2=i_2,\ldots,\xi_m=i_m)=\frac{1}{m!}\sum_{k=1}^{m} N_{i_k} \Big/ \binom{M-1}{m-1} N . \qquad (19.1)$$

Samples are now taken from the chosen sub-populations on the basis of the uniform distribution as before. All of the definitions given in **18** retain their meaning if one takes into account the modifications brought about by application of (19.1) instead of (18.5).

Consider, say, the quantity A defined in (18.14). Define the r.v.

$$\bar{X}_p = N \sum_{i=1}^{m} \frac{N_{\xi_i}\bar{\eta}_{\xi_i}}{\sum_{l=1}^{m} N_{\xi_l}} . \qquad (19.2)$$

We have

$$E(\bar{X}_p)=N E\left[E\left(\sum_{i=1}^{m}\frac{N_{\xi_i}\bar{\eta}_{\xi_i}}{\sum_{l=1}^{m} N_{\xi_l}}\,\middle|\,\xi_1,\ldots,\xi_m\right)\right]=N E\left(\sum_{i=1}^{m}\frac{N_{\xi_i}a_{\xi_i}}{\sum_{l=1}^{m} N_{\xi_l}}\right)$$

$$=\frac{N}{m!\binom{M-1}{m-1}N}\sum_{i_1,\ldots,i_m}^{l}\frac{\sum_{k=1}^{m} N_{i_k}\sum_{k=1}^{m} N_{i_k}a_{i_k}}{\sum_{k=1}^{m} N_{i_k}},$$

where the summation is to be taken over all combinations $i_1,\ldots,i_m$ of m elements taken from a total of M. Here, each summand $N_{i_k}a_{i_k}$ occurs $m!\binom{M-1}{m-1}$ times. Hence

$$E(\bar{X}_p)=\sum_{i=1}^{M} N_i a_i = A .$$

To calculate the variance $E[(\bar{X}_p - A)^2]$ of (19.2) we use I, Theorem 17.2 and obtain first

$$E(\bar{X}_p^2) = N^2 E\left(\frac{\sum_{i,j=1}^{m} N_{\xi_i} N_{\xi_j} \bar{\eta}_{\xi_i} \bar{\eta}_{\xi_j}}{\left(\sum_{i=1}^{m} N_{\xi_i}\right)^2}\right)$$

$$= N^2 E\left[\frac{1}{\left(\sum_{k=1}^{m} N_{\xi_k}\right)^2}\left(E\left(\sum_{i,j=1}^{m} N_{\xi_i} N_{\xi_j} \bar{\eta}_{\xi_i} \bar{\eta}_{\xi_j} \middle| \xi_1, \ldots, \xi_m\right)\right)\right]$$

$$= N^2 E\left[\frac{1}{\left(\sum_{k=1}^{m} N_{\xi_k}\right)^2}\left(\sum_{\substack{i,j=1\\ i\neq j}}^{m} N_{\xi_i} N_{\xi_j} a_{\xi_i} a_{\xi_j} + \sum_{i=1}^{m} N_{\xi_i}^2 \left(a_{\xi_i}^2 + \frac{\sigma_{\xi_i}^2}{n_{\xi_i}} \frac{N_{\xi_i} - n_{\xi_i}}{N_{\xi_i} - 1}\right)\right)\right]$$

$$= N^2 \sum_{i_1, \ldots, i_m} \frac{\sum_{k=1}^{m} N_{i_k}}{m! \binom{M-1}{m-1} N} \frac{1}{\left(\sum_{k=1}^{m} N_{i_k}\right)^2}\left(\sum_{\substack{k,l=1\\ k\neq l}}^{m} N_{i_k} N_{i_l} a_{i_k} a_{i_l} + \sum_{l=1}^{m} N_{i_l}^2 \left(\frac{\sigma_{i_l}^2}{n_{i_l}} \frac{N_{i_l} - n_{i_l}}{N_{i_l} - 1} + a_{i_l}^2\right)\right).$$

Using (19.1) leads to

$$E[(\bar{X}_p - A)^2] = N \sum_{i_1, \ldots, i_m} \frac{\left(\sum_{k=1}^{m} N_{i_k}\right)^{-1}}{\binom{M-1}{m-1}} \left(\left(\sum_{k=1}^{m} N_{i_k} a_{i_k}\right)^2 + \sum_{k=1}^{m} N_{i_k}^2 \frac{\sigma_{i_k}^2}{n_{i_k}} \frac{N_{i_k} - n_{i_k}}{N_{i_k} - 1}\right) - A^2 .$$

The summation is now extended over all combinations of m elements taken from M.

It is obvious that a still more general two-stage procedure can be constructed by choosing a general discrete distribution in place of (19.1) and furthermore, one need not base the sample choice in the subpopulations on a uniform distribution. The main ideas are clear enough from **19**.

We remark finally that questions connected with stratified samples, as in **17**, can also be treated within this framework. All of these investigations for two-stage choices can also be extended without difficulty to infinite populations. See also V, p. 280ff.

Chapter III

Introduction to the Theory of Hypothesis Testing

1. The foundations of the Neyman-Pearson theory of testing. As already mentioned, the procedures discussed in Chapter II for the testing of an hypothesis possess without doubt a certain intuitiveness and are rather convincing. However, we have already pointed out that it is desirable to develop a general test theory which depends on few basic assumptions. In particular, we have given no clear definition of the notion of a "test". We also want to develop criteria for deciding when one of two tests can be viewed as the "better" one.

Neyman and E. S. Pearson deserve credit for creating a clear and fruitful basis for the testing of hypotheses. In a series of fundamental papers[1], they provided the basis for many further investigations in this field. The core of their approach was the recognition that for the test of a hypothesis on the basis of a sample one must not only consider this hypothesis but also possible alternative ones. Precisely what this means will become clear from the definition of the power function of a test, which we will give soon. One can claim that with the theory of tests and that of its dual, confidence regions, both of which are due to Neyman, a new era began for mathematical statistics[2].

Let $(R,\mathcal{S})$ be a measurable space and Γ a set containing at least two elements. Let a probability measure P_γ over $(R,\mathcal{S})$ uniquely correspond to $\gamma\in\Gamma$. Unless the contrary is stated, we will always assume this uniqueness assumption is fulfilled—in this, as well as the following chapters. The set of all P_γ with $\gamma\in\Gamma$ will also be written as P_Γ and γ will be called the *parameter* of this set of probability distributions. This notation will be retained in the sequel. Thus $(R,\mathcal{S})$ *will always be a measurable space,* Γ *always a set of parameters and* P_Γ *the corresponding set of probability measures over* $(R,\mathcal{S})$. Let $\Gamma_0\subset\Gamma$ and assume that

---

[1] Neyman, J. and E. S. Pearson, Biometrika 20A, 175—240 and 263—294 (1928). Philos. Trans. Roy. Soc. London, Ser. A, 231, 289—337 (1933).

[2] A standard reference for the theory of tests and confidence regions, which treats numerous details, is the book by Lehmann l.c. Not.[11]: Testing Statistical Hypotheses, J. Wiley, New York 1959.

both Γ_0 and $\Gamma - \Gamma_0$ are non-empty. Further let α be a real number with $0 \leqslant \alpha \leqslant 1$.

Definition. Each $\mathscr{S}$-measurable map ϕ of R into $[0,1]$ is called a *test.* ϕ is called *a test for the hypothesis* Γ_0 *vs. the alternative hypothesis* $\Gamma - \Gamma_0$ with *level of significance* α, if

$$E(\phi; P_\gamma) \leqslant \alpha \quad (1.1)$$

[3]

for all $\gamma \in \Gamma_0$.

We will then also say that ϕ is a test for the problem $(\alpha, \Gamma_0, \Gamma - \Gamma_0)$ or, for the problem $(\alpha, P_{\Gamma_0}, P_{\Gamma - \Gamma_0})$. If the level of significance is not important we will speak, briefly, of the problem $(\Gamma_0, \Gamma - \Gamma_0)$ or $(P_{\Gamma_0}, P_{\Gamma - \Gamma_0})$. For practical purposes one tries to choose ϕ for given α in such a way that $\sup_{\gamma \in \Gamma} E(\phi; P_\gamma) = \alpha$.[4] For our considerations this will play a lesser role.

Γ, or P_Γ, is called the set of *admissible hypotheses* and Γ_0, or P_{Γ_0}, the *null hypothesis.* If Γ_0, resp., $\Gamma - \Gamma_0$ consists of only one point, then it will be called *simple*, otherwise *composite.* If, in particular, ϕ has the form c_M, $M \in \mathscr{S}$, then M is called a *critical region.* Since in this case $E(\phi; \gamma) = \int_R c_M \, dP_\gamma = P_\gamma(M)$,[5] a critical region M which defines a test c_M for the problem $(\alpha, \Gamma_0, \Gamma - \Gamma_0)$ must fulfill the condition

$$P_\gamma(M) \leqslant \alpha \quad (1.2)$$

for all $\gamma \in \Gamma_0$.

Let us shed some light on these notions with an example. We consider again the test procedure of II.3 for the mean a of a normal distribution with known variance σ^2. The hypothesis $a = a_0$ is to be rejected on the basis of a sample $x_1, \ldots, x_n$ if $\left| \frac{\bar{x} - a_0}{\sigma} \sqrt{n} \right| \geqslant \kappa_\alpha$, where κ_α has been chosen according to II. (3.3) and σ is a given positive real number. The sample space R is thus R_n and $\mathscr{S}$ is to be replaced by $\mathfrak{B}_n$. Which set is to be taken for Γ was not explicitly stated on p. 126ff. We will take R_1 for Γ. P_a is then for each $a \in R_1$ given by the density

$$(x_1, \ldots, x_n) \to \prod_{i=1}^{n} \frac{1}{\sqrt{2\pi}\,\sigma} e^{-\sum_{i=1}^{n} (x_i - a)^2 / 2\sigma^2} \quad (1.3)$$

which follows from the independence assumptions on p. 126. The null hypothesis Γ_0 is for this example the set $\{a_0\}$.

Writing

$$M = \{x \in R_n : |(\bar{x} - a)\sqrt{n}/\sigma| \geqslant \kappa_\alpha\}, \quad (1.4)$$

---

[3] We will also write $E(\phi; \gamma)$ for $E(\phi; P_\gamma)$.

[4] This means that one exploits the given level of significance "as far as possible".

[5] We will sometimes write $P(M; \gamma)$ for $P_\gamma(M)$.

then from II. (3.3) for the test $\phi = c_M$

$$E(\phi; a_0) = P_{a_0}(M) = \alpha. \tag{1.5}$$

We have thus defined a test in the sense of our definition for the problem $(\alpha, \{a_0\}, R_1 - \{a_0\})$ by means of the critical region M given by (1.4). The practical procedure for the test, as described on p. 127, can also be formulated as follows: if the sample point $(x_1, \ldots, x_n)$ lies in the critical region M, then the hypothesis $a = a_0$ is rejected, otherwise accepted. According to (1.4), the "correct" null hypothesis[6] will be rejected with probability α.

Obviously, each test of the form c_M with $M \in \mathscr{S}$ can be analogously interpreted. In the general case, a test ϕ which fulfills (1.1) can be interpreted as follows: For each sample x from the sample space R, $\phi(x)$ determines the probability of rejection of Γ_0. The probability of accepting Γ_0 is $1 - \phi(x)$. Condition (1.1) guarantees that the correct hypothesis Γ_0 is rejected "on the average" with at most probability α. Thus, if one has obtained the sample $x \in R$, then in practice one must carry out another experiment whose possible results can be realized with probability $\phi(x)$, resp., $1 - \phi(x)$. Such a test ϕ is thus also called a *randomized* test. The case of a critical region can be understood as a special case: If $x \in M$, the probability of rejecting Γ_0 is always equal to 1, for $x \notin M$, always 0. The essential reason why the set of all tests ϕ is easier to handle than the subset of critical regions is the fact that the set of all tests is convex. Hence, if ϕ_1, ϕ_2 are two tests and β a real number with $0 < \beta < 1$, then $\beta\phi_1 + (1-\beta)\phi_2$ is also a test. However, if $\phi_1 = c_{M_1}$, $\phi_2 = c_{M_2}$, then $\beta c_{M_1} + (1-\beta) c_{M_2}$ is not in general an indicator function.

We remark that the emphasis in practical applications of statistical tests lies in the rejection of an hypothesis. The smaller the level of significance in (1.1)—and this can be chosen by the statistician—the less probable that one will reject the correct null hypothesis. Accepting the hypothesis, however, means only that it does not contradict what has been observed. This corresponds precisely to the scientific method, which prescribes supporting an hypothesis until it is refuted by new data.

2. The power function. The decisive idea of the Neyman-Pearson theory consists, as we have already indicated, in studying the probability (as function of γ in the space of admissible hypotheses Γ) that the sample lies in a critical region. To this end we introduce

Definition. Let ϕ be a test for the problem $(\alpha, \Gamma_0, \Gamma - \Gamma_0)$. The function g defined for each $\gamma \in \Gamma$ by

$$g_\phi(\gamma) = E(\phi; \gamma) \tag{2.1}$$

---

[6] This terminology is due to the idea that the random experiment which delivers the sample $(x_1, \ldots, x_n)$ is the result of n trials each of which has the same probability distribution with parameter a. The correct value of a is unknown and the null hypothesis, which is to be tested, assumes that $a = a_0$. See also II, p. 127.

is called the *power function* of the test ϕ. If ϕ is understood we will simply write g instead of g_{ϕ}.

We now give an example of the calculation of a power function which will also be important later. Let $\xi_1, \ldots, \xi_n$ be $n \geqslant 1$ independent r.v.'s with ξ_i $N(a_i, 1)$-distributed. The joint distribution of $(\xi_1, \ldots, \xi_n)$ is given for $(x_1, \ldots, x_n) \in R_n$ by the density

$$\left(\frac{1}{\sqrt{2\pi}}\right)^n e^{-1/2\left(\sum_{i=1}^{n}(x_i - a_i)^2\right)}. \tag{2.2}$$

The sample space is naturally R_n. The parameter of the probability distribution P_a given by (2.2) is $a = (a_1, \ldots, a_n)$. Let the set of admissible hypothesis be R_n and the null hypothesis simple and given by $a_1 = a_2 = \cdots = a_n = c_0$, where c_0 is a given real number. We write $c = (c_0, \ldots, c_0)$ and give a level of significance α with $0 < \alpha \leqslant 1$ [7].

We then want to consider the test problem $(\alpha, \{c\}, R_n - \{c\})$. To this end we define a critical region M by

$$M = \left\{x\colon \sum_{i=1}^{n} (x_i - c_0)^2 \geqslant d(\alpha)\right\}.$$

$d(\alpha)$ is to be so determined that $P_c(M) = \alpha$. If the null hypothesis is true, then $\sum_{i=1}^{n} (\xi_i - c_0)^2$ is χ^2-distributed with n degrees of freedom. By I. (28.2), $d(\alpha)$ is uniquely determined by

$$2^{-n/2}(\Gamma(n/2))^{-1} \int_{d(\alpha)}^{\infty} e^{-y/2} y^{(n/2)-1} \, dy = \alpha .$$

The power function g of this test is defined for each $a \in R_n$ by $P_a(M)$. It can be given explicitly if we find the distribution of the r.v.

$$\zeta_n = \sum_{i=1}^{n} \xi_i^2 \tag{2.3}$$

for each $a \in R_n$.

We now proceed by first determining the density of $\sqrt{\zeta_n}$. For $x \in R_n$ and fixed $a \in R_n$ let

$$\frac{a' x}{|a|\,|x|} = \cos \vartheta(x) \tag{2.4}$$

with $0 \leqslant \vartheta(x) \leqslant \pi$. Further let $r(x) = |x|$ for $x \in R_n$. We now transform the density (2.2) by introducing polar coordinates[8]:

$$x_i = r b_i \sin \vartheta, \quad 1 \leqslant i \leqslant n-1$$

$$x_n = r \cos \vartheta, \quad 0 \leqslant \vartheta \leqslant \pi, \quad 0 \leqslant r < \infty, \quad \sum_{i=1}^{n-1} b_i^2 = 1 .$$

---

[7] The case $\alpha = 0$ is trivial and need not be considered.

[8] Strictly speaking, the assumptions of Theorem 12.2 of I are not fulfilled everywhere since the Jacobian vanishes for $r = 0$, $\vartheta = 0$ and $\vartheta = \pi$. However, it is easy to see that the exceptional sets have measure 0.

The absolute value of the corresponding Jacobian is $r^{n-1}\sin^{n-2}\vartheta$. However, since we are only interested in the marginal distribution of $\sqrt{\zeta_n}$, we will not write down the transformed density, but immediately carry out the necessary integrations. Note that the b_i describe the surface $\mathfrak{O}_{n-1}$ of the $(n-1)$-dimensional unit sphere. Thus, taking (2.4) into account, we have for the density $r\to k(r)$ of $\sqrt{\zeta_n}$:

$$k(r)=\int\limits_{\mathfrak{O}_{n-1}}\int\limits_0^\pi \frac{1}{(\sqrt{2\pi})^n} e^{-(r^2+|a|^2-2r|a|\cos\vartheta)/2} r^{n-1}\sin^{n-2}\vartheta\, d\vartheta\, do_{n-1}$$

$$=O_{n-1}\frac{1}{(\sqrt{2\pi})^n} l^{-(r^2+|a|^2)/2} r^{n-1}\int\limits_0^\pi e^{r|a|\cos\vartheta}\sin^{n-2}\vartheta\, d\vartheta,$$

for $r>0$.

Here, do_{n-1} stands for the surface element and O_{n-1} the content of the surface of the $(n-1)$-dimensional sphere. For $r\leqslant 0$, the density $k(r)$ is naturally zero.

With the transformation $r^2=z$ we get the density of ζ_n from $r\to k(r)$:

$$\gamma_n(z,a)=\begin{cases}\pi^{-1/2}2^{-n/2}\left[\Gamma\left(\frac{n-1}{2}\right)\right]^{-1} e^{-(z+|a|^2)/2} z^{(n-2)/2}\cdot\int\limits_0^\pi e^{\sqrt{z}|a|\cos\vartheta}\sin^{n-2}\vartheta\, d\vartheta, & z>0\\ 0, & z\leqslant 0.\end{cases}$$

Because of

$$\frac{1}{\Gamma\left(\frac{n-1}{2}\right)}\int\limits_0^\pi e^{\sqrt{z}|a|\cos\vartheta}\sin^{n-2}\vartheta\, d\vartheta=\sqrt{\pi}\sum_{j=0}^\infty\frac{(|a|^2z)^j\Gamma(j+\frac{1}{2})}{(2j)!\,\Gamma\left(j+\frac{n}{2}\right)} \quad [9]$$

we finally get for $\gamma_n(z,a)$

$$\begin{cases}2^{-n/2}e^{-(z+|a|^2)/2}z^{(n-2)/2}\sum\limits_{j=0}^\infty\frac{(|a|^2z)^j\Gamma(j+\frac{1}{2})}{(2j)!\,\Gamma\left(j+\frac{n}{2}\right)}\ [10], & z>0\\ 0, & z\leqslant 0.\end{cases}\tag{2.5}$$

(2.5) is called the *non-central χ^2-distribution* with n degrees of freedom. We call $|a|^2$ the parameter of (2.5). It is easy to convince oneself that for $a=0$, (2.5) goes into the

[9] For the evaluation of this integral see for example N. Hofreiter and W. Gröbner, Integraltafel, Zweiter Teil: Bestimmte Integrale, 2. Aufl. Springer-Verlag, Wien 1961.

[10] This can be also be written as

$$\sum_{j=0}^\infty\frac{1}{j!}\left(\frac{|a|^2}{2}\right)^j e^{-|a|^2/2}\frac{z^{(2j+n-2)/2}}{2^{(2j+n)/2}\Gamma\left(j+\frac{n}{2}\right)}e^{-z/2}.$$

But for $z\geqslant 0$, $\dfrac{z^{(2j+n-2)/2}}{2^{(2j+n)/2}\Gamma\left(j+\frac{n}{2}\right)}e^{-z/2}$ is the density of a χ^2-distribution with $2j+n$ degrees of freedom. This shows that the non-central χ^2-distribution can be quite easily obtained by induction.

density I. (28.2) of the χ^2-distribution with n degrees of freedom. The power function of the test based on the critical region M is then obtained as follows: if ξ_i is $N(a_i, 1)$-distributed, then $\xi_i - c_0$ is $N(a_i - c_0, 1)$-distributed so that $\sum_{i=1}^{n} (\xi_i - c_0)^2$ is distributed according to (2.5) were a is to be replaced by $a-c$. We thus get for the power function g

$$g(a) = \int_{d(\alpha)}^{\infty} \gamma_n(z, a-c)\,dz, \qquad a \in R_n.$$

g thus has the property of being constant on every sphere with center c.

We now proceed to a classification of tests by means of the power function.

We argue as follows: The condition (1.1) determines the "average" probability of rejecting the correct null hypothesis.

If the null hypothesis is false, then it is desirable that it be rejected with probability "as large as possible". This leads to the

Definition. Let ϕ_1 and ϕ_2 be two tests over the sample space $(R, \mathscr{S})$ for the problem $(\alpha, \Gamma_0, \Gamma - \Gamma_0)$. ϕ_1 is said to be *at least as good as* ϕ_2 for $\Gamma - \Gamma_0$ if

$$g_{\phi_1}(\gamma) \geqslant g_{\phi_2}(\gamma) \tag{2.6}$$

for all $\gamma \in \Gamma - \Gamma_0$.

Based on this is the following important

Definition. Let Φ_α be the set of all tests for the problem $(\alpha, \Gamma_0, \Gamma - \Gamma_0)$ and K a non-empty subset of Φ_α. A test $\phi \in \Phi_\alpha$ is said to be *most powerful* in $\Gamma - \Gamma_0$ w.r.t. K, if ϕ is at least as good for $\Gamma - \Gamma_0$ as any other test from K.

If a test $\phi \in \Phi_\alpha$ is most powerful in $\Gamma - \Gamma_0$ w.r.t. Φ_α then for the sake of brevity we will call ϕ simply most powerful. We also introduce here the notion of an unbiased test:

Definition. A test $\phi \in \Phi_\alpha$ is *unbiased* if for its power function g

$$g(\gamma) \geqslant \alpha \tag{2.7}$$

for all $\gamma \in \Gamma - \Gamma_0$.

The notions of unbiasedness and most powerful are connected by

Theorem 2.1. *Let $\phi \in \Phi_\alpha$ be a most powerful test in $\Gamma - \Gamma_0$ w.r.t. $K \subseteq \Phi$. If K contains an unbiased test, then ϕ is also unbiased. If $K = \Phi$ then ϕ is always unbiased.*

Proof. First one easily sees that for each α with $0 \leqslant \alpha \leqslant 1$ there exists an unbiased test $\psi \in \Phi_\alpha$. Indeed, choose $\psi(x) = \alpha$ for all $x \in R$. Clearly, $\psi \in \Phi_\alpha$ and the power function g_ψ fulfills (2.7) so that ψ is unbiased.

We call ψ a *trivial test*. If $K=\Phi_\alpha$, then $g_\phi(\gamma)\geqslant g_\psi(\gamma)$ for all $\gamma\in\Gamma-\Gamma_0$ and hence also $g_\phi(\gamma)\geqslant\alpha$ for all $\gamma\in\Gamma-\Gamma_0$. However, if $K\neq\Phi_\alpha$ then the assumption assures the existence of an unbiased test. The proof follows exactly the same reasoning.

3. Simple hypotheses. If the set Γ of admissible hypotheses consists of only two elements so that the null hypothesis Γ_0 as well as the alternative $\Gamma-\Gamma_0$ are simple, then a most powerful test can always be constructed. Our next goal is to show this and we will need the so-called Fundamental Lemma of Neyman and Pearson which is given in

Theorem 3.1.[11]

I F. *Let μ be a measure over $(R,\mathcal{S})$ and f_0, f_1 μ-integrable functions with f_0 nonnegative. The map $A\to\int_A f_0\,d\mu$, $A\in\mathcal{S}$, will be denoted by ν. Note that $\nu(R)\geqslant 0$. Let κ be a nonnegative real number with $0\leqslant\kappa\leqslant\nu(R)$. For $k\geqslant-\infty$ set*

$$M_k=\{x: f_1(x)>k\,f_0(x)\}$$

where $-\infty\cdot 0$ is defined, say, as $-\infty$, and for $k>-\infty$ set

$$M_{k+}=\{x: f_1(x)\geqslant k\,f_0(x)\}\,.$$

Then for each κ we can always find a $k\geqslant\infty$ for which

$$\nu(M_k)\leqslant\kappa\leqslant\nu(M_{k+})\ ^{12}. \tag{3.1}$$

II F. *Assume the assumptions of* I F *are fulfilled. Let Φ denote the totality of $\mathcal{S}$-measurable maps from R into $[0,1]$ and Φ_κ the set $\{\phi\in\Phi: \int_R f_0\,\phi\,d\mu\leqslant\kappa\}$. Let $\phi^*\in\Phi_\kappa$ for a $\kappa\geqslant 0$ have the following properties:*

$$\int_R \phi^* f_0\,d\mu=\kappa \tag{3.2}$$

and assume that there is a $k\geqslant 0$ with

$$\phi^*(x)=\begin{cases}1 & x\in M_k\\ 0 & x\in R-M_{k+}\end{cases} \tag{3.3}$$

[11] The first version of this fundamental theorem is in J. Neyman and E. S. Pearson, Statist. Res. Mem. Univ. London 1, 1—37 (1936). Further investigations are: G. B. Dantzig and A. Wald, Ann. Math. Statist. 22, 87—93 (1951); H. Chernoff and H. Scheffé, Ann. Math. Statist. 23, 213—225 (1952); S. Karlin, Mathematical Methods and Theory of Games, Programming and Economics, II, Pergamon Press-Addison Wesley Publishing Company, Oxford-London-New York-Paris 1959, 207 ff.

[12] If $k=-\infty$ define $\nu(M_k)=\nu(M_{k+})$.

Then

$$\int_R \phi^* f_1 \, d\mu = \sup_{\phi \in \Phi_\kappa} \int_R \phi f_1 \, d\mu. \tag{3.4}$$

We emphasize that the definition of ϕ^* *in* $M_{k+} - M_k$ *is of no importance.*

III F. *Let* $Q = \{x: f_1(x) \geqslant 0\}$ *with* $\nu(Q) > 0$. *Let* κ *be a given real number with*

$$0 \leqslant \kappa \leqslant \nu(Q). \tag{3.5}$$

For this κ *choose a* $k \geqslant 0$ *for which* (3.1) *holds. Assume*

a) $\nu(M_{k+} - M_k) = 0$ *or*

b) $\nu(M_{k+} - M_k) > 0$.

Define in case a)

$$\phi^*(x) = \begin{cases} 1 & x \in M_k \\ \text{an arbitrary value from } [0,1] & x \in M_{k+} - M_k \\ 0 & x \in R - M_{k+} \end{cases} \tag{3.6}$$

and in case b)

$$\phi^*(x) = \begin{cases} 1 & x \in M_k \\ \dfrac{\kappa - \nu(M_k)}{\nu(M_{k+}) - \nu(M_k)} & x \in M_{k+} - M_k \\ 0 & x \in R - M_{k+}. \end{cases} \tag{3.7}$$

Then in both cases

$$\int_R \phi^* f_0 \, d\mu = \kappa \tag{3.8}$$

and

$$\int_R \phi^* f_1 \, d\mu = \sup_{\phi \in \Phi_\kappa} \int_R \phi f_1 \, d\mu. \tag{3.9}$$

The following result also holds: Let $\bar{\phi} \in \Phi_\kappa$ *and*

$$\int_R \bar{\phi} f_1 \, d\mu = \sup_{\phi \in \Phi_\kappa} \int_R \phi f_1 \, d\mu. \tag{3.10}$$

Then (3.8) *is fulfilled with* $\bar{\phi}$ *in place of* ϕ^* *provided that*

$$k \neq 0. \tag{3.11}$$

Moreover, $\bar{\phi}(x) = \phi^*(x)$ μ*-a.e. for* $x \in M_k \cup (R - M_{k+})$. *In general, nothing can be said about the behavior of* $\bar{\phi}$ *in* $M_{k+} - M_k$, ϕ^* *is thus "essentially" uniquely determined by* $\phi^* \in \Phi_\kappa$ *and* (3.9).

Supplement to III F. *If* $\nu(Q)\geqslant 0$ *but* $\nu(R)>0$ *and one chooses* κ *with* $0\leqslant\kappa\leqslant\nu(R)$, *then all of the statements of* III F *are correct, even without* (3.11), *provided one allows* $k\geqslant -\infty$ *and replaces* Φ_κ *by*

$$\Phi'_\kappa=\left\{\phi\in\Phi_\kappa:\int_R \phi f_0\,d\mu=\kappa\right\}.$$

Proof of I F. Since $M_{k''}\subseteq M_{k'}$ for $k''>k'$, we have $\nu(M_{k''})\leqslant\nu(M_{k'})$, i.e., the map ψ defined by $k\to\nu(M_k)$ is nonincreasing. Let ε_n be a positive monotone null sequence. Then for $k>-\infty$, $M_k=\bigcup_{n=1}^{\infty} M_{k+\varepsilon_n}$. Hence, from the continuity of the measure ν (see p. 5), $\nu(M_k)=\lim_{n\to\infty}\nu(M_{k+\varepsilon_n})$, i.e., ψ is right-continuous. One shows similarly that $\psi(-\infty)=\lim_{k\to-\infty}\psi(k)$ and $\psi(k-0)=\nu(M_{k+})$, $k>-\infty$ (see p. 31). We have $\psi(-\infty)=\nu(R)$, and from the μ-integrability of f_1 we also have $\lim_{k\to\infty}\psi(k)=0$. For each κ with $0\leqslant\kappa\leqslant\nu(R)$ there thus exists a k with $\psi(k)\leqslant\kappa\leqslant\psi(k-0)$. Here it may be necessary in the cases $\kappa=0$, resp., $\kappa=\nu(R)$ to allow the values $k=\infty$, resp., $k=-\infty$ and then define $\psi(\infty)=\psi(\infty-0)$, resp., $\psi(-\infty)=\psi(-\infty-0)$. The claim is proved.

Proof of II F.[13] Let $\phi\in\Phi_\kappa$. Then

$$\int_R \phi^* f_1\,d\mu-\int_R \phi f_1\,d\mu=\int_R \phi^*(1-\phi)f_1\,d\mu-\int_R \phi(1-\phi^*)f_1\,d\mu.$$

From (3.3) and the fact that $M_{k+}-M_k=\{x: f_1(x)=k f_0(x)\}$ we also have

$$\int_R \phi^*(1-\phi)f_1\,d\mu\geqslant k\int_R \phi^*(1-\phi)f_0\,d\mu$$

and

$$\int_R \phi(1-\phi^*)f_1\,d\mu\leqslant k\int_R \phi(1-\phi^*)f_0\,d\mu.$$

Hence

$$\int_R \phi^* f_1\,d\mu-\int_R \phi f_1\,d\mu\geqslant k\left(\int_R \phi^*(1-\phi)f_0\,d\mu-\int_R \phi(1-\phi^*)f_0\,d\mu\right)$$
$$=k\left(\int_R \phi^* f_0\,d\mu-\int_R \phi f_0\,d\mu\right)\geqslant 0$$

where the last inequality follows from (3.2) and $\phi\in\Phi_\kappa$.

Proof of III F. Since $Q=M_{0+}$ (up to ν-null sets), one can always find a nonnegative k according to (3.5) and I F which satisfies (3.1).

[13] $k=\infty$ requires (also for III F) a trivial special argument.

In both cases a) and b) ϕ^* fulfills (3.8). Hence, (3.9) also follows from II F. We write the relevant string of inequalities again:

$$\begin{aligned}\int_R \phi^* f_1\,d\mu - \int_R \phi f_1\,d\mu &= \int_R \phi^*(1-\phi) f_1\,d\mu - \int_R \phi(1-\phi^*) f_1\,d\mu\\ &\geqslant k\int_R \phi^*(1-\phi) f_0\,d\mu - k\int_R \phi(1-\phi^*) f_0\,d\mu\\ &= k\Big(\int_R \phi^* f_0\,d\mu - \int_R \phi f_0\,d\mu\Big)\\ &\geqslant 0\,.\end{aligned}$$

Now let $\bar\phi\in\Phi_\kappa$ and assume $\bar\phi$ fulfills (3.10) and that (3.11) holds. If $\bar\phi\notin\Phi'_\kappa$ (Φ'_κ is defined on p. 168), then it would follow from the above inequalities that

$$0=\int_R \phi^* f_1\,d\mu - \int_R \bar\phi f_1\,d\mu \geqslant k\Big(\kappa - \int_R \bar\phi f_0\,d\mu\Big)>0$$

which is a contradiction. Hence we must have

$$0=\int_R (\phi^*-\bar\phi) f_1\,d\mu = k\int_R (\phi^*-\bar\phi) f_0\,d\mu$$

or

$$\int_R (\phi^*-\bar\phi)(f_1-k f_0)\,d\mu=0 \tag{3.12}$$

which also holds trivially for $k=0$. Now if

$$\mu(\{x\colon \phi^*(x)>\bar\phi(x)\}\cap M_k)>0$$

hold, or

$$\mu(\{x\colon \phi^*(x)<\bar\phi(x)\}\cap(R-M_{k+}))>0$$

or both, then we would have from the definition of M_k and M_{k+} that $\int_R (\phi^*-\bar\phi)(f_1-k f_0)\,d\mu>0$, which contradicts (3.12). If $k=0$, then one cannot in general conclude that $\phi\in\Phi'_\kappa$ (see p. 173). In any case, $\int_R \phi^* f_1\,d\mu = \int_Q f_1\,d\mu$ holds and this case is not very interesting.

Proof of the Supplement. If $\phi\in\Phi'_\kappa$, then exactly as in the proof of III F,

$$\int_R \phi^* f_1\,d\mu - \int_R \phi f_1\,d\mu \geqslant k\Big(\int_R \phi^* f_0\,d\mu - \int_R \phi f_0\,d\mu\Big)\,. \tag{3.13}$$

But now from (3.8) and the fact that $\phi\in\Phi'_\kappa$ it follows that the right side of (3.13) vanishes provided we define $-\infty\cdot 0=0$. Since (3.12) holds unchanged for all $k>-\infty$ provided $\bar\phi\in\Phi'_\kappa$, we also obtain the remaining parts of III F. They are trivial for $k=-\infty$.

Remark. Theorem 3.1 refers to a maximization problem characterized, say, by (3.8) and (3.9). By modifying Theorem 3.1 appropriately, one obtains the solution of the corresponding minimization problem.

Theorem 3.1 can be generalized to finitely many μ-integrable functions. We will not go into complete generality[14] and choose instead the following formulation:

Theorem 3.2. *Let $(R, \mathscr{S})$ be a measurable space and μ an arbitrary measure on $\mathscr{S}$. Let $f_0, \ldots, f_k$, $k \geqslant 1$, be μ-integrable functions on R and $b_0, \ldots, b_{k-1}$ given real numbers. The totality of all $\mathscr{S}$-measurable maps ϕ from R into $[0, 1]$ satisfying*

$$\int_R \phi f_i \, d\mu = b_i, \qquad i = 0, \ldots, k-1 \tag{3.14}$$

will be denoted by $\Phi_{b_0, \ldots, b_{k-1}}$. If it is possible to determine real numbers $l_0, \ldots, l_{k-1}$ in such a way that $\int_R c_M f_i \, d\mu = b_i$, $i = 0, \ldots, k-1$ for $M = \left\{x : f_k(x) \geqslant \sum_{i=0}^{k-1} f_i(x) l_i\right\}$, then

$$\int_R c_M f_k \, d\mu \geqslant \int_R \phi f_k \, d\mu \tag{3.15}$$

for all $\phi \in \Phi_{b_0, \ldots, b_{k-1}}$.

If $\bar{\phi} \in \Phi_{b_0, \ldots, b_{k-1}}$ and $\int_R \bar{\phi} f_k \, d\mu \geqslant \int_R \phi f_k \, d\mu$ for all $\phi \in \Phi_{b_0, \ldots, b_{k-1}}$, then $\bar{\phi}(x) = c_M(x)$ μ-a.e. for

$$\left\{x : f_k(x) \neq \sum_{i=0}^{k-1} l_i f_i(x)\right\}.$$

The *proof* of (3.15) goes exactly as that of II F, Theorem 3.1. The claim following (3.15) is shown as III F of Theorem 3.1. An analogous theorem again holds for the corresponding minimization problem.

Using Theorem 3.1, we now turn to the construction of a most powerful test for a null hypothesis vs. a simple alternative. We prove

Theorem 3.3. *Let P_0, P_1 be two probability measures over $(R, \mathscr{S})$ with $P_0 \neq P_1$ and α, $0 \leqslant \alpha \leqslant 1$, a given level of significance. Then there always exists a most powerful test for the problem (α, P_0, P_1).*

Proof. Let f_0 (f_1) be the density of P_0 (P_1) w.r.t. a suitable measure μ. Such a measure μ always exists (see Theorem XIII). Now apply Theorem 3.1, III F for $\kappa = \alpha$. A most powerful test is thus always representable

[14] See Dantzig and A. Wald loc. cit.[11].

in the form (3.6) or (3.7). Further one sees that a most powerful test ϕ^* always satisfies

$$\int_R \phi^* f_0 \, d\mu = \alpha \tag{3.16}$$

provided $E(\phi^*; P_1) < 1$ and is "essentially" uniquely determined. Indeed, from $E(\phi^*; P_1) < 1$ it follows that (3.6) or (3.7) with $k > 0$ holds. Under the assumptions of Theorem 3.3 it follows from (3.6) or (3.7) and from (3.8) (with $\kappa = \alpha$) that

$$\int_R \phi^* f_1 \, d\mu = \int_{\{f_1 \geqslant k f_0\}} \phi^* (f_1 - k f_0) \, d\mu + \int_{\{f_1 \geqslant k f_0\}} \phi^* k f_0 \, d\mu$$

thus

$$\int_R \phi^* f_1 \, d\mu = k\alpha + \int_R \phi^* (f_1 - k f_0)^+ \, d\mu \tag{3.17}$$

where $(f_1 - k f_0)^+ = \max(0, f_1 - k f_0)$.

The use of randomized tests thus has the advantage that one can always "completely exploit" the given level of significance. This is naturally not the case for the critical regions. If the null hypothesis is given, say, by $B_1(p)$ and the corresponding probability measure is denoted by P_p, then one has—no matter how a critical region M is chosen—only $0, p, 1-p, 1$ as possible values of $P_p(M)$. For all error probabilities differing from these, one must be satisfied with $P_p(M) < \alpha$. (Cf. also p. 162.)

We now present several properties of the power function of a most powerful test of a simple null hypothesis P_0 vs. a simple alternative P_1. Let $0 \leqslant \alpha \leqslant 1$ and Φ_α the set of all tests ϕ for the problem (α, P_0, P_1). Denote by ϕ_α^* a most powerful test for this problem and for $0 \leqslant \alpha \leqslant 1$ set

$$g(\alpha) = E(\phi_\alpha^*; P_1). \tag{3.18}$$

Then we have

Theorem 3.4[15]. *g is nondecreasing and concave*[16] *in $0 \leqslant \alpha \leqslant 1$ with $0 \leqslant g \leqslant 1$. Consequently, g is continuous in $0 < \alpha < 1$ and $\alpha \to g(\alpha)/\alpha$ nonincreasing. Further*

$$\lim_{\alpha \to 1-0} g(\alpha)/\alpha = 1 \tag{3.19}$$

and

$$1 \leqslant g(\alpha)/\alpha \leqslant 1/\alpha, \qquad 0 < \alpha \leqslant 1. \tag{3.20}$$

[15] See L. Schmetterer, Sankhyā 25, 207—210 (1963). A much deeper result has been given by W. Sendler, Z. Wahrscheinlichkeitstheorie und Verw. Gebiete 18, 183—196 (1971).

[16] This means that $-g$ is convex.

Proof. It follows immediately from the definition of g that g is nondecreasing. Let α_1, α_2 be real numbers with $0<\alpha_i<1$, $i=1,2$, and t_1, t_2 positive numbers with $t_1+t_2=1$. Naturally, $0<t_1\alpha_1+t_2\alpha_2<1$. Let $\phi^*_{\alpha_1}\in\Phi_{\alpha_1}$ and $\phi^*_{\alpha_2}\in\Phi_{\alpha_2}$ be most powerful tests. We then have $0\leqslant t_1\phi^*_{\alpha_1}+t_2\phi^*_{\alpha_2}\leqslant 1$ and also $E(t_1\phi^*_{\alpha_1}+t_2\phi^*_{\alpha_2}; P_0)=t_1 E(\phi^*_{\alpha_1}; P_0)+t_2 E(\phi^*_{\alpha_2}; P_0)\leqslant t_1\alpha_1+t_2\alpha_2$ so that $t_1\phi^*_{\alpha_1}+t_2\phi^*_{\alpha_2}\in\Phi_{t_1\alpha_1+t_2\alpha_2}$. From the definition of a most powerful test we thus have $E(t_1\phi^*_{\alpha_1}+t_2\phi^*_{\alpha_2}; P_1)\leqslant E(\phi^*_{t_1\alpha_1+t_2\alpha_2}; P_1)$ or, from (3.18)

$$t_1 g(\alpha_1)+t_2 g(\alpha_2)\leqslant g(t_1\alpha_1+t_2\alpha_2). \tag{3.21}$$

The continuity of g in $0<\alpha<1$ now follows easily from the monotonicity and (3.21). If the positive numbers t_2 and ε are small enough, so that $\alpha+t_2\varepsilon<1$ and $\alpha-\varepsilon t_1>0$, then

$$(t_1-1)g(\alpha)+t_2 g(\varepsilon)\leqslant g(\alpha+t_2\varepsilon)-g(\alpha). \tag{3.22}$$

On the other hand, from (3.21) we also have

$$t_1 g(\alpha+\varepsilon t_2)+t_2 g(\alpha-\varepsilon t_1)\leqslant g(\alpha),$$

so that

$$g(\alpha+\varepsilon t_2)-g(\alpha)\leqslant -t_2 g(\alpha-\varepsilon t_1)+(1-t_1)g(\alpha+\varepsilon t_2). \tag{3.23}$$

Because $0\leqslant g\leqslant 1$, the right continuity of g then follows for $t_2\to 0$ from (3.22) and (3.23). Similarly follows the left continuity. From the monotonicity and continuity of g in $0<\alpha<1$ follows the existence of $g(0+0)$ and naturally $g(0+0)\geqslant 0$. This implies with (3.19) the monotonicity of $\alpha\to g(\alpha)/\alpha$: choose real numbers γ, δ with $0<\delta<\gamma<1$ and set $\alpha_2=\gamma$, $t_2=\delta/\gamma$ in (3.21). Then letting $\alpha_2\to 0+0$ we easily find that $\alpha\to g(\alpha)/\alpha$ is nonincreasing.

Since for $\alpha=1$ one surely gets a most powerful test by choosing $\phi^*(x)=1$ for all $x\in R$, (3.19) follows trivially. (3.20) is likewise trivial.

We now consider the question of whether the upper bound in (3.20) can always be attained by suitable choice of P_0 and P_1. To this end we give the

Definition. Two arbitrary probability measures P_0 and P_1 over $(R,\mathscr{S})$ are called *orthogonal* if there exists at least one $M\in\mathscr{S}$ with $P_0(M)=1$ and $P_1(M)=0$.

Naturally there then exists an $N\in\mathscr{S}$ with $P_0(N)=0$ and $P_1(N)=1$. It suffices to choose $N=R-M$.

The following result is now easy to prove.

Theorem 3.5. *If P_0 and P_1 are orthogonal measures, then* $g(\alpha)=1$, $0\leqslant\alpha\leqslant 1$.

Proof. Let f_i be the R.-N. densities of P_i, $i=0,1$, w.r.t. the measure $(P_0+P_1)/2$. If $\beta \leqslant \alpha$, then

$$\phi^*(x) = \begin{cases} 1 & \text{for } f_1(x) > 0 \\ \beta & \text{for } f_1(x) = 0 \end{cases}$$

defines a most powerful test for the problem (α, P_0, P_1) for each α, $0 \leqslant \alpha \leqslant 1$. We also have $E(\phi^*; P_1) = 1$.

Thus we have also given an example which shows that a most powerful test need not satisfy (3.16) when the power function for the alternative assumes the value 1.

In applications, small values of α are of especial interest. Consideration of $\lim_{\alpha \to 0+0} g(\alpha)/\alpha$, which must always exist by Theorem 3.4 gives us an idea "how much better a most powerful test is than the trivial test" (see p. 166). The following theorem shows that for suitable choice of P_0 and P_1, a most powerful test can be "almost as bad" as the trivial test.

Theorem 3.6. *Let f be a density over R_1 with the following properties: f is symmetric about zero; f is continuous and strictly monotone decreasing for $x \geqslant 0$. Let σ be a real number >1 and assume the function $x \to f(x/\sigma)/f(x)$ is strictly monotone increasing for $x \geqslant 0$. If P_0 is given by the density f and P_1 by the density $x \to f(x/\sigma)/\sigma$ then $\lim_{\alpha \to 0+0} g(\alpha)/\alpha$ is finite or infinite according as $x \to f(x/\sigma)/f(x)$ is bounded or not.*

Proof. From the assumptions follows the existence of $\lim_{x \to \infty} f(x/\sigma)/f(x)$ as finite or infinite value. For each α, $0<\alpha<1$, let $\eta(\alpha)$ be the only solution of the equation $\int_{\eta(\alpha)}^{\infty} f(x)\,dx = \alpha/2$. Then a most powerful test for (α, P_0, P_1) is given by the critical region

$$M = \{x : \eta(\alpha) \leqslant x < \infty\} \cup \{x : -\infty < x \leqslant -\eta(\alpha)\}.$$

Hence,

$$g(\alpha)/\alpha = \frac{1}{\sigma} \int_{\eta(\alpha)}^{\infty} f(x/\sigma)\,dx \left[\int_{\eta(\alpha)}^{\infty} f(x)\,dx \right]^{-1}$$

so that

$$\lim_{\alpha \to 0+0} g(\alpha)/\alpha = \lim_{\alpha \to 0+0} \frac{1}{\sigma} f(\eta(\alpha)/\sigma)/f(\eta(\alpha))$$

which completes the proof.

The Cauchy distribution (see p. 84), whose density f is $(1/\pi)\cdot(1/(1+x^2))$ for each $x\in R_1$, gives an example of a density satisfying all the conditions of Theorem 3.6. It is easy to see that in this case $\lim_{\alpha\to 0+0} g(\alpha)/\alpha=\sigma$, and σ can be chosen arbitrarily close to 1.

It is also easy to see that for $P_0\neq P_1$ we must always have $\lim_{\alpha\to 0+0} g(\alpha)/\alpha>1$. Let P_0 and P_1 be defined over $(R,\mathscr{S})$ and μ a dominant measure for P_0 and P_1. Denote the corresponding R.-N. densities by f_i, $i=0,1$. Since $P_0\neq P_1$ we surely have $\mu(E)>0$ for $E=\{x: f_1(x)>f_0(x)\}$.

Since

$$E=\bigcup_{n=1}^{\infty}\{x: f_1(x)-f_0(x)\geqslant 1/n\}$$

there exists an $\varepsilon>0$ and an $E_1\subseteq E$ such that $f_1(x)\geqslant f_0(x)+\varepsilon$ for $x\in E_1$ and $\mu(E_1)=\delta>0$.

If $f_0(x)=0$ for $x\in E_1$ μ-a.e. we consider the test

$$\phi_\alpha(x)=\begin{cases}1 & x\in E_1\\ \alpha & x\in R-E_1\end{cases}\qquad\text{with } 0<\alpha<1.$$

Then

$$E(\phi_\alpha;P_1)=\alpha P_1(R-E_1)+P_1(E_1)=\alpha+(1-\alpha)P_1(E_1)>\alpha.$$

But if $\int_{E_1} f_0\,d\mu=\alpha_1>0$, then let

$$\phi_{\alpha_1}(x)=\begin{cases}1 & x\in E_1\\ 0 & x\in R-E_1\end{cases}$$

so that

$$E(\phi_{\alpha_1};P_1)=\int_{E_1} f_1\,d\mu\geqslant\alpha_1+\varepsilon\delta>\alpha_1.$$

In any case, since $\alpha\to g(\alpha)/\alpha$ does not increase, we thus have $\lim_{\alpha\to 0+0} g(\alpha)/\alpha>1$.[17]

The following theorem gives another simple property of a most powerful test for a problem of the form (α,P_0,P_1).

Theorem 3.7. *Let α be a real number with $0\leqslant\alpha\leqslant 1$ and P_i, $i=0,1$, probability measures over $(R,\mathscr{S})$. Let ϕ^* be a most powerful test for the problem (α,P_0,P_1) and assume $E(\phi^*,P_1)=\beta<1$. Then $1-\phi^*$ is a most powerful test for the problem $(1-\beta,P_1,P_0)$.*

---

[17] Without explicitly giving non-trivial tests, one can also argue as follows: From $\lim_{\alpha\to 0+0} g(\alpha)/\alpha=1$ we have from Theorem 3.4, $g(\alpha)=\alpha$, $0\leqslant\alpha\leqslant 1$. Hence, from the definition of g, $\int_R \phi f_1\,d\mu\leqslant\int_R \phi f_0\,d\mu$ for each test $\phi\in\Phi_\alpha$ and $0\leqslant\alpha\leqslant 1$. For the test $\phi=c_E$ we thus have $\mu(E)=0$ which contradicts the assumption.

Proof. $1-\phi^*$ is naturally a test and $E(1-\phi^*;P_1)=1-\beta$. Assume there existed a most powerful test ψ for the problem $(1-\beta,P_1,P_0)$ with $E(\psi;P_0)>E(1-\phi^*;P_0)=1-\alpha$. Then $1-\psi$ would also be a test. Since

$$E(1-\psi;P_0)<\alpha \tag{3.24}$$

$1-\psi$ would also be a test for the problem (α,P_0,P_1) and because $E(1-\psi;P_1)\geqslant\beta$ even a most powerful one. Since, however, $\beta<1$, (3.24) contradicts Theorem 3.1 III F.

Remark for applications: If, for the sake of simplicity we consider a critical region M for a test problem of the form (α,P_0,P_1), then with probability $\leqslant\alpha$ the null hypothesis will be rejected although it is the correct one. This false decision is called an *error of the first kind*. The so-called *error of the second kind* involves not rejecting the null hypothesis when it is false. Then, according to assumption, P_1 is correct. Denoting by β the probability of committing an error of the second kind, we have $\beta=1-P_1(M)$. We give another example to illustrate these ideas. Assume we have a population which is distributed according to a $B_1(p)$-distribution whose parameter p is unknown. From this population we take a sample of size n and want to decide on the basis of this sample whether $p=p_0$ or $=p_1$ [18] with $p_0<p_1$. The sample space is thus R_n and the test problem has the form (α,P_0,P_1), where P_i can be identified with $B_n(p_i)$, $i=0,1$. (See I.**32**.) We interpret the population as the totality of items produced in some process and p the unknown fraction of defectives in the batch. We assume that the units produced are classified only as "good" or "defective". The null hypothesis $p=p_0$ corresponds to the maximum allowable fraction of defectives in the batch (acceptable quality level) and the alternative $p=p_1$ to the fraction of defectives which the consumer is not any more willing to accept. Which of these situations is present is to be checked by means of a sample. This procedure includes the risk that a batch corresponding to the allowable standard is erroneously rejected, that is, that the hypothesis $p=p_0$ is rejected although it is correct. This risk is assumed by the producer and the probability of such an error is thus called the *producer's risk*. The danger to the consumer is that he receives a delivery described as acceptable which does not fulfill the allowable standard, that is, the hypothesis $p=p_0$ is accepted although $p=p_1$. The probability β of this error is therefore called the *consumer's risk*.

4. Composite hypotheses. We first consider the case in which the null hypothesis is composite but the alternative simple. Under these assumptions there exists under some conditions a most powerful test.

We have

Theorem 4.1. *Let P_Γ be a set of probability measures over $(R,\mathscr{S})$ dominated by a σ-finite measure μ, where the set Γ of admissible hypotheses can be completely arbitrary. Let $0\leqslant\alpha\leqslant1$, $\gamma_1\in\Gamma$ and consider the test problem $(\alpha,\Gamma-\{\gamma_1\},\{\gamma_1\})$. For this problem there always exists a most powerful test ϕ^*.*

This is only a special case of the forthcoming Theorem 4.2 and therefore no proof is given here.

[18] Practically speaking, $p\leqslant p_0$, resp., $p\geqslant p_1$ is a more reasonable requirement but we want to consider only simple hypotheses here.

However, this helps little in actually determining a most powerful test for a composite null hypothesis vs. a simple alternative since Theorem 4.1 (resp. Theorem 4.2) merely proves the existence of such a test without allowing its explicit determination.

We now present an example which shows which problems have to be solved in the determination of such a most powerful test[19]. Let the nullhypothesis be given by $m \geqslant 1$ discrete probability distributions P_i defined over a finite sample space $R=\{x_1,\dots,x_k\}$, $k \geqslant 1$. For $i=1,\dots,m$ let them be given by the probabilities $p_{i1},\dots,p_{ik}$. Let the alternative hypothesis be of the same form with probabilities $p_{m+11},\dots,p_{m+1k}$. Denoting by ϕ_j the value of a test ϕ at the j-th mass point, we find that the determination of a most powerful test for the problem $(\alpha, P_i, 1 \leqslant i \leqslant m; P_{m+1})$ leads to the determination of k real numbers ϕ_j satisfying

$$0 \leqslant \phi_j \leqslant 1\,, \qquad 1 \leqslant j \leqslant k$$

$$\sum_{j=1}^{k} p_{ij}\phi_j \leqslant \alpha\,, \qquad 1 \leqslant i \leqslant m$$

$$\sum_{j=1}^{k} p_{m+1j}\phi_j \to \text{Maximum}.$$

One must thus solve a so-called *linear program.* A number of practically effective methods for solving such programs have been developed.[20] The solution of this linear program is of course closely connected with Theorem 3.2. This relation will become much clearer later on.

There is a so-called *dual program* to each linear program which is in this example of the following form:

Determine $m+k$ real numbers $y_1,\dots,y_{m+k}$ satisfying

$$\begin{aligned} &\sum_{i=1}^{m} y_i p_{ij} + y_{m+j} \geqslant p_{m+1j}\,, \qquad 1 \leqslant j \leqslant k\,, \\ &y_l \geqslant 0\,, \qquad 1 \leqslant l \leqslant m+k\,, \\ &\alpha(y_1+\dots+y_m)+y_{m+1}+\dots+y_{m+k} \to \text{Minimum}. \end{aligned} \tag{4.1}$$

It is very easy to see that there always exist real numbers $\phi_1,\dots,\phi_k$ resp. $y_1,\dots,y_{m+k}$ which satisfy the side conditions of the given linear program resp. of its dual program (so-called *feasible solutions*). Therefore both programs admit solutions as is well known. If $y_1^*,\dots,y_{m+k}^*$ is a solution of the dual program and $\phi_1^*,\dots,\phi_k^*$ one of the given program then it is known (see p. 182) that

$$\left.\begin{aligned} \sum_{j=1}^{k} p_{m+1j}\phi_j^* &= \alpha(y_1^*+\dots+y_m^*)+y_{m+1}^*+\dots+y_{m+k}^* \\ &= y_1^* \sum_{j=1}^{k} p_{1j}\phi_j^* + \dots + y_m^* \sum_{j=1}^{k} p_{mj}\phi_j^* + \phi_1^* y_{m+1}^* + \dots + \phi_k^* y_{m+k}^* \end{aligned}\right\}. \tag{4.2}$$

[19] The first systematic investigation of the connection between linear programs and test theory is in E. W. Barankin, Univ. California Publ. Statist. 1, 161—214 (1949—1953).

[20] See for instance S. Vajda, Theory of Games and Linear Programming, John Wiley, New York 1956.

The inequalities $0 \leqslant \phi_i^* \leqslant 1$, $y_{m+i}^* \geqslant 0$, $1 \leqslant i \leqslant k$ together with (4.2) imply the relation

$$\alpha(y_1^* + \cdots + y_m^*) \leqslant \sum_{i=1}^{m} y_i^* \sum_{j=1}^{k} p_{ij} \phi_j^* .$$

From $\sum_{j=1}^{k} p_{ij} \phi_j^* \leqslant \alpha$, $1 \leqslant i \leqslant m$, it follows that

$$\sum_{i=1}^{m} y_i^* \sum_{j=1}^{k} p_{ij} \phi_j^* = \alpha(y_1^* + \cdots + y_m^*) . \tag{4.3}$$

Adding all inequalities (4.1) one gets that

$$\sum_{l=1}^{m+k} y_l \geqslant 1 . \tag{4.4}$$

If the power $\sum_{j=1}^{k} p_{m+1j} \phi_j^*$ of the test $\phi^* = (\phi_1^*, \ldots, \phi_k^*)$ is <1 then from (4.2) and (4.4) it follows that

$$y_1^* + \cdots + y_m^* \neq 0 . \tag{4.5}$$

In this case, therefore,

$$u_i = y_i^* \Big/ \left(\sum_{j=1}^{m} y_j^* \right), \qquad 1 \leqslant i \leqslant m \tag{4.6}$$

can be considered as a probability distribution on $\Gamma_0 = \{1, \ldots, m\}$.

Furthermore, $\sum_{i=1}^{m} p_{ij} u_i$, $1 \leqslant j \leqslant k$, is a probability distribution $\mathfrak{P}_u$ over R. We will show that ϕ^* is a most powerful test for the test problem $(\alpha, \mathfrak{P}_u, P_{m+1})$.

For this purpose let us note that part of (4.2) can be rewritten as

$$\sum_{j=1}^{k} p_{m+1j} \phi_j^* = \sum_{j=1}^{k} \left(\sum_{i=1}^{m} y_i^* p_{ij} + y_{m+j}^* \right) \phi_j^* .$$

This implies together with (4.1): Whenever $\phi_j^* > 0$ then

$$p_{m+1j} = \sum_{i=1}^{m} y_i^* p_{ij} + y_{m+j}^* . \tag{4.7}$$

From (4.3) and again from (4.2) it follows that $\phi_j^* = 1$ whenever $y_{m+j}^* > 0$.

Assuming that (4.5) holds and using the notation introduced by (4.6) we find from (4.7) that

$$\phi_j^* = 1 \quad \text{whenever } p_{m+1j} > k_\alpha \sum_{i=1}^{m} p_{ij} u_i \quad \text{where } k_\alpha = (y_1^* + \cdots + y_m^*) .$$

Furthermore, $p_{m+1} < k_\alpha \sum_{i=1}^{m} p_{ij} u_i$ implies $\phi_j^* = 0$ as follows from the derivation of (4.7).

These facts together with (4.3) imply our claim by an application of Theorem 3.1 IIF. Note that ϕ^* is trivially most powerful for $(\alpha, \mathfrak{P}_u, P_{m+1})$ if the power of ϕ^* w.r.t. the problem $(\alpha, \Gamma_0, \Gamma - \Gamma_0)$ is 1, where $\Gamma - \Gamma_0 = \{m+1\}$.

Conversely if there exists a probability distribution $(u_1,\dots,u_m)$ on Γ_0 such that the linear program

$$\sum_{j=1}^{k} \phi_j \sum_{i=1}^{m} p_{ij} u_i \leqslant \alpha, \quad 0 \leqslant \phi_j \leqslant 1, \quad 1 \leqslant j \leqslant k$$

$$\sum_{j=1}^{k} p_{m+1j} \phi_j \to \text{Maximum}$$

has a solution $\phi^* = (\phi_1^*, \dots, \phi_k^*)$ which also satisfies

$$\sum_{j=1}^{k} p_{ij} \phi_j^* \leqslant \alpha, \quad 1 \leqslant i \leqslant m, \tag{4.8}$$

then ϕ^* is a most powerful test for the problem $(\alpha, \Gamma_0, \Gamma - \Gamma_0)$. This can be easily seen.

Such a distribution $(u_1,\dots,u_m)$ is called *least favorable* (see the definition on p. 182). This terminology is motivated by the following considerations: Let $(v_1,\dots,v_m)$ be any probability distribution on Γ_0 and define $\mathfrak{P}_v$ in a similar manner to $\mathfrak{P}_u$. Let ϕ^* be a most powerful test for the problem $(\alpha, \Gamma_0, \Gamma - \Gamma_0)$. Then the power of a most powerful test for the problem $(\alpha, \mathfrak{P}_v, P_{m+1})$ is greater or equal to $\sum_{j=1}^{k} p_{m+1j} \phi_j^*$. This follows from (4.8) which implies that ϕ^* is a test for $(\alpha, \mathfrak{P}_v, P_{m+1})$.

We now will treat the ideas presented in the above example from a more general point of view.[21] We start with the following

Definition. A test ϕ_0 for the problem $(\alpha, \Gamma_0, \Gamma - \Gamma_0)$ is called a *maximintest* if

$$\inf_{\gamma \in \Gamma - \Gamma_0} g_{\phi_0}(\gamma) = \sup_{\phi \in \Phi_\alpha} \inf_{\gamma \in \Gamma - \Gamma_0} g_\phi(\gamma) \tag{4.9}$$

where Φ_α is again the set of all tests for the problem $(\alpha, \Gamma_0, \Gamma - \Gamma_0)$.

If $\Gamma - \Gamma_0$ contains only one element, then a maximintest is obviously a most powerful test for the problem $(\alpha, \Gamma_0, \Gamma - \Gamma_0)$.

It can be proved that a maximintest always exists under very weak conditions:

Theorem 4.2. *Let P_Γ be a set of probability measures over $(R, \mathscr{S})$ dominated by a σ-finite measure μ. Let $0 \leqslant \alpha \leqslant 1$ and consider the test problem $(\alpha, \Gamma_0, \Gamma - \Gamma_0)$. For this problem there always exists a maximintest.*

Proof. Let f_γ, for every $\gamma \in \Gamma$, be the R.-N.-density of P_γ w.r.t. μ. Then a sequence $\{\phi_i\}$ exists with $\phi_i \in \Phi_\alpha$ such that

$$\lim_{i \to \infty} \inf_{\gamma \in \Gamma - \Gamma_0} g_{\phi_i}(\gamma) = \sup_{\phi \in \Phi_\alpha} \inf_{\gamma \in \Gamma - \Gamma_0} g_\phi(\gamma). \tag{4.10}$$

---

[21] We follow essentially a paper by O. Krafft und H. Witting, Z. Wahrscheinlichkeitstheorie und Verw. Gebiete 7, 289—302 (1967). See also V. Baumann, Z. Wahrscheinlichkeitstheorie und Verw. Gebiete 11, 41—60 (1968).

By Theorem XI there exists a ϕ_0 with $0 \leqslant \phi_0 \leqslant 1$ (μ-a.e.) and a subsequence $\{\phi_{i_k}\}$ from $\{\phi_i\}$ such that

$$\int_R \phi_{i_k} f_\gamma d\mu \to \int_R \phi_0 f_\gamma d\mu \tag{4.11}$$

for all $\gamma \in \Gamma$. Thus, ϕ_0 is a test. Since $\phi_{i_k} \in \Phi_\alpha$ we have from (4.11) $\int_R \phi_0 f_\gamma d\mu \leqslant \alpha$ for $\gamma \in \Gamma_0$, i. e.,

$$\phi_0 \in \Phi_\alpha . \tag{4.12}$$

Moreover, for every $\gamma_1 \in \Gamma - \Gamma_0$ and $k \geqslant 1$

$$\inf_{\gamma \in \Gamma - \Gamma_0} g_{\phi_{i_k}}(\gamma) \leqslant \int_R \phi_{i_k} f_{\gamma_1} d\mu .$$

This yields together with (4.10) and (4.11)

$$\sup_{\phi \in \Phi_\alpha} \inf_{\gamma \in \Gamma - \Gamma_0} g_\phi(\gamma) \leqslant \lim_{k \to \infty} \int_R \phi_{i_k} f_{\gamma_1} d\mu = \int_R \phi_0 f_{\gamma_1} d\mu$$

for every $\gamma_1 \in \Gamma - \Gamma_0$.

It follows that

$$\sup_{\phi \in \Phi_\alpha} \inf_{\gamma \in \Gamma - \Gamma_0} g_\phi(\gamma) \leqslant \inf_{\gamma \in \Gamma - \Gamma_0} g_{\phi_0}(\gamma) . \tag{4.13}$$

Because of (4.12) equality must hold in (4.13).

Theorem 4.2 is closely related to a certain generalized linear program: Let P_Γ, Γ_0, $\Gamma - \Gamma_0$, α, μ and f_γ have the same meaning as in that Theorem.[22] Let $\mathfrak{C}_0$ and $\mathfrak{C}_1$ be a σ-algebra of subsets of Γ_0 and of $\Gamma - \Gamma_0$, resp. Furthermore, suppose that the mapping $(\gamma, x) \to f_\gamma(x)$ is $\mathfrak{C}_0 \otimes \mathscr{S}$-measurable on $\Gamma_0 \times R$ and $\mathfrak{C}_1 \otimes \mathscr{S}$-measurable on $(\Gamma - \Gamma_0) \times R$. *This last assumption will be made throughout the remainder of this section.*[23]

We consider the following linear program which is obviously equivalent to the problem of finding a maximintest:

$$-\int_R f_\gamma \phi d\mu + s \leqslant 0, \qquad \gamma \in \Gamma - \Gamma_0 \tag{4.14a}$$

$$\int_R f_\gamma \phi d\mu \leqslant \alpha, \qquad \gamma \in \Gamma_0 \tag{4.14b}$$

$$0 \leqslant \phi(x) \leqslant 1, \qquad x \in R \tag{4.14c}$$

$$s \geqslant 0 \tag{4.14d}$$

$$s \to \text{Maximum}.$$

Here ϕ is, of course, supposed to be $\mathscr{S}$-measurable.

---

[22] To avoid trivial complications we now assume $0 < \alpha < 1$.

[23] Necessary and sufficient conditions for the existence of product measurable densities can be found in J. Pfanzagl, Sankhyā, Ser. A. 31, 13—18 (1969).

This problem corresponds exactly to the linear program treated on p. 176 where Γ_0 is finite, $\Gamma-\Gamma_0$ contains only one element and R is a finite sample space.

Again we have a dual program which is determined by the following conditions

$$\int_{\Gamma_0} f_\gamma(x)\,d\lambda(\gamma) - \int_{\Gamma-\Gamma_0} f_\gamma(x)\,d\nu(\gamma) + v(x) \geqslant 0, \qquad \mu\text{-a.e.} \tag{4.15a}$$

$$\nu(\Gamma-\Gamma_0) \geqslant 1 \tag{4.15b}$$

$$v(x) \geqslant 0, \qquad \mu\text{-a.e.} \tag{4.15c}$$

$$\alpha\lambda(\Gamma_0) + \int_R v\,d\mu \to \text{Minimum}.$$

where $\lambda \not\equiv 0$ is a bounded measure over $(\Gamma_0, \mathfrak{C}_0)$, ν a bounded measure over $(\Gamma-\Gamma_0, \mathfrak{C}_1)$ and v a $\mathscr{S}$-measurable function on R. Note that the existence of the integrals listed in (4.15a) follows from the measurability assumptions concerning the mapping $(\gamma, x) \to f_\gamma(x)$ and the second part of Theorem X.

Pairs (ϕ, s) or triples (λ, ν, v) which satisfy the conditions (4.14) and (4.15), resp., are called *feasible solutions.*

We now show

Lemma 4.1. *Let* (ϕ, s) *and* (λ, ν, v) *be feasible solutions. Then*

$$s \leqslant \alpha\lambda(\Gamma_0) + \int_R v\,d\mu. \tag{4.16}$$

If (ϕ_0, s_0) *and* (λ_0, ν_0, v_0) *are feasible solutions such that equality holds in* (4.16) *then* (ϕ_0, s_0) *is a solution of the linear program,* (λ_0, ν_0, v_0) *of the dual program, i.e.,* $s_0 = \max s$ *and* $\alpha\lambda_0(\Gamma_0) + \int_R v_0\,d\mu = $ *Minimum.*

Proof. (4.14d), (4.15b) and (4.14a) imply

$$\begin{aligned} s &\leqslant \nu(\Gamma-\Gamma_0)s \leqslant \nu(\Gamma-\Gamma_0) \inf_{\gamma\in\Gamma-\Gamma_0} \int_R f_\gamma \phi\,d\mu \\ &\leqslant \int_{\Gamma-\Gamma_0} \int_R f_\gamma(x)\phi(x)\,d\mu(x)\,d\nu(\gamma). \end{aligned} \tag{4.17}$$

Further, by Theorem X and (4.15a) we obtain

$$\begin{aligned} &\int_{\Gamma-\Gamma_0} \int_R f_\gamma(x)\phi(x)\,d\mu(x)\,d\nu(\gamma) \\ &= \int_R \int_{\Gamma-\Gamma_0} f_\gamma(x)\,d\nu(\gamma)\,\phi(x)\,d\mu(x) \end{aligned}$$

$$\leqslant \int_R \left[\int_{\Gamma_0} f_\gamma(x) d\lambda(\gamma) + v(x) \right] \phi(x) d\mu(x) \tag{4.18}$$
$$= \int_{\Gamma_0} \int_R f_\gamma(x) \phi(x) d\mu(x) d\nu(\gamma) + \int_R v \phi \, d\mu$$
$$\leqslant \alpha \lambda(\Gamma_0) + \int_R v \, d\mu .$$

(4.17) together with (4.18) yields (4.16).

The second claim of Lemma 4.1 is an immediate consequence of (4.16) and the definition of a maximum and a minimum.

We are now going to modify somewhat the conditions (4.15). If (λ, ν, v) is a feasible solution of the dual program then $(\lambda, \nu, \bar{v})$ is also such a solution where

$$\begin{aligned} \bar{v}(x) &= \max\left(0, \int_{\Gamma - \Gamma_0} f_\gamma(x) d\nu(\gamma) - \int_{\Gamma_0} f_\gamma(x) d\lambda(\gamma)\right) \\ &= \left(\int_{\Gamma - \Gamma_0} f_\gamma(x) d\nu(\gamma) - \int_{\Gamma_0} f_\gamma(x) d\lambda(\gamma)\right)^+ . \end{aligned} \tag{4.19}$$

Moreover $\alpha \lambda(\Gamma_0) + \int_R \bar{v} \, d\mu \leqslant \alpha \lambda(\Gamma_0) + \int_R v \, d\mu.$

These two statements follow from (4.15a) and (4.15c). Therefore, the dual program admits the following formulation: Solve the problem

$$\alpha \lambda(\Gamma_0) + \int_R \left(\int_{\Gamma - \Gamma_0} f_\gamma(x) d\nu(\gamma) - \int_{\Gamma_0} f_\gamma(x) d\lambda(\gamma)\right)^+ d\mu(x) \to \text{Minimum}$$

where λ is a bounded measure over $(\Gamma_0, \mathfrak{C}_0)$ and ν one over $(\Gamma - \Gamma_0, \mathfrak{C}_1)$ (with $\nu(\Gamma - \Gamma_0) \geqslant 1$). This formulation shows that one can assume without loss of generality that ν is a probability measure, i.e.,

$$\nu(\Gamma - \Gamma_0) = 1 . \tag{4.20}$$

If one writes $\lambda(\Gamma_0) = k$ then the following formulation of the dual program is obtained which is equivalent to the formulation on p. 180.

Let Λ be the set of all probability measures λ over $(\Gamma_0, \mathfrak{C}_0)$, Ξ the set of all probability measures ν over $(\Gamma - \Gamma_0, \mathfrak{C}_1)$ and k an arbitrary positive number. Let

$$h(k, \lambda, \nu) = \alpha k + \int_R \left(\int_{\Gamma - \Gamma_0} f_\gamma(x) d\nu(\gamma) - k \int_{\Gamma_0} f_\gamma(x) d\lambda(\gamma)\right)^+ d\mu(x). \tag{4.21}$$

Find a triple (k_0, λ_0, ν_0) where $k_0 > 0$, $\lambda_0 \in \Lambda$, $\nu_0 \in \Xi$ such that

$$\inf_{k>0, \lambda \in \Lambda, \nu \in \Xi} h(k, \lambda, \nu) = h(k_0, \lambda_0, \nu_0) .$$

Remark. It is very easy to see that $x \to \int_{\Gamma - \Gamma_0} f_\gamma(x) d\lambda(\gamma)$ and $x \to \int_{\Gamma_0} f_\gamma(x) d\lambda(\gamma)$ are R.-N.-densities of probability measures over $(R, \mathscr{S})$ w.r.t. μ. Remembering the formula (3.17) one recognizes at once that the right side of (4.21) defines the power of a most powerful test.

A more precise result is given in the following

Lemma 4.2. *The equality*

$$\inf_{\gamma\in\Gamma-\Gamma_0} g_{\phi_0}(\gamma)=h(k_0,\lambda_0,\nu_0) \tag{4.22}$$

where $\phi_0\in\Phi_\alpha$ [24] *holds iff*

$$\phi_0(x)=\begin{cases}1 & \int_{\Gamma_1} f_\gamma(x)\,d\nu_0(\gamma)>k_0\int_{\Gamma_0} f_\gamma(x)\,d\lambda_0(\gamma)\\ 0 & \int_{\Gamma_1} f_\gamma(x)\,d\nu_0(\gamma)<k_0\int_{\Gamma_0} f_\gamma(x)\,d\lambda_0(\gamma)\end{cases} \tag{4.23}$$

$$E(\phi_0;\gamma)=\inf_{\gamma_1\in\Gamma-\Gamma_0} E(\phi_0;\gamma_1),\qquad \gamma\in\Gamma_1\ \nu_0\text{-a.e.} \tag{4.24}$$

and

$$E(\phi_0;\gamma)=\alpha,\qquad \gamma\in\Gamma_0\ \lambda_0\text{-a.e.} \tag{4.25}$$

Proof. Suppose that (4.22) holds. Then the equality sign only appears in the strings of inequalities (4.17) and (4.18) if we put $\lambda=k_0\lambda_0$, $\nu=\nu_0$, $\phi=\phi_0$ and $v=v_0$ where

$$v_0(x)=\Big(\int_{\Gamma-\Gamma_0} f_\gamma(x)\,d\nu_0(\gamma)-k_0\int_{\Gamma_0} f_\gamma(x)\,d\lambda_0(\gamma)\Big)^+,\qquad x\in R.$$

At first this implies, together with (4.20), the relation (4.24). Next, it follows together with (4.15a) that

$$\int_{\Gamma-\Gamma_0} f_\gamma(x)\,d\nu_0(\gamma)=k_0\int_{\Gamma_0} f_\gamma(x)\,d\lambda_0(\gamma)+v_0(x)\ \mu\text{-a.e.}$$

whenever $\phi_0(x)>0$. Finally the last equality in (4.18) yields the relation (4.25) as well as $\phi_0(x)=1$ (μ-a.e.) whenever $v_0(x)>0$ (see footnote [22]). An easy argument now leads to (4.23). Conversely it is easy to see that the conditions (4.23), (4.24) and (4.25) together with (4.20) imply equality in (4.18) and (4.17), i. e. (4.22) holds.

Lemma 4.2 and the following Lemma 4.3 suggest the following

Definition. *A pair of probability measures* (λ_0,ν_0) where λ_0 is defined on $(\Gamma_0,\mathfrak{C}_0)$ and ν_0 on $(\Gamma-\Gamma_0,\mathfrak{C}_1)$ is called *least favorable* if

$$\sup_{\phi\in\Phi_\alpha}\ \inf_{\gamma\in\Gamma-\Gamma_0} g_\phi(\gamma)=h(k_0,\lambda_0,\nu_0)$$

for some real k_0.

We introduce the following notation: Let Ω be any set of indices and let P_Ω be a family of probability measures on $(R,\mathscr{S})$ dominated by a σ-finite measure μ. Let f_ρ be the R.-N.-density of P_ρ w.r.t. μ, $\rho\in\Omega$. Let $\mathfrak{C}$ be a σ-algebra of subsets of Ω. Assume that $(\rho,x)\to f_\rho(x)$ is an $(\mathfrak{C}\times\mathscr{S})$-

[24] Φ_α is defined on p. 178.

measurable mapping. Then let $K(\Omega)$ denote the set of all probability measures on $(R,\mathscr{S})$, of the form $A\to\int\limits_{A}\int\limits_{\Omega} f_\rho(x)\,d\eta(\rho)\,d\mu(x)$, $A\in\mathscr{S}$ where η is any probability measure on $(\Omega,\mathfrak{C})$.

Using this terminology we have

Lemma 4.3. *Suppose $P_0\in K(\Gamma_0)$ and $P_1\in K(\Gamma-\Gamma_0)$ where $K(\Gamma_0)$ and $K(\Gamma-\Gamma_0)$ are defined by means of Λ and Ξ resp. Let ϕ^* be a most powerful test for the problem (α, P_0, P_1). Then $g_{\phi^*}(P_1)\geqslant \sup_{\phi\in\Phi_\alpha}\ \inf_{\gamma\in\Gamma-\Gamma_0} g_\phi(\gamma)$.*

Proof. This follows immediately from Lemma 4.2 and the remark on p. 181.

We now formulate a special case of Lemma 4.2 which concerns most powerful tests as

Theorem 4.3. *Suppose $\gamma_1\in\Gamma$ and let be $\Gamma_0=\Gamma-\{\gamma_1\}$. Let be $P_0\in K(\Gamma_0)$ and assume that ϕ^* is a most powerful test for the problem $(\alpha, P_0, P_{\gamma_1})$. Then ϕ^* is also most powerful for $(\alpha, \Gamma_0, \{\gamma_1\})$, provided $E(\phi^*;\gamma)\leqslant\alpha$ for all $\gamma\in\Gamma_0$ holds.*

We now give an example for Theorem 4.3 which will also be relevant to the theory of most powerful tests for composite hypotheses. Assume we have a population distributed according to $N(a,1)$. On the basis of a sample of size n we want to construct a most powerful test for the composite null hypothesis $\Gamma_0=\{a_{-1},a_0\}$ versus the alternative $a=a_1$, $a_{-1}<a_0<a_1$. We thus have a problem of the form $(\alpha, P_{\Gamma_0}, P_{a_1})$, where the P_{a_i} for $i=-1,0,1$, are probability measures over the sample space $(R_n,\mathfrak{B}_n)$ defined by the densities

$$(x_1,\dots,x_n)\to(1/\sqrt{2\pi})^n\, e^{-\left[\sum\limits_{j=1}^{n}(x_j-a_i)^2\right]/2}.$$

We now choose in Γ_0 the probability measure λ given by $\lambda(a_0)=1$, $\lambda(a_{-1})=0$. We thus replace the problem $(\alpha, P_{\Gamma_0}, P_{a_1})$ by $(\alpha, P_{a_0}, P_{a_1})$. According to Theorem 3.3, a most powerful test ϕ^* for the problem $(\alpha, P_{a_0}, P_{a_1})$ is given by

$$\phi^*(x)=\begin{cases}1 & \text{for } e^{-\left[\sum\limits_{j=1}^{n}(x_j-a_1)^2\right]/2}\geqslant k\,e^{-\left[\sum\limits_{j=1}^{n}(x_j-a_0)^2\right]/2}\\ 0 & \text{for } e^{-\left[\sum\limits_{j=1}^{n}(x_j-a_1)^2\right]/2}< k\,e^{-\left[\sum\limits_{j=1}^{n}(x_j-a_0)^2\right]/2}\end{cases}$$

with suitable $k\geqslant0$. Note that

$$\left\{x: e^{-\left[\sum\limits_{j=1}^{n}(x_j-a_1)^2\right]/2}=k\,e^{-\left[\sum\limits_{j=1}^{n}(x_j-a_0)^2\right]/2}\right\}$$

is a null set. ϕ^* is thus defined by a critical region M, which can be given in the form

$$M=\left\{x: \sum_{j=1}^{n}(x_j-a_0)^2-\sum_{j=1}^{n}(x_j-a_1)^2\geqslant 2\log k\right\}. \tag{4.26}$$

By assumption,

$$a_1>a_0\,. \tag{4.27}$$

In this case one also gets for M

$$M=\left\{x:(\bar{x}-a_0)\sqrt{n} \geqslant \frac{\log k}{\sqrt{n}(a_1-a_0)} + (a_1-a_0)\sqrt{n}/2\right\}. \tag{4.28}$$

The determination of k for a given level of significance α, $0 \leqslant \alpha \leqslant 1$ is very easy. If $\xi_1, \ldots, \xi_n$ are independently $N(a_0, 1)$-distributed, then by II, Theorem 3.2 $(\bar{\xi}-a_0)\sqrt{n}$ is $N(0,1)$-distributed. Determining κ'_α uniquely from

$$\frac{1}{\sqrt{2\pi}} \int_{\kappa'_\alpha}^{\infty} e^{-t^2/2}\, dt = \alpha \tag{4.29}$$

we then have by (4.28) $P_{a_0}(M)=\alpha$, provided k satisfies

$$\frac{\log k}{\sqrt{n}(a_1-a_0)} + (a_1-a_0)\sqrt{n}/2 = \kappa'_\alpha .$$

M can also be defined by

$$\{x:(\bar{x}-a_0)\sqrt{n} \geqslant \kappa'_\alpha\} \tag{4.30}$$

instead of using (4.28).

We now show that

$$P_{a_{-1}}(M) \leqslant \alpha. \tag{4.31}$$

M is obviously also defined by $\{x: (\bar{x}-a_{-1})\sqrt{n} \geqslant \kappa'_\alpha + (a_0-a_{-1})\sqrt{n}\}$ and since $a_0 - a_{-1} > 0$, (4.31) is proved. Thus, according to Theorem 4.3 M defines a most powerful test for $(\alpha, P_{\Gamma_0}, P_{a_1})$. Note that this holds for any $a_{-1} < a_0$. It is crucial for our further discussions that the M defined by (4.30) no longer depends on a_1. Hence we have the following result: Let P_a, $a \in R_1$, be the probability measure over $(R_n, \mathfrak{B}_n)$ whose density is given by

$$x \to \frac{1}{(\sqrt{2\pi})^n} e^{-\left[\sum_{i=1}^{n} (x_i-a)^2\right]/2}. \tag{4.32}$$

Let $\Gamma_1 = \{a: a \geqslant a_0\}$. Then the critical region M given by (4.30) defines a most powerful test in $\Gamma_1 - \{a_0\}$ for the problem $(\alpha, \{a_0\}, \Gamma_1 - \{a_0\})$. From (4.31) it also follows that M even defines a most powerful test in $\Gamma_1 - a_0$ for the problem

$$(\alpha, (R_1 - \Gamma_1) \cup \{a_0\}, \Gamma_1 - \{a_0\})\,[25].$$

If we assume that

$$a_1 < a_0 \tag{4.33}$$

then from (4.26) it follows that for the problem $(\alpha, P_{a_0}, P_{a_1})$ with (4.33) a most powerful critical region is given by

$$M_1=\left\{x:(\bar{x}-a_0)\sqrt{n} \leqslant \frac{\log k}{\sqrt{n}(a_1-a_0)} - (a_0-a_1)\sqrt{n}/2\right\}.$$

From (4.29) it follows that

$$(1/\sqrt{2\pi}) \cdot \int_{-\infty}^{-\kappa'_\alpha} e^{-t^2/2}\, dt = \alpha$$

[25] See p. 185ff.

and hence k must now be calculated from

$$\frac{\log k}{\sqrt{n}(a_1-a_0)} - (a_0-a_1)\sqrt{n}/2 = -\kappa'_\alpha .$$

The critical region M_1 is thus given by

$$\{x:(\bar{x}-a_0)\sqrt{n} \leqslant -\kappa'_\alpha\} \tag{4.34}$$

and is again independent of a_1. (4.34) then yields a most powerful test in $R_1-\Gamma_1$ for the problem $(\alpha,\{a_0\},R_1-\Gamma_1)$. Naturally, this result can also be extended by applying Theorem 4.3.

If $0<\alpha<1$, then the critical regions M and M_1 defined by (4.30) and (4.34), resp., are not identical.

We now consider all of R_1 as the set of admissible hypotheses a and $\{a_0\}$ as null hypothesis. We will see that M and M_1 in $R_1-\{a_0\}$ are no longer most powerful for the problem $(\alpha,\{a_0\},R_1-\{a_0\})$, $0<\alpha<1$.[26]

The power function g over R_1 for c_M is obtained in exactly the same way as the inequality (4.31). For each $a\in R_1$, M can also be defined by

$$M=\{x:(\bar{x}-a)\sqrt{n} \geqslant \kappa'_\alpha+(a_0-a)\sqrt{n}\}$$

instead of using (4.30). Thus, with $k(a)=\kappa'_\alpha+(a_0-a)\sqrt{n}$ one gets

$$g(a) = \frac{1}{\sqrt{2\pi}}\int_{k(a)}^{\infty} e^{-t^2/2}\,dt, \qquad a\in R_1 .$$

For the power function g_1 of M_1 one gets quite analogously for $a\in R_1$ with $-\kappa'_\alpha+(a_0-a)\sqrt{n}=k_1(a)$

$$g_1(a) = \frac{1}{\sqrt{2\pi}}\int_{-\infty}^{k_1(a)} e^{-t^2/2}\,dt = \frac{1}{\sqrt{2\pi}}\int_{-k_1(a)}^{\infty} e^{-t^2/2}\,dt .$$

Now, $k(a)>-k_1(a)$ for $a<a_0$ so that also $g(a)<g_1(a)$. Hence c_M cannot be most powerful in $a<a_0$. In the same way one sees that c_{M_1} for $a>a_0$ cannot be most powerful. Thus, for the problem $(\alpha,\{a_0\},R_1-\{a_0\})$ there can exist no most powerful test, since the most powerful test in Γ_1-a_0 is given by c_M and that in $R_1-\Gamma_1$ by c_{M_1}.

Thus there exist most powerful tests for the "*one-sided*" *alternatives* $a<a_0$ or $a>a_0$ but not for the "two-sided" alternative $a\neq a_0$. The test defined by (1.4) is thus not most powerful for this *two-sided alternative*. Its application must thus be "justified" in some other way. We will return to this point in **6**.

5. Hypotheses with monotone likelihood ratios. A more detailed analysis of the example considered at the end of **4** shows that the existence of a most powerful test for a "one-sided" hypothesis is essentially connected with the fact that the quotient

$$\frac{e^{(-1/2)\sum_{j=1}^{n}(x_j-a_1)^2}}{e^{(-1/2)\sum_{j=1}^{n}(x_j-a_0)^2}} = e^{-n\bar{x}(a_0-a_1)-(a_1^2-a_0^2)/2}$$

[26] Essentially, the following considerations represent only an illustration of the uniqueness claim of Theorem 3.1.

dependes on $(x_1, \ldots, x_n)$ for $a_1 \neq a_0$ only via $\bar{x}$ and, indeed, monotonically.

This remark can be generalized.

We first give the following

Definition. Let $\Gamma \subseteq R_1$ and P_Γ a set of probability measures over $(R, \mathcal{S})$ dominated by a σ-finite measure μ. For each $\gamma \in \Gamma$, we denote the R.-N.-density by f_γ. We will say that P_Γ possesses *monotone likelihood ratios* if there exists an $\mathcal{S}$-measurable map T from R into R_1 and for each pair γ_0, γ_1 from Γ with $\gamma_0 < \gamma_1$ a *non-decreasing function* H_{γ_0, γ_1} (which can also assume the value ∞), such that

$$f_{\gamma_1}/f_{\gamma_0} = H_{\gamma_0, \gamma_1} \circ T, \quad \tfrac{1}{2}(P_{\gamma_0} + P_{\gamma_1})\text{-a.e.} \tag{5.1}$$

For the sake of simplicity we now restrict ourselves to the case $\Gamma = R_1$ and show

Theorem 5.1. *Let P_{R_1} be a set of probability measures with monotone likelihood ratios, γ_0 an arbitrary real number and $I_{\gamma_0} = (-\infty, \gamma_0]$. For each α with $0 \leqslant \alpha \leqslant 1$, there exists a most powerful test ϕ^* in $R_1 - I_{\gamma_0}$ for the problem $(\alpha, I_{\gamma_0}, R_1 - I_{\gamma_0})$. Moreover, for the power function g of ϕ^* we have*

$$g(\gamma') \leqslant g(\gamma'') \tag{5.2}$$

for $\gamma' < \gamma''$. Equality holds in (5.2) *iff $g(\gamma') = 1$ or $g(\gamma'') = 0$.*

We first give a theorem which will often be applied in the sequel.

Theorem 5.2.[27] *Let P_Γ be given over $(R, \mathcal{S})$ and be dominated by a σ-finite measure. Then there also exists a measure λ dominating P_Γ such that $P_\gamma(N) = 0$ for all $\gamma \in \Gamma$ implies $\lambda(N) = 0$. The set P_Γ and λ will be called equivalent.*

Proof. If Γ is a countable set, we have already proved the existence of such a measure in Theorem XIII. Theorem 5.2 is thus only interesting for Γ uncountable. Let f_γ for each $\gamma \in \Gamma$ be the R.-N.-density of P_γ w.r.t. μ and

$$S_\gamma = \{x : f_\gamma(x) > 0\}. \tag{5.3}$$

Further let

$$\mathfrak{M} = \left\{M : M \in \mathcal{S},\ M \subseteq \bigcup_{i=1} S_{\gamma_i} \text{ for some sequence } \{\gamma_i\} \in \Gamma\right\}. \tag{5.4}$$

First assume that μ is bounded. Let $s = \sup_{M \in \mathfrak{M}} \mu(M)$. By the definition of the supremum, there exists a sequence $\{M_i\}$ with $M_i \in \mathfrak{M}$ such that

[27] P.R. Halmos and L.J. Savage, Ann. Math. Statist. 20, 225—241 (1949).

$s = \lim_{i\to\infty} \mu(M_i)$. However, by definition there exists for each M_i a countably infinite number of $S_{\gamma_j}^{(i)}$, $j=1,2,\dots$ such that $M_i \subseteq \bigcup_{j=1}^{\infty} S_{\gamma_j}^{(i)}$. We order the countable set of all $S_{\gamma_j}^{(i)}$ $i,j=1,2,\dots$ in some way into a sequence $S_1, S_2, \dots$. Then also $U = \bigcup_{i=1}^{\infty} S_i \in \mathfrak{M}$ and since $U \supseteq M_i$, $1 \leqslant i \leqslant \infty$ we also have

$$s = \mu(U). \tag{5.5}$$

According to definition (5.3), there corresponds to each S_i a measure $P_i \in P_\Gamma$; however, the P_i are not necessarily different from one another. Let

$$\sum_{i=1}^{\infty} g_i P_i = \lambda, \qquad g_i > 0, \qquad \sum_{i=1}^{\infty} g_i = 1. \tag{5.6}$$

We now claim that this λ has the properties of the theorem. It is naturally trivial that $P_\gamma(N)=0$ for all $\gamma \in \Gamma$ implies $\lambda(N)=0$. Conversely, $\lambda(N)=0$ implies

$$P_i(N)=0, \qquad i=1,2,\dots. \tag{5.7}$$

Let P_{γ_0} be an arbitrary measure $\in P_\Gamma$. Since $P_{\gamma_0}(N - S_{\gamma_0})=0$ we can assume that $N \subseteq S_{\gamma_0}$. If we had $P_{\gamma_0}(N-U)>0$ then $\mu(N-U)>0$ would also hold which would imply that $\mu(U \cup (N-U)) = \mu(U) + \mu(N-U) > \mu(U)$ with $U \cup (N-U) \subseteq \bigcup_{i=1}^{\infty} S_i \cup S_{\gamma_0}$ and this would contradict (5.5). Hence $P_{\gamma_0}(N-U)=0$ and we need only investigate $P_{\gamma_0}(N \cap U)$. It is sufficient to show that $\mu(N \cap U)=0$ since this implies $P_{\gamma_0}(N \cap U)=0$. Now,

$$\mu(N \cap U) = \mu\left(N \cap \bigcup_{i=1}^{\infty} S_i\right) \leqslant \sum_{i=1}^{\infty} \mu(N \cap S_i). \tag{5.8}$$

But for each $i=1,2,\dots$ $\mu(N \cap S_i)=0$. Indeed, if this were not true then a contradiction to (5.7), namely $P_i(N \cap S_i)>0$ would follow from the definition of S_i. Hence $\mu(N \cap U)=0$ from (5.8) so that $P_{\gamma_0}(N \cap U)=0$. From $\lambda(N)=0$ we thus have for arbitrary $\gamma_0 \in \Gamma$, $P_{\gamma_0}(N)=0$ which was to be proved. We note that λ is a probability measure.

Now, the more general statement of the theorem follows easily taking into account the definition of a σ-finite measure (see p. 5).

We proceed to the *proof of Theorem* 5.1. From the assumptions and Theorem 5.2 follows the existence of a probability measure λ which is equivalent to P_Γ. Let $M_k = \{x : T(x) > k\}$ for $-\infty \leqslant k \leqslant \infty$. Exactly as in the proof of Theorem 3.1 I F, one sees that the map ψ given by $k \to \lambda(M_k)$ is nonincreasing and right continuous. Thus, for each α with $0 \leqslant \alpha \leqslant 1$, one can find a $k(\alpha)$ and a $c(\alpha)$ with $0 \leqslant c(\alpha) \leqslant 1$ such that for the test defined by

$$\phi_\alpha(x) = \begin{cases} 1 & T(x) > k(\alpha) \\ c(\alpha) & T(x) = k(\alpha) \\ 0 & T(x) < k(\alpha) \end{cases}$$

we have $\int_R \phi_\alpha d\lambda = \alpha$.

The mapping $\alpha \to \phi_\alpha(x)$ is continuous at every α in $0 < \alpha < 1$ for each $x \in R$ up to a null set.

Indeed, let $\{x: T(x) \geqslant k\} = M_{k+}$. If for an α in $0 < \alpha < 1$ we determine $k = k(\alpha)$ so that $\psi(k-0) \geqslant \alpha \geqslant \psi(k)$ then for $x \in M_k$ or $x \in R - M_{k+}$ we also have $\phi_\alpha(x) = \phi_{\alpha'}(x)$ for all $\alpha' \in (\psi(k), \psi(k-0))$. However, if $x \in M_{k+} - M_k$ and $\lambda(M_{k+} - M_k) \neq 0$ then $|\phi_\alpha(x) - \phi_{\alpha'}(x)|$ is arbitrarily small if $|\alpha - \alpha'|$ is sufficiently small. If $\alpha = \psi(k-0)$ then $c(\alpha)$ must equal 1. An easy modification of the following argument then yields $\phi_{\alpha'}(x) = \phi_\alpha(x)$ if $\alpha' > \alpha$ and α' is sufficiently close to α.

If $\alpha = \psi(k)$ and k is a point of continuity of ψ then one can assume without loss of generality that ψ decreases through k. Then sufficiently small changes in k correspond to sufficiently small changes in α. For $x \in M_k$, resp., $x \in R - M_{k+}$ we thus have $\phi_\alpha(x) = \phi_{\alpha'}(x)$ if α' is close enough to α. Since k is a point of continuity of ψ, the set $M_{k+} - M_k$ has λ-measure 0. But since ψ is continuous from the right, the continuity of $\alpha \to \phi_\alpha(x)$ as given in the above statement is shown.

From $\lambda(N) = 0$ we now have $P_\gamma(N) = 0$ for each $\gamma \in R_1$ and since $|\phi_\alpha(x)| \leqslant 1$ for all $x \in R$ and all α, $0 \leqslant \alpha \leqslant 1$ the map $\alpha \to E(\phi_\alpha; \gamma)$ is continuous for each $\gamma \in R_1$ (cf. Theorem VII). But now for each γ, $E(\phi_\alpha; \gamma) \to 0$ for $\alpha \to 0$ and $E(\phi_\alpha; \gamma) \to 1$ for $\alpha \to 1$ and also $E(\phi_0; \gamma) = 0$ and $E(\phi_1; \gamma) = 1$. On the basis of Bolzano's Theorem for continuous functions one can thus choose an α for given $\gamma_0 \in R_1$ and for each α' with $0 \leqslant \alpha' \leqslant 1$, such that the test ϕ_α defined above fulfills $E(\phi_\alpha; \gamma_0) = \alpha'$. Hence, for each α with $0 \leqslant \alpha \leqslant 1$, there exists a suitable $k \geqslant -\infty$ and c with $0 \leqslant c \leqslant 1$ such that the test ϕ^* defined by

$$\phi^*(x) = \begin{cases} 1 & T(x) > k \\ c & T(x) = k \\ 0 & T(x) < k \end{cases} \tag{5.9}$$

fulfills the condition

$$E(\phi^*; \gamma_0) = \alpha . \tag{5.10}$$

In case $\alpha = 0$ we choose $c = 1$ in (5.9) if $H_{\gamma_0, \gamma_1}(k) = \infty$, otherwise $c = 0$.

By assumption H_{γ_0,γ_1} is nondecreasing for $\gamma_0<\gamma_1$. Hence,

$$\{x: H_{\gamma_0,\gamma_1}(T(x))>H_{\gamma_0,\gamma_1}(k)\}\subseteq\{x: T(x)>k\}$$

and likewise

$$\{x: H_{\gamma_0,\gamma_1}(T(x))<H_{\gamma_0,\gamma_1}(k)\}\subseteq\{x: T(x)<k\}.$$

Therefore, ϕ^* fulfills, in particular, the condition

$$\phi^*(x)=\begin{cases}1 & H_{\gamma_0,\gamma_1}(T(x))>H_{\gamma_0,\gamma_1}(k)\\ 0 & H_{\gamma_0,\gamma_1}(T(x))<H_{\gamma_0,\gamma_1}(k)\end{cases}. \tag{5.11}$$

Moreover, ϕ^* satisfies (5.10). Taking (5.1), (5.10) and (5.11) into account and applying Theorem 3.1 II F. one sees that ϕ^* is a most powerful test for the problem $(\alpha,\{\gamma_0\},\{\gamma_1\})$.[28] ϕ^* is, however, also defined by (5.9) and hence independent of γ_1. Hence, we have shown that ϕ^* is a most powerful test in $R_1-I_{\gamma_0}$ for the problem $(\alpha,\{\gamma_0\},R_1-I_{\gamma_0})$. We now have to prove (5.2). Let γ' and γ'' be chosen arbitrarily and $\gamma'<\gamma''$. Let

$$E(\phi^*;\gamma')=\beta. \tag{5.12}$$

If $\beta=0$, we have nothing to prove. However, if $\beta>0$, ϕ^* is most powerful for the problem $(\beta,\{\gamma'\},\{\gamma''\})$. Indeed, we have just shown (with other notation) that a most powerful test for $(\beta,\{\gamma'\},\{\gamma''\})$ is given by one of the form (5.9). Since, however, ϕ^* possesses by (5.12) precisely the prescribed level of significance β for the null hypothesis γ', ϕ^* is most powerful for $(\beta,\{\gamma'\},\{\gamma''\})$ and we then have $g(\gamma')\leqslant g(\gamma'')$ for the power function. If $\beta<1$, then the "$<$" sign holds as we saw on p. 174. This proves (5.2). In particular, $E(\phi^*;\gamma)\leqslant\alpha$ for all $\gamma\in I_{\gamma_0}$ so that ϕ^* is most powerful for the problem $(\alpha,I_{\gamma_0},R_1-I_{\gamma_0})$ which completely proves Theorem 5.1.

We assumed in Theorem 5.1 that the set of admissible parameters is R_1. The arguments do not change if the set of admissible parameters is a subset of R_1. (5.9) gives rise to the following

Definition. Let T be an $\mathscr{S}$-measurable function. If k and c are real numbers with $0\leqslant c\leqslant 1$, and if the test ϕ is defined according to

$$\phi(x)=\begin{cases}1 & T(x)>k\\ c & T(x)=k\\ 0 & T(x)<k\end{cases}$$

then T is called a *test statistic* for ϕ.

[28] To justify this conclusion also in the case $\alpha=0$ and $c=1$, one must define $0\cdot\infty=0$.

The most important example of a set of probability distributions with monotone likelihood ratio are certain *exponential distributions*. We define them by means of their R.-N.-densities w.r.t. a σ-finite measure μ over $(R, \mathscr{S})$: Let T and $h \geqslant 0$ be defined over R and $\mathscr{S}$-measurable. Let Q be a nonincreasing or nondecreasing function defined over a set $\Gamma \subseteq R_1$. Then let for each $\gamma \in \Gamma$

$$f_\gamma(x) = C(\gamma)\, e^{Q(\gamma)T(x)} h(x), \qquad x \in R, \tag{5.13}$$

where $C(\gamma)$ is positive for each $\gamma \in \Gamma$ and must naturally be so determined that

$$C(\gamma) \int_R e^{Q(\gamma)T(x)} h(x)\, d\mu(x) = 1 .$$

The representation of the densities in the form (5.13) possibly might not correspond to our basic assumption (see p. 160) that the relation between the set of parameters and the corresponding set of probability measures be one-to-one. Indeed, if $Q(\gamma_1) = Q(\gamma_2)$ holds for $\gamma_1 \neq \gamma_2$ then obviously (5.13) defines the same density for γ_1 and γ_2 so that in this case also $C(\gamma_1) = C(\gamma_2)$. It is thus expedient to transform the parameter and replace Γ by $Q(\Gamma) = \Gamma'$. We assume that Γ' contains at least two different elements in order to exclude trivialities. The densities (5.13) are now written for each $\gamma' \in \Gamma'$ as

$$f_{\gamma'}(x) = C'(\gamma')\, e^{\gamma' T(x)} h(x), \qquad x \in R, \tag{5.14}$$

where the meaning of $C'(\gamma')$ is obvious. But now $f_{\gamma'_1} \neq f_{\gamma'_2}$ holds for $\gamma'_1 \neq \gamma'_2$, $\gamma'_1, \gamma'_2 \in \Gamma'$. It is easy to see that one can take Γ' as convex.[29] Otherwise let $\gamma'_1, \gamma'_2 \in \Gamma'$ $\gamma' \notin \Gamma'$ and $\gamma'_1 < \gamma' < \gamma'_2$. There always exists a β with $0 < \beta < 1$, such that $\gamma' = \gamma'_1(1-\beta) + \beta\gamma'_2$. The function $x \to e^{\gamma' T(x)} h(x)$ is nonnegative and μ-integrable. This follows immediately from Hölder's inequality (see p. 10):

$$\begin{aligned}\int_R e^{\gamma' T(x)} h(x)\, d\mu(x) &= \int_R \left(e^{\gamma'_1 T(x)} h(x)\right)^{1-\beta} \left(e^{\gamma'_2 T(x)} h(x)\right)^{\beta} d\mu(x) \\ &\leqslant \left(\int_R e^{\gamma'_1 T(x)} h(x)\, d\mu(x)\right)^{1-\beta} \left(\int_R e^{\gamma'_2 T(x)} h(x)\, d\mu(x)\right)^{\beta} .\end{aligned}$$

We now show that a class with monotone likelihood ratios is defined by means of (5.14). Indeed, for all x with $h(x) \neq 0$

$$f_{\gamma'_1}(x) / f_{\gamma'_0}(x) = \frac{C'(\gamma'_1)}{C'(\gamma'_0)}\, e^{(\gamma'_1 - \gamma'_0)T(x)}$$

so that with $H_{\gamma'_0, \gamma'_1}(y) = C'(\gamma'_1) e^{(\gamma'_1 - \gamma'_0)y} / C'(\gamma'_0)$, $\gamma'_0 < \gamma'_1$, $y \in R_1$, the condition (5.1) is fulfilled.

The set of densities (4.32) with $a \in R_1$ is, as one can immediately see, of the form (5.14) for $n \geqslant 1$.

The set of $B_n(p)$, $0 < p < 1$, (see p. 89), whose R.-N.-densities w.r.t. the uniform distribution with mass points $0, 1, \ldots, n$ are given by $(n+1)\binom{n}{r}(1-p)^n e^{r \log[p/(1-p)]}$ for $r = 0, 1, \ldots, n$ represent other examples from this class.

A further class of discrete distributions of the form (5.14) is given by the set of Poisson distribution (see p. 91) with $a > 0$. As final examples we quote the gamma distribution (p. 82) and the beta distributions (p. 86). Thus we have found a large number of examples of most powerful tests for "one-sided" alternatives. In particular, the one-sided test corresponding to the "two-sided" test in II.7 is most powerful for all alternatives of the form $\sigma_0^2 < \sigma^2$ or $\sigma^2 < \sigma_0^2$.

---

[29] From this it naturally does not necessarily follow that the set of corresponding probability measures $P_{\Gamma'}$ is convex.

The arguments leading to the proof of (5.2) also provide the following *remark*: Let $\gamma' < \gamma''$ be two arbitrary hypotheses and ϕ^* an arbitrary test of the form (5.9). Then ϕ^* is most powerful for $(\{\gamma'\}, \{\gamma''\})$, provided that the level of significance is >0.

Taking into account this remark a converse of Theorem 5.1 can be given. If the admissible parameter region is R_1 (or a subset) and the set P_{R_1} of admissible hypotheses is dominated by a σ-finite measure, then there exists a most powerful test for a composite alternative and every level of significance α iff P_{R_1} possesses monotone likelihood ratios. We restrict ourselves for the sake of simplicity to the case where R_1 is the set of admissible hypotheses and prove

Theorem 5.3[30]. *Let R_1 be the space of admissible hypotheses and the set P_{R_1} of probability measures be dominated by a σ-finite measure. Let λ be a probability measure equivalent to P_{R_1}. Denote the R.-N.-densities w.r.t. λ by p_γ for $\gamma \in R_1$. Let there exist a set $\mathfrak{K}$ of tests ϕ with the following properties: Each test $\phi \in \mathfrak{K}$ is most powerful for each null hypothesis γ_0 vs. an arbitrary alternative γ_1 with $\gamma_0 < \gamma_1$, provided $E(\phi; \gamma_0) > 0$. If $E(\phi; \gamma_0) = 0$, then there also exists a $\phi_0 \in \mathfrak{K}$ with $E(\phi_0; \gamma_0) = 0$ such that ϕ_0 is most powerful for each problem $(\{\gamma_0\}, \{\gamma_1\})$ with $\gamma_0 < \gamma_1$. In case there exists a $\phi \in \mathfrak{K}$ with $E(\phi; \gamma_0) = 1$ for a $\gamma_0 \in R_1$, then there also exists a ϕ_1 with $E(\phi_1; \gamma_0) = 1$ such that $1 - \phi_1$ is most powerful for $(\{\gamma_0\}, \{\gamma_1\})$ for each γ_1 with $\gamma_1 < \gamma_0$. For each $\gamma_0 \in R_1$ and each α with $0 \leqslant \alpha \leqslant 1$ there always exists a test $\phi \in \mathfrak{K}$ such that*

$$E(\phi; \gamma_0) = \alpha. \tag{5.15}$$

Then there exists an $\mathscr{S}$-measurable function T and for each $(\gamma_0, \gamma_1) \in R_2$, $\gamma_0 < \gamma_1$, a nondecreasing function H_{γ_0, γ_1} such that

$$p_{\gamma_1} / p_{\gamma_0} = H_{\gamma_0, \gamma_1} \circ T \tag{5.16}$$

holds $(1/2)(P_{\gamma_0} + P_{\gamma_1})$-a.e. and even λ-a.e. if we define the quotient on the left for $\{x : p_{\gamma_0}(x) = 0\}$ appropriately.

To prove this theorem we need the following

Lemma 5.1. *Let $(R, \mathscr{S}, v)$ be a probability space and $\mathfrak{C}$ a subset of $\mathscr{S}$ with the following properties: If $A, B \in \mathfrak{C}$ then $A \subseteq B$ or $B \supseteq A$ v-a.e.*[31] *Then there exists a real-valued $\mathscr{S}$-measurable function T defined everywhere in R with*

$$A = \{x : T(x) \leqslant v(A)\} \tag{5.17}$$

v-a.e. for each $A \in \mathfrak{C}$.

---

[30] J. Pfanzagl, Z. Wahrscheinlichkeitstheorie und Verw. Gebiete 1, 109—115 (1963).

[31] $A \subseteq B$ v-a.e. means that the set of elements of A which do not belong to B form a v-null set. $A = B$ v-a.e. means $v(A - B) + v(B - A) = 0$.

Proof. One can always assume that $R \in \mathfrak{C}$ since one can otherwise add R to $\mathfrak{C}$ without destroying the ordering given in $\mathfrak{C}$. We have $\mathfrak{M} = \{\nu(A): A \in \mathfrak{C}\} \subseteq [0,1]$, thus a subset of R_1. Hence, there exists a countable set $\mathfrak{G} = \{A_i: A_i \in \mathfrak{C}\}$ such that the set $\mathfrak{D} = \{\nu(A_i): A_i \in \mathfrak{G}\}$ is dense in $\mathfrak{M}$, i.e., for each $\varepsilon > 0$ and $A \in \mathfrak{C}$ there is an $A_i \in \mathfrak{G}$ such that $|\nu(A) - \nu(A_i)| < \varepsilon$. We define

$$T(x) = \inf\{\nu(A_i): x \in A_i, A_i \in \mathfrak{G}\}. \tag{5.18}$$

T is defined everywhere in R. We will show that T possesses the property claimed in the theorem.

First, it is easy to show that T is $\mathscr{S}$-measurable. Indeed, for an arbitrary real c,

$$\{x: T(x) < c\} = \bigcup_i \{A_i: A_i \in \mathfrak{G}, \nu(A_i) < c\}. \tag{5.19}$$

If $x_0 \in \{x: T(x) < c\}$, then there exists according to (5.18) an $A_i \in \mathfrak{G}$ with $x_0 \in A_i$ and $\nu(A_i) < c$. Thus x_0 belongs to the set on the right in (5.19). On the other hand, if x_0 is an element of this set, there exists an $A_i \in \mathfrak{G}$ with $x_0 \in A_i$ and $\nu(A_i) < c$ so that $T(x_0) \leqslant \nu(A_i) < c$ and thus $x_0 \in \{x: T(x) < c\}$. From the countability of $\mathfrak{G}$ and the fact that $\mathfrak{G} \subseteq \mathscr{S}$ follows the $\mathscr{S}$-measurability of T. In particular, choosing $c = \nu(A)$ with $A \in \mathfrak{C}$ in (5.19), we find that

$$\{x: T(x) < \nu(A)\} = \bigcup_i \{A_i: A_i \in \mathfrak{G}, \nu(A_i) < \nu(A)\}.$$

From $\nu(A_i) < \nu(A)$ however, it follows by assumption that $A_i \subseteq A$ ν-a.e. so that

$$\{x: T(x) < \nu(A)\} \subseteq A \qquad \nu\text{-a.e.} \tag{5.20}$$

In particular, if $A = A_j \in \mathfrak{G}$ then we even have

$$\{x: T(x) \leqslant \nu(A_j)\} \subseteq A_j. \tag{5.21}$$

Indeed, it is well known that

$$\{x: T(x) \leqslant \nu(A_j)\} = \bigcap_n \{x: T(x) < \nu(A_j) + 1/n\}$$

and hence also

$$\{x: T(x) \leqslant \nu(A_j)\} = \bigcap_n \bigcup_i \{A_i: A_i \in \mathfrak{G}, \nu(A_i) < \nu(A_j) + 1/n\}.$$

However, each set $\{A_i: A_i \in \mathfrak{G}, \nu(A_i) < \nu(A_j) + 1/n\}$ contains A_j since $A_j \in \mathfrak{G}$ which yields (5.21). On the other hand, from $A_j \in \mathfrak{G}$ it also follows that

$$A_j \subseteq \{x: T(x) \leqslant \nu(A_j)\}. \tag{5.22}$$

Indeed, if $x_0 \in A_j$, then from Definition (5.18), $T(x_0) \leqslant \nu(A_j)$. (5.21) and (5.22) show that (5.17) is satisfied for $A_j \in \mathfrak{G}$.

Now let $A \in \mathfrak{C}$ and let I be an index set such that

$$\nu(A) = \inf_{i \in I} \{\nu(A_i) \colon A_i \in \mathfrak{G}\}. \tag{5.23}$$

Then $A_i \supseteq A$ ν-a.e. for each $i \in I$ and also, for each two indices $i, j \in I$ either $A_i \supseteq A_j$ or $A_j \supseteq A_i$ ν-a.e. Hence, $\bigcap_{i \in I} A_i = A$ ν-a.e. and there exists a subsequence i_j of elements from I such that $\nu(A_{i_j})$ is nonincreasing and converges to $\nu(A)$ monotonically (see p. 5).

Therefore,

$$\bigcap_{i \in I} (-\infty, \nu(A_i)] = (-\infty, \nu(A)].$$

and thus

$$\{x \colon T(x) \leqslant \nu(A)\} = T^{-1}(-\infty, \nu(A)] = T^{-1}\Big(\bigcap_{i \in I} (-\infty, \nu(A_i)]\Big)$$

$$= \bigcap_{i \in I} \{x \colon T(x) \leqslant \nu(A_i)\}.$$

Since (5.17) has already been proved for all $A_i \in \mathfrak{G}$, we get $\{x \colon T(x) \leqslant \nu(A)\} = \bigcap_{i \in I} A_i = A$ ν-a.e., i.e., (5.17) holds for all $A \in \mathfrak{C}$ which fulfill (5.23).

If $\nu(A)$ for $A \in \mathfrak{C}$ cannot be represented in the form (5.23), then, because $\mathfrak{D}$ is dense in $\mathfrak{M}$, there must exist a set J of indices such that

$$\nu(A) = \sup_{i \in J} \{\nu(A_i) \colon A_i \in \mathfrak{G}\}. \tag{5.24}$$

Further one can assume that there is no $A_j \in \mathfrak{G}$ for which $A = A_j$ ν-a.e. since otherwise (5.17) is already proved. One now concludes as above that $A_i \subset A$ ν-a.e., $i \in J$. Moreover, for each $i \in J$

$$\nu(A_i) < \nu(A) \tag{5.25}$$

and $\bigcup_{i \in J} A_i = A$ ν-a.e. From (5.25)

$$\{x \colon T(x) \leqslant \nu(A_i)\} \subseteq \{x \colon T(x) < \nu(A)\},$$

so that

$$\bigcup_{i \in J} \{x \colon T(x) \leqslant \nu(A_i)\} \subseteq \{x \colon T(x) < \nu(A)\}.$$

This leads to

$$\bigcup_{i \in J} A_i \subseteq \{x \colon T(x) < \nu(A)\}$$

and finally to $A \subseteq \{x \colon T(x) < \nu(A)\}$ ν-a.e. Taking (5.20) into consideration, we will also have proved (5.17) in this case if we can show that

$\{x: T(x)=v(A)\}=\emptyset$. If $T(x_0)=v(A)$ held, then we would have from (5.18)

$$v(A)=\inf\{v(A_i): x_0 \in A_i, A_i \in \mathfrak{G}\}$$

which contradicts the assumption that $v(A)$ cannot be represented as an infimum of this type.

We need another lemma which will also be useful later.

Lemma 5.2. *Let G be a nondecreasing right continuous function defined on $[a,b]$ which maps $[a,b]$ into $[c,d]$ so that $G(a)=c$ and $G(b)=d$. $a=-\infty$ and $b=\infty$ are also allowed. Then there always exists a function u_G which maps $[c,d]$ into $[a,b]$ and has the following properties: u_G is nondecreasing; for each $x\in[a,b]$*

$$u_G(G(x)) \leqslant x \tag{5.26}$$

and for each $y\in[c,d]$

$$G(u_G(y)) \geqslant y. \tag{5.27}$$

Proof. Let

$$u_G(y)=\inf\{x\in(a,b]: G(x-0)\leqslant y\leqslant G(x)\} \tag{5.28}$$

for $c<y<d$ and

$$u_G(c)=a, \qquad u_G(d)=b. \tag{5.29}$$

The function defined by (5.28) and (5.29) is nondecreasing. Let

$$c<y_1<y_2<d. \tag{5.30}$$

Then from $G(x_i-0)\leqslant y_i\leqslant G(x_i)$, $i=1,2$, it follows that we always have $x_1\leqslant x_2$. From $x_1>x_2$ it follows, in fact, that $G(x_1-0)\geqslant G(x_2)$, which contradicts (5.30).

If $x\in(a,b]$ (5.26) follows immediately from (5.28). For $x=a$, (5.26) follows from (5.29). (5.27) is trivial for $y=c$. If

$$c<y<d \tag{5.31}$$

then $G(x)\geqslant y$ for $G(x)<y$ would imply $x<u_G(y)$ by the right continuity of G and (5.28). This proves (5.27).

From this one obtains another property of u_G: $u_G(y)\leqslant x$ implies $y\leqslant G(x)$ and conversely. The first statement follows from $G(u_G(y))\leqslant G(x)$ and (5.27). Since u_G is nondecreasing, the converse follows in the same way from (5.26).

We proceed to the *proof of Theorem* 5.3. By Theorem 5.2 the existence of a measure λ equivalent to P_{R_1} is assured. Let it be of the form

$$\lambda=\sum_{i=1}^{\infty}\beta_i P_{\gamma_i}, \qquad \beta_i>0, \qquad \sum_{i=1}^{\infty}\beta_i=1, \qquad \beta_i\in R_1. \tag{5.32}$$

For each $\gamma \in R_1$ we define

$$Q_\gamma = \{x: p_\gamma(x) > 0\}, \qquad Q_{\gamma,l} = \bigcup_{\gamma_i < \gamma} Q_{\gamma_i} - Q_\gamma$$

and

$$Q_{\gamma,r} = \bigcup_{\gamma_i > \gamma} Q_{\gamma_i} - Q_\gamma .$$

Further let $A_\phi = \{x: \phi(x) = 0\}$ and $Z_\phi = \{x: \phi(x) = 1\}$ for each $\phi \in \mathfrak{R}$. We first make the following general observation: If $x \notin Q_{\gamma,l} \cup Q_{\gamma,r}$, then either $x \in Q_\gamma$ or x belongs to a set of λ-measure 0. Indeed, if $x \notin Q_\gamma$ then x belongs to $\bigcap_{i=1}^{\infty} (R - Q_{\gamma_i})$ and this is a P_{γ_i}-null set for $i = 1, 2, \ldots$ and thus also *a* λ-null set.

Let γ^* be an arbitrary real number and assume

$$E(\phi; \gamma^*) = 1 \tag{5.33}$$

for a $\phi \in \mathfrak{R}$, thus, in particular

$$P_{\gamma^*}(A_\phi) = 0 . \tag{5.34}$$

From (5.33), for each γ_i with $\gamma_i > \gamma^*$ we also have $E(\phi; \gamma_i) = 1$. Hence A_ϕ is (λ-a.e.) not contained in $Q_{\gamma^*,r}$ and from the above observation

$$A_\phi \subseteq Q_{\gamma^*,l} \qquad \lambda\text{-a.e.} \tag{5.35}$$

But if

$$E(\phi; \gamma^*) = 0 \tag{5.36}$$

then also $E(\phi; \gamma_i) = 0$ for each γ_i with $\gamma_i < \gamma^*$. Hence $A_\phi \supseteq \{x: p_{\gamma^*}(x) > 0\}$ and $A_\phi \supseteq \{x: p_{\gamma_i}(x) > 0\}$ for each $\gamma_i < \gamma^*$ (λ-a.e.) and thus

$$A_\phi \supseteq Q_{\gamma^*,l} \cup Q_{\gamma^*} \qquad \lambda\text{-a.e.} \tag{5.37}$$

We now show that if $\phi \in \mathfrak{R}$, then for arbitrary real γ^* with $0 < E(\phi; \gamma^*) < 1$

$$Q_{\gamma^*,r} \subseteq Z_\phi \qquad \lambda\text{-a.e.} \tag{5.38}$$

$$Q_{\gamma^*,l} \subseteq A_\phi \qquad \lambda\text{-a.e.} \tag{5.39}$$

and

$$A_\phi = Q_{\gamma^*,l} \cup (A_\phi \cap Q_{\gamma^*}) \qquad \lambda\text{-a.e.} \tag{5.40}$$

In fact, by Theorem 3.1. III F $\{x: p_{\gamma_i}(x) > k_i p_{\gamma^*}(x)\} \supseteq Z_\phi$ (λ-a.e.) for arbitrary γ_i with $\gamma_i > \gamma^*$ and suitable real k_i so that (5.38) follows. Conversely, $A_\phi \supseteq \{x: p_{\gamma^*}(x) < k_i p_{\gamma_i}(x)\}$ (λ-a.e.) for suitable real k_i and for each γ_i for which $\gamma_i < \gamma^*$ and $E(\phi; \gamma_i) > 0$. Because $E(\phi; \gamma^*) < 1$, we always have $k_i \neq 0$ so that $Q_{\gamma_i} - Q_{\gamma^*} \subseteq A_\phi$ (λ-a.e.). However, this holds trivially also for $E(\phi; \gamma_i) = 0$. This implies (5.39). From (5.38) we get $A_\phi \cap Q_{\gamma^*,r} = \emptyset$ λ-a.e. As we observed above, this proves (5.40).

If $E(\phi;\gamma^*)=0$ and ϕ is most powerful for $(\{\gamma^*\},\{\gamma\})$ for each γ with $\gamma^*<\gamma$, then likewise $\{x: P_{\gamma_i}(x)>k_i p_{\gamma^*}(x)\}\subseteq Z_\phi$ [32] for each γ_i with $\gamma_i>\gamma^*$ and suitable real k_i. This together with (5.37) implies (5.40) again. If $E(\phi;\gamma^*)=1$ and $1-\phi$ is most powerful for $(\{\gamma^*\},\{\gamma\})$ for each γ with $\gamma<\gamma^*$, then we easily see that $A_\phi\supseteq Q_{\gamma^*,l}$ and $A_\phi\cap Q_{\gamma^*}=\emptyset$ so that (5.40) again results by means of (5.35). Now let

$$\mathfrak{C}=\{A: A=Q_{\gamma^*,l}\cup(\{x: p_\gamma(x)\leqslant k p_{\gamma^*}(x)\}\cap Q_{\gamma^*}), 0\leqslant k\leqslant\infty, \gamma,\gamma^*\in R_1, \gamma^*<\gamma\}.$$

By assumption, there exists for an arbitrary $A\in\mathfrak{C}$ of the form

$$Q_{\gamma^*,l}\cup(\{x: p_\gamma(x)\leqslant k_0 p_{\gamma^*}(x)\}\cap Q_{\gamma^*}) \tag{5.41}$$

a $\phi\in\mathfrak{R}$ with

$$E(\phi;\gamma^*)=1-P_{\gamma^*}(A), \tag{5.42}$$

so that ϕ is most powerful for the problem $(1-P_{\gamma^*}(A),\{\gamma^*\},\{\gamma\})$. If $P_{\gamma^*}(A)=0$, one can also assume that $1-\phi$ is most powerful for each γ with $\gamma<\gamma^*$. From (5.41) we have $P_{\gamma^*}(A)=P_{\gamma^*}(\{x: p_\gamma(x)\leqslant k_0 p_{\gamma^*}(x)\})$ so that from Theorem 3.1 III F,

$$\phi(x)=\begin{cases}1 & p_\gamma(x)>k_0 p_{\gamma^*}(x)\\ 0 & p_\gamma(x)<k_0 p_{\gamma^*}(x)\end{cases}.$$

Now

$$\{x: p_\gamma(x)<k_0 p_{\gamma^*}(x)\}\cap Q_{\gamma^*}=\{x: p_\gamma(x)<k_0 p_{\gamma^*}(x)\}.$$

Either $\{x: p_\gamma(x)=k_0 p_{\gamma^*}(x)\}$ has P_{γ^*}-measure 0 or ϕ vanishes on this set. Otherwise (5.42) would not be fulfilled. Hence

$$A_\phi=A \qquad P_{\gamma^*}\text{-a.e.} \tag{5.43}$$

From (5.40) and (5.41), however, $A_\phi\cap Q_{\gamma^*,l}=A\cap Q_{\gamma^*,l}$ so that from (5.43) we also have

$$A_\phi=A \qquad \lambda\text{-a.e.} \tag{5.44}$$

We now show that for each two elements $A_1, A_2\in\mathfrak{C}$

$$A_1\subseteq A_2 \quad \text{or} \quad A_2\subseteq A_1 \qquad \lambda\text{-a.e.} \tag{5.45}$$

We assume that A_1 is given in the form (5.41), where k_0 is to be replaced by k_1. As we have shown, there exists a ϕ such that

$$A_2=A_\phi \qquad \lambda\text{-a.e.} \tag{5.46}$$

If ϕ is most powerful for $(\{\gamma^*\},\{\gamma\})$, then there exists a k_2 such that

$$\{x: p_\gamma(x)<k_2 p_{\gamma^*}(x)\}\subseteq A_2\subseteq\{x: p_\gamma(x)\leqslant k_2 p_{\gamma^*}(x)\} \qquad \lambda\text{-a.e.}$$

[32] Here, and occasionally later, we will supress the fact that certain relations hold only λ-a.e.

and therefore P_{γ^*}-a.e. From the definition of A_1 we have

$$A_1 = \{x: p_\gamma(x) \leqslant k_1 p_{\gamma^*}(x)\} \qquad P_{\gamma^*}\text{-a.e.}$$

which proves (5.45) P_{γ^*}-a.e.

From (5.46) and (5.40) we have $A_2 = P_{\gamma^*,l} \cup (A_2 \cap P_{\gamma^*})$ λ-a.e., and since A_1 is of the form (5.41) with $k_0 = k_1$ we finally have that (5.45) also holds λ-a.e. However, if $E(\phi;\gamma^*)=0$, resp., $E(\phi;\gamma^*)=1$, then from (5.37), resp., (5.35), (5.45) is fulfilled trivially.

According to Lemma 5.1 p. 191 there exists a λ-a.e. defined, $\mathscr{S}$-measurable function T which satisfies

$$A = \{x: T(x) \leqslant \lambda(A)\} \qquad \lambda\text{-a.e.} \tag{5.47}$$

for all $A \in \mathfrak{C}$.

Let

$$A_k = Q_{\gamma^*,l} \cup (\{x: p_{\bar\gamma}(x) \leqslant k p_{\gamma^*}(x)\} \cap Q_{\gamma^*}) \tag{5.48}$$

for arbitrary but fixed $\gamma^*, \bar\gamma$ with $\gamma^* < \bar\gamma$ and $0 \leqslant k \leqslant \infty$.

By the definition of Q_{γ^*} we also have

$$A_k = Q_{\gamma^*,l} \cup (\{x: p_{\bar\gamma}(x)/p_{\gamma^*}(x) \leqslant k\} \cap Q_{\gamma^*}). \tag{5.49}$$

It follows immediately that the map $k \to \lambda(A_k)$ is nondecreasing. On the basis of the remark following Lemma 5.2 there follows the existence of a nondecreasing function u from $[\lambda(A_0), \lambda(A_\infty)]$ into $[0,\infty]$ such that $y \leqslant \lambda(A_k)$ iff $u(y) \leqslant k$. This remains true for $k=\infty$ if we extend the definition of u to the interval $[0,1]$ by setting $u(y)=0$ for $0 \leqslant y < \lambda(A_0)$ and $u(y)=\infty$ for $\lambda(A_\infty) < y \leqslant 1$. Since $A_k \in \mathfrak{C}$, we obtain from (5.47)

$$A_k = \{x: u(T(x)) \leqslant k\} \qquad \lambda\text{-a.e.} \tag{5.50}$$

From (5.49) and (5.50) there follows $Q_{\gamma^*,l} \subseteq \{x: u(T(x))=0\}$. Thus, defining $p_{\bar\gamma}(x)/p_{\gamma^*}(x)=0$ for $x \in Q_{\gamma^*,l}$ we even have for $x \in Q_{\gamma^*,l} \cup Q_{\gamma^*}$

$$\{x: p_{\bar\gamma}(x)/p_{\gamma^*}(x) \leqslant k\} = \{x: u(T(x)) \leqslant k\} \qquad \lambda\text{-a.e.} \tag{5.51}$$

For $x \in Q_{\gamma^*,r}$ we define $p_{\bar\gamma}(x)/p_{\gamma^*}(x) = \infty$. Then, also for this x we can write $p_{\bar\gamma}(x)/p_{\gamma^*}(x) = u(T(x))$. Indeed, if for $x \in Q_{\gamma^*,r}$ we had $T(x) < \lambda(A_\infty)$, then (5.50) would imply for a suitable $k < \infty$ that $x \in A_k$, i.e., $x \notin Q_{\gamma^*,r}$ λ-a.e. Hence we finally get together with (5.51)

$$p_{\bar\gamma}(x)/p_{\gamma^*}(x) = u(T(x)) \qquad \lambda\text{-a.e.} \tag{5.52}$$

With $H_{\gamma^*,\bar\gamma} = u$ the theorem is proved.

It is of some interest to note that the statement of Theorem 5.3 remains true if (5.15) is assumed to hold for one fixed α only, provided most powerful tests exist for arbitrary large sample sizes[33].

[33] See J. Pfanzagl, Sankhyā Ser. A 30, 147—156 (1968).

6. Locally most powerful tests. Our considerations in **5** have shown that for alternatives comprising more than one element there exist in general no most powerful tests. If the set P_Γ is dominated by a σ-finite measure, the existence of most powerful tests is essentially equivalent to the requirement that the R.-N.-densities possess monotone likelihood ratios. From both a practical and theoretical point of view it is of interest to investigate the power function of a test "in the neighborhood" of the null hypothesis. Indeed, one hardly considers alternatives which are "too far" from the null hypothesis[34]. Here it is necessary to introduce the idea of a distance in the set of admissible hypotheses Γ. Up to now, however, we have in general made no assumptions on the structure of this set.

Consider as an example a set of normal distributions whose densities for an n with $n \geqslant 1$ and for $-\infty < a < \infty$ are defined on p. 184. Then a natural definition for the distance between two hypotheses a_1, a_2 is $|a_1 - a_2|$. Since exactly one normal distribution P_a corresponds to each a, one can also use this distance to define a distance δ for the distributions themselves by simply defining $\delta(P_{a_1}, P_{a_2}) = |a_1 - a_2|$.

One can always introduce a distance in the set of all probability measures over an arbitrary measure space $(R, \mathcal{S})$

$$d(P_1, P_2) = \sup_{A \in S} |P_1(A) - P_2(A)|$$

for each pair of measures P_1 and P_2. It's easy to see that d possesses all the properties of a metric: $d(P_1, P_2) = 0$ iff $P_1 = P_2$. $d(P_1, P_2) = d(P_2, P_1)$. Moreover, the *triangle inequality* holds: $d(P_1, P_3) \leqslant d(P_1, P_2) + d(P_2, P_3)$. This follows easily from the triangle inequality for real numbers and from the definition of the supremum.

We remark incidentally that from $\delta(P_a, P_{a_0}) \to 0$ for $a \to a_0$ we also have $d(P_a, P_{a_0}) \to 0$, and conversely. This can be obtained immediately from the following fact: Let f_i be the R.-N.-densities of P_i, $i = 1, 2$, w.r.t. an appropriate measure μ, for example $P_1 + P_2$. Then

$$d(P_1, P_2) = \tfrac{1}{2} \int_R |f_1 - f_2| \, d\mu . \tag{6.1}$$

This follows from

$$\int_R (f_1 - f_2) \, d\mu = 0 \quad \text{and} \quad d(P_1, P_2) = \int_B |f_1 - f_2| \, d\mu ,$$

where $B = \{x : f_1(x) - f_2(x) \geqslant 0\}$.

The example of the normal distributions just mentioned, for which a distance δ (the usual distance in R_1) can be introduced is typical for numerous other examples. Considering, say, the normal distribution given by (1.3) with $-\infty < a < \infty$, $0 < \sigma < \infty$, we can define $\delta(P_{a_1,\sigma_1}, P_{a_2,\sigma_2}) = \sqrt{(a_1 - a_2)^2 + (\sigma_1 - \sigma_2)^2}$. Again, from $\delta(P_{a,\sigma}, P_{a_0,\sigma_0}) \to 0$ for $(a, \sigma) \to (a_0, \sigma_0)$ there follows $d(P_{a,\sigma}, P_{a_0,\sigma_0}) \to 0$, and conversely.

These considerations are only an illustration of a general theorem which is originally due to Scheffé[35]: *Let $f_1, f_2, \ldots$ be a sequence of densities (w.r.t. a σ-finite measure μ) which converges to a density f (w.r.t. μ). Then $\int_R |f_n - f| \, d\mu \to 0$.*

[34] In practical work, however, alternatives which are "too close" to the null hypothesis are likewise uninteresting.

[35] H. Scheffé, Ann. Math. Statist. 18, 434—438 (1947).

These remarks suggest the following definition:

Let $\mathfrak{P}$ be a set of admissible hypotheses for a test. Suppose a distance can be introduced in $\mathfrak{P}$ such that $\mathfrak{P}$ can be mapped one-to-one and continuously in both directions onto a subset Γ of R_n, $n \geqslant 1$. The test problem $(\Gamma_0, \Gamma - \Gamma_0)$ is called *parametric*[36] when both the null hypothesis Γ_0 and $\Gamma - \Gamma_0$ are open or closed subsets of R_n.

The definition of a locally most powerful test which we give below belongs to the framework of parametric theories.

Definition. Let the set Γ of admissible hypotheses of a test problem be a subset of R_k, $k \geqslant 1$. Let $\gamma_0 \in \Gamma$, let $\Gamma - \{\gamma_0\}$ be open, $0 \leqslant \alpha \leqslant 1$, and Φ_α the set of all tests for the problem $(\alpha, \{\gamma_0\}, \Gamma - \{\gamma_0\})$. ϕ^* is called locally most powerful (l.m.p.) for this problem if for each $\phi \in \Phi_\alpha$ there exists an open ball $K_\phi \subset R_k$ with center γ_0 such that $K_\phi \cap (\Gamma - \{\gamma_0\})$ is nonempty and for the power functions

$$g_{\phi^*}(\gamma) \geqslant g_\phi(\gamma) \tag{6.2}$$

holds for all $\gamma \in K_\phi \cap (\Gamma - \{\gamma_0\})$.

The definition of a l.m.p. test w.r.t. a subset of Φ_α is obvious.

All we need for this definition is the notion of "open" or "closed" for R_n. It can thus be generalized considerably, but we shall not consider this generalization.

One can prove the existence of l.m.p. tests under quite general assumptions. We have

Theorem 6.1. *Let $\Gamma = [\gamma_0, \infty)$* [37] *and let P_Γ be defined over $(R, \mathscr{S})$ and dominated by a σ-finite measure μ. The R.-N.-densities of P_γ w.r.t. μ are designated by f_γ. For all γ in $[\gamma_0, \gamma_1)$, $\gamma_0 < \gamma_1$, and each $x \in R$, with the exception of at most a μ-null set*[38], *let the map $\gamma \to f_\gamma(x)$ be differentiable (at γ_0 from the right). Further let there exist a μ-integrable function ψ such that for $\gamma_0 \leqslant \gamma < \gamma_1$*

$$\left| \frac{\partial}{\partial \gamma} f_\gamma \right| \leqslant \psi \tag{6.3}$$

holds μ-a.e. For each k with $-\infty \leqslant k \leqslant \infty$ assume

$$\mu \left\{ x : \frac{\partial}{\partial \gamma} f_{\gamma_0}(x) = k f_{\gamma_0}(x) \right\} = 0 . \tag{6.4}$$

---

[36] This terminology is not restricted to test problems. It will be used analogously for confidence regions (see IV) and theory of estimation (see also V). See also VII.**1** [2].

[37] The choice of endpoint for this interval is not important. ∞ can be replaced by an arbitrary real number $> \gamma_0$.

[38] This μ-null set may depend on γ.

Each test ϕ^* *of the form*

$$\phi^*(x)=\begin{cases} 1 & \frac{\partial}{\partial\gamma} f_{\gamma_0}(x) > k\, f_{\gamma_0}(x) \\ \text{an arbitrary number in } [0,1] & \frac{\partial}{\partial\gamma} f_{\gamma_0}(x) = k\, f_{\gamma_0}(x) \\ 0 & \frac{\partial}{\partial\gamma} f_{\gamma_0}(x) < k\, f_{\gamma_0}(x) \end{cases} \tag{6.5}$$

with $-\infty \leqslant k \leqslant \infty$ *is l.m.p. for the problem* $(\{\gamma_0\}, \Gamma-\{\gamma_0\})$.[39] *Conversely, if* $\bar{\phi}$ *is l.m.p. for the problem* $(\{\gamma_0\}, \Gamma-\{\gamma_0\})$ *then* $\bar{\phi}=\phi^*$ μ*-a.e. for a suitable k.*

Proof. Let Φ_α be the set of all tests for the problem $(\alpha, \{\gamma_0\}, \Gamma-\{\gamma_0\})$ and let $\Phi'_\alpha=\{\phi\colon \phi\in\Phi_\alpha, E(\phi;\gamma_0)=\alpha\}$. Assume $0<\alpha\leqslant 1$ and let $\phi_1\in\Phi_\alpha$ with

$$E(\phi_1;\gamma_0)=\alpha_1<\alpha. \tag{6.6}$$

Then there always exists a $\phi\in\Phi'_\alpha$ satisfying the condition $\phi_1\leqslant\phi$ such that for the power functions

$$g_\phi(\gamma)\geqslant g_{\phi_1}(\gamma) \tag{6.7}$$

holds for all $\gamma\in\Gamma-\{\gamma_0\}$. This is almost obvious and goes as follows: let $\beta=(1-\alpha)/(1-\alpha_1)$. From (6.6), $\beta<1$. Now we define for $x\in R$

$$\phi(x)=(1-\beta)+\beta\,\phi_1(x).$$

Then an easy calculation shows that ϕ has the claimed property. It is thus enough to show that the test defined by (6.5) belongs to Φ'_α for suitable choice of k and is l.m.p. for Φ'_α. The existence of a suitable k for which $\phi^*\in\Phi'_\alpha$ follows from Theorem 3.1 III F. and the Supplement to III F. From the latter we also find that for $\phi\in\Phi'_\alpha$,

$$\int_R \phi^* \frac{\partial}{\partial\gamma} f_{\gamma_0}\, d\mu \geqslant \int_R \phi \frac{\partial}{\partial\gamma} f_{\gamma_0}\, d\mu.$$

Because of (6.3) (see Theorem VIII) we then also have

$$\frac{\partial}{\partial\gamma} E(\phi^*;\gamma_0) \geqslant \frac{\partial}{\partial\gamma} E(\phi;\gamma_0). \tag{6.8}$$

If the ">" sign holds in (6.8) then the difference of the power functions $g_{\phi^*}-g_\phi$ is locally (strictly) increasing at γ_0. Since $g_{\phi^*}(\gamma_0)-g_\phi(\gamma_0)=0$ we have $g_{\phi^*}(\gamma)>g_\phi(\gamma)$ in a suitable interval that depends in general on ϕ.

[39] We define here $0\cdot\infty=0$ and $-\infty\cdot 0=-\infty$.

However, if equality holds in (6.8), then ϕ must, according to the Supplement to III F be of the form

$$\phi(x)=\begin{cases}1 & \frac{1}{\partial\gamma} f_{\gamma_0}(x) > k f_{\gamma_0}(x)\\ 0 & \frac{\partial}{\partial\gamma} f_{\gamma_0}(x) < k f_{\gamma_0}(x)\end{cases}$$

μ-a.e. Hence, from (6.4), $\phi=\phi^*$ μ-a.e. But then $g_{\phi^*}(\gamma)=g_\phi(\gamma)$ for $\gamma\in\Gamma-\{\gamma_0\}$.

Now let $\bar{\phi}$ be a l.m.p. test for $(\alpha,\{\gamma_0\},\Gamma-\{\gamma_0\})$. There exists then for each $\phi\in\Phi'_\alpha$ an interval (γ_0,γ_ϕ) within which for the power functions

$$g_{\bar{\phi}}(\gamma)\geqslant g_\phi(\gamma). \tag{6.9}$$

From (6.3) we get easily that $\lim\limits_{\gamma\to\gamma_0} g_\phi(\gamma)=g_\phi(\gamma_0)=\alpha$ and thus from (6.9) also $\lim\limits_{\gamma\to\gamma_0} g_{\bar{\phi}}(\gamma)=g_{\bar{\phi}}(\gamma_0)=\alpha$. This along with (6.9) yields for the right derivatives at γ_0

$$\frac{\partial}{\partial\gamma} g_{\bar{\phi}}(\gamma_0)\geqslant\frac{\partial}{\partial\gamma} g_\phi(\gamma_0) \tag{6.10}$$

for each $\phi\in\Phi'_\alpha$. Using (6.3) we get from the Supplement to III F because of (6.4) that $\bar{\phi}$ is μ-a.e. of the form (6.5) which was to be proved.

Remark. The awkward condition (6.4) can be replaced by allowing only those levels of significance which are provided by the P_{γ_0}-measure of critical regions which can be defined by the test statistic $x\to\left[\frac{\partial}{\partial\gamma} f_{\gamma_0}(x)\right]\Big/ f_{\gamma_0}(x)$. In addition to this, one must assume the existence of a set $N\in\mathscr{S}$, such that $\mu(N)=0$ and $\{x: f(x,\gamma)=0\}\subseteq N$ for all $\gamma\in[\gamma_0,\gamma_1)$.

This theorem thus assures the existence of l. m. p. tests for "one-sided" alternatives. However, as the arguments on p. 183ff. show, there exists in general—even under rather strong assumptions—no l.m.p. test for "two-sided" alternatives. Now note that the trivial test, which is always $\equiv\alpha$, belongs to Φ'_α. Hence, by Theorem 2.1, each l.m.p. test ϕ^* is unbiased in a suitable interval of the form $[\gamma_0,\gamma_1]$. Such tests are also called *locally unbiased.* From this it follows for the power function g_{ϕ^*} under the assumptions of Theorem 6.1 that $\frac{\partial}{\partial\gamma} g_{\phi^*}(\gamma_0)\geqslant 0$. These remarks suggest the possibility of constructing tests for simple null hypotheses and two-sided alternatives, which, at least for the class of all locally unbiased tests, are l. m. p.

We can now actually prove

Theorem 6.2. *Let Γ be, say, all of R_1. Let P_Γ, μ and f_γ have meanings analogous to those in Theorem 6.1. For all γ in $(\gamma_0-\delta, \gamma_0+\delta)$, $\delta>0$, and each $x\in R$, with the possible exception of a μ-null set, let the map $\gamma\to f_\gamma(x)$ be twice differentiable. Assume $\frac{\partial}{\partial\gamma} f_{\gamma_0}$ is μ-integrable and let there exist a μ-integrable function ψ defined over R such that in $(\gamma_0-\delta, \gamma_0+\delta)$*

$$\left|\frac{\partial^2}{\partial\gamma^2} f_\gamma(x)\right| \leqslant \psi(x) \tag{6.11}$$

μ-a.e.

Moreover, assume for all real l_1, l_2 that

$$\mu\left\{x: \frac{\partial^2}{\partial\gamma^2} f_{\gamma_0}(x) = l_1 \frac{\partial}{\partial\gamma} f_{\gamma_0}(x) + l_2 f_{\gamma_0}(x)\right\} = 0. \tag{6.12}$$

Let Ψ_α be the set of all locally unbiased tests for the problem $(\alpha, \{\gamma_0\}, \Gamma-\{\gamma_0\})$. Then each test of the form

$$\phi^*(x) = \begin{cases} 1 & \frac{\partial^2}{\partial\gamma^2} f_{\gamma_0}(x) > l_1 \frac{\partial}{\partial\gamma} f_{\gamma_0}(x) + l_2 f_{\gamma_0}(x) \\ \text{an arbitrary number in } [0,1] & \frac{\partial^2}{\partial\gamma^2} f_{\gamma_0}(x) = l_1 \frac{\partial}{\partial\gamma} f_{\gamma_0}(x) + l_2 f_{\gamma_0}(x) \\ 0 & \frac{\partial^2}{\partial\gamma^2} f_{\gamma_0}(x) < l_1 \frac{\partial}{\partial\gamma} f_{\gamma_0}(x) + l_2 f_{\gamma_0}(x) \end{cases} \tag{6.13}$$

with $-\infty < l_1, l_2 < \infty$[40]*, is l.m.p. w.r.t. Ψ_α if α is given by*

$$\alpha = E(\phi^*; \gamma_0) \tag{6.14}$$

and if l_1, l_2 can be determined such that

$$\int_R \frac{\partial}{\partial\gamma} f_{\gamma_0} \phi^* \, d\mu = 0. \tag{6.15}$$

Furthermore, for each test $\bar{\phi}$, l.m.p. w.r.t. Ψ_α, $\bar{\phi} = \phi^*$ μ-a.e.

The *proof* runs essentially like that of Theorem 6.1 but one must also call on Theorem 3.2. One can again restrict oneself to the subset Ψ'_α of Ψ_α for whose elements ϕ we have $E(\phi; \gamma_0) = \alpha$. Since all ϕ are locally

[40] One can also allow $l_1 = \pm\infty$, $l_2 = \pm\infty$.

unbiased, the power function g_ϕ has a local minimum at γ_0. From (6.11) all $\phi \in \Psi'_\alpha$ thus satisfy the conditions

$$\int_R f_{\gamma_0} \phi \, d\mu = \alpha, \qquad \int_R \frac{\partial}{\partial \gamma} f_{\gamma_0} \phi \, d\mu = 0 .$$

ϕ^* thus satisfies, because of (6.14) and (6.15)

$$\int_R \frac{\partial^2}{\partial \gamma^2} f_{\gamma_0} \phi^* \, d\mu = \max_{\phi \in \Psi'_\alpha} \int_R \frac{\partial^2}{\partial \gamma^2} f_{\gamma_0} \phi \, d\mu .$$

If

$$\int_R \frac{\partial^2}{\partial \gamma^2} f_{\gamma_0} \phi^* \, d\mu > \int_R \frac{\partial^2}{\partial \gamma^2} f_{\gamma_0} \phi \, d\mu$$

then we have because of (6.11) for the power function

$$\frac{\partial^2}{\partial \gamma^2} (g_{\phi^*}(\gamma_0) - g_\phi(\gamma_0)) > 0 .$$

Since

$$\frac{\partial}{\partial \gamma} (g_{\phi^*}(\gamma_0) - g_\phi(\gamma_0)) = 0$$

$g_{\phi^*} - g_\phi$ has a local minimum at γ_0 which essentially completes the proof. The further steps in the proof are exactly analogous to those of Theorem 6.1 and we will not repeat them here.

Consider an example. Let P_{R_1} be the set of all n-dimensional normal distributions whose densities are given on p. 184 with $-\infty < a < \infty$. We are interested in the test problem $(\alpha, \{a_0\}, R_1 - \{a_0\})$. The densities fulfill the assumptions of Theorem 6.2. This can be seen at once by using the relation

$$e^{-\frac{1}{2}\sum_{i=1}^{n}(x_i - a)^2} = e^{-\frac{1}{2}\sum_{i=1}^{n}(x_i - \bar{x})^2} e^{-\frac{n}{2}(\bar{x} - a)^2} . \tag{6.16}$$

An easy calculation shows that the test

$$\phi^*(x) = \begin{cases} 1 & n^2(\bar{x} - a_0)^2 - n > l_2 + l_1 n(\bar{x} - a_0) \\ 0 & n^2(\bar{x} - a_0)^2 - n < l_2 + l_1 n(\bar{x} - a_0) \end{cases}$$

corresponds to formula (6.13) for our example. Another easy calculation shows that if l_1, l_2 are chosen such that

$$l_1 = 0 \quad \text{and} \quad \sqrt{1 + l_2/n} = \kappa_\alpha ,$$

where κ_α is defined as in II. (3.3), then

$$\phi^*(x) = \begin{cases} 1 & \sqrt{n}\,|\bar{x} - a_0| > \kappa_\alpha \\ 0 & \sqrt{n}\,|\bar{x} - a_0| < \kappa_\alpha \end{cases}$$

and ϕ^* fulfills (6.14) and (6.15). We have thus arrived at a certain justification of the test procedure in II.3.

Let Δ_α be the set of all tests ϕ for the problem $(\alpha, \{\gamma_0\}, R_1 - \{\gamma_0\})$ whose power function g_ϕ is twice differentiable at γ_0 and satisfies

$$g_\phi(\gamma_0) = \alpha, \qquad \frac{d}{d\gamma} g_\phi(\gamma_0) = 0. \tag{6.17}$$

A test $\phi^* \in \Delta_\alpha$ which satisfies $\dfrac{d^2}{d\gamma^2} g_{\phi^*}(\gamma_0) \geqslant \dfrac{d^2}{d\gamma^2} g_\phi(\gamma_0)$ for all $\phi \in \Delta_\alpha$ is called, according to Neyman, *a test of type A*. Theorem 6.2 states that under the conditions given there, there always exists a type A test. The power function of a type A test has maximum curvature at γ_0.

Let us briefly consider the case in which the parameter γ is multi-dimensional. For simplicity, let R_2 be the set of admissible hypotheses and let $\gamma_0 \in R_2$. Let $\Delta_\alpha^{(1)}$ be the set of all tests ϕ for the problem $(\alpha, \{\gamma_0\}, R_2 - \{\gamma_0\})$ whose power function g_ϕ has the properties: g_ϕ possesses all first and second order derivatives at γ_0,

$$g_\phi(\gamma_0) = \alpha \tag{6.18}$$

and with $\gamma = (\gamma_1, \gamma_2)$,

$$\frac{\partial g_\phi(\gamma_0)}{\partial \gamma_i} = 0, \qquad i = 1, 2 \tag{6.19}$$

holds; further

$$\frac{\partial^2 g_\phi(\gamma_0)}{\partial \gamma_1 \partial \gamma_2} = 0 \tag{6.20}$$

and

$$\frac{\partial^2 g_\phi(\gamma_0)}{\partial \gamma_1^2} = \frac{\partial^2 g_\phi(\gamma_0)}{\partial \gamma_2^2}. \tag{6.21}$$

Denote by $\Delta_\alpha^{(2)}$ the set of all tests ϕ for the same problem whose power functions fulfill (6.18), (6.19) and also

$$\begin{vmatrix} \dfrac{\partial^2 g_\phi(\gamma_0)}{\partial \gamma_1^2} & \dfrac{\partial^2 g_\phi(\gamma_0)}{\partial \gamma_1 \partial \gamma_2} \\ \dfrac{\partial^2 g_\phi(\gamma_0)}{\partial \gamma_1 \partial \gamma_2} & \dfrac{\partial^2 g_\phi(\gamma_0)}{\partial \gamma_2^2} \end{vmatrix} > 0. \tag{6.22}$$

A test ϕ^* is said to be *of type C*[41] for the problem $(\alpha, \{\gamma_0\}, R_2 - \{\gamma_0\})$ if it is either in $\Delta_\alpha^{(1)}$ and its power function g_{ϕ^*} satisfies

$$\frac{\partial^2 g_{\phi^*}(\gamma_0)}{\partial \gamma_1^2} \geqslant \frac{\partial^2 g_\phi(\gamma_0)}{\partial \gamma_1^2} \tag{6.23}$$

[41] J. Neyman and E. S. Pearson, Statist. Res. Mem. Univ. London 2, 25—57 (1938).

for all $\phi \in \Delta_\alpha^{(1)}$, or if it is in $\Delta_\alpha^{(2)}$ and for all $\phi \in \Delta_\alpha^{(2)}$ fulfills

$$\begin{aligned}\frac{\partial^2 g_{\phi^*}(\gamma_0)}{\partial \gamma_1^2} : \frac{\partial^2 g_\phi(\gamma_0)}{\partial \gamma_1^2} &= \frac{\partial^2 g_{\phi^*}(\gamma_0)}{\partial \gamma_1 \partial \gamma_2} : \frac{\partial^2 g_\phi(\gamma_0)}{\partial \gamma_1 \partial \gamma_2} \\ &= \frac{\partial^2 g_{\phi^*}(\gamma_0)}{\partial \gamma_2^2} : \frac{\partial^2 g_\phi(\gamma_0)}{\partial \gamma_2^2} \\ &= \lambda_\phi\end{aligned} \tag{6.24}$$

for $\lambda_\phi \geq 1$.

It is of interest to give a geometric interpretation of this definition by which the analogy to type A tests becomes more clear. By means of the set of all real u, v, with

$$\frac{\partial^2}{\partial \gamma_1^2} g_\phi(\gamma_0) u^2 + 2 \frac{\partial^2}{\partial \gamma_1 \partial \gamma_2} g_\phi(\gamma_0) u v + \frac{\partial^2}{\partial \gamma_2^2} g_\phi(\gamma_0) v^2 = 1$$

we obtain a conic section which represents because of (6.19) the *Dupin indicatrix* of the surface defined by g_ϕ at the point γ_0. From (6.20) and (6.21), we see that for $\phi \in \Delta_\alpha^{(1)}$, the indicatrix is a circle and from (6.19), γ_0 is a *umbilical point* of the surface. The power function g_{ϕ^*} is characterized according to (6.23) by the fact that its indicatrix is contained in that of every other power function g_ϕ, $\phi \in \Delta_\alpha^{(1)}$. If $\phi \in \Delta_\alpha^{(2)}$, then we consider only these power functions g_ϕ which possess an *elliptic point* at γ_0. It follows from (6.24) that only those power functions are considered whose indicatrices at γ_0 are similar ellipses. The indicatrix of g_ϕ at γ_0 is characterized by the fact that it is contained in all other indicatrices at γ_0. The power functions g_{ϕ^*} with $\phi \in \Delta_\alpha^{(1)}$ thus represent a limiting case of those with $\phi \in \Delta_\alpha^2$. The principal axes of the Dupin indicatrix coincide with the principal radii of curvature. These are minimal at γ_0 for g_{ϕ^*} w.r.t. $\Delta_\alpha^{(2)}$ or, in other words, the *Gaussian total curvature* is maximum in $\Delta_\alpha^{(2)}$ for g_{ϕ^*}.

This can serve as the basis of a somewhat different definition: Let Δ_α now be the set of all tests whose power functions possess all first and second order derivatives at γ_0 and which satisfy (6.18) and (6.19). A test $\phi^* \in \Delta_\alpha$ is called of *type D* if its power function g_{ϕ^*} has maximal total curvature at γ_0 w.r.t. all other power functions $\phi \in \Delta_\alpha$. We will not go into the construction of such tests here.[42]

Let us finally make a remark on the case of one-dimensional parameters: The step which led from Theorem 6.1 to Theorem 6.2 consisted essentially of shrinking the set of tests considered. If we consider higher derivatives of the power function we can naturally carry out further steps of this type.

Considering the set of all "one-sided" tests ϕ whose power functions satisfy

$$g_\phi(\gamma_0) = \alpha, \qquad \frac{\partial}{\partial \gamma} g_\phi(\gamma_0) = 0, \qquad \frac{\partial^2}{\partial \gamma^2} g_\phi(\gamma_0) = 0$$

---

[42] See St. L. Isaacson, Ann. Math. Statist. 22, 217—234 (1951). For further generalizations see J. Neyman, Bull. Soc. Math. France 63, 246—266 (1935) and H. K. Nandi, Sankhyā 11, 13—22 (1951).

and assuming the existence of $\frac{\partial^3 g_\phi}{\partial \gamma^3}$ in an interval of the form $[\gamma_0, \gamma_1)$ we can, under reasonable assumptions, show the existence of l.m.p. "one-sided" tests w.r.t. this set. Adding the condition $\frac{\partial^3 g_\phi(\gamma_0)}{\partial \gamma^3}=0$ and assuming the existence of $\frac{\partial^4 g_\phi}{\partial \gamma^4}$ in an interval of the form $(\gamma_0-\delta, \gamma_0+\delta)$, $\delta>0$, one can show the existence of l.m.p. "two-sided" tests w.r.t. this set of tests, and so forth.

The considerations of **5** lead to the question whether one can, under certain conditions, construct tests for the problem $(\alpha, \{\gamma_0\}, R_1-\{\gamma_0\})$ which are most powerful even in $R_1-\{\gamma_0\}$ w.r.t. the set of all unbiased tests for this problem. It turns out that the answer is yes but we will not go into this here. We do remark that these considerations are especially important for the "two-sided" tests for the parameter of the normal distribution considered in Chapter II: All of the tests treated there are most powerful in the set of all unbiased tests (see Lehmann, loc. cit.[2]).

7. Sufficient transformations. We now discuss a notion which is of great importance not only in test theory but also in all parts of mathematical statistics. In **I.18** we defined the conditional probability and the following material will be closely connected to that on p. 59ff. We had there a probability space $(R, \mathcal{S}, P)$ and a measurable space $(Q, \mathfrak{Q})$, as well as a $(\mathcal{S}, \mathfrak{Q})$-measurable map T from R into Q. We then defined for each $A \in \mathcal{S}$ the conditional probability $P(A|T)$ of A "under the hypothesis T". This probability depends, however, in general on the underlying measure P. If one considers another probability measure over $(R, \mathcal{S})$, one obtains another conditional probability. However, it sometimes happens that for a set P_Γ of probability measures there exists a map T such that the conditional probability "under the hypothesis T" does not depend on the probability measure P_γ, $\gamma \in \Gamma$. We will now make these ideas precise and introduce first the following terminology: Let $A \in \mathcal{S}$. Instead of $P_\gamma(A|T)$ we will also write $P(A|T;\gamma)$. We use an analogous notation for the conditional expectation.

Definition. Let P_Γ be a set of probability measures over $(R, \mathcal{S})$, $(Q, \mathfrak{Q})$ a measurable space and T an $(\mathcal{S}, \mathfrak{Q})$-measurable map from R into Q. The σ-algebra $T^{-1}(\mathfrak{Q})$ is denoted by $\mathcal{S}_0$.[43]

If for each $A \in \mathcal{S}$ there exists an $\mathcal{S}_0$-measurable function f_A defined over R such that

$$f_A = P_\gamma(A|T) \quad P_\gamma\text{-a.e.} \tag{7.1}$$

[43] See p. 60.

for all $\gamma \in \Gamma$, then T is called a *sufficient transformation* (or *statistic*[44]) for the set Γ or for P_Γ.

Note that by p. 60 $P_\gamma(A|T)$ is only determined up to P_γ-null sets. Thus f_A is a version of the conditional probability $P_\gamma(A|T)$ for each $\gamma \in \Gamma$.

From the definition we immediately see that if T is sufficient for Γ then it is also so for each subset of Γ. Whether T is sufficient or not depends obviously on the σ-algebra $\mathscr{S}_0 = T^{-1}(\mathfrak{Q})$ induced by T. Every other map T_1 which doesn't have to transform into the same set Q, but which induces the same σ-algebra $\mathscr{S}_0 \subseteq \mathscr{S}$, is sufficient or not according as T is and is thus in this sense equivalent to T. Definition (7.1) shows that T and T_1 can also be viewed as equivalent if for the σ-algebra $\mathscr{S}_0^{(1)}$ induced by T_1 we have $\mathscr{S}_0 = \mathscr{S}_0^{(1)}$ P_Γ-a.e.[45]. Basically, the reference to a transformation T is unnecessary and one can work directly with the sub-σ-algebras of $\mathscr{S}$. Below, we will deal briefly with this point. If T is a sufficient transformation then a statement analogous to (7.1) also holds for the conditional expectation. Indeed, instead of (7.1) one can also write $f_A = E_\gamma(c_A|T)$. The rest follows without difficulty from I, Theorem 20.1. For each r.v. ξ whose expectation exists there thus exists an $\mathscr{S}_0$-measurable function f_ξ defined over R such that for all $\gamma \in \Gamma$

$$f_\xi = E(\xi|T;\gamma), \qquad P_\gamma\text{-a.e.} \tag{7.2}$$

We will often express this situation briefly by saying that $E(\xi|T)$ does not depend on $\gamma \in \Gamma$.

It is obvious that one can also start from $P_T(A|T;\gamma)$ in defining a sufficient transformation. (See p. 60.) If T is sufficient for Γ then there also exists a $\mathfrak{Q}$-measurable function $f_{A,T}$ defined over Q such that for all $\gamma \in \Gamma$

$$f_{A,T} = P_T(A|T;\gamma) \qquad P_{T,\gamma}\text{-a.e.}$$

and conversely. Further by I. (18.11), $f_{A,T} \circ T = f_A$. (Also see I, Theorem 18.3.)

One can naturally use the definition of a conditional probability, "given a σ-algebra" (p. 55) to introduce the idea of a *sufficient σ-algebra* instead of a sufficient transformation. However, we will make little use of this general notion although it is sometimes simpler to prove theorems directly for sufficient σ-algebras.

---

[44] For a more precise terminology see I [22].

[45] More precisely, this means that for each $A \in \mathscr{S}_0$ there is a $B \in \mathscr{S}_0^{(1)}$ such that $P_\gamma((A-B)\cup(B-A))=0$ for all $\gamma \in \Gamma$ and likewise when the roles of $\mathscr{S}_0$ and $\mathscr{S}_0^{(1)}$ are interchanged.

The following almost trivial remark illuminates the role of sufficient transformations in test theory: Let ϕ be a test for the problem $(\alpha, \Gamma_0, \Gamma-\Gamma_0)$, T sufficient for Γ and $\psi=E(\phi|T)$. Since $E(\psi;\gamma)=E(\phi;\gamma)$ for all $\gamma\in\Gamma$, one can always restrict oneself to tests which are functions of T if the power function is the criterion for the behavior of the tests.

We consider two important but rather trivial examples: Let T be such that $\mathscr{S}_0$ coincides with $\mathscr{S}$. Then $f_A=c_A$ P_Γ-a.e. In fact, c_A is $\mathscr{S}$-measurable and $\int_B c_A\,dP_\gamma=P_\gamma(A\cap B)$ for each $\gamma\in\Gamma$ and all $B\in\mathscr{S}$. T is thus sufficient. Note that in this example P_Γ is an arbitrary set of probability measures. We thus want to denote such sufficient transformations as *trivial*. The triviality of these sufficient transformations becomes expecially clear when one considers the measurable space $(R_n, \mathfrak{B}_n)$ and on it an arbitrary set P_Γ of probability measures. For the identity map T which sends x into $x\in R_n$, the set of all inverse images of measurable sets coincides with $\mathfrak{B}_n$ and T is thus sufficient for P_Γ. There thus always exists a sufficient transformation for the sample space $(R_n, \mathfrak{B}_n)$, with arbitrary P_Γ.[46] More generally, the identity map T from R into R, viewed as $(\mathscr{S},\mathscr{S})$-measurable transformation, is a trivial sufficient transformation.

Another example is obtained when $\mathscr{S}_0$ consists only of $\emptyset$ and R. This is precisely the case when T is constant, i.e., if for some element t of a set Q and all $x\in R$ the equation $T(x)=t$ holds. If T is sufficient, then f_A must be $\mathscr{S}_0$-measurable, thus also a constant. According to (7.1), $P_\gamma(A)$ is then also constant for each $A\in\mathscr{S}$ and all $\gamma\in\Gamma$. P_Γ thus consists of only one element and the definition of a sufficient transformation becomes uninteresting.

The construction of nontrivial sufficient transformations is especially easy to carry through when P_Γ is dominated by a σ-finite measure. We have first

Theorem 7.1. *Let P_Γ be a set of probability measures defined over the measurable space $(R, \mathscr{S})$ which is dominated by a σ-finite measure and let λ be a probability measure equivalent to P_Γ of the form given below. For each $\gamma\in\Gamma$ let f_γ denote the R.-N.-density of P_γ w.r.t. λ. Let $\mathscr{S}_0$ be a sub-σ-algebra of $\mathscr{S}$. Then $\mathscr{S}_0$ is sufficient for Γ iff each f_γ, $\gamma\in\Gamma$, is $\mathscr{S}_0$-measurable (λ-a.e.)*[47].

---

[46] In the statistical literature the existence of a sufficient transformation for a set of probability measures over $(R_n, \mathfrak{B}_n)$ is often proved. What is meant is the existence of a nontrivial sufficient transformation. See also J.L. Denny, Fundamenta Math. 55, 95—99 (1964).

[47] Thus for all real α the inverse image of $(-\infty, \alpha)$ under f_γ belongs to $\mathscr{S}_0$ up to a λ-null set.

Proof. By Theorem 5.2, a measure λ with the given properties exists. We first show that this condition is necessary. There exists a countable set $\{\gamma_1, \gamma_2, \ldots\} \subseteq \Gamma$ such that

$$\lambda = \sum_{i=1}^{\infty} p_i P_{\gamma_i}, \quad p_i > 0, \quad \sum_{i=1}^{\infty} p_i = 1\,. \tag{7.3}$$

For each $A \in \mathscr{S}$ and B in $\mathscr{S}_0$

$$\lambda(A \cap B) = \sum_{i=1}^{\infty} p_i P_{\gamma_i}(A \cap B)$$

and, since $\mathscr{S}_0$ is sufficient,

$$\lambda(A \cap B) = \sum_{i=1}^{\infty} p_i \int_B f_A \, dP_{\gamma_i}$$

where f_A is a version of the conditional probability $P_\gamma(A|\mathscr{S}_0)$ for each $\gamma \in \Gamma$. Further,

$$\sum_{i=1}^{\infty} p_i \int_B f_A f_{\gamma_i} d\lambda = \int_B f_A \sum_{i=1}^{\infty} f_{\gamma_i} p_i \, d\lambda,$$

where the exchange of integral and summation is justified by Theorem VI. Finally, (7.3) yields

$$\lambda(A \cap B) = \int_B f_A \, d\lambda$$

i.e.,

$$f_A = \lambda(A|\mathscr{S}_0) \qquad \lambda\text{-a.e.} \tag{7.4}$$

Moreover, for each $B \in \mathscr{S}_0$ and $\lambda \in \Gamma$

$$P_\gamma(B) = \int_B E_\lambda(f_\gamma|\mathscr{S}_0) \, d\lambda.\,^{48} \tag{7.5}$$

Since f_A is $\mathscr{S}_0$-measurable by definition, we have from (7.5) by applying I, Theorem 6.2

$$\int_R f_A \, dP_\gamma = \int_R f_A E_\lambda(f_\gamma|\mathscr{S}_0) \, d\lambda. \tag{7.6}$$

We can also write $f_A = E_\lambda(c_A|\mathscr{S}_0)$ so that from (7.6)

$$\int_R f_A \, dP_\gamma = \int_R E_\lambda(c_A|\mathscr{S}_0) E_\lambda(f_\gamma|\mathscr{S}_0) \, d\lambda,$$

---

[48] $E_\lambda(f_\gamma|\mathscr{S}_0)$ denotes the conditional expectation w.r.t. the measure λ.

and because $E_\lambda(f_\gamma|\mathcal{S}_0)$ is $\mathcal{S}_0$-measurable, we get from I, Theorem 20.2

$$\int_R f_A \, dP_\gamma = \int_R E_\lambda(c_A E_\lambda(f_\gamma|\mathcal{S}_0)|\mathcal{S}_0)\, d\lambda$$

and hence also

$$\int_R f_A \, dP_\gamma = \int_R c_A E_\lambda(f_\gamma|\mathcal{S}_0)\, d\lambda \tag{7.7}$$

which follows simply from the definition of the conditional expectation.

Moreover, for each $\gamma \in \Gamma$

$$\int_A f_\gamma \, d\lambda = \int_R f_A \, dP_\gamma, \tag{7.8}$$

for both integrals are equal to $P_\gamma(A)$.

(7.8) and (7.7) yield: for each $A \in \mathcal{S}$

$$\int_A f_\gamma \, d\lambda = \int_A E_\lambda(f_\gamma|\mathcal{S}_0)\, d\lambda$$

so that

$$f_\gamma = E_\lambda(f_\gamma|\mathcal{S}_0) \qquad \lambda\text{-a.e.}$$

i. e., f_γ is $\mathcal{S}_0$-measurable (λ-a.e.).

Now let conversely f_γ be $\mathcal{S}_0$-measurable for each $\gamma \in \Gamma$. We will show that $\lambda(A|\mathcal{S}_0)$ is for each $A \in \mathcal{S}$ a version of the conditional probability $P_\gamma(A|\mathcal{S}_0)$ for each $\gamma \in \Gamma$. Write $\lambda(A|\mathcal{S}_0) = f_A$. For each $B \in \mathcal{S}_0$ and $\gamma \in \Gamma$

$$P_\gamma(A \cap B) = \int_B c_A f_\gamma \, d\lambda = \int_B E_\lambda(c_A f_\gamma|\mathcal{S}_0)\, d\lambda.$$

From the assumed $\mathcal{S}_0$-measurability of f_γ we have

$$P_\gamma(A \cap B) = \int_B f_\gamma E_\lambda(c_A|\mathcal{S}_0)\, d\lambda = \int_B f_\gamma f_A \, d\lambda = \int_B f_A \, dP_\gamma$$

and the first and last members of this chain of equalities yield the claim.

We apply this to the important case in which T is an $(\mathcal{S}, \mathfrak{Q})$-measurable map from R into Q and $\mathcal{S}_0$ is identified with $T^{-1}(\mathfrak{Q})$. If T is sufficient for Γ, then there exists according to I, Theorem 18.3 for each $\gamma \in \Gamma$ a $\mathfrak{Q}$-measurable map g_γ from Q into R_1 such that

$$f_\gamma = g_\gamma \circ T.$$ [49]

f_γ is thus a function of T for each $\gamma \in \Gamma$ if T is sufficient, and conversely. This remark can be expressed as the important Fisher-Neyman-Halmos-Savage criterion.

[49] We have shown this only for the case $T(R) = Q$. However, see I [23].

Theorem 7.2[50]. *Let P_Γ be dominated by a σ-finite measure μ and T an $(\mathcal{S},\mathfrak{Q})$-measurable map from R into Q. T is sufficient iff there exists a nonnegative $\mathcal{S}$-measurable map h defined over R and for each $\gamma \in \Gamma$ a $\mathfrak{Q}$-measurable g_γ from Q into R_1 such that the R.-N.-density k_γ of P_γ w.r.t. μ is given by $h(g_\gamma \circ T)$.*

Note that the μ-integrability of h is not required.

Proof. Let T be sufficient and λ have the same meaning as in Theorem 7.1. As we have shown above, we then have for the R.-N.-density f_γ of P_γ w.r.t. λ

$$f_\gamma = g_\gamma \circ T \tag{7.9}$$

for each $\gamma \in \Gamma$, where g_γ is a $\mathfrak{Q}$-measurable function. Since from $\mu(N)=0$, we have $P_\gamma(N)=0$ for all $\gamma \in \Gamma$ and hence also $\lambda(N)=0$, λ is absolutely continuous w.r.t. μ. Hence, there exists an $\mathcal{S}$-measurable R.-N.-density h of λ w.r.t. μ and therefore from (7.9) after applying I, Theorem 6.2

$$k_\gamma = h(g_\gamma \circ T). \tag{7.10}$$

Conversely, let a representation of the form (7.10) hold for $k_\gamma, \gamma \in \Gamma$. Then for each $A \in \mathcal{S}$

$$\lambda(A) = \int_A h \sum_{i=1}^{\infty} p_i (g_{\gamma_i} \circ T)\, d\mu,$$

where p_i and γ_i have the meaning of Theorem 7.1. Again let $\mathcal{S}_0 = T^{-1}(\mathfrak{Q})$. Then $k = \sum_{i=1}^{\infty} p_i (g_{\gamma_i} \circ T)$ is a nonnegative $\mathcal{S}_0$-measurable function and hk is the R.-N.-density of λ w.r.t. μ.

Now let

$$g_\gamma^*(x) = \begin{cases} ((g_\gamma \circ T)(x))/k(x) & k(x) > 0 \\ 0 & \text{otherwise.} \end{cases}$$

Then g_γ^* is, as quotient of two $\mathcal{S}_0$-measurable functions, resp., as constant function on the $\mathcal{S}_0$-measurable set $\{x: k(x)=0\}$ likewise $\mathcal{S}_0$-measurable.

We show now that g_γ^* is a version of the R.-N.-density of P_γ w.r.t. λ for each $\gamma \in \Gamma$.

By I, Theorem 6.2 we have for the R.-N.-densities

$$\frac{dP_\gamma}{d\mu} = \frac{dP_\gamma}{d\lambda}\frac{d\lambda}{d\mu} \quad \mu\text{-a.e.}, \quad \gamma \in \Gamma$$

---

[50] See J. Neyman, Giorn. Ital. Attuari 6, 320—334 (1953) as well as P.R. Halmos and L.J. Savage, l.c.[27]. Generalizations are in R.R. Bahadur, Ann. Math. Statist. 25, 423—462 (1954).

or

$$h(g_\gamma \circ T) = \frac{dP_\gamma}{d\lambda} hk \qquad \lambda\text{-a.e.}$$

If $h(x) > 0$ and $k(x) > 0$, then $dP_\gamma(x)/d\lambda = g_\gamma^*(x)$. The sets $\{x: h(x)=0\}$ and $\{x: k(x)=0\}$ are, however, λ-null sets since $hk = d\lambda/d\mu$. Moreover, each μ-null set is a λ-null set. Hence, we actually have $dP_\gamma/d\lambda = g_\gamma^*$ λ-a.e., so that $dP_\gamma/d\lambda$ is $\mathscr{S}_0$-measurable (λ-a.e.) for each $\gamma \in \Gamma$. The claim then follows from Theorem 7.1.

Theorem 7.2 immediately leads to many examples of nontrivial sufficient transformations.

We recognize at once that the set of all probability measures whose densities are given by (5.14) possess the function T defined there as sufficient transformation.

The examples on p. 190 then also provide concrete examples of sufficient transformations.

The set of normal distributions in R_n whose densities for $-\infty < a < \infty$ and $0 < \sigma^2 < \infty$ are given by (1.3) possess the map $(x_1, \ldots, x_n) \to (\bar{x}, s^2)$ as sufficient transformation for the parameter set $\{(a, \sigma^2), -\infty < a < \infty, 0 < \sigma^2 < \infty\}$. Indeed,

$$e^{-\sum\limits_{i=1}^{\infty} (x_i - a)^2/2\sigma^2} = e^{-(n(\bar{x}-a)^2 + (n-1)s^2)/2\sigma^2}.$$

Another example, which is of interest for the so-called *Behrens-Fisher Problem*, is obtained as follows: Let $\xi_1, \ldots, \xi_n$, $n \geqslant 1$, be independent, identically $N(a_1, \sigma_1^2)$-distributed r.v.'s $-\infty < a_1 < \infty$, $0 < \sigma_1^2 < \infty$. Further let $\eta_1, \ldots, \eta_m$, $m \geqslant 1$, be independent, identically distributed normal r.v.'s which are also independent of all the ξ_i, $1 \leqslant i \leqslant n$. The distribution of η_i, $1 \leqslant i \leqslant m$ is denoted by $N(a_2, \sigma_2^2)$, $-\infty < a_2 < \infty$, $0 < \sigma_2^2 < \infty$. The density of the joint distribution of $(\xi_1, \ldots, \xi_n, \eta_1, \ldots, \eta_m)$ is obviously, for each $(x_1, \ldots, x_n, y_1, \ldots, y_m) \in R_{n+m}$ given by

$$(2\pi)^{-\frac{1}{2}(n+m)} \sigma_1^{-n} \sigma_2^{-m} e^{-\frac{1}{2}\sum\limits_{i=1}^{n} (x_i - a_1)^2/\sigma_1^2} e^{-\frac{1}{2}\sum\limits_{j=1}^{m} (y_j - a_2)^2/\sigma_2^2}.$$

One sees at once that the map $(x_1, \ldots, x_n, y_1, \ldots, y_m) \to (\bar{x}, \bar{y}, s_x^2, s_y^2)$ is sufficient for

$$\Gamma = \{(a_1, a_2, \sigma_1^2, \sigma_2^2), -\infty < a_1 < \infty, -\infty < a_2 < \infty, 0 < \sigma_1^2 < \infty, 0 < \sigma_2^2 < \infty\}.$$

More generally, one can show that for the set of probability measures with monotone likelihood ratios the function T, which exists by (5.1), is sufficient. In fact, let $\Gamma \subseteq R_1$ and P_Γ be dominated by a σ-finite measure μ for which we assume that it is a probability measure equivalent to P_Γ. Then we have, in a symbolism already used, $\mu = \sum\limits_{i=1}^{\infty} p_i P_{\gamma_i}$, Then $1 = \sum\limits_{i=1}^{\infty} p_i \frac{dP_{\gamma_i}}{d\mu}$, μ-a.e. so that with $dP_\gamma/d\mu = f_\gamma$ we have

$$f_\gamma = f_\gamma \Big/ \sum_{i=1}^{\infty} p_i f_{\gamma_i} \qquad \mu\text{-a.e.}$$

for an arbitrary $\gamma\in\Gamma$. Then, because of (5.1)

$$f_\gamma = 1\Big/\Big(\sum_{\gamma_i\leqslant\gamma} p_i H_{\gamma_i,\gamma}\circ T + \sum_{\gamma_i>\gamma} p_i (H_{\gamma_i,\gamma}\circ T)^{-1}\Big) \quad \mu\text{-a.e.} \tag{7.11}$$

where $H_{\gamma_i,\gamma}(y)=1$ should hold for $\gamma_i=\gamma$ and all $y\in R_1$. Indeed, the set $\left\{x: \sum_{i=1}^{\infty} p_i f_{\gamma_i}(x)=0\right\}$ is a μ-null set and thus need not be considered. However, if x doesn't belong to this set, then the correctness of (7.11) is easy to check.

Further examples of sufficient transformations are given in VI. See, for example VI, Theorems 3.1, 3.3.

Sufficient transformations play an important role in mathematical statistics. We will illustrate this later or on several occasions. See also p. 208. Much of this importance is due to Theorem 7.2. If $\gamma_0\in\Gamma$ is an arbitrary but fixed element, then for each $\gamma\in\Gamma$

$$f_\gamma/f_{\gamma_0} = g_\gamma\circ T/(g_{\gamma_0}\circ T) \tag{7.12}$$

$\frac{1}{2}(P_\gamma+P_{\gamma_0})$-a.e. Hence, this density ratio depends only on γ and T when γ_0 is fixed. More intuitively: Any two samples x' and x'' from the sample space R for which $T(x')=T(x'')$ yield the same value f_γ/f_{γ_0}. To be more precise, we have simply the following situation: Let T map R into Q and let $M_t=\{x: T(x)=t\}$, $t\in Q$. Then $M_t\cap M_{t'}=\emptyset$ for $t\neq t'$ and $\bigcup_{t\in Q} M_t=R$. According to (7.12), it is sufficient for "knowledge" of the map $\gamma\to f_\gamma(x)/f_{\gamma_0}(x)$ for each sample x to consider the set $\mathfrak{M}=\{M_t: t\in Q\}$. However, $\mathfrak{M}$ contains in general "fewer" elements than R. It is thus advantageous to choose sufficient transformations which define sets M_t which are as "large as possible". To this end we have the

Definition. Let χ be a map from R into some set M. χ is called a *minimal transformation* for Γ if for *each transformation* T mapping R into Q and *sufficient* for Γ there exists a map m_T from Q into M such that

$$\chi = m_T\circ T \qquad P_\Gamma\text{-a.e.} \tag{7.13}$$

Each map χ_0 over R which is a function of χ, i. e., which can be written in the form $\chi_0=q\circ\chi$ by means of a given map q, is likewise minimal since with $q\circ m_T=n_T$ we have immediately from (7.13) $\chi_0=n_T\circ T$, P_Γ-a.e.

If we consider the concept of a sufficient σ-algebra, then the corresponding definition obviously runs as follows:

A sub-σ-algebra $\mathfrak{S}_0$ of $\mathscr{S}$ is called *minimal* if $\mathfrak{S}_0 \subseteq \mathscr{S}_0$ P_Γ-a.e. for each sufficient σ-algebra $\mathscr{S}_0$, i. e., each set of $\mathfrak{S}_0$ is equal to a set of $\mathscr{S}_0$ with the exception of at most a P_Γ-null set. Each σ-algebra contained in $\mathfrak{S}_0$ is naturally also minimal.

Obviously, there always exists a minimal transformation. One obtains one by choosing an arbitrary element $l \in M$ and defining $\chi(x) = l$ for all $x \in R$; m_T is then simply the mapping $t \to m_T(t) = l$ from Q into M, and (7.13) is fulfilled.

Likewise, there always exists a minimal σ-algebra containing only the sets $\emptyset$ and R.

A comparison between the definitions of the minimal transformation and the minimal σ-algebra causes no difficulties if one considers, say, only minimal transformations χ which map R into R_n, $n \geqslant 1$, and for which the corresponding map m_T is $\mathfrak{Q}$-measurable. It then follows from I, Theorem 18.3[51], that χ is necessarily $\mathscr{S}_0$-measurable or, more exactly, $(\mathscr{S}_0, \mathfrak{B}_n)$-measurable, where $\mathscr{S}_0 = T^{-1}(\mathfrak{Q})$ is the σ-algebra induced by T. Considering now the set of all transformations T sufficient for Γ and the corresponding set of all induced sufficient σ-algebras $\mathscr{S}_0$, we see that their intersection $\mathfrak{S}_0$ is a σ-algebra and χ must be $\mathfrak{S}_0$-measurable. But $\mathfrak{S}_0$ is also included in all $\mathscr{S}_0$ and is thus minimal.

From the given definitions it does not follow that a minimal transformation or σ-algebra must also be sufficient.

Of especial interest are therefore minimal transformations which are also sufficient. If the set P_Γ is dominated by a σ-finite measure, then a result in this direction can easily be obtained from Theorem 7.1. For the sake of convenience, we express it for σ-algebras.

Theorem 7.3. *Let P_Γ be a set of probability measures over $(R, \mathscr{S})$ dominated by a σ-finite measure μ. We can assume that μ is equivalent to P_Γ. For each $\gamma \in \Gamma$ denote the R.-N.-density of P_γ w.r.t. μ by f_γ. Let α be a real number and $\mathfrak{S}_0$ the σ-algebra generated by all sets of the form $\{x: f_\gamma(x) < \alpha\}$, $\gamma \in \Gamma$, $\alpha \in R_1$. Then $\mathfrak{S}_0$ is sufficient and minimal for Γ.*

The *proof* is simple: By Theorem 7.1, $\mathfrak{S}_0$ is sufficient. Let $\mathscr{S}_0$ be an arbitrary (for Γ) sufficient σ-algebra. Then again by Theorem 7.1, f_γ is also $\mathscr{S}_0$-measurable (μ-a.e.) for each γ. Hence $\mathfrak{S}_0 \subseteq \mathscr{S}_0$ μ-a.e., which is easy to show. Since, however, P_Γ and μ are equivalent, $\mathfrak{S}_0 \subseteq \mathscr{S}_0$ P_Γ-a.e., which was to be proved.

Theorem 7.3 now leads without difficulty in concrete cases to the construction of sufficient and minimal transformations. For the sake of illustration, we limit ourselves to R_n as sample space and assume that we are starting from n independent, identically distributed r.v.'s.

[51] Actually, I, Theorem 18.3 yields this only for $n=1$, but I, Theorem 18.3 can easily be extended to the case where the function f named there has range R_n with $n>1$.

We first have the following **definition** which we give in quite general form: Let $\mathfrak{F}$ be a (non-empty) set of functions defined over R. The *smallest natural number* r for which there exist r elements $\psi_1,\ldots,\psi_r \in \mathfrak{F}$ such that each $\psi \in \mathfrak{F}$ can be represented with real $c_1,\ldots,c_r$ as

$$\psi = \sum_{i=1}^{r} c_i \psi_i, \tag{7.14}$$

is called *the dimension of* $\mathfrak{F}$. If *no such natural* r *exists, we define* $r = \infty$. The set $\{\psi_1,\ldots,\psi_r\}$ is called *a basis* of $\mathfrak{F}$.

We also have the

Definition. Let P_Γ be a set of probability measures over $(R_n, \mathfrak{B}_n)$, $n \geqslant 1$ and T a map from R_n into an R_r, $r \geqslant 1$. T is called *locally trivial* if there exists an $x_0 \in R$, a neighborhood $U(x_0)$ of x_0 and a measurable map g from R_r into R_n such that for all $x \in U(x_0)$

$$I(x) = (g \circ T)(x) \tag{7.15}$$

where I is the identity map which carries every $x \in R_n$ into itself.

If, in particular, T is a $(\mathfrak{B}_n, \mathfrak{B}_r)$-measurable transformation and (7.15) holds for all $x \in R_n$, then $I^{-1}(\mathfrak{B}_n) = \mathfrak{B}_n$ implies that also $T^{-1}(\mathfrak{B}_r) = \mathfrak{B}_r$. But then T is trivial (see p. 208). This justifies the definition. We now proceed to the announced specialization of Theorem 7.3 and give it in somewhat extended form as

Theorem 7.4. *Let* $P_\Gamma^{(1)}$ *be a set of probability measures over* $(R_1, \mathfrak{B}_1)$ *which are dominated by a* σ*-finite measure* μ. *The R.-N.-density of* $P_\gamma^{(1)}$ *will be denoted for each* $\gamma \in \Gamma$ *by* $f_\gamma^{(1)}$ *and we assume that* $f_\gamma^{(1)} > 0$.[52] *Let* $\mathfrak{F}$ *be the set of all* $\log f_\gamma^{(1)}$, $\gamma \in \Gamma$ *and all functions constant over* R_1. *Let* P_γ *be the set of all product measures over* $(R_n, \mathfrak{B}_n)$, $n \geqslant 2$ *whose densities* f_γ *w.r.t.* $\underbrace{\mu \times \cdots \times \mu}_{n}$ *for each* $\gamma \in \Gamma$ *are given by* $(x_1,\ldots,x_n) \to \prod_{i=1}^{n} f_\gamma^{(1)}(x_i)$.
If $\mathfrak{F}$ *is of finite dimension* r *and* $1, \psi_1,\ldots,\psi_{r-1}$ *is a basis for* $\mathfrak{F}$, *then the map defined for each* $(x_1,\ldots,x_n) \in R_n$ *by*

$$T(x_1,\ldots,x_n) = \left(\sum_{i=1}^{n} \psi_1(x_i), \ldots, \sum_{i=1}^{n} \psi_{r-1}(x_i)\right) \tag{7.16}$$

is a minimal and sufficient transformation from R_n *into* R_{r-1} *for* P_Γ. *If, moreover, the function* $x \to f_\gamma^{(1)}(x)$ *is continuously differentiable in* R_1 *for*

---

[52] The assumption that the densities are >0 in all of R_1 is made only for convenience. It is enough, for example, that the f_γ be >0 for all $\gamma \in \Gamma$ in a fixed open interval and vanish for all γ outside of this fixed interval.

each $\gamma \in \Gamma$ *and* $n < r$, *where* r *can also equal* ∞, *then each transformation defined by* (7.16) *is locally trivial*[53].

Proof. The first part follows almost immediately from Theorem 7.3. Indeed, assume $\mathfrak{S}_0^{(1)}$ is the σ-algebra generated by all sets of the form $\{x: \psi_j(x) < \alpha\}$, $j = 1, \ldots, r-1$, α arbitrary and real. For each $\gamma \in \Gamma$, there are real numbers $c_0(\gamma), \ldots, c_{r-1}(\gamma)$ with

$$\log f_\gamma^{(1)} = c_0(\gamma) + \sum_{j=1}^{r-1} c_j(\gamma)\,\psi_j. \tag{7.17}$$

Hence, the σ-algebra generated by all sets of the form $\{x: f_\gamma^{(1)}(x) < \alpha\}$, $\gamma \in \Gamma$, α arbitrary real, coincides with $\mathfrak{S}_0^{(1)}$. Denoting by $\mathfrak{S}_0$ the $\otimes$-product of n copies of $\mathfrak{S}_0^{(1)}$, we have that the σ-algebra generated by all sets of the form $\{(x_1, \ldots, x_n): f_\gamma(x_1, \ldots, x_n) < \alpha\}$, $\gamma \in \Gamma$, α arbitrary real, coincides precisely with $\mathfrak{S}_0$.

In fact, from (7.17) according to the definition of f_γ:

$$\log f_\gamma(x_1, \ldots, x_n) = n\,c_0(\gamma) + \sum_{i=1}^{n} \sum_{j=1}^{r-1} c_j(\gamma)\,\psi_j(x_i).$$

$\mathfrak{S}_0$ is thus a minimal and sufficient σ-algebra for Γ. Writing $(c_1(\gamma), \ldots, c_{r-1}(\gamma)) = c(\gamma)$ and $(y_1, \ldots, y_{r-1}) = y \in R_{r-1}$ and denoting the function $y \to e^{n c_0(\gamma)}\, e^{(c(\gamma))' y}$ by g_γ, we then have

$$f_\gamma = g_\gamma \circ T.$$

Therefore, T is a minimal and sufficient $(\mathfrak{S}_0, \mathfrak{B}_{r-1})$-measurable transformation.

Regarding the second part of the theorem we remark that because of $n < r$, there exist at least $n+1$ linearly independent functions $\psi_0, \ldots, \psi_n \in \mathfrak{F}$ with $\psi_0(x) = 1$ for all $x \in R_1$. From this it follows that the Jacobian of the transformations $\tau_j: (x_1, \ldots, x_n) \to \sum_{i=1}^{n} \psi_j(x_i)$, $1 \leq j \leq n$, does not identically vanish. Actually, for each $x = (x_1, \ldots, x_n) \in R_n$

$$\begin{vmatrix} \dfrac{\partial \tau_1(x)}{\partial x_1} & \cdots & \dfrac{\partial \tau_1(x)}{\partial x_n} \\ \cdots & \cdots & \cdots \\ \dfrac{\partial \tau_n(x)}{\partial x_1} & \cdots & \dfrac{\partial \tau_n(x)}{\partial x_n} \end{vmatrix} = \begin{vmatrix} \dfrac{\partial \psi_1(x_1)}{\partial x_1} & \cdots & \dfrac{\partial \psi_1(x_n)}{\partial x_n} \\ \cdots & \cdots & \cdots \\ \dfrac{\partial \psi_n(x_1)}{\partial x_1} & \cdots & \dfrac{\partial \psi_n(x_n)}{\partial x_n} \end{vmatrix}.$$

Assume this determinant vanishes identically and that there exists an $(n-1)$-th order determinant which does not vanish identically. That is,

[53] Essentially due to E. B. Dynkin, Uspehi Mat. Nauk 6, 68—90 (1951). See also B. O. Koopman, Trans. Amer. Math. Soc. 39, 399—409 (1936).

there exists an $x=(x_1,\dots,x_{n-1})\in R_{n-1}$ such that (if necessary after a modification of the notation)

$$\begin{vmatrix} \frac{\partial\psi_1(x_1)}{\partial x_1} & \dots & \frac{\partial\psi_1(x_{n-1})}{\partial x_{n-1}} \\ \dots & \dots & \dots \\ \frac{\partial\psi_{n-1}(x_1)}{\partial x_1} & \dots & \frac{\partial\psi_{n-1}(x_{n-1})}{\partial x_{n-1}} \end{vmatrix} \neq 0 .$$

But then, for arbitrary $x_n\in R_1$

$$0=\begin{vmatrix} \frac{\partial\psi_1(x_1)}{\partial x_1} & \dots & \frac{\partial\psi_1(x_{n-1})}{\partial x_{n-1}} & \frac{\partial\psi_1(x_n)}{\partial x_n} \\ \dots & \dots & \dots & \dots \\ \frac{\partial\psi_{n-1}(x_1)}{\partial x_1} & \dots & \frac{\partial\psi_{n-1}(x_{n-1})}{\partial x_{n-1}} & \frac{\partial\psi_{n-1}(x_n)}{\partial x_n} \\ \frac{\partial\psi_n(x_1)}{\partial x_1} & \dots & \frac{\partial\psi_n(x_{n-1})}{\partial x_{n-1}} & \frac{\partial\psi_n(x_n)}{\partial x_n} \end{vmatrix} =$$

$$=(-1)^{n-1}\frac{\partial\psi_1(x_n)}{\partial x_n}\begin{vmatrix} \frac{\partial\psi_2(x_1)}{\partial x_1} & \dots & \frac{\partial\psi_2(x_{n-1})}{\partial x_{n-1}} \\ \dots & \dots & \dots \\ \frac{\partial\psi_n(x_1)}{\partial x_1} & \dots & \frac{\partial\psi_n(x_{n-1})}{\partial x_{n-1}} \end{vmatrix} +\dots+$$

$$+\frac{\partial\psi_n(x_n)}{\partial x_n}\begin{vmatrix} \frac{\partial\psi_1(x_1)}{\partial x_1} & \dots & \frac{\partial\psi_1(x_{n-1})}{\partial x_{n-1}} \\ \dots & \dots & \dots \\ \frac{\partial\psi_{n-1}(x_1)}{\partial x_1} & \dots & \frac{\partial\psi_{n-1}(x_{n-1})}{\partial x_{n-1}} \end{vmatrix} .$$

Thus, there exist n real numbers $\beta_1,\dots,\beta_n$ with $\beta_n\neq 0$, such that

$$\sum_{i=1}^{n}\beta_i\frac{\partial\psi_i(x_n)}{\partial x_n}=0 \quad \text{for all } x_n\in R_1 .$$

This means that the ψ_i must satisfy a relation of the form $\sum_{i=1}^{n}\beta_i\psi_i=C$, where C is a real number. Since $\beta_n\neq 0$, this contradicts the linear independence of $1,\dots,\psi_n$.

On the other hand, if a determinant of order $m-1, 2\leqslant m\leqslant n$, does not vanish identically, then we repeat the argument with $m-1$ and m instead of $n-1$ and n.

Hence, there exists an $x_0 \in R_n$ with $\frac{\partial(\tau_1, \ldots, \tau_n)}{\partial(x_1, \ldots, x_n)}(x_0) \neq 0$ which implies that the map $T_1 = (\tau_1, \ldots, \tau_n)$ is invertible in a neighborhood $U(x_0)$ of x_0, i.e., there exists a map g from R_n into R_n such that $I = g \circ T_1$ holds in $U(x_0)$. If r is finite, then because $n \leqslant r-1$, we also have $I = \bar{g} \circ T$ in $U(x_0)$ for the transformation T defined by (7.16), where $\bar{g}$ is a suitable map from R_{r-1} into R_n. This still holds when $r = \infty$, which proves Theorem 7.4.

For the important question of the extent to which one can determine global properties of sufficient transformations from local ones, see for example Barankin and Katz[54].

The most important examples for the application of Theorem 7.4 are given by distributions of the exponential type: Let μ be a σ-additive measure defined over $(R_1, \mathfrak{B}_1)$ and $T_1, \ldots, T_k$, $k \geqslant 1$, linearly independent functions over R_1 which are not μ-a.e. constant. For all $\gamma = (\gamma_1, \ldots, \gamma_k) \in R_k$ let the functions $C(\gamma) e^{\gamma_1 T_1 + \cdots + \gamma_k T_k}$ be μ-integrable, where $C(\gamma) > 0$ is so chosen that

$$\int_{R_k} e^{\gamma_1 T_1 + \cdots + \gamma_k T_k} d\mu = \frac{1}{C(\gamma)}.$$

Then the transformation $(T_1, \ldots, T_k)$ is sufficient for the set P_Γ with $\Gamma = R_k$ which is defined by the R.-N.-densities $C(\gamma) e^{\gamma_1 T_1 + \cdots + \gamma_k T_k}$. This result is related to the problem of characterizing the family of exponential distributions by the existence of a sufficient transformation. See f.i. J. L. Denny[55].

General investigations for the case of a nondominated set P_Γ have been initiated especially by D. L. Burkholder.[56]

8. The notion of completeness. We begin with the following important

Definition[57]. A set of measures P_Γ over a measure space $(R, \mathscr{S})$ is called *complete* if for each $\mathscr{S}$-measurable function f for which $E(f; \gamma)$ exists for each $\gamma \in \Gamma$ and which fulfills

$$E(f; \gamma) = 0 \tag{8.1}$$

for each $\gamma \in \Gamma$, we always have $f(x) = 0$ P_Γ-a.e. P_Γ is *boundedly complete* when (8.1) for all $\gamma \in \Gamma$ and $|f| \leqslant M$ for an $M \geqslant 0$ imply $f(x) = 0$ P_Γ-a.e.

If $\Gamma^* \subseteq \Gamma$, then from the completeness of P_{Γ^*} follows that of P_Γ if P_{Γ^*} and P_Γ are equivalent to the same measure. However, if $\Gamma^* \supset \Gamma$,

[54] E. W. Barankin and M. Katz, Sankhyā 21, 217—246 (1959) and E. W. Barankin and A. P. Maitra, Sankhyā 25, 217—244 (1963).

[55] J. L. Denny, Proc. Nat. Acad. Sci. U.S.A. 57, 1184—1187 (1967). Ann. Math. Statist. 41, 401—411 (1970). See also O. Barndorff-Nielsen and Karl Pedersen, Math. Scand. 22, 197—202 (1968).

[56] D. L. Burkholder, Ann. Math. Statist. 32, 1191—1200 (1961) and Ann. Math. Statistics 33, 596—599 (1962). See also T. S. Pitcher, loc. cit. VII[7].

[57] The importance of this definition in mathematical statistics was clearly presented for the first time in E. L. Lehmann and H. Scheffé, Sankhyā 10, 305—340 (1950).

then the completeness of $P_{\Gamma*}$ does in general not imply the completeness of P_Γ. Naturally, a similar remark holds also for bounded completeness.

We consider some examples for this definition:

1. From a well-known theorem of Müntz[58] we have that for each function f integrable in $[0,1]$, $\int_0^1 f(x)x^p dx=0$ for all prime numbers p implies $f=0$ a.e.. But then, if Γ consists of 0 and all primes, and P_Γ is the set of all probability measures over $(R_1, \mathfrak{B}_1)$ whose densities for $\gamma\in\Gamma$ are given by

$$\begin{array}{ll}(\gamma+1)x^\gamma & 0\leqslant x\leqslant 1\\ 0 & \text{otherwise},\end{array}$$

we have that P_Γ is complete.

2. If $\Gamma=R_1$ and P_Γ is the set of all $N(\gamma,1)$ with $\gamma\in\Gamma$, then P_Γ is complete. Assume namely that

$$\int_{-\infty}^{+\infty} f(x)e^{\frac{-(x-\gamma)^2}{2}}dx=0 \tag{8.2}$$

for $\gamma\in\Gamma$.

f itself is not necessarily integrable but, by assumption, the function g given for each $x\in R_1$ by $f(x)e^{-x^2/2}$ is integrable. Now, from (8.2),

$$\int_{-\infty}^{+\infty} f(x)e^{-x^2/2}e^{x\gamma}dx=0 \quad \text{for } -\infty<\gamma<\infty$$

and by a known theorem on the Laplace transform of integrable functions it follows that g vanishes a.e. and since $e^{-x^2/2}>0$ for $-\infty<x<\infty$, we also have $f=0$ a.e.

3. Let $\Gamma=\{1,2,\dots,l\}$ with $l\geqslant 1$, and assume $x_1,\dots,x_k$, $k\leqslant l$, are the mass points of a set P_Γ of discrete probability distributions over R_1:

$$P_\gamma(\xi=x_i)=p_{i\gamma}, \quad 1\leqslant i\leqslant k, \quad \sum_{i=1}^k p_{i\gamma}=1, \quad \gamma\in\Gamma.$$

Let $|p_{i\gamma}|_{1k}^{1k}\neq 0$. Then P_Γ is complete. Indeed, if g is an arbitrary function over R_1, then

$$\sum_{i=1}^k g(x_i)p_{i\gamma}=0, \quad \gamma=1,\dots,l$$

implies

$$g(x_i)=0, \quad 1\leqslant i\leqslant k, \quad \text{i.e.,} \quad g=0 \quad P_\Gamma\text{-a.e.}$$

It is easy to show that a set P_Γ of measures can be boundedly complete without being complete.

We do not dwell on this point but consider instead a further example. Let $\Gamma=(0,\infty)$ and P_γ be given for each $\gamma\in\Gamma$ over $(R_2,\mathfrak{B}_2)$ by the density

$$\begin{array}{ll}e^{-\frac{x_1}{\gamma}-x_2\gamma}, & x_1\geqslant 0,\ x_2\geqslant 0\\ 0 & \text{otherwise}.\end{array}$$

[58] See H. Steinhaus—L. Kaczmarz: Theorie der Orthogonalreihen, Monografje Matematyczne VI, Warschau 1935.

Let J_0 be the Bessel function of order zero. It is well known that J_0 is bounded. Then one easily sees that

$$\int_0^\infty \int_0^\infty J_0(\sqrt{x_1 x_2}) e^{-\frac{x_1}{\gamma} - \gamma x_2} dx_1 dx_2 = \tfrac{4}{5} \quad [59]$$

so that

$$\int_0^\infty \int_0^\infty (J_0(\sqrt{x_1 x_2}) - \tfrac{4}{5}) e^{-\frac{x_1}{\gamma} - \gamma x_2} dx_1 dx_2 = 0$$

for all $\gamma \in \Gamma$, i.e., P_Γ is not even boundedly complete. On the other hand it is easy to see that the two families of measures given for each $\gamma \in \Gamma$ by the densities

$$x_1 \to \gamma^{-1} e^{-\frac{x_1}{\gamma}}, \qquad x_1 \geqslant 0$$

and

$$x_2 \to \gamma e^{-\gamma x_2}, \qquad x_2 \geqslant 0$$

resp. are complete.

A refinement of the definition of completeness suggests itself: Let P_Γ be a set of probability measures over $(R, \mathscr{S})$ and f an $\mathscr{S}$-measurable function. Let $p \geqslant 1$ *be a real number and assume* $E(|f|^p; \gamma)$ *exists for each* $\gamma \in \Gamma$. Let $E(f; \gamma) = 0$[60] for each $\gamma \in \Gamma$. Then if this always implies $f = 0$ P_Γ-a.e., P_Γ is called *p-complete.*

Completeness can then also be considered as 1-*completeness*, and bounded completeness as ∞-*completeness.*

Let P_Γ have the same meaning as before and T be a $(\mathscr{S}, \mathfrak{Q})$-measurable transformation from R into Q. It can now happen that although P_Γ is not complete, the totality of measures $P_{T,\Gamma}$ induced by T in $(Q, \mathfrak{Q})$ is complete. In this case, we want to call T itself *complete* for Γ.

These considerations can be extended easily to σ-algebras. Let $\mathscr{S}_0$ be a sub-σ-algebra of $\mathscr{S}$. $\mathscr{S}_0$ is called *p-complete* for Γ if for each $\mathscr{S}_0$-measurable function f for which $E(|f|^p; \gamma)$ exists $E(f, \gamma) = 0$ for each $\gamma \in \Gamma$ implies $f = 0$ P_Γ-a.e. The connection between the definition of a complete transformation and that of a complete σ-algebra is clear.

We comment on the connection between the notions of "minimal and sufficient" and "boundedly complete". For the sake of simplicity we restrict ourselves to σ-algebras and prove

Theorem 8.1. *Let* $\mathscr{S}_0$ *be sufficient and boundedly complete and* $\mathscr{S}_{00}$ *a minimal, sufficient* σ*-algebra. Then,* $\mathscr{S}_0 = \mathscr{S}_{00}$ P_Γ*-a.e.*

In fact, $\mathscr{S}_{00} \subseteq \mathscr{S}_0$ P_Γ-a.e. Let $A \in \mathscr{S}_0$. Then $E(c_A | \mathscr{S}_{00})$ exists "independently" of $\gamma \in \Gamma$ and is $\mathscr{S}_{00}$-measurable, hence, also $\mathscr{S}_0$-measurable.

[59] D. Voelker und G. Doetsch, Die zweidimensionale Laplace-Transformation. Verlag Birkhäuser, Basel 1950, 208.

[60] An application of Hölder's inequality shows that these expectations always exist.

Thus,

$$\int_R (c_A - E(c_A|\mathscr{S}_{00}))\,dP_\gamma = 0 \quad \text{for each } \gamma\in\Gamma$$

and

$$|c_A - E(c_A|\mathscr{S}_{00})| \leqslant 2, \quad P_\Gamma\text{-a.e.}$$

Hence,

$$c_A = E(c_A|\mathscr{S}_{00}) \; P_\Gamma\text{-a.e.}, \quad \text{i.e.,} \quad \mathscr{S}_0 \subseteq \mathscr{S}_{00} \; P_\Gamma\text{-a.e.},$$

which was to be proved.

9. Similar tests[61]. Let $\alpha \geqslant 0$ be a level of significance and consider a simple null-hypothesis. It follows from the remarks on p. 171 that for the construction of a test for such a hypothesis one can always "completely exploit" the given level of significance, i.e., one can always construct at least one nontrivial test whose level of significance is exactly equal to α. If the null hypothesis is composite, then this is no longer always possible, and as we will show, even incorrect. This suggests giving a special name to those tests which also exploit the entire level of significance for a composite null hypothesis. We do this in a

Definition. Let Γ_0 be a (composite) null hypothesis, α a level of significance with $0 \leqslant \alpha \leqslant 1$ and ϕ a test over a sample space $(R, \mathscr{S})$ for the problem (α, Γ_0). ϕ is called *similar* if

$$E(\phi; \gamma) = \alpha \tag{9.1}$$

for all $\gamma \in \Gamma_0$.

Trivially, we have

Theorem 9.1. *For each Γ_0 and each α, $0 \leqslant \alpha \leqslant 1$, there exists at least one similar test.*

To *prove* this, choose the test defined by $\phi(x) = \alpha$ for all $x \in R$.

Calling a critical region $M \in \mathscr{S}$ similar if the indicator function c_M is a similar test, we obtain from Theorem 9.1 that for the trivial cases $\alpha = 0$ and $\alpha = 1$ critical regions always exist. Less trivial is

Theorem 9.2. *Let Γ_0 be a finite set and P_γ non-atomic for each $\gamma \in \Gamma_0$. Then, for each α with $0 \leqslant \alpha \leqslant 1$, there exists a similar critical region.*

Proof. Let $\Gamma_0 = \{\gamma_1, \ldots, \gamma_m\}$, $m \geqslant 1$. By Theorem II the map

$$A \to (P_{\gamma_1}(A), \ldots, P_{\gamma_m}(A)), \quad A \in \mathscr{S},$$

---

[61] Introduced by J. Neyman and E. S. Pearson, Philos. Trans. Roy. Soc. London l.c.[1].

has a convex range in R_m. Since $P_\gamma(\emptyset)=0$ and $P_\gamma(R)=1$ for each $\gamma\in\Gamma_0$, the elements $(0,\ldots,0)$ and $(1,\ldots,1)$ belong to this range. Thus, for each α with $0\leqslant\alpha\leqslant 1$;

$$(1-\alpha)(0,\ldots,0)+\alpha(1,\ldots,1)=(\alpha,\ldots,\alpha)$$

also belongs to this range, i.e., there exists an $A\in\mathscr{S}$ with $P_\Gamma(A)=\alpha$ for each $\gamma\in\Gamma_0$.

In general, however, the trivial test is the only similar test. Indeed, we have the obvious

Theorem 9.3. *If the set P_{Γ_0} is boundedly complete, then there exists (up to P_γ-null sets) for the level of significance α with $0\leqslant\alpha\leqslant 1$, exactly one similar test ϕ.*

Proof. Let ϕ be a similar test. Then $E(\phi-\alpha;\gamma)=0$ for $\gamma\in\Gamma_0$, hence, from $|\phi-\alpha|\leqslant 1$, $\phi(x)=\alpha$ P_{Γ_0}-a.e.

Corollary. *Under the assumptions of Theorem 9.3 there exists no similar critical region for $0<\alpha<1$* [62].

The theory of similar tests becomes more interesting when we combine it with the theory of sufficient transformations.

Definition. Let ϕ be a test for the problem (α,Γ_0) and T *a sufficient transformation* from R into Q for Γ_0 and $\psi=E(\phi|T;\gamma)$, $\gamma\in\Gamma_0$, where ψ "does not depend" on γ and is viewed as a function over Q. If T can be so chosen that $\psi=\alpha$ P_{T,Γ_0}-a.e., we will say that ϕ possesses *Neyman structure*.

Since $E(\phi;\gamma)=E(\psi;\gamma)$ for all $\gamma\in\Gamma_0$, a test that has Neyman structure is similar. We now have

Theorem 9.4[63]. *Let T be sufficient for Γ_0 and $(\mathscr{S},\mathfrak{Q})$-measurable. A similar test ϕ for the problem (α,Γ_0), $0<\alpha<1$ has Neyman structure if P_{T,Γ_0} is boundedly complete. If this condition is not satisfied, then there always exists a similar test which does not possess Neyman structure.*

Proof. Let ψ have the same meaning as above. If ϕ is similar, then

$$E(\psi;P_{T,\gamma})-\alpha=E(\psi-\alpha;P_{T,\gamma})=0 \quad \text{for all } \gamma\in\Gamma_0\,.$$

Since $0\leqslant\phi\leqslant 1$, we have $0\leqslant\psi\leqslant 1$ P_{T,Γ_0}-a.e. Hence, also $|\psi-\alpha|\leqslant 1$ P_{T,Γ_0}-a.e. and therefore, $\psi=\alpha$ P_{T,Γ_0}-a.e.

---

[62] The first known example of this is in W. Feller, Statist. Res. Mem. Univ. London 2, 117—125 (1938). See also H. Kellerer, Z. Wahrscheinlichkeitstheorie und Verw. Gebiete 1, 240—246 (1963).

[63] E. L. Lehmann and H. Scheffé, l. c.[57].

If, conversely, P_{T,Γ_0} is not boundedly complete, then there exists a $\mathfrak{Q}$-measurable function χ from Q into R_1 such that

$$E(\chi; P_{T,\gamma})=0 \tag{9.2}$$

for all $\gamma\in\Gamma_0$, $|\chi|\leqslant M$, $M>0$ and for at least one $\gamma_0\in\Gamma_0$, $\chi(t)\neq 0$ holds in a subset of Q which possesses positive P_{T,γ_0}-measure.

Choosing $0<c\leqslant\min(\alpha/M,(1-\alpha)/M)$ and $\phi=c\chi\circ T+\alpha$, we have $0\leqslant\phi\leqslant 1$, which follows immediately. Further

$$E(\phi;\gamma)=c\,E(\chi\circ T;\gamma)+\alpha$$

for each $\gamma\in\Gamma_0$ so that (9.1) holds because of (9.2) for all $\gamma\in\Gamma_0$. On the other hand, $E(\phi|T;\gamma_0)=\alpha+c\chi$ P_{T,Γ_0}-a.e. and is thus $\neq\alpha$ on a set of positive P_{T,γ_0}-measure. ϕ is thus similar, but does not have Neyman structure.

These facts can be used to construct tests which are most powerful in the set of all similar tests. We will only give an example of this. The general construction is formally simple but measurability difficulties can occur[64].

Let $\xi_1,\dots,\xi_n$, $n\geqslant 2$, be independent sample variables each of which is $N(a,\sigma^2)$-distributed. Let $\sigma_0>0$ and the set of admissible parameters be given say by

$$\Gamma=\{(a,\sigma^2):\,-\infty<a<\infty,\ \sigma_0^2\leqslant\sigma^2\}\,.$$

Further let

$$\Gamma_0=\{(a,\sigma^2):\,-\infty<a<\infty,\ \sigma^2=\sigma_0^2\}\,.$$

We know that the map $(\bar\xi, s_\xi^2)$ is sufficient for Γ. We can thus limit ourselves to tests of the form $\phi(\bar\xi, s_\xi^2)$. $\bar\xi$ is complete for Γ_0 as follows from II, Theorem 3.2 and Example 2, p. 219. From this it results that all similar tests for the problem (α,Γ_0) must satisfy the condition

$$E(\phi(\bar\xi, s_\xi^2)\,|\,\bar\xi;\sigma_0^2)=\alpha\,. \tag{9.3}$$

One can now construct a test most powerful in Γ in the set of these tests by maximizing $E(\phi(\bar\xi, s_\xi^2)\,|\,\bar\xi=x;\sigma^2)$ for each $x\in R_1$ and each $\sigma^2>\sigma_0^2$. This provides in this case one and the same test since the distribution of s_ξ^2, which is independent of $\bar\xi$, possesses monotone likelihood ratios (see II, Theorem 4.3). One constructs the test for each $x\in R_1$ according to Theorem 5.1. In this way we arrive at a "one-sided" test which corresponds to the two-sided test considered in II.**8**. The latter can also be obtained in this way.

One arrives quite analogously at the test considered in II.**4**, whose critical region in the case of a "two-sided" test is given by the set of all $x\in R_n$ which satisfy II. (4.16).

We digress a bit and calculate the power function $(a,\sigma^2)\to g(a,\sigma^2)$ of this test. To this end, we derive the density of the *non-central t-distribution*.

Let ξ be $N(0,1)$-distributed and η be χ^2-distributed with n degrees of freedom, $n\geqslant 1$, independently of ξ. Let c be a real number different from 0. We are interested in the distribution of the r.v.

$$u=(\xi+c)\sqrt{\eta/n} \tag{9.4}$$

---

[64] For an analysis see Lehmann, l.c.[2], 134ff. and especially G. Noelle, Z. Wahrscheinlichkeitstheorie und Verw. Gebiete 11, 208—229 (1969).

which is the t-distribution with n degrees of freedom for $c=0$. The density of (ξ, η) is

$$\frac{1}{\sqrt{2\pi}} \frac{1}{2^{n/2}\Gamma\left(\frac{n}{2}\right)} e^{-y/2} y^{n/2-1} e^{-x^2/2}$$

for $-\infty < x < \infty$, $y > 0$. We proceed by means of the transformation

$$\begin{aligned} &(x+c)/\sqrt{y/n} = u \\ &\sqrt{y/n} = v \qquad -\infty < u < \infty\,, \quad v > 0 \end{aligned} \tag{9.5}$$

to the density of $(\boldsymbol{u}, \boldsymbol{v})$. The absolute value of the Jacobian of (9.5) is $2nv^2$ and the sought for density is

$$\frac{1}{\sqrt{\pi}} \frac{1}{2^{(n-1)/2}\Gamma\left(\frac{n}{2}\right)} n^{n/2} e^{-nv^2/2} e^{-(uv-c)^2/2} v^n\,, \qquad -\infty < u < \infty\,, \qquad v > 0\,. \tag{9.6}$$

We now turn to the marginal distribution of $\boldsymbol{u}$ by integrating (9.6) over v and making the transformation $w = v\sqrt{u^2+n}$ for fixed u. We get for $-\infty < u < \infty$

$$\frac{1}{\sqrt{\pi}\sqrt{n}\,\Gamma\left(\frac{n}{2}\right) 2^{(n-1)/2}} \left(\frac{u^2}{n}+1\right)^{-(n+1)/2} e^{-\frac{1}{2}\frac{uc^2}{u^2+n}} \int_0^\infty e^{-\frac{1}{2}\left(w - \frac{uc}{\sqrt{u^2+n}}\right)^2} w^n dw\,. \tag{9.7}$$

The distribution of the r. v. (9.4) given by this density is called the *non-central t-distribution with n degrees of freedom*. For (9.7) we write $\tau(u, c)$.

Let us return to our example. We have

$$\frac{\bar{\xi}-a_0}{\boldsymbol{s}_\xi}\sqrt{n} = \frac{(\bar{\xi}-a)+(a-a_0)}{\boldsymbol{s}_\xi}\sqrt{n}$$

for each real a. Hence, $(\bar{\xi}-a_0)\sqrt{n}/\boldsymbol{s}_\xi$ is distributed by a non-central t-distribution with $n-1$ degrees of freedom and $c = \frac{(a-a_0)}{\sigma}\sqrt{n}$, if a and σ^2 are the correct parameter values. Hence for each real a and $0 < \sigma^2 < \infty$

$$g(a, \sigma) = 1 - \int_{-t_\alpha}^{+t_\alpha} \tau\left(u, \frac{a-a_0}{\sigma}\sqrt{n}\right) du\,,$$

where we read the definition of t_α from II. (4.15). The power function of this test thus depends on σ[65].

The Behrens-Fisher problem has earned a certain amount of fame[66].

---

[65] G. B. Dantzig, Ann. Math. Statist. 11, 186—192 (1940) proved that there exists no (non-trivial) test for the mean of a normal distribution with given sample size whose power function is independent of σ.

Further examples for similar tests are also found in VI.

[66] W. U. Behrens, Landwirtschaftliche Jahrbücher 48, 807—837 (1929).

It concerns the *problem of two samples* and the question of the existence of similar critical regions. We start with the example treated on p. 212. The set Γ considered there represents the admissible hypotheses. The null hypothesis Γ_0 is of the form

$$\{(a_1, a_2, \sigma_1^2, \sigma_2^2): a_1 = a_2, \ -\infty < a_1 < \infty, \ 0 < \sigma_1^2 < \infty, \ 0 < \sigma_2^2 < \infty\}.$$

Comparing this problem with II.**6**, we see that the former should be described as a test of the difference between the means of two normal distributions with unknown and arbitrary variances. The question arises whether there exists for each α a similar critical region for the problem $(\alpha, \Gamma_0, \Gamma - \Gamma_0)$. Important advances have recently been made in this direction by Linnik and others[67].

10. The maximum likelihood ratio test[68]. Although the construction of a test according to Theorem 3.3 for the case of a simple null-hypothesis and a simple alternative can be carried out in a rather satisfying fashion and this even for certain composite hypotheses, e.g. in the case of monotone likelihood ratios[69], we have until now given no construction principle for a test in the general case of a composite hypothesis. We will now describe such a principle. The extent to which this principle can be justified will not be gone into here. Later we will make further remarks on such a justification.

We first define the notion of a likelihood function: Let P_Γ be dominated by a σ-finite measure μ. For each $\gamma \in \Gamma$ we denote the R.-N.-density of P_γ by $x \to f(x, \gamma)$. For each $x \in R$ the map $\gamma \to f(x, \gamma)$ is defined over Γ. *This map is called the likelihood function.*

Assume we have a test problem $(\Gamma_0, \Gamma - \Gamma_0)$. Let Γ_0 as well as $\Gamma - \Gamma_0$ be composite hypotheses. Let the set of probability measures P_Γ over $(R, \mathscr{S})$ be dominated by a σ-finite measure μ. The R.-N.-densities are denoted by f_γ, $\gamma \in \Gamma$. For each $x \in R$, consider the quotient $\sup_{\gamma \in \Gamma_0} f_\gamma(x) / \sup_{\gamma \in \Gamma} f_\gamma(x)$, provided it is defined. Denoting it by $l(x)$, we have $0 \leqslant l(x) \leqslant 1$, P_Γ-a.e., if l is $\mathscr{S}$-measurable[70]. Then for each λ with $0 \leqslant \lambda \leqslant 1$, we define a critical region by $\{x: l(x) \leqslant \lambda\}$. This is only determined P_Γ-a.e. A test based on such a critical region is called a *maximum likelihood ratio test* (MLRT).

We present several examples.

---

[67] Ju. V. Linnik, Statistical problems with nuisance parameters, Translations of Mathematical Monographs Vol. 20, Amer, Math. Soc., Providence, R.I., 1968.

[68] Due to J. Neyman and E. S. Pearson, Biometrika, l. c.[1].

[69] See **5**.

[70] See V, Lemma 3.2.

Let $\xi_1,\ldots,\xi_n$, $\eta_1,\ldots,\eta_m$ for $n\geqslant 1$ and $m\geqslant 1$ be independent sample variables. Let the first n r.v.'s be Poisson distributed with mean a_1 and the following m Poisson with mean $a_2, 0<a_1<\infty, 0<a_2<\infty$. As a sample space one can choose for example R_{n+m} Γ is given by $\{(a_1,a_2), 0<a_i<\infty, i=1,2\}$. The set P_Γ consists of $(n+m)$-dimensional product measures (n Poisson distributions with mean a_1, m with mean a_2). For μ one can take any measure in R_{n+m} which has all lattice points with nonnegative coordinates as mass points. The null hypothesis will be taken as $\{(a_1,a_2): a_2=q a_1\}$, $q>0$. In the terminology of II, we thus want to test the ratio between the means of two Poisson distributions. We use the fact that $\sum_{i=1}^{n}\xi_i$ and $\sum_{i=1}^{m}\eta_i$ are, according to I, Theorem 33.1, again Poisson distributed with means $n a_1$ and $m a_2$.

Now for positive integers x and y and $0<a_1<\infty$, $0<a_2<\infty$

$$\max_{(a_1,a_2)} \frac{e^{-na_1}}{x!}(n a_1)^x \frac{e^{-ma_2}}{y!}(m a_2)^y = \frac{1}{x!\,y!}x^x y^y e^{-(x+y)}$$

and

$$\max_{a_2=qa_1} \frac{e^{-na_1}}{x!}(n a_1)^x \frac{e^{-ma_2}}{y!}(m a_2)^y = \frac{e^{-(x+y)}(x+y)^{x+y}}{x!\,y!(1+p)^{x+y}}p^y$$

with $p=mq/n$. These are found in the usual way. We remark that for $x\neq 0$, $y\neq 0$, the investigation of maxima at the boundaries need not be carried out since $a_1^x=a_2^y=0$ for $a_1=a_2=0$ and $e^{-na_1}, e^{-ma_2}\to 0$ for $a_1, a_2\to\infty$. For $x=0$, $y=0$ and $0<a_i<\infty$ $(i=1,2)$, the given expressions coincide with the maxima at the boundaries for $a_1=a_2=0$ if one sets, for example $x^x=1$ for $x=0$ etc.

Hence

$$l(x,y) = \left(\frac{x+y}{1+p}\right)^{x+y}\left(\frac{p}{y}\right)^y x^{-x}.$$

A critical region is defined by $0\leqslant l(x,y)\leqslant\lambda$. The determination of λ such that a given level of significance is guaranteed causes practical difficulties. We will not go into a detailed discussion here[71].

Another example is a simple case of the *analysis of variance*[72].

For $i=1,\ldots,k$ and $j=1,\ldots,l$ let ξ_{ij} be independent $N(a+a_i+b_j,\sigma^2)$-distributed r.v.'s. The set of admissible hypotheses is $k+l$-dimensional and is given by $-\infty<a_i<\infty$ with $\sum_{i=1}^{k}a_i=0$, $-\infty<b_j<\infty$ with $\sum_{j=1}^{l}b_j=0$, $-\infty<a<+\infty$ and $0<\sigma^2<\infty$. We consider a null hypothesis H_0 given by $a_1=\cdots=a_k=0$, $-\infty<b_j<+\infty$ with $\sum_{j=1}^{l}b_j=0$, $-\infty<a<\infty$ and $0<\sigma^2<\infty$ as well as a null hypothesis H_0' given by $b_1=\cdots=b_l=0$, $-\infty<a_i<\infty$ with $\sum_{i=1}^{k}a_i=0$, $-\infty<a<\infty$ and $0<\sigma^2<\infty$. In the set of admissible hypotheses the likelihood function is given for each $(x_{11},\ldots,x_{kl})\in R_{kl}$ by

$$(a,a_1,\ldots,a_k,b_1,\ldots,b_l,\sigma^2)\to\frac{1}{(2\pi\sigma^2)^{kl/2}}e^{-\frac{1}{2\sigma^2}\sum_{i,j=1}^{k,l}(x_{ij}-a-a_i-b_j)^2}.$$

Here, since

$$\sum_{i=1}^{k}a_i=0 \tag{10.1}$$

[71] Such a discussion is given by P. Hoel, Ann. Math. Statist. 16, 362—368 (1945).

[72] For details see H. Scheffé, The Analysis of Variance, John Wiley & Sons—Chapman & Hall, New York-London 1959.

and

$$\sum_{j=1}^{l} b_j = 0 \tag{10.2}$$

we must replace, for example, a_k by $-\sum_{i=1}^{k-1} a_i$ and b_l by $-\sum_{j=1}^{l-1} b_j$.

Set the derivatives of the logarithm of the likelihood function w.r.t. σ^2, a, a_i and b_j equal to zero to obtain

$$-\frac{kl}{2}\frac{1}{\sigma^2}+\frac{1}{2\sigma^4}\sum_{i,j=1}^{k,l}(x_{ij}-a-a_i-b_j)^2=0 \tag{10.3}$$

$$\frac{1}{\sigma^2}\sum_{i,j=1}^{k,l}(x_{ij}-a-a_i-b_j)=0 \tag{10.4}$$

$$\frac{1}{\sigma^2}\left(\sum_{j=1}^{l}(x_{ij}-a-a_i-b_j)-\sum_{j=1}^{l}(x_{kj}-a-a_k-b_j)\right)=0,\quad i=1,\dots,k-1 \tag{10.5}$$

$$\frac{1}{\sigma^2}\left(\sum_{i=1}^{k}(x_{ij}-a-a_i-b_j)-\sum_{i=1}^{k}(x_{il}-a-a_i-b_l)\right)=0,\quad j=1,\dots,l-1. \tag{10.6}$$

We introduce some notation which will also be applied to other cases in the sequel:

$$\bar{x}_{i.}=\frac{1}{l}\sum_{j=1}^{l}x_{ij},\qquad \bar{x}_{.j}=\frac{1}{k}\sum_{i=1}^{k}x_{ij},\qquad \bar{x}=\frac{1}{kl}\sum_{i,j=1}^{k,l}x_{ij}.$$

The equations (10.1) to (10.6) can be solved uniquely. Denote their solutions by $\hat{a},\hat{a}_1,\dots,\hat{a}_{k-1},\hat{a}_k,\hat{b}_1,\dots,\hat{b}_{l-1},\hat{b}_l,\hat{\sigma}^2$.

Because of (10.1) and (10.2) we have from (10.4) that $\hat{a}=\bar{x}$.

From (10.5)

$$\hat{a}_i-\hat{a}_k=\bar{x}_{i.}-\bar{x}_{k.},\qquad i=1,\dots,k-1.$$

From (10.6)

$$\hat{b}_j-\hat{b}_l=\bar{x}_{.j}-\bar{x}_{.l},\qquad j=1,\dots,l-1.$$

From this we conclude that

$$\hat{a}_i=\bar{x}_{i.}-\bar{x},\quad \hat{b}_j=\bar{x}_{.j}-\bar{x},\quad i=1,\dots,k-1,\quad j=1,\dots,l-1 \tag{10.7}$$

so that from (10.1) and (10.2) we obtain also $\hat{a}_k=\bar{x}_{k.}-\bar{x}$, $\hat{b}_l=\bar{x}_{.l}-\bar{x}$. From (10.3)

$$\hat{\sigma}^2=\frac{1}{kl}\sum_{i,j=1}^{k,l}(x_{ij}-\bar{x}_{i.}-\bar{x}_{.j}+\bar{x})^2.$$

If H_0 is true we get for the likelihood function

$$(a,\sigma^2,b_1,\dots,b_l)\to\frac{1}{(2\pi\sigma^2)^{kl/2}}e^{-\frac{1}{2\sigma^2}\sum_{i,j=1}^{k,l}(x_{ij}-a-b_j)^2},$$

where b_l is to be replaced again by $-\sum_{j=1}^{l-1} b_j$.

Differentiating the logarithm of the likelihood function w.r.t. σ^2, a and b_j for $j=1,\dots,l-1$, one obtains as before for the solutions of the corresponding equations: $\hat{a}_0=\bar{x}$, $\hat{b}_{j,0}=\bar{x}_{.j}-\bar{x}$, which hold for $j=1,\dots,l$ by (10.2). Also

$$\hat{\sigma}_0^2=\frac{1}{kl}\sum_{i,j=1}^{k,l}(x_{ij}-\bar{x}_{.j})^2.$$

We get

$$l(x_{11},\dots,x_{kl}) = \frac{e^{-\frac{kl}{2}}\left(\sum_{i,j=1}^{k,l}(x_{ij}-\bar{x}_{.j})^2\right)^{-kl/2}}{e^{-\frac{kl}{2}}\left(\sum_{i,j=1}^{k,l}(x_{ij}-\bar{x}_{i.}-\bar{x}_{.j}+\bar{x})^2\right)^{-kl/2}}$$

and a critical region of the form

$$\left\{(x_{11},\dots,x_{kl}): 0 \leqslant \left(\sum_{i,j=1}^{k,l}(x_{ij}-\bar{x}_{i.}-\bar{x}_{.j}+\bar{x})^2\right)^{kl/2} \Big/ \left(\sum_{i,j=1}^{k,l}(x_{ij}-\bar{x}_{i.})^2\right)^{kl/2} \leqslant \lambda\right\}.$$

The same critical region is also described by

$$\left\{(x_{11},\dots,x_{kl}): \sum_{i,j=1}^{k,l}(x_{ij}-\bar{x}_{.j})^2 \Big/ \sum_{i,j=1}^{k,l}(x_{ij}-\bar{x}_{i.}-\bar{x}_{.j}+\bar{x})^2 \geqslant \lambda'\right\} \tag{10.8}$$

with suitable λ'. One verifies immediately the correctness of the identity

$$\sum_{i,j=1}^{k,l}(x_{ij}-\bar{x}_{.j})^2 = \sum_{i,j=1}^{k,l}(x_{ij}-\bar{x}_{i.}-\bar{x}_{.j}+\bar{x})^2 + \sum_{i,j=1}^{k,l}(\bar{x}_{i.}-\bar{x})^2.$$

Hence, we can write (10.8) with $\lambda'-1=\lambda''$ as

$$\left\{(x_{11},\dots,x_{kl}): \sum_{i,j=1}^{k,l}(\bar{x}_{i.}-\bar{x})^2\left(\sum_{i,j=1}^{k,l}(x_{ij}-\bar{x}_{i.}-\bar{x}_{.j}+\bar{x})^2\right)^{-1} \geqslant \lambda''\right\}. \tag{10.9}$$

It is also easy to check that the following identity obtains[73]:

$$\sum_{i,j=1}^{k,l} x_{ij}^2 = kl\bar{x}^2 + \sum_{i,j=1}^{k,l}(\bar{x}_{i.}-\bar{x})^2 + \sum_{i,j=1}^{k,l}(\bar{x}_{.j}-\bar{x})^2 + \sum_{i,j=1}^{k,l}(x_{ij}-\bar{x}_{i.}-\bar{x}_{.j}+\bar{x})^2.$$

The left side is a quadratic form of rank kl. The first summand on the right has rank 1, the second at most rank $k-1$, the third at most rank $l-1$ and the fourth at most rank $kl-k-l+1$.

Since the ranks on the left and right must coincide, each form has exactly the rank given. If H_0 as well as H_0' are true and $a=0$, then by II, Theorem 4.1, the forms on the right, after dividing them by σ^2 and replacing x_{ij} by the r.v. ξ_{ij}, are independently χ^2-distributed with $1, k-1, l-1$ and $(k-1)(l-1)$ degrees of freedom, resp.

We now define r.v.'s ζ_{ij} by $\zeta_{ij}=\xi_{ij}-a-a_i-b_j$, $i=1,\dots,k$, $j=1,\dots,l$.

The ζ_{ij} are independently $N(0,\sigma^2)$-distributed. One immediately calculates that for arbitrary a, a_i, b_j:

$$\sum_{i,j=1}^{k,l}(\xi_{ij}-\bar{\xi}_{i.}-\bar{\xi}_{.j}+\bar{\xi})^2 = \sum_{i,j=1}^{k,l}(\zeta_{ij}-\bar{\zeta}_{i.}-\bar{\zeta}_{.j}+\bar{\zeta})^2.$$

The distribution of

$$\sum_{i,j=1}^{k,l}(\xi_{ij}-\bar{\xi}_{i.}-\bar{\xi}_{.j}+\bar{\xi})^2/\sigma^2$$

does thus not depend on the parameters a, a_i, b_j. Likewise, one sees that when H_0 holds, then

$$\sum_{i,j=1}^{k,l}(\bar{\xi}_{i.}-\bar{\xi})^2 = \sum_{i,j=1}^{k,l}(\bar{\zeta}_{i.}-\bar{\zeta})^2.$$

---

[73] Frequent use is made of such decompositions in the analysis of variance. For the underlying algebraic relations, see H. B. Mann, Ann. Math. Statist. 31, 1—15 (1960).

Hence,

$$\frac{(l-1)\sum_{i,j=1}^{k,l}(\bar{\xi}_{i.}-\bar{\xi})^2}{\sum_{i,j=1}^{k,l}(\xi_{ij}-\bar{\xi}_{i.}-\bar{\xi}_{.j}+\bar{\xi})^2}$$

is F-distributed with $((k-1),(k-1)(l-1))$ degrees of freedom when H_0 is correct. A critical region for H_0 with given level of significance α is thus given by

$$\left\{(x_{11},\dots,x_{kl})\colon \frac{(l-1)\sum_{i,j=1}^{k,l}(\bar{x}_{i.}-\bar{x})^2}{\sum_{i,j=1}^{k,l}(x_{ij}-\bar{x}_{i.}-\bar{x}_{.j}+\bar{x})^2}\geqslant F_\alpha\right\}.$$

F_α is determined in the notation of I. (30.2) by

$$\int_{F_\alpha}^{\infty} k_{k-1,(k-1)(l-1)}(F)\,dF=\alpha.$$

The null hypothesis H_0' is handled completely analogously. One shows that if H_0' is correct, then a critical region is given by

$$\left\{(x_{11},\dots,x_{kl})\colon \frac{(k-1)\sum_{i,j=1}^{k,l}(\bar{x}_{.j}-\bar{x})^2}{\sum_{i,j=1}^{k,l}(x_{ij}-\bar{x}_{i.}-\bar{x}_{.j}+\bar{x})^2}\geqslant F_\alpha'\right\}.$$

Replacing x_{ij} in the quotient above by the r.v. ξ_{ij}, then this quotient is F-distributed with $((l-1),(k-1)(l-1))$ degrees of freedom when H_0' holds. This determines F_α' for given level of significance α.

This inference is used often in applications, for which the observed results of an experiment are subject to constant "row" and "column" effects and the question arises whether these effects actually influence the results. Assume, for example, that several types of wheat are grown with several types of fertilizer in such a way that each wheat type is grown with each type of fertilizer. For the harvest results, which can assumed to be approximately normally distributed, we want to determine whether there exists a type—or fertilizer—effect. This can be done by an analysis of variance[74].

As another example of the MLRT, consider n independently and discretely distributed r.v.'s $\xi_1,\dots,\xi_n$ with the same distribution given by $P(\xi_i=j)=p_j$, $i=1,\dots,n$, $j=1,\dots,k$, $\sum_{j=1}^{k} p_j=1$. The joint distribution of $\sum_{i=1}^{n} c_{\{\xi_i=j\}}$, $j=1,\dots,k$ is given by $\frac{n!}{x_1!x_2!\dots x_k!}p_1^{x_1}\dots p_k^{x_k}$ with $0\leqslant x_i\leqslant n$, where x_i is an integer, $1\leqslant i\leqslant k$, with $\sum_{i=1}^{k} x_i=n$, according to I.**37**. Let the set of admissible hypotheses be given by $0<p_i<1$, $i=1,\dots,k$, $\sum_{j=1}^{k} p_j=1$ and the null hypo-

[74] For a detailed analysis of this model see A. N. Kolmogorov, Proc. Second All-Union Congress Math. Statistics, Sept. 27—Oct. 2, 1948, Acad. Sci. Uzbekistan Soviet Socialist. Republic, Tashkent 1949, 240—268.

thesis by $p_i = p_i^0$, $i=1,\dots,k$, $\sum_{i=1}^{k} p_i^0 = 1$. Considering in place of the likelihood function its logarithm, we find that one can define a critical region by means of

$$\log l(x_1,\dots,x_k) = \sum_{i=1}^{k} x_i \log\left(1+\frac{x_i - p_i^0 n}{p_i^0 n}\right). \tag{10.10}$$

Formally expending the logarithm in a power series, we get by writing $x_i = p_i^0 n + (x_i - p_i^0 n)$:

$$\sum_{i=1}^{k} x_i \log\left(1+\frac{x_i - p_i^0 n}{p_i^0 n}\right) = \sum_{i=1}^{k} [p_i^0 n + (x_i - p_i^0 n)]$$

$$\cdot\left[\left(\frac{x_i - p_i^0 n}{p_i^0 n}\right) - \frac{1}{2}\left(\frac{x_i - p_i^0 n}{p_i^0 n}\right)^2 + \frac{1}{3}\left(\frac{x_i - p_i^0 n}{p_i^0 n}\right)^3 - + \cdots\right]$$

$$= \sum_{i=1}^{k} (x_i - p_i^0 n) + \frac{1}{2}\frac{(x_i - p_i^0 n)^2}{p_i^0 n} + \text{terms of higher order}.$$

Since, however, $\sum_{i=1}^{k} x_i = n$, the first term vanishes. Restricting ourselves to the 2nd term, we get for suitable choice of $d(\alpha)$ a critical region for level of significance α given by

$$\left\{x : \sum_{i=1}^{k} \frac{(x_i - p_i^0 n)^2}{p_i^0 n} \geqslant d(\alpha)\right\}. \tag{10.11}$$

The so-called χ^2-*method of Pearson* corresponds to this practically important procedure[75]. It can also be used for continuous sample variables. In this case one must apply a grouping procedure. If we assume that the sample variables are distributed with a density f, we choose (in principle completely arbitrary)[76] real numbers $-\infty < x_1 < x_2 < \cdots < x_k < \infty$ and set

$$\int_{-\infty}^{x_1} f(x)\,dx = p_1, \quad \int_{x_{i-1}}^{x_i} f(x)\,dx = p_i, \quad 2 \leqslant i \leqslant k-1, \quad \int_{x_k}^{\infty} f(x)\,dx = p_k.$$

One groups the sample values according to which of the intervals $(-\infty, x_1], \dots, (x_{i-1}, x_i]$, $\dots, (x_k, \infty]$ they belong to. Thus, we have the previously considered case.

It's easy to see that the r. v. $\sum_{i=1}^{k} \frac{(\xi_i - p_i^0 n)^2}{p_i^0 n}$ is χ^2-distributed with $k-1$ degrees of freedom when $n \to \infty$[77].

11. Most stringent tests and invariant tests. Wald[78] has introduced a test classification which has proved to be important and which generalizes the notion of a most powerful test. Let $(\alpha, \Gamma_0, \Gamma - \Gamma_0)$ be a test problem and Φ_α the set of all tests for this problem. For each $\gamma \in \Gamma - \Gamma_0$ let

$$G(\gamma) = \sup_{\phi \in \Phi_\alpha} g_\phi(\gamma).$$

[75] K. Pearson, Philos. Mag. 50, Ser. 5, 157—175 (1900).

[76] For the *grouping problem* see H. B. Mann and A. Wald, Ann. Math. Statist. 13, 306—317 (1942) and H. Witting, Arch. Math. 10, 468—479 (1959).

[77] For this and further important results see H. Cramér, l.c. I[58]. The first formulation of such results is in R. A. Fisher, J. Roy. Statist. Soc. 85, 87—94 (1922). Also see W. G. Cochran, Ann. Math. Statist. 23, 315—345 (1952).

[78] A. Wald, Trans. Amer. Math. Soc. 54, 462—482 (1943).

A test $\phi_0 \in \Phi_\alpha$ is called *most stringent* if it satisfies the condition

$$\sup_{\gamma \in \Gamma - \Gamma_0} [G(\gamma) - g_{\phi_0}(\gamma)] \leqslant \sup_{\gamma \in \Gamma - \Gamma_0} [G(\gamma) - g_\phi(\gamma)] \tag{11.1}$$

for each $\phi \in \Phi_\alpha$.

Obviously this concept is closely related to the ideas introduced in **4**, especially to the notation of a maximintest.

If there exists a test $\phi_0 \in \Phi_\alpha$ for which $G(\gamma) = g_{\phi_0}(\gamma)$ for each $\gamma \in \Gamma - \Gamma_0$, then ϕ_0 is naturally most stringent. We thus have the trivial

Theorem 11.1. *If for the problem* $(\alpha, \Gamma_0, \Gamma - \Gamma_0)$, *a most powerful (w.r.t.* $\Gamma - \Gamma_0$*) test exists, then it is most stringent.*

An almost trivial, but practically useful criterion is given by

Theorem 11.2. *Let* Φ_α *be the set of all tests for the test problem* $(\alpha, \Gamma_0, \Gamma - \Gamma_0)$. *Let H be an arbitrary index set and let* $\Gamma_\eta, \eta \in H$ *be pairwise disjoint (non empty) subsets of* $\Gamma - \Gamma_0$ *with* $\bigcup_{\eta \in H} \Gamma_\eta = \Gamma - \Gamma_0$. *Set* $G(\gamma) = \sup_{\phi \in \Phi_\alpha} g_\phi(\gamma)$ *for each* $\gamma \in \Gamma - \Gamma_0$ *and let G be constant on* Γ_η *for each* $\eta \in H$. *Further let there exist a* $\phi_0 \in \Phi_\alpha$ *for which*

$$\inf_{\gamma \in \Gamma_\eta} g_{\phi_0}(\gamma) = \sup_{\phi \in \Phi_\alpha} \inf_{\gamma \in \Gamma_\eta} g_\phi(\gamma) \tag{11.2}$$

for each $\eta \in H$. *Then* ϕ_0 *is most stringent for the above test problem.*

Proof. We have $\sup_{\gamma \in \Gamma - \Gamma_0} (G(\gamma) - g_{\phi_0}(\gamma)) = \sup_{\eta \in H} \sup_{\gamma \in \Gamma_\eta} (G(\gamma) - g_{\phi_0}(\gamma))$. Since G is constant on Γ_η for each arbitrary $\gamma_\eta \in \Gamma_\eta$ we also have $\sup_{\gamma \in \Gamma - \Gamma_0} (G(\gamma) - g_{\phi_0}(\gamma)) = \sup_{\eta \in H} (G(\gamma_\eta) - \inf_{\gamma \in \Gamma_\eta} g_{\phi_0}(\gamma))$. Further

$$\sup_{\eta \in H} (G(\gamma_\eta) - \inf_{\gamma \in \Gamma_\eta} g_{\phi_0}(\gamma)) \leqslant \sup_{\eta \in H} (G(\gamma_\eta) - \inf_{\gamma \in \Gamma_\eta} g_\phi(\gamma)) \tag{11.3}$$

for each $\phi \in \Phi_\alpha$ because of (11.2). The right side of (11.3) is, however,

$$\sup_{\eta \in H} \sup_{\gamma \in \Gamma_\eta} (G(\gamma) - g_\phi(\gamma)) = \sup_{\gamma \in \Gamma - \Gamma_0} (G(\gamma) - g_\phi(\gamma))$$

which completes the proof.

Consider an example for this theorem. Assume we have a set of k-dimensional normal distributions, $k \geqslant 1$, with meanvectors $a \in R_k$ and given positive definite covariance matrix B^{-1}. Let $a_0 \in R_k$ and $a = a_0$ be a simple hypothesis on the mean. Γ_0 is thus to be identified with $\{a_0\}$ and $\Gamma - \Gamma_0$ with $R_k - \{a_0\}$. Denote by Φ'_α the set of tests ϕ for the problem $\{\alpha, \Gamma_0, \Gamma - \Gamma_0\}$ which fulfill

$$E(\phi; a_0) = \alpha. \tag{11.4}$$

Limiting ourselves to such tests causes no loss of generality. We can likewise assume $\alpha > 0$.

We now claim that, for suitable choice of $d(\alpha)$,

$$K=\{x:(x-a_0)'B(x-a_0)\geqslant d(\alpha)\} \tag{11.5}$$

defines a critical region in R_k which yields a most stringent test c_K for the problem $\{\alpha,\Gamma_0,\Gamma-\Gamma_0\}$. Here, $d(\alpha)$ must be so determined that $E(c_K;a_0)=\alpha$, which is always uniquely possible. (See VI, p. 369.)

Since B^{-1} and B are positive definite, one can simplify the problem a bit by assuming that B is the $k\times k$ unit matrix. Indeed, there exists an orthogonal matrix C such that $C'BC=\Lambda$, where Λ is a diagonal matrix of the form

$$\begin{pmatrix} \lambda_1 & 0 & \dots & 0 \\ 0 & \lambda_2 & \dots & 0 \\ \dots & \dots & \dots & \dots \\ 0 & 0 & \dots & \lambda_k \end{pmatrix} \quad \text{with } \lambda_i>0,\ 1\leqslant i\leqslant k.$$

Thus, writing

$$y=Cx \tag{11.6}$$

and

$$b=Ca \tag{11.7}$$

and, in particular, $b_0=Ca_0$, we see that (11.5) goes into $\{y:(y-b_0)'\Lambda(y-b_0)\geqslant d(\alpha)\}$. Denoting by $\Lambda^{\frac{1}{2}}$ the diagonal matrix

$$\begin{pmatrix} \lambda_1^{1/2} & 0 & \dots & 0 \\ 0 & \lambda_2^{1/2} & \dots & 0 \\ \dots & \dots & \dots & \dots \\ 0 & 0 & \dots & \lambda_k^{1/2} \end{pmatrix}$$

and setting

$$z=\Lambda^{\frac{1}{2}}y \tag{11.8}$$

and

$$c=\Lambda^{\frac{1}{2}}b \tag{11.9}$$

and, in particular, $c_0=\Lambda^{\frac{1}{2}}b_0$, we see that (11.5) goes finally into

$$\{z:|z-c_0|^2\geqslant d(\alpha)\}\,.$$

Applying the transformations (11.6) and then (11.8) to the density

$$|B|^{1/2}(2\pi)^{-k/2}e^{-(1/2)[(x-a)'B(x-a)]}x\in R_k\,,$$

and taking (11.7) and (11.9) into consideration, one gets a new density given for all $z\in R_k$ by

$$(2\pi)^{-k/2}e^{-\frac{1}{2}|z-c|^2}\,. \tag{11.10}$$

In addition, the inverse of $\Lambda^{\frac{1}{2}}C$ exists. Thus, essentially by means of a change of notation we find that instead of starting from the given set of normal distributions, we can start from the set of all normal distributions, with the $k\times k$-unit matrix as covariancematrix, whose densities for $z\in R_k$ are given by (11.10). Writing x for z and a for c again, we have in place of (11.5)

$$K=\{x:|x-a_0|^2\geqslant d(\alpha)\}\,. \tag{11.11}$$

Now let for each real $l>0$

$$E_l=\{a:|a-a_0|^2=l\}\,. \tag{11.12}$$

Obviously, $\bigcup_{l>0} E_l = R_k - \{a_0\} = \Gamma - \Gamma_0$. For each $a \in \Gamma - \Gamma_0$ let $G(a) = \sup_{\phi \in \Phi'_\alpha} g_\phi(a)$. We will show that G is constant on each E_l.

Let a and $a^* \in E_l$ so that

$$|a - a_0| = |a^* - a_0|. \tag{11.13}$$

Further let $\phi_a^{(i)} \in \Phi'_\alpha$, $i = 1, 2 \ldots$ be a sequence of tests for which

$$G(a) = \lim_{i \to \infty} E(\phi_a^{(i)}; a). \tag{11.14}$$

Likewise, let $\phi_{a^*}^{(i)} \in \Phi'_\alpha$, $i = 1, 2 \ldots$ be a sequence of tests with

$$G(a^*) = \lim_{i \to \infty} E(\phi_{a^*}^{(i)}; a^*). \tag{11.15}$$

Choose an orthogonal transformation $\mathfrak{O}$ such that

$$y - a_0 = \mathfrak{O}(x - a_0) \tag{11.16}$$

as well as

$$(a^* - a_0)'(x - a_0) = (a - a_0)'(y - a_0) \tag{11.17}$$

hold. For each $y \in R_k$, define a test ψ_i by means of

$$\psi_i(y) = \phi_{a^*}^{(i)}(\mathfrak{O}^{-1}(y - a_0) + a_0), \quad 1 \leqslant i. \tag{11.18}$$

The density of the normal distribution corresponding to the null hypothesis with mean-vector a_0 does not change when the transformation (11.16) is applied. From (11.18)

$$E(\psi_i; a_0) = (2\pi)^{-k/2} \int_{R_k} \psi_i(y) e^{-\frac{1}{2}|y - a_0|^2} dy = (2\pi)^{-k/2} \int_{R_k} \phi_{a^*}(x) e^{-\frac{1}{2}|x - a_0|^2} dx = \alpha,$$

i.e.,

$$\psi_i \in \Phi'_\alpha. \tag{11.19}$$

Further

$$\begin{aligned} E(\psi_i; a) &= (2\pi)^{-k/2} \int_{R_k} \psi_i(y) e^{-\frac{1}{2}|y - a|^2} dy \\ &= (2\pi)^{-k/2} \int_{R_k} \psi_i(y) e^{-\frac{1}{2}(|y - a_0|^2 + 2(y - a_0)'(a_0 - a) + |a_0 - a|^2)} dy. \end{aligned}$$

Taking (11.17) and (11.18) into consideration, we find that (11.16) takes the integral into

$$\int_{R_k} \phi_{a^*}^{(i)}(x) e^{-\frac{1}{2}(|x - a_0|^2 + 2(a^* - a_0)'(x - a_0) + |a_0 - a^*|^2)} dx.$$

We also used (11.13) here.

Thus, $E(\psi_i; a) = E(\phi_{a^*}^{(i)}; a^*)$. (11.14), (11.15) and (11.19) thus lead to

$$G(a^*) \leqslant G(a).$$

Reversing the roles of a^* and a, one gets in the same way

$$G(a) \leqslant G(a^*).$$

Hence, $G(a) = G(a^*)$ which was to be shown.

We still need to prove that the test c_K belonging to the critical region (11.11) satisfies, for each $l > 0$, the condition

$$\inf_{a \in E_l} g_{c_K}(a) = \sup_{\phi \in \Phi'_\alpha} \inf_{a \in E_l} g_\phi(a). \tag{11.20}$$

It is enough to show that g_{c_K} is constant on each E_l and

$$\int_{E_l} g_{c_K}\, do_l \geqslant \int_{E_l} g_\phi\, do_l \tag{11.21}$$

for each $\phi \in \Phi'_\alpha$ and all $l>0$, where do_l is the element of integration on E_l.

This first statement follows essentially as the analogous one for G. (11.21) can be shown with the Neyman-Pearson Fundamental Lemma. We proceed as follows: For each $\phi \in \Phi'_\alpha$

$$\int_{E_l} g_\phi\, do_l = \frac{1}{(\sqrt{2\pi})^k}\int_{E_l} \phi(x) \int_{R_k} e^{-\frac{1}{2}|x-a|^2}\, dx\, do_l = \frac{1}{(\sqrt{2\pi})^k}\int_{R_k} \phi(x) \int_{E_l} e^{-\frac{1}{2}|x-a|^2}\, do_l\, dx,$$

where the exchange of integration order is justified by Theorem X.

(11.21) is thus to be proved by applying Theorem 3.1 to

$$f_1(x) = (2\pi)^{-k/2} e^{-\frac{1}{2}|x-a_0|^2}, \qquad x \in R_k$$

and

$$f_2(x) = (2\pi)^{-k/2} \int_{E_l} e^{-\frac{1}{2}|x-a|^2}\, do_l, \qquad x \in R_k$$

provided we can show that (11.11) can also be defined by

$$\left\{x\colon \left(\int_{E_l} e^{-\frac{1}{2}|x-a|^2}\, do_l \big/ e^{-\frac{1}{2}|x-a_0|^2}\right) \geqslant \gamma(\alpha)\right\} \tag{11.22}$$

for suitable choice of $\gamma(\alpha)$. The quotient in (11.22) is, from (11.12), equal to

$$e^{-l/2} \int_{E_l} e^{-(x-a_0)'(a-a_0)}\, do_l\,.$$

We proceed exactly as in the arguments on p. 163. For $y \neq a_0$ let

$$\cos\vartheta(a) = \frac{(y-a_0)'(a-a_0)}{|y-a_0|\,|a-a_0|} \tag{11.23}$$

with $0 \leqslant \vartheta(a) \leqslant \pi$. We introduce polar coordinates with $a_0 = (a_1^{(0)}, \ldots, a_k^{(0)})$

$$a_i - a_i^{(0)} = l\alpha_i \sin\vartheta\,, \qquad 1 \leqslant i \leqslant k-1\,, \qquad \sum_{i=1}^{k-1} \alpha_i^2 = 1\,,$$

$$a_k - a_k^{(0)} = l\cos\vartheta\,, \qquad 0 \leqslant \vartheta \leqslant \pi\,.$$

We then obtain from (11.23)

$$\int_{E_l} e^{(x-a_0)'(a-a_0)}\, do_l = l^{k-1}\, O_{k-1} \int_0^\pi e^{|y-a_0|\sqrt{l}\cos\vartheta} \sin^{k-2}\vartheta\, d\vartheta\,,$$

where O_{k-1} is the surface area of the $(k-1)$-dimensional unit sphere. It suffices to show that the map $I(z)\colon z \to \int_0^\pi e^{z\sqrt{l}\cos\vartheta} \sin^{k-2}\vartheta\, d\vartheta$ is strictly monotone increasing in $0<z<\infty$. Applying Theorem VIII one easily sees that

$$I'(z) = \sqrt{l} \int_0^\pi e^{z\sqrt{l}\cos\vartheta} \cos\vartheta \sin^{k-2}\vartheta\, d\vartheta\,.$$

Splitting the integration into sections from 0 to $\frac{\pi}{2}$ and $\frac{\pi}{2}$ to π and replacing ϑ in the second integral by $\pi-\vartheta$, one gets

$$I'(z) = \sqrt{l} \int_0^{\pi/2} \cos\vartheta \sin^{k-2}\vartheta \left[e^{z\sqrt{l}\cos\vartheta} - e^{-z\sqrt{l}\cos\vartheta}\right] d\vartheta > 0$$

for $z>0$ which completes the demonstration.

The critical region K given by (11.11) thus actually defines a most stringent test.

It has turned out that the definition of a most stringent test is closely connected with another idea, referred to as the *invariance principle.* We will formulate it here only for the theory of testing.

Let $\mathfrak{G}$ be an arbitrary multiplicatively written group and $(R,\mathcal{S})$ a measure space. We assume that one can "apply" the elements of $\mathfrak{G}$ to $(R,\mathcal{S})$, i.e., the following should hold: A map into R is defined over $\mathfrak{G}\times R$: If $\mathfrak{g}\in\mathfrak{G}$ and $x\in R$ and y is the image of $(\mathfrak{g}, x)$, then one writes simply $\mathfrak{g}x=y$. For $\mathfrak{g}_1,\mathfrak{g}_2\in\mathfrak{G}$ and $x\in R$ assume that we always have

$$(\mathfrak{g}_1\mathfrak{g}_2)x=\mathfrak{g}_1(\mathfrak{g}_2 x). \tag{11.24}$$

If $\mathfrak{e}$ is the unit element of $\mathfrak{G}$, let

$$\mathfrak{e}x=x \tag{11.25}$$

for each $x\in R$.

We require further that the map $x\to\mathfrak{g}x$ is for each $\mathfrak{g}\in\mathfrak{G}$ an $(\mathcal{S},\mathcal{S})$-measurable transformation from R into itself, i.e., the inverse image of each $M\in\mathcal{S}$ is again in $\mathcal{S}$ under this mapping.

It follows from (11.24) and (11.25) that this inverse image is given by $\mathfrak{g}^{-1}M$. It also follows from this that for each $\mathfrak{g}\in\mathfrak{G}$ and all $M\in\mathcal{S}$ we also have $\mathfrak{g}M\in\mathcal{S}$. Further the map $x\to\mathfrak{g}x$ is an one-to-one correspondence for each $\mathfrak{g}\in\mathfrak{G}$. Let P_Γ be a set of probability measures defined on $(R,\mathcal{S})$. For each $\gamma\in\Gamma$ and each $\mathfrak{g}\in\mathfrak{G}$ we define a probability measure $P_{\bar{\mathfrak{g}}\gamma}$ by

$$P_{\bar{\mathfrak{g}}\gamma}(A)=P_\gamma(\mathfrak{g}^{-1}A) \tag{11.26}$$

for each $A\in\mathcal{S}$.

Replacing A by $\mathfrak{g}A$ one sees that (11.26) is equivalent to

$$P_{\bar{\mathfrak{g}}\gamma}(\mathfrak{g}A)=P_\gamma(A). \tag{11.27}$$

It follows easily that $P_{\bar{\mathfrak{g}}\gamma}$ is actually a probability measure. The notation chosen here for the probability measure $P_{\bar{\mathfrak{g}}\gamma}$ defined by (11.26) becomes more meaningful provided we require $\bar{\mathfrak{g}}\Gamma\subseteq\Gamma$. [79]

We then say that P_Γ is *invariant* w.r.t. $\mathfrak{G}$. Then

$$P_{\overline{(\mathfrak{g}_1\mathfrak{g}_2)}\gamma}(A)=P_\gamma((\mathfrak{g}_1\mathfrak{g}_2)^{-1}A)=P_\gamma(\mathfrak{g}_2^{-1}\mathfrak{g}_1^{-1}A)=P_{\bar{\mathfrak{g}}_1\gamma}(\mathfrak{g}_1^{-1}A)=P_{\bar{\mathfrak{g}}_1\bar{\mathfrak{g}}_2\gamma}(A)$$

for each $\gamma\in\Gamma$, each $A\in\mathcal{S}$ and $\mathfrak{g}_1,\mathfrak{g}_2\in\mathfrak{G}$. Thus $\overline{\mathfrak{g}_1\mathfrak{g}_2}=\bar{\mathfrak{g}}_1\bar{\mathfrak{g}}_2$.

Further $P_{\overline{\mathfrak{g}^{-1}}\gamma}(A)=P_\gamma(\mathfrak{g}A)$. Defining $(\bar{\mathfrak{g}})^{-1}$ by means of $(\bar{\mathfrak{g}})^{-1}\bar{\mathfrak{g}}=\bar{\mathfrak{e}}$, one sees that $(\bar{\mathfrak{g}})^{-1}$ always exists and coincides with $\overline{\mathfrak{g}^{-1}}$. Indeed,

$$P_{\bar{\mathfrak{e}}\gamma}(A)=P_\gamma(A)=P_\gamma(\mathfrak{g}^{-1}\mathfrak{g}A)=P_{\bar{\mathfrak{g}}\gamma}(\mathfrak{g}A)=P_{\overline{\mathfrak{g}^{-1}}\bar{\mathfrak{g}}\gamma}(A).$$

The set $\bar{\mathfrak{g}}$ of all $\bar{\mathfrak{g}}$ is thus likewise a group[80]. Then we call $\mathfrak{G}$ *admissible* for the test problem.

Retaining the notation introduced we give a

---

[79] Strictly speaking, $\bar{\mathfrak{g}}\gamma$ is for the time being not at all defined for $\gamma\in\Gamma$; only $P_{\bar{\mathfrak{g}}\gamma}$ is defined.

[80] $\bar{\mathfrak{G}}$ is thus a *homomorphic image* of $\mathfrak{G}$.

Definition. *A test problem* (Γ_0, Γ) is *invariant* w.r.t. an *admissible group* $\mathfrak{G}$ if

$$\bar{\mathfrak{g}}\Gamma_0 = \Gamma_0 \tag{11.28}$$

and

$$\bar{\mathfrak{g}}(\Gamma - \Gamma_0) = \Gamma - \Gamma_0 \tag{11.29}$$

for each $\mathfrak{g} \in \mathfrak{G}$.

A test ϕ is *invariant* w.r.t. $\mathfrak{G}$ if $\phi(\mathfrak{g}x) = \phi(x)$ for all $\mathfrak{g} \in \mathfrak{G}$ and each $x \in R$.

This definition of invariance is naturally not restricted to tests and can be applied in an analogous way to arbitrary maps from R into some set. It we have an invariant test problem, then the use of invariant tests suggests itself.

A simple example of an invariant test problem is obtained as follows: Let $\mathfrak{G}$ be the additive group of real numbers which we apply as follows to R_n, $n \geqslant 1$: Let $\mathfrak{g}x = (x_1 + \mathfrak{g}, \ldots, x_n + \mathfrak{g})$ for $x \in R_n$ and each real $\mathfrak{g} \in \mathfrak{G}$. Let f_i, $i = 1, 2$ be densities in R_n. Consider the set of all probability measures over $(R_n, \mathfrak{B}_n)$ whose densities are given for each $x \in R_n$ by $f_i(x_1 + \gamma, \ldots, x_n + \gamma)$, $i = 1, 2$, $-\infty < \gamma < \infty$. Define Γ_0 as the set of all pairs of the form $(\gamma, 1)$ and $\Gamma - \Gamma_0$ as the set of all pairs of the form $(\gamma, 2)$, $\gamma \in R_1$, where (γ, i) is associated with the measure $P_{(\gamma, i)}$ with density $x \to f_i(x_1 + \gamma, \ldots, x_n + \gamma)$ $(i = 1, 2)$. We naturally require that $P_{\Gamma_0} \cap P_{\Gamma - \Gamma_0} = \emptyset$. Now for all $A \in \mathfrak{B}_n$

$$\begin{aligned} P_{\bar{\mathfrak{g}}(\gamma,1)}(A) &= P_{(\gamma,1)}(-\mathfrak{g} + A) = \int_{-\mathfrak{g}+A} f_1(x_1 + \gamma, \ldots, x_n + \gamma)\, dx_1 \ldots dx_n \\ &= \int_A f_1(x_1 + \gamma - \mathfrak{g}, \ldots, x_n + \gamma - \mathfrak{g})\, dx_1 \ldots dx_n = P_{(\gamma - \mathfrak{g}, 1)}(A). \end{aligned}$$

Thus, $\bar{\mathfrak{g}}\Gamma_0 = \Gamma_0$. Quite analogously one gets $\bar{\mathfrak{g}}(\Gamma - \Gamma_0) = \Gamma - \Gamma_0$, so that $\mathfrak{G}$ leaves the test problem invariant. Each function defined over R_n of the form $x \to \phi(x_1 - x_n, \ldots, x_{n-1} - x_n)$, $0 \leqslant \phi \leqslant 1$, is an invariant test, which one sees at once.

We turn to the connection between most stringent tests and the invariance principle. We remark that (11.27) can also be written as

$$\int_R c_{\mathfrak{g}A}\, dP_{\bar{\mathfrak{g}}\gamma} = \int_R c_A\, dP_\gamma$$

or, because $c_{\mathfrak{g}A}(x) = c_A(\mathfrak{g}^{-1}x)$ for $x \in R$, also as

$$\int_R c_A(\mathfrak{g}^{-1}x)\, dP_{\bar{\mathfrak{g}}\gamma}(x) = \int_R c_A\, dP_\gamma. \tag{11.30}$$

Thus, from (11.26) or (11.30) it follows in the usual way (see p. 8) for each P_γ-integrable function f, that

$$\int_R f(\mathfrak{g}^{-1}x)\, dP_{\bar{\mathfrak{g}}\gamma}(x) = \int_R f(x)\, dP_\gamma(x). \tag{11.31}$$

We can now prove

Lemma 11.1. *Let* $(\alpha, \Gamma_0, \Gamma - \Gamma_0)$ *be an invariant test problem w.r.t. a group* $\mathfrak{G}$. *Let* Φ_α *be the set of all tests for this problem.*

Again let $G(\gamma) = \sup_{\phi \in \Phi_\alpha} g_\phi(\gamma)$. *Then* G *is invariant w.r.t. the group* $\bar{\mathfrak{G}}$ *generated by* $\mathfrak{G}$.

Proof. If $\phi \in \Phi_\alpha$, then for each $\mathfrak{g} \in \mathfrak{G}$ the map $x \to \phi(\mathfrak{g}^{-1}x)$ also belongs to Φ_α. This follows from an application of (11.31) to ϕ and the assumed validity of (11.28). Further

$$G(\gamma) = \sup_{\phi \in \Phi_\alpha} \int_R \phi(x) dP_\gamma(x) = \sup_{\phi \in \Phi_\alpha} \int_R \phi(\mathfrak{g}x) dP_\gamma(x)$$
$$= \sup_{\phi \in \Phi_\alpha} \int_R \phi(x) dP_{\bar{\mathfrak{g}}\gamma}(x) = G(\bar{\mathfrak{g}}\gamma),$$

for each $\bar{\mathfrak{g}} \in \bar{\mathfrak{G}}$ which was to be shown.

In almost the same way we prove

Lemma 11.2. *Assume we have a test problem $(\alpha, \Gamma_0, \Gamma - \Gamma_0)$ invariant w.r.t. a group $\mathfrak{G}$. Let ϕ be a test for this problem which fulfills the following condition: For each $\mathfrak{g} \in \mathfrak{G}$ let $\phi(\mathfrak{g}x) = \phi(x)$ $P_{\Gamma - \Gamma_0}$-a.e., where the exceptional set can depend on $\mathfrak{g}$. Then the power function g_ϕ is invariant w.r.t. $\bar{\mathfrak{G}}$ on $\Gamma - \Gamma_0$.*

The same method of proof also yields a

Corollary. *Let the assumptions of Lemma 11.2 be fulfilled for Γ_0 instead of $\Gamma - \Gamma_0$. If there exists a $\gamma \in \Gamma_0$ such that $\{\bar{\mathfrak{g}}\gamma : \bar{\mathfrak{g}} \in \bar{\mathfrak{G}}\} = \Gamma_0$,*[81] *then ϕ is similar w.r.t. Γ_0.*

Let $\mathfrak{G}$ be a group and $\mathfrak{S}$ a σ-algebra of subsets of $\mathfrak{G}$. For each $B \in \mathfrak{S}$ and $\mathfrak{g} \in \mathfrak{G}$ assume that $\mathfrak{g}B$ and $B\mathfrak{g}$ are in $\mathfrak{S}$. A measure ν on $\mathfrak{S}$ is called *invariant* if

$$\nu(\mathfrak{g}B) = \nu(B) = \nu(B\mathfrak{g}) \tag{11.32}$$

for all $B \in \mathfrak{S}$ and each $\mathfrak{g} \in \mathfrak{G}$. For all $\mathfrak{S}$-measurable and ν-integrable maps f from $\mathfrak{G}$ into R_1 we then get

$$\int_{\mathfrak{G}} f(\mathfrak{g}^{-1}\mathfrak{g}_1) d\nu(\mathfrak{g}_1) = \int_{\mathfrak{G}} f(\mathfrak{g}_1 \mathfrak{g}^{-1}) d\nu(\mathfrak{g}_1) = \int_{\mathfrak{G}} f(\mathfrak{g}_1) d\nu(\mathfrak{g}_1) \tag{11.33}$$

for all $\mathfrak{g} \in \mathfrak{G}$.

With this terminology we can now prove

Theorem 11.3. *Let $(\alpha, \Gamma_0, \Gamma - \Gamma_0)$ be a test problem invariant w.r.t. a group $\mathfrak{G}$. Let P_Γ be dominated by a probability measure μ. If $N \in \mathscr{S}$ and $\mu(N) = 0$, then assume $\mu(\mathfrak{g}N) = 0$ also holds for each $\mathfrak{g} \in \mathfrak{G}$. Let an invariant probability measure ν exist on $(\mathfrak{G}, \mathfrak{S})$. Let the map $(\mathfrak{g}, x) \to \mathfrak{g}x$ defined over $\mathfrak{G} \times R$ be $(\mathscr{S}, \mathfrak{S} \otimes \mathscr{S})$-measurable and assume there exists an invariant test ϕ_0 which is most powerful w.r.t. the subset of all invariant tests in Φ_α on $\Gamma - \Gamma_0$. Then ϕ_0 is most stringent for the given test problem*[82].

---

[81] The group is then called *transitive*.

[82] The notion of an invariant test is viewed somewhat more generally in this theorem: ϕ is called *invariant* if there exists a μ-null set M such that $\phi(\mathfrak{g}x) = \phi(x)$ for all $\mathfrak{g} \in \mathfrak{G}$ and each $x \in R - M$.

Proof. We will use Theorem 11.2. Let

$$\Gamma_\eta = \{\gamma_\eta : \gamma_\eta = \bar{\mathfrak{g}}\gamma, \bar{\mathfrak{g}} \in \mathfrak{G}\}$$

for each $\gamma \in \Gamma - \Gamma_0$. For $\eta \neq \eta_1$ we have either $\Gamma_\eta = \Gamma_{\eta_1}$ or $\Gamma_\eta \cap \Gamma_{\eta_1} = \emptyset$. Hence, $\Gamma - \Gamma_0 = \bigcup_{\eta \in H} \Gamma_\eta$ with $\Gamma_\eta \cap \Gamma_{\eta'} = \emptyset$ for $\eta, \eta' \in H, \eta \neq \eta'$, if the index set H is suitably chosen. By Lemma 11.1, G is constant on each Γ_η. It is thus enough to show that for each $\eta \in H$,

$$\inf_{\gamma \in \Gamma_\eta} g_{\phi_0}(\gamma) = \sup_{\phi \in \Phi_\alpha} \inf_{\gamma \in \Gamma_\eta} g_\phi(\gamma).$$

Now for each $\varepsilon > 0$ there exists a test $\phi_\varepsilon \in \Phi_\alpha$ such that

$$\inf_{\gamma \in \Gamma_\eta} g_{\phi_\varepsilon}(\gamma) + \varepsilon \geqslant \sup_{\phi \in \Phi_\alpha} \inf_{\gamma \in \Gamma_\eta} g_\phi(\gamma). \tag{11.34}$$

$(\mathfrak{g}, x) \to \phi_\varepsilon(\mathfrak{g}x)$ is $\mathfrak{S} \otimes \mathscr{S}$-measurable and $\int_R \int_{\mathfrak{G}} \phi_\varepsilon(\mathfrak{g}x)\, d\nu(\mathfrak{g})\, d\mu(x)$ is thus meaningful. Thus, the $\mathscr{S}$-measurable map

$$\psi : x \to \int_{\mathfrak{G}} \phi_\varepsilon(\mathfrak{g}x)\, d\nu(\mathfrak{g})$$

exists μ-a.e. (see Theorem X). The exceptional set is denoted by N. ψ is a test for the problem $(\alpha, \Gamma_0, \Gamma - \Gamma_0)$. Indeed, $0 \leqslant \psi \leqslant 1$ μ-a.e. Moreover, for $\gamma \in \Gamma_0$

$$\begin{aligned} E(\psi; P_\gamma) &= \int_R \int_{\mathfrak{G}} \phi_\varepsilon(\mathfrak{g}x)\, d\nu(\mathfrak{g})\, dP_\gamma(x) = \int_{\mathfrak{G}} \int_R \phi_\varepsilon(\mathfrak{g}x)\, dP_\gamma(x)\, d\nu(\mathfrak{g}) \\ &= \int_{\mathfrak{G}} \int_R \phi_\varepsilon(x)\, dP_{\bar{\mathfrak{g}}\gamma}(x)\, d\nu(\mathfrak{g}) \leqslant \alpha, \end{aligned}$$

since the test problem is invariant w.r.t. $\mathfrak{G}$ and $\phi_\varepsilon \in \Phi_\alpha$.

In addition, for each $\mathfrak{g}_1 \in \mathfrak{G}$ and all $x \notin \mathfrak{g}_1^{-1} N$

$$\psi(\mathfrak{g}_1 x) = \int_{\mathfrak{G}} \phi_\varepsilon(\mathfrak{g}(\mathfrak{g}_1 x))\, d\nu(\mathfrak{g}) = \int_{\mathfrak{G}} \phi_\varepsilon(\mathfrak{g}x)\, d\nu(\mathfrak{g})$$

when we apply (11.33). But then for all $x \notin N \cup \mathfrak{g}_1^{-1} N$

$$\psi(\mathfrak{g}_1 x) = \psi(x) \tag{11.35}$$

and, by assumption, $\mu(N \cup \mathfrak{g}_1^{-1} N) = 0$, so that also $P_{\Gamma - \Gamma_0}(N \cup \mathfrak{g}_1^{-1} N) = 0$. We can thus apply Lemma 11.2 and obtain for each $\gamma \in \Gamma_\eta$

$$\begin{aligned} \inf_{\gamma \in \Gamma_\eta} g_\psi(\gamma) &= \int_R \psi(x)\, dP_\gamma(x) = \int_{\mathfrak{G}} \int_R \phi_\varepsilon(\mathfrak{g}x)\, dP_\gamma(x)\, d\nu(\mathfrak{g}) \\ &= \int_{\mathfrak{G}} \int_R \phi_\varepsilon(x)\, dP_{\overline{\mathfrak{g}^{-1}}\gamma}(x)\, d\nu(\mathfrak{g}) \geqslant \int_{\mathfrak{G}} \inf_{\gamma \in \Gamma_\eta} E(\phi_\varepsilon; \gamma)\, d\nu(\mathfrak{g}) = \inf_{\gamma \in \Gamma_\eta} g_{\phi_\varepsilon}(\gamma). \end{aligned} \tag{11.36}$$

Now define say $\psi(x)=0$ for $x\in N$ and note that $(g,x)\to\psi(gx)$ is $\mathfrak{S}\otimes\mathcal{S}$-measurable. Since the map $(g,x)\to\psi(x)$ is also $\mathfrak{S}\otimes\mathcal{S}$-measurable,

$$\int_{\mathfrak{G}}\int_{R} |\psi(gx)-\psi(x)|\,d\mu(x)\,d\nu(g)$$

exists and, by (11.35), this integral vanishes. (See Theorem X.) Therefore, we also have

$$\int_{R}\int_{\mathfrak{G}} |\psi(gx)-\psi(x)|\,d\nu(g)\,d\mu(x)=0$$

so that we must have μ-a.e.

$$\int_{\mathfrak{G}} |\psi(gx)-\psi(x)|\,d\nu(g)=0\,.$$

Then for all x up to a μ-null set M

$$\psi(gx)=\psi(x) \qquad \nu\text{-a.e.} \tag{11.37}$$

Hence,

$$\psi_1(x)=\int_{\mathfrak{G}} \psi(gx)\,d\nu(g) \tag{11.38}$$

is defined μ-a.e. ψ_1 is invariant, for

$$\int_{\mathfrak{G}} \psi(g g_1 x)\,d\nu(g)=\int_{\mathfrak{G}} \psi(x)\,d\nu(g)=\int_{\mathfrak{G}} \psi(gx)\,d\nu(g)=\psi_1(x) \qquad \mu\text{-a.e.}$$

Thus, $\psi_1(g_1 x)$ is defined for all $g_1\in\mathfrak{G}$ and all $x\in R-M$ and we have there $\psi_1(x)=\psi_1(g_1 x)$. Trivially, by (11.38)

$$g_{\psi_1}(\gamma)=g_{\psi}(\gamma) \tag{11.39}$$

for $\gamma\in\Gamma-\Gamma_0$.

(11.39) and (11.36) together with (11.34) now allow us to conclude by means of the assumptions on ϕ_0 that

$$\inf_{\gamma\in\Gamma_n} g_{\phi_0}(\gamma)+\varepsilon \geqslant \inf_{\gamma\in\Gamma_n} g_{\psi_1}(\gamma)+\varepsilon \geqslant \inf_{\gamma\in\Gamma_n} g_{\phi_\varepsilon}(\gamma)+\varepsilon \geqslant \sup_{\phi\in\Phi_\alpha}\inf_{\gamma\in\Gamma_n} g_{\phi}(\gamma)$$

and since ε was arbitrarily small, the theorem is proved.

One thus sees that the core of Theorem 11.2 (p. 231) is essentially group-theoretic. This is also demonstrated by the example considered there.

For the application of Theorem 11.3, the existence of an invariant measure on the group $\mathfrak{G}$ is crucial.

It is known that such measures exist iff the group is compact[83]. Without going into details, we point out that this assumption is quite restrictive. Theorem 11.3 can actually be generalized and one gets the so-called *Hunt-Stein theorem*[84].

---

[83] See for example A. Weil, L'intégration dans les groupes topologiques et ses applications, Actualités scientifiques et industrielles 869—1145, Hermann & Cie, 2nd ed., Paris 1953.

[84] See e.g. E. L. Lehmann, l.c.[2], 335. Also O. Wesler, Ann. Math. Statist. 30, 1—20 (1959).

12. Introduction to asymptotic theory. Up to now we have always considered a fixed sample space. We have thus almost always assumed that an arbitrary, but fixed sample size was under consideration. In applications the idea is prevalent that one can judge the correctness of a null hypothesis on the parameter of a set of probability measures with "increasing reliability" with increasing sample size. This idea corresponds roughly to the following procedure: Let $(R^{(n)}, \mathscr{S}^{(n)})$ for $n=1,2,\ldots$ be a measurable space and $P_\Gamma^{(n)}$ a set of probability measures on $(R^{(n)}, \mathscr{S}^{(n)})$, where Γ is the set of admissible hypotheses and does not depend on n. We can naturally drop the last assumption and assume, for example, that a set of admissible hypotheses Γ_n is associated with each n and $\Gamma_1 \subseteq \Gamma_2 \subseteq \cdots$ holds.[85] We will, however, not consider such generalizations. Let the test problem be of the form $(\Gamma_0, \Gamma - \Gamma_0)$, so that we now have to compare sequences $\{\phi_n\}$ of tests. Here, ϕ_n is an $\mathscr{S}^{(n)}$-measurable function defined over $R^{(n)}$ for each $n \geqslant 1$. Naturally, we are primarily interested in the behavior of ϕ_n for large n. In particular, we want to find reasonable criteria for deciding when a sequence of tests $\{\phi_n^*\}$ is "asymptotically better" then a sequence of tests $\{\phi_n\}$. These problems are closely connected with problems of the asymptotic theory of estimation which we treat in V.**4** and **5**. We will only consider certain of the more important aspects here. We first present a definition which is based on the idea that for "infinitely large sample size" one never rejects the "true" null hypothesis and never accepts the "false" one. We use the notation above.

Definition. A sequence of tests $\{\phi_n\}$ is called *a consistent sequence of tests* for the test problem $(\Gamma_0, \Gamma - \Gamma_0)$ if $\lim_{n\to\infty} E^{(n)}(\phi_n; \gamma) = 0$ [86] for $\gamma \in \Gamma_0$ and $\lim_{n\to\infty} E^{(n)}(\phi; \gamma) = 1$ for $\gamma \in \Gamma - \Gamma_0$.

In most practically important cases, we don't have a completely arbitrary sequence of sample spaces $(R^{(n)}, \mathscr{S}^{(n)})$ but rather $R^{(n)}$ is R_n and $\mathscr{S}^{(n)}$ the σ-algebra $\mathfrak{B}_n$. As sample variables ξ_i, $i=1,2,\ldots$, we take independent r.v.'s which all possess the same probability distribution $P_\gamma, \gamma \in \Gamma$, independent of i. For each $\gamma \in \Gamma, P_\gamma^{(n)}$ over $(R_n, \mathfrak{B}_n)$ is then given by $\prod_{i=1}^{n} P_{\gamma,i}$ with $P_{\gamma,i} = P_\gamma$, $1 \leqslant i \leqslant n$, $n \geqslant 1$. We recall that one can always give a probability space over which the r.v.'s $\xi_1, \xi_2, \ldots$ are defined and over which they are independent in the sense of the Definition, p. 46, so that the marginal distribution of ξ_i for $i=1,2,\ldots$ is precisely P_γ. In fact, it is enough to take the probability space $\left(R_\infty, B_\infty, \prod_{i=1}^{\infty} P_{\gamma,i}\right)$.

---

[85] See for example J. Neyman and E. Scott, Econometrica 16, 1—32 (1948).

[86] $E^{(n)}(\phi_n; \gamma)$ naturally means $\int_{R^{(n)}} \phi_n \, d P_\gamma^{(n)}$.

Under these more special assumptions, at least when Γ_0 and $\Gamma - \Gamma_0$ are countable, one can always construct a consistent sequence of tests[87]. We consider only the case when Γ_0 and $\Gamma - \Gamma_0$ are simple, which allows the essential idea to be presented and which is due to Kakutani[88]. We first prove

Lemma 12.1. *Let P_1 and P_2 be arbitrary probability measures over $(R, \mathcal{S})$. Denote by μ a probability measure which dominates P_1 and P_2. Let f_1 and f_2 be the corresponding R.-N.-densities and let $\rho(P_1, P_2) = \int_R \sqrt{f_1 f_2}\, d\mu$. Then*

$$0 \leqslant \rho(P_1, P_2) \leqslant 1\,. \tag{12.1}$$

$\rho(P_1, P_2) = 1$ iff $P_1 = P_2$ and $\rho(P_1, P_2) = 0$ iff P_1 and P_2 are orthogonal measures.

Proof. (12.1) follows immediately from Schwarz' inequality. $\rho(P_1, P_2) = 1$ holds iff equality holds in Schwarz' inequality. This is true iff $f_1 = f_2$ μ-a.e. Now let P_1 and P_2 be orthogonal. Then there exists an $M \in \mathcal{S}$ with $P_1(M) = 1$, $P_2(M) = 0$. This implies

$$\begin{aligned}\rho(P_1, P_2) &= \int_R c_M \sqrt{f_1 f_2}\, d\mu + \int_R c_{R-M} \sqrt{f_1 f_2}\, d\mu \\ &\leqslant \left(\int_R f_1\, d\mu\right)^{1/2} \left(\int_R c_M\, f_2\, d\mu\right)^{1/2} \\ &\quad + \left(\int_R c_{R-M}\, f_1\, d\mu\right)^{1/2} \left(\int_R f_2\, d\mu\right)^{1/2} \\ &= 0\,.\end{aligned}$$

We now prove

Theorem 12.1. *Let P_{γ_0} and P_{γ_1} be two probability measures over $(R_1, \mathfrak{B}_1)$. Consider the sequence of sample spaces $\{(R_n, \mathfrak{B}_n)\}$ and on them the product measures $P_{\gamma_i}^{(n)} = \prod_{j=1}^{n} P_{\gamma_i, j}$ with $P_{\gamma_i, j} = P_{\gamma_i}$, $1 \leqslant j \leqslant n$, $i = 0, 1$. Then there always exists a consistent sequence of tests for the problem $(\{\gamma_0\}, \{\gamma_1\})$.*

Proof. Let μ be a probability measure dominating P_{γ_0} and P_{γ_1}. Denote the R.-N.-densities of P_{γ_i} w.r.t. μ by f_{γ_i}, $i = 0, 1$. Then the R.-N.-density of $P_{\gamma_i}^{(n)}$ w.r.t. $\prod_{j=1}^{n} \mu_j$ with $\mu_j = \mu$, $1 \leqslant j \leqslant n$, is given by $(x_1, \ldots, x_n) \to \prod_{j=1}^{n} f_{\gamma_i}(x_j)$.

---

[87] See A. Berger, Ann. Math. Statist. 22, 289—293 (1951) and Ch. Kraft, Univ. California Publ. Statist. 2, 125—141 (1953—1958).

[88] S. Kakutani, Ann. of Math. II. Ser. 49, 214—224 (1948).

Hence,

$$\rho(P_{\gamma_0}^{(n)}, P_{\gamma_1}^{(n)}) = \int_{R_n} \left(\prod_{j=1}^{n} f_{\gamma_0}(x_j) f_{\gamma_1}(x_j) \right)^{1/2} d\mu(x_1) \ldots d\mu(x_n)$$

$$= \prod_{j=1}^{n} \int_{R_1} \sqrt{f_{\gamma_0}(x_j) f_{\gamma_1}(x_j)}\, d\mu(x_j)$$

and so

$$\rho(P_{\gamma_0}^{(n)}, P_{\gamma_1}^{(n)}) = (\rho(P_{\gamma_0}, P_{\gamma_1}))^n. \tag{12.2}$$

Since $P_{\gamma_0} \neq P_{\gamma_1}$, we have from (12.2) by Lemma 12.1

$$\rho(P_{\gamma_0}^{(n)}, P_{\gamma_1}^{(n)}) \to 0 \quad \text{for } n \to \infty .$$

Now choose a sequence of real numbers $\{k_n\}$ and real numbers c_1, c_2 with $0 < c_1 \leqslant c_2$ such that for $n = 1, 2, \ldots$

$$0 < c_1 \leqslant k_n \leqslant c_2 \tag{12.3}$$

holds. For $n \geqslant 1$, let

$$\phi_n(x_1, \ldots, x_n) = \begin{cases} 1 & \prod_{j=1}^{n} (f_{\gamma_1}(x_j))^{\frac{1}{2}} > k_n \prod_{j=1}^{n} (f_{\gamma_0}(x_j))^{\frac{1}{2}} \\ 0 & \prod_{j=1}^{n} (f_{\gamma_1}(x_j))^{\frac{1}{2}} \leqslant k_n \prod_{j=1}^{n} (f_{\gamma_0}(x_j))^{\frac{1}{2}} \end{cases}.$$

Then

$$E(\phi_n; \gamma_0) = \int_{R_n} \phi_n(x_1, \ldots, x_n) \prod_{j=1}^{n} f_{\gamma_0}(x_j)\, d\mu(x_1) \ldots d\mu(x_n)$$

$$\leqslant \frac{1}{k_n} \int_{R_n} \left(\prod_{j=1}^{n} f_{\gamma_0}(x_j) \prod_{j=1}^{n} f_{\gamma_1}(x_j) \right)^{\frac{1}{2}} d\mu(x_1) \ldots d\mu(x_n) = \frac{1}{k_n} \rho(P_{\gamma_0}^{(n)}, P_{\gamma_1}^{(n)})$$

and because of (12.3) we thus have $E(\phi_n; \gamma_0) \leqslant \frac{1}{c_1} \rho(P_{\gamma_0}^{(n)}; P_{\gamma_1}^{(n)}) \to 0$ for $n \to \infty$.

Similarly,

$$E(1 - \phi_n; \gamma_1) \leqslant k_n \int_{R_n} \left(\prod_{j=1}^{n} f_{\gamma_0}(x_j) \prod_{j=1}^{n} f_{\gamma_1}(x_j) \right)^{\frac{1}{2}} d\mu(x_1) \ldots d\mu(x_n)$$

$$\leqslant c_2 \rho(P_{\gamma_0}^{(n)}, P_{\gamma_1}^{(n)}) \to 0 \quad \text{for } n \to \infty .$$

This completes the proof.

If $\xi_1, \xi_2, \ldots$ is a sequence of r.v.'s whose distribution fulfills, say, the assumptions of V, Theorem 3.2, then one can prove the following result on the MLRT: For $k \geqslant 2$ let a k-dimensional interval Q_k be the set of admissible hypotheses and the interval Q_l, $1 \leqslant l < k$, the null hypothesis. Then, in the notation of p. 225, the sequence of tests defined by the statistic $x^{(n)} \to l(x^{(n)})$, $x^{(n)} \in R_n$, $n = 1, 2, \ldots$, is consistent for $(\alpha, Q_l, Q_k - Q_l)$[89].

If we have two sequences of tests $\{\phi_n^{(j)}\}$, $j = 1, 2$ for a test problem, and want to compare the power functions for large n, the consideration of

$$\lim_{n \to \infty} g_{\phi_n^{(1)}}(\gamma) / g_{\phi_n^{(2)}}(\gamma) \quad \text{for } \gamma \in \Gamma - \Gamma_0 \tag{12.4}$$

provided it exists, suggests itself. However, if both are consistent sequences of tests for the given problem, as occurs in most applications, then the limit (12.4) always has the value 1. Such a comparison criterium is thus of little use. A useful concept has been given by Pitman. Assume given a test problem of the form $(\alpha, \{\gamma_0\}, \Gamma - \{\gamma_0\})$. Let $\gamma_i \in \Gamma - \{\gamma_0\}$, $i = 1, 2, \ldots$ and $\lim_{i \to \infty} \gamma_i = \gamma_0$. Let $\{\phi_n^{(j)}\}, j = 1, 2$ be two sequences of tests which satisfy

$$\lim_{n \to \infty} E(\phi_n^{(1)}; \gamma_0) = \lim_{n \to \infty} E(\phi_n^{(2)}; \gamma_0) = \alpha \,. \tag{12.5}$$

Assume there exist two increasing sequences of natural numbers $\{n_i^{(j)}\}$, $j = 1, 2$ for which $\lim_{n_i^{(j)} \to \infty} g_{\phi_{n_i^{(j)}}}(\gamma_i)$, $j = 1, 2$ exist and are equal. Moreover, assume they are different from 0 and 1. It can happen that

$$\lim_{i \to \infty} n_i^{(2)} / n_i^{(1)} \tag{12.6}$$

exists for every possible choice of the sequences $\{n_i^{(j)}\}, j = 1, 2$ and has the same value. This limit is called the *asymptotic relative efficiency* or the *Pitman efficiency* of $\{\phi_n^{(1)}\}$ w.r.t. $\{\phi_n^{(2)}\}$ for the sequence $\{\gamma_i\}$. We write for it

$$\mathrm{re}(\{\phi_n^{(1)}\}, \{\phi_n^{(2)}\}, \alpha, \{\gamma_i\}) \,.$$

The considerations in **3** suggest using tests ϕ_n over measurable spaces $(R^{(n)}, \mathscr{S}^{(n)})$ which are defined as follows. Let α be a real number with $0 \leqslant \alpha \leqslant 1$. For $n = 1, 2, \ldots$, let T_n be an $\mathscr{S}^{(n)}$-measurable function defined over $R^{(n)}$. Assume we have the test problem $(\alpha, \{\gamma_0\}, \Gamma - \{\gamma_0\})$ and set

$$\phi_n(x^{(n)}) = \begin{cases} 1 & T_n(x^{(n)}) > k_{n,\alpha} \\ c_n & T_n(x^{(n)}) = k_{n,\alpha} \\ 0 & T_n(x^{(n)}) < k_{n,\alpha} \end{cases}$$

[89] See for details A. Wald, l.c.[78].

Here, $\{c_n\}$ is a sequence of real numbers with $0 \leqslant c_n \leqslant 1$, $n \geqslant 1$ and $\{k_{n,\alpha}\}$ is a sequence of real numbers which are so chosen that

$$E(\phi_n; \gamma_0) = \alpha$$

for $n = 1, 2, \ldots$ or that

$$\lim_{n\to\infty} E(\phi_n; \gamma_0) = \alpha . \tag{12.7}$$

We can view the test statistics T_n for $n = 1, 2, \ldots$ as r.v.'s over the probability space $(R^{(n)}, \mathscr{S}^{(n)}, P_\gamma^{(n)})$, $\gamma \in \Gamma$.

The following arguments become more understandable when we view them against the background of CAN-estimates, which we treat in V, p. 336ff.

We first have

Lemma 12.2[90]. *Let* $\Gamma \subseteq R_1$, *and for the sake of simplicity,* $\gamma_0 = 0$. *Let the* T_n *have the above meaning. Let F be a continuous, strictly monotone distribution function,* $\{a_n\}$ *a sequence of positive numbers and* $\{\eta_n\}$ *a sequence of functions over* Γ *with the following property: For each sequence* $\{\gamma_n\}$ *with* $\gamma_n \in \Gamma, n \geqslant 1$, *and* $\lim_{n\to\infty} \gamma_n = 0$ *there holds*

$$\lim_{n\to\infty} P_{\gamma_n}\left(\frac{T_n - \eta_n(\gamma_n)}{a_n} \leqslant x\right) \to F(x) \tag{12.8}$$

for $-\infty < x < \infty$. *Define by*

$$\omega_n = \left\{x^{(n)} : \frac{T_n(x^{(n)}) - \eta_n(0)}{a_n} > k_n\right\} \tag{12.9}$$

for $n = 1, 2, \ldots$ *a sequence of critical regions with asymptotic level of significance* $\alpha, 0 < \alpha < 1$, i.e., *let*

$$\lim_{n\to\infty} k_n = k \tag{12.10}$$

with

$$k = F^{-1}(1-\alpha) . \tag{12.11}$$

Denote the power function of ω_n *by* $g_n, n \geqslant 1$. *Then*

$$g_n(\gamma_n) \to \beta \tag{12.12}$$

holds for a real c, where c is defined by

$$\beta = 1 - F(k - c) \tag{12.13}$$

iff

$$\frac{\eta_n(\gamma_n) - \eta_n(0)}{a_n} \to c . \tag{12.14}$$

---

[90] See J. L. Hodges Jr. and E. L. Lehmann, Proc. Fourth Berkeley Sympos. Math. Statist. and Prob. Vol. I, pp. 307—317, Univ. California Press, Berkeley Calif., 1961.

For the simple *proof* note that

$$g_n(\gamma_n)=1-P_{\gamma_n}\left(\frac{T_n-\eta_n(0)}{a_n}\leqslant k_n\right)$$
$$=1-P_{\gamma_n}\left(\frac{T_n-\eta_n(\gamma_n)}{a_n}\leqslant k_n-\frac{\eta_n(\gamma_n)-\eta_n(0)}{a_n}\right).$$

Since F is continuous and strictly monotone, the claim follows from the assumptions.

We also have the similar

Lemma 12.3. *Let the sequences* $\{T_n^{(i)}\}, \{a_n^{(i)}\}$ *and* $\{\eta_n^{(i)}\}$ *fulfill for* $i=1,2,$ *the assumptions of Lemma 12.2 for the same* α. *Let* $\gamma_n\to 0$ *and for a real* β *with* $0<\beta<1$ *(in obvious notation)*

$$\lim_{n\to\infty} g_n^{(2)}(\gamma_n)=\beta\,. \tag{12.15}$$

Let β *satisfy* (12.13) *with* $c>0$. *With each n we associate a natural number* $r_n\geqslant n$ *such that for the power function* $g_n^{(1)}$[91] *of the critical region defined by*

$$\left\{x^{(r_n)}:\frac{T_{r_n}^{(1)}(x^{(r_n)})-\eta_{r_n}^{(1)}(\gamma_n)}{a_{r_n}^{(1)}}>k_{r_n}^{(1)}\right\}$$

$$\lim_{n\to\infty} g_n^{(1)}(\gamma_n)=\beta \tag{12.16}$$

holds. Then

$$\frac{\eta_n^{(2)}(\gamma_n)-\eta_n^{(2)}(0)}{\eta_{r_n}^{(1)}(\gamma_n)-\eta_{r_n}^{(1)}(0)}\cdot\frac{a_{r_n}^{(1)}}{a_n^{(2)}}\to 1 \tag{12.17}$$

for $n\to\infty$.

Proof. Writing for the moment $\gamma_n=\delta_{r_n}$ for $n\geqslant 1$, we then have from $\lim\limits_{n\to\infty}\delta_{r_n}=0$ by (12.16) and Lemma 12.2

$$\frac{\eta_{r_n}^{(1)}(\delta_{r_n})-\eta_{r_n}^{(1)}(0)}{a_{r_n}^{(1)}}\to c$$

and, likewise, from (12.15)

$$\frac{\eta_n^{(2)}(\gamma_n)-\eta_n^{(2)}(0)}{a_n^{(2)}}\to c\,.$$

This implies (12.17).

These results hold, in particular, when one defines

$$F(x)=(\sqrt{2\pi})^{-1}\int_{-\infty}^{x} e^{-t^2/2}\,dt \quad \text{for each } x\in R_1\,.$$

[91] Note that $g_n^{(1)}$ is somewhat differently defined as $g_n^{(2)}$.

The results we have obtained allow us to define the Pitman efficiency for a wide class of sequences of tests. It will turn out that in the comparison of two sequences of tests from this class it has the same value for each sequence $\{\gamma_n\}$ satisfying (12.15) and (12.16). Hence, the reference to the sequence $\{\gamma_n\}$ can be suppressed. In fact, we have

Theorem 12.2. *Let the assumptions of Lemma 12.3 be fulfilled. Let Γ be an open interval (containing zero) and the functions $\eta_n^{(i)}$ for $n=1,2,\dots$ and $i=1,2$, m-times continuously differentiable there, $m\geqslant 1$. Denote by $\eta_{n,s}^{(i)}(\gamma)$ for $n\geqslant 1$ and $i=1,2$ the s-th derivative at the point γ. For each natural s with $0<s<m$ let*

$$\eta_{n,s}^{(i)}(0)=0, \qquad n\geqslant 1.^{92} \tag{12.18}$$

Further let

$$\eta_{n,m}^{(i)}(0)>0, \qquad n\geqslant 1. \tag{12.19}$$

Let there exist a $\delta>0$ such that

$$0<\lim_{n\to\infty} \eta_{n,m}^{(i)}(0)\,n^{-\delta}(a_n^{(i)})^{-1}<\infty. \tag{12.20}$$

Also assume that for each null sequence $\{\gamma_n\}$ with $\gamma_n\in\Gamma$

$$\lim_{n\to\infty} \eta_{n,m}^{(i)}(\gamma_n)/\eta_{n,m}^{(i)}(0)=1. \tag{12.21}$$

The conditions (12.18)—(12.21) *are to hold for $i=1,2$. Then*

$$\lim_{n\to\infty}\frac{n}{r_n}=\lim_{n\to\infty}\left(\frac{\eta_{n,m}^{(1)}(0)\,a_n^{(2)}}{\eta_{n,m}^{(2)}(0)\,a_n^{(1)}}\right)^{1/\delta}. \tag{12.22}$$

This limit is thus independent of the null sequence $\{\gamma_n\}$ and can therefore be denoted by $\mathrm{re}(\{\phi_n^{(1)}\},\{\phi_n^{(2)}\},\alpha)$.

Proof. Let $\gamma_n\to 0$. Then, from (12.18) and the extended Mean Value Theorem there follows the existence of sequences $\{\gamma_n^{(j)}\}$ with $0<\gamma_n^{(j)}<\gamma_n$, $n\geqslant 1$, $1\leqslant j\leqslant m$, such that

$$\frac{\eta_n^{(2)}(\gamma_n)-\eta_n^{(2)}(0)}{\eta_{r_n}^{(1)}(\gamma_n)-\eta_{r_n}^{(1)}(0)}=\frac{\eta_{n,1}^{(2)}(\gamma_n^{(1)})}{\eta_{r_n,1}^{(1)}(\gamma_n^{(1)})}=\frac{\eta_{n,1}^{(2)}(\gamma_n^{(1)})-\eta_{n,1}^{(2)}(0)}{\eta_{r_n,1}^{(1)}(\gamma_n^{(1)})-\eta_{r_n,1}^{(1)}(0)}$$

$$=\frac{\eta_{n,2}^{(2)}(\gamma_n^{(2)})}{\eta_{r_n,2}^{(1)}(\gamma_n^{(2)})}=\frac{\eta_{n,2}^{(2)}(\gamma_n^{(2)})-\eta_{n,2}^{(2)}(0)}{\eta_{r_n,2}^{(1)}(\gamma_n^{(2)})-\eta_{r_n,2}^{(1)}(0)}=\dots=\frac{\eta_{n,m}^{(2)}(\gamma_n^{(m)})}{\eta_{r_n,m}^{(1)}(\gamma_n^{(m)})}.$$

All of these quotients make sense because of (12.19). Thus, from Lemma 12.3

$$1=\lim_{n\to\infty}\frac{\eta_{n,m}^{(2)}(\gamma_n^{(m)})}{\eta_{r_n,m}^{(1)}(\gamma_n^{(m)})}\,\frac{a_{r_n}^{(1)}}{a_n^{(2)}}.$$

---

[92] If $m=1$, this condition is omitted.

Thus, after a simple calculation which takes (12.20) and (12.21) into consideration, we get

$$1 = \lim_{n\to\infty} \left(\frac{n}{r_n}\right)^{\delta} \lim_{n\to\infty} \frac{\eta^{(2)}_{n,m}(0)}{n^{\delta} a^{(2)}_n} \bigg/ \lim_{n\to\infty} \frac{\eta^{(1)}_{r_n,m}(0)}{r_n^{\delta} a^{(1)}_{r_n}}$$

$$= \lim_{n\to\infty} \left(\frac{n}{r_n}\right)^{\delta} \lim_{n\to\infty} \frac{\eta^{(2)}_{n,m}(0)}{n^{\delta} a^{(2)}_n} \bigg/ \lim_{n\to\infty} \frac{\eta^{(1)}_{n,m}(0)}{n^{\delta} a^{(1)}_n}$$

whence (12.22) easily follows.

This concept of Pitman[93] has proved to be quite useful. We will give some examples in VII, p. 466. This definition is, however, basically arbitrary. Pitman fixes the asymptotic level of significance α and the asymptotic power β and lets the alternative γ_n converge for $n\to\infty$ to the null hypothesis. With the same justification, one can base comparisons of sequences of tests on fixed α and fixed alternative γ and compares the sequences of power functions at the point γ.[94] One can also hold γ and β fixed and compare the sequences of levels of significance. This possibility has been analyzed by Bahadur and we treat this approach here.

Let $\{\phi_n\}$ be a sequence of tests for a problem of testing of the form $(\{\gamma_0\}, \Gamma - \{\gamma_0\})$, and let γ be an arbitrary, but fixed element from $\Gamma - \{\gamma_0\}$. We consider those sequences of tests which fulfill the following conditions:

Let

$$b(\{\phi_n\}, \gamma_0) = \lim_{n\to\infty} 2n^{-1} \log(1/E(\phi_n; \gamma_0)) \tag{12.23}$$

exist as finite or infinite limit which is necessarily $\geqslant 0$. Since we will always be referring to γ_0 in the sequel, we will also write $b(\{\phi_n\})$ in place of $b(\{\phi_n\}, \gamma_0)$. Now let $\{\phi_n^{(i)}\}, u=1,2,$ be sequences of tests for which $b(\{\phi_n^{(i)}\})$ exists. Further assume that

$$0 < b(\{\phi_n^{(i)}\}) < \infty, \qquad i=1,2. \tag{12.24}$$

We then denote as the *Bahadur efficiency*[95] of $\{\phi_n^{(1)}\}$ w.r.t. $\{\phi_n^{(2)}\}$ the quotient

$$b(\{\phi_n^{(1)}\})/b(\{\phi_n^{(2)}\}). \tag{12.25}$$

(12.24) implies that $E(\phi_n^{(i)}; \gamma_0) \neq 0$ at least for all sufficiently large n. On the other hand, however, $\lim_{n\to\infty} E(\phi_n^{(i)}; \gamma_0) = 0$ must hold. Indeed, we

---

93 See J. G. Pitman, Lecture Notes on Nonparametric Inference. Columbia University, New York 1949. See also G. E. Noether, Ann. Math. Statist. 26, 64—68 (1955).

94 See, however, the developments on p. 243. Also J. L. Hodges jr. and E. L. Lehmann, Ann. Math. Statist. 27, 324—335 (1956).

95 R. R. Bahadur, Ann. Math. Statist. 31, 276—295 (1960).

obviously have from Definition (12.23) and from (12.24) for two suitable real numbers c_1, c_2 with $0<c_1<c_2<\infty$ and all sufficiently large n

$$e^{-c_2 n} < E(\phi_n^{(i)}; \gamma_0) < e^{-c_1 n}. \tag{12.26}$$

The sequence of levels of significance thus converges exponentially to 0. The quotient (12.25) can likewise, in analogy to the Pitman efficiency, be represented as the limit of a quotient of sample sizes. Intuitively, this limit indicates for which sequence of tests the corresponding sequence of levels of significance goes faster to zero. In fact, let $\varepsilon>0$ be given and define $N^{(i)}(\varepsilon)$ to be the smallest natural number with $E(\phi^{(i)}_{N^{(i)}(\varepsilon)}; \gamma_0) \leqslant \varepsilon$, $i=1,2$. From (12.26), $\lim_{\varepsilon\to 0} N^{(i)}(\varepsilon)=\infty$. Hence, for each $\varepsilon>0$ there exists by (12.23) a $\delta^{(i)}(\varepsilon)>0$ with $\lim_{\varepsilon\to 0}\delta^{(i)}(\varepsilon)=0$ such that

$$\frac{b(\{\phi_n^{(i)}\})+\delta^{(i)}(\varepsilon)}{2} N^{(i)}(\varepsilon) \geqslant \log(1/E(\phi^{(i)}_{N^{(i)}(\varepsilon)}; \gamma_0) \geqslant \frac{1}{\varepsilon} > \log(1/E(\phi^{(i)}_{N^{(i)}(\varepsilon)-1}; \gamma_0)$$

$$\geqslant \frac{b(\{\phi_n^{(i)}\})-\delta^{(i)}(\varepsilon)}{2}(N^{(i)}(\varepsilon)-1)$$

for $i=1,2$.

But then

$$\frac{b(\{\phi_n^{(1)}\})}{b(\{\phi_n^{(2)}\})}\underline{\lim}_{\varepsilon\to 0}\frac{N^{(1)}(\varepsilon)}{N^{(2)}(\varepsilon)} \geqslant 1 \geqslant \frac{b(\{\phi_n^{(1)}\})}{b(\{\phi_n^{(2)}\})}\overline{\lim}_{\varepsilon\to 0}\frac{N^{(1)}(\varepsilon)}{N^{(2)}(\varepsilon)}$$

so that

$$\lim_{\varepsilon\to 0}\frac{N^{(2)}(\varepsilon)}{N^{(1)}(\varepsilon)} = \frac{b(\{\phi_n^{(1)}\})}{b(\{\phi_n^{(2)}\})}.$$

We have not motivated the definition of $b(\{\phi_n\})$ according to (12.23). We now make this a bit clearer. We will again specialize to independent, one-dimensional sample variables $\xi_1, \xi_2, \ldots$ with the same distribution. Consider, then, the sequence of sample spaces $(R_n, \mathfrak{B}_n)$ and let each ξ_i, $i=1,2,\ldots$ be distributed by $P_\gamma^{(1)}$, $\gamma\in\Gamma$. Let the set $P_\Gamma^{(1)}$ of probability measures, defined over $(R_1, \mathfrak{B}_1)$, be dominated by a σ-finite measure μ. For each $\gamma\in\Gamma$ denote the R.-N.-density of $P_\gamma^{(1)}$ w.r.t. μ by $x\to f(x,\gamma)$.

For $n\geqslant 2$, $P_\Gamma^{(n)}$ is the set of all product measures of the form $\prod_{i=1}^n P_{\gamma,i}^{(1)}$, $\gamma\in\Gamma$, $P_{\gamma,i}^{(1)}=P_\gamma^{(1)}$, $1\leqslant i\leqslant n$. Let γ_0 and γ_1 be two elements from Γ. For $x\in R_1$ set

$$q(x;\gamma_1,\gamma_0)=\log[f(x,\gamma_1)/f(x,\gamma_0)] \tag{12.27}$$

provided this expression is defined. Assume that

$$E(q^2(\xi;\gamma_1,\gamma_0);\gamma_1) \tag{12.28}$$

exists for all $\gamma_0, \gamma_1 \in \Gamma$, where ξ is, of course, a r.v. distributed by $P_{\gamma_1}^{(1)}$. The existence of (12.28) implies that for each pair (γ_0, γ_1)

$$\{x: f(x,\gamma_1)=0\} = \{x: f(x,\gamma_0)=0\} \quad \mu\text{-a.e.}$$

Write

$$H(\gamma_1,\gamma_0) = E(q(\xi;\gamma_1,\gamma_0);\gamma_1) \tag{12.29}$$

and choose two arbitrary, but fixed elements to which we will refer until further notice γ_0, γ_1 from Γ with

$$\gamma_0 \neq \gamma_1 . \tag{12.30}$$

In this connection we will thus write H instead of $H(\gamma_1,\gamma_0)$, $q(\xi)$ instead of $q(\xi,\gamma_1,\gamma_0)$, etc.

With this terminology we prove

Theorem 12.3. *Let all of the assumptions above be fulfilled. Moreover, assume*

$$E(e^{tq(\xi)};\gamma_0) \tag{12.31}$$

exists for all real t belonging to an interval I which contains $t=1$ as interior point. Let a be a real number, $-\infty<a<\infty$, and for $n=1,2,\dots$

$$r_n = \exp(nH + \sqrt{n}\,a\sigma) \tag{12.32}$$

where σ^2 denotes the variance of $q(\xi)$ w.r.t. $P_{\gamma_1}^{(1)}$. Let A_n be the subset

$$\left\{x: \prod_{i=1}^{n} f(x_i,\gamma_1) \geqslant r_n \prod_{i=1}^{n} f(x_i,\gamma_0)\right\} \tag{12.33}$$

of R_n and define for $n=1,2,\dots$ with $\xi^{(n)}=(\xi_1,\dots,\xi_n)$

$$E(c_{A_n}\circ\xi^{(n)};\gamma_0) = \alpha_n . \tag{12.34}$$

Then

$$\lim_{n\to\infty} n^{-1}\log 1/\alpha_n = H \tag{12.35}$$

no matter how a is chosen in (12.32).

Moreover, likewise for all real a,

$$\underline{\lim}_{n\to\infty} E(c_{A_n}\circ\xi^{(n)};\gamma_1) > 0 . \tag{12.36}$$

If $\{\phi_n\}$ is a sequence of tests for the problem $(\{\gamma_0\},\{\gamma_1\})$ such that ϕ_n is defined for $n\geqslant 1$ over R_n and if for the corresponding sequence of power functions

$$\underline{\lim}_{n\to\infty} g_{\phi_n}(\gamma_1) > 0 , \tag{12.37}$$

then

$$\overline{\lim}_{n\to\infty} n^{-1}\log(1/E(\phi_n\circ\xi^{(n)};\gamma_0)) \leqslant H . \tag{12.38}$$

Proof. Since $P^{(1)}_{\gamma_1} \neq P^{(1)}_{\gamma_0}$, one has $0 < \sigma^2$, and from the existence of (12.28), $\sigma^2 < \infty$. Moreover, as we will show in the proof of V, Theorem 3.6, $0 < H < \infty$. Thus, $0 < r_n < \infty$ for all $n \geq 1$. A_n can also be defined as

$$\left\{x: \sum_{i=1}^{n} q(x_i) \geq nH + \sqrt{n}\, a\sigma\right\}. \tag{12.39}$$

Hence,

$$E(1 - c_{A_n} \circ \xi^{(n)}; \gamma_1) = P_{\gamma_1}\left(\left(\sum_{i=1}^{n} q(\xi_i) - nH\right)\Big/(\sigma\sqrt{n}) < a\right).$$

Since H is given by (12.29), from I, Theorem 39.1

$$\lim_{n\to\infty} E(c_{A_n} \circ \xi^{(n)}; \gamma_1) = \frac{1}{\sqrt{2\pi}} \int_a^\infty e^{-t^2/2}\, dt > 0. \tag{12.40}$$

This proves (12.36). Because of (12.34) and the fact that A_n is also given by (12.39), we obviously have

$$\alpha_n = P_{\gamma_0}\left(\sum_{i=1}^{n} q(\xi_i) \geq nH + \sqrt{n}\, a\sigma\right). \tag{12.41}$$

We now apply I, Theorem 39.5 using the terminology introduced on p. 109.

Let $\psi(t) = E(e^{tq(\xi)}; \gamma_0)$ for $t \in I$. Using (12.27) and (12.29), we find $\psi'(1)/\psi(1) = H$. Also, $\chi(1) = e^{-H}$. We have

$$\lim_{n\to\infty} n^{-1} \log P_{\gamma_0}\left(\sum_{i=1}^{n} q(\xi_i) \geq nH + \sqrt{n}\, a\sigma\right) = -H.$$

From (12.41) follows (12.35). Now let $\{\phi_n\}$ be an arbitrary test sequence which fulfills (12.37). We can thus always choose an $\varepsilon > 0$ and a $b > 0$ such that

$$g_{\phi_n}(\gamma_1) > \frac{1}{\sqrt{2\pi}} \int_b^\infty e^{-t^2/2}\, dt + \varepsilon \tag{12.42}$$

for all sufficiently large n. Choose $a = b$ in (12.32) and use (12.39) to (12.41). Since the sequence of critical regions $\{A_n\}$ is given by (12.33), one can apply Theorem 3.3. Comparing (12.40) with (12.42) leads to $E(\phi_n \circ \xi^{(n)}; \gamma_0) > \alpha_n$ for large enough n. Hence, from (12.35) there follows (12.38).

Thus, if $\lim\limits_{n\to\infty} b(\{\phi_n \circ \xi^{(n)}\}; \gamma_0)$ exists, then the limit, under assumption (12.37), is equal to at most $2H$ and this value is actually assumed for the sequence of most powerful tests for the problem $(\{\gamma_0\}, \{\gamma_1\})$. This result motivates the definition of Bahadur efficiency.

We now give a class of sequences of tests for which the limit (12.23) exists. The following arguments will also provide further insights. We proceed from general measurable spaces $(R^{(n)}, \mathscr{S}^{(n)})$, $n=1,2,\ldots$, consider a set of probability measures $P_\Gamma^{(n)}$ over $(R^{(n)}, \mathscr{S}^{(n)})$ and show

Theorem 12.4. *For $n \geq 1$ define $\mathscr{S}^{(n)}$-measurable functions h_n (r.v.'s) over $(R^{(n)}, \mathscr{S}^{(n)}, P_\Gamma^{(n)})$. Let $\gamma_0 \in \Gamma$. For each real y, define*

$$P_{\gamma_0}^{(n)}(h_n \leq y) = F_n(y).$$

Let there exist a positive c such that for each sequence of positive numbers b_n with $b_n/\sqrt{n} \to y$, $0 < y < \infty$,

$$\lim_{n\to\infty} 2n^{-1} \log(1 - F_n(b_n)) = -c y^2. \tag{12.43}$$

Define a function d over Γ and assume for $\gamma_1 \in \Gamma$ with $\gamma_1 \neq \gamma_0$

$$d(\gamma_1) > 0 \tag{12.44}$$

holds. Assume $h_n/\sqrt{n}$ converges in $P_{\gamma_1}^{(n)}$-probability to $d(\gamma_1)$ for $n \to \infty$, $\{k_n\}$ is a sequence of positive numbers and set $\alpha_n = 1 - F_n(k_n)$ for $n \geq 1$. Consider the sets $M_n = \{x^{(n)}: h_n(x^{(n)}) \geq k_n\}$ as critical regions for the test problem $(\alpha_n, \{\gamma_0\}, \{\gamma_1\})$. Let

$$0 < \varliminf_{n\to\infty} g_{cM_n}(\gamma_1) \leq \varlimsup g_{cM_n}(\gamma_1) < 1 \tag{12.45}$$

hold for the power functions g_{cM_n}. Then

$$\lim_{n\to\infty} 2n^{-1} \log 1/\alpha_n = c d^2(\gamma_1). \tag{12.46}$$

Moreover, $c d^2(\gamma_1) \leq 2H$.

Proof. For $n \to \infty$

$$k_n/\sqrt{n} \to d(\gamma_1). \tag{12.47}$$

Indeed, from the convergence in probability of $h_n/\sqrt{n}$ to $d(\gamma_1)$ it follows for each $\varepsilon > 0$ that

$$\lim_{n\to\infty} P_{\gamma_1}^{(n)}(\sqrt{n}(d(\gamma_1) - \varepsilon) < h_n < \sqrt{n}(d(\gamma_1) + \varepsilon)) = 1. \tag{12.48}$$

If there existed a sequence of natural numbers n_i with $k_{n_i}/\sqrt{n_i} \to d_1$ and if $d_1 < d(\gamma_1)$ held, then for sufficiently small $\varepsilon > 0$ and all large enough n_i,

$$P_{\gamma_1}^{(n_i)}(h_{n_i} < k_{n_i}) \leq P_{\gamma_1}^{(n_i)}(h_{n_i} < \sqrt{n_i}(d(\gamma_1) - \varepsilon))$$

thus

$$\lim_{i\to\infty} P_{\gamma_1}^{(n_i)}(h_{n_i} < k_{n_i}) = 0$$

which contradicts the last inequality in (12.45). One treats the case $d_1 > d(\gamma_1)$ similarly. Thus, we have demonstrated (12.47). (12.46) follows without difficulty from (12.43). The last inequality of the theorem then follows from Theorem 12.3.

The following result, which we will not prove, is closely connected with the preceeding.

Except where noted, we will retain the notation above. Let F be a continuous d.f. and assume $F_n \to F$ weakly for $n \to \infty$. Moreover, for a $c > 0$ let

$$\log(1-F(x)) = -\frac{cx^2}{2}(1+o(1)) \quad \text{for } x \to \infty . \tag{12.49}$$

These assumptions replace (12.43), otherwise h_n is to fulfill the conditions of Theorem 12.4. Let $\{\alpha_n\}$ be an arbitrary sequence of real numbers in $(0, 1)$. We define M_n now by

$$\{x^{(n)}: 1 - F(h_n(x^{(n)})) \leq \alpha_n\} .$$

If (12.45) holds for these critical regions, we then conclude that (12.46) holds. Note, that in general, $P_{\gamma_0}^{(n)}(M_n) \neq \alpha_n$, but the sequence of distributions of $F \circ h_n$ still converges to the uniform distribution, which is easy to see.

We now treat briefly the *connection between the efficiencies of Pitman and Bahadur* and use the notation of Lemma 12.3:

Let the assumptions of Theorem 12.2 be satisfied with $\delta = 1/2$. Let F now be the distribution function of $N(0, 1)$. For all $\gamma \in \Gamma$ with $\gamma \neq 0$ and $i = 1, 2$, let there exist

$$\lim_{n\to\infty} \frac{\eta_n^{(i)}(\gamma) - \eta_n^{(i)}(0)}{a_n^{(i)}} \frac{1}{\sqrt{n}} = d^{(i)}(\gamma) . \tag{12.50}$$

We further assume that

$$\lim_{n\to\infty} \lim_{\gamma\to 0} \frac{\eta_n^{(1)}(\gamma) - \eta_n^{(1)}(0)}{\eta_n^{(2)}(\gamma) - \eta_n^{(2)}(0)} \frac{a_n^{(2)}}{a_n^{(1)}} = \lim_{\gamma\to 0} \lim_{n\to\infty} \frac{\eta_n^{(1)}(\gamma) - \eta_n^{(1)}(0)}{\eta_n^{(2)}(\gamma) - \eta_n^{(2)}(0)} \frac{a_n^{(2)}}{a_n^{(1)}} . \tag{12.51}$$

The Pitman efficiency is then given by

$$\lim_{n\to\infty} \lim_{\gamma\to 0} \left(\frac{\eta_n^{(1)}(\gamma) - \eta_n^{(1)}(0)}{\eta_n^{(2)}(\gamma) - \eta_n^{(2)}(0)} \frac{a_n^{(2)}}{a_n^{(1)}} \right)^2$$

because of (12.21) and (12.22). Hence from (12.51) and (12.50)

$$\mathrm{re}(T_n^{(1)}, T_n^{(2)}) = \lim_{\gamma\to 0} \left(\frac{d^{(1)}(\gamma)}{d^{(2)}(\gamma)} \right)^2 .$$

Denoting $\frac{T_n^{(i)} - \eta_n^{(i)}(0)}{a_n^{(i)}}$ by $h_n^{(i)}$ for $n \geqslant 1$, $i=1,2$, and assuming that $h_n^{(i)}/\sqrt{n}$ converges stochastically to $d^{(i)}(\gamma)$ (when γ is the true parameter), then we see that the Pitman efficiency is represented by the limit of the Bahadur efficiency.

This last assumption is often fulfilled, for example, under the following additional and almost always satisfied assumptions: Let $\gamma \to a_n^{(i)}(\gamma)$ for $n \geqslant 1$, $i=1,2$, be maps into the positive numbers, where $a_n^{(i)}(0) = a_n^{(i)}$ is assumed to hold. Let

$$\lim_{n\to\infty} P_\gamma\left(\frac{T_n^{(i)} - \eta_n^{(i)}(\gamma)}{a_n^{(i)}(\gamma)} \leqslant x\right) = \frac{1}{\sqrt{2\pi}} \int_{-\infty}^{x} e^{-t^2/2}\, dt$$

for each real x and

$$\lim_{n\to\infty} \frac{a_n^{(i)}(\gamma)}{a_n^{(i)}(0)} \Big/ \sqrt{n} = 0$$

for each $\gamma \in \Gamma$. Now for $\gamma \neq 0$

$$h_n^{(i)}/\sqrt{n} - \frac{\eta_n^{(i)}(\gamma) - \eta_n^{(i)}(0)}{a_n^{(i)}(0)\sqrt{n}} = \frac{1}{\sqrt{n}} \frac{a_n^{(i)}(\gamma)}{a_n^{(i)}(0)} \frac{T_n^{(i)} - \eta_n^{(i)}(\gamma)}{a_n^{(i)}(\gamma)}.$$

But now the claim follows from (12.50) according to I, Theorem 40.2.

13. Sequential tests. We briefly go into an important generalization of the original Neyman-Pearson concept. It is related to the material of **12** by the fact that one also considers sequences of tests. We give only a rough outline. We consider the sequence of sample spaces $\{(R_n, \mathfrak{B}_n)\}$, as defined on p. 240 and the corresponding sequence $\{P_\Gamma^{(n)}\}$ and the test problem $(\Gamma_0, \Gamma - \Gamma_0)$. For each $n = 1, 2, \ldots$, we define three pairwise disjoint sets $M_i^{(n)} \in \mathfrak{B}_n$, $1 \leqslant i \leqslant 3$, which can also be empty, and which fulfill

$$\bigcup_{i=1}^{3} M_i^{(n)} = R_n. \tag{13.1}$$

The $M_i^{(n)}$ define a *sequential test* which is carried out in practice as follows: Let $x_1 \in R_1$ be a sample value. One determines whether x_1 belongs to $M_1^{(1)}$ to $M_2^{(1)}$ or to $M_3^{(1)}$. In this first case one accepts the null hypothesis Γ_0, in the second one decides at the earliest on the basis of the next sample value, and in the third one rejects the null hypothesis. If $x_1 \in M_2^{(1)}$, one takes the additional sample value x_2 and investigates whether $(x_1, x_2) \in M_1^{(2)}$ or $\in M_2^{(2)}$ or $\in M_2^{(3)}$. In the first case, one accepts

Γ_0 again, in the third, one rejects it and in the second one continues the procedure, and so on.

The sample size is thus no longer fixed, but depends on the sample values. The sample size is hence also a r.v.

Let α and $\beta, 0 \leqslant \alpha, \beta \leqslant 1$, be two levels of significance. Then, the $M_i^{(n)}$ must also satisfy another condition in addition to (13.1). The error of first kind first and the error of second kind are supposed to be $\leqslant \alpha$ and $\leqslant \beta$ respectively. The essential problem naturally consists in choosing the $M_i^{(n)}$ in such a way that they lead under the given conditions to tests with "optimal" properties. Test procedures based on fixed sample size m can also be viewed as sequential tests. If, say, $M \in \mathfrak{B}_m$ is the critical region of a test based on a sample of size m, then define a sequential test according to

$$M_2^{(j)} = R_j, \qquad 1 \leqslant j \leqslant m-1, \qquad M_2^{(m)} = \emptyset, \qquad M_3^{(m)} = M.$$

The development of these ideas is due essentially to Wald[96]. The important problem of finding a construction principle for a sequential test with prescribed level of significance can be considered solved by a construction analogous to Theorem 3.3, at least in the case of a simple null hypothesis and a simple alternative. Indeed, one obtains in this way a test which is optimal in a certain sense. The test possesses for given level of significance the smallest average sample size [97].

The basic idea of a sequential test is obviously not limited to test theory. See VII.**9**.

[96] The fundamental paper is A. Wald. Ann. Math. Statist. 16, 117—186 (1945). Also see A. Wald, Sequential Analysis, John Wiley & Sons—Chapman & Hall, New York-London 1947. See also G.A. Barnard, Suppl. J. Roy. Statist. Soc. 8, 1—21 (1946).

[97] For details see A. Wald and J. Wolfowitz, Ann. Math. Statist. 19, 326—339 (1948).

Chapter IV

The Theory of Confidence Sets

1. Construction of confidence intervals. We first demonstrate the notion of a confidence interval with an example. Let $\xi_1, \ldots, \xi_n$ be sample variables from an $N(a, \sigma_0^2)$-distributed population, where σ_0 is a given positive number and $-\infty < a < \infty$. In II, Theorem 3.2, we showed that $(\bar{\xi} - a)\sqrt{n}/\sigma_0$ is $N(0,1)$-distributed when a is the true parameter value[1]. If we define κ_α for given α with $0 < \alpha < 1$ according to II.(3.3), then we get

$$P\left(-\kappa_\alpha \leqslant \frac{\bar{\xi} - a}{\sigma_0}\sqrt{n} \leqslant \kappa_\alpha; a\right) = 1 - \alpha.\,^2 \tag{1.1}$$

In place of (1.1) one can also write

$$P(\bar{\xi} - \sigma_0 \kappa_\alpha/\sqrt{n} \leqslant a \leqslant \bar{\xi} + \sigma_0 \kappa_\alpha/\sqrt{n}; a) = 1 - \alpha. \tag{1.2}$$

It would be wrong to express the content of (1.2) as follows: The true mean always lies with probability $1-\alpha$ in the interval $[\bar{\xi} - \sigma_0 \kappa_\alpha/\sqrt{n}, \bar{\xi} + \sigma_0 \kappa_\alpha/\sqrt{n}]$. a is *not* a r.v., but an (unknown) real number. (1.2) means rather that the totality of intervals whose boundaries are the two r.v.'s $\bar{\xi} - \sigma_0 \kappa_\alpha/\sqrt{n}$ and $\bar{\xi} + \sigma_0 \kappa_\alpha/\sqrt{n}$ cover the true parameter value a with probability $\beta = 1 - \alpha$.

Each interval of the form $[\bar{x} - \sigma_0 \kappa_\alpha/\sqrt{n}, \bar{x} + \sigma_0 \kappa_\alpha/\sqrt{n}]$ is called a *confidence interval* for the parameter a corresponding to *confidence level* β[3] and determined by the realization $\bar{x}$ of $\bar{\xi}$.

---

[1] For this terminology see III[6].

[2] Since the normal distribution is continuous, one can delete the equality signs within the large brackets without altering the meaning of the expression.

[3] The notion of a confidence interval is due to J. Neyman. See J. Neyman, Ann. Math. Statist. 6, 111—116 (1935); Actualités scientifiques et industrielles 739, 25—57, Hermann & Cie, Paris 1938. See also J. Neyman, Biometrika 32, 128—150 (1941) and J. Neyman, Philos. Trans. Roy. Soc. London, Ser. A 236, 333—380 (1937).

In practice one naturally proceeds by constructing the interval $[\bar{x}-\sigma_0\kappa_\alpha/\sqrt{n}, \bar{x}+\sigma_0\kappa_\alpha/\sqrt{n}]$ by means of a sample $x_1,\dots,x_n$ in such a way that one can assume it to contain the true parameter value "in most cases". With the frequency interpretation this can be made more precise as follows: By (1.2) one can assume that the true parameter value a will on the average be covered 100β times by the given confidence intervals for each 100 samples of size n. Hence, confidence levels near 1 are of practical interest.

We now formulate the problem of the construction of confidence intervals in somewhat more general form. Let a sample space $(R,\mathscr{S})$ be given and a set P_Γ of probability measures over $\mathscr{S}$. Let Γ be an interval of R_1 and β a real number with $0\leqslant\beta\leqslant 1$. We seek functions h_1,h_2 defined over R with $h_1\leqslant h_2$ such that $\{x:\gamma\in[h_1(x),h_2(x)]\}\in\mathscr{S}$ for each $\gamma\in\Gamma$ and such that for each γ

$$P_\gamma(\{x:\gamma\in[h_1(x),h_2(x)]\})\geqslant\beta. \tag{1.3}$$

In the example above, $(R_n,\mathfrak{B}_n)$ is the sample space, $\Gamma=R_1$ and P_{R_1} the set of all probability measures whose densities are given by III. (1.3) with $\sigma=\sigma_0$. h_1 is to be identified with $\bar{\xi}-\sigma_0\kappa_\alpha/\sqrt{n}$ and h_2 with $\bar{\xi}+\sigma_0\kappa_\alpha/\sqrt{n}$.

We now prove a theorem which allows the construction of confidence intervals in some special cases.

Theorem 1.1. *Let P_Γ be a set of probability distributions over $(R,\mathscr{S})$, where Γ is an interval of R_1. Let there exist a function T defined over $R\times\Gamma$ such that the map $x\to T(x,\gamma)$ is $\mathscr{S}$-measurable for each γ and the maps $\gamma\to T(x,\gamma)$ are strictly monotone (all in the same sense) for each $x\in R$. Let $T(R\times\Gamma)=A\subseteq R_1$ and assume $a=T(x,\gamma)$ is solvable for each $a\in A$ and each $x\in R$. The r.v.'s T_γ defined for each γ by $x\to T(x,\gamma)$ are assumed to have the same distribution function w.r.t. P_γ, independent of γ. Then a confidence interval for γ can always be constructed.*

Proof. For given β, $0\leqslant\beta\leqslant 1$, we can determine two real numbers $\varepsilon_1(\beta)$, $\varepsilon_2(\beta)$, which by assumption do not depend on γ, such that

$$P_\gamma(\varepsilon_1(\beta)\leqslant T_\gamma\leqslant\varepsilon_2(\beta))\geqslant\beta. \tag{1.4}$$

Naturally, $\varepsilon_1(\beta)$ and $\varepsilon_2(\beta)$ are, in general, not uniquely determined by (1.4). The inequality

$$P_\gamma(\{x:\varepsilon_1(\beta)\leqslant T(x,\gamma)\leqslant\varepsilon_2(\beta)\})\geqslant\beta \tag{1.5}$$

is equivalent to (1.4).

Because of the monotonicity in γ assumed for T, the equations

$$\varepsilon_1(\beta)=T(x,\gamma) \tag{1.6}$$

and

$$\varepsilon_2(\beta)=T(x,\gamma) \tag{1.7}$$

can be uniquely solved for γ for each x if one chooses $\varepsilon_1(\beta)\in A$ and $\varepsilon_2(\beta)\in A$ [4].

If we now assume that $\gamma\to T(x,\gamma)$ increases monotonically for each x and denote the solution of (1.6) by $T_2(x,\beta)$ and that of (1.7) by $T_1(x,\beta)$, then we have $T_2(x,\beta)\geqslant T_1(x,\beta)$.

If for a γ with fixed x

$$\varepsilon_1(\beta)\leqslant T(x,\gamma)\leqslant\varepsilon_2(\beta) \tag{1.8}$$

then

$$T_1(x,\beta)\leqslant\gamma\leqslant T_2(x,\beta). \tag{1.9}$$

Conversely, each γ which satisfies (1.9) also satisfies (1.8). Thus, due to the connection between (1.8) and (1.9) we get

$$\{x:\varepsilon_1(\beta)\leqslant T(x,\gamma)\leqslant\varepsilon_2(\beta)\}$$
$$=\{x:\gamma\in[T_1(x,\beta),T_2(x,\beta)]\}$$

and from (1.5), $[T_1(x,\beta),T_2(x,\beta)]$ represents a confidence interval with confidence level β.

Theorem 1.1 is illustrated by the example previously given on p. 255. Another example, which, as we will see, is of additional interest, can be constructed with reference to II, Theorem 4.4. We again seek a confidence interval for the unknown mean a of a normal distribution on the basis of a sample of size n. Note here that the function $T(x,a)=(\bar{x}-a)\sqrt{n}/s$ is defined for all $x\in R_n$ and $a\in R_1$ and fulfills all assumptions of Theorem 1.1. The r.v. $(\bar{\xi}-a)\sqrt{n}/s$ is t-distributed with $n-1$ degrees of freedom when a is the true parameter value. This distribution naturally depends not on a. One must now determine $\varepsilon_1(\beta)$ and $\varepsilon_2(\beta)$ from

$$\int_{\varepsilon_1(\beta)}^{\varepsilon_2(\beta)} h_{n-1}(t)\,dt=\beta \tag{1.10}$$

(see I.(29.3)) and sees at once that there exist infinitely many pairs $\varepsilon_1(\beta)$ and $\varepsilon_2(\beta)$ which satisfy (1.10). Then

$$P(\varepsilon_1(\beta)\leqslant(\bar{\xi}-a)\sqrt{n}/s\leqslant\varepsilon_2(\beta);a)=\beta. \tag{1.11}$$

Solving the equations $\varepsilon_1(\beta)=(\bar{x}-a)\sqrt{n}/s$ and $\varepsilon_2(\beta)=(\bar{x}-a)\sqrt{n}/s$, one obtains a confidence interval $[\bar{x}-s\varepsilon_2(\beta)/\sqrt{n},\bar{x}+s\varepsilon_1(\beta)/\sqrt{n}]$ for a which satisfies according to (1.11), the condition

$$P(a\in[\bar{\xi}-s\varepsilon_2(\beta)/\sqrt{n},\bar{\xi}+s\varepsilon_1(\beta)/\sqrt{n};a)=\beta$$

for each $a\in R_1$. Note that the variance σ^2 of the normal distribution plays no role in this example. We will return to this point later.

2. Confidence sets. Although one can construct confidence intervals for a one-dimensional parameter in many practical examples, this construction still depends on certain monotonicity assumptions. (See Theorem 1.1 and the forthcoming Theorem 2.1.). These are not in general satisfied. Moreover, when the parameter belongs to an arbitrary set, the

[4] If $\inf A\notin A$ or $\sup A\notin A$ the cases $\beta=0$ or $\beta=1$ can cause (trivial) difficulties.

definition of a confidence interval may become meaningless. We thus introduce the more general notion of confidence sets and show at the same time that, at least in principle, one can construct confidence sets under completely general assumptions. We start again from a measurable space $(R,\mathscr{S})$ and a set of probability measures P_Γ, where Γ is now an arbitrary set. Let $K \subseteq R \times \Gamma$, $K_\gamma = \{x:(x,\gamma)\in K\}$ and $K(x)=\{\gamma:(x,\gamma)\in K\}$. We assume that $K_\gamma \in \mathscr{S}$ for each γ. Further for each $x \in R$ assume

$$K(x) \neq \emptyset \tag{2.1}$$

or, somewhat more generally, $\{x: K(x)=\emptyset\} \subseteq N \in \mathscr{S}$ with

$$P_\gamma(N)=0 \tag{2.2}$$

for all $\gamma \in \Gamma$.

Let β be a real number with $0<\beta<1$ and

$$\beta \leqslant \inf_{\gamma\in\Gamma} P_\gamma(K_\gamma). \tag{2.3}$$

The fact that

$$K_\gamma = \{x:\gamma\in K(x)\} \tag{2.4}$$

motivates the following

Definition. Under the given condition $K(x)$ is called a *confidence set* for γ with *confidence level* β.

Indeed, because of (2.3) and (2.4)

$$P_\gamma\{x:\gamma\in K(x)\} \geqslant \beta. \tag{2.5}$$

Conditions (2.1) or (2.2) mean, briefly speaking, that to each or almost each sample x there corresponds a confidence set. Hence, the only essential problem in the construction of confidence sets is to assure that (2.3) is fulfilled for given β. In practice, one naturally chooses β in such a way that, if possible, the equality sign holds in (2.3) for at least one $\gamma \in \Gamma$.

If Γ is an interval of R_1 and all the $K(x)$ are intervals, we are led to the confidence intervals considered in **1**. The construction of Theorem 1.1 is covered by this definition. One need only identify K_γ with $\{x:\varepsilon_1(\beta) \leqslant T(x,\gamma) \leqslant \varepsilon_2(\beta)\}$ and $K(x)$ with $\{\gamma: T_1(x,\beta) \leqslant \gamma \leqslant T_2(x,\beta)\}$. More generally, the construction of confidence intervals relies primarily on the following

Theorem 2.1. *Let P_Γ have the same meaning as in Theorem 1.1. Let T be an $\mathscr{S}$-measurable function defined for all $x\in R$, i.e., a r.v. For given β, $0 \leqslant \beta \leqslant 1$, choose $\varepsilon_1(\beta,\gamma)$ and $\varepsilon_2(\beta,\gamma)$, $\gamma\in\Gamma$, such that for all γ*

$$P(\varepsilon_1(\beta,\gamma) \leqslant T \leqslant \varepsilon_2(\beta,\gamma);\gamma) \geqslant \beta. \tag{2.6}$$

Assume also that $\gamma \to \varepsilon_i(\beta, \gamma)$, $i=1,2$ *are both strictly monotone functions in the same sense. Let* $T(R)=A$ *and assume* $\varepsilon_i(\beta,\gamma)=a$, $a \in A$ *are always solvable for* γ. *Then it is possible to construct a confidence interval for* γ *with confidence level* β.

Proof. Assume that $\gamma \to \varepsilon_i(\beta,\gamma)$ are, say, strictly increasing. For each $a \in A$ the equations

$$\varepsilon_1(\beta,\gamma)=a \tag{2.7}$$

and

$$\varepsilon_2(\beta,\gamma)=a \tag{2.8}$$

are uniquely solvable in γ. The solution of (2.7) is denoted by $T_2(a,\beta)$ and that of (2.8) by $T_1(a,\beta)$. Since $\varepsilon_1(\beta,\gamma) \leqslant \varepsilon_2(\beta,\gamma)$, $T_1(a,\beta) \leqslant T_2(a,\beta)$. If

$$\varepsilon_1(\beta,\gamma) \leqslant a \leqslant \varepsilon_2(\beta,\gamma) \tag{2.9}$$

holds for a γ, then also

$$T_1(a,\beta) \leqslant \gamma \leqslant T_2(a,\beta), \tag{2.10}$$

and conversely. Hence,

$$P(\varepsilon_1(\beta,\gamma) \leqslant T \leqslant \varepsilon_2(\beta,\gamma); \gamma) = P(T_1(T,\beta) \leqslant \gamma \leqslant T_2(T,\beta); \gamma).$$

Hence, from (2.6), $[T_1(T,\beta), T_2(T,\beta)]$ is a confidence interval for γ with confidence level β.

Fig. 10 shows how to find the confidence interval $[T_1, T_2]$ graphically under the assumptions of Theorem 2.1 with $\Gamma=[a,b]$.

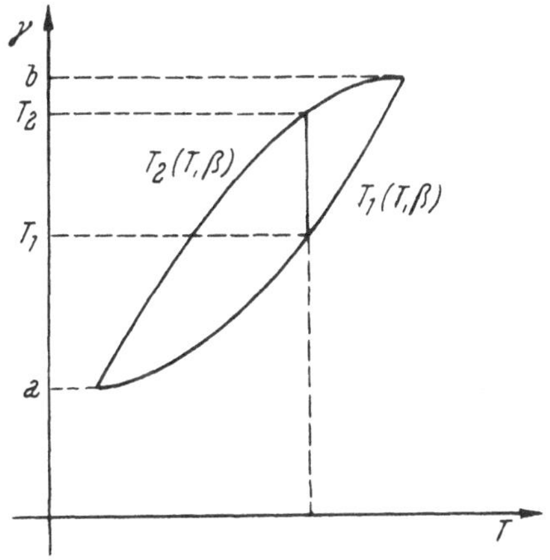

Fig. 10

Consider two practically important examples. The first one concerns the *binomial distribution*. The sample space is $(R_n, \mathfrak{B}_n)$ and the sample variables possess the distribution given by I. (32.1) with $0<p<1$ so that the set Γ is the interval (0, 1). The r.v. T is now the r.v. ζ_n which is $B_n(p)$-distributed. For given $\beta, 0 \leqslant \beta \leqslant 1$, we now determine the $\varepsilon_i(\beta, p)$, $i=1,2$, as functions of p in $0<p<1$ in such a way that $\varepsilon_1(\beta, p)$ is the largest and $\varepsilon_2(\beta, p)$ the smallest integer for which

$$\sum_{r=0}^{\varepsilon_1(\beta,p)} \binom{n}{r} p^r(1-p)^{n-r} \leqslant (1-\beta)/2\,, \tag{2.11}$$

resp.,

$$\sum_{r=\varepsilon_2(\beta,p)}^{n} \binom{n}{r} p^r(1-p)^{n-r} \leqslant (1-\beta)/2\,. \tag{2.12}$$

From (2.11) and (2.12)

$$P(\varepsilon_1(\beta,p)+1 \leqslant \zeta_n \leqslant \varepsilon_2(\beta,p)-1\,;\, p) \geqslant \beta\,.$$

We check that the functions $p \to \varepsilon_i(\beta, p)$, $i=1,2$ never decrease. By I, Theorem 32.3, in the notation used there, $p \to S_k(p)$ is a monotone decreasing function. Hence, $p \to \varepsilon_1(\beta, p)$ can never decrease since $\varepsilon_1(\beta, p)$ is for each p the largest integer satisfying (2.11). On the other hand, $p \to 1-S_k(p)$ increases. Since $\varepsilon_2(\beta, p)$ is for each p the smallest integer satisfying (2.12), $p \to \varepsilon_2(\beta, p)$ can also never decrease[5]. Thus, the assumptions for the construction of confidence intervals are satisfied[6]. For the case $n=50$, $\beta=0.99$ and a sample value 11 for ζ_n, the procedure is represented graphically in the following figure (Fig. 11.).

As second example we consider the construction of a *confidence interval for the parameter a of a Poisson distribution*. (See I. **33**.) One proceeds exactly as in the binomial case. If $\xi_1, \ldots, \xi_n$ are n independent, Poisson-distributed (parameter a) r.v.'s, $0<a<\infty$, then by I, Theorem 33.1, $\zeta_n = \sum_{i=1}^{n} \xi_i$ is also Poisson with mean na. For given β the functions $a \to \varepsilon_i(\beta, a)$, $i=1,2$, are then determined as follows: Let $\varepsilon_1(\beta, a)(\varepsilon_2(\beta, a))$ be the largest (smallest) integer fulfilling

$$\sum_{k=0}^{\varepsilon_1(\beta,a)} \frac{(na)^k}{k!} e^{-na} \leqslant \frac{1-\beta}{2} \quad \text{resp.} \quad \sum_{k=\varepsilon_2(\beta,a)}^{\infty} \frac{(na)^k}{k!} e^{-na} \leqslant \frac{1-\beta}{2}\,.$$

Then

$$P(\varepsilon_1(\beta,a)+1 \leqslant \zeta_n \leqslant \varepsilon_2(\beta,a)-1\,;\, a) \geqslant \beta$$

and one uses I, Theorem 33.2, proceeding exactly as before[7].

---

[5] However, in Theorem 2.1 strict monotonicity is required. If this is not the case, the uniqueness of the construction is lost.

[6] Confidence intervals for the binomial parameter were first given by J. Clopper and E.S. Pearson, Biometrika 26, 404—413 (1934). Also see O. Bunke, Wiss. Z. Humboldt-Univ. Berlin, Math.-Natur. Reihe 9, 335—363 (1959/60).

[7] Also see E. Ricker, J. Amer. Statist. Assoc. 32, 349—356 (1937).

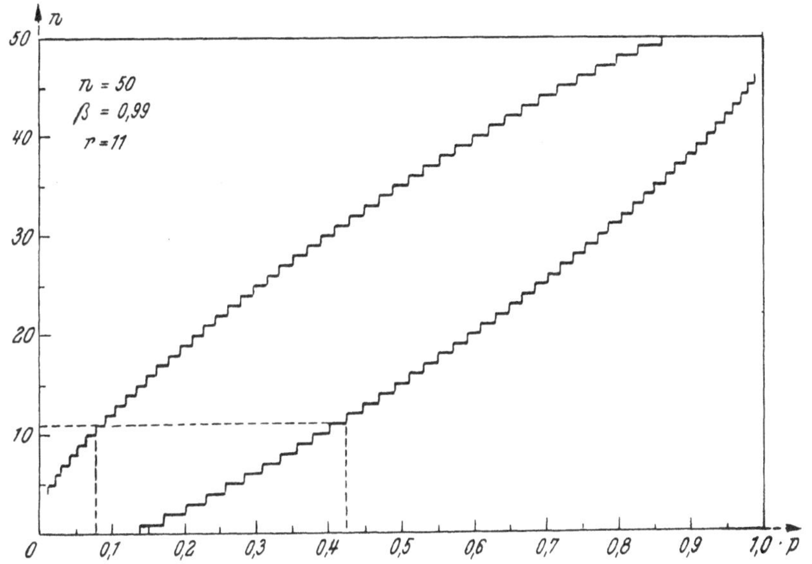

Fig. 11

The previous examples and arguments indicate that there is a close connection between the theory of tests and the theory of confidence sets. We make this precise in a

Duality principle. Let P_Γ be a set of probability measures over $(R, \mathscr{S})$ and assume given a confidence region $K(x)$ for $\gamma \in \Gamma$ with given confidence level β, $0 < \beta < 1$. Then, in the notation of p. 258, $R - K_\gamma$ *is for each γ a critical region for the test problem $(\alpha, \{\gamma\}, \Gamma - \{\gamma\})$ with $\alpha = 1 - \beta$. Conversely, let $M(\gamma)$ be a critical region for each $\gamma \in \Gamma$ for the test problem $(\alpha, \{\gamma\}, \Gamma - \{\gamma\})$. If we define $K = \bigcup_{\gamma \in \Gamma} (R - M(\gamma)) \times \{\gamma\}$ then $K(x) = \{\gamma : (x, \gamma) \in K\}$ is a confidence set for γ with confidence level $\beta = 1 - \alpha$ if* (2.1) *or* (2.2) *is fulfilled.*

The *proof* is obvious.

Essentially, this principle allows us to give a dual notion and a dual theorem from the theory of confidence sets corresponding to each notion and theorem from the theory of tests. We will demonstrate this with several examples which will suffice to show how one can systematically transfer the content of Chapter III to the theory of confidence sets.

We must, however, point out that conditions (2.1) and (2.2) need not necessarily be satisfied. We give a simple example for which these are not fulfilled. Let $\Gamma = \{0, 1\}$ and let the measures P_Γ over $(R_1, \mathfrak{B}_1)$

be defined by their densities f_0 and f_1. Let f_0 be the density of the uniform distribution over $(-1/2, 1/2)$. Let ε be real with

$$0<\varepsilon<\tfrac{1}{2} \tag{2.13}$$

and

$$u=\frac{1-\varepsilon}{\varepsilon}+\varepsilon. \tag{2.14}$$

Set

$$f_1(x)=\begin{cases} 0 & |x|\geqslant\tfrac{1}{2} \\ \varepsilon & -\tfrac{1}{2}<x\leqslant-\varepsilon,\ \varepsilon\leqslant x<\tfrac{1}{2} \\ u+\dfrac{u-\varepsilon}{\varepsilon}x & -\varepsilon\leqslant x\leqslant 0 \\ u-\dfrac{u-\varepsilon}{\varepsilon}x & 0\leqslant x\leqslant\varepsilon \end{cases}$$

and let

$$\frac{\alpha}{2}=\frac{u-2\varepsilon}{u-\varepsilon}\varepsilon. \tag{2.15}$$

From (2.13) and (2.14) we have $0<\alpha<1$.

For the test problem $(\alpha, \{0\}, \{1\})$ we define with $c=2\varepsilon$ a critical region

$$M(0)=\{x: f_1(x)\geqslant c f_0(x)\}.$$

It is easy to check that $P_0(M(0))=\alpha$.

For the problem $(\alpha, \{1\}, \{0\})$ we define with

$$\frac{1}{c_1^2}=(1+\varepsilon-4\varepsilon^2+\varepsilon^3)/\varepsilon \tag{2.16}$$

a critical region

$$M(1)=\{x: f_0(x)\geqslant c_1 f_1(x)\}.$$

Because of (2.13), the right side of (2.16) is actually positive. One again checks easily that $P_1(M(1))=\alpha$. Also, $\dfrac{1}{c_1}>c$ holds. Hence, the set $\left\{x: c f_0(x)\leqslant f_1(x)\leqslant\dfrac{1}{c_1}f_0(x)\right\}$ is not a P_γ-null set for any $\gamma\in\Gamma$ and belongs to no $R-M(\gamma)$.

We point out that one can also introduce the notion of a randomized confidence interval in analogy to that of the randomized test. We will not go into this here.

The example considered on p. 257 illustrates the following general problem: Let $\Gamma=\Gamma_1\times\Gamma_2$. For given confidence level β, $0<\beta<1$, a confidence set $K(x)$ is to be constructed for $\gamma_1\in\Gamma_1$ such that

$$P_{(\gamma_1,\gamma_2)}(\{x: \gamma_1\in K(x)\})\geqslant\beta \tag{2.17}$$

for each $\gamma_1 \in \Gamma_1$ and all $\gamma_2 \in \Gamma_2$. If the equality sign holds in (2.17) for all $\gamma_2 \in \Gamma_2$, then $R - K(\gamma_1)$ is a similar critical region for the test problem

$$(1-\beta, \{\gamma_1\} \times \Gamma_2, (\Gamma_1 - \{\gamma_1\}) \times \Gamma_2).$$

Conversely, if for each $\gamma_1 \in \Gamma_1$ one has constructed a similar critical region $M(\gamma_1)$ for this problem, then

$$K = \bigcup_{\gamma_1 \in \Gamma_1} (R - M(\gamma_1)) \times (\{\gamma_1\} \times \Gamma_2)$$

yields a confidence set $K(x)$ which even fulfills (2.17) with the equality sign and which can also be denoted as similar. Hence, this problem leads to the theory of similar tests whenever conditions (2.1) or (2.2) are satisfied.

We now present an almost trivial construction principle for confidence sets. First, an illustrative example.

We saw on p. 257 how one can obtain, on the basis of a sample of size n, a confidence interval for the mean a of a normal distribution $N(a, \sigma^2)$, $-\infty < a < \infty$, $0 < \sigma^2 < \infty$, which, briefly speaking, does not depend on the parameter σ^2. Using II, Corollary to Theorem 4.3 and Theorem 1.1, one easily sees how to construct a confidence interval for σ^2, under the same assumptions, which is independent of a. The question of how one can exploit this to construct a two-dimensional confidence set (simultaneous confidence set) for (a, σ^2) with given confidence level now arises. An almost obvious answer is given in

Theorem 2.2. *Let P_Γ be a set of probability measures over $(R, \mathscr{S})$ and let Γ for $k \geq 2$ be of the form $\Gamma = \prod_{i=1}^{k} \Gamma_i$. If for each confidence level β', $0 < \beta' < 1$, one can construct confidence sets $K_i(x)$ for each $\gamma_i \in \Gamma_i$ such that*

$$P_{(\gamma_1, \ldots, \gamma_i, \ldots, \gamma_k)}(\{x : \gamma_i \in K_i(x)\}) \geq \beta' \tag{2.18}$$

for all $\gamma_j \in \Gamma_j$ with $i \neq j$, then one can also construct a simultaneous confidence set for $(\gamma_1, \ldots, \gamma_k) \in \Gamma$ for each confidence level β.

Proof. Let β, $0 < \beta < 1$, be given. For $i = 1, \ldots, k$ in (2.18), choose

$$\beta' = 1 - (1-\beta)/k. \tag{2.19}$$

Then $K(x) = \prod_{i=1}^{k} K_i(x)$ is for each $x \in R$ a confidence set with confidence level β. Indeed,

$$P_{(\gamma_1, \ldots, \gamma_k)}\left(\left\{x : (\gamma_1, \ldots, \gamma_k) \in \prod_{i=1}^{k} K_i(x)\right\}\right) = P_{(\gamma_1, \ldots, \gamma_k)}\left(\bigcap_{i=1}^{k} \{x : \gamma_i \in K_i(x)\}\right),$$

so that

$$P_{(\gamma_1, \ldots, \gamma_k)}(\{x : (\gamma_1, \ldots, \gamma_k) \in K(x)\}) = P_{(\gamma_1, \ldots, \gamma_k)}\left(\bigcap_{i=1}^{k} K_{\gamma_i}\right).$$

Hence, from (2.18) and (2.19) we get

$$P_{(\gamma_1,\ldots,\gamma_k)}(\{x:(\gamma_1,\ldots,\gamma_k)\in K(x)\})\geq 1-k\frac{1-\beta}{k}=\beta.$$

An illustration of this theorem is in VI, p. 361 ff.

Since $1-\frac{1-\beta}{k}>\beta$ for $0\leq\beta<1$, this construction is associated in general with a "loss of information", i.e., optimality properties possessed by the confidence sets $K_i(x)$ are in general not valid for $K(x)$.

Optimality principles for confidence sets can be defined by means of the duality principle in a way parallel to those for tests. It thus makes sense to speak of *unbiased*, *most accurate* (= most powerful[8]) and *invariant* confidence sets, etc. We will go into this briefly in the next section.

3. Unbiased and most accurate confidence sets. (2.3) says that a confidence set $K(x)$ covers the true parameter γ with probability at least as large as β. Nothing is, however, said about the probability of a "false" parameter being covered or, in other words, how $P_\gamma(\{x:\gamma'\in K(x)\})$ behaves for $\gamma\neq\gamma'$. It is for example desirable that the false parameter be covered with probability "as small as possible". More precisely, we have

Definition. Let $K(x)$ be a confidence set for $\gamma\in\Gamma$ with confidence level β, $0\leq\beta\leq 1$. Then, we call the function

$$k(\gamma,\gamma')=P_\gamma(\{x:\gamma'\in K(x)\})$$

defined over $\Gamma\times\Gamma$ the *accuracy function* of $K(x)$.

The accuracy function is obviously the dual of the power function of a test.

We now define the notion of an unbiased confidence set.

Definition. A confidence set $K(x)$ with confidence level β for $\gamma\in\Gamma$ is called *unbiased* when its accurateness function k has the following properties:

$$k(\gamma,\gamma)\geq\beta \qquad \text{for } \gamma\in\Gamma$$
$$k(\gamma,\gamma)\geq k(\gamma,\gamma') \quad \text{for } \gamma,\gamma'\in\Gamma \quad \text{with } \gamma\neq\gamma'.$$

The definition corresponds exactly to that of an unbiased test. The transfer of the definition of a most powerful test to confidence sets is almost obvious.

[8] *Most selective* or *shortest* are also used instead of accurate.

Definition. Let $K(x)$ be a confidence set for $\gamma \in \Gamma$ with confidence level β. It is called *most accurate w.r.t. a set $\mathfrak{K}_\beta$ of confidence sets $K^*(x)$* with the same confidence level β if for the accuracy function k of $K(x)$ and the accuracy function k^* of $K^*(x)$

$$k(\gamma,\gamma') \leqslant k^*(\gamma,\gamma') \tag{3.1}$$

holds for all $\gamma,\gamma' \in \Gamma$, $\gamma \neq \gamma'$ and each $K^*(x) \in \mathfrak{K}_\beta$. If $\mathfrak{K}_\beta$ is the set of all confidence sets with confidence level β, then we call simply $K(x)$ *most accurate.*

The theorems of III on unbiased and most powerful tests can be immediately carried over. Certain difficulties may be caused by the verification of (2.1) or (2.2). We have

Theorem 3.2. *Let $\Gamma = \{\gamma_0, \gamma_1\}$ and f_0, f_1 be the R.-N.-densities of $P_{\gamma_0}, P_{\gamma_1}$ w.r.t. a dominating measure. Let a $c \geqslant 0$ with*

$$P_{\gamma_0}(\{x : f_1(x) \leqslant c f_0(x)\}) = P_{\gamma_1}(\{x : f_1(x) \geqslant c f_0(x)\})$$

be given. Then there exists a most accurate confidence set for $\gamma \in \Gamma$, for each confidence level β with $P_{\gamma_0}(\{x : f_1(x) \leqslant c f_0(x)\}) \leqslant \beta \leqslant 1$.

Proof. Construct by III, Theorem 3.3 most powerful critical regions $M(\gamma_0)$ and $M(\gamma_1)$ for the test problems $(1-\beta, \{\gamma_0\}, \{\gamma_1\})$, resp., $(1-\beta, \{\gamma_1\}, \{\gamma_0\})$[9]. Defining $K = (R - M(\gamma_0)) \times \{\gamma_0\} \cup ((R - M(\gamma_1)) \times \{\gamma_1\})$, we see immediately that the corresponding sets $K(x)$ yield a most accurate confidence set for $\gamma \in \Gamma$.

4. Bayesian confidence sets. If Γ consists of more than two elements, then there exist in general no accurate confidence sets. Theorem 5.1 and 6.1 of III can naturally also be formulated for confidence sets. We have already assumed on occasion in the theory of tests[10] that measures have been defined over a suitable σ-algebra of subsets of Γ. This leads to the Bayesian viewpoint. If, briefly speaking, the parameter Γ possesses a probability distribution, then the latter is referred to as the *a priori distribution.*

In practice one chooses the a priori distribution of the parameter in such a way that it reflects "our knowledge about the parameter available before the experiment as well as possible". The Bayesian viewpoint allows the formulation of new optimality principles which are still related in many ways to the ideas already mentioned[11]. We first have a

---

[9] For simplicity we have assumed that for each $1-\beta$, $0 \leqslant \beta \leqslant 1$, such regions exist.

[10] See the arguments following III, Theorem 4.2 as well as the appendix p. 481 ff.

[11] See also V, p. 330.

Definition. Let $\mathfrak{S}$ be a σ-algebra of subsets of Γ which contains the set $A(\gamma)=\Gamma-\{\gamma\}$ for each $\gamma\in\Gamma$. Let ν be an arbitrary measure over $\mathfrak{S}$. Let $\mathfrak{K}$ be any set of confidence sets $K^*(x)$ for $\gamma\in\Gamma$. *$K(x)$ is called a Bayes solution corresponding to ν w.r.t. $\mathfrak{K}$ or also ν-most accurate w.r.t. $\mathfrak{K}$ if for all $\gamma\in\Gamma$,*

$$\int_{A(\gamma)} k(\gamma,\gamma')d\nu(\gamma')=\inf_{\mathfrak{K}}\int_{A(\gamma)} k^*(\gamma,\gamma')d\nu(\gamma'). \tag{4.1}$$

Here, k and k^* are the accuracy functions of $K(x)$, resp., $K^*(x)\in\mathfrak{K}$. From (3.1), each confidence set which is most accurate w.r.t. a set $\mathfrak{K}_\beta$ is also ν most accurate with respect to this set.

The following theorem, which is obviously closely connected with III, Theorem 4.2 and the considerations following this Theorem, treats the construction of ν-most powerful confidence sets. The assumptions of our formulation can be generalized[12].

Theorem 4.1. *Let Γ be an open cube (or a bounded open set) of R_k, $k\geqslant 1$. Let $\mathfrak{S}$ be the set of Borel sets of Γ and ν the measure over $(\Gamma,\mathfrak{S})$ with constant density $1/L(\Gamma)$, where L is Lebesgue measure. Let the set P_Γ defined over $(R,\mathscr{S})$ be dominated by a σ-finite measure μ. For each $\gamma\in\Gamma$ denote the R.-N.-density of P_γ by $x\to f(x,\gamma)$ and let the map $(x,\gamma)\to f(x,\gamma)$ be $\mathscr{S}\otimes\mathfrak{S}$-measurable. Then $g(x)=\int_\Gamma f(x,\gamma)d\nu(\gamma)$ exists μ-a.e. and we assume that $g(x)>0$ μ-a.e. For all $\gamma\in\Gamma$ and all $x\in R$ let $h(x,\gamma)=f(x,\gamma)/g(x)$ up to a set of μ-measure 0. Let β be real with*

$$0<\beta<1 \tag{4.2}$$

and for each $\gamma\in\Gamma$ choose a real number $c(\gamma)$ such that

$$P_\gamma(\{x:h(x,\gamma)>c(\gamma)\})\leqslant\beta\leqslant P_\gamma(\{x:h(x,\gamma)\geqslant c(\gamma)\}).$$

Assume that for each $\gamma\in\Gamma$ the set $K_\gamma\in\mathscr{S}$ can be so determined that

$$\{x:h(x,\gamma)>c(\gamma)\}\subseteq K_\gamma\subseteq\{x:h(x,\gamma)\geqslant c(\gamma)\}, \tag{4.3}$$

$$P_\gamma(K_\gamma)=\beta \tag{4.4}$$

and such that $K(x)$ fulfills (2.1) or (2.2) with $K=\bigcup_{\gamma\in\Gamma}K_\gamma\times\{\gamma\}$. Then $K(x)$ is ν-most accurate w.r.t. the set $\mathfrak{K}_\beta$ of all confidence sets for confidence level β, and each ν-most powerful confidence set $\in\mathfrak{K}_\beta$ is (up to μ-null sets) given in this way.

Proof. The existence of $c(\gamma)$ for each $\gamma\in\Gamma$ follows from III, Theorem 3.1 I F. Because of (4.2), one can always choose

$$c(\gamma)>0. \tag{4.5}$$

[12] See R. Borges, Z. Wahrscheinlichkeitstheorie und Verw. Gebiete 1, 47—69 (1962).

Now note that

$$\int_{A(\gamma)}\int_{K_\gamma} f(x,\gamma')\,d\mu(x)\,d\nu(\gamma') = \int_{K_\gamma}\int_{A(\gamma)} f(x,\gamma')\,d\nu(\gamma')\,d\mu(x)$$
$$= \int_{K_\gamma}\int_{\Gamma} f(x,\gamma')\,d\nu(\gamma')\,d\mu(x).$$

Hence,

$$\int_{A(\gamma)}\int_{K_\gamma} f(x,\gamma')\,d\mu(x)\,d\nu(\gamma') = \int_{K_\gamma} g(x)\,d\mu(x). \tag{4.6}$$

Now we apply III, Theorem 3.1 II F. (See the remark on p. 170.) Since condition (4.3) can also be written as

$$\{x: f(x,\gamma) > c(\gamma)g(x)\} \subseteq K_\gamma \subseteq \{x: f(x,\gamma) \geqslant c(\gamma)g(x)\}$$

from (4.4) and (4.6) it follows that $K(x)$ is ν-most accurate.

But if $K^*(x) \in \mathfrak{K}_\beta$ and is ν-most accurate, then from (4.5) and an application of III, Theorem 3.1 III F shows that K^*_γ fulfills (4.4) for each $\gamma \in \Gamma$ and also a relation of the form (4.3).

Chapter V

Theory of Estimation

1. Unbiased estimates[1]. We first superficially sketch the problem we will treat in this chapter. In the previous chapter we dealt with the question of how one can acquire more precise information on the value of an unknown parameter on the basis of a sample. Although one tries to construct confidence sets which are "as small as possible", one cannot be guided in such a construction by the idea of "exactly" determining the parameter. To work out this concept is the goal of the theory of estimation. If $(R, \mathcal{S})$ is a sample space and Γ a set of parameters of a class of probability measures P_Γ over $(R, \mathcal{S})$, then one seeks a map h of R into Γ such that $h(x)$ for a sample $x \in R$ is "approximately" equal to the true parameter value. We are primarily concerned with the case in which Γ is a subset of R_1 or where we have to estimate a mapping d from Γ into R_1. For the sake of simplified formulation we agree that: *Γ will always be a non-empty set of parameters of a class of probability measures and d a map from Γ into R_1, unless something else is specifically said. Further conditions can be also imposed on Γ as well as d.*

There is a number of possible ways to make the statement of the problem of estimation precise. The first is to require that h be equal "in the mean" to the true parameter value. In this connection we have the

Definition. Let P_Γ be a non-empty set of probability measures over $(R, \mathcal{S})$. An $\mathcal{S}$-measurable mapping h of R into R_1 is called an *unbiased estimate* for d if

$$E(h; \gamma) = d(\gamma) \tag{1.1}$$

for each $\gamma \in \Gamma$.

In particular, this definition requires that h be P_γ-integrable for each γ. It can immediately be extended to the case where d is a map from Γ into R_n, $n \geqslant 2$. If $d = (d_1, \ldots, d_n)$, where d_i, $1 \leqslant i \leqslant n$, is a map

[1] This concept was given its first clear treatment in F. N. David and J. Neyman, Statist. Res. Mem. Univ. London 2, 105—116 (1938).

from Γ into R_1, then an $\mathscr{S}$-measurable map $h=(h_1, \dots, h_n)$ from R into R_n is called an unbiased estimate for d if

$$E(h_i; \gamma) = d_i(\gamma), \quad 1 \leqslant i \leqslant n \tag{1.2}$$

for each $\gamma \in \Gamma$. We make the following

Definition. *The expectation $E(M)$ of a matrix M of random variables will be understood as the matrix of the expectations of these r.v.'s*, whose existence is assumed.

Under this agreement, (1.2) can be written in the form (1.1). From (1.2) we have for each n-tuple of real numbers $\alpha_1, \dots, \alpha_n$

$$E(\alpha_1 h_1 + \cdots + \alpha_n h_n; \gamma) = \sum_{i=1}^{n} \alpha_i d_i(\gamma). \tag{1.3}$$

On the other hand, if (1.3) holds for each n-tuple of real numbers, then $h=(h_1, \dots, h_n)$ is an unbiased estimate for $d=(d_1, \dots, d_n)$. To each n-tuple $(\alpha_1, \dots, \alpha_n)$ there corresponds a linear mapping of R_n into R_1 given by $(x_1, \dots, x_n) \to \sum_{i=1}^{n} \alpha_i x_i$. Conversely, each such linear map L satisfying $\alpha L(x) + \beta L(y) = L(\alpha x + \beta y)$ for each $x, y \in R_n$ and each $\alpha, \beta \in R_1$ can be written in this form. Thus, (1.3) says that the problem of unbiased estimation consists of finding an n-dimensional random variable h which satisfies

$$E(L \circ h; \gamma) = L \circ d$$

for each γ and all linear maps L. In this form the problem can, however, be generalized considerably.

We consider some examples of unbiased estimates. Let $(\xi_1, \dots, \xi_n)$ be an n-dimensional r.v. The sample space is thus $(R_n, \mathfrak{B}_n)$. Over this sample space we look at the set $\mathfrak{P}$ of all probability measures, so that

$$E(\xi_1; P) = \cdots = E(\xi_n; P) = a(P), \quad P \in \mathfrak{P}. \tag{1.4}$$

Hence all the ξ_i, $1 \leqslant i \leqslant n$, have the same expectation, which we denote by $a(P)$, indicating the dependence of $P \in \mathfrak{P}$. A trivial application of I, Theorem 16.3 then yields: $\bar{\xi} = (\xi_1 + \cdots + \xi_n)/n$ is an unbiased estimate for the mean a.

Another example results from a somewhat more precise formulation of Theorem 4.2 of II: Let $\xi_1, \dots, \xi_n$, $n \geqslant 2$ be independent r.v.'s with the same one-dimensional distribution P. The joint distribution of $(\xi_1, \dots, \xi_n)$ is given by the product measure $\underbrace{P \times \cdots \times P}_{n}$. We consider all P for which $E(\xi_i) = a(P)$ and $E(\xi_i - a(P))^2 = \sigma^2(P)$ exist. Then

$$s^2 = \sum_{i=1}^{n} (\xi_i - \bar{\xi})^2/(n-1)$$

is an unbiased estimate for σ^2. It is easy to see that s is an unbiased estimate for σ iff P is a Dirac measure. An application of Schwarz' inequality yields $E(s) \leqslant (E(1)E(s^2))^{1/2} = \sigma$. Equality holds iff $s = \lambda$ a.e. where λ is a nonnegative real number and the claim follows.

For example, if $\xi_1, \ldots, \xi_n$ are independently $N(a, \sigma^2)$-distributed, $-\infty < a < \infty, 0 < \sigma < \infty$, then, according to the Corollary to II, Theorem 4.3, $\frac{s^2}{\sigma^2}(n-1)$ is distributed by the Helmert-Pearson law with $n-1$ degrees of freedom and has the density g_{n-1} given in I. (28.2). Hence,

$$E\left(\frac{s}{\sigma}\sqrt{n-1}\right) = \int_0^{\infty} \sqrt{y}\, g_{n-1}(y)\,dy = 2^{-(n-1)/2}[\Gamma((n-1)/2)]^{-1} \int_0^{\infty} y^{\frac{n}{2}-1} e^{-\frac{y}{2}} dy$$

$$= \sqrt{2}\,\Gamma(n/2)[\Gamma((n-1)/2)]^{-1}$$

$$\text{or } E(s;\sigma) = \sigma \sqrt{\frac{2}{n-1}}\,\Gamma(n/2)[\Gamma((n-1)/2)]^{-1}.$$

Let us treat two more examples: Let P_Γ be a set of probability measures over $(R, \mathscr{S})$ dominated by a σ-additive measure μ. For each γ, let the R.-N.-densities be given by $x \to f(x, \gamma)$ and let these have the form

$$f(x,\gamma) = \sum_{i=1}^{l} h_i(x) d_i(\gamma), \qquad x \in R, \qquad \gamma \in \Gamma \tag{1.5}$$

with $\mathscr{S}$-measurable h_i, although the h_i need not be densities. However, let $\int_R h_i^2\, d\mu$ exist, $1 \leqslant i \leqslant l$. Assume the h_i as well as the d_i are linearly independent, more exactly, from $\sum_{i=1}^{l} \lambda_i h_i = 0$ (λ_i real) μ-a.e., $\lambda_1 = \cdots = \lambda_l = 0$ should follow. Likewise, from $\sum_{i=1}^{l} \lambda_i d_i = 0$ for all $\gamma \in \Gamma$ should follow $\lambda_1 = \cdots = \lambda_l = 0$. Then, an unbiased estimate k for d exists iff d has the form

$$\sum_{i=1}^{l} \alpha_i d_i,$$

where the α_i are arbitrary real numbers.

Let k be an unbiased estimate for d. Then $E(k;\gamma) = \int_R k \sum_{i=1}^{l} h_i d_i(\gamma)\,d\mu$ exists for all $\gamma \in \Gamma$. Since the d_i are linearly independent,

$$E(k;\gamma) = \sum_{i=1}^{l} d_i(\gamma) \int_R k h_i\, d\mu = \sum_{i=1}^{l} \alpha_i d_i(\gamma)$$

with $\alpha_i = \int_R k h_i\, d\mu$, which proves the necessity.

Now let $d = \sum_{i=1}^{l} \alpha_i d_i$. We claim that for arbitrary α_i one can determine constants c_i such that $\sum_{i=1}^{l} c_i h_i$ is an unbiased estimate for d. In this connection we have to show that

$$\sum_{i=1}^{l} d_i(\gamma) \int_R \sum_{j=1}^{l} c_j h_j h_i\, d\mu = \sum_{i=1}^{l} \alpha_i d_i(\gamma) \tag{1.6}$$

can be fulfilled by suitable choice of the c_i, which, naturally, must be independent of γ. Because the d_i are linearly independent, (1.6) holds iff the system of equations

$$\sum_{j=1}^{l} c_j \int_R h_i h_j\, d\mu = \alpha_i, \qquad i = 1, \ldots, l \tag{1.7}$$

is solvable. We remark that all the integrals in (1.7) make sense because of Schwarz' inequality and the assumed existence of $\int_R h_i^2\,d\mu$. For arbitrary α_i (1.7) is always solvable if the determinant

$$\begin{vmatrix} \int_R h_1 h_1\,d\mu \ldots \int_R h_1 h_l\,d\mu \\ \cdots\cdots\cdots\cdots\cdots\cdots \\ \int_R h_l h_1\,d\mu \ldots \int_R h_l h_l\,d\mu \end{vmatrix} \tag{1.8}$$

is different from zero. This holds due to the linear independence of the h_i. The quadratic from

$$\int_R (u_1 h_1 + \cdots + u_l h_l)^2\,d\mu = \sum_{i,j=1}^{l} u_i u_j \int_R h_i h_j\,d\mu$$

in the real variables u_j, $j=1,\ldots,l$ with determinant (1.8) is obviously positive semi-definite. If there existed real numbers $u_1^{(0)},\ldots,u_l^{(0)}$, however, not all zero, such that

$$\int_R (u_1^{(0)} h_1 + \cdots + u_l^{(0)} h_l)^2\,d\mu = 0$$

were valid, then, according to I, Theorem 17.6, this would contradict the linear independence. Hence (1.8) is $\neq$ zero.

This proof is well known in the theory of integral equations with degenerate kernel. (1.8) is the so-called *Gramian determinant* of the h_i [2].

We now treat an example related to II.**13** [3]. We will use the terminology of quality control introduced there. Let N be the size of the batch to be checked and M the (unknown) number of defective pieces in the batch.

First consider the case where the pieces are destroyed by the checking process. k_α, α and n have the meaning introduced in II.**13**. Let $p=M/N$ and the r.v. η^* have the following discrete distribution (see also p. 144)

$$P(\eta^* = pN - r) = \binom{Np}{r}\binom{N-Np}{n-r}\Big/\binom{N}{n}, \qquad r \leqslant k_\alpha,$$

$$P(\eta^* = 0) = \sum_{r=k_\alpha+1}^{n} \binom{Np}{r}\binom{N-Np}{n-r}\Big/\binom{N}{n}.$$

Here let $n<N$. η^* can be viewed as the number of defective pieces accepted. Setting

$$P(p,r,n) = \binom{Np}{r}\binom{N-Np}{n-r}\Big/\binom{N}{n} \quad \text{for } 0 \leqslant p \leqslant 1,$$

we have

$$d(p) = E\left(\frac{\eta^*}{N}; p\right) = \sum_{r \leqslant k_\alpha} \left(p - \frac{r}{N}\right) P(p,r,n). \tag{1.9}$$

Now

$$P(p,r,n) = \binom{n}{r} \frac{(N-n)!}{N!} N^n p\left(p - \frac{1}{N}\right)\cdots\left(p - \frac{r-1}{N}\right)(1-p)\cdots\left(1-p-\frac{n-r-1}{N}\right). \tag{1.10}$$

[2] See also Theorem 1 in H. Teicher, Ann. Math. Statist. 34, 1265—1269 (1963).

[3] Due to A. N. Kolmogorov, Izv. Akad. Nauk SSSR Ser. Mat. 14, 303—326 (1950).

Substituting this into (1.9) we see that $E\left(\frac{\eta^*}{N}\right)$ is a polynomial of degree $n+1$ in p. We now consider the r.v. η (see I.**34**), which represents the sample variable and is hypergeometrically distributed. Let h be an arbitrary function on R_1. Then

$$E(h\circ\eta)=\sum_{r=0}^{n} h(r)\binom{n}{r}\frac{(N-n)!}{N!}N^n p\left(p-\frac{1}{N}\right)\dots\left(p-\frac{r-1}{N}\right)(1-p)\dots\left(1-p-\frac{n-r-1}{N}\right). \tag{1.11}$$

(1.11) is an n-degree polynomial in p. Hence, the polynomials at (1.9) and (1.11), for any choice of h, are always different. So that for $n<N$ there exists no unbiased estimate for d which is a function of η.

We now turn to the case in which the articles are not destroyed by checking. Define a r.v. ξ by

$$P(\xi=r)=\binom{Np}{r}\binom{N-Np}{n-r}\Big/\binom{N}{n}, \qquad r\leqslant k_\alpha,$$

$$P(\xi=pN)=\sum_{r=k_\alpha+1}^{n}\binom{Np}{r}\binom{N-Np}{n-r}\Big/\binom{N}{n}.$$

ξ can be interpreted as the number of unusable pieces found. We recall that the sample plan prescribes control if the sample contains more than k_α unusable pieces. Then all pN defectives will be found. Note that in the previously considered case in which checking causes destruction, the r.v. ξ has no practical significance since the realization of ξ can, in some instances, lead to complete destruction of the production.

Now let b be some function of the parameter p which latter assumes only the values $0, 1/N, \dots, N/N=1$ and let k be a function over R_1. We claim that one can always choose k such that $k\circ\xi$ is an unbiased estimate for b. Indeed,

$$E(k\circ\xi)=\sum_{r\leqslant k_\alpha} k(r)P(p,r,n)+k(pN)\left(1-\sum_{r\leqslant k_\alpha}P(p,r,n)\right).$$

For $p=0, 1/N, \dots, 1$ we must show that

$$E[k\circ\xi;p]=b(p) \tag{1.12}$$

can be fulfilled. Using (1.10) one easily sees that $1-\sum_{r\leqslant k_\alpha}P(p,r,n)$ vanishes for $p\leqslant\frac{k_\alpha}{N}$. Thus, for $p=0,\dots,\frac{k_\alpha}{N}$, (1.12) is identical to the system of equations

$$\sum_{r\leqslant k_\alpha} k(r)P\left(\frac{l}{N},r,n\right)=b\left(\frac{l}{N}\right), \qquad l=0,1,\dots,k_\alpha.$$

The determinant of this system is different from zero because $P(p,r,n)$ vanishes for $p<\frac{r}{N}$ and the elements of the main diagonal are $\neq 0$. For $m>k_\alpha$ one determines $k(m)$ according to (1.12) by means of

$$k(m)=\frac{b\left(\frac{m}{N}\right)-\sum_{r\leqslant k_\alpha}k(r)P\left(\frac{m}{N},r,n\right)}{1-\sum_{r\leqslant k_\alpha}P\left(\frac{m}{N},r,n\right)}.$$

This makes sense since $1-\sum_{r\leq k_\alpha} P\left(\frac{m}{N}, r, n\right)\neq 0$ for $m>k_\alpha$.

A large number of further examples of unbiased estimates are also provided by the considerations in II.**14** when one formulates them in the terminology of this chapter.

We return to the general case. Let h be an unbiased estimate for d, that is, (1.1) holds for all $\gamma\in\Gamma$. Let h_1 be an unbiased estimate differing from h. Thus, there exists at least one γ such that $P_\gamma(\{x: h_1(x)\neq h(x)\})>0$. However, in this case there are infinitely many different unbiased estimates for d. Indeed, let $0<\alpha<1$, then $\alpha h+(1-\alpha)h_1$ is also unbiased for d.

However, if there are infinitely many estimates for d, we have the problem of which one to choose in order to estimate d "as well as possible". The quality of the approximation is to be judged "in the mean". It must be related to the behavior of the estimate which is a r.v. and not to each of its realizations. The following two definitions serve this purpose.

Definition. Let H_{γ_0} be the class of all unbiased estimates for d according to (1.1) such that $E(h^2;\gamma_0)$ exists for all $h\in H_{\gamma_0}$ and $\gamma_0\in\Gamma$. We will call $h_0\in H_{\gamma_0}$ *locally minimal at* γ_0 if

$$E[(h_0-d(\gamma_0))^2;\gamma_0]\leq E[(h-d(\gamma_0))^2;\gamma_0] \tag{1.13}$$

for all $h\in H_{\gamma_0}$.

There can, of course, exist unbiased estimates for h for which $E(h^2;\gamma_0)$ does not exist. Such estimates are to be excluded from this definition. An analogous remark pertains to the following

Definition. Let H be the *set of all unbiased estimates* for d with the property: *For all* $h\in H$ *and all* $\gamma\in\Gamma$, $E(h^2;\gamma)$ *exists.* $h_0\in H$ is called *uniformly minimal* if

$$E[(h_0-d(\gamma))^2;\gamma]\leq E[(h-d(\gamma))^2;\gamma] \tag{1.14}$$

for all $\gamma\in\Gamma$ *and each* $h\in H$.

We have now

Theorem 1.1[4]**.** *Let* P_Γ *be a set of probability measures on* $(R,\mathscr{S})$ *and assume that the above defined class H of unbiased estimates for d is not empty. Let* V_2 *be the set of all unbiased estimates* v *for zero, (the mapping which sends each* $\gamma\in\Gamma$ *into zero) such that* $E(v^2;\gamma)$ *exists for each* $\gamma\in\Gamma$. *Then* $h_0\in H$ *is uniformly minimal iff*

$$E(vh_0;\gamma)=0 \tag{1.15}$$

for each $\gamma\in\Gamma$ *and all* $v\in V_2$.

[4] This result is essentially due to C. R. Rao, Sankhyā 12, 27—42 (1952).

Proof. From the assumptions it follows, after an application of Schwarz' inequality, that the left side of (1.15) always makes sense.

First let $h_0 \in H$ be uniformly minimal and assume $E(v_0 h_0; \gamma_0) \neq 0$ for a $\gamma_0 \in \Gamma$ and a $v_0 \in V_2$. Then $h_0 + \lambda v_0 \in H$ for each real λ. If $E(v_0^2; \gamma_0)$ were zero, then $E(v_0 h_0; \gamma_0)$ would also be zero. Hence, $E(v_0^2; \gamma_0) \neq 0$. Choosing $\lambda_0 = -E(v_0 h_0; \gamma_0)/E(v_0^2; \gamma_0)$ we have

$$E[(h_0 + \lambda_0 v_0)^2; \gamma_0] = E(h_0^2; \gamma_0) - [E(v_0 h_0; \gamma_0)]^2 / E(v_0^2; \gamma_0)$$

which, in the light of I, Theorem 17.2, contradicts the assumption that h_0 is uniformly minimal.

Conversely, if (1.15) holds for an $h_0 \in H$, each $\gamma \in \Gamma$ and all $v \in V_2$ and if h is an arbitrary element from H, then $h_0 - h \in V_2$. Thus, for each γ

$$E(h_0(h_0 - h); \gamma) = 0 \quad \text{or} \quad E(h_0^2; \gamma) = E(h_0 h; \gamma).$$

An application of Schwarz' inequality yields

$$E(h_0^2; \gamma) \leqslant (E(h_0^2; \gamma))^{\frac{1}{2}} (E(h^2; \gamma))^{\frac{1}{2}}. \tag{1.16}$$

Either $E(h_0^2; \gamma) = 0$ and we are through, or $E(h_0^2; \gamma) \leqslant E(h^2; \gamma)$ from (1.16), which we wanted to prove.

Remark. It is easy to see that $2E(v h_0) = \dfrac{d}{d\lambda} E[(h_0 + \lambda v)^2]\big|_{\lambda=0}$. (1.15) can therefore be interpreted as the vanishing of a differential. This allows considerable generalization of Theorem 1.1, by replacing the variances in the definition of uniformly minimal estimates, say, by p-th absolute moments ($p \geqslant 1$). [5]

We also have

Theorem 1.2. *Assume P_Γ and H have the same meaning as in Theorem 1.1. Then there exists at most one uniformly minimal estimate for d.*

Proof. For any r.v. k over $(R, \mathscr{S})$ whose second moment exists for a P_γ we write for abbreviation $\left(\int_R k^2 \, dP_\gamma\right)^{\frac{1}{2}} = \|k\|_\gamma$. One sees immediately that $\|\lambda k\|_\gamma = |\lambda| \, \|k\|_\gamma$ for each real λ. Since for any r.v.'s k_1, k_2 with finite second moment w.r.t. P_γ we always have (Schwarz' inequality)

$$\left|\int_R k_1 k_2 \, dP_\gamma\right| \leqslant \left(\int_R k_1^2 \, dP_\gamma\right)^{\frac{1}{2}} \left(\int_R k_2^2 \, dP_\gamma\right)^{\frac{1}{2}},$$

[5] See L. Schmetterer, Ann. Math. Statist. 31, 1154—1163 (1960) and Publ. Math. Inst. Hungar. Acad. Sci. Ser. A. 6, 295—300 (1961). See also A. Kramli, Studia Sci. Math. Hung. 2, 159—161 (1967).

then also

$$\int_R k_1^2\,dP_\gamma + 2\int_R k_1 k_2\,dP_\gamma + \int_R k_2^2\,dP_\gamma \leqslant \int_R k_1^2\,dP_\gamma$$
$$+2\left(\int_R k_1^2\,dP_\gamma\right)^{\frac{1}{2}}\left(\int_R k_2^2\,dP_\gamma\right)^{\frac{1}{2}} + \int_R k_2^2\,dP_\gamma,$$

i.e.,

$$\|k_1+k_2\|_\gamma \leqslant \|k_1\|_\gamma + \|k_2\|_\gamma. \tag{1.17}$$

Assume h_1 and h_2 are two different [6] uniformly minimal estimates $\in H$. Then $\|h_1\|_\gamma = \|h_2\|_\gamma$ for each $\gamma \in \Gamma$. Moreover, $(h_1+h_2)/2 \in H$, so that

$$\tfrac{1}{2}(\|h_1\|_\gamma + \|h_2\|_\gamma) \leqslant \|(h_1+h_2)/2\|_\gamma. \tag{1.18}$$

Applying (1.17) we get

$$\|(h_1+h_2)/2\|_\gamma \leqslant \tfrac{1}{2}\|h_1\|_\gamma + \tfrac{1}{2}\|h_2\|_\gamma. \tag{1.19}$$

Hence, according to (1.18) and (1.19) we have for all $\gamma \in \Gamma$

$$\|h_1+h_2\|_\gamma = \|h_1\|_\gamma + \|h_2\|_\gamma. \tag{1.20}$$

From (1.20)

$$\int_R h_1 h_2\,dP_\gamma = \left(\int_R h_1^2\,dP_\gamma\right)^{\frac{1}{2}}\left(\int_R h_2^2\,dP_\gamma\right)^{\frac{1}{2}}. \tag{1.21}$$

The equality sign holds, however, in Schwarz' inequality iff P_γ-a.e. $h_1 = \lambda h_2$ for a real number λ. But if $\|h_1\|_\gamma = \|h_2\|_\gamma \neq 0$, then with (1.21), $\lambda = 1$. If, on the other hand $\|h_1\|_\gamma = \|h_2\|_\gamma = 0$, then $h_1 = h_2 = 0$ P_γ-a.e.

We remark that if we assign the real number $\|k_1 - k_2\|_\gamma$ as distance between two r.v.'s k_1 and k_2 with finite second moments w.r.t. P_γ, then this distance satisfies the triangle inequality, as an easy application of (1.17) shows. In addition, from $\|k_1+k_2\|_\gamma = \|k_1\|_\gamma + \|k_2\|_\gamma$ follows $k_1 = \lambda k_2$ for real λ. Such distance functions are called *strictly convex* and for all such distances (e.g. for the p-th absolute moments with $p>1$) one can prove an analogue to Theorem 1.2.[7]

Obviously, one can obtain analogous results in the case of locally minimal estimates. For example, the analogue to Theorem 1.1 is as follows:

Let P_Γ and d have the same meaning as in Theorem 1.1. and assume H_{γ_0} is non-empty for a $\gamma_0 \in \Gamma$. Take V_{2,γ_0} as the set of all unbiased estimates v of zero for which $E(v^2;\gamma_0)$ exists. h_0 is then locally minimal at γ_0 iff (1.15) holds for all γ_0 and all $v \in V_{2,\gamma_0}$.

For locally minimal estimates one can prove the following existence theorem:[8]

---

[6] See p. 273.

[7] See L. Schmetterer, Mitteilungsbl. Math. Statist. 9, 147—152 (1957).

[8] See E. W. Barankin, Ann. Math. Statist. 20, 477—501 (1949) and L. Schmetterer, loc. cit.[5].

Let P_Γ be dominated by a measure P_{γ_0} with $\gamma_0 \in \Gamma$ and assume H_{γ_0} is non-empty. f_γ will denote the R.-N. density of P_γ w.r.t. P_{γ_0} and we assume that f_γ^2 is P_{γ_0}-integrable for each $\gamma \in \Gamma$. Then there always exists a locally minimal estimate at γ_0.

This result is especially easy to apply when each P_γ from P_Γ dominates all other measures from P_Γ. Then each P_γ can assume the rôle of P_{γ_0}.

We proceed now to the study of the structure of the set of all uniformly minimal estimates. We will answer questions such as the following: Let P_Γ be, as usual, a set of probability measures over $(R, \mathscr{S})$ and let d_i, $1 \leqslant i \leqslant 2$ be maps from Γ into R_1. Let uniformly minimal estimates h_i for d_i exist. According to Theorem 1.2, the h_i are (essentially) uniquely determined. Then is λh_i for arbitrary real λ a uniformly minimal estimate for λd_i and is $h_1 + h_2$ uniformly minimal for $d_1 + d_2$? If the answers are affirmative, then the uniformly minimal estimates form a linear space. We have, in fact

Theorem 1.3. *Let P_Γ be a set of probability measures on $(R, \mathscr{S})$ and d_i, $1 \leqslant i \leqslant 2$, mappings of Γ into R_1. If uniformly minimal estimates h_i for d_i exist, then for λd_i, λ arbitrary and real, and for $d_1 + d_2$ there exist uniformly minimal estimates given by λh_i and $h_1 + h_2$, resp.*

Proof. This theorem follows almost immediately from Theorem 1.1. We show for example the second claim. Since h_i is uniformly minimal, we have for all $v \in V_2$ and all $\gamma \in \Gamma$

$$E(v h_i; \gamma) = 0. \tag{1.22}$$

However, from (1.22) we have $E v((h_1 + h_2); \gamma) = 0$ for all $v \in V_2$ and all $\gamma \in \Gamma$ which proves the claim.

We also have

Theorem 1.4[9]. *Let $h_1, \ldots, h_k$, $k \geqslant 1$ be a set of uniformly minimal and bounded estimates. Then each polynomial of the form $\sum_{i_1, \ldots, i_k} \alpha_{i_1 \ldots i_k} h_1^{i_1} \ldots h_k^{i_k}$, $\alpha_{i_1 \ldots i_k}$ real, is a uniformly minimal estimate.*

Proof. First it is clear that the map $h(x) = c$, $x \in R$, c arbitrary and real, is a uniformly minimal estimate and hence, it is sufficient, because of Theorem 1.3, to show that the product of two uniformly minimal and bounded estimates is again uniformly minimal. Let h_1 and h_2 be such estimates. Then, according to Theorem 1.1 for all $v \in V_2$ and all $\gamma \in \Gamma$

$$E(v h_1; \gamma) = 0. \tag{1.23}$$

---

[9] R.R. Bahadur, Sankhyā 18, 211—224 (1957).

Since, however, h_1 is bounded, $v h_1 \in V_2$ because of (1.23). Hence, for all $v \in V_2$ and all γ we necessarily also have

$$E(v h_1 h_2; \gamma) = 0. \tag{1.24}$$

But (1.24) is also sufficient to insure that $h_1 h_2$ is a uniformly minimal estimate.

Finally we have

Theorem 1.5[10]. *Let h_i be a sequence of uniformly minimal estimates and h an $\mathcal{S}$-measurable mapping from R into R_1 such that $E(h^2; \gamma)$ exists for all $\gamma \in \Gamma$. Let, further $\int_R |h - h_i|^2 \, dP_\gamma \to 0$ for $i \to \infty$ and each $\gamma \in \Gamma$. Then h is also a uniformly minimal estimate.*

Proof. For all $v \in V_2$, all $\gamma \in \Gamma$ and each $i = 1, 2, \ldots$ we have $E(v h_i; \gamma) = 0$. Thus, $E(v h; \gamma) = E(v(h - h_i); \gamma)$. But

$$\left|(E(v(h-h_i)); \gamma)\right| \leqslant (E(v^2; \gamma))^{\frac{1}{2}} (E((h-h_i)^2; \gamma))^{\frac{1}{2}} \to 0 \quad \text{for } i \to \infty$$

for all $v \in V_2$ and all $\gamma \in \Gamma$. This implies that $E(v h; \gamma) = 0$ for all $v \in V_2$ and all $\gamma \in \Gamma$.

The most important aids to the effective construction of uniformly minimal estimates are two theorems due to Blackwell and to Lehmann and Scheffé.

Theorem 1.6[11]. *Let P_Γ be a set of probability measures over $(R, \mathcal{S})$. The class H of unbiased estimates for d is assumed to be non-empty. We recall that $E(h^2; \gamma)$ exists for all $h \in H$ and all $\gamma \in \Gamma$. Let T be a transformation which is sufficient for P_Γ and h an arbitrary element from H. The conditional expectation $E(h \mid T)$ is independent of γ*[12] *and is an unbiased estimate for d. Moreover,*

$$E[(E(h \mid T) - d(\gamma))^2; \gamma] \leqslant E[(h - d(\gamma))^2; \gamma] \tag{1.25}$$

for all $\gamma \in \Gamma$. Equality holds in (1.25) *for all $\gamma \in \Gamma$ iff $h = E(h \mid T)$* P_Γ-a.e.

Proof. According to I. (20.5), we always have $E(E(h \mid T)) = E(h)$ so that $E(h \mid T)$ is unbiased for d. From I, Theorem 17.2 it is sufficient in place of (1.25) to show that

$$E[(E(h \mid T))^2; \gamma] \leqslant E(h^2; \gamma) \tag{1.26}$$

---

[10] R.R. Bahadur, loc. cit.[9] and L. Schmetterer, loc. cit.[5].

[11] D. Blackwell, Ann. Math. Statist. 18, 105—110 (1947). See also A.N. Kolmogorov, loc. cit.[3].

[12] For this terminology see III, p. 206.

for each $\gamma \in \Gamma$. But $E(h^2;\gamma)=E(E(h^2|T);\gamma)$ and if we can prove that

$$(E(h|T))^2 \leqslant E(h^2|T)^{13} \tag{1.27}$$

P_γ-a.e. then we will have also shown (1.25). But we know that one can P_γ-a.e. apply Schwarz' inequality to $E(h|T)$. (See I, p. 64.) Then $(E(h|T))^2 \leqslant E(h^2|T)E(1|T)$ P_γ-a.e. and hence also (1.27). If equality holds in (1.25) for a γ then we must have:

$$E((E(h|T))^2;\gamma) = E(h^2;\gamma). \tag{1.28}$$

But, according to I, Theorem 20.2:

$$E(hE(h|T)) = E(E(hE(h|T)|T)) = E(E(h|T)E(h|T)).$$

Thus, with (1.28)

$$E(h-E(h|T))^2=0.$$

Then $h=E(h|T)$ P_γ-a.e. and the last claim is proved.

Theorem 1.6 shows that one need consider only functions of T in the search for uniformly minimal estimates. If T is the trivial sufficient transformation, then, naturally, nothing is won. We have, however,

Theorem 1.7[14]. *Assume the assumptions of the previous theorem are fulfilled and that, in addition, the set of measures $P_{\gamma,T}$, $\gamma \in \Gamma$ induced by T is complete. Then $E(h|T)$ is a uniformly minimal estimate no matter how $h \in H$ is chosen.*

Proof. We first show that for $h_1, h_2 \in H$ we always have

$$E(h_1|T) = E(h_2|T) \tag{1.29}$$

P_Γ-a.e. Indeed, since both $E(h_1|T)$ and $E(h_2|T)$ are unbiased, we obtain

$$\int_Q (E(h_1|T)) - E(h_2|T))\, dP_{\gamma,T}=0$$

for all $\gamma \in \Gamma$ where, $Q=T(R)$.

From the completeness of the $P_{\gamma,T}$ follows (1.29) $P_{\Gamma,T}$-a.e. Thus all the $E(h|T)$ coincide for $h \in H$ P_Γ-a.e. and then, according to Theorem 1.6, $E(h|T)$ is a uniformly minimal estimate.

Using the examples of III, pp. 212—213 it is quite easy to give examples of the application of Theorems 1.6 and 1.7. Also see VII, Theorems 1.8 and 1.9.

One can give a different interpretation to the content of Theorem 1.6. We have first a

---

[13] Here and in the following lines we have occasionally suppressed the reference to γ.

[14] E. L. Lehmann and H. Scheffé, Sankhyā 10, 305—340 (1950).

Definition. Let $(R,\mathscr{S})$, P_Γ and d have the same meaning as in Theorem 1.6. Let K be the *set of all $\mathscr{S}$-measurable functions h over R for which $E(h^2;\gamma)$ exists for all $\gamma\in\Gamma$. A subset C of K is called complete (w.r.t. K)* if for each $h\in K-C$ there exists an $h_1\in C$ such that

$$E((h_1-d(\gamma))^2;\gamma)\leqslant E((h-d(\gamma))^2;\gamma) \tag{1.30}$$

for all $\gamma\in\Gamma$ and for at least one γ the inequality sign holds.

We remark that we have not required here that h or h_1 be unbiased for d.

The definition can immediately be generalized, for example, to p-th moments $(p\geqslant 1)$.[15]

A further generalization consists of choosing a function G defined on $R_1\times\Gamma$, assuming for it the necessary measurability and integration properties and replacing (1.30) by $E(G(h_1,\gamma);\gamma)\leqslant E(G(h;\gamma);\gamma)$. (See p. 330 and p. 477.)

We have now

Theorem 1.8. *Let T be a sufficient transformation for P_Γ. Then the set $C=\{E(h|T):h\in K\}$ is complete.*

The *proof* follows immediately from (1.26) and the last claim of Theorem 1.6.

We give a further

Definition. Let $(R,\mathscr{S})$ and P_Γ have their previous meanings and let *K be a set of $\mathscr{S}$-measurable functions h* defined over R *for which $E(h^2;\gamma)$ exists for all $\gamma\in\Gamma$.* $h_0\in K$ is called *admissible* w.r.t. K if there is no $h\in K$ for which

$$E[(h-d(\gamma))^2;\gamma]\leqslant E[(h_0-d(\gamma))^2;\gamma] \tag{1.31}$$

for all $\gamma\in\Gamma$ and

$$E[(h-d(\gamma_0))^2;\gamma_0]<E[(h_0-d(\gamma_0))^2;\gamma_0] \tag{1.32}$$

for at least one $\gamma_0\in\Gamma$.

We have the trivial

Theorem 1.9. *Each (w.r.t. K) complete class C must contain all (w.r.t. K) admissible estimates. If K is, in particular, the set of all unbiased estimates for d and there exists a uniformly minimal estimate h in K, then h is admissible.*[16]

---

[15] More precisely, we are speaking here of absolute moments. Thus, we consider $E(|h-d(\gamma)|^p;\gamma)$.

[16] See also p. 477.

Let us illustrate these notions with another general example which summarizes many known facts about the sampling theory of finite populations[17].

Let $Z=\{1,2,\dots,N\}$, $N\geqslant 2$ and Γ be the set of all N-tuples of pairs of the form $\{(1,y_1),\dots,(N,y_N)\}$ with $y_j\in R_1$ for $j\in Z$. Furthermore, let $Q_1=Z\times R_1$ and for $n\geqslant 1$ $Q_{n+1}=Q_n\times Q_1$. We define a sample space $R=\bigcup_{n=1}^{\infty} Q_n$ and let $\mathscr{S}$ be the set of all subsets of R. Each element x of R thus has the form

$$x=\{(i_1,x_{i_1}),\dots,(i_n,x_{i_n})\} \quad \text{with } i_j\in Z\,, \quad x_{i_j}\in R_1 \quad \text{for } j\leqslant n\,, \quad n\geqslant 1\,.$$

Each (i_j,x_{i_j}), $1\leqslant j\leqslant n$, will be called a component of x.

Let $Z_1=Z$ and $Z_{n+1}=Z_n\times Z_1$ for $n\geqslant 1$. We choose finitely or countably many elements from $\bigcup_{n=1}^{\infty} Z_n$ and denote their number by m. Thus m can equal ∞. The i-th element chosen we denote (slightly abusing notation) by $(k_1,k_2,\dots,k_{j_i})$, $1\leqslant j_i$. To each $\gamma\in\Gamma$ we assign a probability measure P_γ over $(R,\mathscr{S})$: Let P_γ be a discrete distribution with mass points $x_\gamma^{(i)}\in R$, $i=1,\dots,m$, so that

$$P_\gamma(\{x_\gamma^{(i)}\})=p_i \tag{1.33}$$

independently of γ with $p_i>0$ and $\sum_{i=1}^{m} p_i=1$. The $x_\gamma^{(i)}$ are defined as follows:

If

$$\gamma=\{(1,y_1^{(\gamma)}),\dots,(N,y_N^{(\gamma)})\}\,,$$

then

$$x_\gamma^{(i)}=\{(k_1,y_{k_1}^{(\gamma)}),(k_2,y_{k_2}^{(\gamma)}),\dots,(k_{j_i},y_{k_{j_i}}^{(\gamma)})\}\,.$$

Thus, each component of $x_\gamma^{(i)}$ is a component of γ but not all components of γ need appear as components of $x_\gamma^{(i)}$. Which components of γ are components of $x_\gamma^{(i)}$ depends not on γ, but only on the choice of the m elements from $\bigcup_{n=1}^{\infty} Z_n$.[18] If $(j,y_j^{(\gamma)})$ is a component of $x_\gamma^{(i)}$ we will write $y_j^{(\gamma)}\kappa\, x_\gamma^{(i)}$, otherwise $y_j^{(\gamma)}\not\kappa\, x_\gamma^{(i)}$.

We now consider the problem of unbiased estimation of the function d defined for each γ by $d(\gamma)=(y_1^{(\gamma)}+\cdots+y_N^{(\gamma)})/N$.

In this connection we define functions $Y_1,\dots,Y_N$ from R into R_1 as follows: For all $x_\gamma^{(i)}$, $i=1,\dots,m$, $\gamma\in\Gamma$ let

$$Y_j(x_\gamma^{(i)})=\begin{cases} y_j^{(\gamma)}, & y_j^{(\gamma)}\kappa\, x_\gamma^{(i)} \\ 0, & y_j^{(\gamma)}\not\kappa\, x_\gamma^{(i)}\,, \end{cases}$$

$Y_j(x)=$ arbitrary for $x\neq x_\gamma^{(i)}$, $j=1,\dots,N$. Further, we define functions[19] $c_1,\dots,c_N$ from R into R_1 which for $j=1,\dots,N$ satisfy the conditions

$$c_j(x_\gamma^{(i)})=0 \quad \text{for } y_j^{(\gamma)}\not\kappa\, x_\gamma^{(i)}\,, \quad 1\leqslant i\leqslant m \tag{1.34}$$

and for each i

$$c_j(x_{\gamma_1}^{(i)})=c_j(x_{\gamma_2}^{(i)}) \quad \text{for all } \gamma_1,\gamma_2\in\Gamma\,. \tag{1.35}$$

---

[17] See L. Schmetterer, Abh. Deutsch. Akad. Wiss. Berlin, Kl. Math. Phys., Tech. 1964, Nr. 4, 117—120 and J. Roy and I. M. Chakravarti, Ann. Math. Statist. 31, 392—398 (1960).

[18] From the sampling theory standpoint the case in which all components of γ appear in $x_\gamma^{(i)}$ is trivial.

[19] The definition of the c_j, $1\leqslant j\leqslant N$ in the elements of R that are different from $x_\gamma^{(i)}$ is of no consequence.

The c_j at the mass points $x_\gamma^{(i)}$ thus also depend only on the choice of the m elements from $\bigcup_{n=1}^{\infty} Z_n$ and not on γ. We now consider estimates for d of the form $h=\sum_{j=1}^{N} Y_j c_j$. Since h depends linearly on the Y_j, one calls such estimates linear.

If $E(h;\gamma)$ exists we obtain

$$E(h;\gamma)=\sum_{i=1}^{m}\sum_{j=1}^{N} Y_j(x_\gamma^{(i)})c_j(x_\gamma^{(i)})p_i=\sum_{j=1}^{N} y_j^{(\gamma)}\sum_{i=1}^{m} c_j(x_\gamma^{(i)})p_i$$

since if $Y_j(x_\gamma^{(i)})\neq y_j^{(\gamma)}$ we have $c_j(x_\gamma^{(i)})=0$ according to the definition of the Y_j and (1.34). Hence,

$$E(h;\gamma)=\sum_{j=1}^{N} y_j E(c_j;\gamma)\,. \tag{1.36}$$

From (1.33) and (1.35) it follows that $E(c_j;\gamma)$ does not depend on γ and we thus write $E(c_j)$ instead of $E(c_j;\gamma)$. Since almost all of the definitions given in this example do not depend on γ we will frequently omit the reference to γ.

Now it follows from (1.36) that a linear estimate for d is unbiased iff

$$E(c_j)=1/N\,,\quad j=1,\ldots,N\,. \tag{1.37}$$

Let $\mathfrak{C}$ be the set of all mappings $(c_1,\ldots,c_N)$ from R into R_N possessing the following properties[20]: (1.34), (1.35) and (1.37) hold and for $j=1,\ldots,N$, c_j has finite second moment. Let H be the set of all unbiased linear estimates h for d obtained by means of the elements of $\mathfrak{C}$.

We first note that, in general, there exist no uniformly minimal estimates for d in H. Indeed, let $h^*=\sum_{j=1}^{N} Y_j c_j^*$ be such an estimate. Then, for all real $y_1,\ldots,y_N$ and all $(c_1,\ldots,c_N)\in\mathfrak{C}$ we must have

$$\sum_{j,k=1}^{N} y_j y_k E(c_j^* c_k^*)\leqslant \sum_{j,k=1}^{N} y_j y_k E(c_j c_k)\,. \tag{1.38}$$

If one chooses $y_k=0$ for $k\neq j$, $y_j=1$, then it follows, in particular, that c_j^* must possess minimal variance for each $j=1,\ldots,N$. Now let

$$M(j)=\{i: y_j \kappa x^{(i)}\} \tag{1.39}$$

and

$$P_\gamma(M(j))=w_j \tag{1.40}$$

for $j=1,\ldots,N$. One immediately sees that the definition of $M(j)$ is independent of γ. Because of (1.37), $M(j)$ is always $\neq\emptyset$. Using Schwarz' inequality, we get

$$1/N^2=\Big|\sum_{i\in M(j)} c_j(x^{(i)})p_i\Big|^2\leqslant \sum_{i\in M(j)} c_j^2(x^{(i)})p_i \sum_{i\in M(j)} p_i$$

and equality holds iff $c_j(x^{(i)})\sqrt{p_i}=\lambda\sqrt{p_i}$ for real λ and $i\in M(j)$.

Since $p_i\neq 0$, all $c_j(x^{(i)})$ are constant for $i\in M(j)$. Hence, for $i\in M(j)$ and $j=1,\ldots,N$

$$c_j(x^{(i)})=1/N w_j \tag{1.41}$$

This result can be succinctly formulated if we introduce "indicator functions" for the $x^{(i)}$. For $j=1,\ldots,N$ let

$$v_j(x^{(i)})=\begin{cases}1 & i\in M(j)\\ 0 & i\notin M(j)\end{cases}.$$

[20] We naturally assume that this set is non-empty.

Then, (1.41) says that for $i=1,\dots,m$ one necessarily has

$$c_j(x^{(i)})=v_j(x^{(i)})/(N w_j). \tag{1.42}$$

It is easy to see that $\sum_{j=1}^{N} Y_j v_j/N w_j \in H$. Let $v=\sum_{j=1}^{N} Y_j e_j$ be a not identically vanishing unbiased estimate for zero, where $(e_1,\dots,e_N)$ possesses all the properties of the elements of $\mathfrak{C}$ except for (1.37). Instead of (1.37)

$$E(e_j)=0 \tag{1.43}$$

must hold for $j=1,\dots,N$. If no such v exists, then everything is trivial. If $h^*=\sum_{j=1}^{N} Y_j v_j/N w_j$ were a uniformly minimal estimate for d, then we would have to have, for all real $y_1,\dots,y_N$, according to Theorem 1.1 [21]

$$\sum_{j,k=1}^{N} y_j y_k E\left(e_j \frac{v_k}{N w_k}\right)=0$$

or

$$E(e_j v_k)=0, \quad j,k=1,\dots,N. \tag{1.44}$$

We now define, in analogy to (1.39), for all $j\neq k$, $j,k=1,\dots,N$

$$M(j,k)=\{i: y_j \kappa x^{(i)},\ y_k \kappa x^{(i)}\}. \tag{1.45}$$

(1.43) means the same as

$$\sum_{i\in M(j)} e_j(x^{(i)})\,p_i=0.$$

We assume for simplicity that $M(j)$ is a finite set. Then one easily sees that $\varepsilon_j=e_j(x^{(i)})$ for $i\in M(j)$ can always be chosen such that all partial sums $\sum \varepsilon_i p_i$ are different from zero if i does not run through all of $M(j)$. Let $M(j,k)\neq\emptyset$ and $M(j,k)\neq M(j)$. Then (1.44) is identical to

$$\sum_{i\in M(j,k)} e_j(x^{(i)})\,p_i=\sum_{i\in M(j,k)} \varepsilon_i p_i=0$$

which contradicts the construction of the ε_i.

But the arguments which led to (1.42) imply almost immediately that h^* is admissible in H. Otherwise there would have to exist a $\sum_{j=1}^{N} Y_j c_j \in H$ such that for all real $y_1,\dots,y_N$

$$\sum_{j,k=1}^{N} y_j y_k\left(E(c_j c_k)-E\left(\frac{v_j v_k}{N^2 w_j w_k}\right)\right)\leqslant 0.$$

Thus, we would have to have $E(c_j^2)\leqslant E\left(\frac{v_j^2}{N^2 w_j^2}\right)$ for $j=1,\dots,N$ which would imply $c_j=v_j/(N w_j)$ as we previously demonstrated.

We now want to show how to construct a complete class $\mathfrak{H}$ w.r.t. H. Let $\{j_1,\dots,j_u\}$ be a non-empty subset of Z and let $N(j_1,\dots,j_u)=\{i: y_{l_1}\kappa x^{(i)},\ l_1\in\{j_1,\dots,j_u\};\ y_{l_2}\not\kappa x^{(i)},\ l_2\in Z-\{j_1,\dots,j_u\}\}$.

---

[21] Theorem 1.1 referred, however, to the totality of all unbiased estimates and not just to the linear estimates. The theorem holds trivially for linear estimates h if one limits the class of estimates of zero to the linear ones.

Let $\mathfrak{D}$ be a subset of $\mathfrak{C}$ characterized by the following conditions: For $j=1,\dots,N$ and $i\in N(j_1,\dots,j_u)$

$$c_j(x^{(i)})=c_j^{(j_1,\dots,j_u)}$$

where the $c_j^{(j_1,\dots,j_u)}$ are real numbers and $\{j_1,\dots,j_u\}$ runs through all non-empty subsets $\neq Z$ of Z.

If $j\notin\{j_1,\dots,j_u\}$, then $c_j^{(j_1,\dots,j_u)}=0$ because of (1.34).

The subset of H corresponding to $\mathfrak{D}$ will be denoted by $\mathfrak{H}$. We will show that $\mathfrak{H}$ is complete. Let

$$w_{(j_1,\dots,j_u)}=\sum_{i\in N(j_1,\dots,j_u)} p_i \tag{1.46}$$

and let $h=\sum_{j=1}^{N} Y_j c_j$ be an element of $H-\mathfrak{H}$. We fix this h and define

$$b_j^{(j_1,\dots,j_u)}=\begin{cases} 0 & w_{(j_1,\dots,j_u)}=0\\ \sum\limits_{k\in N(j_1,\dots,j_u)} c_j(x^{(k)})\,p_k/w_{(j_1,\dots,j_u)}, & w_{(j_1,\dots,j_u)}>0\,.\end{cases}$$

For $j=1,\dots,N$, let

$$a_j(x^{(i)})=b_j^{(j_1,\dots,j_u)}\qquad i\in N(j_1,\dots,j_u)\,. \tag{1.47}$$

Then, $\sum_{j=1}^{N} Y_j a_j\in\mathfrak{H}$ for $E(a_j)=\sum_{i=1}^{m} a_j(x^{(i)})\,p_i=\sum_{\{j_1,\dots,j_u\}}\sum_{i\in N(j_1,\dots,j_u)} p_i a_j(x^{(i)})$, and thus $E(a_j)=1/N$ because of (1.46), (1.47) and the fact that $h\in H$.

Furthermore, for $j,k=1,\dots,N$

$$E(a_j c_k)=E(a_j a_k)\,. \tag{1.48}$$

Indeed,

$$E(a_j a_k)=\sum_{\{j_1,\dots,j_u\}}\sum_{i\in N(j_1,\dots,j_u)} a_j(x^{(i)})\,p_i\sum_{l\in N(j_1,\dots,j_u)} c_k(x^{(l)})\,p_l/w_{j_1,\dots,j_u}$$

which equals $E(a_j c_k)$ because of (1.46). We now have

$$\sum_{j,k=1}^{N} y_j y_k(E(c_j c_k)-E(a_j a_k))\geqslant 0 \tag{1.49}$$

for all real $y_1,\dots,y_N$.

In fact, from (1.48)

$$E(c_j c_k)-E(a_j a_k)=E[(c_j-a_j)(c_k-a_k)]$$

and thus, $(E(c_j c_k)-E(a_j a_k))_{1N}^{1N}$ is a covariance matrix. But, according to I, Theorem 17.5, this matrix is always positive semi-definite which implies (1.49). However, equality in (1.49) for all real $y_1,\dots,y_N$ cannot hold because then $a_j=c_j$ for $j=1,\dots,N$ would obtain.

We now specialize as follows: Let $n\leqslant N$. Choose exactly those elements $\{k_1,\dots,k_n\}\in Z_n$ for which

$$k_i\neq k_j\quad\text{for } i\neq j\,. \tag{1.50}$$

There are $N!/(N-n)!$ such choices. Let $x_\gamma^{(i)}=\{(k_1,y_{k_1}^{(\gamma)}),\dots,(k_n,y_{k_n}^{(\gamma)})\}$ and $P_\gamma(x_\gamma^{(i)})=1/[N(N-1)\dots(N-n+1)]$ for $i=1,\dots,N!/(N-n)!$. This brings us back to the case considered in II.**12**. P_γ is the joint distribution of the r.v.'s there.

Modifying this example by dropping (1.50), we obtain N^n mass points $x_\gamma^{(i)}$ for each γ. Defining $P_\gamma(x_\gamma^{(i)})=1/N^n$ for $i=1,\dots,N^n$, we have a case of sampling "with replacement", i.e., the corresponding r.v.'s $\xi_1,\dots,\xi_n$ are independent.

Locally minimal (uniformly minimal) estimates deliver the smallest variance about the estimated function d at a point (everywhere in Γ). Under certain assumptions, one can obtain a lower bound for this minimal variance. This turns out to be independent of the existence of locally or uniformly minimal estimates. The corresponding theorem is called the Cramér-Fréchet-Rao theorem[22]. We first show

Theorem 1.10. *Let Γ be an open set of R_1 and P_Γ a set of probability measures on $(R,\mathcal{S})$ dominated by a σ-finite measure μ. For each $\gamma\in\Gamma$, let the R.-N.-densities be denoted by $x\to f(x,\gamma)$ and let $f(x,\gamma)\neq 0$ for each γ μ-a.e.*[23]*. For each $\gamma\in\Gamma$ let the mapping $x\to\partial f(x,\gamma)/\partial\gamma$ be defined μ-a.e., and let*

$$\frac{d}{d\gamma}\int\limits_R f(x,\gamma)\,d\mu(x)=\int\limits_R \frac{\partial}{\partial\gamma}f(x,\gamma)\,d\mu(x). \tag{1.51}$$

Furthermore, let $\int\limits_R \left(\frac{\partial \log f(x,\gamma)}{\partial\gamma}\right)^2 f(x,\gamma)\,d\mu(x)$ exist for each $\gamma\in\Gamma$ and let ψ be a function defined on Γ and everywhere differentiable there. Assume h is an unbiased estimate for ψ such that $E(h^2;\gamma)$ exists for all $\gamma\in\Gamma$. Then $\int\limits_R h\frac{\partial f}{\partial\gamma}\,d\mu$ exists for all $\gamma\in\Gamma$ and we assume that

$$\frac{d}{d\gamma}\int\limits_R h(x)f(x,\gamma)\,d\mu(x)=\int\limits_R h\frac{\partial f}{\partial\gamma}\,d\mu \tag{1.52}$$

for all $\gamma\in\Gamma$. Let ϕ be an arbitrary map from Γ into R_1. Then, for each $\gamma\in\Gamma$

$$(\psi'(\gamma))^2\leqslant E[(h-\phi(\gamma))^2;\gamma]\,E\left[\left(\frac{\partial\log f}{\partial\gamma}\right)^2;\gamma\right]. \tag{1.53}$$

---

[22] See M. Fréchet, Rev. Inst. Internat. Statist. 11, 182—205 (1943), C. R. Rao, Bull. Calcutta Math. Soc. 37, 81—91 (1945) and H. Cramér, Skand. Aktuarietidskr. 29, 85—94 (1946). Of many generalizations we mention E. W. Barankin, loc. cit.[8], C. R. Seth, Ann. Math. Statist. 20, 1—27 (1949), D. G. Chapman and H. Robbins, Ann. Math. Statist. 22, 581—586 (1951), J. Kiefer, Ann. Math. Statist. 23, 627—629 (1952), D. A. S. Fraser and L. Guttman, Ann. Math. Statist. 23, 629—632 (1952). Also see L. N. Bol'šev, Teor. Verojatnost. i Primenen 6, 319—326 (1961).

[23] One can replace this assumption by the requirement that $f(x,\gamma)=0$ for each γ in a set independent of γ.

Proof. By assumption $\partial \log f(x,\gamma)/\partial\gamma$ exists μ-a.e. for each $\gamma\in\Gamma$. Thus, $\int_R h\left|\frac{\partial f}{\partial\gamma}\right| d\mu = \int_R h\sqrt{f}\left|\frac{\partial \log f}{\partial\gamma}\right|\sqrt{f}\,d\mu$ and by Schwarz' inequality

$$\int_R \left|h\frac{\partial f}{\partial\gamma}\right| d\mu \leqslant [E(h^2;\gamma)]^{1/2}\left[E\left(\left(\frac{\partial \log f}{\partial\gamma}\right)^2;\gamma\right)\right]^{1/2},$$

whence the existence of $\int_R h\frac{\partial f}{\partial\gamma}\,d\mu$ for each $\gamma\in\Gamma$ follows. Since $\int_R f(x,\gamma)\,d\mu(x)=1$ for $\gamma\in\Gamma$, we have from (1.51)

$$\int_R \frac{\partial}{\partial\gamma} f(x,\gamma)\,d\mu(x)=0.$$

Hence, from $\int_R h(x)f(x,\gamma)\,d\mu(x)=\psi(\gamma)$ and (1.52) we obtain

$$\psi'(\gamma)=\int_R (h-\phi(\gamma))\frac{\partial f}{\partial\gamma}\,d\mu.$$

Because $\frac{\partial f}{\partial\gamma}=\frac{\partial \log f}{\partial\gamma}\cdot f$, the result follows from an application of Schwarz' inequality.

Corollary (Cramér-Fréchet-Rao inequality). *Let the hypotheses in Theorem* 1.10 *hold. Then, for arbitrary* $\gamma_0\in\Gamma$ *either* a) $\psi'(\gamma_0)=0$ *and equality holds in* (1.53) *for* $\gamma=\gamma_0$ *or* b) *we have*

$$E[(h-\phi(\gamma_0)^2);\gamma_0]\geqslant(\psi'(\gamma_0))^2\Big/\left[E\left(\left(\frac{\partial \log f}{\partial\gamma}\right)^2;\gamma_0\right)\right], \tag{1.54}$$

where

$$E\left(\left(\frac{\partial \log f}{\partial\gamma}\right)^2;\gamma_0\right)=\int_R\left(\frac{\partial \log f(x,\gamma_0)}{\partial\gamma}\right)^2 f(x,\gamma_0)\,d\mu(x).$$

If in case b) *equality holds in* (1.54), *then there is a real number* $\lambda(\gamma_0)\neq 0$ *with*

$$h(x)-\phi(\gamma_0)=\lambda(\gamma_0)\frac{\partial \log f(x,\gamma_0)}{\partial\gamma} \tag{1.55}$$

μ*-a.e. provided h is not constant* μ*-a.e.*

For the *proof* it suffices to consider the case where either $\psi'(\gamma_0)\neq 0$ or for $\gamma=\gamma_0$ in (1.53) equality does not hold. In both cases it follows from (1.53) for $\gamma=\gamma_0$ that $E\left(\left(\frac{\partial \log f}{\partial \gamma}\right)^2; \gamma_0\right)>0$. This proves (1.54). If equality in (1.54) holds in case b) (and h is not constant μ-a.e.), then necessarily $\psi'(\gamma_0)\neq 0$. Then there exists a real number $\lambda(\gamma_0)$ for which

$$(h(x)-\phi(\gamma_0))\sqrt{f(x,\gamma_0)}=\lambda(\gamma_0)\frac{\partial \log f(x,\gamma_0)}{\partial \gamma}\sqrt{f(x,\gamma_0)} \quad \mu\text{-a.e.}$$

Thus (1.55) holds and, since h is not constant μ-a.e., $\lambda(\gamma_0)\neq 0$.

If the hypotheses of Theorem 1.10 are not satisfied, then (1.54) need not hold. (See VII, p. 420.)

Inequality (1.54) leads to the following

Definition. If h is an *unbiased estimate* for ψ and *if equality in* (1.54) *holds for a* $\gamma_0\in\Gamma$, then h is called *locally efficient* at γ_0. If *equality holds in* (1.54) *for all* $\gamma\in\Gamma$, then h is said to be *efficient* in Γ.

The arbitrary function ϕ enters into this definition. This dependence on ϕ is, however, only apperant, at least under the assumptions of Theorem 1.10 and if

$$\psi'(\gamma)\neq 0, \qquad \gamma\in\Gamma. \tag{1.56}$$

Indeed, if equality holds in (1.54) for some γ, and thereby (1.55) also holds for $\gamma_0=\gamma$, then $E(h-\phi(\gamma);\gamma)=0$, i.e., $\phi(\gamma)=\psi(\gamma)$[24]. Hence, instead of (1.55) we have

$$h(x)-\psi(\gamma)=\lambda(\gamma)\frac{\partial \log f(x,\gamma)}{\partial \gamma} \quad \mu\text{-a.e.} \tag{1.57}$$

with $\lambda(\gamma)\neq 0$.

From (1.57)

$$E((h-\psi(\gamma))^2;\gamma)=(\lambda(\gamma))^2 E\left(\left(\frac{\partial \log f}{\partial \gamma}\right)^2;\gamma\right). \tag{1.58}$$

On the other hand,

$$E((h-\psi(\gamma))^2;\gamma)=(\psi'(\gamma))^2/E\left(\left(\frac{\partial \log f}{\partial \gamma}\right)^2;\gamma\right)$$

and with (1.58) this yields

$$(\lambda(\gamma))^2=(\psi'(\gamma))^2\bigg/\left[E\left(\left(\frac{\partial \log f}{\partial \gamma}\right)^2;\gamma\right)\right]^2. \tag{1.59}$$

[24] This result is of course related to I, Theorem 17.1.

Thus, if h is efficient in Γ, we have under the assumption (1.56)

$$\frac{1}{\lambda(\gamma)}(h(x)-\psi(\gamma)) = \frac{\partial \log f(x,\gamma)}{\partial \gamma} \tag{1.60}$$

μ-a.e. for all $\gamma \in \Gamma$. Here, $\lambda(\gamma)$ is given by (1.59) and never vanishes.

These ideas are also of interest for the asymptotic theory and then lead to the concept of *asymptotic efficiency* (see **4**).

Let us now assume for simplicity that Γ is all of R_1, that (1.60) holds for all $x \in R$ and all $\gamma \in R_1$ and that, for all $\gamma \in R_1$

$$\Lambda(\gamma) = \int_{-\infty}^{\gamma} \frac{dt}{\lambda(t)} \quad \text{and} \quad \Phi(\gamma) = \int_{-\infty}^{\gamma} \frac{\psi(t)}{\lambda(t)}\, dt$$

are defined, and that for some function k defined over R, $\lim_{\gamma \to -\infty} \log f(x,\gamma) = k(x)$ for all $x \in R$. Then

$$f(x,\gamma) = e^{h(x)\Lambda(\gamma)-\Phi(\gamma)}\, e^{k(x)}$$

for all $x \in R$ and all $\gamma \in R_1$. We thus have a distribution of exponential type. According to III, Theorem 7.2, h is a sufficient transformation for the set P_Γ. This can be viewed as a sort of converse to Theorems 1.6 and 1.7. There is obviously a close connection between the notion of an efficient estimate in Γ and that of a uniformly minimal estimate. If H is a class of unbiased estimates for ψ and if, for all $h \in H$, the assumptions of Theorem 1.10 are fulfilled along with the remaining hypotheses of that theorem, then an estimate efficient in Γ is also uniformly minimal.

However, a connection between sufficient transformations and uniformly minimal estimates—representing a converse to Theorems 1.6 and 1.7—exists under much more general assumptions. We present here one aspect of this connection and prove

Theorem 1.11[25]. *Let P_Γ have the usual meaning and let C be the set of all indicator functions c_A with $A \in \mathscr{S}$ which are uniformly minimal estimates. The set $\mathscr{S}_0$ of all A for which $c_A \in C$ is a σ-algebra. Moreover, all $\mathscr{S}_0$-measurable functions h defined over R for which $E(h^2;\gamma)$ exists for $\gamma \in \Gamma$ are uniformly minimal.*

Proof. The set C is non-empty. It contains at least c_R and $c_\emptyset$. Let $c_A, c_B \in C$. Then $c_A \cdot c_B = c_{A \cap B} \in C$ according to Theorem 1.4 and hence if $A, B \in \mathscr{S}_0$, then $A \cap B \in \mathscr{S}_0$. From Theorem 1.3, if $c_A \in C$, then $c_R - c_A = c_{R-A} \in C$ i.e., $A \in \mathscr{S}_0$ implies $R - A \in \mathscr{S}_0$. We need now only show that

$$A_i \in \mathscr{S}_0, \quad A_i \cap A_j = \emptyset, \quad i \neq j, \quad i,j = 1,2,\ldots \tag{1.61}$$

[25] R.R. Bahadur, loc. cit.[9].

implies that $\bigcup_{i=1}^{\infty} A_i \in \mathscr{S}_0$. This follows from Theorem 1.3 and Theorem 1.5. Indeed, $c_{A_i} \in C$ implies that $\sum_{i=1}^{n} c_{A_i} \in C$ for $n=1,2,\ldots$. Moreover, for each $x \in R$ $\sum_{i=1}^{n} c_{A_i}(x) \to \sum_{i=1}^{\infty} c_{A_i}(x) = c_{\bigcup_{i=1}^{\infty} A_i}(x)$ because of (1.61). Thus from $\left|\sum_{i=1}^{n} c_{A_i}\right| \leq 1$, $n=1,2,\ldots$ we have $\int_R \left|\sum_{i=1}^{n} c_{A_i} - c_{\bigcup_{j=1}^{\infty} A_j}\right|^2 dP_\gamma \to 0$ for each $\gamma \in \Gamma$ (see Theorem VI). Hence $c_{\bigcup_{i=1}^{\infty} A_i} \in C$ which was to be proved.

Now let h be an arbitrary $\mathscr{S}_0$-measurable function over R for which $E(h^2;\gamma)$ exists for all $\gamma \in \Gamma$. Then, there exists a sequence $\sum_{i=1}^{k_n} \alpha_i c_{A_i}$ with $A_i \in \mathscr{S}_0$, α_i real, $k_n \to \infty$ such that

$$\sum_{i=1}^{k_n} \alpha_i c_{A_i}(x) \to h(x) \tag{1.62}$$

for all $x \in R$ and for which

$$\left|h - \sum_{i=1}^{k_n} c_{A_i}\right|^2 \leq 4h^2. \tag{1.63}$$

Such a sequence can be constructed as follows: Let, for $n=1,2,\ldots$

$$A_{-n2^n-1} = \{x : h(x) < -n\}, \qquad A_{n2^n} = \{x : h(x) \geq n\}$$

and

$$A_k = \{x : k/2^n \leq h(x) < (k+1)/2^n\}, \qquad k = -n2^n, \ldots, n2^n - 1.$$

By the definition of $\mathscr{S}_0$-measurability, all the $A_i \in \mathscr{S}_0$ and, moreover, $A_i \cap A_j = \emptyset$, $i \neq j$, $i,j = -n2^n - 1, \ldots, n2^n$. Now consider the sequence $d_n = \sum_{k=0}^{n2^n} \frac{k}{2^n} c_{A_k} + \sum_{k=-n2^n-1}^{-1} \frac{k+1}{2^n} c_{A_k}$. Then, $\lim_{n\to\infty} d_n(x) = h(x)$ for each $x \in R$, for if n is sufficiently large, x belongs to exactly one A_k with $-n2^n - 1 \leq k \leq n2^n$, and $|h(x) - d_n(x)| < 1/2^n$ there. In addition, $|h| \geq |d_n|$ follows from the construction of d_n, so that $|h - d_n| \leq 2|h|$, or $|h - d_n|^2 \leq 4h^2$. Then d_n satisfies (1.62) and (1.63). Since $E(h^2;\gamma)$ exists for all $\gamma \in \Gamma$

$$\int_R |h - d_n|^2 dP_\gamma \to 0 \tag{1.64}$$

for each $\gamma \in \Gamma$. (See Theorem VI.) But d_n is uniformly minimal. Now apply Theorem 1.5 which shows that h is uniformly minimal.

Remark. Let v be an $\mathscr{S}_0$-measurable unbiased estimate of zero for which $E(v^2;\gamma)$ exists for each $\gamma\in\Gamma$. Then v is uniformly minimal by Theorem 1.11 and is thus equal to zero P_Γ-a.e. Thus, $\mathscr{S}_0$ is 2-complete. It is, moreover, easy to show that $\mathscr{S}_0$ is also minimal. It is more difficult to prove that under some more conditions $\mathscr{S}_0$ is also sufficient[26].

We consider some more examples for Theorem 1.10. Let $\xi_1,\ldots,\xi_n$, $n\geqslant 1$ be independent, identically distributed r.v.'s with density given by $x\to f(x,\gamma)$. The variable x is assumed to vary in R_1 and γ will be assumed to belong to an open interval denoted by Γ. Let $f(x,\gamma)\neq 0$ for almost all $x\in R_1$ and each $\gamma\in\Gamma$. Furthermore, let $\frac{\partial}{\partial\gamma}f(x,\gamma)$ exist for almost all x and each γ and assume that

$$\frac{d}{d\gamma}\int_{R_1} f(x,\gamma)\,dx=\int_{R_1}\frac{\partial}{\partial\gamma}f(x,\gamma)\,dx$$

and that $E\left(\left(\frac{\partial\log f}{\partial\gamma}\right)^2;\gamma\right)$ exists and is non-zero for each $\gamma\in\Gamma$. The joint density of $\xi_1,\ldots,\xi_n$ is then given by $(x_1,\ldots,x_n)\to\prod_{i=1}^n f(x_i,\gamma)$. Let ψ be a function defined and differentiable on Γ and h an unbiased estimate for ψ defined over R_n. Let $\int_{R_n} h^2(x_1,\ldots,x_n)\prod_{i=1}^n f(x_i,\gamma)\,dx=E(h^2;\gamma)$ exist for each $\gamma\in\Gamma$ and assume that for each γ

$$\frac{d}{d\gamma}\int_{R_n} h(x_1,\ldots,x_n)\prod_{i=1}^n f(x_i;\gamma)\,dx=\int_{R_n} h(x_1,\ldots,x_n)\frac{\partial}{\partial\gamma}\prod_{i=1}^n f(x_i;\gamma)\,dx.$$

Under these assumptions

$$E[(h-\psi(\gamma))^2;\gamma]\geqslant(\psi'(\gamma))^2\Big/\left(nE\left[\left(\frac{\partial\log f}{\partial\gamma}\right)^2;\gamma\right]\right). \tag{1.65}$$

All we need to shows is that

$$E\left[\left(\frac{\partial\log\prod_{i=1}^n f(\xi_i,\gamma)}{\partial\gamma}\right)^2;\gamma\right]=nE\left[\left(\frac{\partial\log f}{\partial\gamma}\right)^2;\gamma\right].$$

This follows from I, Theorem 17.4 and the fact that $E\left(\frac{\partial\log f(\xi_i;\gamma)}{\partial\gamma}\right)=0$, $1\leqslant i\leqslant n$.

We specialize further:

[26] R.R. Bahadur loc. cit.[9] and L. Schmetterer loc. cit.[5].

Let $\xi_1, \ldots, \xi_n$, $n \geqslant 2$ be independent r.v.'s, distributed according to $N(0, \sigma^2)$, $0 < \sigma^2 < \infty$. One easily sees that the r.v.

$$s' = \phi(n)\left(\sum_{i=1}^{n} \xi_i^2\right)^{1/2} \quad \text{with } \phi(n) = 2^{-1/2}\Gamma\left(\frac{n}{2}\right)[\Gamma(n+1)/2]^{-1}$$

is unbiased for σ (Cf. the calculations on p. 270). Moreover, we have for the joint density of $\xi_1, \ldots, \xi_n$

$$\left(\frac{1}{\sqrt{2\pi}\,\sigma}\right)^n e^{-\sum_{i=1}^{n} x_i^2/(2\sigma^2)} = \left(\frac{1}{\sqrt{2\pi}\,\sigma}\right)^n e^{-\phi^2(n)\sum_{i=1}^{n} x_i^2/(2\sigma^2\phi^2(n))} \tag{1.66}$$

for all $(x_1, \ldots, x_n) \in R_n$. Hence, the transformation $(x_1, \ldots, x_n) \to \phi(n)\left(\sum_{i=1}^{n} x_i^2\right)^{1/2}$ is sufficient for the set of normal distributions with densities given by (1.66) with $0 < \sigma < \infty$. It is easy to show that $E[(s' - \sigma)^2; \sigma] = \sigma^2[\phi^2(n)n - 1]$ and that with $f(x, \sigma) = (\sqrt{2\pi}\,\sigma)^{-1} \cdot e^{-x^2/2\sigma^2)}$, $x \in R_1$

$$\int_{-\infty}^{\infty} \left(\frac{\partial \log f(x, \sigma)}{\partial \sigma}\right)^2 f(x, \sigma)\, dx = 2/\sigma^2 .$$

However, an efficient estimation function would, for $0 < \sigma < \infty$, have to possess minimal variance $\sigma^2/2n$. But $\phi^2(n)n - 1 > 1/2n$ for $n \geqslant 2$.

Finally, we mention as a further example:

If $\xi_1, \ldots, \xi_n$ are independent, $N(a, 1)$, $-\infty < a < \infty$ distributed r.v.'s, then it is easy to check that $(\xi_1 + \cdots + \xi_n)/n$ is efficient for a in $-\infty < a < \infty$.

We turn now to the unbiased estimation of n-dimensional parameters. Let P_Γ again be a set of probability measures over $(R, \mathscr{S})$ and $d = (d_1, \ldots, d_n)$ a map from Γ into R_n, $n \geqslant 2$. We have already indicated (p. 269) that one can reduce the unbiased estimation of multi-dimensional maps to the one-dimensional case. One also uses this to define the notions of locally and uniformly minimal estimates. For example: Let H be the class of those unbiased estimates $h = (h_1, \ldots, h_n)$ for d for which $E(h_i^2; \gamma)$ exists for $i = 1, \ldots, n$ and each $\gamma \in \Gamma$. h is called a uniformly minimal estimate for d if $\sum_{i=1}^{n} u_i h_i$ is a uniformly minimal estimate for $\sum_{i=1}^{n} u_i d_i$ for all real n-tuples $u = (u_1, \ldots, u_n)$.

Denoting the covariance matrix of $h \in H$ (w.r.t. the probability measure P_γ) by $(\sigma_{ij}^{(h)}(\gamma))_{1n}^{1n}$, we have that h_0 is uniformly minimal iff, for all n-tuples of real numbers u, each $\gamma \in \Gamma$ and all $h \in H$

$$\sum_{i,j=1}^{n} u_i u_j \sigma_{ij}^{(h_0)}(\gamma) \leqslant \sum_{i,j=1}^{n} u_i u_j \sigma_{ij}^{(h)}(\gamma). \tag{1.67}$$

Using the notation just introduced we prove

Theorem 1.12. *Let h_0 be a uniformly minimal estimate for d and h an arbitrary unbiased estimate $\in H$ for d. Let $|\sigma_{ij}^{(h_0)}(\gamma)|_{1n}^{1n} \neq 0$ for each $\gamma \in \Gamma$. Then, for each $\gamma \in \Gamma$ the concentration ellipsoid corresponding to h_0 lies completely in that corresponding to h.*

Proof. By definition, (1.67) holds for all $\gamma \in \Gamma$. But since $|\sigma_{ij}^{(h_0)}|_{1n}^{1n} \neq 0$ the quadratic form in u_i on the left side of (1.67) is positive definite and hence also the quadratic form on the right. The inverse matrix of $(\sigma_{ij}^{(h_0)}(\gamma))_{1n}^{1n}$ will be written as $(\sigma_{(h_0)}^{ij}(\gamma))_{1n}^{1n}$. An analogous meaning is given to $(\sigma_{(h_0)}^{ij}(\gamma))_{1n}^{1n}$. It is sufficient to show that for each n-tuple $(u_1, \ldots, u_n)$

$$\sum_{i,j=1}^{n} u_i u_j \sigma_{(h_0)}^{ij}(\gamma) \geqslant \sum_{i,j=1}^{n} u_i u_j \sigma_{(h)}^{ij}(\gamma) \tag{1.68}$$

follows from (1.67). Let u and v be n-tuples of real numbers and λ, μ real numbers. Write $(\sigma_{ij}^{(h_0)})_{1n}^{1n} = S_{h_0}$ and $(\sigma_{ij}^{(h)})_{1n}^{1n} = S_h$, where we have suppressed the dependence on γ. Since S_{h_0} is positive definite and symmetric,

$$(\mu S_{h_0}^{-1} u + \lambda v)' S_{h_0} (\mu S_{h_0}^{-1} u + \lambda v) \geqslant 0.$$

The quadratic form in μ and λ, $\mu^2 u' S_{h_0}^{-1} u + 2\mu\lambda u'v + \lambda^2 v' S_{h_0} v$ is thus always nonnegative. Hence, for all n-tuples u, v

$$(u'v)^2 \leqslant (u' S_{h_0}^{-1} u)(v' S_{h_0} v). \tag{1.69}$$

Hence,

$$\sup_{v \in R_n} \frac{(u'v)^2}{v' S_{h_0} v} \leqslant u' S_{h_0}^{-1} u. \tag{1.70}$$

But $\dfrac{(u'v)^2}{v' S_{h_0} v} = u' S_{h_0}^{-1} u$ for $v = S_{h_0}^{-1} u$ and so equality holds in (1.70). This argument naturally holds for each positive definite symmetric matrix and hence also for S_h. Now, from (1.67) $\dfrac{(u'v)^2}{v' S_h v} \leqslant \dfrac{(u'v)^2}{v' S_{h_0} v}$ and thus, also $\sup\limits_{v \in R_n} \dfrac{(u'v)^2}{v' S_h v} \leqslant \sup\limits_{v \in R_n} \dfrac{(u'v)^2}{v' S_{h_0} v}$. From (1.70) this is the same as (1.68).

The generalization of the Cramer-Fréchet-Rao inequality is formulated in an analogous way. Taking $\phi(\gamma) = \psi(\gamma) = \gamma$ for the sake of simplicity in Theorem 1.10, we can also interpret (1.54) as follows: Let C be an arbitrary positive number. The closed interval defined by $\{u : u^2 [E((h-\gamma)^2; \gamma)]^{-1} \leqslant C\}$ contains the closed interval

$$\left\{u : u^2 E\left[\left(\frac{\partial \log f}{\partial \gamma}\right)^2; \gamma\right] \leqslant C\right\}.$$

Now Theorem 1.10 and its corollary can be expressed for n-dimensional parameters.

Theorem 1.13. *Let P_Γ have the usual meaning but assume that Γ is an open set in $R_n, n \geqslant 2$. Let μ be a σ-finite measure which dominates P_Γ. Denote again by $x \to f(x,\gamma)$ the R.-N.-densities for each $\gamma=(\gamma_1,\ldots,\gamma_n)$, assume that $f(x,\gamma) \neq 0$ for each γ μ-a.e. and that the maps $x \to \frac{\partial f(x,\gamma)}{\partial \gamma_i}$ exist μ-a.e. for each γ and $i=1,\ldots,n$. For each γ and i let*

$$\frac{\partial}{\partial \gamma_i} \int_R f(x,\gamma)\, d\mu(x) = \int_R \frac{\partial}{\partial \gamma_i} f(x,\gamma)\, d\mu(x). \tag{1.71}$$

Moreover, assume $v_{ij}(\gamma)=E\left[\left(\frac{\partial \log f}{\partial \gamma_i}\,\frac{\partial \log f}{\partial \gamma_j}\right); \gamma\right]$ exists for each γ and $i,j=1,\ldots,n$, and let $h=(h_1,\ldots,h_n)$ be an unbiased estimate for γ and for each γ and $i,j=1,\ldots,n$, let

$$\frac{\partial}{\partial \gamma_i} \int_R h_j(x) f(x,\gamma)\, d\mu(x) = \int_R h_j(x) \frac{\partial}{\partial \gamma_i} f(x,\gamma)\, d\mu(x). \tag{1.72}$$

Assume $\sigma_{ij}(\gamma)=E[(h_i-\gamma_i)(h_j-\gamma_j); \gamma]$ exists for each γ and denote the matrix $(\sigma_{ij}(\gamma))_{1n}^{1n}$ by $K(\gamma)$. We take $K(\gamma)$ as positive definite for each $\gamma \in \Gamma$. Denoting the matrix $(v_{ij}(\gamma))_{1n}^{1n}$ by $V(\gamma)$ we then have: For each positive number c the ellipsoid $u' K^{-1}(\gamma) u \leqslant c$ completely contains the ellipsoid $u' V(\gamma) u \leqslant c$, where u is an n-dimensional column vector.

Proof. Consider for fixed γ the $2n$ r.v.'s $h_1,\ldots,h_n, \frac{\partial \log f}{\partial \gamma_1}, \ldots, \frac{\partial \log f}{\partial \gamma_n}$. It follows easily from the hypotheses that the covariance matrix $L(\gamma)$ of these $2n$ r.v.'s exists. We have

$$E(h_j; \gamma) = \gamma_j, \quad j=1,\ldots,n. \tag{1.73}$$

Differentiating (1.73) w.r.t. γ_i for $i=1,\ldots,n$, we obtain by means of (1.71) and (1.72)

$$E\left[(h_j-\gamma_j)\frac{\partial \log f}{\partial \gamma_i}; \gamma\right] = \begin{cases} 0 & j \neq i \\ 1 & j=i \end{cases}. \tag{1.74}$$

We now omit in our notation the dependence of these expressions on γ. Since L is positive semi-definite, $z' L z \geqslant 0$ for each $2n$-tuple of real numbers z. Let I_n denote the n-dimensional identity matrix $\begin{pmatrix} 1 & & 0 \\ & \ddots & \\ 0 & & 1 \end{pmatrix}$ and 0 the $n \times n$ matrix all of whose elements are zero. Set $M = \begin{pmatrix} K^{-1} & 0 \\ 0 & I_n \end{pmatrix}$. If we perform the transformation $z=Mw$, then, because of $|M| \neq 0$, w runs through all real $2n$-tuples if z does and vice versa. Thus, we also have $w'M'LMw \geqslant 0$ for all $w \in R_{2n}$. Now $M' = M$ and it is easy to verify that $M'LM = \begin{pmatrix} K^{-1} & K^{-1} \\ K^{-1} & V \end{pmatrix}$ and this matrix is positive semi-definite. Choosing $w_i = u_i$, $1 \leqslant i \leqslant n$, $w_{i+n} = -u_i$, $1 \leqslant i \leqslant n$, we obtain

$$w'\begin{pmatrix} K^{-1} & K^{-1} \\ K^{-1} & V \end{pmatrix} w = u'(V - K^{-1})u \geqslant 0.$$

But then $u'Vu \geqslant u'K^{-1}u$ for each $u \in R_n$, which was to be proved.

We remark that a particular consequence of this theorem is the fact that the concentration ellipsoid of h is contained in $u'Vu \leqslant n+2$ for each γ.

2. Consistent sequences of estimates. Up to this point we have investigated the problem of estimation for the case of a fixed sample space. In practice, however, the idea that one can estimate the parameter "better and better" the larger one takes the sample size plays an important role. This concept, as in the theory of testing, leads to the study of sequences of estimates. We present first the important

Definition. Let $(R^{(n)}, \mathscr{S}^{(n)})$ for $n = 1, 2, \ldots$, be a measurable space. Let a set of probability measures $P_\Gamma^{(n)}$ be defined on each $(R^{(n)}, \mathscr{S}^{(n)})$, whereby Γ is assumed to be independent of n. Let h_n be a r.v. over $(R^{(n)}, \mathscr{S}^{(n)})$ for $n = 1, 2, \ldots$, i.e., an $\mathscr{S}^{(n)}$-measurable function. $\{h_n\}$ is called a *consistent sequence of estimates for* d[27] or, abbreviating, *consistent for* d, if for each $\gamma \in \Gamma$, each $\varepsilon > 0$ and each $\delta > 0$

$$P_\gamma^{(n)}(|h_n - d(\gamma)| < \varepsilon) > 1 - \delta \tag{2.1}$$

for $n \geqslant N(\varepsilon, \delta, \gamma)$.

This definition says that for each γ the sequence of r.v.'s $\{h_n\}$ converges stochastically to $d(\gamma)$.

$\{h_n\}$ is called *uniformly consistent* for d if $N(\varepsilon, \delta, \gamma)$ depends only on ε and δ and not on $\gamma \in \Gamma$.

In general, we will deal with the most important special case of this definition in which we take $R^{(n)} = R_n$, $n \geqslant 1$ (where R_n is as usual the n-dimensional Euclidean space) and independent r.v.'s as the sample variables. More precisely:

Let $\xi_1, \xi_2, \ldots$ be a sequence of independent r.v.'s of the same dimension (which we take as one for simplicity) possessing the same distribution $P_{\gamma, \xi_i} = P_\gamma$, $\gamma \in \Gamma$[28]. For every $n \geqslant 1$ let h_n be a function on R_n.

---

[27] If $\Gamma \subseteq R_1$ and $d(\gamma) = \gamma$ for all $\gamma \in \Gamma$, then we also say that h_n is consistent for $\gamma \in \Gamma$.

[28] See also III, p. 240ff.

Then $\{h_n\}$ is called a consistent sequence of estimates for d if for each $\gamma\in\Gamma$, $\varepsilon>0$ and $\delta>0$

$$P[|h_n(\xi_1,\ldots,\xi_n)-d(\gamma)|<\varepsilon;\gamma]>1-\delta \tag{2.2}$$

for $n\geqslant N(\varepsilon,\delta,\gamma)$.

We will see in the sequel that we will rarely make use of the independence of the ξ_i.

Let us consider some examples: From I, Theorem 38.2 we immediately have: Let $\xi_1,\xi_2,\ldots$ be a sequence of independent r.v.'s with the same distribution for which $E(\xi_1)=a$, $-\infty<a<\infty$ exists. Then $\bar{\xi}_n=(\xi_1+\cdots+\xi_n)/n$, $n\geqslant 1$ is a consistent sequence of estimates for a.

If also $\sigma^2=E[(\xi_1-a)^2;a]$ exists and $0<\sigma^2<c_0$ holds, then one can immediately transform such consistency statements by means of Čebyšev's inequality into statements about uniform consistency. Indeed, we then have $P(|\bar{\xi}_n-a|>\varepsilon;a)\leqslant\delta$ for $n\geqslant\frac{c_0}{\varepsilon^2\delta}$ and this no longer depends on a.

One can always obtain a consistent sequence of estimates according to this model provided one has solved the problem of unbiased estimates. If one has a consistent sequence $\{h_n\}$ of estimates for d, then it is immediately possible to give infinitely many other consistent sequences of estimates for d. One need only choose an arbitrary null-sequence of real numbers a_n. Then the sequence $\{h_n+a_n\}$ is also consistent for d. This follows immediately from (2.1).

The definition of a consistent sequence of estimates can be extended without difficulty to mappings d of Γ into R_k, $k\geqslant 2$. It is merely necessary to interpret the absolute value in (2.2) as the Euclidean distance in R_k, where $\{h_n\}$ is naturally a sequence of k-dimensional r.v.'s.

The notion of a consistent sequence of tests is only a special case of a consistent sequence of estimates. This is shown in

Theorem 2.1. *$\phi_1,\phi_2,\ldots$ is a consistent sequence of tests for the problem $(\Gamma_0,\Gamma-\Gamma_0)$ iff $\phi_1,\phi_2,\ldots$ is a consistent sequence of estimates for d with $d(\gamma)=0$ for $\gamma\in\Gamma_0$ and $d(\gamma)=1$ for $\gamma\in\Gamma-\Gamma_0$.*

Proof. Let $P_\gamma^{(n)}$ denote the distribution of $(\xi_1,\ldots,\xi_n)$ for each $\gamma\in\Gamma$ and $n=1,2,\ldots$. If $\{\phi_n\}$ is a consistent sequence of tests, then

$$E[\phi_n(\xi_1,\ldots,\xi_n);\gamma]\to 0,\qquad n\to\infty \tag{2.3}$$

for $\gamma\in\Gamma_0$ and

$$E[\phi_n(\xi_1,\ldots,\xi_n);\gamma]\to 1,\qquad n\to\infty \tag{2.4}$$

for $\gamma\in\Gamma-\Gamma_0$.

Let $M_n=\{(x_1,\ldots,x_n)\in R_n:\phi_n(x_1,\ldots,x_n)\geqslant\varepsilon\}$ for given $\varepsilon>0$. For given ε one can choose $N(\varepsilon)$[29] such that $\int\limits_{R_n}\phi_n\,dP_\gamma^{(n)}<\varepsilon^2$, $n\geqslant N(\varepsilon)$, $\gamma\in\Gamma_0$. Then

[29] We do not exclude the possibility that $N(\varepsilon)$ depends on γ.

$\int\limits_{M_n} \phi_n dP_\gamma^{(n)} < \varepsilon^2$ and thus also $\varepsilon P_\gamma^{(n)}(M_n) < \varepsilon^2$, i.e., $P_\gamma^{(n)}(\phi_n(\xi_1, \ldots, \xi_n) < \varepsilon) > 1 - \varepsilon$ so that $\{\phi_n\}$ is consistent for d over Γ_0. In order to show that $\{\phi_n\}$ is also consistent for d over $\Gamma - \Gamma_0$, we need only take into consideration the fact that (2.4) is equivalent to $E(1 - \phi_n(\xi_1, \ldots, \xi_n); \gamma) = 0$. One then proceeds as above.

On the other hand, if $\{\phi_n\}$ is a consistent sequence of estimates for d, then for each $\gamma \in \Gamma_0$, $\varepsilon > 0$ and $\delta > 0$

$$P_\gamma^{(n)}(\phi_n(\xi_1, \ldots, \xi_n) < \varepsilon) > 1 - \delta \tag{2.5}$$

for $n \geqslant N(\varepsilon, \delta, \gamma)$. Since $0 \leqslant \phi_n \leqslant 1$, we then have because of (2.5)

$$\int\limits_{R_n} \phi_n dP_\gamma^{(n)} = \int\limits_{M_n} \phi_n dP_\gamma^{(n)} + \int\limits_{R_n - M_n} \phi_n dP_\gamma^{(n)} \leqslant \delta \cdot 1 + \varepsilon \cdot 1$$

for $n \geqslant N(\varepsilon, \delta, \gamma)$. (2.3) then follows. (2.4) goes through analogously for $\gamma \in \Gamma - \Gamma_0$.

3. The Maximum Likelihood Principle. The form of mappings d for which there exist consistent sequences of estimates can be determined under very general assumptions[30]. However, for practical applications, the problem of construction of consistent sequences of estimates is more important than an existence theorem. We will now present such a construction principle which has retained its importance although recently, its limitations have been recognized and other useful construction schemes have been discovered. We are referring to the so-called *Maximum Likelihood Principle* (MLP), whose beginnings go back to Gauss. Its modern formulation was given by Fisher[31]. (The notion of a consistent sequence of estimates is likewise due essentially to Fisher.) Roughly speaking, the MLP is as follows: Let Γ be a subset of R_k and $\xi_1, \ldots, \xi_n$ be r.v.'s with the joint density $(x_1, \ldots, x_n) \to f(x_1, \ldots, x_n, \gamma)$, $\gamma \in \Gamma$. Let a mapping $(x_1, \ldots, x_n) \to \gamma^*(x_1, \ldots, x_n)$ from R_n into R_k have the property

$$\max_{\gamma \in \Gamma} f(x_1, \ldots, x_n, \gamma) = f(x_1, \ldots, x_n, \gamma^*(x_1, \ldots, x_n)).$$

In short, we maximize the likelihood function. The r.v. $\gamma^*(\xi_1, \ldots, \xi_n)$ is called the ML-estimate. Its great importance will soon become clear.

Let ξ be now a r.v. of arbitrary dimension $\geqslant 1$, which we take as 1 for simplicity. Let Γ be a k-dimensional, open cube, $k \geqslant 1$. Let the distribution P_γ of ξ be given by the density $x \to f(x, \gamma)$, $\gamma \in \Gamma$ w.r.t. a σ-finite

30 L. Le Cam and L. Schwartz, Ann. Math. Statist. 31, 140—150 (1960). See also J. L. Doob, Colloques internationaux Centre National de la Recherche Scientifique, no. 13, 23—27 Centre National de la Recherche Scientifique, Paris 1949.

31 R.A. Fisher, Messenger of Math. 41, 150—160 (1912).

measure μ. *For each* γ *and all* $x \in R_1$ *assume* $f(x,\gamma) \neq 0$[32]. *Moreover, assume that for each* x[33], *all first and second order derivatives of the likelihood function* $\gamma \to f(x,\gamma)$ *w.r.t.* γ_j *exist and are continuous. Also let*

$$\frac{\partial}{\partial \gamma_i} \int_{R_1} f(x,\gamma)\, d\mu(x) = \int_{R_1} \frac{\partial}{\partial \gamma_i} f(x,\gamma)\, d\mu(x) \tag{3.1}$$

and

$$\frac{\partial^2}{\partial \gamma_i \partial \gamma_j} \int_{R_1} f(x,\gamma)\, d\mu(x) = \int_{R_1} \frac{\partial^2}{\partial \gamma_i \partial \gamma_j} f(x,\gamma)\, d\mu(x) \tag{3.2}$$

for all γ *and* $i,j=1,\ldots,k$. *We also require the existence of* $E\left[\frac{\partial^2 \log f(\xi,\gamma)}{\partial \gamma_i \partial \gamma_j}; \gamma\right]$ *for* $\gamma \in \Gamma$, $i,j=1,\ldots,k$. *Let* $A(x,\gamma)$ *denote, for all* x *and* $\gamma \in \Gamma$, *the* $k \times k$ *matrix with elements* $\frac{\partial \log f(x,\gamma)}{\partial \gamma_i} \frac{\partial \log f(x,\gamma)}{\partial \gamma_j}$. *Then the matrix* $E(A(\xi,\gamma);\gamma)$ *is defined for each* γ. *Assume that*

$$|E(A(\xi,\gamma);\gamma)| \neq 0 \tag{3.3}$$

for $\gamma \in \Gamma$.

We now prove

Theorem 3.1. *Let* $\xi_1, \xi_2, \ldots$ *be a sequence of independent r.v.'s all distributed like* ξ *where* ξ *is a r.v. having the properties described above. The density of* $\xi^{(n)} = (\xi_1, \ldots, \xi_n)$, $n \geq 1$ *w.r.t.* $\mu \times \cdots \times \mu$ *(n times), given for each* $x^{(n)} = (x_1, \ldots, x_n)$ *and* $\gamma \in \Gamma$ *by* $\prod_{i=1}^{n} f(x_i,\gamma)$ *will be denoted by* $f^{(n)}(x^{(n)},\gamma)$. $B(x^{(n)},\gamma)$ *will stand for the* $k \times k$ *matrix with elements* $\frac{\partial^2}{\partial \gamma_i \partial \gamma_j} \log f^{(n)}(x^{(n)},\gamma)$, $i,j=1,\ldots,k$ *and we will write*

$$\begin{pmatrix} \dfrac{\partial \log f^{(n)}(x^{(n)},\gamma)}{\partial \gamma_1} \\ \vdots \\ \dfrac{\partial \log f^{(n)}(x^{(n)},\gamma)}{\partial \gamma_k} \end{pmatrix} = \frac{\partial \log f^{(n)}(x^{(n)},\gamma)}{\partial \gamma}$$

---

[32] For the following one can replace the assumption $f(x,\gamma) \neq 0$ for all $x \in R_1$ by $f(x,\gamma) \neq 0$ for all $x \in R_1$ up to a μ-null set without difficulty. Moreover, R_1 can also be replaced by an arbitrary Borel set M not depending on γ, i.e., it is sufficient to require $f(x,\gamma) \neq 0$ for each $\gamma \in \Gamma$ and all $x \in M$ and $f(x,\gamma) = 0$ for each $\gamma \in \Gamma$ and all $x \in R_1 - M$.

[33] One can again allow exceptional sets of μ-measure zero here.

and assume: For each $\eta>0$ *and* $\gamma\in\Gamma$ *there exists a natural number* $n(\eta,\gamma)=n(\eta)$ *for which there is a* $q(\gamma)$ *with* $0<q(\gamma)<1$ *and a closed sphere* $K(\gamma)$ *with center* γ *and radius* $\rho(\gamma)>0$ *such that for each* $\tilde{\gamma}$ *and* $\tilde{\tilde{\gamma}}\in K(\gamma)$ *and* $n\geqslant n(\eta)$

$$P_\gamma\left(\left|\frac{\partial\log f^{(n)}(\xi^{(n)},\tilde{\gamma})}{\partial\gamma}-\frac{\partial\log f^{(n)}(\xi^{(n)},\tilde{\tilde{\gamma}})}{\partial\gamma}-B(\xi^{(n)},\gamma)(\tilde{\gamma}-\tilde{\tilde{\gamma}})\right| \leqslant q(\gamma)|B(\xi^{(n)},\gamma)(\tilde{\gamma}-\tilde{\tilde{\gamma}})|\right)\geqslant 1-\eta\,. \tag{3.4}$$

Then we claim that for each $\gamma\in\Gamma$, $\varepsilon>0$ *and* $\delta>0$ *there exists a natural number* $N(\delta,\varepsilon)$ *with the following property: For each* $n\geqslant N(\delta,\varepsilon)$[34] *one can find a* $\mathfrak{B}_n$*-measurable mapping* $\hat{\gamma}_n$ *from* R_n *into* Γ *and a set* $F_n\in\mathfrak{B}_n$ *such that for all* $x^{(n)}\in F_n$

$$\frac{\partial\log f^{(n)}(x^{(n)},\hat{\gamma}_n(x^{(n)}))}{\partial\gamma}=0\,^{35}, \tag{3.5}$$

$$P(\xi^{(n)}\in F_n;\gamma)\geqslant 1-\delta$$

and

$$P(|\hat{\gamma}_n(\xi^{(n)})-\gamma|<\varepsilon;\gamma)\geqslant 1-\delta\,. \tag{3.6}$$

Before proceeding to the proof we need

Lemma 3.1. *From the existence of* $E\left(\frac{\partial^2\log f(\xi,\gamma)}{\partial\gamma_i\,\partial\gamma_j};\gamma\right)$ *and the validity of* (3.2) *follows the existence of* $E\left(\frac{\partial\log f(\xi,\gamma)}{\partial\gamma_i}\,\frac{\partial\log f(\xi,\gamma)}{\partial\gamma_j};\gamma\right)$. *Moreover,*

$$E\left(\frac{\partial^2\log f(\xi,\gamma)}{\partial\gamma_i\,\partial\gamma_j};\gamma\right)=-E\left(\frac{\partial\log f(\xi,\gamma)}{\partial\gamma_i}\,\frac{\partial\log f(\xi,\gamma)}{\partial\gamma_j};\gamma\right). \tag{3.7}$$

Proof. For all $x\in R_1$

$$f(x,\gamma)\frac{\partial^2}{\partial\gamma_i\,\partial\gamma_j}\log f(x,\gamma)-\frac{\partial^2}{\partial\gamma_i\,\partial\gamma_j}f(x,\gamma) =-\frac{1}{f^2(x,\gamma)}\,\frac{\partial f(x,\gamma)}{\partial\gamma_i}\,\frac{\partial f(x,\gamma)}{\partial\gamma_j}\,f(x,\gamma)\,.$$

---

[34] $N(\delta,\varepsilon)$ may also depend on γ but we suppress this. This will pertain to analogous statements in the following proof.

[35] "0" is here the k-dimensional zero vector.

The function on the left is μ-integrable and hence, so is the function on the right. But

$$\int_{R_1} \frac{1}{f^2(x,\gamma)} \frac{\partial f(x,\gamma)}{\partial \gamma_i} \frac{\partial f(x,\gamma)}{\partial \gamma_j} f(x,\gamma)\, d\mu(x) = E\left[\frac{\partial \log f(\xi,\gamma)}{\partial \gamma_i} \frac{\partial \log f(\xi,\gamma)}{\partial \gamma_j}; \gamma\right]$$

and since $\int_{R_1} \frac{\partial^2}{\partial \gamma_i \partial \gamma_j} f(x,\gamma)\, d\mu(x) = 0$ follows from (3.2), we also have (3.7).

Proof of Theorem 3.1[36]. Choose a $\gamma \in \Gamma$ and hold it fixed. All probabilities and expectations refer to this γ and we thus suppress it in the sequel. From the definition of $f^{(n)}$ we have for each $(x_1, \dots, x_n) \in R_n$

$$\frac{\partial^2 \log f^{(n)}(x^{(n)},\gamma)}{\partial \gamma_i \partial \gamma_j} = \sum_{l=1}^{n} \frac{\partial^2 \log f(x_l,\gamma)}{\partial \gamma_i \partial \gamma_j}.$$

Thus from I, Theorem 38.2 we have the following result: Let $\varepsilon_1 > 0$ and $\delta_1 > 0$ be given. Then there exists an $n(\varepsilon_1, \delta_1)$ such that

$$P\left(\left|\frac{1}{n} \frac{\partial^2 \log f^{(n)}(\xi^{(n)},\gamma)}{\partial \gamma_i \partial \gamma_j} - E\left(\frac{\partial^2 \log f(\xi,\gamma)}{\partial \gamma_i \partial \gamma_j}\right)\right| < \varepsilon_1\right) > 1 - \delta_1 \tag{3.8}$$

for $n \geqslant n(\varepsilon_1, \delta_1)$ and $i,j = 1, \dots, k$. Let

$$E_{ij}^{(n)} = \left\{x^{(n)} : \left|\frac{1}{n} \frac{\partial^2 \log f^{(n)}(x^{(n)},\gamma)}{\partial \gamma_i \partial \gamma_j} - E\left(\frac{\partial^2 \log f(\xi,\gamma)}{\partial \gamma_i \partial \gamma_j}\right)\right| \geqslant \varepsilon_1\right\}$$

$i,j = 1, \dots, k$. Then, $P_{\xi^{(n)}}\left(\bigcup_{i,j=1}^{k} E_{ij}^{(n)}\right) \leqslant \delta_1 k^2$ from (3.8). Now $R_n - \bigcup_{i,j=1}^{k} E_{ij}^{(n)} = \bigcap_{i,j=1}^{k} (R_n - E_{ij}^{(n)})$. But on this set

$$\left|\frac{1}{n} \frac{\partial^2 \log f^{(n)}(x^{(n)},\gamma)}{\partial \gamma_i \partial \gamma_j} - E\left(\frac{\partial^2 \log f(\xi,\gamma)}{\partial \gamma_i \partial \gamma_j}\right)\right| < \varepsilon_1, \quad i,j = 1, \dots, k$$

so that

$$P\left(\left|\frac{1}{n} \frac{\partial^2 \log f^{(n)}(\xi^{(n)},\gamma)}{\partial \gamma_i \partial \gamma_j} - E\left(\frac{\partial^2 \log f(\xi,\gamma)}{\partial \gamma_i \partial \gamma_j}\right)\right| < \varepsilon_1,\ i,j = 1, \dots, k\right) > 1 - k^2 \delta_1 \tag{3.9}$$

for $n \geqslant n(\varepsilon_1, \delta_1)$.

[36] This proof is somewhat related to that of H. Cramér, Loc. cit. I,[58] 500ff. For the case considered here of a multi-dimensional parameter, K. C. Chanda, Biometrika 41, 56—61 (1954) has carried out Cramér's proof in detail.

For each $x^{(n)} \in R_n$

$$\frac{\partial \log f^{(n)}(x^{(n)},\gamma)}{\partial \gamma_i} = \sum_{l=1}^{k} \frac{\partial \log f(x_l,\gamma)}{\partial \gamma_i}.$$

Moreover, from (3.1)

$$E\left(\frac{\partial \log f(\xi,\gamma)}{\partial \gamma_i}\right)=0, \quad i=1,\ldots,k. \tag{3.10}$$

Thus, applying I, Theorem 38.2 again, we obtain

$$P\left(\frac{1}{n}\left|\frac{\partial \log f^{(n)}(\xi^{(n)},\gamma)}{\partial \gamma_i}\right|<\varepsilon_1\right)>1-\delta_1 \tag{3.11}$$

for $i=1,\ldots,k$ and $n \geqslant n_2(\varepsilon_1,\delta_1)$. Proceeding exactly as before, we get from (3.11)

$$P\left(\frac{1}{n}\left|\frac{\partial \log f^{(n)}(\xi^{(n)},\gamma)}{\partial \gamma_i}\right|<\varepsilon_1,\ i=1,\ldots,k\right)>1-k\delta_1 \tag{3.12}$$

for $n \geqslant n_2(\varepsilon_1,\delta_1)$.

Taking (3.4) into consideration we arrive at the following result: Let F_n be the set of all $x^{(n)} \in R_n$ for which

$$\left|\frac{\partial \log f^{(n)}(x^{(n)},\tilde{\gamma})}{\partial \gamma} - \frac{\partial \log f^{(n)}(x^{(n)},\tilde{\tilde{\gamma}})}{\partial \gamma} - B(x^{(n)},\gamma)(\tilde{\gamma}-\tilde{\tilde{\gamma}})\right| \leqslant q(\gamma)\,|B(x^{(n)},\gamma)(\tilde{\gamma}-\tilde{\tilde{\gamma}})| \tag{3.13}$$

for all $\tilde{\gamma},\tilde{\tilde{\gamma}} \in K(\gamma)$ and

$$\left|\frac{1}{n}\frac{\partial^2 \log f^{(n)}(x^{(n)},\gamma)}{\partial \gamma_i \partial \gamma_j} - E\left(\frac{\partial^2 \log f(\xi,\gamma)}{\partial \gamma_i \partial \gamma_j}\right)\right|<\varepsilon_1 \tag{3.14}$$

for $i,j=1,\ldots,k$, and, finally

$$\left|\frac{1}{n}\frac{\partial \log f^{(n)}(x^{(n)},\gamma)}{\partial \gamma_i}\right|<\varepsilon_1 \tag{3.15}$$

for $i=1,\ldots,k$.

If $n(\delta_1,\varepsilon_1,\eta)=\max(n_1(\varepsilon_1,\delta_1),n_2(\varepsilon_1,\delta_1),n(\eta))$, then, as in the proof of (3.9) we have from (3.4), (3.9) and (3.12)

$$P(\xi^{(n)} \in F_n)>1-\eta-\delta_1 k-\delta_1 k^2 \tag{3.16}$$

for $n \geqslant n(\delta_1,\varepsilon_1,\eta)$.

Now consider a fixed $x^{(n)} \in F_n$ for which (3.13), (3.14) and (3.15) hold. We will thus suppress reference to $x^{(n)}$ (and also to the r.v. $\xi^{(n)}$) in the notation of the sequel and will write $\frac{\partial \log f^{(n)}(\gamma)}{\partial \gamma}$ instead of $\frac{\partial \log f^{(n)}(x^{(n)},\gamma)}{\partial \gamma}$ or $E(B(\gamma))$ instead of $E(B(\xi^{(n)},\gamma))$, etc.

From

$$E\left[\frac{1}{n}\,\frac{\partial^2 \log f^{(n)}(\gamma)}{\partial\gamma_i\,\partial\gamma_j}\right]=E\left(\frac{\partial^2 \log f(\gamma)}{\partial\gamma_i\,\partial\gamma_j}\right),\qquad i,j=1,\dots,k \tag{3.17}$$

and Lemma 3.1 we obtain

$$\frac{1}{n}\,E(B(\gamma)) = -E(A(\gamma)). \tag{3.18}$$

Because of (3.10) we can view $E(A(\gamma))$ as the covariance matrix of the k-dimensional r. v.'s $\left(\frac{\partial \log f(\xi,\gamma)}{\partial\gamma_1},\dots,\frac{\partial \log f(\xi,\gamma)}{\partial\gamma_k}\right)$. According to I, Theorem 17.5, $E(A(\gamma))$ is always positive semi-definite and from (3.3) positive definite. Since a determinant is a continuous function of its elements, one can, because of (3.18), for suitable choice of ε_1, according to (3.14), always take $B(\gamma)/n$ as negative definite. Since from (3.14), (3.17) and (3.18) the elements of $B(\gamma)/n$ differ for all $n \geqslant n(\delta_1,\varepsilon_1,\eta)$ by at most the fixed amount ε_1 from the elements of $-E(A(\gamma))$, we have for suitable choice of $\lambda>0$, depending only on γ and not on $\tilde{\gamma},\tilde{\tilde{\gamma}}\in R_k$ or on n

$$\left|\frac{B(\gamma)}{n}(\tilde{\gamma}-\tilde{\tilde{\gamma}})\right| \geqslant \lambda\,|\tilde{\gamma}-\tilde{\tilde{\gamma}}|. \tag{3.19}$$

We now show that for $x^{(n)}\in F_n$ there exists a solution $\gamma^*(x^{(n)})$ of

$$\frac{\partial \log f^{(n)}(x^{(n)},\gamma)}{\partial\gamma}=0 \tag{3.20}$$

which lies in $K(\gamma)$ and is uniquely determined there. Indeed, if (3.20) were satisfied for $\gamma^*(x^{(n)})$ and $\gamma^{**}(x^{(n)})$, with $\gamma^*,\gamma^{**}\in K(\gamma)$ and $\gamma^*\neq\gamma^{**}$, then from (3.13) we would have

$$|B(\gamma)(\gamma^*-\gamma^{**})| \leqslant q(\gamma)\,|B(\gamma)(\gamma^*-\gamma^{**})|$$

and hence, because $q(\gamma)<1$ also $B(\gamma)(\gamma^*-\gamma^{**})=0$. Then $\gamma^*=\gamma^{**}$ follows from (3.19) which is a contradiction[37]. The solution is thus unique.

We now demonstrate the existence of a solution $\gamma^*(x^{(n)})$ to (3.20). Since $B(\gamma)$ is negative definite, its inverse $B^{-1}(\gamma)$ exists. We now define a sequence $c_0,c_1,\dots$, with $c_i\in R_k$, whereby $c_0=\gamma$ and for $l\geqslant 0$

$$c_{l+1}=c_l-B^{-1}(\gamma)\,\frac{\partial \log f^{(n)}(c_l)}{\partial\gamma}. \tag{3.21}$$

---

[37] We have modified here and in the sequel an argument of H. Hornich, Monatsh. Math. 54, 130—134 (1950).

We show that c_l for $l \geqslant 1$ belongs to $K(\gamma)$. Considering (3.21) for l and $l-1, l\geqslant 1$, one obtains by subtraction and multiplication by $B(\gamma)$

$$B(\gamma)(c_{l+1}-c_l) = B(\gamma)(c_l - c_{l-1}) - \frac{\partial \log f^{(n)}(c_l)}{\partial \gamma} + \frac{\partial \log f^{(n)}(c_{l-1})}{\partial \gamma}. \quad (3.22)$$

We now carry through a proof by induction and show in addition that

$$|B(\gamma)(c_{l+1}-c_l)| \leqslant q(\gamma)|B(\gamma)(c_l-c_{l-1})|, \qquad l\geqslant 1. \quad (3.23)$$

According to (3.15)

$$\left|\frac{1}{n}\frac{\partial \log f^{(n)}(\gamma)}{\partial \gamma}\right| < k\varepsilon_1. \quad (3.24)$$

Choosing ε_1 sufficiently small, with $q=q(\gamma), \rho=\rho(\gamma)$ and the positive number λ appearing in (3.19), we have

$$k\varepsilon_1 \leqslant \rho\lambda(1-q). \quad (3.25)$$

From (3.21) follows $B(\gamma)(c_1-c_0) = -\dfrac{\partial \log f^{(n)}(c_0)}{\partial \gamma}$ when one sets $l=0$, and thus from (3.24) and (3.25)

$$\left|\frac{1}{n} B(\gamma)(c_1-c_0)\right| \leqslant \rho\lambda(1-q). \quad (3.26)$$

Hence from (3.19) $|c_1-c_0|<\rho$, i.e., $c_1 \in K(\gamma)$. Applying (3.13) with $\tilde{\gamma}=c_0$ and $\tilde{\tilde{\gamma}}=c_1$ to (3.22) for $l=1$, we obtain (3.23) for $l=1$. Assume now that for $1\leqslant l\leqslant m$ we have proved that

$$c_l \in K(\gamma) \quad (3.27)$$

and that (3.23) holds. We will show that both (3.27) and (3.23) are valid for $l=m+1$. First, we have for each $l\geqslant 0$

$$|B(\gamma)(c_{l+1}-c_0)| \leqslant |B(\gamma)(c_{l+1}-c_l)| + |B(\gamma)(c_l-c_{l-1})| + \cdots + |B(\gamma)(c_1-c_0)|$$

and hence from (3.23) for $l\leqslant m$ we get

$$|B(\gamma)(c_{l+1}-c_0)| \leqslant |B(\gamma)(c_1-c_0)|\,(1-q^{l+1})/(1-q)$$

and thus, according to (3.26)

$$\left|\frac{1}{n} B(\gamma)(c_{l+1}-c_0)\right| \leqslant \rho\lambda(1-q^{l+1}). \quad (3.28)$$

Using (3.28) for $l=m$ and applying (3.19) we find $c_{m+1}\in K(\gamma)$. But then from (3.22) for $l=m+1$ and an application of (3.13) there follows (3.23) for $l=m+1$.

As we have just shown,

$$\left|\frac{1}{n}B(\gamma)(c_s-c_r)\right| \leqslant (q^{s-1}+\cdots+q^r)\left|\frac{B(\gamma)}{n}(c_1-c_0)\right|$$

holds for all $s>r\geqslant 0$.

Because of (3.26), the fact that $q<1$ and (3.19), $|c_s-c_r|$ becomes arbitrary small for sufficiently large r. Hence, there exists a k-dimensional vector γ^* with $c_r\to\gamma^*$ for $r\to\infty$ and

$$|\gamma^*-\gamma|\leqslant\rho \tag{3.29}$$

since $K(\gamma)$ is compact.

Since the likelihood function $\gamma\to f^{(n)}(x^{(n)},\gamma)$ is continuously differentiable,

$$\frac{\partial\log f^{(n)}(c_l)}{\partial\gamma}\to\frac{\partial\log f^{(n)}(\gamma^*)}{\partial\gamma}\quad\text{for } l\to\infty$$

and since $B(\gamma)(c_{l+1}-c_l)=-\dfrac{\partial\log f^{(n)}(c_l)}{\partial\gamma}$ we have for $l\to\infty$

$$\frac{\partial\log f^{(n)}(\gamma^*)}{\partial\gamma}=0. \tag{3.30}$$

In order to emphasize the dependence of this construction on n, we write γ_n^* instead of γ^*.

The mapping $x^{(n)}\to\gamma_n^*(x^{(n)})$ defined over F_n is measurable; this follows from (3.21) by induction.

Now define

$$\hat{\gamma}_n(x^{(n)})=\begin{cases}\gamma_n^*(x^{(n)}) & x^{(n)}\in F_n\\ 0 & x^{(n)}\in R_n-F_n\end{cases}. \tag{3.31}$$

If $\varepsilon>0$ and $\delta>0$ are given and we choose $\rho<\varepsilon$ and then ε_1 and δ_1 sufficiently small, so that, in particular, (3.25) and $\eta+\delta_1(k+k^2)<\delta$ hold, then we can choose $N(\delta,\varepsilon)$ in such a way that $\hat{\gamma}_n$ fulfills (3.30) on F_n for $n\geqslant N(\delta,\varepsilon)$ and, moreover, because of (3.29) and (3.16)

$$P_\gamma(|\hat{\gamma}_n(\xi^{(n)})-\gamma|<\varepsilon)>1-\delta \tag{3.32}$$

holds.

One now proceedes in the usual way: choose strictly decreasing null-sequences $\{\delta_i\}$ and $\{\varepsilon_i\}$. Then there exists a strictly increasing sequence of positive integers n_i such that for $n_i\leqslant n\leqslant n_{i+1}-1$, $i\geqslant 1$, $\hat{\gamma}_n$ is defined according to (3.31), fulfills (3.30) and (3.32) holds for $\varepsilon=\varepsilon_i$ and $\delta=\delta_i$.

We also remark that the construction of γ_n^* depends on the chosen $\gamma \in \Gamma$. Hence, one cannot, without further restrictions, claim that γ_n^* is a consistent sequence of estimates for $\gamma \in \Gamma$. But this theorem is still one of the bases for the construction of such consistent sequences of estimates as we will see in the sequel. Condition (3.4) was essential for the proof of Theorem 3.1. We want to find a convenient sufficient condition for this to hold. We maintain all previous notation and prove

Theorem 3.2. *Let all hypotheses of Theorem* 3.1 *hold with the exception of* (3.4). *Moreover, let the map* $\gamma \to f(x, \gamma)$ *in* Γ *be thrice differentiable for all* $x \in R_1$. *Let there exist a function* ϕ *defined over* R_1 *such that for each* $x \in R_1$, *all* $\gamma \in \Gamma$ *and all* $h, i, j = 1, \ldots, k$

$$\left| \frac{\partial^3 \log f(x, \gamma)}{\partial \gamma_h \partial \gamma_i \partial \gamma_j} \right| \leqslant \phi(x). \tag{3.33}$$

Let $E(\phi; \gamma)$ *exist for all* $\gamma \in \Gamma$. *Then* (3.4) *holds.*

Proof. From (3.33), for each $x^{(n)} \in R_n$ and all $\gamma \in \Gamma$

$$\frac{1}{n} \left| \frac{\partial^3 \log f^{(n)}(x^{(n)}, \gamma)}{\partial \gamma_h \partial \gamma_i \partial \gamma_j} \right| \leqslant (\phi(x_1) + \cdots + \phi(x_n))/n. \tag{3.34}$$

Denoting $E(\phi; \gamma)$ by $M(\gamma)$, we have from I, Theorem 38.2 for each $\gamma_0 \in \Gamma$ and given $\varepsilon_1, \delta_1 > 0$

$$P_{\gamma_0}\left(\left| \frac{\phi(\xi_1) + \cdots + \phi(\xi_n)}{n} - M(\gamma_0) \right| < \varepsilon_1 \right) > 1 - \delta_1 \tag{3.35}$$

for all sufficiently large n. If $M(\gamma_0) + \varepsilon_1 = m(\gamma_0)$, say, then from (3.34) and (3.35), for all sufficiently large n, there follows the existence of a set $E_n \in \mathfrak{B}_n$ such that for all $\gamma \in \Gamma$ and $x^{(n)} \in E_n$

$$\left| \frac{1}{n} \frac{\partial^3 \log f^{(n)}(x^{(n)}, \gamma)}{\partial \gamma_h \partial \gamma_i \partial \gamma_j} \right| \leqslant m(\gamma_0), \qquad h, i, j = 1, \ldots, k \tag{3.36}$$

with

$$P(\xi^{(n)} \in E_n; \gamma_0) > 1 - \delta_1 \tag{3.37}$$

for each $\gamma_0 \in \Gamma$.

Let for all $x^{(n)} \in R_n$ and $\gamma \in \Gamma$ and all $i = 1, \ldots, k$

$$\frac{\partial^2 \log f^{(n)}(x^{(n)}, \gamma)}{\partial \gamma_i \partial \gamma} = \begin{bmatrix} \dfrac{\partial^2 \log f^{(n)}(x^{(n)}, \gamma)}{\partial \gamma_i \partial \gamma_1} \\ \cdots\cdots\cdots\cdots \\ \dfrac{\partial^2 \log f^{(n)}(x^{(n)}, \gamma)}{\partial \gamma_i \partial \gamma_k} \end{bmatrix}.$$

Then by the Mean Volume Theorem for all γ_0, $\tilde{\gamma}$ and $\tilde{\tilde{\gamma}}$ from Γ, all $x^{(n)} \in R_n$ and $1 \leqslant i \leqslant k$

$$\frac{1}{n}\frac{\partial \log f^{(n)}(x^{(n)},\tilde{\tilde{\gamma}})}{\partial \gamma_i} - \frac{1}{n}\frac{\partial \log f^{(n)}(x^{(n)},\tilde{\gamma})}{\partial \gamma_i} - \frac{1}{n}\left(\frac{\partial^2 \log f^{(n)}(x^{(n)},\gamma_0)}{\partial \gamma_i \partial \gamma}\right)'(\tilde{\tilde{\gamma}}-\tilde{\gamma})$$
$$= \frac{1}{n}\left(\frac{\partial^2 \log f^{(n)}(x^{(n)},\tilde{\gamma}+\vartheta_i(x^{(n)})(\tilde{\tilde{\gamma}}-\tilde{\gamma}))}{\partial \gamma_i \partial \gamma}\right)'(\tilde{\tilde{\gamma}}-\tilde{\gamma})$$
$$-\frac{1}{n}\left(\frac{\partial^2 \log f^{(n)}(x^{(n)},\gamma_0)}{\partial \gamma_i \partial \gamma}\right)'(\tilde{\tilde{\gamma}}-\tilde{\gamma}), \qquad 0<\vartheta_i(x^{(n)})<1.$$

Another application of the Mean Value Theorem of calculus yields

$$\frac{1}{n}\frac{\partial \log f^{(n)}(x^{(n)},\tilde{\tilde{\gamma}})}{\partial \gamma_i} - \frac{1}{n}\frac{\partial \log f^{(n)}(x^{(n)},\tilde{\gamma})}{\partial \gamma_i}$$
$$-\frac{1}{n}\left(\frac{\partial^2 \log f^{(n)}(x^{(n)},\gamma_0)}{\partial \gamma_i \partial \gamma}\right)'(\tilde{\tilde{\gamma}}-\tilde{\gamma})$$
$$=\frac{1}{n}(\bar{\gamma}-\gamma_0)'\left(\frac{\partial^3 \log f^{(n)}(x^{(n)},\gamma_0+\vartheta_{jh}(x^{(n)})(\bar{\gamma}-\gamma_0))}{\partial \gamma_i \partial \gamma_j \partial \gamma_h}\right)_{1k}^{1k}(\tilde{\tilde{\gamma}}-\tilde{\gamma})$$

with $\bar{\gamma}=\tilde{\gamma}+\vartheta_i(x^{(n)})(\tilde{\tilde{\gamma}}-\tilde{\gamma})$ and $0<\vartheta_{jh}(x^{(n)})<1$. We now choose a closed ball $K(\gamma_0)$ with γ_0 as center and radius $\rho>0$ which we will determine later. Let $\tilde{\gamma},\tilde{\tilde{\gamma}}\in K(\gamma_0)$, i.e.

$$|\tilde{\gamma}-\gamma_0| \leqslant \rho, \quad |\tilde{\tilde{\gamma}}-\gamma_0| \leqslant \rho. \tag{3.38}$$

Then from (3.36) and (3.38)

$$\left|\frac{1}{n}\frac{\partial \log f^{(n)}(x^{(n)},\tilde{\tilde{\gamma}})}{\partial \gamma} - \frac{1}{n}\frac{\partial \log f^{(n)}(x^{(n)},\tilde{\gamma})}{\partial \gamma} - \frac{1}{n}B(x^{(n)},\gamma_0)(\tilde{\tilde{\gamma}}-\tilde{\gamma})\right|$$
$$\leqslant m(\gamma_0)k^2|\tilde{\tilde{\gamma}}-\tilde{\gamma}|\rho \tag{3.39}$$

and this holds because of (3.37) and a previous argument for all $x^{(n)}$ belonging to a set from $\mathfrak{B}_n$ whose $P_{\gamma_0,\xi(n)}$-measure is arbitrarily close to 1 if n is sufficiently large. Furthermore, in analogy to (3.19), one can also assume that for all these $x^{(n)}$

$$\left|\frac{B(x^{(n)},\gamma_0)}{n}(\tilde{\tilde{\gamma}}-\tilde{\gamma})\right| \geqslant \lambda|\tilde{\tilde{\gamma}}-\tilde{\gamma}|$$

with $\lambda>0$. Choosing q arbitrarily from $(0,1)$ and then ρ according to $0<\rho<q\lambda/(m(\gamma_0)k^2)$, the claim follows from (3.39).

We now augment Theorem 3.2, retaining the same notation.

Theorem 3.3.[38] *Assume the assumptions of Theorem 3.2 are satisfied. Then for each $\gamma_0 \in \Gamma$ and each $\delta > 0$ the set of all $x^{(n)}$ for which*

$$\left(\frac{\partial^2 \log f^{(n)}(x^{(n)}, \hat{\gamma}_n(x^{(n)}))}{\partial \gamma_i \partial \gamma_j}\right)_{1k}^{1k}$$

is negative definite possesses $P_{\gamma_0, \xi^{(n)}}$-measure $> 1-\delta$ for all sufficiently large n.

Proof. From the assumptions of Theorem 3.2 follow these of Theorem 3.1. Hence, for given $\varepsilon > 0$ and $\delta_1 > 0$

$$P_{\gamma_0}(|\hat{\gamma}_n(\xi^{(n)}) - \gamma_0| < \varepsilon) > 1 - \delta_1 \tag{3.40}$$

holds for $n \geqslant N(\delta_1, \varepsilon)$. From this and the considerations on p. 299 it follows that for all sufficiently large n, one can prove the existence of a set $E_n \in \mathfrak{B}_n$ such that for $x^{(n)} \in E_n$, all the inequalities (3.36), all those of (3.14) with small enough ε_1 for $\gamma = \gamma_0$ and

$$|\hat{\gamma}_n(x^{(n)}) - \gamma_0| < \varepsilon \tag{3.41}$$

hold so that, for $\delta > 0$

$$P(\xi^{(n)} \in E_n; \gamma_0) > 1 - \delta. \tag{3.42}$$

Applying the Mean Value Theorem again, we have for $i,j = 1, \ldots, k$ from (3.36)

$$\frac{1}{n}\left|\frac{\partial^2 \log f^{(n)}(x^{(n)}, \hat{\gamma}_n(x^{(n)}))}{\partial \gamma_i \partial \gamma_j} - \frac{\partial^2 \log f^{(n)}(x^{(n)}, \gamma_0)}{\partial \gamma_i \partial \gamma_j}\right| \leqslant \sqrt{k}\, m(\gamma_0)\, |\hat{\gamma}_n(x^{(n)}) - \gamma_0|. \tag{3.43}$$

The right side of (3.43) for $x^{(n)} \in E_n$ can, because of (3.41), be made uniformly arbitrarily small, for example, smaller than ε_1. Hence, from (3.14) for $i,j = 1, \ldots, k$ and all $x^{(n)} \in E_n$

$$\left|\frac{1}{n}\frac{\partial^2 \log f^{(n)}(x^{(n)}, \hat{\gamma}_n(x^{(n)}))}{\partial \gamma_i \partial \gamma_j} - E\left(\frac{\partial^2 \log f(\xi, \gamma_0)}{\partial \gamma_i \partial \gamma_j}; \gamma_0\right)\right| < 2\varepsilon_1. \tag{3.44}$$

However, since on the right side of the difference in (3.44) we have, according to (3.7), the elements of the negative definite matrix $-E(A(\gamma_0))$, the theorem follows from the fact that all the sub-determinants of the matrix $E(A(\gamma_0))$ are continuous functions of their elements.

Hence, at least for $x^{(n)} \in E_n$, $\hat{\gamma}_n(x^{(n)})$ is a relative maximum for the mapping $\gamma \to f^{(n)}(x^{(n)}, \gamma)$, i.e., $\hat{\gamma}_n$ is in fact a ML estimate.

From Theorem 3.1 we have the following theorem, which is important for applications.

---

[38] See V. S. Huzurbazar, Ann. Eugenics 14, 185—200 (1948).

Theorem 3.4. *Assume the assumptions of Theorem 3.2 are fulfilled and that equation (3.20) for each $x^{(n)}$, $n \geqslant 1$ (or only $\mu \times \cdots \times \mu$-a.e.) has exactly one solution $\gamma_n^+(x^{(n)})$ in Γ. Then $\{\gamma_n^+\}$ is a consistent sequence of estimates for $\gamma \in \Gamma$.*

Proof. The maps $x^{(n)} \to \hat{\gamma}_n(x^{(n)})$ defined in Theorem 3.1 must coincide with γ_n^+ at least on F_n because of (3.30) and (3.31). Thus from (3.6) γ_n^+ fulfills for each $\gamma \in \Gamma$ the condition

$$P_\gamma(|\gamma_n^+(\xi^{(n)}) - \gamma| < \varepsilon) > 1 - \delta$$

for $n \geqslant N(\varepsilon, \delta)$, where we have not excluded the possibility that $N(\varepsilon, \delta)$ might also depend on γ.

From Theorem 3.3 follows the quite similar

Theorem 3.5. *Let the assumptions of Theorem 3.3 be fulfilled and assume that the mapping $\gamma \to f^{(n)}(x^{(n)}, \gamma)$ possesses exactly one maximum $\gamma_n^+(x^{(n)})$ in Γ for each $x^{(n)} \in R_n$, $n = 1, 2, \ldots$. Then γ_n^+ is a consistent sequence of estimates for $\gamma \in \Gamma$ if γ_n^+ is measurable.*

The *proof* is analogous to that of Theorem 3.4.

We consider as simple application of Theorem 3.4 a Poisson distribution with parameter $a > 0$. If we choose some measure μ over $(R_1, \mathfrak{B}_1)$ with $\mu(\{i\}) = p_i > 0$, $i = 0, 1, \ldots$ and $\sum_{i=0}^{\infty} p_i = 1$, then all Poisson distributions are absolutely continuous w.r.t. μ. One can then easily see that all the assumptions of Theorem 3.4 are satisfied. For nonnegative integers $x_1, \ldots, x_n$ we then have $f^{(n)}(x_1, \ldots, x_n, a) = (x_1! \ldots x_n!)^{-1} e^{-na} a^{n\bar{x}} (p_{x_1} \cdots p_{x_n})^{-1}$. The likelihood equation is then $\bar{x} n/a - n = 0$ and this has as its only solution $\hat{a}(x_1, \ldots, x_n) = \bar{x}$. This depends trivially on $(x_1, \ldots, x_n)$ only via $\bar{x}$ which is naturally not surprising.

Assume, more generally, that T is a sufficient transformation for a set of probability measures P_Γ over $(R, \mathscr{S})$, where Γ is an open cube $\subseteq R_k$, $k \geqslant 1$. Let P_Γ be dominated by a σ-finite measure and let the R.-N.-densities be denoted by $x \to f(x, \gamma)$ for each $\gamma \in \Gamma$. According to III, Theorem 7.2, for suitable g and h we have on $R \times \Gamma$

$$f(x, \gamma) = h(x) g(T(x), \gamma).$$

Under appropriate assumptions, the likelihood equation is, for each $x \in R$, given by

$$\frac{\partial \log g(T(x), \gamma)}{\partial \gamma} = 0$$

and each of its solutions is a function depending only on T.

Remark. If we use the strong instead of the weak law of large numbers in the proof of Theorem 3.1, we can even show that $\hat{\gamma}_n \to \gamma$ with probability 1.[39] However, it is then necessary to modify assumption (3.4) appropriately: one must replace the "weak" formulation by the "strong" one. This generalization is not very important in practice. For applications, the results of Theorems 3.1—3.5 are sufficient. A shortcoming of Theorems 3.1 and 3.3 lies, however, in the fact that they hold only "locally". Only the "global" assumptions of Theorem 3.4 and 3.5 allow global results. When such global assumptions are not fulfilled, a sequence of ML estimates is not necessarily consistent[40]. This is due naturally to the local character of the notion of a "relative maximum". The supremum of a function over a set is, however, a global notion. One can thus expect the consideration of $\sup_{\gamma\in\Gamma} f^{(n)}(x^{(n)},\gamma)$ (maintaining the notation already introduced) to provide more comprehensive results. We first prove several lemmas.

Lemma 3.2. *Let* $(R,\mathscr{S})$ *be a measurable space and* C *a compact set from* R_k, $k\geqslant 1$. *Let* u *be a function defined on* $R\times C$ *such that the map* $x\to u(x,t)$ *is* $\mathscr{S}$*-measurable for each* $t\in C$ *and the map* $t\to u(x,t)$ *is continuous for each* $x\in R$. *Set* $u^*(x)=\sup_{t\in C} u(x,t)$ *for* $x\in R$. *Then,* u^* *is* $\mathscr{S}$*-measurable.*

Proof. Choose a countable subset D of C which is dense in C[41]. Since $t\to u(x,t)$ is continuous for each x, $\sup_{t\in C} u(x,t)=\sup_{t\in D} u(x,t)$ so that

$$u^*(x)=\sup_{t\in D} u(x,t). \tag{3.45}$$

Hence, u^* is $\mathscr{S}$-measurable. (See p. 7.)

Lemma 3.3. *Let the hypotheses of Lemma* 3.2 *be satisfied. Then one can always define an* $\mathscr{S}$*-measurable mapping* $x\to t(x)$ *from* R *into* C *such that* $\sup_{t\in C} u(x,t)=u(x,t(x))$ *for all* $x\in R$.

Proof[42]. We will make frequent use of the following known property of compact subsets C of a Euclidean space: from each sequence of elements in C one can choose a convergent subsequence (whose limit naturally belongs to C). We order the elements of R_k as follows: let $x,y\in R_k$ and $x\neq y$. Then let $x<y$ iff there exists an i with $1\leqslant i\leqslant k$ for which $x_1=y_1,\ x_2=y_2,\dots,x_{i-1}=y_{i-1},\ x_i<y_i$ holds. Set

[39] A more precise formulation can be given with the help of the statement of Theorem 3.8.

[40] See L. Le Cam and Ch. Kraft, Ann. Math. Statist. 27, 1174—1177 (1956) and also R. R. Bahadur, Sankhyā 20, 207—210 (1958).

[41] See p. 192. One can for example choose the set of all k-tuples in C with rational components. Note that only the existence of a dense set in C is used in the proof and not the compactness of C.

[42] Cf. H. Richter, Math. Ann. 150, 85—90 and 440—441 (1963), M. Sion, Trans. Amer. Math. Soc. 96, 237—246 (1960).

$M_x=\{t: t\in C, u(x,t)=u^*(x)\}$. From the compactness of C and the continuity assumptions it follows that M_x is never empty and is even compact. Now order the elements of M_x as above. There always exists an element $t(x)\in M_x$ with $t(x)>t$ for all $t\in M_x$ with $t(x)\neq t$. This is shown as follows. Let i with $1\leqslant i\leqslant k$ be the smallest index for which there are at least two elements $\tilde{t},\tilde{\tilde{t}}\in M_x$ with $\tilde{t}_j=\tilde{\tilde{t}}_j$, $1\leqslant j\leqslant i-1$, $\tilde{t}_i<\tilde{\tilde{t}}_i$. Set

$$t_i^* = \sup_{t\in M_x} t_i\,. \tag{3.46}$$

Furthermore, let $t^{(n)}\in M_x$ for $n\geqslant 1$ and $\lim_{n\to\infty} t_i^{(n)}=t_i^*$. A subsequence $\{t^{(n_k)}\}$ can be so chosen that $\lim_{k\to\infty} t^{(n_k)}=\bar{t}$ exists. Then

$$\bar{t}_i = t_i^* \tag{3.47}$$

and for all $t\in M_x$ we have $t_j=\bar{t}_j$, $1\leqslant j\leqslant i-1$. Finally, from (3.46) and (3.47)

$$t_i\leqslant \bar{t}_i\,.$$

If the $<$-sign holds here for all $t\neq\bar{t}$, then we can identify $t(x)$ with $\bar{t}$. Otherwise consider the (compact) subset of M_x with $t_i=t_i^*$ and repeat the construction for it, etc.

One can now simplify the problem somewhat by defining a function v over $R\times C$ by $v(x,t)=u^*(x)-u(x,t)$. Since u^* is $\mathscr{S}$-measurable by Lemma 3.2, v satisfies the same assumptions as u in addition to the condition

$$\inf_{t\in C} v(x,t)=0\,. \tag{3.48}$$

for every $x\in R$. Moreover, $M_x=\{t: t\in C, v(x,t)=0\}$.

We now show that the map $x\to t(x)=(t_1(x),\dots,t_k(x))$ is $\mathscr{S}$-measurable. Let α_1 be an arbitrary real number. It is easy to see that

$$\{x: t_1(x)\geqslant\alpha_1\} = \bigcup_{\substack{t\in C\\ t_1\geqslant\alpha_1}} \{x: v(x,t)=0\}\,.$$

We next show that

$$\bigcup_{\substack{t\in C\\ t_1\geqslant\alpha_1}} \{x: v(x,t)=0\} = \bigcap_{n=1}^{\infty} \bigcup_{\substack{t\in C\\ t_1>\alpha_1-\frac{1}{n}}} \{x: v(x,t)<1/n\}\,. \tag{3.49}$$

The set on the left is obviously contained in that on the right. Conversely, if $x\in \bigcap_{n=1}^{\infty} \bigcup_{\substack{t\in C\\ t_1>\alpha_1-\frac{1}{n}}} \{x: v(x,t)<1/n\}$, then there exists for each $n\geqslant 1$ a $t^{(n)}\in C$ with $t_1^{(n)}>\alpha_1-\dfrac{1}{n}$ and $v(x,t^{(n)})<1/n$. Hence, there exists a convergent subsequence $\{t^{(n_i)}\}$ which converges to a limit $t_1 \in C$ for which $t_1 \geqslant \alpha_1$.

Because of continuity, $\lim_{i\to\infty} v(x, t^{(n_i)}) = v(x,t) = 0$ from (3.48). Now let D be a countable subset of C which is dense there. For each $n \geqslant 1$

$$\bigcup_{\substack{t\in C \\ t_1 > \alpha_1 - \frac{1}{n}}} \{x : v(x,t) < 1/n\} = \bigcup_{\substack{t\in D \\ t_1 > \alpha_1 - \frac{1}{n}}} \{x : v(x,t) < 1/n\}. \tag{3.50}$$

We need only show that the set on the left in (3.50) is contained in that on the right. If, however, for an $x \in R$ and a $t \in C$ with $t_1 > \alpha_1 - \frac{1}{n}$ the inequality $v(x,t) < 1/n$ holds, then there is a $t^{(0)} \in D$ such that $|t - t^{(0)}|$ is arbitrarily small, in particular, $t_1^{(0)} > \alpha_1 - \frac{1}{n}$. Then from continuity we also have $v(x, t^{(0)}) < 1/n$. By assumption, the set on the right in (3.50) is $\mathscr{S}$-measurable. Thus, $\{x : t_1(x) \geqslant \alpha_1\}$ also belongs to $\mathscr{S}$ by (3.49) so that $x \to t_1(x)$ is $\mathscr{S}$-measurable. Now let α_2 be another arbitrary real number. From the definition of the map $x \to t(x)$

$$\{x : t_2(x) \geqslant \alpha_2\} = \bigcup_{\substack{t\in C \\ t_2 \geqslant \alpha_2}} \{x : v(x,t) = 0,\ t_1(x) = t_1\}. \tag{3.51}$$

The set on the right is also equal to the set

$$\bigcup_{\substack{t\in C \\ t_2 \geqslant \alpha_2}} \{x : v(x,t) = 0,\ t_1(x) \leqslant t_1\}.$$

We now proceed as before. We have

$$\bigcup_{\substack{t\in C \\ t_2 \geqslant \alpha_2}} \{x : v(x,t) = 0,\ t_1(x) \leqslant t_1\} = \bigcap_{n=1}^{\infty} \bigcup_{\substack{t\in C \\ t_2 > \alpha_2 - \frac{1}{n}}} \left\{x : v(x,t) < \frac{1}{n},\ t_1(x) < t_1 + \frac{1}{n}\right\}. \tag{3.52}$$

Furthermore

$$\bigcup_{\substack{t\in C \\ t_2 > \alpha_2 - \frac{1}{n}}} \left\{x : v(x,t) < \frac{1}{n},\ t_1(x) < t_1 + \frac{1}{n}\right\} = \bigcup_{\substack{t\in D \\ t_2 > \alpha_2 - \frac{1}{n}}} \left\{x : v(x,t) < \frac{1}{n},\ t_1(x) < t_1 + \frac{1}{n}\right\}. \tag{3.53}$$

Since $x \to t_1(x)$ is $\mathscr{S}$ measurable, the set on the right in (3.53) is also $\mathscr{S}$-measurable and we have—because of (3.52)—also demonstrated the measurability of $x \to t_2(x)$. If the measurability of $x \to t_i(x)$ for $1 \leqslant i \leqslant k-1$ obtains, then the measurability of $x \to t_k(x)$ follows as in the demonstration just given. The lemma is proved.

It is easy to see that, in general, not every map $x \to t(x)$ satisfying the condition $u^*(x) = u(x, t(x))$ for all $x \in R$ is $\mathscr{S}$-measurable. Let $u(x,t)$ for instance be constant on $R \times C$. Assume there exists a set $M \subset R$ which does not belong to $\mathscr{S}$. Defining

$$t(x) = \begin{cases} t^{(0)}, & x \in M \\ t^{(1)}, & x \in R - M \end{cases} \quad \text{with } t^{(0)} \neq t^{(1)} \text{ and } t^{(0)}, t^{(1)} \in C,$$

we see that $x \to t(x)$ is not $\mathscr{S}$-measurable and $u^*(x) = u(x, t(x))$.

Finally, we prove

Lemma 3.4. *Let $(R, \mathscr{S}, P)$ be a probability space and h a P-integrable function. Then*

$$\int_R h\, dP \leqslant \log \int_R e^h\, dP \tag{3.54}$$

where $\int_R e^h\, dP = \infty$ if the integral does not exist. The equality sign holds iff h is P-a.e. constant.

Proof. If $\int_R e^h\, dP = \infty$ there is nothing to prove. We will not consider this case further.

For all $x \in R_1$, $e^x \geqslant 1 + x$, where equality holds iff $x = 0$. From this follows

$$e^{\left(h - \int_R h\, dP\right)} \geqslant 1 + h - \int_R h\, dP$$

and integrating,

$$\int_R e^{\left(h - \int_R h\, dP\right)} dP \geqslant \int_R \left(1 + h - \int_R h\, dP\right) dP = 1.$$

Taking logarithms yields (3.54). Furthermore, we have equality iff $h - \int_R h\, dP = 0$ P-a.e., so that h is constant P-a.e.

We arrive now at the present goal of our considerations. All undefined symbols have the same meaning as in Theorem 3.1.

Theorem 3.6[43]**.** *Let Γ be an open cube*[44] *of R_k, $k \geqslant 1$. Assume that for each $x \in R_1$ the mapping $\gamma \to f(x, \gamma)$*[45] *is continuous and that $K_\rho(\gamma)$ is the closed ball with radius $\rho > 0$ and center γ. For $x \in R_1$, set*

$$f^*(x, \gamma, \rho) = \sup_{\tilde{\gamma} \in K_\rho(\gamma)} f(x, \tilde{\gamma}).$$ [46]

---

[43] A. Wald, Ann. Math. Statist. 20, 595—601 (1949), J. Wolfowitz, Ann. Math. Statist. 20, 601—602 (1949). See also J. L. Doob, Trans. Amer. Math. Soc. 36, 759—775 (1934).

[44] Γ can in principle be any open set.

[45] It is enough to require that the mapping be continuous for all x up to a μ-null set.

[46] For each $\gamma \in \Gamma$ there is a sufficiently small ρ such that $K_\rho(\gamma) \subset \Gamma$.

Let

$$\log^+ x = \begin{cases} \log x & x \geqslant 1 \\ 0 & x < 1 \end{cases}.$$

Assume

$$E[\log^+ f^*(\xi, \gamma, \rho); \gamma_0] \tag{3.55}$$

exists for all $\gamma, \gamma_0 \in \Gamma$ *and all sufficiently small* $\rho > 0$ *and that*

$$E[\log f(\xi, \gamma_0); \gamma_0] \tag{3.56}$$

exists for each $\gamma_0 \in \Gamma$.

Let Γ_0 *be an arbitrary compact subset of* Γ *and* $\gamma_0 \notin \Gamma_0$. *Then for each* $\varepsilon > 0$ *and* $\delta > 0$

$$P\left(\left| \frac{\sup\limits_{\gamma \in \Gamma_0} f^{(n)}(\xi^{(n)}, \gamma)}{f^{(n)}(\xi^{(n)}, \gamma_0)} \right| < \varepsilon; \gamma_0 \right) > 1 - \delta \tag{3.57}$$

for $n \geqslant N(\varepsilon, \delta)$.

Proof. First it follows from Lemma 3.2 that $x \to f^*(x, \gamma, \rho)$ is measurable and $f^*(\xi, \gamma, \rho)$ is thus a r.v. Now choose a $\gamma_0 \in \Gamma$ and hold it fixed. If probabilities and expectations are taken w.r.t. γ_0, we will suppress it in the notation. We now show that for each $\gamma \neq \gamma_0$

$$E[\log f(\xi, \gamma)] < E[\log f(\xi, \gamma_0)] \tag{3.58}$$

$E[\log f(\xi, \gamma)]$ always exists in the sense that it is finite or $-\infty$. Namely, $\log^+ f(x, \gamma) \leqslant \log^+ f^*(x, \gamma, \rho)$ for all $x \in R_1$, $\gamma \in \Gamma$ and $\rho > 0$. Hence, $E[\log^+ f(\xi, \gamma)]$ is finite, since (3.55) is. On the other hand, $\log f(x, \gamma) - \log^+ f(x, \gamma) \leqslant 0$ for all $x \in R_1$ and $\gamma \in \Gamma$ which shows the existence of $E[\log f(\xi, \gamma)]$ in the given sense. If, however, $E[\log f(\xi, \gamma)] = -\infty$, then (3.58) holds trivially, since (3.56) is finite. Now let $E[\log f(\xi, \gamma)]$ be finite. Then

$$P\{x : f(x, \gamma) = 0\} = 0 \tag{3.59}$$

since otherwise $\log f(x, \gamma) = -\infty$ on a set of positive P-measure and $E[\log f(\xi, \gamma)]$ could not be finite. Since further $P\{x : f(x, \gamma) \neq f(x, \gamma_0)\} > 0$ (see the agreement on p. 160), $\log f(x, \gamma) - \log f(x, \gamma_0)$ is non-constant P-a.e. Thus, according to Lemma 3.4

$$\int_{R_1} (\log f(x, \gamma) - \log f(x, \gamma_0)) dP < \log \int_{R_1} f(x, \gamma)(f(x, \gamma_0))^{-1} dP. \tag{3.60}$$

From (3.59)

$$\int_{R_1} f(x, \gamma)(f(x, \gamma_0))^{-1} dP = \int_{\{x : f(x, \gamma) > 0\}} f(x, \gamma)(f(x, \gamma_0))^{-1} dP \tag{3.61}$$

so that

$$\int_{R_1} f(x, \gamma)(f(x, \gamma_0))^{-1} dP = \int_M f(x, \gamma) d\mu \tag{3.62}$$

with

$$M = \{x: f(x,\gamma)>0, f(x,\gamma_0)>0\}.$$

Thus we finally have

$$\int_{R_1} f(x,\gamma)(f(x,\gamma_0))^{-1}\, dP \leqslant 1. \tag{3.63}$$

Inserting this in (3.60) makes the right side there <0. This shows (3.58).

We now prove that for each $\gamma\in\Gamma$

$$\lim_{\rho\to 0} E[\log f^*(\xi,\gamma,\rho)] = E[\log f(\xi,\gamma)]. \tag{3.64}$$

Since the mapping $y\to\log^+ y$ is continuous, we have from the continuity of $\gamma\to f(x,\gamma)$ for each $x\in R_1$

$$\lim_{\rho\to 0} \log^+ f^*(x,\gamma,\rho) = \log^+ f(x,\gamma). \tag{3.65}$$

But the map $\rho\to f^*(x,\gamma,\rho)$ is monotone decreasing by the definition of the supremum. Hence, from the existence of (3.55) and from (3.65) there follows for each $\gamma\in\Gamma$

$$\lim_{\rho\to 0} E[\log^+ f^*(\xi,\gamma,\rho)] = E[\log^+ f(\xi,\gamma)]. \tag{3.66}$$

Furthermore, as one easily shows, for each $x\in R_1$, each $\gamma\in\Gamma$ and each $\rho>0$

$$\log f(x,\gamma) - \log^+ f(x,\gamma) \leqslant \log f^*(x,\gamma,\rho) - \log^+ f^*(x,\gamma,\rho) \leqslant 0. \tag{3.67}$$

On the other hand, again for reasons of continuity, for all $x\in R_1$, and all $\gamma\in\Gamma$

$$\lim_{\rho\to 0} \log f^*(x,\gamma,\rho) = \log f(x,\gamma) \tag{3.68}$$

which also holds when this limit is $-\infty$.

Thus, if $\lim\limits_{\rho\to 0} (E[\log f^*(\xi,\gamma,\rho)] - E[\log^+ f^*(\xi,\gamma,\rho)])$ is finite, it is equal to $E[\log f(\xi,\gamma)] - E[\log^+ f(\xi,\gamma)]$ because of (3.68) and (3.65). If, however, it is $-\infty$, then from (3.67), $E[\log f(\xi,\gamma)] - E[\log^+ f(\xi,\gamma)]$ is also $-\infty$. This together with (3.66) yields (3.64).

From (3.64) and (3.58) follows: for each $\gamma\in\Gamma_0$ there is a $\rho_\gamma>0$ with

$$E[\log f^*(\xi,\gamma,\rho_\gamma)] < E[\log f(\xi,\gamma_0)]. \tag{3.69}$$

By the Borel covering theorem (see p. 3) there are finitely many $\gamma_1,\ldots,\gamma_l$ from Γ_0 for which $\Gamma_0 \subseteq \bigcup\limits_{i=1}^{l} K_{\rho_{\gamma_i}}(\gamma_i)$. Naturally, for each $x^{(n)}=(x_1,\ldots,x_n)\in R_n$ with $\rho_{\gamma_i}=\rho_i$

$$0 \leqslant \sup_{\gamma\in\Gamma_0} f^{(n)}(x^{(n)},\gamma) \leqslant \sum_{j=1}^{l} \prod_{i=1}^{n} f^*(x_i,\gamma_j,\rho_j). \tag{3.70}$$

It is thus enough to show that for each $\varepsilon_1>0, \delta_1>0$ and γ_j there exists an $n(\varepsilon_1,\delta_1,\gamma_j)$ such that

$$P\left(\prod_{i=1}^{n} f^*(\xi_i,\gamma_j,\rho_j)/f^{(n)}(\xi^{(n)},\gamma_0)<\varepsilon_1\right)>1-\delta_1 \tag{3.71}$$

for $n\geqslant n(\varepsilon_1,\delta_1,\gamma_j)$, $j=1,\ldots,l$.

Since l is independent of n, from (3.71) we conclude—with an argument already used several times—that there exists an $E_n\in\mathfrak{B}_n$ such that for all $x^{(n)}\in E_n$ the inequalities

$$\prod_{i=1}^{n} f^*(x_i,\gamma_j,\rho_j)/f^{(n)}(x^{(n)},\gamma_0)<\varepsilon_1\,,\quad j=1,\ldots,l \tag{3.72}$$

are satisfied and the $P_{\xi^{(n)}}$-measure of E_n is larger than $1-\delta$, provided $n\geqslant N(\varepsilon,\delta)$. Then the claim follows from (3.70) with $\varepsilon=\varepsilon_1 l$.

But (3.71) is equivalent to

$$P\left(\left(\sum_{i=1}^{n}\log f^*(\xi_i,\gamma_j,\rho_j)-\sum_{i=1}^{n}\log f(\xi_i,\gamma_0)\right)<\log\varepsilon_1\right)>1-\delta_1\,. \tag{3.73}$$

An application of I, Theorem 38.2 and the supplement to it yields, however, the stochastic convergence of

$$\frac{1}{n}\sum_{i=1}^{n}\log f^*(\xi_i,\gamma_j,\rho_j)-\frac{1}{n}\sum_{i=1}^{n}\log f(\xi_i,\gamma_0)$$

to

$$E[\log f^*(\xi,\gamma_j,\rho_j)]-E[\log f(\xi,\gamma_0)]$$

and because of (3.69), this is equivalent to (3.73).

From Theorem 3.6 follows easily

Theorem 3.7. *We retain the notation of Theorem 3.6 and assume all its assumptions hold. Choose for each $n\geqslant 1$ a measurable function $\hat\Theta_n$ over R_n such that for each $x^{(n)}\in R_n$*

$$\sup_{\gamma\in\Gamma_0} f^{(n)}(x^{(n)},\gamma)=f^{(n)}(x^{(n)},\hat\Theta_n(x^{(n)}))\,.$$

Such a choice is always possible according to Lemma 3.3. Then, $\{\hat\Theta_n\}$ is a consistent sequence of estimates for $\gamma\in\Gamma_0$.

Proof. We must show that for all $\gamma\in\Gamma_0$, each $\varepsilon>0$ and $\delta>0$

$$P(|\hat\Theta_n(\xi^{(n)})-\gamma|<\varepsilon;\gamma)>1-\delta \tag{3.74}$$

for all $n \geqslant n(\varepsilon, \delta)$[47]. Let $\gamma_0 \in \Gamma_0$. Then by the definition of $\hat{\Theta}_n$

$$f^{(n)}(x^{(n)}, \hat{\Theta}_n(x^{(n)}))/f^{(n)}(x^{(n)}, \gamma_0) \geqslant 1 \tag{3.75}$$

$P_{\gamma_0, \xi^{(n)}}$-a.e.

If (3.74) were not true, we could choose a sequence of natural numbers $n_1 < n_2 < \cdots$ and give two positive numbers ε_0 and δ_0 for which

$$P(|\hat{\Theta}_{n_i}(\xi^{(n_i)}) - \gamma_0| < \varepsilon_0; \gamma_0) \leqslant 1 - \delta_0, \quad i = 1, 2, \ldots. \tag{3.76}$$

This can also be formulated as follows: For $i = 1, 2, \ldots$ there exist sets $E_{n_i} \in \mathfrak{B}_{n_i}$ such that for all $x^{(n_i)} \in E_{n_i}$

$$|\hat{\Theta}_{n_i}(x^{(n_i)}) - \gamma_0| \geqslant \varepsilon_0 \tag{3.77}$$

and

$$P(E_{n_i}; \gamma_0) > \delta_0. \tag{3.78}$$

Now for all $x^{(n_i)} \in E_{n_i}$ we have from (3.77)

$$\sup_{\substack{|\gamma - \gamma_0| \geqslant \varepsilon_0 \\ \gamma \in \Gamma_0}} f^{(n_i)}(x^{(n_i)}, \gamma) \geqslant f^{(n_i)}(x^{(n_i)}, \hat{\Theta}_{n_i}(x^{(n_i)}))$$

and hence according to (3.75) we would have for all $x^{(n_i)} \in E_{n_i}$ with the possible exception of a $P_{\gamma_0, \xi^{(n_i)}}$-null set

$$\sup_{\substack{|\gamma - \gamma_0| \geqslant \varepsilon_0 \\ \gamma \in \Gamma_0}} f^{(n_i)}(x^{(n_i)}, \gamma)/f^{(n_i)}(x^{(n_i)}, \gamma_0) \geqslant 1.$$

$\Gamma_0 \cap \{\gamma: |\gamma - \gamma_0| \geqslant \varepsilon_0\}$ is compact. Thus, if n_i is large enough, the set E_{n_i} would have to have arbitrarily small $P_{\gamma_0, \xi^{(n_i)}}$-measure by (3.57), which contradicts (3.78). Hence, (3.74) and the consistency of the sequence $\{\hat{\Theta}_n\}$ for Γ_0 is proved.

The "strong" form of Theorem 3.7 is proved in exactly the same way by applying I, Theorem 38.5, instead of I, Theorem 38.2 in the proof of Theorem 3.6. We formulate this result as

Theorem 3.8. *Assume all of the hypotheses of Theorem 3.7 are fulfilled. Let* $P_{\gamma, \infty} = \prod_{i=1}^{\infty} P_\gamma^{(i)}$ *with* $P_\gamma^{(i)} = P_\gamma$, $i = 1, 2, \ldots$.

The sequence of r.v.'s $\xi_1, \xi_2, \ldots$ *will be taken w.r.t. the probability space* $(R_\infty, \mathfrak{B}_\infty, P_{\gamma, \infty})$, $\gamma \in \Gamma$. *Then for each* $\gamma \in \Gamma_0$

$$\lim_{n \to \infty} \hat{\Theta}_n(\xi^{(n)}) = \gamma, \quad P_{\gamma, \infty}\text{-a.e.} \tag{3.79}$$

Let us mention the connection between Theorems 3.1 and 3.7. For the sake of simplicity we use the "strong" form of these theorems. Let Γ

---

[47] We do not exclude the possibility that $n(\varepsilon, \delta)$ depends on γ.

be an arbitrary open cube in R_k, $k \geqslant 1$, $\Gamma_0 \subset \Gamma$ a closed and bounded, thus compact cube and $\Gamma^* \subset \Gamma_0$ the corresponding open cube[48]. If all the assumptions of Theorem 3.1 are satisfied—condition (3.4) even in the "strong" version—and $\hat{\Theta}_n$ has the same meaning as in Theorem 3.7, then $\hat{\Theta}_n$ is also a ML estimate for Γ^* in the sense of Theorem 3.1. More precisely, let $\gamma_0 \in \Gamma^*$. Then, according to Theorem 3.8 the set of all sequences $x_1, x_2, \ldots$ from R_∞ such that for each $\varepsilon > 0$ and sufficiently large n

$$|\hat{\Theta}_n(x_1, \ldots, x_n) - \gamma_0| < \varepsilon \tag{3.80}$$

holds, has $P_{\gamma_0, \infty}$-measure 1. However, if ε is small enough, then

$$\{\gamma : |\gamma - \gamma_0| < \varepsilon\} \subseteq \Gamma^* . \tag{3.81}$$

Then necessarily for all of these sequences

$$\frac{\partial f^{(n)}(x^{(n)}, \hat{\Theta}_n(x^{(n)}))}{\partial \gamma} = 0$$

or also

$$\frac{\partial \log f^{(n)}(x^{(n)}, \hat{\Theta}_n(x^{(n)}))}{\partial \gamma} = 0 .$$

We know, however, that equation (3.20) possesses exactly one solution in a sufficiently small sphere about γ_0 and this solution must therefore coincide with $\hat{\Theta}_n(x^{(n)})$ $P_{\gamma_0, \infty}$-a.e. One can easily give sufficient conditions which guarantee that the assumptions of Theorem 3.2 as well as Theorem 3.6. hold. Except for the necessary differentiability conditions, one can take, for example, the following: Let $\phi_1, \phi_2, \phi_3, \phi_4$ and ϕ_5 be nonnegative, μ-integrable functions defined over R_1. Assume that for all $\gamma \in \Gamma$ and $x \in R_1$

$$|f(x,\gamma) \log \sup_{\gamma \in \Gamma} f(x,\gamma)| \leqslant \phi_1(x), \quad \left|\frac{\partial f(x,\gamma)}{\partial \gamma}\right| \leqslant \phi_2(x), \quad \left|\frac{\partial^2 f(x,\gamma)}{\partial \gamma_i \partial \gamma_j}\right| \leqslant \phi_3(x),$$

$$\left|\frac{\partial^2 \log f(x,\gamma)}{\partial \gamma_i \partial \gamma_j}\right| \leqslant \phi_4(x) \quad \text{and} \quad \left|\frac{\partial^3 \log f(x,\gamma)}{\partial \gamma_i \partial \gamma_j \partial \gamma_h}\right| \leqslant \phi_5(x) \quad \text{for } i,h,j = 1, \ldots, k$$

(see Theorem VII).

Furthermore, let there exist real numbers M_1, M_2 such that for all $\gamma \in \Gamma$

$$|E(\phi_4; \gamma)| \leqslant M_1 \quad \text{and} \quad |E(\phi_5; \gamma)| \leqslant M_2 .$$

---

[48] One sees immediately that these assumptions can be generalized. Γ can be an arbitrary open set and Γ^* an open, bounded subset of Γ with closure belonging to Γ. The *closure* is the *smallest closed set* containing Γ^*, i.e., the intersection of all closed subsets of R_k containing Γ^*.

These are joined in addition by condition (3.3). Note that $f(x,\gamma)\neq 0$ μ-a.e., which follows from the μ-integrability of ϕ_1.

Theorem 3.8 admits a generalization to a larger class of consistent estimates than the ML estimates.[49]

Only rarely is one fortunate enough to obtain a simply constructed sequence of ML estimates for a parameter as in the example on p. 306. It is, however, desirable in many practical situations to know the probability distribution of ML estimates—at least approximately, for large n. We now show that at least under the assumptions of Theorem 3.2[50], the sequence of ML estimates is asymptotically distributed according to a k-dimensional normal distribution with mean vector γ and covariance matrix $(E(A(\xi,\gamma);\gamma))^{-1}$. More precisely, we have in the notation of Theorem 3.1

Theorem 3.9. *Assume all of the hypotheses of Theorem 3.2 are fulfilled and let* $E(\phi;\gamma)\leqslant M$ *for all* $\gamma\in\Gamma$, *where* M *is a real, positive number. Let* $\hat{\gamma}_n$ *have the same meaning as in* (3.31). *Then, for all* $y\in R_k$ *and each* $\gamma\in\Gamma$

$$P(\sqrt{n}(\hat{\gamma}_n(\xi^{(n)})-\gamma)\leqslant y;\gamma)\to\frac{|E(A(\xi,\gamma);\gamma)|^{1/2}}{(\sqrt{2\pi})^k}\int_{-\infty}^{y}e^{-(1/2)x'E(A(\xi,\gamma);\gamma)x}\,dx \tag{3.82}$$

for $n\to\infty$.

Proof. Choose $\gamma\in\Gamma$ fixed. Probabilities or expectations taken w.r.t. this γ will not show this explicitly.

From (3.30) for all $x^{(n)}\in F_n$ (see p. 299) we have by an application of the Mean Value Theorem of calculus

$$-\frac{1}{n}\frac{\partial\log f^{(n)}(x^{(n)},\gamma)}{\partial\gamma}=\frac{1}{n}B(x^{(n)},\gamma)(\hat{\gamma}_n(x^{(n)})-\gamma)+r_n \tag{3.83}$$

where r_n is a k-dimensional vector depending on the third derivative w.r.t. γ. Recall that for all $x^{(n)}\in F_n$ and arbitrary small $\rho>0$

$$|\hat{\gamma}_n(x^{(n)})-\gamma|\leqslant\rho, \tag{3.84}$$

provided n is large enough. Thus, in complete analogy to (3.39), we find for r_n

$$|r_n|\leqslant 2^{-1}k^2M|\hat{\gamma}_n(x^{(n)})-\gamma|\rho. \tag{3.85}$$

[49] See f.i. J. Pfanzagl, Metrika 14, 249—272 (1969).

[50] A careful analysis of the assumptions used to prove asymptotic normality of ML estimates is given by L. Le Cam, Ann. Math. Statist. 41, 802—828 (1970). See also P. J. Huber, Proc. 5th. Berkeley Sympos. Math. Stat. Probab. Vol. I, 221—233 (1967).

Because of (3.3) we obtain from (3.83) and (3.18) after slight rearrangement

$$\left.\begin{aligned}&(E(A(\xi,\gamma)))^{-1}\frac{1}{\sqrt{n}}\frac{\partial\log f^{(n)}(x^{(n)},\gamma)}{\partial\gamma}\\&=\sqrt{n}(\hat{\gamma}_n(x^{(n)})-\gamma)+\sqrt{n}[E(A(\xi,\gamma))]^{-1}\\&\times\left\{\frac{1}{n}[E(B(\xi^{(n)},\gamma))-B(x^{(n)},\gamma)](\hat{\gamma}_n(x^{(n)})-\gamma)-r_n\right\}.\end{aligned}\right\}\quad(3.86)$$

It now suffices to apply I, Theorems 38.3, 40.1 and 40.2 repeatedly. We indicate this procedure. Let $\eta_1^{(n)}$ denote the k-dimensional r. v.

$$(E(A(\xi,\gamma)))^{-1}\frac{1}{\sqrt{n}}\frac{\partial\log f(\xi^{(n)},\gamma)}{\partial\gamma}$$

and $\eta_2^{(n)}$ the r. v. on the right side of (3.86). Since (3.86) is at least valid for $x^{(n)}\in F_n$, (3.86) and (3.16) state that $\eta_1^{(n)}=\eta_2^{(n)}+\eta_3^{(n)}$, where $\eta_3^{(n)}$ converges stochastically to 0.[51]

From (3.85) and (3.14) follows:

$$\eta_2^{(n)}=[I+C_n(\xi^{(n)})]\sqrt{n}(\hat{\gamma}_n(\xi^{(n)})-\gamma)$$

where I is the $k\times k$ identity matrix and $C_n(\xi^{(n)})$ is a $k\times k$ matrix of r.v.'s all of whose elements converge stochastically to 0 for $n\to\infty$.

Further, it follows from (3.10) and (3.3) by means of I, Theorem 39.4, that the sequence $\frac{1}{\sqrt{n}}\sum_{i=1}^{n}\frac{\partial\log f(\xi_i,\gamma)}{\partial\gamma}$, $n=1,2,\dots$ is asymptotically normally distributed with mean vector 0[52] and covariance matrix $E(A(\xi,\gamma))$. Hence, $\eta_1^{(n)}$ and $\eta_1^{(n)}-\eta_3^{(n)}$ possess, according to the supplement to Theorem 39.4, an asymptotic normal distribution with mean vector 0 and covariance matrix $(E(A(\xi,\gamma)))^{-1}$. Since, however,

$$\sqrt{n}(\hat{\gamma}_n(\xi^{(n)})-\gamma)=(I+C_n(\xi^{(n)}))^{-1}(\eta_1^{(n)}-\eta_3^{(n)})\,^{53},$$

we are finished.

Specializing to the important case for which Γ is a one-dimensional set we obtain as special case of Theorem 3.9:

$\hat{\gamma}_n(\xi^{(n)})-\gamma$ is asymptotically normally distributed according to $N\left(0,\frac{1}{n}\left[E\left(\frac{\partial\log f(\xi,\gamma)}{\partial\gamma}\right)^2;\gamma\right]^{-1}\right)$.

[51] All these statements are related to probabilities w.r.t. γ.

[52] 0 represents here the k-dimensional null vector.

[53] Actually, we have suppressed a step here since $(I+C_n(\xi^{(n)}))^{-1}$ is only defined with probability arbitrarily close to 1 for sufficiently large n.

Referring to the example of the Poisson distribution on p. 306, we find that $(\bar{\xi}_n - a)\sqrt{n}/\sqrt{a}$ is asymptotically $N(0,1)$-distributed—a result we already know (see p. 108).

We now combine all of the results of Theorems 3.1—3.9 and obtain an important result for which we choose the "strong" formulation. We retain the previous notation and get

Theorem 3.10. *Let Γ be an open cube of R_k, $k \geqslant 1$ and Γ^* an arbitrary open set whose closure Γ_0*[54] *is compact and contained in Γ. Let all of the assumptions of Theorems 3.6 and 3.9 be fulfilled in Γ (in the strong formulation) and let the map $\gamma \to E(A(\xi,\gamma);\gamma)$ be continuous in Γ. Then there exists a consistent sequence $\{\hat{\gamma}_n\}$ of estimates for $\gamma \in \Gamma$ which possesses the following properties: for each $\gamma \in \Gamma^*$, $\hat{\gamma}_n$ is a solution of (3.20) with $P_{\gamma,\infty}$-probability one; moreover, $\hat{\gamma}_n$ is, with $P_{\gamma,\infty}$-probability one, the only solution of (3.20) in a closed ball with center γ and radius >0 which can be chosen independently of $\gamma \in \Gamma^*$ provided n is sufficiently large*[55]. *Further, $\sqrt{n}(\hat{\gamma}_n(\xi^{(n)}) - \gamma)$ is (w.r.t. $P_{\gamma,\infty}$) asymptotically normally distributed with mean vector 0 and covariance matrix $(E(A(\xi,\gamma)))^{-1}$.*

This theorem can be proved under weaker conditions and can also be augmented by statements about uniform convergence[56]. We will not pursue this further here.

Wald's method of proof has turned out to be quite important. A whole series of new results rest on the extension of this method. We give some of these here. They are also of interest in so far as they establish a connection between the MLP and the *Bayes' approach*, which we have already mentioned. Since we have already considered Wald's method and related topics in some detail on the previous pages, we will often be brief in the sequel. We prefer the strong version of the results from now on.

We beginn with

Lemma 3.5. *Let ξ be a r.v. with probability distribution P_ξ and $\xi_1, \xi_2, \ldots$ a sequence of independent r.v.'s possessing the same distribution P_ξ. Take $P_\infty = \prod_{i=1}^{\infty} P_{\xi_i}$ with $P_{\xi_i} = P_\xi$. Let Γ be an arbitrary open set of R_k, $k \geqslant 1$ and $\Gamma_0 \subset \Gamma$ a compact cube. Assume u is a function defined on $R_1 \times \Gamma$ such that $x \to u(x,\gamma)$ is measurable for each $\gamma \in \Gamma$ and $\gamma \to u(x,\gamma)$ is continuous for each $x \in R_1$. Let $E(u(\xi,\gamma))$ exist for each $\gamma \in \Gamma$ and let $\gamma \to E(u(\xi,\gamma))$*

---

[54] See[48] for remarks on this formulation.

[55] This follows from the additional continuity assumption and (3.3). Thus, $\gamma \to |E(A(\xi,\gamma);\gamma)|$ possesses a positive lower bound in Γ_0.

[56] L. Le Cam, Univ. California Publ. Statist. 1, 277—329 (1953) and l.c.[50].

be continuous in Γ. Finally, let $E\left(\sup_{\gamma\in K_\rho(\gamma_0)} u(\xi,\gamma)\right)$ *and* $E\left(\inf_{\gamma\in K_\rho(\gamma_0)} u(\xi,\gamma)\right)$[57] *exist for each $\gamma_0\in\Gamma$ and for sufficiently small ρ.*

Then

$$\overline{\lim}\sup_{\gamma\in\Gamma_0}\left(\frac{1}{n}\sum_{i=1}^{n} u(\xi_i,\gamma)-E(u(\xi,\gamma))\right)\leqslant 0, \tag{3.87}$$

and

$$\underline{\lim}\inf_{\gamma\in\Gamma_0}\left(\frac{1}{n}\sum_{i=1}^{n} u(\xi_i,\gamma)-E(u(\xi,\gamma))\right)\geqslant 0 \tag{3.88}$$

P_∞-a.e. It is easily seen that (3.87) and (3.88) are equivalent to

$$\lim_{n\to\infty}\sup_{\gamma\in\Gamma_0}\left|\frac{1}{n}\sum_{i=1}^{n} u(\xi_i,\gamma)-E(u(\xi,\gamma))\right|=0.$$

Proof. For each $x\in R_1$ and $\gamma_0\in\Gamma$ set

$$u^*(x,\gamma_0,\rho)=\sup_{\gamma\in K_\rho(\gamma_0)} u(x,\gamma).$$

Write $E(u(\xi,\gamma))=a(\gamma)$. For each $x\in R_1$ and $\gamma_0\in\Gamma$ $u^*(x,\gamma_0,\rho)$ increases monotonically to $u(x,\gamma_0)$ for $\rho\to 0$ and thus also

$$\lim_{\rho\to 0} E(u^*(\xi,\gamma_0,\rho))=a(\gamma_0). \tag{3.89}$$

Hence, for each $\varepsilon>0$ and $\gamma\in\Gamma$ one can choose a $\rho_{\varepsilon,\gamma}>0$ such that

$$a(\gamma)\leqslant E(u^*(\xi,\gamma,\rho_{\varepsilon,\gamma}))\leqslant a(\gamma)+\varepsilon/2. \tag{3.90}$$

Moreover, because of the continuity of $\gamma\to a(\gamma)$ one can also assume that for each $\gamma\in\Gamma$

$$|a(\tilde{\gamma})-a(\gamma)|<\varepsilon/2 \tag{3.91}$$

for all $\tilde{\gamma}$ with $|\tilde{\gamma}-\gamma|<\rho_{\varepsilon,\gamma}$.

Since Γ_0 is compact, one can choose finitely many $\gamma_i\in\Gamma_0$ with, say, $1\leqslant i\leqslant m$, for which

$$\Gamma_0\subseteq\bigcup_{i=1}^{m} K_{\rho_{\varepsilon,\gamma_i}}(\gamma_i). \tag{3.92}$$

From (3.90) and (3.91) for $1\leqslant i\leqslant m$ and $\gamma\in K_{\rho_{\varepsilon,\gamma_i}}(\gamma_i)$ we have

$$a(\gamma)-\varepsilon\leqslant E(u^*(\xi,\gamma_i,\rho_{\varepsilon,\gamma_i}))\leqslant a(\gamma)+\varepsilon. \tag{3.93}$$

Let $(x_1,x_2,\dots,x_n,\dots)\in R_\infty$ and $\gamma\in K_{\rho_{\varepsilon,\gamma_j}}(\gamma_j)$ for j with $1\leqslant j\leqslant m$. Then according to (3.93) for $n\geqslant 1$

$$\frac{1}{n}\sum_{i=1}^{n} u(x_i,\gamma)-a(\gamma)\leqslant\frac{1}{n}\sum_{i=1}^{n} u^*(x_i,\gamma_j,\rho_{\varepsilon,\gamma_j})-E(u^*(\xi,\gamma_j,\rho_{\varepsilon,\gamma_j}))+\varepsilon$$

[57] For the notation see Theorem 3.6.

and thus for each $\gamma \in \Gamma_0$

$$\frac{1}{n}\sum_{i=1}^{n} u(x_i,\gamma)-a(\gamma) \leqslant \sup_{j=1,\ldots,m}\left(\frac{1}{n}\sum_{i=1}^{n} u^*(x_i,\gamma_j,\rho_{\varepsilon,\gamma_j})-E(u^*(\xi,\gamma_j,\rho_{\varepsilon,\gamma_j}))\right)+\varepsilon.$$

Hence, also

$$\sup_{\gamma\in\Gamma_0}\left(\frac{1}{n}\sum_{i=1}^{n} u(x_i,\gamma)-a(\gamma)\right)$$

$$\leqslant \sup_{j=1,\ldots,m}\left(\frac{1}{n}\sum_{i=1}^{n} u^*(x_i,\gamma_j,\rho_{\varepsilon,\gamma_j})-E(u^*(\xi,\gamma_j,\rho_{\varepsilon,\gamma_j}))\right)+\varepsilon.$$

But now according to I, Theorem 38.5, for all elements from R_∞, with at most the exception of a P_∞-null set,

$$\lim_{n\to\infty}\left(\frac{1}{n}\sum_{i=1}^{n} u^*(x_i,\gamma_j,\rho_{\varepsilon,\gamma_j})-E(u^*(\xi,\gamma_j,\rho_{\varepsilon,\gamma_j}))\right)=0.$$

From this (3.87) follows. One proves (3.88) analogously.

Remark. If u is only defined over $R_1\times\Gamma_0$ and all the other assumptions on u are satisfied, then Lemma 3.5 still holds.

We now prove

Theorem 3.11[58]. *Let Γ be an open cube of R_k, $k\geqslant 1$ and $\boldsymbol{\gamma}$ a k-dimensional r.v. with density p which vanishes in $R_k-\Gamma$. Let $\Gamma_0\subset\Gamma$ be a compact cube and $\Gamma^*\subset\Gamma_0$ the corresponding open cube. Take $\gamma_0\in\Gamma^*$ and let p at γ_0 be continuous and positive. Let ξ be a r.v. such that the joint distribution of ξ and $\boldsymbol{\gamma}$ possesses a density and let $f(x\,|\,\gamma)$ for all $x\in R_1$ be the conditional density of ξ under the hypothesis $\boldsymbol{\gamma}=\gamma$, provided it exists*[59]. *For all $\gamma^{(1)},\gamma^{(2)}\in\Gamma_0$, $\gamma^{(1)}\neq\gamma^{(2)}$ let*

$$P_\xi(\{x: f(x\,|\,\gamma^{(1)})-f(x\,|\,\gamma^{(2)})\neq 0\})>0.$$

For each $x\in R_1$[60] *and $\gamma\in\Gamma_0$ assume all first and second derivatives of $\gamma\to f(x\,|\,\gamma)$ exist and are continuous. Moreover, assume*

$$\frac{\partial}{\partial\gamma_i}\int_{R_1} f(x\,|\,\gamma_0)\,dx=\int_{R_1}\frac{\partial}{\partial\gamma_i} f(x\,|\,\gamma_0)\,dx \tag{3.94}$$

[58] Cf. L. Le Cam, loc. cit.[56]. See also P.J. Bickel and J.A. Yahav, Z. Wahrscheinlichkeitstheorie und Verw. Gebiete 11, 257—276 (1969) and the references given there.

[59] See I, p. 59. The map $\gamma\to f(x\,|\,\gamma)$ is thus defined up to a null set (which does not depend on x).

[60] Possibly with the exception of L-null sets.

and

$$\frac{\partial^2}{\partial\gamma_i\partial\gamma_i}\int_{R_1} f(x|\gamma_0)dx=\int_{R_1}\frac{\partial^2}{\partial\gamma_i\partial\gamma_j} f(x|\gamma_0)dx, \qquad i,j=1,\ldots,k. \tag{3.95}$$

Furthermore, let there exist for all $\gamma\in\Gamma_0$, *sufficiently small* ρ *and* $i,j=1,\ldots,k$

$$E\left(\sup_{\tilde\gamma\in K_\rho(\gamma)}\left|\frac{\partial^2\log f(\xi|\tilde\gamma)}{\partial\gamma_i\partial\gamma_j}\right||\gamma_0\right)$$

and let $\gamma\to E\left(\frac{\partial^2\log f(\xi|\gamma)}{\partial\gamma_i\partial\gamma_j}\middle|\gamma_0\right)$ *be continuous. Assume* $E(A(\xi|\gamma_0)|\gamma_0)$ [61] *is positive definite. For each* $\gamma\in\Gamma$, $\rho>0$ *and* $x\in R_1$ *set*

$$f^*(x,\gamma,\rho)=\sup_{\tilde\gamma\in K_\rho(\gamma)} f(x|\tilde\gamma)$$ [62]

and

$$f_*(x,\gamma,\rho)=\inf_{\tilde\gamma\in K_\rho(\gamma)} f(x|\tilde\gamma).$$

For all $\gamma\in\Gamma_0$ *and all sufficiently small* ρ *let there exist* $E(\log f_*(\xi,\gamma,\rho)|\gamma_0)$ *and* $E(\log f^*(\xi,\gamma,\rho)|\gamma_0)$. *Let the map* $\gamma\to E(\log f(\xi|\gamma)|\gamma_0)$ *be continuous in* Γ_0. *For an* $\eta>0$ *and all* $\gamma\in\Gamma-\Gamma_0$ *let*

$$E\left(\log\frac{f(\xi|\gamma_0)}{f(\xi|\gamma)}\middle|\gamma_0\right)\geqslant\eta \text{ [63]}. \tag{3.96}$$

Let $\xi_1,\xi_2,\ldots$ *be a sequence of r.v.'s. For each* $n\geqslant 1$ *and* $x^{(n)}=(x_1,\ldots,x_n)\in R_n$ *let the conditional density of* $\xi^{(n)}=(\xi_1,\ldots,\xi_n)$ *under the hypothesis* $\gamma=\gamma$ *be given by*

$$f^{(n)}(x^{(n)}|\gamma)=\prod_{i=1}^{n} f(x_i|\gamma).$$

Suppose

$$\int_{\Gamma-\Gamma_0} f^{(n)}(x^{(n)}|\gamma)p(\gamma)d\gamma\left[\int_\Gamma f^{(n)}(x^{(n)}|\gamma)p(\gamma)d\gamma\right]^{-1}\to 0 \tag{3.97}$$

as $n\to\infty$ *with the exception of a set of* $P_{\gamma_0,\infty}$*-measure zero.*

Assume there exists a Borel measurable map Θ_n *from* R_n *into* Γ_0 *for* $n=1,2,\ldots$ *which fulfills the condition: for all sequences* $(x_1,x_2,\ldots)\in R_\infty$

[61] We use the notation of Theorem 3.1 although the context is somewhat different.

[62] $K_\rho(\gamma)$ has the same meaning as in Theorem 3.6.

[63] We allow the possibility that the expectation on the left is $+\infty$.

and sufficiently large n there hold with exception of a set of $P_{\gamma_0,\infty}$-measure zero:

$$\sup_{\gamma\in\Gamma_0} f^{(n)}(x^{(n)}|\gamma)=f^{(n)}(x^{(n)}|\hat{\Theta}_n(x^{(n)})) \tag{3.98}$$

$$\frac{\partial \log f^{(n)}(x^{(n)}|\hat{\Theta}_n(x^{(n)}))}{\partial\gamma}=0 \tag{3.99}$$

and

$$\lim_{n\to\infty}\hat{\Theta}_n(x^{(n)})=\gamma_0\,. \tag{3.100}$$

For each $\gamma\in\Gamma$ and $x^{(n)}\in R_n$ for the sake of simplicity there shall exist

$$p^{(n)}(\gamma|x^{(n)})=\frac{f^{(n)}(x^{(n)}|\gamma)p(\gamma)}{\int_\Gamma f^{(n)}(x^{(n)}|\gamma')p(\gamma')d\gamma'}\,. \tag{3.101}$$

Then: the set of elements of R_∞ with

$$\lim_{n\to\infty}\int_\Gamma |p^{(n)}(\gamma|x^{(n)})-|E(A(\xi|\gamma_0)|\gamma_0)|^{1/2}(n/(2\pi))^{k/2}$$
$$\times e^{-\frac{1}{2}n(\gamma-\hat{\Theta}_n(x^{(n)}))'E(A(\xi|\gamma_0)|\gamma_0)(\gamma-\hat{\Theta}_n(x^{(n)}))}|d\gamma=0 \tag{3.102}$$

has $P_{\gamma_0,\infty}$-measure 1 [64].

This theorem says in *summary*: Let "the parameter" of a set of probability measures be a k-dimensional r.v. γ with density p which will be called in this connection the "*a priori*" *distribution* of γ. The so-called "*a posteriori*" *distribution*[65] of γ, whose density for $n\geqslant 1$ and each $x^{(n)}\in R_n$ is given by (3.101), is then, for sufficiently large n and independently of the a priori distribution, approximated by a normal distribution with mean value $\hat{\Theta}_n(x^{(n)})$ and covariance matrix $\frac{1}{n}[E(A(\xi|\gamma_0)|\gamma_0)]^{-1}$.

In practice, this result is often stated in another form: The a posteriori distribution of $(\gamma-\hat{\Theta}_n(x^{(n)}))\sqrt{n}$ is, for sufficiently large n, (approximately) given by a normal distribution with mean 0 and covariance matrix $[E(A(\xi|\gamma_0)|\gamma_0)]^{-1}$.

We proceed to the *proof*[66] of Theorem 3.11. We first show that one can replace the conditional density $p^{(n)}$ at (3.101) by another conditional density which vanishes outside an open subset of Γ^* and for which a result analogous to (3.102) holds.

---

[64] For the meaning of $P_{\gamma_0,\infty}$ see the definition of $P_{\gamma_0,\infty}$ on p. 314. Denoting the probability defined by the conditional density $x\to f(x|\gamma_0)$ by P_{γ_0} we have $P_{\gamma_0,\infty}=\prod_{i=1}^{\infty}P_{\gamma_0,i}$ with $P_{\gamma_0,i}=P_{\gamma_0}$ for $i=1,2,\ldots$.

[65] See IV, p. 265.

[66] L. Le Cam, loc. cit.[56].

Consider the set X of all $(x_1, x_2, \dots) \in R_\infty$ for which $f^{(n)}(x^{(n)} | \gamma_0) \neq 0$ for $n = 1, 2, \dots$. The complement of X has $P_{\gamma_0,\infty}$-measure 0. In the sequel, we will always refer to X. For $\gamma \in \Gamma$ let

$$b(\gamma) = E(\log f(\xi | \gamma) | \gamma_0).$$

Let $\varepsilon > 0$ and

$$2\varepsilon < \eta. \tag{3.103}$$

Further let

$$\Gamma_{1,\varepsilon} = \Gamma_1 = \{\gamma: -b(\gamma_0) + b(\gamma) > -\varepsilon\} \tag{3.104}$$

and

$$\Gamma_{2,\varepsilon} = \Gamma_2 = \{\gamma: -b(\gamma_0) + b(\gamma) \leqslant -2\varepsilon\} \tag{3.105}$$

Naturally, $\gamma_0 \in \Gamma_1$. Since b is continuous in Γ_0, $\Gamma_0 \cap \Gamma_2$ is compact. From (3.96) and (3.103), $\Gamma_1 \subset \Gamma_0$.

We show that the assumptions imply the existence of $E\Big(\sup\limits_{\tilde{\gamma} \in K_\rho(\gamma)} \log f(\xi | \tilde{\gamma}) | \gamma_0\Big)$ for all $\gamma \in \Gamma_0$ and all sufficiently small ρ. Because of the monotonicity of the logarithm it follows from the existence of $E(\log f^*(\xi, \gamma, \rho) | \gamma_0)$ for these γ and ρ. Similarly the existence of $E\Big(\inf\limits_{\tilde{\gamma} \in K_\rho(\gamma)} \log f(\xi | \tilde{\gamma}) | \gamma_0\Big)$ follows from that of $E(\log f_*(\xi, \gamma, \rho) | \gamma_0)$. Now apply Lemma 3.5, use (3.104) and (3.105) and replace $E(\log f(\xi | \gamma_0) | \gamma_0)$ according to I, Theorem 38.5 by $\frac{1}{n} \sum\limits_{i=1}^{n} \log f(x_i | \gamma_0)$: For each $\delta > 0$ and $x \in X$ there is—with the exception of at most a set of $P_{\gamma_0,\infty}$-measure 0—an $N(\delta, x)$ such that for $n \geqslant N(\delta, x)$

$$\inf_{\gamma \in \Gamma_1} \frac{1}{n} \sum_{i=1}^{n} (\log f(x_i | \gamma) - \log f(x_i | \gamma_0)) \geqslant -\varepsilon - \delta \tag{3.106}$$

and

$$\sup_{\gamma \in \Gamma_0 \cap \Gamma_2} \frac{1}{n} \sum_{i=1}^{n} (\log f(x_i | \gamma) - \log f(x_i | \gamma_0)) \leqslant -2\varepsilon + \delta. \tag{3.107}$$

Choose

$$\delta < \varepsilon/3. \tag{3.108}$$

From (3.106) for each $\gamma \in \Gamma_1$

$$\prod_{i=1}^{n} f(x_i | \gamma) \Big/ \prod_{i=1}^{n} f(x_i | \gamma_0) \geqslant e^{-n(\varepsilon + \delta)} \tag{3.109}$$

and from (3.107) for $\gamma \in \Gamma_0 \cap \Gamma_2$

$$\prod_{i=1}^{n} f(x_i|\gamma) \Big/ \prod_{i=1}^{n} f(x_i|\gamma_0) \leqslant e^{-n(2\varepsilon-\delta)} \tag{3.110}$$

for all $n \geqslant N(\delta, x)$.

Now let $\Gamma_* \subseteq \Gamma^*$ be an arbitrary open subset containing γ_0. Using the continuity of b in Γ_0 and (3.58) it can be seen that there exists an $\varepsilon > 0$ satisfying (3.103) and such that $\Gamma - \Gamma_* \subseteq \Gamma_2$.

For $n \geqslant 1$ and $x^{(n)} \in R_n$ define

$$q^{(n)}(\gamma|x^{(n)}) = \begin{cases} 0 & \gamma \in \Gamma - \Gamma_* \\ p^{(n)}(\gamma|x^{(n)}) \Big/ \int\limits_{\Gamma_*} p^{(n)}(\gamma'|x^{(n)})\,d\gamma' & \gamma \in \Gamma_* . \end{cases}$$

This definition makes sense only up to null sets.

We want to show that

$$\lim_{n\to\infty} \int\limits_{\Gamma} |p^{(n)}(\gamma|x^{(n)}) - q^{(n)}(\gamma|x^{(n)})|\,d\gamma = 0 . \tag{3.111}$$

According to (3.101) and the definition of X we have, using the abbreviations

$$f^{(n)}(x^{(n)}|\gamma) = f_n \quad \text{and} \quad f^{(n)}(x^{(n)}|\gamma_0) = f_n^{(0)}$$

for the integrands:

$$\begin{aligned} \int\limits_{\Gamma} |p^{(n)}(\gamma|x^{(n)}) - q^{(n)}(\gamma|x^{(n)})|\,d\gamma &= 2 \int\limits_{\Gamma-\Gamma_*} p^{(n)}(\gamma|x^{(n)})\,d\gamma \\ &= 2 \int\limits_{\Gamma-\Gamma_*} f_n (f_n^{(0)})^{-1} p(\gamma)\,d\gamma \Big[\int\limits_{\Gamma} f_n (f_n^{(0)})^{-1} p(\gamma)\,d\gamma\Big]^{-1} . \end{aligned} \tag{3.112}$$

Here we have used the fact that for two arbitrary probability densities f and g $\int\limits_{R_1} |f(x) - g(x)|\,dx = 2 \int\limits_{\{x: f(x) \geqslant g(x)\}} (f(x) - g(x))\,dx$ always holds.

Moreover, from (3.109)

$$\int\limits_{\Gamma} f_n (f_n^{(0)})^{-1} p(\gamma)\,d\gamma \geqslant \int\limits_{\Gamma_1} f_n (f_n^{(0)})^{-1} p(\gamma)\,d\gamma \geqslant e^{-n(\varepsilon+\delta)} \int\limits_{\Gamma_1} p(\gamma)\,d\gamma . \tag{3.113}$$

Since $\gamma_0 \in \Gamma_1$, $p(\gamma_0) > 0$ and p is continuous at γ_0

$$\int\limits_{\Gamma_1} p(\gamma)\,d\gamma > 0 .$$

We now assume that (3.111) is false. Then from (3.112) on a set E of positive $P_{\gamma_0,\infty}$-measure

$$\overline{\lim_{n\to\infty}} \int\limits_{\Gamma-\Gamma_*} f_n (f_n^{(0)})^{-1} p(\gamma)\,d\gamma \Big[\int\limits_{\Gamma} f_n (f_n^{(0)})^{-1} p(\gamma)\,d\gamma\Big]^{-1} > 0 .$$

Then (3.113) and (3.97) yield

$$\varlimsup_{n\to\infty} e^{n(\varepsilon+\delta)} \int_{\Gamma_0-\Gamma_*} f_n(f_n^{(0)})^{-1} p(\gamma)d\gamma > 0 .$$

From (3.107) and (3.108) we conclude that

$$\int_{\Gamma_0-\Gamma_*} \varlimsup e^{n(\varepsilon+\delta)} f_n(f_n^{(0)})^{-1} p(\gamma)d\gamma > 0$$

using Theorem V and therefore

$$\int_{\Gamma-\Gamma_*} \varlimsup_{n\to\infty} e^{n(\varepsilon+\delta)} f_n(f_n^{(0)})^{-1} p(\gamma)d\gamma > 0 .$$

Hence, for each $x\in E$ there exists a $\gamma\in\Gamma-\Gamma_*$ and an $\alpha>0$ such that for infinitely many n, which can depend on x,

$$e^{n(\varepsilon+\delta)} f_n(f_n^{(0)})^{-1} > \alpha$$

and further $n(\varepsilon+\delta)+\log(f_n/f_n^{(0)})>\log\alpha$ and

$$\frac{1}{n}\sum_{i=1}^{n} \log\frac{f(x_i|\gamma)}{f(x_i|\gamma_0)} > -(\varepsilon+\delta) + \frac{1}{n}\log\alpha .$$

Thus, according to (3.107), $\gamma\notin\Gamma_0\cap\Gamma_2$. Since E possesses positive $P_{\gamma_0,\infty}$-measure, we have from this last inequality by I, Theorem 38.5:

$$E\left(\log\frac{f(\xi|\gamma)}{f(\xi|\gamma_0)}\middle|\gamma_0\right) \geqslant -(\varepsilon+\delta) .$$

From (3.103), (3.108) and (3.96) $\gamma\in\Gamma_0$ and further $\gamma\in\Gamma_0-\Gamma_2$ which contradicts $\gamma\in\Gamma-\Gamma_*\subseteq\Gamma_2$. This proves (3.111).

Thus one can replace $p^{(n)}(\gamma|x^{(n)})$ in (3.102) by $q^{(n)}(\gamma|x^{(n)})$ and by suitable choice of ε we can cause $\gamma\to q^{(n)}(\gamma|x^{(n)})$ to vanish outside of an arbitrary open set containing γ_0.

We now show that we can further reduce the problem by proving: For each $\beta>0$ with $p(\gamma_0)-\beta>0$, all $x\in R_\infty$ up to a $P_{\gamma_0,\infty}$-null set and sufficiently large n an open set $\Gamma_*=\Gamma_\beta$ can be chosen such that

$$\int_{\Gamma_\beta} \left| q^{(n)}(\gamma|x^{(n)}) - \frac{f^{(n)}(x^{(n)}|\gamma)}{\int_{\Gamma_\beta} f^{(n)}(x^{(n)}|\gamma')d\gamma'}\right| d\gamma \leqslant \frac{2\beta}{p(\gamma_0)-\beta} . \tag{3.114}$$

We may assume that $x\in X$ and that Γ_β is so chosen that

$$|p(\gamma)-p(\gamma_0)|<\beta \tag{3.115}$$

for $\gamma\in\Gamma_\beta$.

From (3.109) and (3.115), at least for sufficiently large n,

$$\int_{\Gamma_\beta} f_n(f_n^{(0)})^{-1} p(\gamma)d\gamma > 0 . \tag{3.116}$$

We then obtain

$$\int_{\Gamma_\beta}\left|q^{(n)}(\gamma\,|\,x^{(n)}) - \frac{f^{(n)}(x^{(n)}\,|\,\gamma)}{\int_{\Gamma_\beta} f^{(n)}(x^{(n)}\,|\,\gamma')\,d\gamma'}\right| d\gamma$$

$$\leqslant \int_{\Gamma_\beta}\left|f_n(f_n^{(0)})^{-1}(p(\gamma)-p(\gamma_0))\right| d\gamma \Big/ \left|\int_{\Gamma_\beta} f_n(f_n^{(0)})^{-1}\,p(\gamma)\,d\gamma\right|$$

$$+\int_{\Gamma_\beta} f_n(f_n^{(0)})^{-1}\,p(\gamma_0)\,d\gamma\left|\int_{\Gamma_\beta} f_n(f_n^{(0)})^{-1}(p(\gamma)-p(\gamma_0))\,d\gamma\right|$$

$$\times\left[\int_{\Gamma_\beta} f_n(f_n^{(0)})^{-1}\,p(\gamma_0)\,d\gamma \int_{\Gamma_\beta} f_n(f_n^{(0)})^{-1}\,p(\gamma)\,d\gamma\right]^{-1}.$$

Hence,

$$\int_{\Gamma_\beta}\left|q^{(n)}(\gamma\,|\,x^{(n)}) - \frac{f^{(n)}(x^{(n)}\,|\,\gamma)}{\int_{\Gamma_\beta} f^{(n)}(x^{(n)}\,|\,\gamma')\,d\gamma'}\right| d\gamma$$

$$\leqslant 2\int_{\Gamma_\beta}\left|f_n(f_n^{(0)})^{-1}(p(\gamma)-p(\gamma_0))\right| d\gamma\left[\int_{\Gamma_\beta} f_n(f_n^{(0)})^{-1}\,p(\gamma)\,d\gamma\right]^{-1}. \tag{3.117}$$

Replacing $\int_{\Gamma_\beta} f_n(f_n^{(0)})^{-1}\,p(\gamma)\,d\gamma$ by

$$\int_{\Gamma_\beta} f_n(f_n^{(0)})^{-1}\,p(\gamma_0)\,d\gamma+\int_{\Gamma_\beta} f_n(f_n^{(0)})^{-1}(p(\gamma)-p(\gamma_0))\,d\gamma$$

and noting that from (3.115)

$$\int_{\Gamma_\beta} f_n(f_n^{(0)})^{-1}(p(\gamma)-p(\gamma_0))\,d\gamma \geqslant -\frac{\beta}{p(\gamma_0)}\int_{\Gamma_\beta} f_n(f_n^{(0)})^{-1}\,p(\gamma_0)\,d\gamma$$

we see that the right side of (3.117) is $\leqslant 2\,\dfrac{\beta/p(\gamma_0)}{1-\beta/p(\gamma_0)} = \dfrac{2}{p(\gamma_0)-\beta}$ and from this follows (3.114). With the notation

$$r^{(n)}(\gamma\,|\,x^{(n)}) = f^{(n)}(x^{(n)}\,|\,\gamma)\Big/\int_{\Gamma_\beta} f^{(n)}(x^{(n)}\,|\,\gamma)\,d\gamma \quad \text{for } \gamma\in\Gamma_\beta$$

we must thus finally show that

$$\lim_{n\to\infty}\int_{\Gamma_\beta}\left|r^{(n)}(\gamma\,|\,x^{(n)}) - |E(A(\xi\,|\,\gamma_0)\,|\,\gamma_0)|^{1/2}(n/(2\pi))^{k/2}\right.$$

$$\left.\times\exp\left[-\tfrac{1}{2}n(\gamma-\hat{\Theta}_n(x^{(n)}))'\,E(A(\xi\,|\,\gamma_0)\,|\,\gamma_0)(\gamma-\hat{\Theta}_n(x^{(n)}))\right]\right|d\gamma = 0$$

with $P_{\gamma_0,x}$-measure 1.

We introduce in addition the following notation: Let A be an arbitrary $k\times k$ matrix. Order its elements into a vector with k^2 elements and denote the length of this vector, which obviously does not depend on the ordering chosen, by $\|A\|$.

Since $E(A(\xi|\gamma_0)|\gamma_0)$ is positive definite and symmetric, there exists a $\lambda>0$ such that

$$u' E(A(\xi|\gamma_0)|\gamma_0)u \geqslant \lambda |u|^2 \tag{3.118}$$

for all $u \in R_k$.

By continuity there exists for each $\varepsilon_1>0$ an open sphere K_{ε_1} with center γ_0 for which

$$\|E(B(\xi|\gamma)|\gamma_0)+E(A(\xi|\gamma_0)|\gamma_0)\| < \varepsilon_1 .^{67} \tag{3.119}$$

Hence,

$$\begin{aligned}\|E(A(\xi|\gamma_0)|\gamma_0)\|-\varepsilon_1 &\leqslant \inf_{\gamma\in K_{\varepsilon_1}} \|E(B(\xi|\gamma)|\gamma_0)\| \\ &\leqslant \sup_{\gamma\in K_{\varepsilon_1}} \|E(B(\xi|\gamma)|\gamma_0)\| \leqslant \|E(A(\xi|\gamma_0)|\gamma_0)\|+\varepsilon_1 .\end{aligned} \tag{3.120}$$

From (3.119) we have after applying Lemma 3.5 k^2 times and for all $x \in R_\infty$ up to a $P_{\gamma_0,\infty}$-null set

$$\sup_{\gamma\in K_{\varepsilon_1}} \left\| \frac{1}{n} B(x^{(n)}|\gamma)+E(A(\xi|\gamma_0)|\gamma_0)\right\| < 2\varepsilon_1 \tag{3.121}$$

provided one chooses n large enough. In particular, choosing ε_1 sufficiently small we also have from (3.118)

$$\inf_{\gamma\in K_{\varepsilon_1}} \frac{1}{n} u' B(x^{(n)}|\gamma)u \leqslant -\frac{\lambda}{2}|u|^2 \tag{3.122}$$

for all $u \in R_k$.

We now take $\varepsilon_1>0$ so small that, taking (3.115) under consideration, we may set $\Gamma_\beta = K_{\varepsilon_1}$. Moreover, we choose a fixed $x \in X$ and can assume that (3.109), (3.110), (3.121) and (3.122) are fulfilled for sufficiently large n. One can also assume because of (3.109) that $f^{(n)}(x^{(n)}|\gamma) \neq 0$ for $\gamma \in K_{\varepsilon_1}$. For sufficiently large n, (3.99) is also fulfilled and we can take $\hat{\Theta}_n(x^{(n)}) \in K_{\varepsilon_1}$ because of (3.100). Writing for arbitrary $\gamma \in \Gamma$

$$v=(\gamma-\hat{\Theta}_n(x^{(n)}))\sqrt{n}$$

we have

$$\begin{aligned}&\log f^{(n)}(x^{(n)}|\gamma)-\log f^{(n)}(x^{(n)}|\hat{\Theta}_n(x^{(n)})) \\ &= \frac{1}{2}\frac{v'}{\sqrt{n}} B\left(x^{(n)}|\hat{\Theta}_n(x^{(n)})+\vartheta(\gamma)\frac{v}{\sqrt{n}}\right)\frac{v}{\sqrt{n}}\end{aligned}$$

---

67 For the notation see p. 296. Also cf. 61.

with $0<\vartheta(\gamma)<1$. Hence, for $\gamma\in\Gamma_\varepsilon$

$$r^{(n)}(\gamma\,|\,x^{(n)})=\exp\left[\frac{1}{2}\,\frac{v'B\left(x^{(n)}\,|\,\hat{\Theta}_n(x^{(n)})+\vartheta(\gamma)\dfrac{v}{\sqrt{n}}\right)v}{n}\right]$$

$$\times\left[\int\limits_{K_{\varepsilon_1}}\exp\left[\frac{1}{2}\,\frac{v'B\left(x^{(n)}\,|\,\hat{\Theta}_n(x^{(n)})+\vartheta(\gamma)\dfrac{v}{\sqrt{n}}\right)v}{n}\right]d\gamma\right]^{-1}.$$

Now define for $n\geqslant 1$:

$$V_n=\{v:(\hat{\Theta}_n(x^{(n)})+v/\sqrt{n})\in K_{\varepsilon_1}\}\quad\text{and}\quad U_n^*=\{v:|v|<n^{1/4}\}.$$

$$\delta_n=\sup_{v\in U_n^*}\left\|B(x^{(n)}\,|\,\hat{\Theta}_n(x^{(n)})+\vartheta(\gamma)\,v/\sqrt{n})\frac{1}{n}+E(A(\xi\,|\,\gamma_0)\,|\,\gamma_0)\right\|.$$

For large enough n we also have for $v\in U_n^*$

$$\hat{\Theta}_n(x^{(n)})+\vartheta(\gamma)\,v/\sqrt{n}\in\Gamma_\varepsilon \tag{3.123}$$

and, furthermore, from (3.121)

$$\delta_n\to 0. \tag{3.124}$$

Set

$$\alpha_n=\min(n^{1/4},\delta_n^{-1/4}) \tag{3.125}$$

where, naturally, $\alpha_n=n^{1/4}$ should hold if $\delta_n=0$. Also let

$$U_n=\{v:|v|<\alpha_n\}. \tag{3.126}$$

Introducing v as variable of integration, we then need to show that

$$\int\limits_{V_n}\Bigg\{\Bigg|\exp\left[\frac{1}{2}\,\frac{v'B\left(x^{(n)}\,|\,\hat{\Theta}_n(x^{(n)})+\vartheta(\gamma)\dfrac{v}{\sqrt{n}}\right)v}{n}\right]\times$$

$$\times\left(\int\limits_{V_n}\exp\left[\frac{1}{2}\,\frac{u'B\left(x^{(n)}\,|\,\hat{\Theta}_n(x^{(n)})+\vartheta(\gamma)\dfrac{u}{\sqrt{n}}\right)u}{n}\right]du\right)^{-1}-$$

$$-(\exp[-\tfrac{1}{2}\,v'E(A(\xi\,|\,\gamma_0)\,|\,\gamma_0)v])\times$$

$$\times\left(\int\limits_{R_k}\exp[-\tfrac{1}{2}\,u'E(A(\xi\,|\,\gamma_0)u]\,du\right)^{-1}\Bigg|\Bigg\}\,dv\to 0\quad\text{for } n\to\infty.$$

We abbreviate this integral in a suggestive way:

$$\int_{U_n} \left| e^{-B(v)} \left(\int_{V_n} e^{-B(u)} du \right)^{-1} - e^{-A(v)} \left(\int_{R_k} e^{-A(u)} du \right)^{-1} \right| dv$$
$$+ \int_{V_n - U_n} \left| e^{-B(v)} \left(\int_{V_n} e^{-B(u)} du \right)^{-1} - e^{-A(v)} \left(\int_{R_k} e^{-A(u)} du \right)^{-1} \right| dv .$$

It now suffices to show that

$$\int_{U_n} e^{-A(v)} dv \to \int_{R_k} e^{-A(v)} dv \tag{3.127}$$

$$\int_{U_n} e^{-B(v)} dv - \int_{U_n} e^{-A(v)} dv \to 0 \tag{3.128}$$

$$\int_{V_n - U_n} e^{-B(v)} dv \to 0 \tag{3.129}$$

for $n \to \infty$. Because of (3.126), (3.127) is trivial. For (3.128) we note that from (3.123), (3.124), (3.125) and (3.126) there follows for $v \in U_n$

$$|e^{-B(v)} - e^{-A(v)}| = e^{-A(v)} |e^{-B(v)+A(v)} - 1|$$
$$\leqslant e^{-\frac{1}{2} v' E(A(\xi|\gamma_0)|\gamma_0) v} \left| e^{\frac{k}{2} \delta_n \delta_n^{-1/2}} - 1 \right| = o(e^{-v' E(A(\xi|\gamma_0)|\gamma_0) v}) .$$

But this implies the claim.

For the third assertion note that from (3.122)

$$e^{-B(v)} \leqslant e^{-\frac{\lambda}{2}|v|^2}$$

and since $\int_{R_k - U_n} e^{-\frac{\lambda}{4}|v|^2} dv \to 0$, this claim is also proved, and with it, the theorem.

As example consider the following case: Let γ be a one-dimensional r.v., uniformly distributed over (0,1). Let a sequence $\xi_1, \xi_2, \ldots$ of r.v.'s be given which are independent and are distributed according to the same distribution P_γ under the hypotheses $\gamma = \gamma$:

$$\left.\begin{aligned} P_\gamma(\xi_i = 1 \mid \gamma = \gamma) &= \gamma \\ P_\gamma(\xi_i = 0 \mid \gamma = \gamma) &= 1 - \gamma \end{aligned}\right. \qquad i = 1, 2, \ldots, \qquad 0 < \gamma < 1 .$$

With the notation $\zeta_n = \xi_1 + \cdots + \xi_n$, $n \geqslant 1$, we obtain for each integer z_n with $0 \leqslant z_n \leqslant n$ for the conditional density of γ under the hypotheses $\zeta_n = z_n$, the density of $B(1 + z_n, n - z_n + 1)$:

$$p^{(n)}(\gamma \mid z_n) = \begin{cases} 0 & \gamma \leqslant 0 \\ \gamma^{z_n}(1-\gamma)^{n-z_n} \Big/ \int_0^1 x^{z_n}(1-x)^{n-z_n} dx , & 0 < \gamma < 1 . \\ 0 & \gamma \geqslant 1 \end{cases}$$

Thus, for $0 < \gamma < 1$

$$p^{(n)}(\gamma \mid z_n) = \frac{\Gamma(n+2)}{\Gamma(z_n + 1)\Gamma(n - z_n + 1)} \gamma^{z_n}(1-\gamma)^{n-z_n} .$$

Hence,

$$\lim_{n\to\infty}\int_0^1\left|\frac{\Gamma(n+2)}{\Gamma(z_n+1)\Gamma(n-z_n+1)}\gamma^{z_n}(1-\gamma)^{n-z_n}-(\gamma_0(1-\gamma_0))^{-\frac{1}{2}}\right.$$

$$\left.\times(n/(2\pi))^{\frac{1}{2}}\exp\left\{-\frac{1}{2}n\left(\gamma-\frac{z_n}{n}\right)^2(\gamma_0(1-\gamma_0))^{-1}\right\}\right|d\gamma=0$$

with $P_{\gamma_0,\infty}$-measure 1. In applications, an error estimate is important. One has been given (in slightly altered form) by van der Waerden.[68]

We will use Theorem 3.11 later to derive a result of Bayes type. The definitions necessary in this connection are closely related to IV.**4**. We recall those already made. Let $(R,\mathscr{S})$ be a measurable space, Γ an arbitrary set $\neq\emptyset$ and $\mathfrak{S}$ a σ-algebra of subsets of Γ. Let ν be a measure over $(\Gamma,\mathfrak{S})$ and P_Γ a set of measures over $(R,\mathscr{S})$ dominated by a σ-finite measure μ. The R.-N.-densities w.r.t. μ will be given by $x\to f(x,\gamma)$, $\gamma\in\Gamma$. The map $(x,\gamma)\to f(x,\gamma)$ is to be $(\mathscr{S}\otimes\mathfrak{S})$-measurable and $(y,\gamma)\to w(y,\gamma)$ will be a $(\mathfrak{B}_1\otimes\mathfrak{S})$-measurable map from $R_1\times\Gamma$ into R_1. Let H be the set of all $\mathscr{S}$-measurable functions h over R such that

$$L(h,\gamma)=\int_R w(h(x),\gamma)\,f(x,\gamma)\,d\mu(x)$$

for all $\gamma\in\Gamma$ exists and is ν-integrable. $h^*\in H$ will be called a *Bayes estimate*[69] *(for w), w.r.t.* ν if

$$\int_\Gamma L(h^*,\gamma)\,d\nu(\gamma)=\inf_{h\in H}\int_\Gamma L(h,\gamma)\,d\nu(\gamma).$$

Frequently one assumes that for a function d defined on Γ

$$w(y,\gamma)=|y-d(\gamma)|^2 \tag{3.130}$$

for all $y\in R_1$ and $\gamma\in\Gamma$.

We now consider a further specialization of the given definitions. Let $n\geqslant 1$, $R=R_n$, $\mathscr{S}=\mathfrak{B}_n$, $\Gamma=R_k$ and $\mathfrak{S}=\mathfrak{B}_k$, $k\geqslant 1$. Let the measure ν be given by a density p over R_k and let $\boldsymbol{\gamma}$ be a k-dimensional r.v. with this density. For each $x\in R_n$ let $f(x|\gamma)$ be the conditional density of an n-dimensional r.v. under the hypothesis $\boldsymbol{\gamma}=\gamma$, provided it can be well defined. Let w be given by (3.130) for each $y\in R_1$ and $\gamma\in R_k$. H is then the set of all functions h for which

$$\int_{R_k}\int_{R_n}(h(x)-d(\gamma))^2 f(x|\gamma)\,dx\,p(\gamma)\,d\gamma \tag{3.131}$$

exists. Then we have

---

[68] B. van der Waerden, Ber. Verh. sächs. Akad. Leipzig, Math.-Phys. Kl. 87, 353—364 (1935).

[69] See appendix, p. 480ff.

Theorem 3.12. *In the notation just given, let*

$$q(\gamma|x)=f(x|\gamma)p(\gamma)/\int_{R_k} f(x|\gamma')p(\gamma')d\gamma'.$$

Then the function

$$h^*(x)=\int_{R_k} d(\gamma)q(\gamma|x)d\gamma \tag{3.132}$$

defined for almost all $x\in R_n$, *is a Bayes estimate for d (for w) w.r.t. v. Thus,* h^* *minimizes the expression* (3.131).

Proof. Let $f(x)=\int_{R_k} f(x|\gamma)p(\gamma)d\gamma$ for $x\in R_n$. Writing, for $h\in H$ and $\gamma\in\Gamma$

$$\int_{R_n} (h(x)-d(\gamma))^2 f(x|\gamma)dx=L(h,\gamma),$$

we have according to Fubini's theorem

$$\int_{R_k} L(h,\gamma)p(\gamma)d\gamma=\int_{R_n} f(x)\int_{R_k} (h(x)-d(\gamma))^2 q(\gamma|x)d\gamma\, dx.$$

This implies the existence of (3.132) for almost all $x\in R_n$. I, Theorem 21.1. then delivers the assertion.

In this special case we have thus established the existence of a Bayes estimate. The special form of w is essential. In general, Bayes estimates need not exist. However, if one considers an infinite sequence of sample spaces $\{(R^{(n)};\mathscr{S}^{(n)})\}$ in place of a fixed sample space, the idea that one ought to obtain a Bayes estimate for $n\to\infty$ is quite natural[70]. With this idea in mind, we make the following definition:

Let $(R^{(n)},\mathscr{S}^{(n)})$ for $n=1,2,\ldots,$ be a sample space and $P_\Gamma^{(n)}$ probability measures over $(R^{(n)},\mathscr{S}^{(n)})$ which are dominated by a σ-finite measure $\mu^{(n)}$. $\mathfrak{S}$ is a σ-algebra over Γ and $\{v^{(n)}\}$ is a sequence of measures over $\mathfrak{S}$. The R.-N.-density of $P_\gamma^{(n)}$, w.r.t., $\mu^{(n)}$ will be denoted by $x^{(n)}\to f^{(n)}(x^{(n)},\gamma)$, $\gamma\in\Gamma$. The map $(x^{(n)},\gamma)\to f^{(n)}(x^{(n)},\gamma)$ is assumed to be $\mathscr{S}^{(n)}\otimes\mathfrak{S}$-measurable and let $(y,\gamma)\to w^{(n)}(y,\gamma)$ *be a* $\mathfrak{B}_1\otimes\mathfrak{S}$-*measurable map from* $R_1\times\Gamma$ *into* R_1 *for* $n=1,2,\ldots$. Let $H^{(n)}$ be the set of all $\mathscr{S}^{(n)}$-measurable functions $h^{(n)}$ such that

$$L^{(n)}(h^{(n)},\gamma)=\int_{R^{(n)}} w^{(n)}(h^{(n)}(x^{(n)}),\gamma)f^{(n)}(x^{(n)},\gamma)d\mu^{(n)}(x^{(n)})$$

exists for all $\gamma\in\Gamma$ and such that $\int_\Gamma L^{(n)}(h^{(n)},\gamma)dv^{(n)}(\gamma)$ also exists. $h^{(n)*}\in H^{(n)}$, $n=1,2,\ldots$ will be called an *asymptotic Bayes estimate for* $w^{(n)}$, w.r.t.

[70] The theory of Bayes estimates has been carefully studied recently. We only mention: M. H. de Groot and M. M. Rao, Ann. Math. Statist. 34, 598—611 (1963) and L. Schwartz, Z. Wahrscheinlichkeitstheorie und Verw. Gebiete 4, 10—26 (1965).

$v^{(n)}$ if there exists a null sequence $\{\varepsilon_n\}$ of positive numbers such that for all h

$$\int_\Gamma L^{(n)}(h^{(n)*},\gamma)\,d\,v^{(n)}(\gamma) \leqslant \inf_{h^{(n)}\in H^{(n)}} \int_\Gamma L(h^{(n)},\gamma)\,d\,v^{(n)}(\gamma)+\varepsilon_n\,.$$

We have made no essential restrictions on the sequence $w^{(n)}$ in this definition. It is almost evident that one can prove the existence of asymptotic Bayes estimates only under less general assumptions. We will give a theorem under such assumptions and again come in contact with a property of ML estimates.

We first prove

Lemma 3.6. *Let w be a bounded map from $R_k, k\geqslant 1$ into R_1 which depends on $x\in R_k$ by way of $|x|$ in such a way that $|x|\to w(|x|)$ is nondecreasing. Let A be an arbitrary positive definite $k\times k$ matrix and $a\in R_k$. Then*

$$\int_{R_k} w(v-a)e^{-\frac{1}{2}v'Av}\,d\,v \geqslant \int_{R_k} w(v)e^{-\frac{1}{2}v'Av}\,d\,v\,.$$

This result also holds if w is a convex and (w.r.t. zero) symmetric map from R_k into R_1.

Proof. First, we claim that the relation

$$\int_{|x|\leqslant r} e^{-x'Ax}\,d\,x \geqslant \int_{|x|\leqslant r} e^{-(x+a)'A(x+a)}\,d\,x \tag{3.133}$$

holds where r is an arbitrary positive number. This is quite obvious if $k=1$. If $k>1$ one only has to transform the positive quadratic form $x'Ax$ by means of an orthogonal transformation into the form $y'\Lambda y$, where Λ is a diagonal matrix as considered on p. 98 (3.133) then follows by induction.

The inequality (3.133) is equivalent to

$$\int_{R_k} c_{\{|x|\leqslant r\}}e^{-x'Ax}\,d\,x \geqslant \int_{R_k} c_{\{|x-a|\leqslant r\}}e^{-x'Ax}\,d\,x\,. \tag{3.134}$$

Let f be a bounded function from R_k into R_1 which depends on x by way of $|x|$ in such away that $|x|\to f(|x|)$ is nonincreasing. Then a real number M, sequences $\{a_{k_1},\dots,a_{k_n}\}$ with $a_{k_i}>0, 1\leqslant i\leqslant n$, and sequences $\{r_{k_1},\dots,r_{k_n}\}$ with $0\leqslant r_{k_1}<r_{k_2}<\dots<r_{k_n}, n\geqslant 1$, all exist, such that

$$f(x)=M+\lim_{n\to\infty}\sum_{i=1}^{n} a_{k_i}c_{\{|x|\leqslant r_{k_i}\}}\,. \tag{3.135}$$

(Recall that a monotone function has at most countably many discontinuities. See p. 33.)

(3.134) and (3.135) yield $\int_{R_k} f(x)e^{-x'Ax}\,d\,x \geqslant \int_{R_k} f(x-a)e^{-x'Ax}\,d\,x$ and the claim for nondecreasing functions w as considered in the first statement of the lemma follows.

As far as the second claim of the lemma is concerned one may assume that the integrals in the inequality stated in the lemma are finite. Then

$$\begin{aligned}&\int_{R_k}(w(v-a)-w(v))\,e^{-\frac{1}{2}v'Av}\,dv\\&=\tfrac{1}{2}\int_{R_k}(w(v+a)+w(v-a)-2w(v))\,e^{-\frac{1}{2}v'Av}\,dv\end{aligned}$$

follows immediately from the symmetry of w. The convexity of w implies the assertion.

We proceed to the announced theorem. Retaining the notation of Theorem 3.11 we have

Theorem 3.13[71]**.** *Let all the hypotheses of Theorem* 3.11 *be fulfilled, this time for all* $\gamma\in\Gamma_0$. *Let, however,* $p(\gamma)=0$ *for* $\gamma\in R_k-\Gamma_0$. *For each* $B\in\mathfrak{B}_k$ *set* $v(B)=\int_B p(\gamma)\,d\gamma$ *and let* w *be a bounded and nondecreasing function as considered in Lemma* 3.6. *For* $n=1,2,\ldots$ *let* $h^{(n)}$ *be a measurable map from* R_n *into* R_k. *Furthermore, let* $v=v^{(n)}$ *for* $n\geqslant 1$. *Let* $w^{(n)}(h^{(n)}(x^{(n)}),\gamma)=w((h^{(n)}(x^{(n)})-\gamma)\sqrt{n})$. *Then* $H^{(n)}$ *is the set of all these maps* $h^{(n)}$. $\{\hat{\Theta}_n\}$ *is an asymptotic Bayes estimate for* $w^{(n)}$ *and all* $\gamma\in\Gamma_0$, *w.r.t. the measure* v *in the set of all sequences* $\{h^{(n)}\}$ *with* $h^{(n)}\in H^{(n)}$.

Proof. We have

$$L^{(n)}(h^{(n)},\gamma)=\int_{R_n} w((h^{(n)}(x^{(n)})-\gamma)\sqrt{n})\,f^{(n)}(x^{(n)}|\gamma)\,dx^{(n)}$$

for each $\gamma\in\Gamma_0$. We must show that there exists a sequence of positive numbers $\{\varepsilon_n\}$ with $\varepsilon_n\to 0$ such that

$$\int_{\Gamma_0} L^{(n)}(\hat{\Theta}_n,\gamma)\,p(\gamma)\,d\gamma\leqslant\inf_{h^{(n)}\in H^{(n)}}\int_{\Gamma_0} L^{(n)}(h^{(n)},\gamma)\,p(\gamma)\,d\gamma+\varepsilon_n\,. \tag{3.136}$$

Writing

$$q^{(n)}(\gamma|x^{(n)})=\frac{f^{(n)}(x^{(n)}|\gamma)p(\gamma)}{\int_{\Gamma_0} f^{(n)}(x^{(n)}|\gamma')p(\gamma')\,d\gamma'}$$

for $\gamma\in\Gamma$ and (almost all) $x^{(n)}\in R_n$ and

$$f^{(n)}(x^{(n)})=\int_{\Gamma_0} f^{(n)}(x^{(n)}|\gamma)\,p(\gamma)\,d\gamma \tag{3.137}$$

we obtain, after an application of Fubini's theorem

$$\begin{aligned}&\int_{\Gamma_0} L^{(n)}(h^{(n)},\gamma)\,p(\gamma)\,d\gamma\\&=\int_{R_n} f^{(n)}(x^{(n)})\int_{\Gamma_0} w((h^{(n)}(x^{(n)})-\gamma)\sqrt{n})\,q^{(n)}(\gamma|x^{(n)})\,d\gamma\,dx^{(n)}\,.\end{aligned}$$

---

[71] L. Le Cam, loc. cit.[56].

Now let

$$W_n = \{v : (\hat{\Theta}_n(x^{(n)}) + v/\sqrt{n}) \in \Gamma_0\}.$$

One obtains without difficulty

$$\left.\begin{aligned} &\int_{\Gamma_0} w((h^{(n)}(x^{(n)}) - \gamma)\sqrt{n})q^{(n)}(\gamma|x^{(n)})\,d\gamma \\ &= \int_{W_n} w([h^{(n)}(x^{(n)}) - \hat{\Theta}_n(x^{(n)})]\sqrt{n} - v) \\ &\times q^{(n)}((\hat{\Theta}_n(x^{(n)}) + v/\sqrt{n})|x^{(n)})n^{-\frac{k}{2}}\,dv\,. \end{aligned}\right\} \tag{3.138}$$

But now from Theorem 3.11 and the boundedness of w for each $\gamma \in \Gamma_0$

$$\left.\begin{aligned} &\Big| \int_{W_n} w([h^{(n)}(x^{(n)}) - \hat{\Theta}_n(x^{(n)})]\sqrt{n} - v) \\ &\times q^{(n)}(\hat{\Theta}_n(x^{(n)}) + v/\sqrt{n}|x^{(n)})n^{-\frac{k}{2}}\,dv \\ &- \int_{R_k} w([h^{(n)}(x^{(n)}) - \hat{\Theta}_n(x^{(n)})]\sqrt{n} - v) \\ &\times \frac{|E(A(\xi|\gamma)|\gamma)|^{\frac{1}{2}}}{(2\pi)^{k/2}} e^{-\frac{1}{2}v' E(A(\xi|\gamma)|\gamma)v}\,dv \Big| \end{aligned}\right\} \tag{3.139}$$

tends to 0 for $n \to \infty$ and indeed for all $x \in R_\chi$ up to a $P_{\gamma,\chi}$-null set. Note that in the second integral at (3.139) we are integrating over v, while γ in $E(A(\xi|\gamma)|\gamma)$ is an arbitrary but fixed element of R_k.

In addition, from the boundedness of w follows the boundedness in n, γ and $x^{(n)} \in R_n$ of the expression in (3.139). Denoting this quantity by $\varepsilon_n(x^{(n)}, \gamma)$, we see that $\varepsilon_n(x^{(n)}, \gamma)$ tends to zero as $n \to \infty$ for $\gamma \in \Gamma_0$ and for $P_{\gamma,\chi}$-almost all $x \in R_\chi$. ε_n is also uniformly bounded for $n \geqslant 1$ in the entire domain of definition.

A simple application of Lemma 3.6 yields the inequality

$$\left.\begin{aligned} &\int_{\Gamma_0} w((h^{(n)}(x^{(n)}) - y)\sqrt{n})q^{(n)}(y|x^{(n)})\,dy \\ &\geqslant \int_{R_k} w(v)|E(A(\xi|\gamma)|\gamma)|^{1/2}(2\pi)^{-k/2} e^{-\frac{1}{2}v' E(A(\xi|\gamma)|\gamma)v}\,dv - \varepsilon_n(x^{(n)}, \gamma)\,. \end{aligned}\right\} \tag{3.140}$$

Replacing $h^{(n)}$ by $\hat{\Theta}_n$ in (3.138) one obtains directly

$$\begin{aligned} &\int_{R_k} w(v)|E(A(\xi|\gamma)|\gamma)|^{1/2}(2\pi)^{-k/2} e^{-\frac{1}{2}v' E(A(\xi|\gamma)|\gamma)}\,dv \\ &\geqslant \int_{\Gamma_0} w((\hat{\Theta}_n(x^{(n)}) - y)\sqrt{n})q^{(n)}(y|x^{(n)})\,dy - \eta_n(x^{(n)}, \gamma)\,, \end{aligned} \tag{3.141}$$

where η_n is defined analogously to ε_n and has analogous properties.

Writing $\varepsilon_n + \eta_n = \delta_n$, we obtain from (3.140) and (3.141) $P_{\gamma,\gamma}$-a.e.

$$\int_{\Gamma_0} w((h^{(n)}(x^{(n)}) - y)\sqrt{n})\, q^{(n)}(y|x^{(n)})\, dy$$
$$\geqslant \int_{\Gamma_0} w((\hat{\Theta}_n(x^{(n)}) - y)\sqrt{n})\, q^{(n)}(y|x^{(n)})\, dy - \delta_n(x^{(n)}, \gamma).$$

Then,

$$\int_{R_n}\int_{\Gamma_0} w((h^{(n)}(x^{(n)}) - y)\sqrt{n})\, q^{(n)}(y|x^{(n)})\, dy\, f^{(n)}(x^{(n)}|\gamma)\, dx^{(n)}$$
$$\geqslant \int_{R_n}\int_{\Gamma_0} w((\hat{\Theta}_n(x^{(n)}) - y)\sqrt{n})\, q^{(n)}(y|x^{(n)})\, dy\, f^{(n)}(x^{(n)}|\gamma)\, dx^{(n)}$$
$$- \int_{R_n} \delta_n(x^{(n)}, \gamma)\, f(x^{(n)}|\gamma)\, dx^{(n)}.$$

This leads to

$$\int_{\Gamma_0}\int_{R_n} f^{(n)}(x^{(n)}|\gamma) \int_{\Gamma_0} w((h^{(n)}(x^{(n)}) - y)\sqrt{n})\, q^{(n)}(y|x^{(n)})\, dy\, dx^{(n)}\, p(\gamma)\, d\gamma$$
$$\geqslant \int_{\Gamma_0}\int_{R_n} f^{(n)}(x^{(n)}|\gamma) \int_{\Gamma_0} w((\hat{\Theta}_n(x^{(n)}) - y)\sqrt{n})\, q^{(n)}(y|x^{(n)})\, dy\, dx^{(n)}\, p(\gamma)\, d\gamma$$
$$- \int_{\Gamma_0}\int_{R_n} \delta_n(x^{(n)}, \gamma)\, f^{(n)}(x^{(n)}|\gamma)\, dx^{(n)}\, p(\gamma)\, d\gamma.$$

Considering the map $x^{(n)} \to \delta_n(x^{(n)}, \gamma)$ for each n as a function over R_γ, we obtain by applying Theorem VI

$$\int_{R_n} \delta_n(x^{(n)}, \gamma)\, f^{(n)}(x^{(n)}|\gamma)\, dx \to 0$$

for each $\gamma \in \Gamma_0$ and another application of this theorem yields

$$\lim_{n\to\gamma} \int_{\Gamma_0}\int_{R_n} \delta_n(x^{(n)}, \gamma)\, f^{(n)}(x^{(n)}|\gamma)\, dx^{(n)}\, p(\gamma)\, d\gamma = 0.$$

The claim follows from Fubini's theorem and consideration of (3.137).

Remark. A quite similar result can be established using the second part of Lemma 3.6. The necessary modifications of the assumptions of Theorem 3.13 can easily be ascertained.

4. The notion of asymptotic efficiency. We have already made clear in Theorem 2.1 that the ideas of consistency in estimation and theory of testing are closely connected. We want to elaborate here on this connection. We will often be able to view the results obtained here as extensions of those in III.**12**. On the other hand, we will also refer to Theorem 1.10 and the results of **3**.

We have already encountered in connection with the Cramér-Rao inequality the matrix $E(A(\xi, \gamma); \gamma)$, which played such a prominant rôle in **3**. We pointed out there that efficient estimates are the exception. We will, however, see that there exist sequences of "asymptotically efficient" estimates under general assumptions. We have not yet defined the notion of asymptotic efficiency. For the sake of simplicity we restrict ourselves to the case of a one-dimensional parameter or, somewhat more generally, to mappings of the set of parameters into R_1. The extension to the case of several dimensions is not difficult.

Definition. Let Γ be a set of parameters and d a map from Γ into R_1. Let $(R^{(n)}, \mathcal{S}^{(n)})$ be a measurable space for $n=1,2,\ldots$ and $P_\Gamma^{(n)}$ a set of probability measures over $(R^{(n)}, \mathcal{S}^{(n)})$. Assume $h_n, n \geqslant 1$ is an $\mathcal{S}^{(n)}$-measurable map from $R^{(n)}$ into R_1. $\{h_n\}$ is called a *sequence of CAN*[72]*-estimates for d* if $\{h_n\}$ *is consistent for d and if for each* $\gamma \in \Gamma$ *there exists a sequence of positive numbers* $\{\sigma_n(\gamma)\}$ *with* $\sigma_n(\gamma) \to 0$ such that for each $y \in R_1$

$$P_\gamma^{(n)}(\{x^{(n)}: (h_n(x^{(n)}) - d(\gamma))/\sigma_n(\gamma) \leqslant y\}) \to \frac{1}{\sqrt{2\pi}} \cdot \int_{-\infty}^{y} e^{-x^2/2}\, dx \tag{4.1}$$

for $n \to \infty$.

Examples of CAN-estimates are the ML estimates (see Theorem 3.9).

$\sigma_n^2(\gamma)$ is called the *asymptotic variance* of h_n for the element γ. The sequence of asymptotic variances is naturally not unique. If $\{c_n\}$ is an arbitrary sequence of positive numbers with $\lim_{n\to\infty} c_n = 1$, then $c_n \sigma_n^2(\gamma)$ is also an asymptotic variance for h_n, which follows from I, Theorem 40.3. The converse also holds. Assume that $\sigma_n^{*2}(\gamma)$ were an asymptotic variance of h_n and that $\lim_{n\to\infty} \sigma_n^*(\gamma)/\sigma_n(\gamma) = 1$ did not hold. Then, for example, $\underline{\lim}_{n\to\infty} \sigma_n^*(\gamma)/\sigma_n(\gamma) = a > 1$. Thus, there would exist an $\varepsilon > 0$ with the following properties: $\sigma_n^*(\gamma)/\sigma_n(\gamma) \geqslant a - \varepsilon > 1$ for all sufficiently large n; we have

$$\begin{aligned} &P_\gamma^{(n)}(\{x^{(n)}: (h_n(x^{(n)}) - d(\gamma))/\sigma_n^*(\gamma) \leqslant y\}) \\ &= P_\gamma^{(n)}\left(\left\{x^{(n)}: (h_n(x^{(n)}) - d(\gamma))/\sigma_n(\gamma) \leqslant \frac{y\sigma_n^*(\gamma)}{\sigma_n(\gamma)}\right\}\right) \\ &\to \frac{1}{\sqrt{2\pi}} \int_{-\infty}^{y} e^{-x^2/2}\, dx \end{aligned}$$

for $n \to \infty$; moreover, for all large enough n

$$\begin{aligned} &P_\gamma^{(n)}\left(\left\{x^{(n)}: (h_n(x^{(n)}) - d(\gamma))/\sigma_n(\gamma) \leqslant \frac{y\sigma_n^*(\gamma)}{\sigma_n(\gamma)}\right\}\right) \\ &\geqslant P_\gamma^{(n)}(\{x^{(n)}: (h_n(x^{(n)}) - d(\gamma))/\sigma_n(\gamma) \leqslant y(a-\varepsilon)\}) \end{aligned}$$

---

[72] Consistent, Asymptotically Normal.

and from (4.1) also

$$P_\gamma^{(n)}(\{x^{(n)}: (h_n(x^{(n)}) - d(\gamma))/\sigma_n(\gamma) \leqslant y(a-\varepsilon)\})$$
$$\to \frac{1}{\sqrt{2\pi}} \int_{-\infty}^{y(a-\varepsilon)} e^{-x^2/2} dx .$$

This leads to a contradiction. All other possible cases are taken care of in a similar way.

We remark incidentally that the consistency of $\{h_n\}$ for d follows simply from (4.1). For given $\varepsilon, \delta > 0$, one need only choose $y > 0$ so large that $\frac{1}{\sqrt{2\pi}} \int_{-y}^{y} e^{-\frac{x^2}{2}} dx \geqslant 1 - \delta/2$ and then N so large that both $\sup_{n \geqslant N} \sigma_n(\gamma)\, y < \varepsilon$ and for $n \geqslant N$

$$\left| P_\gamma^{(n)}(\{x^{(n)}: |h_n(x^{(n)}) - d(\gamma)| \leqslant \sigma_n(\gamma)\, y\}) - (2\pi)^{-1/2} \int_{-y}^{y} e^{-x^2/2} dx \right| \leqslant \delta/2$$

hold. In case the sequence of Euclidian spaces $\{R_n\}$ is taken for the sequence of sample spaces $\{R^{(n)}\}$, Γ is a subset of R_1 and $d(\gamma) = \gamma$ for all $\gamma \in \Gamma$, then Theorems 1.10 and 3.9 suggest the following

Definition. Let Γ *be an open set of* R_1 and ξ a r.v. distributed according to $P_\gamma, \gamma \in \Gamma$. Let P_Γ be dominated by a σ-finite measure μ. Denote the corresponding R.-N.-densities by $x \to f(x, \gamma)$. Let $\gamma \to f(x, \gamma)$ be differentiable for all $x \in R_1$ with the exception of a μ-null set and assume that

$$E\left[\left(\frac{\partial \log f(\xi, \gamma)}{\partial \gamma}\right)^2 ; \gamma\right]$$

exists for each $\gamma \in \Gamma$. Let $\xi_1, \xi_2, \ldots$ be a sequence of independent, identically distributed r.v.'s and $\{h_n(\xi_1, \ldots, \xi_n)\}$ *a sequence of CAN-estimates for* $\gamma \in \Gamma$. For each $\gamma \in \Gamma$ let $\{\sigma_n^2(\gamma)\}$ be *a sequence of asymptotic variances of* $\{h_n\}$. The sequence of the $h_n(\xi_1, \ldots, \xi_n)$ is called *locally asymptotically efficient* at if γ_0

$$\overline{\lim_{n \to \infty}}\, \sigma_n^2(\gamma_0) \Bigg/ \frac{1}{n E\left[\left(\frac{\partial \log f(\xi, \gamma_0)}{\partial \gamma}\right)^2 ; \gamma_0\right]} \leqslant 1 . \tag{4.2}$$

The sequence is called *asymptotically efficient in* Γ if (4.2) holds for all $\gamma_0 \in \Gamma$.

An example connected with this definition is given by Theorem 3.9. At least under the assumptions there, the sequence of ML estimates turns out to be asymptotically efficient in Γ.

This definition is related in a natural way to the following one, which we again formulate under general assumptions.

Let $\{h_n\}$ and $\{h'_n\}$ be *two sequences of CAN-estimates* over some sequence of sample spaces for a mapping d from Γ into R_1. For each $\gamma\in\Gamma$ let $\{\sigma_n^2(\gamma)\}$ and $\{\sigma_n'^2(\gamma)\}$ be the *corresponding sequences of asymptotic variances*. $\{h_n\}$ is said to be *more efficient than* $\{h'_n\}$ *in* Γ if $\overline{\lim}_{n\to\infty}\ \sigma_n^2(\gamma)/\sigma_n'^2(\gamma)\leqslant 1$ *for all* $\gamma\in\Gamma$ and *for at least one* γ the "$<$" sign holds.

For a long time it was assumed—without restrictions—that there existed no sequence of CAN-estimates more efficient than the ML estimates.[73] It was even assumed that one always has

$$\underline{\lim}_{n\to\infty}\ \sigma_n^2(\gamma)\bigg/\frac{1}{nE\left[\left(\dfrac{\partial\log f(\xi,\gamma)}{\partial\gamma}\right)^2;\gamma\right]}\geqslant 1$$

and thereby a further important optimality property of ML estimates. Only recently have these questions been exposed to a detailed analysis and to a large extent clarified. Using an idea of Hodges jr., Le Cam[74] showed that the "$<$" sign in (4.2) can even hold for infinitely many γ. Sequences of CAN-estimates which do this are called *super-efficient*. Following Le Cam, we now construct such super-efficient sequences. It is obviously enough to prove

Theorem 4.1. *Let Γ be a subset of R_1 and Γ_0 a compact, countable subset of Γ. Let $(R^{(n)},\mathcal{S}^{(n)})$ for $n=1,2,\ldots$ be measurable spaces and $P_\Gamma^{(n)}$ a set of probability measures over $(R^{(n)},\mathcal{S}^{(n)})$ and $\{h_n\}$ a sequence of CAN-estimates for $\gamma\in\Gamma$ and for each γ let $\{\sigma_n^2(\gamma)\}$ be a sequence of asymptotic variances. Set $\sup_{\gamma\in\Gamma_0}\sigma_n^2(\gamma)=c_n$ and assume $\lim_{n\to\infty}c_n=0$. Then there exists a sequence $\{h_n^*\}$ of CAN-estimates for $\gamma\in\Gamma$ whose asymptotic variances are given for each $\gamma\in\Gamma-\Gamma_0$ and $n\geqslant n(\gamma)$ by*

$$\sigma_n^{*2}(\gamma)=\sigma_n^2(\gamma)$$

and for $\gamma\in\Gamma_0$ and $n\geqslant n(\gamma)$ by

$$\sigma_n^*(\gamma)=\beta\,\sigma_n^2(\gamma)\,.$$

Here, β is an arbitrary positive number <1.

Proof. Let $b_n>0$ for $n=1,2,\ldots$ and assume $\{b_n\}$ converges monotonically to 0 so that for each $\gamma\in\Gamma_0$

$$\lim_{n\to\infty} b_n/\sigma_n(\gamma)=\infty\,. \tag{4.3}$$

[73] See, for example, R.A. Fisher, Philos. Trans. Roy. Soc. London Ser. A 222, 309—368 (1922) and R.A. Fisher, Proc. Cambridge Philos. Soc. 22, 700—725 (1925).

[74] L. Le Cam, l.c.[56].

Since $c_n \to 0$, the existence of such a sequence $\{b_n\}$ is obvious. (Let say, $d_n = \sup_{r \geq n} c_r$ and $b_n = d_n^{1/4}$.) Let $\Gamma_0 = \{\gamma_1, \gamma_2, \ldots\}$ and $4\delta_n = \min_{1 \leq i < j \leq n} |\gamma_i - \gamma_j|$, $n = 2, 3, \ldots$. The sequence $\{\delta_n\}$ is obviously monotone nonincreasing and $\delta_n \to 0$ since Γ_0 contains at least one accumulation point. Let k_r be the smallest natural number for which

$$b_{k_r} < \delta_r \tag{4.4}$$

For $\gamma \in \Gamma$ set

$$I_{r,\gamma} = [\gamma - b_{k_r}, \gamma + b_{k_r}].$$

Then, from (4.4)

$$I_{r,\gamma} \subset [\gamma - \delta_r, \gamma + \delta_r]. \tag{4.5}$$

Moreover, for all $i, j \leq r$ with $i \neq j$

$$I_{r,\gamma_i} \cap I_{r,\gamma_j} = \emptyset. \tag{4.6}$$

Indeed, for all such i, j

$$|\gamma_i - \gamma_j| \geq 4\delta_r.$$

And hence (4.6) follows from (4.5).

Now let $k_r < k_{r+1}$ and define h_n^* for $k_r \leq n < k_{r+1}$ as follows:

$$h_n^*(x^{(n)}) = h_n(x^{(n)}) \quad \text{on} \left\{x^{(n)}: h_n(x^{(n)}) \notin \bigcup_{i \leq r} I_{r,\gamma_i}\right\} \tag{4.7}$$

$$h_n^*(x^{(n)} = \gamma_j + \beta(h_n(x^{(n)}) - \gamma_j) \quad \text{on} \ \{x^{(n)}: h_n(x^{(n)}) \in I_{r,\gamma_j}\}, \quad 1 \leq j \leq r. \tag{4.8}$$

In case $k_r = \cdots = k_{r+m}$ but $k_r < k_{r+m+1}$ for some $m \geq 1$ we apply definitions (4.7) and (4.8) analogously to the h_n^* with $k_r \leq n < k_{r+m+1}$. Assume now that $\gamma \notin \Gamma_0$. Since Γ_0 is compact, there exists an $\eta(\gamma) > 0$ such that

$$\inf_{\gamma_i \in \Gamma_0} |\gamma - \gamma_i| = \eta(\gamma) > 0. \tag{4.9}$$

Let r_0 be the smallest integer with

$$4\delta_{r_0} \leq \eta(\gamma). \tag{4.10}$$

According to (4.4), $b_k < \delta_{r_0}$ for all $k \geq k_{r_0}$. From (4.9) and (4.10)

$$[\gamma - \delta_{r_0}, \gamma + \delta_{r_0}] \cap \bigcup_{i \leq r} I_{r,\gamma_i} = \emptyset \tag{4.11}$$

for all $r \geq r_0$. From the consistency of $\{h_n\}$ for $\gamma \in \Gamma$ we have for each $\delta > 0$ and $\varepsilon > 0$

$$P_\gamma^{(n)}\{x^{(n)}: |h_n(x^{(n)}) - \gamma| < \varepsilon\} \geq 1 - \delta \tag{4.12}$$

provided that $n \geq n(\delta, \varepsilon)$.

Let $E_n = \{x^{(n)}: h_n^*(x^{(n)}) = h_n(x^{(n)})\}$.

Choosing $\gamma \notin \Gamma_0$ and $\varepsilon < \delta_{r_0}$ in (4.12), we have from (4.11), (4.12) and (4.7)

$$P_\gamma^{(n)}(E_n) \geqslant 1 - \delta \tag{4.13}$$

for

$$n \geqslant \max(k_{r_0}, n(\delta, \varepsilon)) . \tag{4.14}$$

Then for each real y and all n which satisfy (4.14)

$$\left| P_\gamma^{(n)}\left(\left\{x^{(n)} : \frac{h_n^*(x^{(n)}) - \gamma}{\sigma_n(\gamma)} \leqslant y\right\}\right) - P_\gamma^{(n)}\left(\left\{x^{(n)} : \frac{h_n(x^{(n)}) - \gamma}{\sigma_n(\gamma)} \leqslant y\right\}\right)\right| < 2\delta$$

and since h_n is a CAN-estimate with asymptotic variance $\sigma_n^2(\gamma)$ at γ, the claim about h_n^* for $\gamma \in \Gamma - \Gamma_0$ is proved.

When $\gamma \in \Gamma_0$, say $\gamma = \gamma_{j_0}$, then $\gamma_{j_0} \in \bigcup_{i \leqslant r} I_{r, \gamma_i}$ for all large enough r. But now

$$P_{\gamma_{j_0}}^{(n)}\{x^{(n)}: |h_n(x^{(n)}) - \gamma_{j_0}| < b_n\} \to 1 \tag{4.15}$$

for $n \to \infty$. Since h_n is a CAN-estimate with asymptotic variance $\sigma_n^2(\gamma_{j_0})$, this follows easily from (4.3). But then it follows from (4.8) that for all large enough n,

$$h_n^*(x^{(n)}) = \gamma_{j_0} + \beta(h_n(x^{(n)}) - \gamma_{j_0})$$

up to a set of arbitrarily small $P_{\gamma_{j_0}}^{(n)}$-measure. However, this implies exactly as above, that

$$(h_n^* - \gamma_{j_0})/(\beta\sigma_n(\gamma_{j_0})) \text{ for } n \to \infty$$

is asymptotically $N(0, 1)$-distributed.

Theorem 4.1 is thus proved and shows in particular the following: Let $h_n, n \geqslant 1$ be an arbitrary CAN-estimate for $\gamma \in \Gamma$ with asymptotic variance $\sigma_n^2(\gamma)$ at γ. Then there always exists a sequence $\{h_n^*\}$ of CAN-estimates which is more efficient than $\{h_n\}$ for a given $\gamma \in \Gamma$.

Applying Theorem 4.1 to the ML estimates, we obtain in the special one-dimensional case of Theorem 3.9 (see p. 317) the following result: Let

$$\inf_{\gamma \in \Gamma} E\left[\left(\frac{\partial \log f(\xi, \gamma)}{\partial \gamma}\right)^2 ; \gamma\right] > 0 .$$

Then one can always find a sequence of CAN-estimates which is more efficient than the ML estimates at least at a countably infinite number of points $\gamma_i \in \Gamma$. Sequences of CAN-estimates which are asymptotically more efficient than the ML estimates in a set of parameters Γ are said, as already mentioned, to be *super-efficient* there. We have thus shown that in countable sets, there always exist super-efficient sequences of CAN-estimates.

Later we will give a theorem with the opposite aim of establishing conditions under which a sequence of CAN-estimates cannot be super-efficient. (See Theorem 5.2.)

The phenomenon of super-efficiency has occasioned several investigations of more detailed properties of ML estimates. We mention here especially the papers of Rao, whose results we can only touch on. Retaining the notation and assumptions of Theorem 3.1, we obtain from the arguments following (3.86) that

$$\frac{1}{\sqrt{n}}\,\frac{\partial \log f^{(n)}(\xi^{(n)},\gamma)}{\partial\gamma}-\sqrt{n}(\hat{\gamma}_n(\xi^{(n)})-\gamma)$$

converges for all $\gamma\in\Gamma$ to zero in probability (that is, to the k-dimensional null vector). On the other hand, it follows from this because of the stochastic convergence of $\frac{1}{n}\,\frac{\partial \log f^{(n)}(\xi^{(n)},\gamma)}{\partial\gamma}$ for each $\gamma\in\Gamma$ to 0 that this ML estimate is consistent for $\gamma\in\Gamma$. This caused Rao to formulate the following definition (we retain the notation of Theorem 3.1 and assume for the sake of simplicity that Γ is one-dimensional): For $n=1,2,\ldots$ let $h^{(n)}$ be measurable functions over R_n. They are called *asymptotically efficient for* $\gamma\in\Gamma$ *in the sense of Rao* if there exist mappings a and b from Γ into R_1 such that

$$\frac{1}{\sqrt{n}}\,\frac{\partial \log f^{(n)}(\xi^{(n)},\gamma)}{\partial\gamma}-a(\gamma)-b(\gamma)\sqrt{n}(h^{(n)}\circ\xi^{(n)}-\gamma)$$

tends to zero in probability for each $\gamma\in\Gamma$. Thus, under the given conditions (consistent) sequences of ML estimates are asymptotically efficient in the sense of Rao. In the set $\mathfrak{F}$ of all sequences of asymptotically efficient estimates in the sense of Rao we will now introduce a measure for "second-order efficiency" (provided it exists): For each real λ consider the sequence of r.v.'s

$$\left\{\frac{\partial \log f^{(n)}(\xi^{(n)},\gamma)}{\partial\gamma}-a(\gamma)\sqrt{n}-b(\gamma)n(h^{(n)}\circ\xi^{(n)}-\gamma)-\lambda n(h^{(n)}\circ\xi^{(n)}-\gamma)^2\right\}.$$

Suppose it possesses a limiting distribution whose variance will be denoted by $\sigma^2(\lambda,\{h^{(n)}\};\gamma)$. Then $\min\limits_{\lambda}\sigma^2(\lambda,\{h^{(n)}\};\gamma)$ is the desired measure. Rao[75] has shown that in some cases the value $\min\limits_{\{h^{(n)}\}\in\mathfrak{F}}\ \min\limits_{\lambda}\sigma^2(\lambda, h^{(n)};\gamma)$ is actually assumed and, indeed, by a consistent sequence of ML

[75] See in connection with the entire question C. R. Rao, J. Roy. Statist. Soc. Ser. B, 24, 46—72 (1962). Proc. Fourth Berkeley Sympos. Math. Statist. and Prob. Vol. I., pp. 531—545, Univ. California Press, Berkeley, Calif., (1960) and Sankhyā 24, Ser. A, 73—101 (1962), as well as Sankhyā 25, Ser. A, 189—206 (1963).

estimates. These considerations suggest the investigation of other construction principles for sequences of CAN-estimates which possess the essential properties of ML estimates as stated, say, in Theorem 3.9. Such an investigation was first suggested by Neyman[76], who was also the first to tackle the problem. Since that time, the theory of so-called *sequences of best asymptotically normal estimates* has been widely developed.

5. Bahadur's concept[77]. Recently, Bahadur has considered problems of asymptotic efficiency from a somewhat different standpoint. We have already mentioned in III.**12** that these problems are closely connected with the corresponding ones for the efficiency of tests. Indeed, Theorem 2.1 has made this connection completely clear. We begin by indicating the weakness of the definition of asymptotic efficiency that has been used up to now: In particular, it is not difficult to give examples of the following type:

Let d be a map from Γ into R_1 and $\{h_n\}$, resp., $\{h_n^*\}$ sequences of CAN-estimates for d defined over a sequence $(R^{(n)}, \mathscr{S}^{(n)})$ of measurable spaces. Take $\{\sigma_n^2\}$, resp., $\{\sigma_n^{*2}\}$ as the corresponding sequences of asymptotic variances. Then

$$\lim_{n\to\infty} \sigma_n^2(\gamma)/\sigma_n^{*2}(\gamma) = 0 \tag{5.1}$$

for each $\gamma \in \Gamma$, but for each $\varepsilon > 0$

$$\lim_{n\to\infty} \frac{P_\gamma^{(n)}(\{x^{(n)}: |h_n(x^{(n)}) - \gamma| \geqq \varepsilon\})}{P_\gamma^{(n)}(\{x^{(n)}: |h_n^*(x^{(n)}) - \gamma| \geqq \varepsilon\})} = \infty\,. \tag{5.2}$$

An especially simple example of this type is due to Basu[78]. It is obviously of the same type as the construction of sequences of super efficient estimates on p. 338 ff.

---

[76] J. Neyman, Proceedings of the Berkeley Symposium on Mathematical Statistics and Probability pp. 239—273, (1949), University of California Press, Berkeley and Los Angeles. Trabajos Estadist. 5, 161—168 (1954). A further (quite incomplete) list of literature: E. W. Barankin and J. Gurland, Univ. California Publ. Statist. 1, 89—129 (1951), T. S. Ferguson, Ann. Math. Statist. 29, 1046—1062 (1958), L. Le Cam, Proceedings of the 3rd Berkeley Symposium on Mathematical Statistics and Probability 1954—1955, Vol. I, pp. 129—156, University of California Press, Berkeley and Los Angeles, (1956), Univ. California Publ. Statist. 3, 37—98 (1960), J. Hájek, Ann. Math. Statist. 33, 1124—1147 (1962).

[77] R. R. Bahadur, Sankhyā 22, 229—253 (1960).

[78] See D. Basu, Sankhyā 17, 193—196 (1956) as well as E. L. Lehmann, Proceedings of the Berkeley Symposium on Mathematical Statistics and Probability 1949, pp. 451—457, University of California Press, Berkeley and Los Angeles where, in a somewhat different connection, a similar example is considered.

Let $\xi_1, \xi_2, \ldots$ be independent, identically $N(a, 1)$-distributed r.v.'s, where, say, $|a| \leqslant m$ with $m>0$ holds. Let $\varepsilon>0$ but otherwise arbitrary. Then, according to II, Theorem 3.2, for $n \geqslant 1$

$$P_a(|\bar{\xi}_n - a| > \varepsilon) = \sqrt{2/\pi} \int_{\varepsilon\sqrt{n}}^{\infty} e^{-y^2/2}\, dy$$

so that

$$P_a(|\bar{\xi}_n - a| > \varepsilon) = o(1/n) \tag{5.3}$$

for $n \to \infty$.

Let $\{b_n\}$ be a sequence of real numbers with

$$P_a(s_n^2(n-1) > b_n) = 1/n, \qquad n \geqslant 1. \tag{5.4}$$

The sequence $\{b_n\}$ can be chosen independently of a (see II, Corollary to Theorem 4.3). Further let

$$c_n(x_1, \ldots, x_n) = \begin{cases} 0 & \sum_{i=1}^{n} (x_i - \bar{x}_n)^2 \leqslant b_n \\ 1 & \sum_{i=1}^{n} (x_i - \bar{x}_n)^2 > b_n \end{cases},$$

and for $n \geqslant 1$ and all $x^{(n)} \in R_n$ set

$$h_n(x^{(n)}) = (1 - c_n(x^{(n)}))\bar{x}_n + n\, c_n(x^{(n)})$$

and

$$h_n^*(x^{(n)}) = \bar{x}_{[\sqrt{n}]}.$$

For each $x^{(n)} \in R_n$ we have

$$\sqrt{n}(h_n(x^{(n)}) - a) = \sqrt{n}(\bar{x}_n - a) + \sqrt{n}\, c_n(x^{(n)})(n - \bar{x}_n).$$

According to I, Theorem 39.1, $\bar{\xi}_n$ is a CAN-estimate for a with asymptotic variance $1/n$ which does not depend on a. Since, however,

$$P_a(c_n(\xi^{(n)}) = 0) = 1 - \frac{1}{n} \to 1$$

with $\xi^{(n)} = (\xi_1, \ldots, \xi_n)$ for $n \to \infty$, we have from I, Theorem 40.1 that $h_n \circ \xi^{(n)}$ is also a CAN-estimate for a with the same asymptotic variance. The asymptotic variance of $h_n^* \circ \xi^{(n)}$ is $1/\sqrt{n}$. Thus, with an appropriate interpretation, (5.1) is fulfilled. On the other hand, for $\varepsilon>0$

$$\begin{aligned} &P_a(|h_n \circ \xi^{(n)} - a| > \varepsilon) \\ &= P_a(\{x^{(n)} : |(\bar{x}_n - a) + c_n(x^{(n)})(n - \bar{x}_n)| > \varepsilon\} \cap \{x^{(n)} : c_n(x^{(n)}) = 0\}) \\ &\quad + P_a(\{x^{(n)} : |(\bar{x}_n - a) + c_n(x^{(n)})(n - \bar{x}_n)| > \varepsilon\} \cap \{x^{(n)} : c_n(x^{(n)}) = 1\}). \end{aligned}$$

Since $c_n \circ \xi^{(n)}$ and $\bar{\xi}_n$ are independent by II, Theorem 4.3 and I, Theorem 13.1, for $n > m + \varepsilon$ we have

$$P_a(|h_n \circ \xi^{(n)} - a| > \varepsilon) = P_a(c_n(\xi^{(n)}) = 0)\, P_a(|\bar{\xi}_n - a| > \varepsilon) + P_a(c_n(\xi^{(n)}) = 1)$$

since for $n > m + \varepsilon$, the inequality

$$|(\bar{x}_n - a) + c_n(x^{(n)})(n - \bar{x}_n)| > \varepsilon$$

is trivially fulfilled for all $x^{(n)}$ with $c_n(x^{(n)}) = 1$.

From (5.3) and (5.4)

$$P_a(|h_n \circ \xi^{(n)} - a| > \varepsilon) = o(1/n) + 1/n\,. \tag{5.5}$$

Moreover, we also have

$$P_a(|h_n^* \circ \xi^{(n)} - a| > \varepsilon) = o(1/n)\,.$$

This, together with (5.5) yields (5.2).

Essentially, this example only serves to clearly point out the not surprising fact that the behavior of the asymptotic variance of a sequence of CAN-estimates tells nothing about the asymptotic behavior of the probability distribution of the estimates in the complement of a neighborhood of the parameter. This fact has already been expressed by the different character of Theorems I, 39.1 and I, 39.5.

As we already mentioned at the end of **4**, such examples suggest an investigation of the rate of convergence of consistent sequences of estimates. We do this now, following Bahadur. This will lead in a natural way to another notion of "asymptotic efficiency" which is, however, closely connected with the ideas of **4**.

First we have a

Definition. Let $\{(R^{(n)}, \mathcal{S}^{(n)})\}$ be a sequence of measurable spaces. Suppose that h_n is defined for $n \geqslant 1$ over $R^{(n)}$ and is $\mathcal{S}^{(n)}$-measurable. $P_\gamma^{(n)}$ will be a probability measure over $(R^{(n)}, \mathcal{S}^{(n)})$ for $\gamma \in \Gamma$. Let $\{h_n\}$ be a consistent sequence of estimates for d. Let $\varepsilon > 0$. For each $\gamma \in \Gamma$ $\tau(h_n, \varepsilon, \gamma)$ will be called *effective standard deviation* of h_n if

$$\sqrt{2/\pi} \int\limits_{\varepsilon/\tau(h_n,\varepsilon,\gamma)}^{\infty} e^{-x^2/2}\, dx = P_\gamma^{(n)}(\{x^{(n)} : |h^{(n)}(x^{(n)}) - d(\gamma)| \geqslant \varepsilon\})\,. \tag{5.6}$$

Since $y \to \int\limits_y^{\infty} e^{-x^2/2}\, dx$ is strictly monotone decreasing and continuous for $y \in [0, \infty)$, there always exists exactly one $\tau(h_n, \varepsilon, \gamma)$ which satisfies (5.6). Obviously, we always have $0 \leqslant \tau(h_n, \varepsilon, \gamma) \leqslant \infty$.

It is easy to show, say by integrating twice by parts, that

$$\log \int_x^\infty e^{-t^2/2}\,dt = -\frac{x^2}{2}(1+o(1)) \tag{5.7}$$

for $x \to \infty$. From the consistency of $\{h_n\}$ we have according to (5.6)

$$\lim_{n\to\infty} 1/\tau(h_n, \varepsilon, \gamma) \to \infty \quad \text{for each } \varepsilon>0 \quad \text{and } \gamma \in \Gamma .$$

Hence, from (5.7)

$$\frac{2}{\varepsilon^2} \log P_\gamma^{(n)}(\{x^{(n)}: |h_n(x^{(n)}) - d(\gamma)| > \varepsilon\}) = -\frac{1}{\tau^2(h_n, \varepsilon, \gamma)}(1+o(1)) \tag{5.8}$$

for $n \to \infty$.

Now let h_n^* be defined over $(R^{(n)}, \mathscr{S}^{(n)})$ for $n=1,2,\ldots$ and assume $\{h_n^*\}$ is a sequence of ($\mathscr{S}^{(n)}$-measurable) estimates which is consistent for d. For each $\varepsilon>0$ and large enough n let the quotient $\tau^2(h_n^*, \varepsilon, \gamma)/\tau^2(h_n, \varepsilon, \gamma)$ be well defined. Then set

$$e(\{h_n\}, \{h_n^*\}, \gamma) = \overline{\lim_{\varepsilon\to 0}}\; \overline{\lim_{n\to\infty}}\; \tau^2(h_n^*, \varepsilon, \gamma)/\tau^2(h_n, \varepsilon, \gamma) .$$

If

$$e(\{h_n\}, \{h_n^*\}, \gamma) < 1 , \tag{5.9}$$

then obviously for small enough $\varepsilon>0$ and large enough n

$$\tau^2(h_n^*, \varepsilon, \gamma)/\tau^2(h_n, \varepsilon, \gamma) < 1$$

and thus, because of (5.8)

$$P_\gamma^{(n)}(\{x^{(n)}: |h_n(x^{(n)}) - d(\gamma)| < \varepsilon\}) < P_\gamma^{(n)}(\{x^{(n)}: |h_n^*(x^{(n)}) - d(\gamma)| < \varepsilon\}) .$$

The assumption (5.9) thus implies that for the γ under consideration and sufficiently small $\varepsilon>0$, h_n^* is concentrated near $d(\gamma)$ with greater probability when $n\to\infty$ than h_n and is in this sense "asymptotically more efficient" than h_n. We thus call $e(\{h_n\}, \{h_n^*\}, \gamma)$ the *asymptotic relative efficiency* of h_n w.r.t. h_n^* at γ *in the sense of Bahadur.*

This immediately recalls the corresponding notion for tests. (See III.**12.**) We now want to investigate this further. We use the notation of III, Theorem 12.3 and prove

Theorem 5.1. *Let the assumptions of* III, *Theorem* 12.3 *hold, in as far as they apply to the r.v.* $q(\xi)$. *Let* d *be a function defined over* Γ. *For a* $\gamma_0 \in \Gamma$ *let* d *be differentiable with*

$$d'(\gamma_0) \neq 0 . \tag{5.10}$$

Further let

$$\lim_{\gamma\to\gamma_0} H(\gamma,\gamma_0)\Big/\frac{(\gamma-\gamma_0)^2}{2} = I(\gamma_0) \tag{5.11}$$

with $0<I(\gamma_0)<\infty$.

If $\{h_n\circ\xi^{(n)}\}$ *with* $\xi^{(n)}=(\xi_1,\ldots,\xi_n)$ *is a consistent sequence of estimates for d, then*

$$\left.\begin{aligned}&\varliminf_{\varepsilon\to 0}\varliminf_{n\to\infty}(n\varepsilon^2)^{-1}\log(P_{\gamma_0}(|h_n\circ\xi^{(n)}-d(\gamma)|\geq\varepsilon))\\ &\geq -\tfrac{1}{2}I(\gamma_0)/(d'(\gamma_0))^2\end{aligned}\right\} \tag{5.12}$$

or

$$\varliminf_{\varepsilon\to 0}\varliminf_{n\to\infty} n\tau^2(h_n,\varepsilon,\gamma_0)\geq\frac{(d'(\gamma_0))^2}{I(\gamma_0)}. \tag{5.13}$$

Proof. From (5.10) for each sufficiently small $\varepsilon>0$

$$D(\gamma_0,\varepsilon)=\{\gamma:|d(\gamma)-d(\gamma_0)|=\varepsilon\}\neq\emptyset.$$

Such an $\varepsilon>0$ will now be chosen and held fixed. With

$$\delta=v\varepsilon,\qquad 0<v<1 \tag{5.14}$$

define the sets $M_n\subseteq R_n$ for $n\geq 1$ by means of

$$M_n=\{x^{(n)}:|h_n(x^{(n)})-d(\gamma_0)|\geq\delta\}.$$

Assume

$$\gamma_1\in D(\gamma_0,\varepsilon). \tag{5.15}$$

We consider $\{M_n\}$ as a sequence of critical regions for the problem $(\{\gamma_0\},\{\gamma_1\})$. Our first goal is to show the applicability of III, Theorem 12.3 and in particular, of the inequality III, (12.38). It suffices to prove that $\varliminf_{n\to\infty} g_{c_{M_n}}(\gamma_1)>0$. Now from (5.15) and the definition of $D(\gamma_0,\varepsilon)$, for each $x^{(n)}=x\in R_n$

$$|h_n(x)-d(\gamma_0)|\geq|\varepsilon-|h_n(x)-d(\gamma_1)||.$$

Hence,

$$\{x:|\varepsilon-|h_n(x)-d(\gamma_1)||\geq\delta\}\subseteq M_n$$

or, taking (5.14) into account

$$\{x:|h_n(x)-d(\gamma_1)|\leq\varepsilon(1-v)\}\subseteq M_n.$$

Since, however, $\{h_n\circ\xi^{(n)}\}$ is consistent, we even have $\lim_{n\to\infty} g_{c_{M_n}}(\gamma_1)=1$ so that

$$\varliminf_{n\to\infty} n^{-1}\log P_{\gamma_0}(|h_n\circ\xi^{(n)}-d(\gamma_0)|\geq\delta)\geq -H(\gamma_1,\gamma_0). \tag{5.16}$$

Note that the left side of (5.16) is independent of γ_1; we can thus choose $\gamma_1=\gamma_\varepsilon$ in (5.16) in such a way that $\lim_{\varepsilon\to 0}\gamma_\varepsilon=\gamma_0$.

Thus also

$$\varliminf_{n\to\infty}(n\delta^2)^{-1}\log P_{\gamma_0}(|h_n\circ\xi^{(n)}-d(\gamma_0)|\geqslant\delta)$$
$$\geqslant\frac{1}{v^2}\,\frac{H(\gamma_\varepsilon,\gamma_0)(\gamma_\varepsilon-\gamma_0)^2/2}{(d(\gamma_\varepsilon)-d(\gamma_0))^2(\gamma_\varepsilon-\gamma_0)^2/2}.$$

Now use (5.11) to obtain

$$\varliminf_{\delta\to 0}\varliminf_{n\to\infty}(n\delta^2)^{-1}\log P_{\gamma_0}(|h_n\circ\xi^{(n)}-d(\gamma_0)|\geqslant\delta)\geqslant -\frac{1}{v^2}\,\frac{I(\gamma_0)}{2(d'(\gamma_0))^2}.$$

Since this holds for each v with $0<v<1$, we obtain (5.12). (5.13) follows from the definition of $\tau^2(h_n,\varepsilon,\gamma_0)$.

(5.13) recalls the Cramér-Fréchet-Rao inequality. The connection becomes clear if we make regularity assumptions such as in Theorem 3.2. Then

$$I(\gamma_0)=E\left[\left(\frac{\partial\log f(\xi,\gamma_0)}{\partial\gamma}\right)^2;\gamma_0\right].$$

This result, together with the considerations on p. 316ff., suggests that the ML estimates also play an important rôle here.

Indeed, under certain assumptions—which we do not go into here—we can obtain the following result by means of a refinement (in the direction of Wald's method) of previous arguments:

For the sake of simplicity, let $d(\gamma)=\gamma$ for each $\gamma\in\Gamma$ and $\{\hat{\Theta}_n\}$ be a consistent sequence of ML estimates for γ. Then

$$\varlimsup_{\varepsilon\to 0}\varlimsup_{n\to\infty}n\tau^2(\hat{\Theta}_n,\varepsilon,\gamma_0)\leqslant\frac{1}{I(\gamma_0)}.$$

We do not give a proof here. One can be found in Bahadur, loc. cit.[77].

However we will present here a related result with all the details. This result has already been announced on p. 341.

First we prove the following

Lemma 5.1. *Let Γ be an open interval (or more generally an open set) $\subseteq R_1$. Let P_Γ be a class of probability measures defined on $(R_1,\mathfrak{B}_1)$ which are dominated by a σ-finite measure μ. Denote by $x\to f(x,\gamma)$ the R.-N.-density of P_γ w.r.t. μ for every $\gamma\in\Gamma$.*

Suppose that the map $\gamma\to f(x,\gamma)$ is continuous μ-a.e. in R_1. Furthermore assume that $\gamma\to\dfrac{\partial\log f(x,\gamma)}{\partial\gamma}$ exists μ-a.e. in R_1, is continuous and that

$x \to \dfrac{\partial \log f(x,\gamma)}{\partial \gamma}$ *is* $\neq 0$ *on a set of positive* P_γ*-measure for every* $\gamma \in \Gamma$. *Let the mappings* $(x,\gamma) \to f(x,\gamma)$ *and* $(x,\gamma) \to \dfrac{\partial \log f(x,\gamma)}{\partial \gamma}$ *be* $\mathfrak{B}_1 \otimes \mathfrak{B}_1^*$*-measurable where* $\mathfrak{B}_1^*$ *is the class of all Borel sets of* Γ. *Suppose that there exist for every* $\gamma_0 \in \Gamma$ *a neighborhood* $U(\gamma_0)$ *and a function* $x \to A(x,\gamma_0)$ *with*

$$\int_{R_1} A^2(x,\gamma_0)\, dP_{\gamma_0}(x) < \infty \tag{5.17}$$

such that

$$\left|\frac{f(x,\gamma_1)}{f(x,\gamma_2)} - 1\right| \leqslant A(x,\gamma_0)|\gamma_1 - \gamma_2| \tag{5.18}$$

for all $x \in R_1$, μ*-a.e. and all* $\gamma_i \in U(\gamma_0)$, $i = 1, 2$.

Then

$$\left|\frac{\partial \log f(x,\gamma)}{\partial \gamma}\right| \leqslant A(x,\gamma_0) \tag{5.19}$$

for all $\gamma \in U(\gamma_0)$ *and* μ*-a.e. in* R_1. *Moreover, using the notation*

$$\int_{R_1} \left(\frac{\partial \log f(x,\gamma)}{\partial \gamma}\right)^2 dP_\gamma(x) = I(\gamma),$$

we have

$$0 < I(\gamma) < \infty. \tag{5.20}$$

Furthermore, denoting by ξ_γ *(or briefly by* ξ*) a r.v. with distribution* P_γ, *we have*

$$E[(\log f(\xi_{\gamma_0},\gamma) - \log f(\xi_{\gamma_0},\gamma_0));\gamma_0] = -\frac{(\gamma-\gamma_0)^2}{2}[I(\gamma_0) + o(1)] \tag{5.21}$$

and

$$E[(\log f(\xi_{\gamma_0},\gamma) - \log f(\xi_{\gamma_0},\gamma_0))^2;\gamma_0] = (\gamma-\gamma_0)^2(I(\gamma_0) + o(1)) \tag{5.22}$$

as $\gamma \to \gamma_0$.

Proof. (5.19) can easily be derived from the following considerations: Let $\gamma, \gamma_1 \in U(\gamma_0)$. Then

$$\left|\frac{\partial \log f(x,\gamma)}{\partial \gamma}\right| = \left|\frac{\partial f(x,\gamma)}{\partial \gamma}\,\frac{1}{f(x,\gamma)}\right| = \left|\lim_{\gamma_1 \to \gamma} \frac{f(x,\gamma_1) - f(x,\gamma)}{\gamma_1 - \gamma}\,\frac{1}{f(x,\gamma)}\right|$$

$$= \lim_{\gamma_1 \to \gamma}\left|\frac{1}{\gamma_1 - \gamma}\left[\frac{f(x,\gamma_1)}{f(x,\gamma)} - 1\right]\right| \leqslant A(x,\gamma_0), \quad \mu\text{-a.e.}$$

where the last inequality follows from (5.18). The relation (5.20) is an easily seen consequence of the assumptions made. To get (5.21) we show first that

$$E\left[\frac{\partial \log f(\xi,\gamma)}{\partial \gamma};\gamma\right]=0 \tag{5.23}$$

for every $\gamma\in\Gamma$. For this we observe that $\left|\frac{f(x,\gamma_1)-f(x,\gamma)}{\gamma_1-\gamma}\right|\leqslant A(x,\gamma)\,f(x,\gamma)$, μ-a.e. for every $\gamma_1\neq\gamma$ which belongs to $U(\gamma)$ and such that $x\to A(x,\gamma)\,f(x,\gamma)$ is μ-integrable. Therefore, using Theorem VIII we get

$$\begin{aligned}E\left[\frac{\partial \log f(\xi,\gamma)}{\partial \gamma};\gamma\right]&=\int_{R_1}\lim_{\gamma_1\to\gamma}\frac{f(x,\gamma_1)-f(x,\gamma)}{\gamma_1-\gamma}\,\frac{1}{f(x,\gamma)}\,dP_\gamma(x)\\&=\int_{R_1}\lim_{\gamma_1\to\gamma}\frac{f(x,\gamma_1)-f(x,\gamma)}{\gamma_1-\gamma}\,d\mu(x)\\&=\lim_{\gamma_1\to\gamma}\int_{R_1}\frac{f(x,\gamma_1)-f(x,\gamma)}{\gamma_1-\gamma}\,d\mu(x)=0\,.\end{aligned}$$

From (5.23) and (5.19) we obtain

$$\lim_{\gamma\to\gamma_0}\frac{1}{\gamma-\gamma_0}E\left[\frac{\partial \log f(\xi_{\gamma_0},\gamma)}{\partial\gamma};\gamma_0\right]=\lim_{\gamma\to\gamma_0}\int_R\frac{\partial \log f(x,\gamma)}{\partial\gamma}\,\frac{f(x,\gamma_0)-f(x,\gamma)}{\gamma-\gamma_0}\,d\mu(x).$$

This is equal to $\int_{R_1}\frac{\partial \log f(x,\gamma_0)}{\partial\gamma}\,\frac{\partial f(x,\gamma_0)}{\partial\gamma}\,d\mu(x)=-I(\gamma_0)$ as follows from (5.18) and from the continuity assumption. Thus,

$$E\left[\frac{\partial \log f(\xi_{\gamma_0},\gamma)}{\partial\gamma};\gamma_0\right]=-(\gamma-\gamma_0)\left[I(\gamma_0)+o(1)\right] \tag{5.24}$$

holds.

Finally

$$\begin{aligned}E[(\log f(\xi_{\gamma_0},\gamma)-\log f(\xi_{\gamma_0},\gamma_0));\gamma_0]&=\int_{R_1}\int_{\gamma_0}^{\gamma}\frac{\partial \log f(x,\gamma_1)}{\partial\gamma_1}\,d\gamma_1\,f(x,\gamma_0)\,d\mu(x)\\&=\int_{\gamma_0}^{\gamma}\int_{R_1}\frac{\partial \log f(x,\gamma_1)}{\partial\gamma_1}\,f(x,\gamma_0)\,d\mu(x)\,d\gamma_1\end{aligned}$$

as follows from Theorem X. This equals $\int_{\gamma_0}^{\gamma} -(\gamma_1-\gamma_0)[I(\gamma_0)+o(1)]\,d\gamma_1$ because of (5.24), whence (5.21) follows.

(5.22) is an easy consequence of (5.17) and (5.18).

Next, we show

Lemma 5.2. *Suppose that all assumptions of Lemma 5.1 are satisfied. Let $\{\xi_n\}$ be a sequence of independent identically distributed r.v.'s with distribution P_{γ_0} where $\gamma_0 \in \Gamma$. Define*

$$\gamma_n = \gamma_0 + 1/\sqrt{n}, \quad n \geqslant 1. \tag{5.25}$$

We may assume that $\gamma_n \in \Gamma$, $n \geqslant 1$.

Then

$$P_{\gamma_0}\left\{\frac{\sum_{i=1}^{n}[\log f(\xi_i,\gamma_n)-\log f(\xi_i,\gamma_0)]+\frac{1}{2}I(\gamma_0)}{\sqrt{I(\gamma_0)}} \leqslant y\right\} = \Phi(y)+o(1) \tag{5.26}$$

and

$$P_{\gamma_n}\left\{\frac{\sum_{i=1}^{n}[\log f(\xi_i,\gamma_n)-\log f(\xi_i,\gamma_0)]+\frac{1}{2}I(\gamma_0)}{\sqrt{I(\gamma_0)}} \leqslant y\right\} = \Phi(y-\sqrt{I(\gamma_0)})+o(1) \tag{5.27}$$

for every $y \in R_1$ as $n \to \infty$. Here $\Phi(x) = \frac{1}{\sqrt{2\pi}}\int_{-\infty}^{x} e^{-t^2/2}\,dt$ for every $x \in R_1$.

Proof. Set $\xi_{nk} = \sqrt{n}[\log f(\xi_k,\gamma_n)-\log f(\xi_k,\gamma_0)]$, $1 \leqslant k \leqslant n$, $n \geqslant 1$. We want to verify that the r.v.'s ξ_{nk} (or more precisely $\xi_{nk}-E(\xi_{nk})$) satisfy the conditions of I, Theorem 39.2. First it follows immediately from Lemma 5.1 that

$$E(\xi_{nk}) = -\frac{1}{2\sqrt{n}}I(\gamma_0)+o\left(\frac{1}{\sqrt{n}}\right) \tag{5.28}$$

and also that

$$\sigma_n^2 = E(\xi_{nk}-E(\xi_{nk}))^2 = I(\gamma_0)+o(1) \tag{5.29}$$

as $n \to \infty$.

It is more cumbersome to show that (using the notation of p. 107) ϕ_n'', $n \geqslant 1$, is equicontinuous at 0. From (5.19) we have

$$|\log f(x,\gamma_n)-\log f(x,\gamma_0)| \leqslant A(x,\gamma_0)\,|\gamma_n-\gamma_0|, \quad \mu\text{-a.e.} \tag{5.30}$$

(5.30) implies

$$|\phi_n''(t)-\phi_n''(0)|$$
$$\leqslant \int_{R_1} n A^2(x,\gamma_0)\frac{1}{n}\,|e^{it\sqrt{n}[\log f(x,\gamma_n)-\log f(x,\gamma_0)]}-1|\,f(x,\gamma_0)\,d\mu(x)\,.$$

Therefore, it is enough to show that

$$\int_{R_1} A^2(x,\gamma_0)\,|e^{it\sqrt{n}[\log f(x,\gamma_n)-\log f(x,\gamma_0)]}-1|\,f(x,\gamma_0)\,d\mu(x)=o(1) \qquad (5.31)$$

as $t\to 0$ uniformly for all $n\geqslant 1$.

Let K be a positive real number and

$$M=\left\{x:|\log f(x,\gamma_n)-\log f(x,\gamma_0)|\leqslant\frac{K}{\sqrt{n}},\ n=1,2,\ldots\right\}. \qquad (5.32)$$

We split the integral on the right side of (5.31) into two parts: $\int_M+\int_{R_1-M}$. We obtain from (5.32) that

$$\int_M = O\Big(t K e^{tK}\int_{R_1} A^2(x,\gamma_0)\,f(x,\gamma_0)\,d\mu(x)\Big)=O(t K e^{tK}) \qquad (5.33)$$

where the "O" does not depend on n.

Now, assume that $x\in R_1-M$. Then there exists a natural number N such that $|\log f(x,\gamma_N)-\log f(x,\gamma_0)|>\frac{K}{\sqrt{N}}$. Then it follows from (5.30) and (5.25) that

$$A(x,\gamma_0)>\frac{K}{\sqrt{N}\,|\gamma_N-\gamma_0|}=K\,. \qquad (5.34)$$

(5.34) implies

$$P_{\gamma_0}(R_1-M)\leqslant\frac{1}{K}\int_{R_1} A(x,\gamma_0)\,dP_{\gamma_0}(x)\,. \qquad (5.35)$$

It follows from (5.35) that $\int_{R_1-M}\leqslant 2\int_{R_1-M} A^2(x,\gamma_0)\,dP_{\gamma_0}(x)\to 0$ for $K\to\infty$. But (5.35) implies that $\int_M\to 0$ for every fixed K and $n\geqslant 1$ as $t\to 0$. This proves the equicontinuity of ϕ_n'', $n\geqslant 1$, at 0.

Thus, (5.26) is a consequence of I, Theorem 39.2 and I, Theorem 40.1.

Relation (5.27) can be obtained as follows: Set $f_n(x^{(n)},\gamma)=\prod_{i=1}^{n} f(x_i,\gamma)$ for $x^{(n)}=(x_1,\ldots,x_n)\in R_n$, $\xi^{(n)}=(\xi_1,\ldots,\xi_n)$ and

$$g_n(x^{(n)},\gamma_0)=\frac{\log f_n(x^{(n)},\gamma_n)-\log f_n(x^{(n)},\gamma_0)+\frac{1}{2}I(\gamma_0)}{\sqrt{I(\gamma_0)}}\,.$$

Then

$$P_{\gamma_n}\{g_n(\xi^{(n)},\gamma_0)\leqslant y\}=\int\limits_{M_n(y)} f_n(x^{(n)},\gamma_n)\,d\mu^{(n)}(x^{(n)})$$
$$=e^{-\frac{1}{2}I(\gamma_0)}\int\limits_{M_n(y)} e^{g_n(x^{(n)},\gamma_0)\sqrt{I(\gamma_0)}}\,dP^{(n)}_{\gamma_0}(x^{(n)})$$

where $M_n(y)=\{x^{(n)}: g_n(x^{(n)},\gamma_0)\leqslant y\}$, $\mu^{(n)}=\mu\times\cdots\times\mu$ and $P^{(n)}_{\gamma_0}=P_{\gamma_0}\times\cdots\times P_{\gamma_0}$. It follows that $P_{\gamma_n}\{g_n(\xi^{(n)},\gamma_0)\leqslant y\}=e^{-\frac{1}{2}I(\gamma_0)}\int_{-\infty}^{y} e^{z\sqrt{I(\gamma_0)}}\,dG_n(z)$, G_n being the d.f. of $g_n(\xi^{(n)},\gamma_0)$. (5.26) says that G_n converges weakly to Φ. Therefore using (a slight modification of) I, Lemma 23.1 one obtains

$$P_{\gamma_n}\{g_n(\xi^{(n)},\gamma_0)\leqslant y\}\to e^{-\frac{1}{2}I(\gamma_0)}\int_{-\infty}^{y} e^{[z\sqrt{I(\gamma_0)}-z^2/2]}\,dz=\Phi(y-\sqrt{I(\gamma_0)})$$

and (5.27) is proved.

Retaining the notation of Lemma 5.1 and Lemma 5.2 we are going to prove

Theorem 5.2.[79] *Suppose that all assumptions of Lemma 5.1 are satisfied. Let $\xi_1,\xi_2,\ldots$ be a sequence of independent identically distributed r.v.'s with distribution P_γ, $\gamma\in\Gamma$. Assume further that T_n is a $\mathfrak{B}_n$-measurable function on R_n for every $n\geqslant 1$. Denote for each real y and $\gamma\in\Gamma$ the probability $P_\gamma\{\sqrt{n}[T_n(\xi_1,\ldots,\xi_n)-\gamma]\leqslant y\}$ by $F_n(y;\gamma)$. Let σ be a function defined on Γ such that $1/\sigma$ is positive and finite on Γ. Suppose further that $\{F_n(y;\gamma)\}$ converges uniformly on Γ to $\Phi[y/\sigma(\gamma)]$ for each fixed $y\in R_1$. Then $\{T_n(\xi_1,\ldots,\xi_n)\}$ is not superefficient on Γ, that is, $\sigma^2(\gamma)\geqslant 1/I(\gamma)$ for all $\gamma\in\Gamma$.*

Proof. Fix a $\gamma_0\in\Gamma$. The assumptions imply that $F_n(y;\gamma_n)-F_n(y;\gamma_0)\to 0$ for every $y\in R_1$ and this in turn implies that

$$\lim_{n\to\infty} P_{\gamma_n}\{\sqrt{n}[T_n(\xi_1,\ldots,\xi_n)-\gamma_n]\geqslant 0\}=\Phi(0)=\tfrac{1}{2}.$$

It follows from (5.27) that for each real y with

$$y>[I(\gamma_0)]^{1/2} \tag{5.36}$$

and all sufficiently large n

$$P_{\gamma_n}\{\sqrt{n}[T_n(\xi_1,\ldots,\xi_n)-\gamma_n]\geqslant 0\}>P_{\gamma_n}(g_n(\xi^{(n)},\gamma_0)\geqslant y). \tag{5.37}$$

---

[79] See L. Schmetterer, Research Papers in Statistics (Neyman Festschrift, 301—317) John Wiley, New York 1966. Some errata contained in this paper have been corrected here. (See also L. Schmetterer, Symposium on Probability Methods in Analysis (Lecture Notes in Mathematics) 31, 291—295, Springer-Verlag, Berlin-Heidelberg-New York 1967. Previous investigations of the same kind have been made by R. R. Bahadur, Ann. Math. Statist. 35, 1545—1552 (1964) and J. Wolfowitz, Theor. Probab. Appl. 10, 247—260 (1965). Many references can be found in R. R. Bahadur, Ann. Math. Statist. 38, 303—324 (1967).

Now, recall that the set $\{x^{(n)}: \log[f^{(n)}(x^{(n)}, \gamma_n)/f^{(n)}(x^{(n)}, \gamma_0)] \geqslant y[I(\gamma_0)]^{1/2} - \frac{1}{2}I(\gamma_0)\}$ can be viewed as the critical region of a most powerful test for the problem $(\{\gamma_0\}, \{\gamma_n\})$. It follows from III, Theorem 3.3 that the inequality (5.37) also holds when P_{γ_0} is replaced by P_{γ_n}. As $n \to \infty$ it follows from (5.25) and (5.26) that $1-\Phi(1/\sigma(\gamma_0)) \geqslant 1-\Phi(y)$ and therefore

$$y \geqslant 1/\sigma(\gamma_0) \tag{5.38}$$

for each y which satisfies (5.36). But this implies $\sigma^2(\gamma_0) \geqslant 1/I(\gamma_0)$ since $[I(\gamma_0)]^{1/2} < [\sigma(\gamma_0)]^{-1}$ would entail the existence of a y satisfying (5.26) and $y < 1/\sigma(\gamma_0)$ which is a contradiction of (5.38).

Remark. If $\{T_n\}$ is a consistent sequence of ML estimates which are asymptotically normally distributed, then $\sigma^2(\gamma_0) = 1/I(\gamma_0)$. (See p. 317.)

Chapter VI

Theory of Regression and the Sampling Theory of Multidimensional Normal Distributions

1. The theory of regression. Let $\xi_{p+11},\dots,\xi_{p+1n}$, $p\geq 1$, $n\geq p+2$ be r.v.'s possessing the following properties: $E(\xi_{p+1i})$ exists for $1\leq i\leq n$ and

$$E(\xi_{p+1i})=\beta_0+\beta_1 x_{1i}+\dots+\beta_p x_{pi}. \tag{1.1}$$

Further let the covariance matrix of $(\xi_{p+11},\dots,\xi_{p+1n})$ exist and be denoted by $U=(u_{ij})_{1n}^{1n}$. Here, the x_{ji}, $1\leq j\leq p$, $1\leq i\leq n$ are given real numbers and the β_i, $0\leq i\leq p$, as well as the u_{ij}, $1\leq i,\ j\leq n$, real parameters. The β_i satisfy $-\infty<\beta_i<\infty$ and the u_{ij} need only satisfy the trivial restriction that U be positive semi-definite. To be more precise, we should denote the right side of (1.1) by $E(\xi_{p+1i};\beta_0,\dots,\beta_p)$ or even by $E(\xi_{p+1i};\beta_0,\dots,\beta_p,u_{ij},1\leq i,j\leq n)$ but the abbreviated notation should cause no misunderstanding. Our task is to construct *unbiased estimates for each* β_i, $0\leq i\leq p$. In order to bring this problem into the general framework of V.1, we let the sample space be given by $(R_n,\mathfrak{B}_n)$ and the set of joint distributions of $(\xi_{p+11},\dots,\xi_{p+1n})$ be so restricted by the parameters $\beta_0,\dots,\beta_p$; u_{ij}, $1\leq i,j\leq n$ that (1.1) holds and $(u_{ij})_{1n}^{1n}$ is positive semi-definite. To obtain the estimates we make use of Gauss' *method of least squares*, which is closely connected with the MLP.

Before we go into the details we make the following *notational convention*, which will be sufficiently explained by examples: Let $(x_1,\dots,x_n)\to f(x_1,\dots,x_n)$ be a measurable function on R_n and $\zeta_1,\dots,\zeta_n$ some r.v.'s. We then denote the r.v. $f(\zeta_1,\dots,\zeta_n)$ simply by f. It will always be clear from the context to which r.v.'s we are referring. Sometimes we will also apply this convention in the opposite direction. We also introduce for $j=1,\dots,p+1$, the notation $x_j=(x_{j1},\dots,x_{jn})$.

The method of least squares consists of finding for $0\leq i\leq p$ functions $x_{p+1}\to\hat{b}_i(x_{p+1})$[1] such that for each set of real numbers $\beta_0,\dots,\beta_p$

$$\sum_{i=1}^{n}(x_{p+1i}-\beta_0-\beta_1 x_{1i}-\dots-\beta_p x_{pi})^2\geq\sum_{i=1}^{n}(x_{p+1i}-\hat{b}_0(x_{p+1})-\hat{b}_1(x_{p+1})x_{1i}-\dots-\hat{b}_p(x_{p+1})x_{pi})^2$$

[1] The $\hat{b}_j$, provided they exist, will in general also depend on the $x_1,\dots,x_p$. Since, however, we view these n-tuples here as given, we suppress this dependence.

holds. Thus, we want to choose $\hat{b}_0(x_{p+1}), \ldots, \hat{b}_p(x_{p+1})$ such that the function $(\beta_0, \ldots, \beta_p) \to \sum_{i=1}^{n} (x_{p+1\,i} - \beta_0 - \beta_1 x_{1i} - \cdots - \beta_p x_{pi})^2$ is minimized. This explains the method's name.

It will turn out that $\hat{\boldsymbol{b}}_j$ is an unbiased estimate for β_j, $0 \leqslant j \leqslant p$. We will make this precise below.

One applies the method of least squares practically by determining the quantities $\hat{b}_j(x_{p+1})$ by means of a sample $(x_{p+1\,1}, \ldots, x_{p+1\,n})$ and then using them as estimates for β_j, $0 \leqslant j \leqslant p$.

This procedure will be justified by the consideradions below.

The problem posed above appears very frequently in statistical applications. A good example (from A. Linder, loc. cit. II[1]) (for $p=1$) occurs in the study of the connection between the velocity and the stopping distance of autos. The velocity is controlled by the driver. For given squared velocity, one can view the stopping distance as a r.v. ξ_{2i}.

There obviously exists a connection between velocity and stopping distance, which is not of strictly functional form, and which may be represented by the regression curve of stopping distance versus square of velocity. We assume that this regression curve is a straight line. (See the remark following Theorem 1.2.) A natural problem is then the determination of the coefficients of this line when a stopping distance x_{2i} is measured for squared velocity x_{1i}.

Let X denote the matrix

$$\begin{pmatrix} 1\, x_{11} & \ldots & x_{p1} \\ \ldots & \ldots & \ldots \\ 1\, x_{1n} & \ldots & x_{pn} \end{pmatrix}$$

and A the matrix $X'X$. Then we have

Theorem 1.1. *Let $\xi_{p+1\,i}$, $1 \leqslant i \leqslant n$ be n r.v.'s of the previously defined type such that, in particular,* (1.1) *is fulfilled. If the inverse A^{-1} of A exists, then one can make use of the method of least squares described above for the construction of unbiased estimates for β_j, $0 \leqslant j \leqslant p$. The covariance matrix of $(\hat{\boldsymbol{b}}_0, \ldots, \hat{\boldsymbol{b}}_p)$ is given by $A^{-1} X'U X A^{-1}$.*

Proof. Introduce the notation $B = \begin{pmatrix} \beta_0 \\ \vdots \\ \beta_p \end{pmatrix}$. Then the expression

$$\sum_{i=1}^{n} \left(x_{p+1\,i} - \beta_0 - \sum_{j=1}^{p} \beta_j x_{ji} \right)^2$$

assumes the form

$$(x_{p+1} - XB)'(x_{p+1} - XB). \tag{1.2}$$

In order for (1.2) as function of β_i to be a minimum it is necessary that all partial derivatives w.r.t. β_i, $0 \leqslant i \leqslant p$, vanish. Then one obtains a system of linear equations for the $\hat{b}_i(x_{p+1})$. Denoting any solution, provided it exists, by $\hat{b}_0, \ldots, \hat{b}_p$ (where we now even suppress, somewhat inconsistently, the dependence on x_{p+1}) and summarizing these by a column vector $\hat{B}$, we see that $\hat{B}$ must satisfy

$$X' x_{p+1} = X' X \hat{B}$$

or in other notation,

$$X' x_{p+1} = A \hat{B} . \tag{1.3}$$

We show that each vector $\hat{B}$ satisfying (1.3) minimizes (1.2). Indeed, if B_0 is any vector with $p+1$ components, then

$$\begin{aligned} &(x_{p+1} - X B_0)'(x_{p+1} - X B_0) = (x_{p+1} - X \hat{B})'(x_{p+1} - X \hat{B}) \\ &-(X(B_0 - \hat{B}))'(x_{p+1} - X \hat{B}) - (x_{p+1} - X \hat{B})' X (B_0 - \hat{B}) \\ &+(X(B_0 - \hat{B}))' X (B_0 - \hat{B}) . \end{aligned} \tag{1.4}$$

From (1.3), however,

$$(X(B_0 - \hat{B}))'(x_{p+1} - X \hat{B}) = (B_0 - \hat{B})'(X' x_{p+1} - X' X \hat{B}) = 0$$

and the same holds for the transpose matrices so that the two middle terms on the right in (1.4) vanish. There remains

$$\begin{aligned} &(x_{p+1} - X B_0)'(x_{p+1} - X B_0) = (x_{p+1} - X \hat{B})'(x_{p+1} - X \hat{B}) \\ &+(B_0 - \hat{B})' X' X (B_0 - \hat{B}) . \end{aligned} \tag{1.5}$$

Since, however, $(X u)'(X u) = u' X' X u$ is nonnegative for each $(p+1)$-tuple of real numbers u, we are finished.

A^{-1} exists by assumption. Then, (1.3) is uniquely solvable and we get

$$\hat{B} = A^{-1} X' x_{p+1} . \tag{1.6}$$

From (1.6) we obtain with the notation $\xi_{p+1} = (\xi_{p+1\,1}, \ldots, \xi_{p+1\,n})$

$$E(\hat{\boldsymbol{B}}) = E(A^{-1} X' \xi_{p+1}) = A^{-1} X' E(\xi_{p+1}) .$$

According to (1.1)

$$E(\xi_{p+1}) = X B . \tag{1.7}$$

Hence, $E(\hat{\boldsymbol{B}}) = A^{-1} X' X B = A^{-1} A B = B$, i.e., $\hat{\boldsymbol{B}}$ is unbiased for B.

From (1.6) and (1.7) we have for the covariance matrix of $\hat{\boldsymbol{B}}$:

$$E[(\hat{\boldsymbol{B}} - B)(\hat{\boldsymbol{B}} - B)'] = A^{-1} X' E[(\xi_{p+1} - E(\xi_{p+1}))(\xi_{p+1} - E(\xi_{p+1}))'] .$$

$X(A^{-1})' = A^{-1} X' U X A^{-1}$ since A and hence A^{-1} are symmetric.

This proves the theorem. We emphasize that we have obtained these results without any special assumptions (other than (1.1) and the requirement that U exists, which are basic) on the underlying probability distributions[2].

An easy and practically important consequence is the following: *If the r.v.'s* ξ_{p+1i}, $1 \leqslant i \leqslant n$, *are independent*, satisfy (1.1) and *have the same variance* σ^2, $0 < \sigma^2 < \infty$, then *the covariance matrix of* $\hat{\boldsymbol{B}}$ is given by $\sigma^2 A^{-1}$.

We remark that the requirement of the existence of A^{-1} in Theorem 1.1 can be replaced by the assumption that X has rank $p+1$. In particular, we have the well-known

Lemma 1.1[3]. *The matrix $X'X$ has the same rank as X.*

Proof. Let u be a column vector with $p+1$ real components. From $X'Xu=0$ we have, in order, $u'X'Xu=0$, $(Xu)'(Xu)=0$, $Xu=0$ and again $X'Xu=0$. We see that the columns of $X'X$ are linearly independent in the same way as those of X and vice versa, i.e., $X'X$ and X have the same rank.

Before we proceed to further specialization we point out that the decisive equation (1.3) can also be obtained immediately by geometric means[4]. Consider the hyperplane

$$\beta_0 e + \beta_1 x_1 + \cdots + \beta_p x_p, \qquad -\infty < \beta_i < \infty, \qquad 0 \leqslant i \leqslant p$$

spanned by the vectors $e=(1, \ldots, 1)$, $x_1, \ldots, x_p$. (1.2) is obviously equal to the squared distance of the point x_{p+1} from the point $\beta_0 e + \beta_1 x_1 + \cdots + \beta_p x_p$ of the hyperplane. This distance is minimal when $(\beta_0, \ldots, \beta_p)$ is chosen such that

$$x_{p+1} - \beta_0 e - \beta_1 x_1 - \cdots - \beta_p x_p$$

is orthogonal to the hyperplane and thus to e and to all x_i, $1 \leqslant i \leqslant p$. This leads to the condition $X'(x_{p+1} - XB) = 0$.

We now assume as special case that we have n independent r.v.'s ξ_{p+1i}, each of which following a $N(\beta_0 + \beta_1 x_{1i} + \cdots + \beta_p x_{pi}, \sigma^2)$. We repeat that

$$-\infty < \beta_j < \infty, \qquad 0 \leqslant j \leqslant p \tag{1.8}$$

and

$$0 < \sigma^2 < \infty \tag{1.9}$$

are assumed to hold and that the x_{ji}, $1 \leqslant j \leqslant p$, $1 \leqslant i \leqslant n$, are given real numbers.

---

2 From the extensive literature on the method of least squares we indicate only: A. C. Aitken, Proc. Roy. Soc. Edinburgh Sect. A, 55, 42—48 (1935), B. J. van Ijzeren, Statistica Rijswijk 8, 21—45 (1954), O. Kempthorne. The Design and Analysis of Experiments, John Wiley & Sons-Chapman & Hall, New York-London 1952, J. V. Linnik, Die Methode der kleinsten Quadrate in moderner Darstellung, VEB Deutscher Verlag der Wissenschaften, Berlin 1961.

3 See B. J. van Ijzeren, l. c.[2].

4 See A. N. Kolmogorov, Uspehi. Mat. Nauk 1, 57—70 (1946).

We are interested in the distribution of

$$(\xi_{p+1}-X\hat{\boldsymbol{B}})'(\xi_{p+1}-X\hat{\boldsymbol{B}}). \tag{1.10}$$

For each realization x_{p+1} of the r.v. ξ_{p+1},

$$(x_{p+1}-X\hat{B})'(x_{p+1}-X\hat{B}) \tag{1.11}$$

represents the minimal value of (1.2). We call (1.11) or also the r.v. (1.10) the *residual term.*

We replace B_0 in (1.5) by B and write again the relation thus obtained:

$$\begin{aligned}&(x_{p+1}-XB)'(x_{p+1}-XB)\\&=(x_{p+1}-X\hat{B})'(x_{p+1}-X\hat{B})+(\hat{B}-B)'X'X(\hat{B}-B).\end{aligned} \tag{1.12}$$

Using (1.6) and substituting the r.v. ξ_{p+1} for x_{p+1}, we obtain from (1.7) and (1.12)

$$\begin{aligned}\frac{1}{\sigma^2}(\xi_{p+1}-E(\xi_{p+1}))'(\xi_{p+1}-E(\xi_{p+1})) &= \frac{1}{\sigma^2}(\xi_{p+1}-X\hat{\boldsymbol{B}})'(\xi_{p+1}-X\hat{\boldsymbol{B}})\\&+\frac{1}{\sigma^2}(\xi_{p+1}-E(\xi_{p+1}))'XA^{-1}X'(\xi_{p+1}-E(\xi_{p+1})).\end{aligned}$$

The second summand on the right equals $\frac{1}{\sigma^2}(\hat{\boldsymbol{B}}-B)'A(\hat{\boldsymbol{B}}-B)$.

The left side of this equality is by assumption equal to the sum of the squares of n independent, $N(0,1)$-distributed r.v.'s and is thus itself chi-square-distributed with n degrees of freedom. Since the components of $\hat{\boldsymbol{B}}$ depend linearly on the r.v.'s $\xi_{p+1i}-E(\xi_{p+1i})$, $1\leqslant i\leqslant n$, the two summands on the right also represent quadratic forms in these r.v.'s. The first summand on the right in (1.12) is from (1.6) also equal to

$$(x_{p+1}-XB-XA^{-1}X'(x_{p+1}-XB))'(x_{p+1}-XB-XA^{-1}X'(x_{p+1}-XB))$$

and is thus a quadratic form in the n variables $x_{p+1i}-\beta_0-\beta_1x_{1i}-\cdots-\beta_px_{pi}$. This is of at most rank $n-p-1$. Indeed, from (1.3), $X'(x_{p+1}-X\hat{B})=0$ and these are $p+1$ independent linear relations for the given variables. The matrix $X'X$ is, however, a $(p+1)\times(p+1)$ matrix, and hence, the quadratic form $(B-\hat{B})'X'X(B-\hat{B})$ has at most rank $p+1$.

Applying II, Theorem 4.1 we thus get

Theorem 1.2. *Let* ξ_{p+1i}, $1\leqslant i\leqslant n$ *be* $n\geqslant p+2$ *independent r.v.'s, each of which following a* $N(\beta_0+\beta_1x_{1i}+\cdots+\beta_px_{pi},\sigma^2)$, *where the parameters satisfy* (1.8) *and* (1.9). *Let* A^{-1} *exist. Then*

$$\frac{1}{\sigma^2}(\xi_{p+1}-X\hat{\boldsymbol{B}})'(\xi_{p+1}-X\hat{\boldsymbol{B}}) \quad \text{and} \quad \frac{1}{\sigma^2}(\hat{\boldsymbol{B}}-B)'A(\hat{\boldsymbol{B}}-B)$$

are independent and are chi-square distributed with $n-p-1$ *and* $p+1$ *degrees of freedom, resp.*

I. (35.12) is an important case of an example for which the assumptions of Theorem 1.2 are realized. A simple comparison shows that in this case

$$\sigma^2=(d_{p+1\,p+1})^{-1}, \quad \beta_0=a_{p+1}+\sum_{i=1}^{p}\frac{d_{ip+1}}{d_{p+1\,p+1}}a_i, \qquad\qquad (1.13)$$

$$\beta_j=-\frac{d_{jp+1}}{d_{p+1\,p+1}}, \quad 1\leqslant j\leqslant p$$

hold.

Under the assumptions of Theorem 1.2, the estimate $\hat{\boldsymbol{B}}$ possesses an important minimality property, which in its simplest form was discovered by A. Markov[5].

This is given by the so-called *Gauss-Markov theorem:*

Theorem 1.3. *Let the assumptions of Theorem 1.2 be fulfilled and assume for the sake of simplicity that* $\sigma^2=1$. *Let* $p+1$ *real numbers* $c_0,\ldots,c_p$ *be given. If* c *denotes the vector of these* $p+1$ *numbers, then* $c'\hat{\boldsymbol{B}}$ *is a uniformly minimal unbiased estimate for the map* $B\to c'B$ *defined over* R_{p+1}.

Proof. We use V, Theorems 1.1 and 1.2. Let V be the set of all unbiased estimates for 0 whose second moment w.r.t. all n-dimensional normal distributions with density

$$x\to(2\pi)^{-n/2}e^{-\frac{1}{2}(x-XB)'(x-XB)}$$

exists.

For each $v\in V$ we thus have

$$\int_{R_n} v(x)e^{-\frac{1}{2}(x-XB)'(x-XB)}dx=0. \qquad (1.14)$$

For each $B\in R_{p+1}$, one can differentiate the left side of (1.14) w.r.t. β_j, $0\leqslant j\leqslant p$, under the integral sign and we get

$$\int_{R_n} v(x)X'xe^{-\frac{1}{2}(x-XB)'(x-XB)}dx=0$$

---

[5] A. Markov, Wahrscheinlichkeitsrechnung, 2nd ed., Leipzig-Berlin 1912. Also see F.N. David and J. Neyman, l.c. V.[1] and L. Schmetterer, l.c. V.[5], second paper listed. For a somewhat more general formulation, also see H. Scheffé, l.c. III[72], 14. A recent reference is H. Drygas, The Coordinate-Free Approach to Gauss-Markov Estimation, Lecture Notes in Operations Research and Mathematical Systems 40, Springer-Verlag, Berlin-Heidelberg-New York 1970.

from which

$$\int_{R_n} v(x) A^{-1} X' x e^{-\frac{1}{2}(x-XB)'(x-XB)} dx = 0.$$[6]

This vector equality yields, however,

$$\int_{R_n} v(x) c' A^{-1} X' x e^{-\frac{1}{2}(x-XB)'(x-XB)} dx = 0.$$

Writing x_{p+1} instead of x again, one concludes the proof using (1.6) and the mentioned theorems from V.

We want to consider now a problem of the construction of confidence sets and proceed from the assumptions of Theorem 1.2. Let c_{kj}, $0 \leqslant j \leqslant p$, $1 \leqslant k \leqslant l$ be real numbers. Consider for $1 \leqslant k \leqslant l$ the linear functions $\gamma_k(B) = \sum_{j=0}^{p} c_{kj}\beta_j$ of the parameter B over R_{p+1}. For the set of all $(\gamma_1, \ldots, \gamma_l)$ we want to construct confidence sets. We assume to this end that the l vectors $(c_{k0}, \ldots, c_{kp})$ span a space of dimension m. Naturally, $m \leqslant p+1$, but we assume to avoid trivial complications that $m = p+1$. Theorem 1.3 suggests the use of $C\hat{\boldsymbol{B}}$ as an unbiased estimation for $\begin{pmatrix}\gamma_0\\ \vdots \\ \gamma_l\end{pmatrix}$, where $C = (c_{kj})_{1l}^{0p}$. By assumption, C has rank $p+1$.

We now use the technique of the so-called *canonical form of the method of least squares*[7], which formed in essence the basis for the proof of Theorem 1.2. Namely, since X has rank $p+1$, the column vectors of X span a $(p+1)$-dimensional linear space. Introducing $p+1$ mutually orthogonal vectors $(g_{i1}, \ldots, g_{in})$ of length 1 into this space and writing $G = (g_{ij})_{1\,p+1}^{1n}$, we find that there exists a $(p+1) \times (p+1)$ matrix D with

$$|D| \neq 0$$

such that

$$X' = DG. \tag{1.15}$$

G can be extended to an orthogonal $n \times n$ matrix, which we denote by $\begin{pmatrix}G\\E\end{pmatrix}$. Introduce a new r.v.

$$\eta = \begin{pmatrix}G\\E\end{pmatrix} \xi_{p+1} \tag{1.16}$$

with components η_i. According to I, Theorem 27.2, the η_i for $1 \leqslant i \leqslant n$ are independent and normally distributed with variance σ^2.

---

[6] It should cause no difficulty when the symbol 0 denotes both the $(p+1)$-dimensional null-vector and zero itself.

[7] See H. Scheffé, l.c. III[72],

From (1.16), (1.15) and (1.7) it easily follows that

$$E(\eta_i)=0\,, \qquad p+2\leqslant i\leqslant n\,. \tag{1.17}$$

Writing $\eta^{(0)}=\begin{pmatrix}\eta_1\\ \vdots\\ \eta_{p+1}\end{pmatrix}$ we also get

$$E(\eta^{(0)})=D'B\,. \tag{1.18}$$

From (1.6), (1.16) and (1.15), $\hat{\boldsymbol{B}}=D'^{-1}\eta^{(0)}$. Furthermore,

$$(\xi_{p+1}-X\hat{\boldsymbol{B}})'(\xi_{p+1}-X\hat{\boldsymbol{B}})=\xi'_{p+1}\xi_{p+1}-\hat{\boldsymbol{B}}'X'\xi_{p+1}$$
$$=\eta'\eta-\eta^{(0)\prime}D^{-1}DG(G'E')\eta=\sum_{i=p+2}^{n}\eta_i^2\,.$$

Together with (1.17) this shows that

$$S^2/\sigma^2=\frac{1}{\sigma^2}(\xi_{p+1}-X\hat{\boldsymbol{B}})'(\xi_{p+1}-X\hat{\boldsymbol{B}})$$

is chi-square distributed with $n-p-1$ degrees of freedom, independently of $\hat{\boldsymbol{B}}$ (provided σ^2 is the true parameter). This is essentially again Theorem 1.2.

With $(f_{ij})_{11}^{0p}=F=CD'^{-1}$ one thus obtains $C\hat{\boldsymbol{B}}=F\eta^{(0)}$ or from (1.18), for $1\leqslant k\leqslant l$,

$$\left(\sum_{j=0}^{p}c_{kj}\hat{\boldsymbol{b}}_j-\gamma_k(B)\right)\Big/(\boldsymbol{S}|f^{(k)}|)=\frac{(f^{(k)})'(\eta^{(0)}-D'B)}{\boldsymbol{S}|f^{(k)}|} \tag{1.19}$$

with $f^{(k)}=(f_{k0},\dots,f_{kp})$. Thus, if B is the true parameter, then the left and, consequently also the right side of (1.19) is, after multiplication by $\sqrt{n-p-1}$, t-distributed with $n-p-1$ degrees of freedom. Further if κ is a positive number, then for each k and all $x_{p+1}\in R_n$ [8]

$$\left(-\kappa S(x_{p+1})|f^{(k)}|+\sum_{j=0}^{p}c_{kj}\hat{b}_j(x_{p+1}),\ \kappa S(x_{p+1})|f^{(k)}|+\sum_{j=0}^{p}c_{kj}\hat{b}_j(x_{p+1})\right)$$

defines a confidence interval $K_k(x_{p+1})$ for γ_k which is even similar and whose confidence level can easily be given (dependent on κ). To solve our original problem we must thus, according to IV, Theorem 2.2, calculate

$$P_B\left(\bigcap_{i=1}^{l}\{x_{p+1}:\gamma_i(B)\in K_i(x_{p+1})\}\right).$$

[8] For clarity, we now write $S(x_{p+1})$ instead of S.

But if

$$-\kappa \leqslant \left(\sum_{j=0}^{p} c_{kj} \hat{b}_j(x_{p+1}) - \gamma_k(B) \right) \Big/ (S(x_{p+1}) |f^{(k)}|) \leqslant \kappa \tag{1.20}$$

holds for a B and all k with $1 \leqslant k \leqslant l$, then also

$$\begin{aligned} -\kappa &\leqslant \min_k \left(\sum_{j=0}^{p} c_{kj} \hat{b}_j(x_{p+1}) - \gamma_k(B) \right) \Big/ (S(x_{p+1}) |f^{(k)}|) \\ &\leqslant \max_k \left(\sum_{j=0}^{p} c_{kj} \hat{b}_j(x_{p+1}) - \gamma_k(B) \right) \Big/ (S(x_{p+1}) |f^{(k)}|) \leqslant \kappa . \end{aligned} \tag{1.21}$$

On the other hand, the inequalities (1.20) also follow from (1.21). However, if we define $\mathfrak{f} = \{f \in R_{p+1} : |f| > 0\}$, then (1.19) shows that if we set

$$a_k = \left(\sum_{j=0}^{p} c_{kj} \hat{b}_j - \gamma_k(B) \right) \Big/ (S |f^{(k)}|)$$

then

$$\begin{aligned} & P_B \left(-\kappa \leqslant \min_k \boldsymbol{a}_k \leqslant \max_k \boldsymbol{a}_k \leqslant \kappa \right) \\ & \geqslant P_B \left(-\kappa \leqslant -\sup_{f \in \mathfrak{f}} \frac{f'(\eta^{(0)} - D' B)}{S|f|} \leqslant \sup_{f \in \mathfrak{f}} \frac{f'(\eta^{(0)} - D' B)}{S|f|} \leqslant \kappa \right). \end{aligned}$$

From Schwarz' inequality it follows that the latter is equal to

$$P_B \left(\frac{(\eta^{(0)} - D' B)' (\eta^{(0)} - D' B)}{S^2} \leqslant \kappa^2 \right).$$

Now $\dfrac{n-p-1}{n+1} \left(\dfrac{(\eta^{(0)} - D' B)' (\eta^{(0)} - D' B)}{S^2} \right)$ is F-distributed with $((p+1), (n-p-1))$ degrees of freedom. Thus, if we choose κ for given β, $0 < \beta < 1$ such that with $\lambda = \kappa^2 \dfrac{n-p-1}{p+1}$

$$\int_0^{\lambda} k_{p+1, n-p-1}(F) \, dF = \beta ,$$

then the totality of inequalities (1.20) yields a confidence set for $(\gamma_1, \ldots, \gamma_l)$ with confidence level β.

Introducing by (1.16) the n-dimensional r.v. ξ_{p+1} again, one obtains the following dual test:

For the *null hypothesis* $B = B_0$, $0 < \sigma^2 < \infty$, a *similar critical region* with level of significance $1 - \beta$ is defined by

$$\left\{ x_{p+1} : \frac{n-p-1}{p+1} \frac{(B_0 - \hat{B})' A (B_0 - \hat{B})}{(x_{p+1} - X \hat{B})' (x_{p+1} - X \hat{B})} \geqslant \lambda \right\}.$$

The set of admissible hypotheses is given by, say $-\infty < B < \infty, 0 < \sigma^2 < \infty$. One can also obtain this critical region under the same hypotheses by means of the MLRT.

2. Linear constraints. We first want to consider under general assumptions the case in which the parameters β_i are no longer independent but satisfy, briefly speaking, linear conditions. This case is obviously related to the last section of **1**. Let C be a matrix of the form

$$C = (c_{ij})_{1r}^{1p+1}, \quad 1 \leq r < p+1 .$$

Let its rank be r and for $0 \leq i \leq r-1$, let l_i be real numbers which we collect into a column vector l. Assume the parameter B varies in the set

$$\{B : CB = l, -\infty < B < \infty\} . \tag{2.1}$$

We consider again the problem of constructing unbiased estimates for B. We will simplify it somewhat. By assumption we can assume that $|c_{ij}|_{1r}^{1r} \neq 0$. Introduce a one-to-one parameter transformation in the set given by (2.1):

$$\begin{aligned} \delta_0 &= c_{11}\beta_0 + \cdots + c_{1p+1}\beta_p \\ &\cdots\cdots\cdots\cdots\cdots\cdots \\ \delta_{r-1} &= c_{r1}\beta_0 + \cdots + c_{rp+1}\beta_p \\ \delta_r &= \beta_r \\ &\cdots\cdots\cdots\cdots\cdots\cdots \\ \delta_p &= \beta_p \end{aligned} \tag{2.2}$$

Collect the δ_i, $0 \leq i \leq p$ into a column vector Δ and denote the matrix of the transformation (2.2) by C_1. Instead of (2.2) we thus have

$$\Delta = C_1 B . \tag{2.3}$$

With the notation

$$\Delta_1 = \begin{pmatrix} \delta_0 \\ \vdots \\ \delta_{r-1} \end{pmatrix}, \quad \Delta_2 = \begin{pmatrix} \delta_r \\ \vdots \\ \delta_p \end{pmatrix}$$

the set (2.1) then takes the form

$$\{\Delta : \Delta_1 = l, -\infty < \Delta_2 < \infty\} . \tag{2.4}$$

Instead of (1.7) we get

$$E(\xi_{p+1}) = X C_1^{-1} \Delta .$$

$X C_1^{-1}$ is, along with X, an $n \times (p+1)$ matrix. $(C_1^{-1})' A C_1^{-1}$ possesses, along with $X'X = A$, an inverse, and vice versa. It is thus no loss of generality to assume for the sake of simplified notation that

$$E(\xi_{p+1}) = X\Delta, \tag{2.5}$$

where (2.4) holds. Otherwise, the assumptions of Theorem 1.1 should hold. The problem consists of constructing unbiased estimates for Δ_2. We again make use of the *method of least squares* and want to minimize

$$(\delta_r, \dots, \delta_p) \to \sum_{i=1}^{n} (x_{p+1i} - l_0 - \dots - x_{r-1i} l_{r-1} - x_{ri} \delta_r - \dots - x_{pi} \delta_p)^2 .$$

Let

$$\begin{pmatrix} 1\, x_{11} & \dots & x_{r-11} \\ \dots & \dots & \dots \\ 1\, x_{1n} & \dots & x_{r-1n} \end{pmatrix} = X_0, \quad \begin{pmatrix} x_{r1} & \dots & x_{p1} \\ \dots & \dots & \dots \\ x_{rn} & \dots & x_{pn} \end{pmatrix} = X^0 .$$

Further let $\bar{A} = (X_0' X_0 \;\; X_0' X^0)$, $A_0 = X^{0\prime} X_0$, $A^0 = X^{0\prime} X^0$. Then $X = (X_0 \;\; X^0)$ and $A = \begin{pmatrix} & \bar{A} \\ A_0 & A^0 \end{pmatrix}$.

As in the proof of Theorem 1.1 we see, substituting $\begin{pmatrix} l \\ \Delta_2 \end{pmatrix}$ for B and $\begin{pmatrix} l \\ \hat{D}_2 \end{pmatrix}$ for $\hat{B}$, that each solution $\hat{d}_r, \dots, \hat{d}_p$ of the posed extremum problem must satisfy

$$X^{0\prime} x_{p+1} - A_0 l = A^0 \hat{D}_2 . \tag{2.6}$$

Here, $\hat{D}_2 = \begin{pmatrix} \hat{d}_r \\ \vdots \\ \hat{d}_p \end{pmatrix}$. We have again suppressed the dependence on x_{p+1} in our notation. We assume that (2.6) has a unique solution and obtain

$$(A^0)^{-1} X^{0\prime} x_{p+1} - (A^0)^{-1} A_0 l = \hat{D}_2 \tag{2.7}$$

and $\hat{D}_2$ actually solves the extremum problem. In analogy to Theorem 1.1 we now have

Theorem 2.1. *Let $\xi_{p+11}, \dots, \xi_{p+1n}$ be $n \geq 2$ r.v.'s. Assume that (2.5) holds for $r \leq p \leq n-2$ and that Δ satisfies (2.4). Moreover, let the covariance matrix U of the ξ_{p+1i}, $1 \leq i \leq n$, exist. It can otherwise be completely arbitrary. Substituting ξ_{p+1i} for x_{p+1i}, $i = 1, \dots, n$, we have that the $(p+1-r)$-dimensional r.v. $\hat{\boldsymbol{D}}_2$ is unbiased for Δ_2. For the covariance matrix of $\hat{\boldsymbol{D}}_2$ one obtains $(A^0)^{-1} X^{0\prime} U X^0 (A^0)^{-1}$. Under the additional assumption that the ξ_{p+1i} are independent and normally distributed with the same positive variance σ^2 for $i = 1, \dots, n$, the residual term*

$$\frac{1}{\sigma^2} \left(\xi_{p+1} - X \begin{pmatrix} l \\ \hat{\boldsymbol{D}}_2 \end{pmatrix} \right)' \left(\xi_{p+1} - X \begin{pmatrix} l \\ \hat{\boldsymbol{D}}_2 \end{pmatrix} \right)$$

is chi-square distributed with $(n+r-p-1)$ *degrees of freedom. Moreover,*

$$\frac{1}{\sigma^2}(\xi_{p+1}-XA^{-1}X'\xi_{p+1})'(\xi_{p+1}-XA^{-1}X'\xi_{p+1}),$$

$$\frac{1}{\sigma^2}\left(\begin{pmatrix} l \\ \hat{D}_2 \end{pmatrix}-A^{-1}X'\xi_{p+1}\right)' A\left(\begin{pmatrix} l \\ \hat{D}_2 \end{pmatrix}-A^{-1}X'\xi_{p+1}\right)$$

and

$$\frac{1}{\sigma^2}(\hat{D}_2-\Delta_2)'X^{0\prime}X^0(\hat{D}_2-\Delta_2)$$

are independent and chi-square distributed with $n-p-1$, r *and* $p-r+1$ *degrees of freedom, resp.*

Proof. First, an easy calculation using (2.7) and (2.5) yields

$$\begin{aligned} E(\hat{D}_2)&=(A^0)^{-1}(X^0)'E(\xi_{p+1})-(A^0)^{-1}A_0 l \\ &=(A^0)^{-1}(X^0)'X\begin{pmatrix} l \\ \Delta_2 \end{pmatrix}-(A^0)^{-1}A_0 l \\ &=(A^0)^{-1}X^{0\prime}(X_0 l+X^0\Delta_2)-(A^0)^{-1}A_0 l \end{aligned}$$

so that

$$E(\hat{D}_2)=\Delta_2\,.$$

For the covariance of $\hat{D}_2$ one gets again according to (2.7):

$$\begin{aligned} E[(\hat{D}_2-\Delta_2)(\hat{D}_2-\Delta_2)'] &= E[((A^0)^{-1}X^{0\prime}\xi_{p+1}-(A^0)^{-1}A_0 l \\ &\quad -(A^0)^{-1}X^{0\prime}X^0\Delta_2)((A^0)^{-1}X^{0\prime}\xi_{p+1}-(A^0)^{-1}A_0 l-(A^0)^{-1}X^{0\prime}X^0\Delta_2)'] \\ &=(A^0)^{-1}X^{0\prime}E\left[\left(\xi_{p+1}-X\begin{pmatrix} l \\ \Delta_2 \end{pmatrix}\right)\left(\xi_{p+1}-X\begin{pmatrix} l \\ \Delta_2 \end{pmatrix}\right)'\right]X^0(A^0)^{-1} \end{aligned}$$

from which the claim on the covariance follows.

To demonstrate the rest, abbreviate $x_{p+1}-X_0 l-X^0\Delta_2$ by y_{p+1}. From (2.6) we get an analog to (1.12) in the same way as (1.12) was derived from (1.3):

$$\begin{aligned} y'_{p+1}y_{p+1}&=(x_{p+1}-X_0 l-X^0\hat{D}_2)'(x_{p+1}-X_0 l-X^0\hat{D}_2) \\ &\quad +(\hat{D}_2-\Delta_2)'X^{0\prime}X^0(\hat{D}_2-\Delta_2)\,. \end{aligned} \tag{2.8}$$

However, substituting $\begin{pmatrix} l \\ \hat{D}_2 \end{pmatrix}$ for B in (1.12) and expressing $\hat{B}$ by $A^{-1}X'x_{p+1}$, we find

$$\begin{aligned}&\left(x_{p+1}-X\begin{pmatrix} l \\ \hat{D}_2 \end{pmatrix}\right)'\left(x_{p+1}-X\begin{pmatrix} l \\ \hat{D}_2 \end{pmatrix}\right)\\ &=(x_{p+1}-XA^{-1}X'x_{p+1})'(x_{p+1}-XA^{-1}X'x_{p+1}) \qquad (2.9)\\ &+\left(\begin{pmatrix} l \\ \hat{D}_2 \end{pmatrix}-A^{-1}X'x_{p+1}\right)'A\left(\begin{pmatrix} l \\ \hat{D}_2 \end{pmatrix}-A^{-1}X'x_{p+1}\right).\end{aligned}$$

From (2.8) and (2.9)

$$\begin{aligned}y'_{p+1}y_{p+1}&=(x_{p+1}-XA^{-1}X'x_{p+1})'(x_{p+1}-XA^{-1}X'x_{p+1})\\ &+\left(\begin{pmatrix} l \\ \hat{D}_2 \end{pmatrix}-A^{-1}X'x_{p+1}\right)'A\left(\begin{pmatrix} l \\ \hat{D}_2 \end{pmatrix}-A^{-1}X'x_{p+1}\right) \qquad (2.10)\\ &+(\hat{D}_2-\Delta_2)'X^{0\prime}X^0(\hat{D}_2-\Delta_2).\end{aligned}$$

We also have

$$\begin{aligned}\begin{pmatrix} l \\ \hat{D}_2 \end{pmatrix}-A^{-1}X'x_{p+1}&=\begin{pmatrix} 0 \\ \hat{D}_2-\Delta_2 \end{pmatrix}+\begin{pmatrix} l \\ \Delta_2 \end{pmatrix}-A^{-1}X'x_{p+1}\\ &=\begin{pmatrix} 0 \\ (A^0)^{-1}X^{0\prime}y_{p+1} \end{pmatrix}-A^{-1}X'(x_{p+1}-X_0 l-X^0\Delta_2).\end{aligned}$$

Here, 0 represents the r-dimensional null-vector. This leads to

$$\begin{aligned}\begin{pmatrix} l \\ \hat{D}_2 \end{pmatrix}-A^{-1}X'x_{p+1}&=A^{-1}\left[A\begin{pmatrix} 0 \\ (A^0)^{-1}X^{0\prime}y_{p+1} \end{pmatrix}-X'y_{p+1}\right]\\ &=A^{-1}\left[\begin{pmatrix} X_0'X^0(A^0)^{-1}X^{0\prime} \\ X^{0\prime} \end{pmatrix}y_{p+1}-\begin{pmatrix} X_0' \\ X^{0\prime} \end{pmatrix}y_{p+1}\right]\\ &=A^{-1}\left[\begin{pmatrix} X_0'X^0(A^0)^{-1}X^{0\prime}-X_0' \\ 0 \end{pmatrix}y_{p+1}\right].\end{aligned}$$

The last zero now represents a null vector with $(p-r+1)$ rows. For the second summand on the right in (2.10) we thus obtain

$$y'_{p+1}\begin{pmatrix} X_0'X^0(A^0)^{-1}X^{0\prime}-X_0' \\ 0 \end{pmatrix}'A^{-1}\begin{pmatrix} X_0'X^0(A^0)^{-1}X^{0\prime}-X_0' \\ 0 \end{pmatrix}y_{p+1}. \qquad (2.11)$$

We have already shown on p. 358 that the first summand on the right in (2.10) is a quadratic form in the y_{p+1}. The expression (2.11) is a quadratic form in the y_{p+1} of at most rank r. Since $X^{0\prime}X^0$ is a $(p-r+1)\times(p-r+1)$ matrix, the 3rd summand on the right in (2.10) has at most rank $p-r+1$. Since all of the named quadratic forms are positive semi-definite and the quadratic form on the left side of (2.10) has exactly the rank n, the quadratic forms on the right have, in order, ranks $n-p-1, r$, and $p-r+1$. Hence, from II, Theorem 4.1, the last claim follows. The claim preceding it follows immediately from (2.9) and I, Theorem 28.1.

We note explicitly that

$$(\zeta_{p+1}-XA^{-1}X'\xi_{p+1})'(\xi_{p+1}-XA^{-1}X'\xi_{p+1})$$

is the *residual term* which results when B is subject to no constraints, i.e., $-\infty<B<\infty$. By considering the transformation inverse to (2.2) we can formulate the part of Theorem 2.1 dealing with the residual terms as follows: *If one considers the standardized residual term under the assumptions of Theorem* 1.2, *where B is subject to no (linear) constraints, and then the standardized residual term when the parameters* β_i, $0\leqslant i\leqslant p$, *are subject to a number of independent, linear constraints of the type* (2.1), *then both are chi-square distributed. The distribution of the latter residual term possesses as many more degrees of freedom than the first as there are independent linear constraints.*

We now write the residual (1.11) as quotient of determinants. To this end, we introduce the notation:

$$\frac{1}{n}\sum_{l=1}^{n} x_{il}=\bar{x}_i, \quad i=1,\dots,p+1 \tag{2.12}$$

$$\sum_{l=1}^{n}(x_{il}-\bar{x}_i)(x_{jl}-\bar{x}_j)=w_{ij}, \quad i,j=1,\dots,p+1 \tag{2.13}$$

$$a_{0i}=a_{i0}=n\bar{x}_i \tag{2.14}$$

$$a_{00}=n \tag{2.15}$$

$$a_{ij}=\sum_{l=1}^{n} x_{il}x_{jl}, \quad i,j=1,\dots,p+1. \tag{2.16}$$

In this notation one obtains in place of (1.3)

$$\begin{gathered}a_{00}\hat{b}_0+a_{01}\hat{b}_1+\cdots+a_{0p}\hat{b}_p=a_{0p+1}\\ \cdots\cdots\cdots\cdots\cdots\cdots\cdots\cdots\\ a_{p0}\hat{b}_0+a_{p1}\hat{b}_1+\cdots+a_{pp}\hat{b}_p=a_{pp+1}\,.\end{gathered}$$

From the first equation and (2.15) we get $\hat{b}_0=(a_{0p+1}-a_{01}\hat{b}_1-\cdots-a_{0p}\hat{b}_p)/n$. Substituting this in the rest of the equations, we get using (2.13) and (2.16)

$$\sum_{j=1}^{p} w_{ij}\hat{b}_j=w_{ip+1}, \quad i=1,\dots,p. \tag{2.17}$$

Eliminating $\hat{b}_0$ in the same way as in (1.11) as well, we obtain

$$(x_{p+1}-X\hat{B})'(x_{p+1}-X\hat{B})=w_{p+1\,p+1}-2\sum_{k=1}^{p} w_{p+1\,k}\hat{b}_k+\sum_{i,k=1}^{p} w_{ik}\hat{b}_i\hat{b}_k$$

and from (2.17), this becomes

$$(x_{p+1}-X\hat{B})'(x_{p+1}-X\hat{B})=w_{p+1\,p+1}-\sum_{k=1}^{p} w_{p+1\,k}\hat{b}_k. \tag{2.18}$$

Assuming again the existence of A^{-1}, we get from (2.17)

$$\hat{b}_i=\begin{vmatrix} w_{11} \dots w_{1i-1}\, w_{1p+1}\, w_{1i+1} \dots w_{1p} \\ \dots\dots\dots\dots\dots\dots\dots\dots \\ w_{p1} \dots w_{pi-1}\, w_{pp+1}\, w_{pi+1} \dots w_{pp} \end{vmatrix} \left(\begin{vmatrix} w_{11} \dots w_{1p} \\ \dots\dots\dots \\ w_{p1} \dots w_{pp} \end{vmatrix}\right)^{-1} \tag{2.19}$$

$(i=1,\dots,p)$.

From (2.19) and (2.18)

$$(x_{p+1}-X\hat{B})'(x_{p+1}-X\hat{B})=|w_{ij}|_{1\,p+1}^{1\,p+1}(|w_{ij}|_{1\,p}^{1\,p})^{-1}. \tag{2.20}$$

Denote by $M^{(p+1)}$ the expression on the right in (2.20).

Determining the residual term under the conditions $\beta_0=0$, $-\infty<\beta_i<\infty$, $i=1,\dots,p$, we obtain after a simple calculation, that it can likewise be represented by the ratio of two determinants:

$$|a_{ij}|_{1\,p+1}^{1\,p+1}/|a_{ij}|_{1\,p}^{1\,p} \tag{2.21}$$

Denote (2.21) by $M_1^{(p+1)}$.

These results lead to

Theorem 2.2. *Let* $\xi_{p+1\,1},\dots,\xi_{p+1\,n}$ *be* $n\geqslant p+2\geqslant 3$ *independent, normally distributed r.v.'s satisfying the conditions of Theorem* 1.2 *with, however,* $\beta_0=0$, $-\infty<\beta_i<\infty$, $i=1,\dots,p$. *Then*

$$\boldsymbol{M}^{(p+1)}/\boldsymbol{M}_1^{(p+1)} \tag{2.22}$$

is $B((n-p-1)/2,\tfrac{1}{2})$*-distributed.*[9]

Proof. Using Theorem 2.1 one sees that $\boldsymbol{M}^{(p+1)}/\sigma^2$ is chi-square distributed with $n-p-1$ degrees of freedom and that $\boldsymbol{M}_1^{(p+1)}/\sigma^2$ is chi-square distributed with $n-p$ degrees of freedom. Moreover, $(\boldsymbol{M}_1^{(p+1)}-\boldsymbol{M}^{(p+1)})/\sigma^2$ is chi-square distributed with one degree of freedom

[9] The r.v. $\boldsymbol{M}^{(p+1)}$ as well as $\boldsymbol{M}_1^{(p+1)}$ and $\boldsymbol{M}^{(p+1)}/\boldsymbol{M}_1^{(p+1)}$ are undefined on a set of probability zero, i.e., on the set where the denominator in the definition vanishes. Naturally, this does not affect the statement of the theorem. (See also the remark concerning the definition of the t-distribution p. 83.)

independent of $M^{(p+1)}/\sigma^2$. Writing

$$M^{(p+1)}/M_1^{(p+1)} = \frac{M^{(p+1)}/(M_1^{(p+1)} - M^{(p+1)})}{1 + M^{(p+1)}/(M_1^{(p+1)} - M^{(p+1)})}$$

and applying the arguments which led to I. (30.3), we obtain the proof.

The result of Theorem 2.2 can be used under the corresponding assumptions to test $(\Gamma_0, \Gamma - \Gamma_0)$ with $\Gamma_0 = \{\beta_0 = 0, -\infty < \beta_i < \infty, 1 \leqslant i \leqslant p, 0 < \sigma^2 < \infty\}$ and $\Gamma = \{-\infty < B < \infty, 0 < \sigma^2 < \infty\}$ and, in fact, by means of similar critical regions.

3. Sampling theory for populations with multi-dimensional normal distribution. We first prove a theorem which is completely analogous to II, Theorem 3.2 and which performs the same task in the testing of a hypothesis on the mean of a multi-dimensional normal distribution.

Theorem 3.1. *Let $\xi_1, \ldots, \xi_n, n \geqslant 1$ be independent identically distributed k-dimensional r.v.'s, $k \geqslant 1$, which have a normal distribution with positive-definite covariance matrix D^{-1} and mean vector a. Then the k-dimensional r.v. $\bar{\xi} = (\xi_1 + \cdots + \xi_n)/n$ is likewise normally distributed with the density*

$$n^{k/2} |D|^{1/2} (2\pi)^{-k/2} e^{-\frac{1}{2}(x-a)' Dn(x-a)}$$

for all $x \in R_n$.

The *proof* follows from an application of I, Theorem 36.2.

It is immediately clear how this result can be applied for testing a simple null hypothesis $\{a_0\}$ with given covariance matrix D^{-1}. In particular, if

$$M = \{x : (x - a_0)' Dn(x - a_0) \geqslant d(\alpha)\},$$

where α is the level of significance and $d(\alpha)$ is determined by

$$n^{k/2} |D|^{1/2} (2\pi)^{-k/2} \int_M e^{-\frac{1}{2}(x-a_0)' Dn(x-a_0)} dx = \alpha$$

then III, p. 231 ff. says that M defines a most stringent test for the problem $(\alpha, \{a_0\}, R_k - \{a_0\})$. In order to drop the assumption that the covariance matrix is known, we try to obtain a multi-dimensional generalization of the t-distribution, guided, say, by the results of II.**4**. For this purpose it will be expedient initially to look for a generalization of the chi-square distribution.

Consider n independent, k-dimensional, normally distributed r.v.'s $\xi_l = (\xi_{1l}, \ldots, \xi_{kl})$, $1 \leqslant l \leqslant n$ whose density for each $x \in R_k$ is given by

$$|D|^{1/2} (2\pi)^{-k/2} e^{-\frac{1}{2}x' Dx}, \quad n \geqslant k \geqslant 1. \tag{3.1}$$

The matrix $D=(d_{ij})_{1k}^{1k}$ is again assumed to be positive-definite. For $i,j=1,\dots,k$ we write

$$\boldsymbol{a}_{ij}=\sum_{l=1}^{n}\xi_{il}\xi_{jl}$$

and claim

Theorem 3.2. *Under the assumptions made above, the distribution of the $k(k+1)/2$-dimensional r.v. $(\boldsymbol{a}_{11},\boldsymbol{a}_{21},\boldsymbol{a}_{22},\dots,\boldsymbol{a}_{k1},\dots,\boldsymbol{a}_{kk})$ possesses the density*

$$|D|^{n/2}|A|^{1/2(n-k-1)}2^{-kn/2}\pi^{-\frac{1}{4}k(k-1)}\left(\prod_{i=1}^{k}\Gamma(1/2(n-i+1))\right)^{-1}e^{-\frac{1}{2}\sum_{i,j=1}^{k}d_{ij}a_{ij}}$$

for all a_{ij} for which $A=(a_{ij})_{1k}^{1k}$ is positive definite; (3.2)

0 *otherwise.*

Here, $a_{ij}=a_{ji}$, $i,j=1,\dots,k$.

Proof by induction[10]. For $k=1$, the r.v. $\boldsymbol{a}_{11}$ has the form $\sum_{l=1}^{n}\xi_{1l}^2$. Formula (3.2) goes with $|D|=d_{11}$ into the density of a chi-square distribution with n degrees of freedom, which proves the claim for $k=1$. Now assume that the claim holds for a $(k-1)$-dimensional normal distribution. The density of the distribution of $(\xi_1,\dots,\xi_n)$ is, according to (3.1), given in R_{kn} by[11]

$$\begin{aligned}&h(x_{11},\dots,x_{k-1\,1},x_{k1},\dots,x_{1n},\dots,x_{k-1\,n},x_{kn})\\&=|D|^{n/2}(2\pi)^{-nk/2}\exp\left(-\frac{1}{2}\sum_{i=1}^{n}x_i'Dx_i\right).\end{aligned}\tag{3.3}$$

Denoting the marginal density of $(\xi_{k1},\dots,\xi_{kn})$ by h_k, we obtain in R_{kn}

$$\begin{aligned}&h(x_{11},\dots,x_{k-1\,1},x_{k1},\dots,x_{1n},\dots,x_{k-1\,n},x_{kn})\\&=h(x_{11},\dots,x_{k-1\,1},\dots,x_{1n},\dots,x_{k-1\,n}\,|\,x_{k1},\dots,x_{kn})\,h_k(x_{k1},\dots,x_{kn}).\end{aligned}\tag{3.4}$$

Now consider the conditional density

$$\begin{aligned}&(x_{11},\dots,x_{k-1\,1},\dots,x_{1n},\dots,x_{k-1\,n})\\&\to h(x_{11},\dots,x_{k-1\,1},\dots,x_{1n},\dots,x_{k-1\,n}\,|\,x_{k1},\dots,x_{kn})\end{aligned}$$

---

[10] This distribution was first found by J. Wishart, Biometrika 20A, 32—52 (1928). The induction proof here is due essentially to P. L. Hsu, Proc. Cambridge Philos. Soc. 35, 336—338 (1939). For an exhaustive study of the distributions connected with the multidimensional normal distribution see T. W. Anderson, Introduction to Multivariate Statistical Analysis, John Wiley & Sons-Chapman & Hall, New York-London 1958.

[11] Note that here, as was not the case in **1** and **2**, the symbols x_i and ξ_i denote k-dimensional vectors.

provided that $|z_k| = \left(\sum_{l=1}^{n} x_{kl}^2\right)^{1/2} > 0$. We proceed to new variables by means of the following transformations: for $j=1, \ldots, k-1$, let

$$\left.\begin{aligned} y_{ji} &= \sum_{l=1}^{n} c_{li} x_{jl}, \quad 1 \leqslant i \leqslant n-1 \\ y_{jn} &= |z_k|^{-1/2} \sum_{l=1}^{n} x_{kl} x_{jl} \end{aligned}\right\}. \tag{3.5}$$

The c_{li} should be so chosen that (3.5) is an orthogonal transformation.

From this orthogonality requirement it is easy to see that (3.5) takes $\sum_{l=1}^{n} x_{il} x_{jl}$ for $1 \leqslant i, j \leqslant k$ into $\sum_{l=1}^{n} y_{il} y_{jl}$. Since $P(\xi_{k1}^2 + \cdots + \xi_{kn}^2 = 0) = 0$, we obtain in obvious notation from (3.3) and (3.4) for the density of the joint distribution of $(\eta_{11}, \ldots, \eta_{k-11}, \ldots, \eta_{1n}, \ldots, \eta_{k-1n}, \xi_{k1}, \ldots, \xi_{kn})$ in R_{kn}

$$\begin{aligned} |D|^{n/2} (2\pi)^{-nk/2} \exp\Bigg[-\frac{1}{2} \Bigg(& \sum_{i,j=1}^{k-1} d_{ij} \sum_{l=1}^{n-1} y_{il} y_{jl} + d_{kk} |z_k|^2 \\ & + 2|z_k| \sum_{i=1}^{k-1} d_{ik} y_{in} + \sum_{i,j=1}^{k-1} d_{ij} y_{in} y_{jn} \Bigg) \Bigg]. \end{aligned} \tag{3.6}$$

But from this one easily sees that the $(k-1)$-dimensional r.v.'s $(\eta_{1l}, \ldots, \eta_{k-1l})$ for $l=1, \ldots, n-1$ are independent and identically distributed according to a $(k-1)$-dimensional normal distribution. Further, they are also independent of the r.v. $(\eta_{1n}, \ldots, \eta_{k-1n}, \xi_{k1}, \ldots, \xi_{kn})$. Thus, defining for $i,j = 1, \ldots, k-1$ the r.v.'s $\boldsymbol{a}_{ij}^* = \sum_{i=1}^{n-1} \eta_{il} \eta_{jl}$, one can use the induction hypothesis.

With the transformation

$$a_{ij}^* = \sum_{l=1}^{n-1} y_{il} y_{jl}, \quad i,j = 1, \ldots, k-1,$$

one obtains from (3.6) for the density of the joint distribution of $(\boldsymbol{a}_{11}^*, \boldsymbol{a}_{21}^*, \boldsymbol{a}_{22}^*, \ldots, \boldsymbol{a}_{k-11}^*, \ldots, \boldsymbol{a}_{k-1k-1}^*, \eta_{1n}, \ldots, \eta_{k-1n}, \xi_{k1}, \ldots, \xi_{kn})$

$$\left.\begin{aligned} & |D|^{n/2} (|a_{ij}^*|_{1k-1}^{1k-1})^{(n-k-1)/2} (2\pi)^{-\frac{n+k-1}{2}} 2^{-\frac{(k-1)(n-1)}{2}} \pi^{-\frac{1}{4}(k-1)(k-2)} \\ & \times \left(\prod_{i=1}^{k-1} \Gamma((n-i)/2) \right)^{-1} \exp\Bigg[-\frac{1}{2} \Bigg(\sum_{i,j=1}^{k-1} d_{ij} (a_{ij}^* + y_{in} y_{jn}) \\ & \qquad + 2|z_k| \sum_{i=1}^{k-1} d_{ik} y_{in} + d_{kk} |z_k|^2 \Bigg) \Bigg] \end{aligned}\right\} \tag{3.7}$$

provided that

$$(a^*_{ij})^{1k-1}_{1k-1} \text{ is positive-definite, and } 0<|z_k|<\infty, -\infty<y_{in}<\infty,$$
$$0 \qquad \text{otherwise.}$$

Now we introduce polar coordinates for the $x_{k,l}$, $1\leqslant l\leqslant n$ in (3.7). We write them in the notation of (2.16) as follows

$$x_{kl}=a_{kk}^{1/2}\alpha_l\,, \quad l=1,\ldots,n\,, \quad \sum_{i=1}^{n}\alpha_i^2=1\,.$$

Since our ultimate interest is the joint distribution of the r.v.'s $\boldsymbol{a}_{ij}$, we integrate w.r.t. the α_i after introducing the polar coordinates and obtain for the density of the joint distribution of $\boldsymbol{a}^*_{ij}$, η_{in}, $i,j,=1,\ldots,k-1$ and $\boldsymbol{a}_{kk}$

$$|D|^{n/2}2^{-nk/2}\pi^{-\frac{1}{4}k(k-1)}\left(\prod_{i=1}^{k}\Gamma((n-i+1)/2)\right)^{-1}$$
$$\times(|a^*_{ij}|^{1k-1}_{1k-1})^{(n-k-1)/2}a_{kk}^{(n-2)/2}\exp\left[-\frac{1}{2}\left(\sum_{i,j=1}^{k-1}d_{ij}(a^*_{ij}+y_{in}y_{jn})\right.\right. \tag{3.8}$$
$$\left.\left.+2a_{kk}^{1/2}\sum_{i=1}^{k-1}d_{ik}y_{in}+d_{kk}a_{kk}\right)\right],$$

provided it does not vanish. It is now easy to see that the following connection exists between the r.v.'s $\boldsymbol{a}^*_{ij}$ and $\boldsymbol{a}_{ij}$: $\boldsymbol{a}^*_{ij}=\boldsymbol{a}_{ij}-\boldsymbol{a}_{ik}\boldsymbol{a}_{jk}/\boldsymbol{a}_{kk}$, $i,j=1,\ldots,k-1$. This suggests proceeding from (3.8) to a new density by means of the following transformation:

$$\begin{aligned} a^*_{ij} &= a_{ij}-a_{ik}a_{jk}/a_{kk}\,, && i,j=1,\ldots,k-1,\\ y_{in} &= a_{ik}/a_{kk}^{1/2}\,, && i=1,\ldots,k-1,\\ a_{kk} &= a_{kk}\,. \end{aligned}$$

The Jacobian of this transformation is $a_{kk}^{-(k-1)/2}$. Denoting the determinant

$$\begin{vmatrix} a_{11}-a_{1k}^2/a_{kk} & \ldots & a_{1k-1}-a_{1k}a_{k-1k}/a_{kk}\\ a_{21}-a_{2k}a_{1k}/a_{kk} & \ldots & a_{2k-1}-a_{2k}a_{k-1k}/a_{kk}\\ \cdots & \cdots & \cdots\\ a_{k-11}-a_{k-1k}a_{1k}/a_{kk} & \ldots & a_{k-1k-1}-a_{k-1k}^2/a_{kk} \end{vmatrix}$$

by $|A(a_{ij})|$, one gets from (3.8) for the new density (provided it does not vanish)

$$|D|^{n/2}2^{-kn/2}\pi^{-k(k-1)/4}\left(\prod_{i=1}^{k}\Gamma((n-i+1)/2)\right)^{-1}$$
$$+|A(a_{ij})^{(n-k-1)/2}a_{kk}^{(n-k-1)/2}e^{-\frac{1}{2}\sum_{i,j=1}^{k}d_{ij}a_{ij}}$$

if $A(a_{ij})$ is positive definite.

Since

$$|A|=|a_{ij}|_{1k}^{1k}=a_{kk}\begin{vmatrix} a_{11} & a_{12} & \dots a_{1k} \\ \dots & \dots & \dots \\ a_{k-11} & a_{k-12} & \dots a_{k-1k} \\ a_{k1}/a_{kk} & a_{k2}/a_{kk} & \dots 1 \end{vmatrix}$$

we have $|A(a_{ij})|=a_{kk}^{-1}|A|$. Hence, $|a_{ij}^{*}|_{1k-1}^{1k-1}$ is positive definite for $a_{kk}>0$ iff A is, which proves the theorem.

The distribution just derived, whose density is given by (3.2) is called the *Wishart distribution with n degrees of freedom and parameter k.*

An important application of this distribution is given by

Theorem 3.3. *Let the hypotheses of Theorem 3.2 be satisfied for* $n\geqslant k+1$. *Then, the* $\frac{1}{2}k(k+1)$*-dimensional r.v.* $(\mathbf{w}_{11}, \mathbf{w}_{21}, \mathbf{w}_{22}, \dots, \mathbf{w}_{k1}, \dots, \mathbf{w}_{kk})$ *is Wishart distributed with* $n-1$ *degrees of freedom and parameter k independently of the k-dimensional r.v.* $\bar{\xi}=(\bar{\xi}_1, \dots, \bar{\xi}_k)$.

The definition of the $\mathbf{w}_{ij}$ and the $\bar{\xi}_i$, $i,j=1,\dots,k$ is easily taken from the notation introduced in (2.12) and (2.13).

Proof. We proceed from the density (3.1) by means of the transformation

$$\begin{aligned} y_{i1} &= c_{11}x_{i1}+\dots+c_{1n}x_{in} \\ &\dots\dots\dots\dots, \qquad 1\leqslant i\leqslant k \\ y_{in} &= c_{n1}\,x_{i1}+\dots+c_{nn}\,x_{in} \end{aligned} \tag{3.9}$$

to a new density for the r.v.'s η_{ij}, $1\leqslant i\leqslant k$, $1\leqslant j\leqslant n$. Here, the matrix $(c_{ij})_{1n}^{1n}$ should be orthogonal and be determined such that

$$y_{i1}=\sqrt{n}\,\bar{x}_i, \qquad 1\leqslant i\leqslant k. \tag{3.10}$$

The transformation (3.9) then takes $\mathbf{w}_{ij}=\sum_{l=1}^{n} x_{il}x_{jl}-n\bar{x}_i\bar{x}_j$ into $\sum_{l=2}^{n} y_{il}y_{jl}$, $i,j=1,\dots,k$. Since the joint distribution of the η_{ij} is normal and coincides with the joint distribution of all ξ_{ij}, we immediately obtain the claim on the joint distribution of the $\mathbf{w}_{ij}$. (See also p. 391.)

The statement of the theorem concerning independence can be demonstrated as follows: The density of the joint distribution of the ξ_{ij}, $i=1,\dots,k$, $j=1,\dots,n$, is given in R_{nk} by

$$|D|^{n/2}(2\pi)^{-nk/2}\,e^{-\frac{1}{2}\sum_{i,j=1}^{k}\bar{x}_i\bar{x}_j d_{ij} n}\,e^{-\frac{1}{2}\sum_{l=1}^{n}\sum_{i,j=1}^{k} d_{ij}(x_{il}-\bar{x}_i)(x_{jl}-\bar{x}_j)} \tag{3.11}$$

which is easy to see. Applying the transformation given by (3.9) and (3.10) to (3.11), one obtains in R_{nk} the density

$$|D|^{n/2}(2\pi)^{-nk/2}\,e^{-\frac{1}{2}\sum_{i,j=1}^{k} y_{i1}y_{j1}d_{ij}}\,e^{-\frac{1}{2}\sum_{l=2}^{n}\sum_{i,j=1}^{k} d_{ij}y_{il}y_{jl}}. \tag{3.12}$$

The function $(y_{11},\ldots,y_{k1})\to e^{-\frac{1}{2}\sum_{i,j=1}^{k} y_{i1}y_{j1}d_{ij}}$ represents, up to a constant factor, the density of the distribution of $\sqrt{n}\,\bar{\xi}$. Introducing a new variable into (3.12) by means of the transformation $w_{ij}=\sum_{l=2}^{n} y_{il}y_{jl},\ 1\leqslant i,\ j\leqslant k,$ takes (3.12) into a new density—the product of the densities of $\bar{\xi}$ and $(w_{11},w_{21},w_{22},\ldots,w_{k1},\ldots,w_{kk})$. This completes the proof.

It is hardly necessary to mention that the result of Theorem 3.3 does not change when the mean value vector 0 is replaced by an arbitrary k-dimensional mean value vector a.

We now prove the *reproductive property of the Wishart distribution*, obtaining the analog to I, Theorem 28.1.

Theorem 3.4. *Let the joint distribution of the* $\boldsymbol{a}_{ij}^{(n)}$, $1\leqslant j\leqslant i\leqslant k$, *be Wishart with n degrees of freedom and parameter k. Let the joint distribution of the* $\boldsymbol{a}_{ij}^{(m)}$, $1\leqslant j\leqslant i\leqslant k$, *also be Wishart, but with m degrees of freedom. Assume the two distributions are independent. Then the joint distribution of the r.v.'s* $\boldsymbol{a}_{ij}^{(n)}+\boldsymbol{a}_{ij}^{(m)}$, $1\leqslant j\leqslant i\leqslant k$ *is a Wishart distribution with* $n+m$ *degrees of freedom and parameter k.*

Proof. We begin with a distribution with density (3.1) and determine, for $l=1,\ldots,n$, the characteristic function of the joint distribution of $\xi_{il}\xi_{jl}$, $i,j=1,\ldots,k$. This does not depend on l and we denote it by $T\to\phi(T)$, where

$$T=(t_{ij})_{1k}^{1k} \quad \text{with } t_{ij}=t_{ji}, \qquad -\infty<t_{ij}<\infty, \qquad i,j=1,\ldots,k.$$

With $x_l=(x_{1l},\ldots,x_{kl})$ we then get for each T

$$\phi(T)=|D|^{1/2}(2\pi)^{-k/2}\int_{R_k} e^{ix_l'Tx_l'}e^{-\frac{1}{2}x_lDx_l}dx_l=|D|^{1/2}|D-2iT|^{-1/2}$$

which immediately follows quite formally from I. (35.3). Since the matrices here have complex elements, this formal calculation requires some justification, which can be modelled after the justification of I. (28.6). By assumption, ξ_l and ξ_m are independent for $1\leqslant l<m\leqslant k$. Thus, for the characteristic function of a Wishart distribution with n d.f. and parameter k, one gets $|D|^{n/2}|D-2iT|^{-n/2}$ for each $T\in R_{k(k+1)/2}$ which follows from I, Theorem 24.4. Applying this theorem again proves Theorem 3.4.

We now consider some auxiliary results which are also of independent interest.

Lemma 3.1. *Let f be the density of an n-dimensional r.v.* $(\zeta_1,\ldots,\zeta_n)$, $n\geqslant 2$. *Let a density* f^* *be given over* R_{n-k} *such that for the density of the conditional distribution of* $(\zeta_{k+1},\ldots,\zeta_n)$*—given* $x_1,\ldots,x_k$, $1\leqslant k\leqslant n$, *we* *have* $f(x_{k+1},\ldots,x_n|x_1,\ldots,x_k)=f^*(x_{k+1},\ldots,x_n)$ *for all* $(x_{k+1},\ldots,x_n)$ $\in R_{n-k}$ *and all* $(x_1,\ldots,x_k)\in R_k$ *provided the conditional density is defined.*

Then f^* *coincides a.e. with the density of the marginal distribution of* $(\zeta_{k+1},\ldots,\zeta_n)$.

This follows immediately from I. (14.4) and the multi-dimensional analog of I. (18.9).

Lemma 3.2[12]. *Let* ξ *be a* $B(a,b)$*-distributed r.v. Let* η *be independent of* ξ *and* $B(c,d)$*-distributed,* $a,b,c,d>0$. *Moreover, let*

$$a=c+d. \tag{3.13}$$

Then, the r.v. $\zeta=\xi\eta$ *is* $B(c,b+d)$*-distributed.*

Proof. From I. (30.4), taking (3.13) into consideration, we get for the density of the joint distribution of (ξ,η) in $0<x<1, 0<y<1$

$$\Gamma(a+b)(\Gamma(b)\Gamma(c)\Gamma(d))^{-1}x^{a-1}(1-x)^{b-1}y^{c-1}(1-y)^{d-1}.$$

We make the transformation

$$u=x, \qquad z=xy,$$

which sends $0<x<1,\ 0<y<1$ into $0<z<1,\ z<u<1$.

We then get for the density of the marginal distribution of ζ, again with the help of (3.13),

$$\Gamma(a+b)(\Gamma(b)\Gamma(c)\Gamma(d))^{-1}\int_z^1(1-u)^{b-1}z^{c-1}(u-z)^{d-1}du$$

$(0<z<1)$. Letting $u=v(1-z)+z$ in this integral for a fixed z with $0<z<1$, we find for the density of ζ

$$\Gamma(b+c+d)(\Gamma(c)\Gamma(b+d))^{-1}z^{c-1}(1-z)^{b+d-1}, \qquad 0<z<1.$$

We now prove

Theorem 3.5. *Let* $\xi_1,\ldots,\xi_n$ *be independent. k-dimensional r.v.'s,* $n>k$, *each possessing the same normal distribution whose density for all* $x\in R_k$ *is given by* (3.1) *with* D *positive-definite. With* $\xi_l=(\xi_{1l},\ldots,\xi_{kl})$, $1\leqslant l\leqslant n$, *we consider the r.v.'s* $\boldsymbol{a}_{ij}(\boldsymbol{w}_{ij})$ *defined in Theorem 3.2 (Theorem 3.3),* $i,j=1,\ldots,k$. *Then, the r.v.*

$$V=|\boldsymbol{w}_{ij}|_{1k}^{1k}/|\boldsymbol{a}_{ij}|_{1k}^{1k} \tag{3.14}$$

is $B((n-k)/2,k/2)$*-distributed.*

---

[12] C.R. Rao, Sankhyā 9, 343—366 (1949).

Proof. We have often used the fact (see for Example I, p. 96) that an appropriate orthogonal transformation of the form

$$y_l = \mathfrak{O}\, x_l \tag{3.15}$$

takes $x_l' D x_l$ into $y_l' \Lambda y_l$, where

$$\Lambda = \begin{pmatrix} \lambda_1 & 0 & \dots 0 \\ 0 & \lambda_2 & \dots 0 \\ \dots & \dots & \dots \\ 0 & 0 & \dots \lambda_k \end{pmatrix} \qquad \text{with } \lambda_j > 0, \quad 1 \leqslant j \leqslant k.$$

Applying (3.15) for $1 \leqslant l \leqslant n$ to $V = |w_{ij}|_{1k}^{1k} / |a_{ik}|_{1k}^{1k}$, one easily shows that V is invariant w.r.t. this transformation. In determining the distribution of (3.14) one can thus assume that $\xi_{1l}, \dots, \xi_{kl}$ are also independently distributed and that the density of the distribution of $\xi_l = (\xi_{1l}, \dots, \xi_{kl})$, $1 \leqslant l \leqslant n$, is given in R_k by

$$(2\pi)^{-k/2} \prod_{j=1}^{k} \lambda_j^{1/2} e^{-\lambda_j y_{jl}^2/2}. \tag{3.16}$$

We now define for $1 \leqslant m \leqslant k$ the r.v.

$$\boldsymbol{N}^{(m)} = |\boldsymbol{w}_{ij}|_{1m}^{1m} / |\boldsymbol{w}_{ij}|_{1m-1}^{1m-1},$$

where $|\boldsymbol{w}_{ij}|_{10}^{10} = 1$. We define $\boldsymbol{N}_1^{(m)}$ in a similar way, using $\boldsymbol{a}_{ij}$ in place of $\boldsymbol{w}_{ij}$. Obviously,

$$\boldsymbol{V} = \prod_{m=1}^{k} \boldsymbol{N}^{(m)} / \boldsymbol{N}_1^{(m)}. \tag{3.17}$$

This suggests making use of the results of Theorem 2.2. We replace the index $p+1$ chosen there by m. Note that in this notation, only the r.v.'s ξ_{mi}, $1 \leqslant i \leqslant n$, appear in Theorem 2.2. We have, however, already pointed out on p. 359 that the distribution of ξ_{mi} can be viewed as a conditional distribution under the assumptions of Theorem 2.2. One obtains it by starting, for $i = 1, \dots, n$, with m-dimensional normally distributed r.v.'s $(\xi_{1i}, \dots, \xi_{m-1i}, \xi_{mi})$ and then considering the conditional distribution of ξ_{mi} under the hypothesis $\xi_{1i} = x_{1i}, \dots, \xi_{m-1i} = x_{m-1i}$, $1 \leqslant i \leqslant n$. I. (35.12) and (1.13) also show that for all m with $2 \leqslant m \leqslant k$, the assumption $\beta_0 = 0$ of Theorem 2.2 is in each case satisfied since the mean vector of the given k-dimensional normal distribution vanishes. Hence, one can claim that $\boldsymbol{N}^{(m)} / \boldsymbol{N}_1^{(m)}$ for $2 \leqslant m \leqslant k$ is $B((n-m)/2, 1/2)$-distributed. One need only note that the distribution of $\boldsymbol{M}^{(m)} / \boldsymbol{M}_1^{(m)}$ can be viewed as the conditional distribution of $\boldsymbol{N}^{(m)} / \boldsymbol{N}_1^{(m)}$ under the hypothesis

$$\xi_{11} = x_{11}, \dots, \xi_{m-11} = x_{m-11}, \dots, \xi_{1n} = x_{1n}, \dots, \xi_{m-1n} = x_{m-1n}$$

and this distribution does not depend on the hypothesis. Applying Lemma 3.1 then yields the claim. Further it is easy to see that for $m=1$ as well, the r.v.

$$N^{(1)}/N_1^{(1)}=\sum_{i=1}^{n}(\xi_{1i}-\overline{\xi}_1)^2\Big/\sum_{i=1}^{n}\xi_{1i}^2$$

is $B((n-1)/2, 1/2)$-distributed. A repeated application of Lemma 3.2 now leads to our goal because of (3.17) provided one can show that $N^{(j)}/N_1^{(j)}$ for $j=2,\ldots,k$ is independent of $N^{(m)}/N_1^{(m)}$ for $1\leqslant m\leqslant j-1$. However, we have even demonstrated that one can start from the assumption that the k r.v.'s $\xi_{1l},\ldots,\xi_{kl}$ are independently distributed and possess in R_k for $1\leqslant l\leqslant n$ the density given by (3.16). The claim on the independence now follows easily by means of Lemma 3.1.

We now arrive at the goal of our considerations and prove

Theorem 3.6. *Let $\xi_1,\ldots,\xi_n$, $n\geqslant 2$ be independent k-dimensional r.v.'s satisfying the assumptions of Theorem 3.1 with $n>k$. Denote the inverse of $(w_{ij})_{1k}^{1k}$ by $(W_{ij})_{1k}^{1k}$. Then, the r.v.*[13]

$$T^2=(n-1)n\sum_{i,j=1}^{k}W_{ij}(\overline{\xi}_i-a_i)(\overline{\xi}_j-a_j)$$

with $(\overline{\xi}_1,\ldots,\overline{\xi}_k)=\overline{\xi}=(\xi_1+\cdots+\xi_n)/n$ possesses a distribution with the density

$$\left.\begin{array}{ll}\dfrac{1}{n-1}\Gamma(n/2)\left(\Gamma((n-k)/2)\Gamma(k/2)\right)^{-1} & \\ \times(1+y/(n-1))^{-n/2}(y/(n-1))^{-1+k/2}, & y>0\\ 0 & y\leqslant 0\end{array}\right\}\tag{3.18}$$

Proof. Since we can replace ξ_l for $l=1,\ldots,n$ by ξ_l-a, we can assume that the mean value vector vanishes. We now show that

$$V=(1+T^2/(n-1))^{-1}.\tag{3.19}$$

Then the result of Theorem 3.5 yields the claim of Theorem 3.6 by means of a simple transformation. From

$$a_{ij}=\sum_{l=1}^{n}(x_{il}-\overline{x}_i)(x_{jl}-\overline{x}_j)+n\overline{x}_i\overline{x}_j=w_{ij}+n\overline{x}_i\overline{x}_j,\qquad 1\leqslant i,j\leqslant k$$

---

[13] The matrix $(W_{ij})_{1k}^{1k}$ exists iff $|w_{ij}|_{1k}^{1k}\neq 0$. The matrix $(w_{ij})_{1k}^{1k}$ is positive-semi-definite. Under the assumption of Theorem 3.6, however, $P(|w_{ij}|_{1k}^{1k}=0)=0$. Thus we can assume that the inverse exists. We will take these considerations as tacit in the sequel.

there follows

$$|a_{ij}|_{1k}^{1k} = \begin{vmatrix} w_{11}+n\bar{x}_1^2 & \dots & w_{1k}+n\bar{x}_1\bar{x}_k \\ \dots & \dots & \dots \\ w_{k1}+n\bar{x}_k\bar{x}_1 & \dots & w_{kk}+n\bar{x}_k^2 \end{vmatrix} = \begin{vmatrix} 1 & \sqrt{n}\bar{x}_1 & \dots & \sqrt{n}\bar{x}_k \\ 0 & w_{11}+n\bar{x}_1^2 & \dots & w_{1k}+n\bar{x}_1\bar{x}_k \\ \dots & \dots & \dots & \dots \\ 0 & w_{k1}+n\bar{x}_k\bar{x}_1 & \dots & w_{kk}+n\bar{x}_k^2 \end{vmatrix}.$$

Multiplying the first row of the last determinant by $-\sqrt{n}\bar{x}_i$ for $i=1,\dots,k$, and adding the result to the $(i+1)$st row, we find

$$\begin{vmatrix} 1 & \sqrt{n}\bar{x}_1 & \dots & \sqrt{n}\bar{x}_k \\ -\sqrt{n}\bar{x}_1 & w_{11} & \dots & w_{1k} \\ \dots & \dots & \dots & \dots \\ -\sqrt{n}\bar{x}_k & w_{k1} & \dots & w_{kk} \end{vmatrix} = |w_{ij}|_{1k}^{1k}\left(1+n\sum_{i,j=1}^{k} W_{ij}\bar{x}_i\bar{x}_j\right).$$

Whence (3.19) immediately follows.

The distribution with density (3.18) is called the *Hotelling distribution*[14] with $n-1$ *degrees of freedom and parameter k.*

One easily sees that if $k=1$ in (3.18) then one obtains the distribution of the square of a Student-distributed variable with $n-1$ degrees of freedom. The Hotelling distribution is thus a reasonable generalization of the t-distribution[15].

Combining Theorem 3.3 and 3.6, one gets the following result, which is an exact parallel to the derivation of the t-distribution in I.**29**.

Theorem 3.7. *Let $\xi=(\xi_1,\dots,\xi_k)$ be a k-dimensional normally distributed r.v. with mean a and covariance D^{-1}. The $k(k+1)/2$-dimensional r.v. $(\eta_{11},\eta_{12},\eta_{22},\dots,\eta_{1k},\dots,\eta_{kk})$ is assumed to be Wishart-distributed with n degrees of freedom and parameter $n\geqslant k$, independently of ξ. Then the r.v.*

$$\boldsymbol{T}^2 = n\sum_{i,j=1}^{k} \Xi_{ij}(\xi_i-a_i)(\xi_j-a_j)$$

is Hotelling-distributed with n degrees of freedom and parameter k. $|\Xi_{ij}|_{1k}^{1k}$ is the inverse of $|\eta_{ij}|_{1k}^{1k}$, where $\eta_{ij}=\eta_{ji}$, $1\leqslant i,j\leqslant k$ by definition.

---

[14] This distribution was first discovered by H. Hotelling, Ann. Math. Statist. 2, 360—378 (1931). Also see H. Hotelling, Proceedings of 2nd Berkeley Symposium on Mathematical Statistics and Probability 1951, pp. 23—41 University of California Press, Berkeley and Los Angeles.

[15] A comparison with the F-distribution shows that $\frac{n-k}{(n-1)k}\boldsymbol{T}^2$ possesses an F-distribution with $(k, n-k)$ degrees of freedom.

Theorem 3.6 forms the basis for testing the mean of a k-dimensional normal distribution with unknown correlation matrix by means of a sample of size n. More precisely, the set of admissible hypotheses is thus given by $\{-\infty<a_i<\infty,\ -\infty<d_{ij}=d_{ji}<\infty,\ i,j=1,\dots,k,\ (d_{ij})_{1k}^{1k}$ positive definite$\}$[16]. The null hypothesis has the form $a=a^{(0)}$, where $a^{(0)}$ is an element from R_k. The conditions on the d_{ij}, $1\leq i,j\leq k$, are the same.

4. The discriminant function of Fisher and Mahalanobis' distance[17]. We will begin our discussion of these topics with one example: Let $x_1,\dots,x_n$ be a sample of k-dimensional elements from a k-dimensional, normally distributed population. Let its mean value be a and covariance matrix G. Let $y_1,\dots,y_m$ likewise be a sample from a k-dimensional, normally distributed population with mean b and the same covariance matrix G. Assume a and b are unknown and G is known. Our problem is to characterize as precisely as possible the difference between a and b (and thus between the populations) and on the basis of this characterization, to assign further sample values to one or the other population. A series of important contributions to these problems have been made by the Indian School of statistics[18]. We can only treat some of the more important results here.

Let $\xi=(\xi_1,\dots,\xi_k)$ be a k-dimensional normally distributed r.v. with mean $a=(a_1,\dots,a_k)$ and positive definite covariance matrix $G=(g_{ij})_{1k}^{1k}$. Let $\eta=(\eta_1,\dots,\eta_k)$ be a normally distributed r.v. independent of ξ and with mean $b=(b_1,\dots,b_k)$ and the same covariance matrix. Set $a_i-b_i=d_i$, $1\leq i\leq k$. We consider the problem of maximizing the quotient

$$\left(\sum_{i=1}^{k}\lambda_i d_i\right)^2\Bigg/\sum_{i,j=1}^{k}\lambda_i\lambda_j g_{ij} \tag{4.1}$$

as function of $(\lambda_1,\dots,\lambda_k)\in R_k$. Our interest in this problem becomes more intuitive when we note that the numerator of (4.1) is $\left[E\left(\sum_{i=1}^{k}\lambda_i(\xi_i-\eta_i)\right)\right]^2$ and the denominator the variance of $\sum_{i=1}^{k}\lambda_i\xi_i$, or equivalently, of $\sum_{i=1}^{k}\lambda_i\eta_i$. (4.1) remains invariant when λ_i is replaced for $c\neq 0$ by $c\lambda_i$, $1\leq i\leq k$. Thus, we normalize our extremal problem by considering only the set of all $(\lambda_1,\dots,\lambda_k)$ which satisfy

$$\sum_{i=1}^{k}\lambda_i d_i\Bigg/\left(\sum_{i=1}^{k}\lambda_i\lambda_j g_{ij}\right)=1\,. \tag{4.2}$$

---

[16] $(d_{ij})_{1k}^{1k}$ denotes the inverse of the covariance matrix D^{-1}.

[17] R.A. Fisher, Ann. Eugenics 7, 179—188 (1936); P.C. Mahalanobis, Proc. Nat. Inst. Sci. India 2, 49—55 (1936).

[18] We refer only to P.C. Mahalanobis, Sankhyā 9, 237—239 (1949) and papers by C.R. Rao, such as Biometrika 35, 58—79 (1948); Sankhyā, l.c.[12]; Sankhyā 10, 257—268 (1950).

It is easy to show that under this restriction (4.1) assumes the maximum value $\sum_{i,j=1}^{k} G_{ij} d_i d_j$, where $(G_{ij})_{1k}^{1k} = G^{-1}$. The maximization problem has a unique solution $l = (l_1, \ldots, l_k)$, which is given by $l = G^{-1} d$. Here, d denotes the (column) vector of the d_i, $1 \leqslant i \leqslant k$. But then

$$d' G^{-1} d = l' d . \tag{4.3}$$

The r.v.'s $\sum_{i=1}^{k} l_i \xi_i$, resp., $\sum_{i=1}^{k} l_i \eta_i$ are called *linear discriminant functions.*

One can arrive at the idea of a discriminant function in a somewhat different way[19], which makes its "discriminating character" clear. Consider the problem of finding a set $M \in \mathfrak{B}_k$ for which

$$P(\xi \in M) = P(\eta \in R_k - M) \tag{4.4}$$

and these probabilities are maximal w.r.t. all M satisfying (4.4). To this end, consider the set

$$M = \{x : \exp(-\tfrac{1}{2}(x-a)' G^{-1}(x-a)) > \exp(-\tfrac{1}{2}(x-b)' G^{-1}(x-b))\} .$$

We first show that M fulfills (4.4). Indeed, it is easy to show that

$$\begin{aligned} M &= \{x : x' G^{-1} x - 2a' G^{-1} x + a' G^{-1} a < x' G^{-1} x - 2b' G^{-1} x + b' G^{-1} b\} \\ &= \{x : (a-b)' G^{-1}(x - (a+b)/2) > 0\} = \{x : l'(x - (a+b)/2) > 0\} . \end{aligned}$$

Applying I, Theorem 36.2 we find that for $a \neq b$ the r.v. $(a-b)' G^{-1}\left(\xi - \dfrac{a+b}{2}\right)$ is $N(\tfrac{1}{2}(a-b)' G^{-1}(a-b), (a-b)' G^{-1}(a-b))$-distributed and that $(a-b)' G^{-1}\left(\eta - \dfrac{a+b}{2}\right)$ is $N(-\tfrac{1}{2}(a-b)' G^{-1}(a-b), (a-b)' G^{-1}(a-b))$-distributed. This implies that M satisfies (4.4).

Using the original definition of M, we obtain from the Neyman-Pearson lemma

$$\begin{aligned} P(\xi \in M) &= \max_{K \in \mathfrak{B}_k} \{P(\xi \in K) : P(\eta \in K) \leqslant P(\eta \in M)\} , \\ P(\eta \in M) &= \min_{K \in \mathfrak{B}_k} \{P(\eta \in K) : P(\xi \in K) \geqslant P(\xi \in M)\} \end{aligned}$$

and further

$$\begin{aligned} P(\xi \in M) &= \max_{K \in \mathfrak{B}_k} \{P(\xi \in K) : P(\eta \in R_k - K) \geqslant P(\eta \in R_k - M)\} , \\ P(\eta \in R_k - M) &= \max_{K \in \mathfrak{B}_k} \{P(\eta \in R_k - K) : P(\xi \in K) \geqslant P(\xi \in M)\} . \end{aligned}$$

---

[19] Communicated by R. Borges.

Thus, M has the required maximal property. The r.v.'s $l'\xi$ and $l'\eta$ therefore define a set $M\in\mathfrak{B}_k$ with the "best" discriminating properties.

However, this determination of l is in general of no practical value since one does not know a and b. Usually l is estimated from sample values. One proceeds practically as follows: let $x_1,\dots,x_{n_1}$, $n_1\geqslant 2$ be a sample from a k-dimensional, normally distributed population with mean a and covariance matrix G. Let $y_1,\dots,y_{n_2}$, $n_2\geqslant 2$ be an analogous sample from a population with mean b. Set

$$x_l=(x_{1l},\dots,x_{kl}),\qquad 1\leqslant l\leqslant n_1\quad\text{and}\ \sum_{i=1}^{n} x_i/n_1=(\bar{x}_1,\dots,\bar{x}_k);$$

$$y_m=(y_{1m},\dots,y_{km}),\qquad 1\leqslant m\leqslant n_2,\qquad \sum_{j=1}^{n_2} y_j/n_2=(\bar{y}_1,\dots,\bar{y}_k)^{20}.$$

Further we write

$$\delta_i=\bar{x}_i-\bar{y}_i,$$

$$s_{ij}=(n_1+n_2-2)^{-1}\left[\sum_{l=1}^{n_1}(x_{il}-\bar{x}_i)(x_{jl}-\bar{x}_j)+\sum_{m=1}^{n_2}(y_{im}-\bar{y}_i)(y_{jm}-\bar{y}_j)\right],$$

$$i,j=1,\dots,k.$$

Repeating the considerations above with δ_i in place of d_i and s_{ij} instead of g_{ij}, where we naturally assume that $(s_{ij})_{1k}^{1k}$ possesses an inverse $(S_{ij})_{1k}^{1k}$, we obtain for the maximum of

$$\left(\sum_{i=1}^{k}\lambda_i\delta_i\right)^2\Bigg/\sum_{i,j=1}^{k}\lambda_i\lambda_j s_{ij}$$

under a condition analogous to (4.2), the quantity

$$D^2=\sum_{i,j=1}^{k} S_{ij}\delta_i\delta_j. \tag{4.5}$$

Denoting now the uniquely determined solution of the maximum problem by $l_i(x,y)$, $1\leqslant i\leqslant k$, we get in place of (4.5) $D^2=\sum_{i=1}^{k} l_i(x,y)\delta_i$. Hence, the two discriminant functions $e_1=\sum_{i=1}^{k} l_i(x,y)\bar{x}_i$ and $e_2=\sum_{i=1}^{k} l_i(x,y)\bar{y}_i$ correspond to the sample values. Practically speaking, the process of

---

[20] The notation here is so chosen that x_j is a k-dimensional vector but $\bar{x}_j$ is the real number $\frac{1}{n_1}\sum_{i=1}^{n_1} x_{ji}$ and similarly for the y_{ji}.

assigning a new sample $(z_1,\dots,z_k)$ to one of the two populations consists in determining whether

$$\sum_{i=1}^{k} l_i(x,y)z_i > (e_1+e_2)/2 \quad \text{or} \quad < (e_1+e_2)/2 .$$

To justify such a method, we must study more exactly the distribution of $\boldsymbol{D}^2$, viewed as a function of k-dimensional r.v.'s ξ_l, $1 \leqslant l \leqslant n_1$ and η_m, $1 \leqslant m \leqslant n_2$. We prove

Theorem 4.1. *Let $\xi_1,\dots,\xi_{n_1}, \eta_1,\dots,\eta_{n_2}$ be independent k-dimensional r.v.'s. Let ξ_l for $1 \leqslant l \leqslant n_1$ be normally distributed with mean a and positive-definite covariance matrix G. Let η_m, $1 \leqslant m \leqslant n_2$, be similarly distributed, however, with mean vector b. Assume $n_1+n_2-2 \geqslant k \geqslant 1$. If $a=b$, then the density of $\boldsymbol{D}^2$ is given by*

$$\begin{cases} C(n_1,n_2,k)\left(1+\dfrac{n_1 n_2}{n_1+n_2}\,\dfrac{z}{n_1+n_2-2}\right)^{-(n_1+n_2-1)/2} z^{k/2-1} & z>0 \\ 0 & z \leqslant 0 . \end{cases}$$

Here,

$$C(n_1,n_2,k) = \Gamma((n_1+n_2-1)/2)\,[(n_1 n_2/(n_1+n_2)]^{k/2}(n_1+n_2-2)^{k/2} \,.$$
$$[\Gamma((n_1+n_2-k-1)/2)\,\Gamma(k/2)]^{-1} .$$

Further for $n_1+n_2 \geqslant k+4$

$$E(\boldsymbol{D}^2) = k(n_1+n_2-2)(n_1+n_2)((n_1+n_2-k-3)n_1 n_2)^{-1} . \tag{4.6}$$

Using Theorem 3.7, we find the theorem quite easy to prove. First, with $\bar{\xi} = n_1^{-1} \sum_{l=1}^{n_1} \xi_l$ and $\bar{\eta} = n_2^{-1} \sum_{m=1}^{n_2} \eta_m$, $\bar{\xi}-\bar{\eta}$ is, according to I, Theorem 36.2, k-dimensional normally distributed with mean vector zero and covariance matrix $(n_1+n_2)(n_1 n_2)^{-1} G$. From Theorem 3.3, along with the reproductive property of the Wishart distribution, we find that the joint distribution of the r.v.'s

$$(n_1+n_2-2)\boldsymbol{S}_{ij} = \sum_{l=1}^{n_1} (\xi_{il}-\bar{\xi}_i)(\xi_{jl}-\bar{\xi}_j) + \sum_{m=1}^{n_2} (\eta_{im}-\bar{\eta}_i)(\eta_{jm}-\bar{\eta}_j)$$

$1 \leqslant i \leqslant j \leqslant k$ is a Wishart distribution with n_1+n_2-2 degrees of freedom and parameter k. Moreover, this $k(k+1)/2$-dimensional r.v. is distributed independently of $\bar{\xi}-\bar{\eta}$. Hence, by Theorem 3.7, the r.v.

$$\boldsymbol{T}^2 = \frac{n_1 n_2}{n_1+n_2} \sum_{i,j=1}^{k} \boldsymbol{S}_{ij}(\bar{\xi}_i-\bar{\eta}_i)(\bar{\xi}_j-\bar{\eta}_j)$$

is Hotelling distributed with n_1+n_2-2 degrees of freedom. We also have the relation

$$T^2 = \frac{n_1 n_2}{n_1+n_2} D^2 . \tag{4.7}$$

From this, the claim concerning the distribution of D^2 follows immediately. For the rest, we easily calculate, using the transformation $u=1/(1+z)$, that

$$\Gamma((n+1)/2)[\Gamma((n+1-k)/2)\Gamma(k/2)]^{-1}\int_0^\infty (1+z)^{-(n+1)/2} z^{k/2} dz = k/(n-k-1).$$

This is, up to a constant factor, the first moment of a Hotelling distribution with n degrees of freedom and parameter k, $n \geqslant k+2$. For $n=n_1+n_2-2$ one gets by (4.7) the claim (4.6). It is not difficult to generalize this theorem to the case $a \neq b$ but we will not go into this here[21].

We note that the right and hence also the left side of (4.3) vanishes for $a=b$. Thus, in this case D^2 is not unbiased for $l'd$ but $D^2-k(n_1+n_2-2)(n_1+n_2)\cdot((n_1+n_2-k-3)n_1 n_2)^{-1}$ is.

It is possible to justify use of the discriminant function given by (4.3) in another way. The basic idea for what follows is attributed to Mahalanobis.

Let ξ be an n-dimensional ($n \geqslant 1$) r.v. with probability distribution $P_\gamma, \gamma \in \Gamma$, where Γ is an open set of $R_k, k \geqslant 2$. As usual we assume that the correspondence $\gamma \to P_\gamma$ is one-to-one. Let the set P_Γ be dominated by a σ-finite measure μ and let the R.-N.-densities be given by $x \to f(x,\gamma)$ for each $\gamma \in \Gamma$. For each $\gamma=(\gamma_1, \ldots, \gamma_k) \in \Gamma$, assume that for all $x \in R_n$—with the possible exception of a μ-null set—the derivative $\dfrac{\partial \log f(x,\gamma)}{\partial \gamma_i}$, $1 \leqslant i \leqslant k$ exists. For $i,j=1,\ldots,k$, let

$$\int_{R_n} \frac{\partial \log f(x,\gamma)}{\partial \gamma_i} \frac{\partial \log f(x,\gamma)}{\partial \gamma_j} f(x,\gamma) d\mu(x)$$

be defined and denote it for each $\gamma \in \Gamma$ by $G_{ij}(\gamma)$. Assume also that $E\left[\dfrac{\partial \log f(\xi,\gamma)}{\partial \gamma_i}; \gamma\right]=0$ for $1 \leqslant i \leqslant k$ and $\gamma \in \Gamma$. Then $(G_{ij}(\gamma))_{1k}^{1k}$ is the covariance matrix of the k-dimensional r.v. $\left(\dfrac{\partial \log f(\xi,\gamma)}{\partial \gamma_1}, \ldots, \dfrac{\partial \log f(\xi,\gamma)}{\partial \gamma_k}\right)$.

Assume that this is positive-definite for each $\gamma \in \Gamma$. We now consider, following Mahalanobis, the problem of defining a distance in P_Γ between two distributions.

$$ds^2 = \sum_{i,j=1}^{k} G_{ij}(\gamma) d\gamma_i d\gamma_j \tag{4.8}$$

---

[21] See C.R. Rao, l.c.[12].

defines a metric in the set P_Γ. To obtain the distance between two distributions corresponding to the parameters $\gamma^{(1)}$, resp., $\gamma^{(2)}$, we must integrate along a *geodesic* from $\gamma^{(1)}$ to $\gamma^{(2)}$. The geodesics are distinguished from all other curves connecting $\gamma^{(1)}$ and $\gamma^{(2)}$ by the fact that this integral assumes an extremum under the metric at (4.8).

We will not go into a detailed discussion here but consider only as example the set of k-dimensional normal distributions possessing the same positive-definite covariance matrix $G=(g_{ij})_{1k}^{1k}$ and differing from one another only through their different mean vectors a, $-\infty<a<\infty$. For each $x\in R_k$, the density $f(a,x)$ is given by

$$|G^{-1}|^{1/2}(2\pi)^{-k/2}e^{-(x-a)'G^{-1}(x-a)/2}.$$

One then obtains for each $x\in R_k$ and all a

$$\frac{\partial}{\partial a_i}\log f(x,a)=\sum_{l=1}^{k}(x_l-a_l)G_{il}, \qquad 1\leqslant i\leqslant k$$

with $(G_{ij})_{1k}^{1k}=G^{-1}$. Since (in easily understood notation)

$$E[(\xi_i-a_i)(\xi_j-a_j);a]=g_{ij},$$

we get

$$E\left[\frac{\partial\log f(\xi,a)}{\partial a_i}\,\frac{\partial\log f(\xi,a)}{\partial a_j};a\right]=G_{ij}, \qquad 1\leqslant i,\ j\leqslant k.$$

Thus, in place of (4.8) we obtain

$$ds^2=\sum_{i,j=1}^{k}G_{ij}\,da_i\,da_j. \tag{4.9}$$

The *differential equation of the geodesics* is $\dfrac{d^2a_i}{ds^2}=0$, $1\leqslant i\leqslant k$, because the G_{ij} are constant and therefore do not depend on a. Thus, the metric is Euclidean. Each solution of the differential equation is given for all real s by $a_i(s)=\alpha_i+\beta_i s$, $1\leqslant i\leqslant k$, where α_i, β_i are arbitrary real numbers. Hence, if $a^{(i)}\in R_k$, $a^{(1)}\neq a^{(2)}$ and, say, $a_j^{(1)}-a_j^{(2)}=\beta_j^{(0)}(s_1-s_2)$, $1\leqslant j\leqslant k$, then we find for the integral from $a^{(1)}$ to $a^{(2)}$ along a geodesic

$$\int_{s_1}^{s_2}\left(\sum_{i,j=1}^{k}\frac{da_i}{ds}\frac{da_j}{ds}G_{ij}\right)^{1/2}ds=(s_2-s_1)\left(\sum_{i,j=1}^{k}\beta_i^{(0)}\beta_j^{(0)}G_{ij}\right)^{1/2}.$$

Thus, for the square of the distance defined by (4.9) between the two k-dimensional normal distributions given by $a^{(1)}$ and $a^{(2)}$ resp., we obtain the expression $\sum_{i,j=1}^{k}G_{ij}(a_i^{(2)}-a_i^{(1)})(a_j^{(2)}-a_j^{(1)})$ which, in slightly different notation, is the left side of (4.3).

5. Regression theory revisited. We refer to I.**21** and use the notation introduced there. Let $\xi=(\xi_1,\ldots,\xi_{p+1})$ be a $(p+1)$-dimensional r.v., $p\geqslant 1$, all of whose first and second order moments are assumed to exist. The regression function ρ of ξ_{p+1} w.r.t. $(\xi_1,\ldots,\xi_p)$ solves, as we know from I, Theorem 21.1, a minimum problem which we want to consider again from a somewhat different standpoint. In general, ρ is of quite complicated form. We would thus like to replace ρ by a linear approximation and consider the problem of minimization of $E[(\xi_{p+1}-\beta_0-\beta_1\xi_1-\cdots-\beta_p\xi_p)^2]$ when $(\beta_0,\ldots,\beta_p)$ varies throughout R_{p+1}. This naturally suggests an extension to polynomials of arbitrary degree. This extended problem will, however, also be solved by the solution of the mentioned minimum problem. If we consider for example the 3-dimensional r.v. (η_1,η_2,η_3) and try to minimize $E[\eta_3-\beta_0-\beta_1\eta_1-\beta_2\eta_2-\beta_3\eta_1^2-\beta_4\eta_1\eta_2-\beta_5\eta_2^2)]$ as function of $\beta_0,\beta_1,\beta_2,\beta_3,\beta_4,\beta_5$, then we arrive with $p=5$, $\eta_3=\xi_6$, $\eta_1=\xi_1$, $\eta_2=\xi_2$, $\eta_1^2=\xi_3$, $\eta_1\eta_2=\xi_4$, $\eta_2^2=\xi_5$ back at the linear problem. The extention to arbitrary polynomials is trivial.

The formal analogy to the problem considered in **1** is immediate. Denoting the components of the mean vector of ξ by a_i, $1\leqslant i\leqslant p+1$ and the elements of the covariance matrix by σ_{ij}, $1\leqslant i,\ j\leqslant p+1$, we easily obtain as necessary and sufficient condition for solutions of the minimization problem the system of equations

$$\beta_0+\sum_{i=1}^{p} a_i\beta_i=a_{p+1}, \tag{5.1}$$

$$\sum_{j=1}^{p} \beta_j\sigma_{ij}=\sigma_{ip+1}, \qquad 1\leqslant i\leqslant p. \tag{5.2}$$

Assume $|\sigma_{ij}|_{1p}^{1p}\neq 0$. Then the uniquely determined solutions are

$$\beta_i^{(0)}=\frac{\begin{vmatrix}\sigma_{11}\cdots\sigma_{1i-1}\,\sigma_{1p+1}\,\sigma_{1i+1}\cdots\sigma_{1p}\\ \cdots\cdots\cdots\cdots\cdots\cdots\cdots\\ \sigma_{p1}\cdots\sigma_{pi-1}\,\sigma_{pp+1}\,\sigma_{pi+1}\cdots\sigma_{pp}\end{vmatrix}}{|\sigma_{ij}|_{1p}^{1p}}, \qquad i=1,\ldots,p, \tag{5.3}$$

and

$$\beta_0^{(0)}=\frac{\begin{vmatrix}a_{p+1}\; a_1\;\ldots a_p\\ \sigma_{1p+1}\,\sigma_{11}\cdots\sigma_{1p}\\ \cdots\cdots\cdots\cdots\\ \sigma_{pp+1}\,\sigma_{p1}\cdots\sigma_{pp}\end{vmatrix}}{|\sigma_{ij}|_{1p}^{1p}}. \tag{5.4}$$

In what follows we write β_i instead of $\beta_i^{(0)}$, $0\leqslant i\leqslant p$ and from now on β_i will have only this meaning. The β_i are called *regression coefficients* and we have the almost obvious

Theorem 5.1. *If the regression function of a $(p+1)$-dimensional distribution*[22] *is linear, i.e., of the form*

$$(x_1, \dots, x_p) \to \delta_0 + \delta_1 x_1 + \cdots + \delta_p x_p, \quad \delta_i \text{ real and } |\sigma_{ij}|_{1p}^{1p} \neq 0,$$

then, necessarily, $\beta_i = \delta_i$, $0 \leqslant i \leqslant p$.

Naturally, we assume that the first and second order moments exist.

Proof. According to I, Theorem 21.1, we must have

$$E\left[\left(\xi_{p+1} - \delta_0 - \sum_{i=1}^{p} \delta_i \xi_i\right)^2\right] \leqslant E\left[\left(\xi_{p+1} - \beta_0 - \sum_{i=1}^{p} \beta_i \xi_i\right)^2\right].$$

On the other hand, the reverse inequality must hold because of the definition of the β_i, $0 \leqslant i \leqslant p$. By (5.3) and (5.4), however, the β_i, $0 \leqslant i \leqslant p$, are uniquely determined, which completes the proof.

The regression function of a $(p+1)$-dimensional normal distribution is linear. This follows immediately from the inequality I. (35.12).

We now consider the r.v.

$$\eta_{p+1} = \xi_{p+1} - \beta_0 - \sum_{i=1}^{p} \beta_i \xi_i. \tag{5.5}$$

From (5.3) and (5.4)

$$\eta_{p+1} = \begin{vmatrix} \xi_{p+1} - a_{p+1} & \xi_1 - a_1 \dots & \xi_p - a_p \\ \sigma_{1p+1} & \sigma_{11} & \dots \sigma_{1p} \\ \dots & \dots & \dots \\ \sigma_{pp+1} & \sigma_{p1} & \dots \sigma_{pp} \end{vmatrix} (|\sigma_{ij}|_{1p}^{1p})^{-1}. \tag{5.6}$$

Since $E(\xi_i - a_i) = 0$, $1 \leqslant i \leqslant p+1$, we get $E(\eta_{p+1}) = 0$ immediately from (5.6), so that from (5.2)

$$E[(\xi_i - a_i)\eta_{p+1}] = E[(\xi_i - a_i)(\eta_{p+1} - E(\eta_{p+1})] = 0$$

for $1 \leqslant i \leqslant p$, i.e., η_{p+1} and ξ_i are uncorrelated. This follows, of course, directly from the remark contained in I[27].

Hence by means of (5.5) and (5.1) for the variance of η_{p+1}:

$$E(\eta_{p+1}^2) = E[(\xi_{p+1} - a_{p+1})\eta_{p+1}] \tag{5.7}$$

or

$$E(\eta_{p+1}^2) = \frac{|\sigma_{ij}|_{1p+1}^{1p+1}}{|\sigma_{ij}|_{1p}^{1p}}. \tag{5.8}$$

[22] That is the regression function of ξ_{p+1} w.r.t. $(\xi_1, \dots, \xi_p)$, say.

The r.v. $\eta_{p+1}^2 = \left(\xi_{p+1} - \beta_0 - \sum_{i=1}^{p} \beta_i \xi_i\right)^2$ will again be called the *residual term*. Its expectation is nothing other than the minimum of

$$E\left[\left(\xi_{p+1} - \gamma_0 - \sum_{i=1}^{p} \gamma_i \xi_i\right)^2\right] \quad \text{for} \quad -\infty < \gamma_i < \infty\,, \quad 0 \leqslant i \leqslant p\,.$$

We now define the *multiple correlation coefficient* (of ξ_{p+1}, w.r.t. $(\xi_1, \ldots, \xi_p)$) by

$$K_{p+1} = \frac{E[((\xi_{p+1} - a_{p+1}) - \eta_{p+1})(\xi_{p+1} - a_{p+1})]}{(E[(\xi_{p+1} - a_{p+1} - \eta_{p+1})^2]\, E[(\xi_{p+1} - a_{p+1})^2])^{1/2}}. \tag{5.9}$$

Naturally, this definition only makes sense for $\sigma_{p+1\,p+1} \neq 0$. For $p=1$, K_2 coincides with the correlation coefficient of ξ_1 and ξ_2 defined in I, p. 52.

From (5.7) and (5.8) follows

$$E[(\xi_{p+1} - a_{p+1} - \eta_{p+1})(\xi_{p+1} - a_{p+1})] = \sigma_{p+1\,p+1} - \frac{|\sigma_{ij}|_{1\,p+1}^{1\,p+1}}{|\sigma_{ij}|_{1\,p}^{1\,p}}$$

and likewise, $E[(\xi_{p+1} - a_{p+1} - \eta_{p+1})^2]$ has the same value.

Hence, from (5.9)

$$K_{p+1} = (1 - |\sigma_{ij}|_{1\,p+1}^{1\,p+1} (\sigma_{p+1\,p+1} |\sigma_{ij}|_{1\,p}^{1\,p})^{-1})^{1/2}. \tag{5.10}$$

From the definition of η_{p+1}^2

$$E(\eta_{p+1}^2) \leqslant E[(\xi_{p+1} - a_{p+1})^2]$$

or, because of (5.8)

$$|\sigma_{ij}|_{1\,p+1}^{1\,p+1} \leqslant \sigma_{p+1\,p+1} |\sigma_{ij}|_{1\,p}^{1\,p}.$$

Hence, from (5.10) we have $0 \leqslant K_{p+1} \leqslant 1$. The *equality* $K_{p+1} = 1$ *holds iff* $|\sigma_{ij}|_{1\,p+1}^{1\,p+1} = 0$. According to I, Theorem 17.6, there exist under this condition real numbers u_i, $1 \leqslant i \leqslant p+1$, with $\sum_{i=1}^{p+1} u_i^2 > 0$ such that

$$P\left(\sum_{i=1}^{p+1} u_i(\xi_i - a_i) = 0\right) = 1\,.$$

From (5.8) and (5.10) we can also write

$$K_{p+1} = (1 - E(\eta_{p+1}^2)/E[\xi_{p+1} - a_{p+1})^2])^{1/2}.$$

Hence, $K_{p+1} = 0$ implies $E(\eta_{p+1}^2) = E[(\xi_{p+1} - a_{p+1})^2]$ which is equivalent to

$$\sum_{i,j=1}^{p} \beta_i \beta_j \sigma_{ij} - 2 \sum_{i=1}^{p} \beta_i \sigma_{i\,p+1} + \left(\beta_0 - a_{p+1} + \sum_{i=1}^{p} \beta_i a_i\right)^2 = 0\,.$$

From this by means of (5.1) and (5.2), we get $\sum_{i,j=1}^{p} \beta_i \beta_j \sigma_{ij} = 0$. Since, however, $(\sigma_{ij})_{1p}^{1p}$ is positive-definite by assumption, $\beta_i = 0$, $1 \leqslant i \leqslant p$.

Again taking account of (5.2), we get $\sigma_{ip+1} = 0$, i.e., ξ_{p+1} and ξ_i are uncorrelated for $i = 1, \ldots, p$. Summarizing, we have

Theorem 5.2. *Let $\xi_1, \ldots, \xi_{p+1}$, $p \geqslant 1$, be any r.v.'s all of whose first and second order moments exist. Let $E(\xi_i) = a_i$ and $E[(\xi_i - a_i)(\xi_j - a_j)] = \sigma_{ij}$, $i,j = 1, \ldots, p+1$. Assume $|\sigma_{ij}|_{1p}^{1p} \neq 0$ and $\sigma_{p+1p+1} \neq 0$. Let the r.v. η_{p+1} be defined by (5.5). Then the multiple correlation coefficient K_{p+1} defined by (5.9) possesses the following properties: $0 \leqslant K_{p+1} \leqslant 1$; $K_{p+1} = 1$ iff there exist real numbers u_i, $1 \leqslant i \leqslant p+1$, with $\sum_{i=1}^{p+1} u_i^2 \neq 0$ such that $P\left(\sum_{i=1}^{p+1} u_i(\xi_i - a_i) = 0\right) = 1$. Further $K_{p+1} = 0$ iff ξ_{p+1} and ξ_i, $i = 1, \ldots, p$, are uncorrelated.*

We now consider a $(p+2)$-dimensional r.v. $(\xi_1, \ldots, \xi_p, \xi_{p+1}, \xi_{p+2})$ all of whose first and second order moments exist. Denote the mean again by a_i the covariance by σ_{ij}, $i,j = 1, \ldots, p+2$. $(\sigma_{ij})_{1p}^{1p}$ is assumed to be positive-definite. The real numbers $\beta_i^{(j)}$, $0 \leqslant i \leqslant p$, $j = 1, 2$, are assumed to be the uniquely determined solution of the problem of minimizing

$$E\left[\left(\xi_{p+j} - \gamma_0 - \sum_{i=1}^{p} \gamma_i \xi_i\right)^2\right], \qquad -\infty < \gamma_i < \infty, \qquad 0 \leqslant i \leqslant p.$$

We then obtain two r.v.'s

$$\eta_{p+j} = \xi_{p+j} - \beta_0^{(j)} - \sum_{i=1}^{p} \beta_i^{(j)} \xi_i, \qquad j = 1, 2.$$

Let $E(\eta_{p+i}^2) \neq 0$, $i = 1, 2$. We will call

$$E(\eta_{p+1} \eta_{p+2}) (E(\eta_{p+1}^2) E(\eta_{p+2}^2))^{-1/2}$$

the *partial correlation coefficient* of ξ_{p+1} and ξ_{p+2} (w.r.t. $\xi_1, \ldots, \xi_p$). Since η_{p+2} is uncorrelated with ξ_i, $i = 1, \ldots, p$, we obtain from an application of (5.6)

$$E(\eta_{p+1} \eta_{p+2}) = \begin{vmatrix} \sigma_{p+1p+2} & \sigma_{p+11} & \cdots & \sigma_{p+1p} \\ \sigma_{1p+2} & \sigma_{11} & \cdots & \sigma_{1p} \\ \cdots & \cdots & \cdots & \cdots \\ \sigma_{pp+2} & \sigma_{p1} & \cdots & \sigma_{pp} \end{vmatrix} (|\sigma_{ij}|_{1p}^{1p})^{-1}. \tag{5.11}$$

Hence, from (5.8) and the corresponding relation for η_{p+2}, we have for the partial correlation coefficient the expression

$$\frac{\begin{vmatrix} \sigma_{p+1\,p+2} & \sigma_{p+1\,1} & \cdots & \sigma_{p+1\,p} \\ \sigma_{1\,p+2} & \sigma_{11} & \cdots & \sigma_{1p} \\ \cdots & \cdots & \cdots & \cdots \\ \sigma_{pp} & \sigma_{p1} & \cdots & \sigma_{pp} \end{vmatrix}}{\left(\begin{vmatrix} \sigma_{p+1\,p+1} & \sigma_{p+1\,1} & \cdots & \sigma_{p+1\,p} \\ \sigma_{1\,p+1} & \sigma_{11} & \cdots & \sigma_{1p} \\ \cdots & \cdots & \cdots & \cdots \\ \sigma_{p+1} & \sigma_{p1} & \cdots & \sigma_{pp} \end{vmatrix} \begin{vmatrix} \sigma_{p+2\,p+2} & \sigma_{p+2\,1} & \cdots & \sigma_{p+2\,p} \\ \sigma_{1\,p+2} & \sigma_{11} & \cdots & \sigma_{1p} \\ \cdots & \cdots & \cdots & \cdots \\ \sigma_{pp+2} & \sigma_{p1} & \cdots & \sigma_{pp} \end{vmatrix} \right)^{1/2}}. \tag{5.12}$$

If ρ_{ik} is the correlation coefficient of ξ_i and ξ_k, $i \neq k$ (see I, p. 52), then $\sigma_{ik} = (\sigma_{ii}\sigma_{kk})^{1/2}\rho_{ik}$. Using this, one can easily express (5.12) by means of the ρ_{ik}.

It is expedient at this point to use the notation of Yule[23] *for the regression coefficients and the various correlation coefficients* that have been defined up to now. Let $\xi_\mu, \xi_{\nu_1}, \xi_{\nu_2}, \ldots, \xi_{\nu_k}$, $k \geqslant 1$, be any (one-dimensional) r.v.'s. The regression coefficients determined on the basis of the condition $\min\limits_{\substack{-\infty<\gamma_i<\infty \\ 0 \leqslant i \leqslant k}} E\left[\left(\xi_\mu - \gamma_0 - \sum\limits_{i=1}^{k} \gamma_i \xi_{\nu_i}\right)^2\right]$ will be denoted, in order, by $\beta_{\mu 0 \,.\, \nu_1 \ldots \nu_k}$, $\beta_{\mu\nu_1 \,.\, \nu_2 \ldots \nu_k}, \ldots, \beta_{\mu\nu_k \,.\, \nu_1 \ldots \nu_{k-1}}$. The regression coefficients defined according to (5.3) are thus denoted by

$$\beta_{p+1\,i\,.\,1 \ldots (i-1)(i+1) \ldots p}, \qquad 1 \leqslant i \leqslant p.$$

The r.v.

$$\xi_\mu - \beta_{\mu 0\,.\,\nu_1 \ldots \nu_k} - \sum_{i=1}^{k} \beta_{\mu\nu_i\,.\,\nu_1 \ldots \nu_{i-1}\nu_{i+1} \ldots \nu_k} \xi_{\nu_i}$$

will be denoted by $\eta_{\mu\,.\,\nu_1 \ldots \nu_k}$. The r.v. defined by (5.5) is now, for example, $\eta_{p+1\,.\,1 \ldots p}$.

The expectation $E(\eta^2_{\mu\,.\,\nu_1 \ldots \nu_k})$ of the residual is written $\sigma^2_{\mu\,.\,\nu_1 \ldots \nu_k}$. For the left side of (5.8) we thus write $\sigma^2_{p+1\,.\,1 \ldots p}$. The partial correlation coefficient (5.11) is denoted by $\rho_{p+1\,p+2\,.\,1 \ldots p}$. This should make the new notation sufficiently clear. Using it, we get for $p \geqslant 2$ the following relation which will often be used in the sequel:

$$\beta_{p+1\,p\,.\,1 \ldots p-1} = \rho_{p+1\,p\,.\,1 \ldots p-1} \frac{\sigma_{p+1\,.\,1 \ldots p-1}}{\sigma_{p\,.\,1 \ldots p-1}}. \tag{5.13}$$

[23] G. U. Yule, Proc. Roy. Soc. London Ser. A, 79, 182—193 (1907).

This follows easily after writing (5.11) for $p-1$ instead of p, using (5.8) and then comparing with (5.3) for $i=p$.

6. The statistical theory of regression. The practical calculation of the regression and correlation coefficients defined in **5** is only possible when the parameters which appear—means and covariances—of the underlying distribution are known. If they are not, then we have the problem of deriving estimates for the regression and correlation coefficients. The arguments of **5** were to a large extent independent of the underlying distribution. From now on we will always assume that it is normal.

We again use notation similar to that of **3**. Let now $\xi_1, \ldots, \xi_n$ be n independent $(p+1)$-dimensional r.v.'s with $n \geqslant p+2$, each distributed by the same $(p+1)$-dimensional normal distribution. Let its density be given for each $x \in R_{p+1}$ by

$$|D|^{1/2}(2\pi)^{-(p+1)/2} e^{-x'Dx/2} \tag{6.1}$$

where $D=(d_{ij})_{1\,p+1}^{1\,p+1}$ is assumed to be positive-definite. We denote the covariance matrix D^{-1} also by $(\sigma_{ij})_{1\,p+1}^{1\,p+1}$. It will sometimes be convenient—in the spirit of the definition of $\sigma_{p+1\,.\,1\ldots p}^2$ (indicated on p. 389)—to write σ_i^2 instead of σ_{ii}.

Let $\xi_l=(\xi_{1l}, \ldots, \xi_{p+1\,l})$, $1 \leqslant l \leqslant n$. The possibility of obtaining estimates for the regression and correlation coefficients by replacing the σ_{ij} by the estimates $\mathbf{w}_{ij}(n-1)^{-1}$, $1 \leqslant i, j \leqslant p+1$ suggests itself. We recall in this connection the notation introduced by (2.12) and (2.13). $\mathbf{w}_{ij}$ is then the r.v. obtained from w_{ij} by replacing x_{il}(resp. x_{jl}) by $\xi_{il}(\xi_{jl})$ and $\bar{x}_i(\bar{x}_j)$ by $\bar{\xi}_i(\bar{\xi}_j)$, $1 \leqslant i, j \leqslant p+1$, $1 \leqslant l \leqslant n$. If we do this, we get expressions which we have already encountered in a formal context. If we introduce this substitution for example into the right side of (5.3), then the analog to (2.19), or more precisely, to the corresponding r.v.'s $\mathbf{b}_i$, is immediately obvious. Similar results obtain for the comparison of (5.8) and (2.20) etc. One should, however, note that in **2** we have only considered part of the arguments appearing in the estimates as r.v.'s (namely $\xi_{p+1\,1}, \ldots, \xi_{p+1\,n}$). See also the proof of Theorem 3.5.

We also want to apply the notation introduced at the end of **5**. For example, $b_{p+1\,1\,.\,2\ldots p}$ denotes the quotient

$$\begin{vmatrix} w_{1\,p+1} & w_{12} & \cdots & w_{1p} \\ \cdots & \cdots & \cdots & \cdots \\ w_{p\,p+1} & w_{p2} & \cdots & w_{pp} \end{vmatrix} (|w_{ij}|_{1\,p}^{1\,p})^{-1}$$

in analogy to the definition of $\beta_{p+1\,1\,.\,2\ldots p}$.[24]

[24] It would perhaps be more consistent in the sense of the notation used in **1** and **2** to write $\hat{b}_{p+1\,1.2\ldots p}$ instead of $b_{p+1\,1.2\ldots p}$.

Likewise, we write $r_{p+1p.1\dots p-1}$ for the quotient

$$\frac{\begin{vmatrix} w_{11} \dots w_{1p-1}\; w_{1p+1} \\ \dots\dots\dots\dots\dots \\ w_{p1} \dots w_{pp-1}\; w_{pp+1} \end{vmatrix}}{\left[\begin{vmatrix} w_{11} & \dots w_{1p-1} & w_{1p} \\ \dots & \dots & \dots \\ w_{p-11} & \dots w_{p-1p-1} & w_{p-1p} \\ w_{p1} & \dots w_{pp-1} & w_{pp} \end{vmatrix} \begin{vmatrix} w_{11} & \dots w_{1p-1} & w_{1p+1} \\ \dots & \dots & \dots \\ w_{p-11} & \dots w_{p-1p-1} & w_{p-1p+1} \\ w_{p+11} & \dots w_{p+1p-1} & w_{p+1p+1} \end{vmatrix}\right]^{1/2}}$$

in analogy to the definition of $\rho_{p+1p.1\dots p-1}$. It is expedient to call $b_{p+1i.1\dots(i-1)(i+1)\dots p}$ the *sample regression coefficient* $(1\leqslant i\leqslant p)$ and $r_{p+1p.1\dots p-1}$ the *partial sample correlation coefficient*[25], etc. We also use this terminology for the corresponding r.v.'s $\boldsymbol{b}_{p+1i.1\dots(i-1)(i+1)\dots p}$, $\boldsymbol{r}_{p+1p.1\dots p-1}$, etc.

A series of further definitions are to be understood in this sense: Let $\mu, \nu_1, \nu_2, \dots, \nu_k$, $k\geqslant 1$ be different elements from the set $\{1, \dots, p+1\}$. We write $x_i^{(l)}$ instead of x_{il}, $1\leqslant i\leqslant p+1$, $1\leqslant l\leqslant n$ and define

$$x^{(l)}_{\mu.\nu_1\dots\nu_k} = x^{(l)}_\mu - b_{\mu\nu_1.\nu_2\dots\nu_k} x^{(l)}_{\nu_1} - b_{\mu\nu_2.\nu_1\nu_3\dots\nu_k} x^{(l)}_{\nu_2} - \dots - b_{\mu\nu_k.\nu_1\dots\nu_{k-1}} x^{(l)}_{\nu_k}.$$

If $i, j, \nu_1, \dots, \nu_k$ have a meaning similar to that above, we define

$$w_{ij.\nu_1\dots\nu_k} = \sum_{l=1}^{n} (x^{(l)}_{i.\nu_1\dots\nu_k} - \bar{x}_{i.\nu_1\dots\nu_k})(x^{(l)}_{j.\nu_1\dots\nu_k} - \bar{x}_{j.\nu_1\dots\nu_k}) \tag{6.2}$$

where naturally

$$\bar{x}_{\mu.\nu_1\dots\nu_k} = \frac{1}{n}\sum_{l=1}^{n} x^{(l)}_{\mu.\nu_1\dots\nu_k}.$$

Similar definitions in the sequel will be easy to comprehend in the context of these examples.

In order to derive the distribution of $\boldsymbol{b}_{ij.\nu_1\dots\nu_k}$, $\boldsymbol{r}_{ij.\nu_1\dots\nu_k}$, $\boldsymbol{w}_{ij.\nu_1\dots\nu_k}$ and other estimates, one can use the methods we applied in the proof of Theorem 3.5. However, we want to sketch a somewhat different method which allows a deeper understanding of the structure of the Wishart distribution and which forms essentially the basis for the proof of Theorem 3.2[26].

---

[25] This terminology clearly points out the fact that one can calculate these quantities from a sample.

[26] See M. S. Bartlett, Proc. Roy. Soc. Edinburgh Sect. A, 53, 260—283 (1932—1933). For a extension of this method see R.A. Wijsman, Ann. Math. Statist. 28, 415—422 (1957) and A.M. Kshirsagar, Ann. Math. Statist. 30, 239—241 (1959).

Firstly, Theorem 3.3 shows that under the assumptions formulated on p. 390, the joint distribution of the r.v.'s $\boldsymbol{w}_{ij}$, $1 \leqslant i \leqslant j \leqslant p+1$, is Wishart with $n-1$ d.f. and parameter $p+1$, whose density in $R_{(p+1)(p+2)/2}$ is given by

$$W_{p+1}(w_{ij}; n-1) = C_1 |W|^{(n-p-3)/2} e^{-\sum_{i,j=1}^{p+1} d_{ij} w_{ij}/2} \tag{6.3}$$

provided it does not vanish. Here, $w_{ij} = w_{ji}$ for $1 \leqslant i, j \leqslant p+1$ and $W = (w_{ij})_{1\,p+1}^{1\,p+1}$. The symbol C_1 stands for a positive real number whose value becomes obvious by comparing with (3.2). We introduce new variables into (6.3) which are related to the w_{ij} as follows:

$$\begin{aligned} w_{ij.1} &= w_{ij} - w_{i1} w_{j1}/w_{11}, \\ u_{i1} &= w_{11}^{1/2}(b_{i1} - \beta_{i1}), \qquad i,j = 2, \ldots, p+1, \\ w_{11} &= w_{11}. \end{aligned} \tag{6.4}$$

In what follows we will not explicitly state the range of variation of the indices i,j and will always assume that they run through all of the elements of the set $\{1, \ldots, p+1\}$ which do not lie to the right of the "point". Analogous situations will be understood in this context.

We recall that according to our agreement, the relations

$$b_{i1} = w_{i1}/w_{11} \tag{6.5}$$

and

$$\beta_{i1} = \sigma_{i1}/\sigma_{11} \tag{6.6}$$

hold. (Cf. p. 390.) Since no "point" appears, i runs in (6.5) and (6.6) from 1 to $p+1$.

We use the notational convention introduced by rule for the transformation given by (6.2) and the following transformations of variables. This can hardly lead to misunderstanding and turns out to be practically quite useful. Also note that the transformations concerning $w_{ij.1}$ represent an identity in the following sense: applying (6.2), one gets

$$w_{ij.1} = \sum_{l=1}^{n} (x_{i.1}^{(l)} - \bar{x}_{i.1})(x_{j.1}^{(l)} - \bar{x}_{j.1}) \quad \text{where } x_{i.1}^{(l)} = x_i^{(l)} - b_{i1} x_1^{(l)}$$

$(1 \leqslant l \leqslant n)$. Since (6.5) is also an identity in this sense, the claimed identity for the $w_{ij.1}$ follows.

The transformation (6.4) takes the exponent $\sum_{i,j=1}^{p+1} d_{ij} w_{ij}$ appearing in (6.3) into

$$\frac{w_{11}}{\sigma_{11}} + \sum_{i,j=2}^{p+1} d_{ij}(w_{ij.1} + u_{i.1} u_{j.1}). \tag{6.7}$$

Indeed, from (6.5) and (6.6),

$$\sum_{i,j=2}^{p+1} d_{ij} w_{1j} - \sum_{i,j=2}^{p+1} \frac{\sigma_{1i}}{\sigma_{11}} d_{ij} w_{1j} + \sum_{i,j=2}^{p+1} \frac{\sigma_{1j}}{\sigma_{11}} d_{ij} w_{i1} + \frac{w_{11}}{\sigma_{11}^2} \sum_{i,j=1}^{p+1} \sigma_{1i}\sigma_{1j} d_{ij} \quad (6.8)$$

is taken by (6.4) into $\sum_{i,j=2}^{p+1} d_{ij}(w_{ij.1} + u_{i.1} u_{j.1})$. Since, however,

$$\sum_{m=1}^{p+1} \sigma_{1m} d_{im} = \begin{cases} 0 & 1 \neq i \\ 1 & 1 = i, \end{cases} \quad (6.9)$$

(6.8) is identical to

$$\sum_{\substack{i,j=1\\ i+j\geq 3}}^{p+1} w_{ij} d_{ij} + \frac{w_{11}}{\sigma_{11}^2} \sum_{i,j=2}^{p+1} \sigma_{1i}\sigma_{1j} d_{ij}. \quad (6.10)$$

From (6.9) for $i \neq 1$ it easily follows that

$$\sum_{i,m=2}^{p+1} \sigma_{1i}\sigma_{1m} d_{1m} + \sigma_{11} \sum_{i=2}^{p+1} \sigma_{1i} d_{1i} = 0. \quad (6.11)$$

Now using (6.9) for $i=1$, one finds the identity $\sum_{m=2}^{p+1} \sigma_{1m} d_{1m} = 1 - \sigma_{11} d_{11}$. Hence, together with (6.11), this says that

$$\sum_{i,m=2}^{p+1} \sigma_{li}\sigma_{1m} d_{1m}/\sigma_{11}^2 = d_{11} - 1/\sigma_{11}.$$

so that in place of (6.10) one can write $\sum_{i,j=1}^{p+1} d_{ij} w_{ij} - w_{11}/\sigma_{11}$ which implies, as claimed, that $\sum_{i,j=1}^{p+1} d_{ij} w_{ij}$ goes into (6.7). Exactly as at the end of the proof of Theorem 3.2, one obtains

$$|W| = w_{11} |w_{ij.1}|_{2\ p+1}^{2\ p+1}.$$

The Jacobian of the transformation inverse to (6.4) is given by $\Delta_1 = w_{11}^{p/2}$. We thus get from (6.3) for the density of the joint distribution of the r.v.'s $\boldsymbol{w}_{ij.1}$, $\boldsymbol{b}_{i.1}$, $\boldsymbol{w}_{11}$ in $R_{(p+1)(p+2)/2}$, provided it does not vanish,

$$\begin{aligned} &W_{p+1}(w_{ij}; n-1)\Delta_1 \\ &= C_2 w_{11}^{\frac{n-1}{2}-1} e^{-w_{11}/(2\sigma_{11})} (|w_{ij.1}|_{2\ p+1}^{2\ p+1})^{(n-p-3)/2} \exp\left(-1/2 \sum_{i,j=2}^{p+1} d_{ij} w_{ij.1}\right) \\ &\quad \times \exp\left(-1/2 \sum_{i,j=2}^{p+1} d_{ij} u_{i.1} u_{j.1}\right). \end{aligned} \quad (6.12)$$

C_2 is a positive normalizing factor which is easy to determine. In the symbol $W_{p+1}(w_{ij}; n-1)$, the w_{ij} are to be replaced by the new variables according to (6.4). The equality (6.12) says that the distribution of $\boldsymbol{w}_{11}$ is independent of the joint distribution of the $\boldsymbol{w}_{ij\,.\,1}$, and both of these distributions are independent of the joint distribution of the $\boldsymbol{u}_{i\,.\,1}$. The r.v. $\boldsymbol{w}_{11}/\sigma_{11}$ (in the alternate notation $\boldsymbol{w}_{11}/\sigma_1^2$) is χ^2-distributed with $n-1$ degrees of freedom. The joint distribution of the $\boldsymbol{w}_{ij\,.\,1}$ is Wishart with $n-2$ d.f. and parameter p. The joint distribution of the $\boldsymbol{u}_{i\,.\,1}$ is p-dimensional normal, which we will denote by $N_p(u_{i\,.\,1})$. Thus, in easily understood symbolism we can write

$$W_{p+1}(w_{ij}; n-1)\,\Delta_1 = W_1(w_{11}; n-1)\,W_p(w_{ij\,.\,1}; n-2)\,N_p(u_{i\,.\,1}). \quad (6.13)$$

Continuing this procedure, we consider the distribution whose density is given by $W_p(w_{ij\,.\,1}; n-2)$ and introduce new variables by means of the transformation

$$\left.\begin{aligned} w_{ij\,.\,12} &= w_{ij\,.\,1} - w_{i2\,.\,1}\,w_{j2\,.\,1}/w_{22\,.\,1} \\ u_{i\,.\,21} &= w_{22\,.\,1}^{1/2}(b_{i2\,.\,1} - \beta_{i2\,.\,1}) \\ w_{22\,.\,1} &= w_{22\,.\,1}\,. \end{aligned}\right\} \quad (6.14)$$

Denoting the Jacobian of the transformation inverse to (6.14) by Δ_2, we obtain instead of (6.13)

$$W_p(w_{ij\,.\,1}; n-2)\,\Delta_2 = W_1(w_{22\,.\,1}; n-2)\,W_{p-1}(w_{ij\,.\,12}; n-3)\,N_{p-1}(u_{i\,.\,21}).$$

Thus, in symbolic notation we get

$$\begin{aligned} W_{p+1}(w_{ij}; n-1)\,\Delta_1\,\Delta_2 &= W_1(w_{11}; n-1)\,W_1(w_{22\,.\,1}; n-2) \\ &\quad\times W_{p-1}(w_{ij\,.\,12}; n-3)\,N_{p-1}(u_{i\,.\,21})\,N_p(u_{i\,.\,1}). \end{aligned} \quad (6.15)$$

By induction we arrive at

$$\left.\begin{aligned} & W_{p+1}(w_{ij}; n-1)\prod_{i=1}^{p}\Delta_1 \\ & = \prod_{k=1}^{p+1} W_1(w_{kk\,.\,1\ldots k-1}; n-k)\prod_{k=0}^{p-1} N_{p-k}(u_{i\,.\,k+1\ldots 1}). \end{aligned}\right\} \quad (6.16)$$

This says, in particular, that $\boldsymbol{w}_{11}, \boldsymbol{w}_{22\,.\,1}, \ldots, \boldsymbol{w}_{p+1\,p+1\ldots p}$ are mutually independent. The $p-k+1$ r.v.'s $(\boldsymbol{u}_{i,k\ldots 1})$[27] are for $1 \leqslant k \leqslant p$ independent of all these r.v.'s and are also themselves mutually independent. For $1 \leqslant j \leqslant p+1$, the $\boldsymbol{w}_{jj\,.\,1\ldots j-1}/\sigma_{j\,.\,1\ldots j-1}^2$ are χ^2-distributed with

[27] The symbol $(\boldsymbol{u}_{i\,.\,k\ldots 1})$ stands for the $(p-k+1)$-dimensional r.v. $(\boldsymbol{u}_{p+1\,.\,k\ldots 1}, \ldots, \boldsymbol{u}_{k+1\,.\,k\ldots 1})$.

$n-j$ degrees of freedom, and the $(\boldsymbol{u}_{i.p-k\ldots 1})$ are $(k+1)$-dimensionally normally distributed for $0 \leqslant k \leqslant p-1$.

From these considerations we easily derive the distribution of $\boldsymbol{b}_{p+1p.1\ldots p-1}$. As we have just shown, the density of the joint distribution of $\boldsymbol{w}_{pp.1\ldots p-1}$ and $\boldsymbol{u}_{p+1.p\ldots 1}$ for $-\infty < x < \infty$, $0 < y < \infty$ is given by

$$\left.\begin{aligned} &C(n,p)\exp(-y/(2\sigma^2_{p.1\ldots p-1}))\left[(y/\sigma^2_{p.1\ldots p-1})^{\frac{n-p}{p}-1}\right] \\ &\times\exp\left(-\frac{1}{2}x^2/\sigma^2_{p+1.1\ldots p}\right). \end{aligned}\right\} \tag{6.17}$$

Here,

$$C(n,p) = \left[\sigma^2_{p.1\ldots p-1}2^{(n-p)/2}\Gamma((n-p)/2)(2\pi)^{1/2}\sigma_{p+1.1\ldots p}\right]^{-1}.$$

Continuing the sequence of transformation equations (6.4) and (6.14) (arriving finally at (6.16)), we see that between the r.v.'s $\boldsymbol{u}_{p+1.1\ldots p}$, $\boldsymbol{b}_{p+1p.1\ldots p-1}$ and $\boldsymbol{w}_{pp.1\ldots p-1}$ the following relation exists

$$\boldsymbol{u}_{p+1.1\ldots p} = \boldsymbol{w}^{1/2}_{pp.1\ldots p-1}(\boldsymbol{b}_{p+1p.1\ldots p-1} - \beta_{p+1p.1\ldots p-1}). \tag{6.18}$$

Hence, to obtain from (6.17) the distribution of

$$\boldsymbol{b}_{p+1p.1\ldots p-1} - \beta_{p+1p.1\ldots p-1},$$

we make the transformation

$$x = u^{1/2}(z - \beta_{p+1p.1\ldots p-1}),$$
$$y = u$$

which takes $-\infty < x < \infty$, $y > 0$ into $u > 0$, $-\infty < z < \infty$ and possesses the Jacobian $\sqrt{u}$. With the abbreviations

$$\beta_{p+1p.1\ldots p-1} = \varepsilon_p, \quad \sigma^2_{p.1\ldots p-1} = \tau^2_p, \quad \sigma^2_{p+1.1\ldots p} = \tau^2_{p+1}$$

one easily obtains (integrating over u) for the density of $\boldsymbol{b}_{p+1p.1\ldots p-1} - \varepsilon_p$:

$$\left.\begin{aligned} &\tau_p\tau^{-1}_{p+1}\Gamma((n-p+1)/2)\left[\Gamma((n-p)/2)(\Gamma(1/2)\right]^{-1} \\ &\times\left(1+\frac{\sigma^2_p(z-\varepsilon_p)^2}{\tau^2_{p+1}}\right)^{-\frac{n-p+1}{2}}, \qquad -\infty < z < \infty. \end{aligned}\right\} \tag{6.19}$$

The densities of $\boldsymbol{b}_{p+1i.1\ldots(i-1)(i+1)\ldots p}$, $1 \leqslant i \leqslant p-1$, are obtained similarly. We then get without difficulty

$$E(\boldsymbol{b}_{p+1i.1\ldots(i-1)(i+1)\ldots p} - \beta_{p+1i.1\ldots(i-1)(i+1)\ldots p}) = 0$$

$1 \leqslant i \leqslant p$, i.e., $\boldsymbol{b}_{p+1i.1\ldots(i-1)(i+1)\ldots p}$ is unbiased for $\beta_{p+1i.1\ldots(i-1)(i+1)\ldots p}$ in analogy to the result of Theorem 1.1.

One can only use the distribution given by (6.19) to test *a simple hypothesis on* $\beta_{p+1\,p.1\ldots p-1}$ when τ_p and τ_{p+1} are *given* positive numbers. It is thus of practical importance to be able to derive another distribution which allows one to test *an hypothesis on a regression coefficient* when—briefly speaking—the matrix D (or equivalently, *the covariance matrix*) in (6.1) *is unknown*. We will find a new application of the t-distribution. Limiting ourselves again for the sake of simplicity to the regression coefficient $\beta_{p+1\,p.1\ldots p-1}$, we shall obtain a similar test for the null hypothesis:

$$\{\beta_{p+1\,p.1\ldots p-1}=\gamma_0,\ \gamma_0 \text{ real, } D \text{ positive-definite, otherwise arbitrary}\}.$$

According to (6.16) and (6.18), $w^{1/2}_{pp.1\ldots p-1}(b_{p+1\,p.1\ldots p-1}-\beta_{p+1\,p.1\ldots p-1})$ is $N(0,\sigma^2_{p+1.1\ldots p})$-distributed and $w_{p+1\,p+1.1\ldots p}/\sigma^2_{p+1.1\ldots p}$ chi-square with $n-p-1$ degrees of freedom, independently of the former. Hence, if the null hypothesis is true,

$$\frac{w^{1/2}_{pp.1\ldots p-1}(b_{p+1\,p.1\ldots p-1}-\gamma_0)\,(n-p-1)^{1/2}}{w^{1/2}_{p+1\,p+1.1\ldots p}}$$

is t-distributed with $n-p-1$ degrees of freedom.

This also allows a *test of the hypothesis*

$$\rho_{p+1\,p.1\ldots p-1}=0 \tag{6.20}$$

by means of the *test statistic* $r_{p+1\,p.1\ldots p-1}$, even if the *covariance matrix* of the distribution defined by (6.1) *is unknown*. To show this note that (6.20) implies, according to (5.13), $\beta_{p+1\,p.1\ldots p-1}=0$. However, under this hypothesis

$$\frac{w^{1/2}_{pp.1\ldots p-1}\,b_{p+1\,p.1\ldots p-1}(n-p-1)^{1/2}}{w^{1/2}_{p+1\,p+1.1\ldots p}}$$

is—t-distributed with $n-p-1$ degrees of freedom.

On the other hand, an easy calculation shows that

$$\frac{w^{1/2}_{pp.1\ldots p-1}\,b_{p+1\,p.1\ldots p-1}}{w^{1/2}_{p+1\,p+1.1\ldots p}}=\frac{r_{p+1\,p.1\ldots p-1}}{\sqrt{1-r^2_{p+1\,p.1\ldots p-1}}}.$$

This method also allows the derivation of the distribution of the *multiple sample correlation coefficient* provided

$$\beta_{p+1\,i.1\ldots(i-1)(i+1)\ldots p}=0 \tag{6.21}$$

holds for $1\leqslant i\leqslant p$. In analogy to (5.10), this coefficient is given by

$$\hat{K}_{p+1}=(1-|w_{ij}|^{1\,p+1}_{1\,p+1}(w_{p+1\,p+1}|w_{ij}|^{1\,p}_{1\,p})^{-1})^{1/2}. \tag{6.22}$$

Furthermore, from (5.3) and (5.10), (6.21) for $1 \leqslant i \leqslant p$ also implies $K_{p+1}=0$. Later we will derive the distribution of $\hat{K}_{p+1}$ without the assumption (6.21).

We illustrate the results just obtained in the practically important case $p=1$.

We thus start with a two-dimensional normal distribution with density

$$(|d_{ij}|_{12}^{12})^{1/2}(2\pi)^{-1}e^{-(1/2)\sum_{i,j=1}^{2} y_i y_j d_{ij}}, \qquad (y_1, y_2) \in R_2 \tag{6.23}$$

where $(d_{ij})_{12}^{12}$ is assumed to be positive-definite. Let the covariance matrix $(\sigma_{ij})_{12}^{12}$ also be written as $\begin{pmatrix} \sigma_1^2 & \sigma_{12} \\ \sigma_{21} & \sigma_2^2 \end{pmatrix}$. Then

$$d_{11}=(\sigma_1^2(1-\rho_{12}^2))^{-1}, \tag{6.24}$$

$$d_{21}=d_{12}=-\rho_{12}/(\sigma_1\sigma_2(1-\rho_{12}^2)), \tag{6.25}$$

$$d_{22}=(\sigma_2^2(1-\rho_{12}^2))^{-1} \tag{6.26}$$

where ρ_{12} is the correlation coefficient defined on p. 52.

Hence, instead of (6.23) we can also write

$$(2\pi\sigma_1\sigma_2(1-\rho_{12}^2)^{1/2})^{-1}\exp\left(-\frac{1}{2}\left(\frac{y_1^2}{\sigma_1^2}-\frac{2\rho_{12}y_1y_2}{\sigma_1\sigma_2}+\frac{y_2^2}{\sigma_2^2}\right)(1-\rho_{12}^2)^{-1}\right). \tag{6.27}$$

From (6.27) there follows

Theorem 6.1. *If the joint distribution of the r.v.'s η_1 and η_2 is two-dimensional normal, then the vanishing of the correlation coefficient ρ_{12} implies the independence of η_1 and η_2.*

Let $\xi_1, \ldots, \xi_n$, $n \geqslant 3$ be two-dimensional independent r.v.'s all possessing the same distribution whose density in R_2 is given by (6.27). We again write $\xi_l=(\xi_{1l}, \xi_{2l})$. The density of (w_{11}, w_{12}, w_{22}) is then (provided it does not vanish) given in R_3 by

$$W_2(w_{ij}; n-1)=(|d_{ij}|_{12}^{12})^{(n-1)/2}(2^{n-1}\pi^{1/2}\Gamma((n-1)/2)\Gamma((n-2)/2))^{-1}$$
$$\times(|w_{ij}|_{12}^{12})^{(n-4)/2}e^{-1/2\sum_{i,j=1}^{2} d_{ij}w_{ij}}$$

with $w_{ij}=w_{ji}$.

The transformation equations (6.4) now have the form

$$\left.\begin{aligned} w_{22.1} &= w_{22}-w_{12}w_{21}/w_{11} \\ w_{2.1} &= w_{11}^{1/2}(b_{21}-\beta_{21}) \\ w_{11} &= w_{11} \end{aligned}\right\} \tag{6.28}$$

b_{21} is the sample regression coefficient given by $b_{21}=w_{21}/w_{11}$. The number $\beta_{21}=\sigma_{12}/\sigma_1^2$ solves the problem of minimizing $E[(\xi_{2l}-\delta\xi_{1l})^2]$ for $-\infty<\delta<\infty$ (which naturally does not depend on l with $1\leqslant l\leqslant n$).

Specializing (5.8) to the case $p=1$, we get

$$\sigma_{2.1}^2=\sigma_1^2\sigma_2^2(1-\rho_{12}^2)/\sigma_1^2=\sigma_2^2(1-\rho_{12}^2)$$

so that

$$|d_{ij}|_{12}^{12}=(\sigma_1^2\sigma_2^2(1-\rho_{12}^2))^{-1}=(\sigma_1^2\sigma_{2.1}^2)^{-1}$$

The symbolic relation

$$W_2(w_{ij};n-1)\,w_{11}^{1/2}=W_1(w_{11};n-1)\,W_1(w_{22.1};n-2)\,N_1(u_{2.1}) \tag{6.29}$$

to which (6.28) leads us, has the following detailed form:

$$\begin{aligned} W_2(w_{ij};n-1)\,w_{11}^{1/2}=&\Big((2^{(n-1)/2}\Gamma((n-1)/2)\sigma_1^2)^{-1}e^{-\frac{1}{2}w_{11}\sigma_1^2}(w_{11}/\sigma_1^2)^{\frac{n-1}{2}-1}\Big)\\ &\times\Big((2^{(n-2)/2}\Gamma((n-2)/2)\sigma_{2.1}^2)^{-1}e^{-\frac{1}{2}w_{22.1}/\sigma_{2.1}^2}(w_{22.1}/\sigma_{2.1}^2)^{\frac{n-2}{2}-1}\Big)\\ &\times\Big(((2\pi)^{\frac{1}{2}}\sigma_{2.1})^{-1}e^{-\frac{1}{2}u_{2.1}/\sigma_{2.1}^2}\Big). \end{aligned}$$

We bring together the results obtained on p. 395 and 396 for the distribution of the sample regression coefficient in the case $p=1$ in

Theorem 6.2[28]. *Let $\xi_1,\dots,\xi_n$ for $n\geqslant 3$ be independent, two-dimensional, identically distributed r.v.'s with density in R_2 given by (6.27). Then $\boldsymbol{b}_{21}-\beta_{21}$ has a distribution with density*

$$\sigma_1(\sigma_{2.1})^{-1}\Gamma(n/2)\,(\Gamma((n-1)/2)\,\Gamma(1/2))^{-1}(1+\sigma_1^2(z-\beta_{21})^2/\sigma_{2.1}^2)^{-n/2}\,,$$
$$-\infty<z<\infty\,.$$

Moreover, the r.v.

$$\frac{\mathbf{w}_{11}^{1/2}(\boldsymbol{b}_{21}-\beta_{21})}{\mathbf{w}_{22.1}^{1/2}}(n-2)^{1/2}$$

is t-distributed with $n-2$ degrees of freedom.

Naturally, these theorems still hold if one assumes that the distribution of ξ_l, $1\leqslant l\leqslant n$, has an arbitrary mean vector a. Hence, for level of significance α, one gets for *the null hypothesis*

$$\Big\{-\infty<a<\infty,\ 0<\sigma_1^2<\infty,\ 0<\sigma_2^2<\infty,\ -\infty<\sigma_{12}<\infty,$$
$$\sigma_1^2\sigma_2^2-\sigma_{12}>0,\ \beta_{21}=\frac{\sigma_{12}}{\sigma_1^2}=\gamma_0, \gamma_0 \text{ fixed}\Big\}$$

[28] First found by V. Romanovskij, Bull. Acad. Sci. Leningrad (6) 20, 643—648 (1926) and K. Pearson, Proc. Roy. Soc. London Ser. A, 112, 1—14 (1926).

a *similar critical region* in R_{2n} of the form

$$\{(x_{1l}, x_{2l}),\ 1 \leqslant l \leqslant n : |w_{11}^{1/2}(b_{21}-\gamma_0)(n-2)^{1/2}/w_{22.1}| \geqslant \tau_\alpha\},$$

where τ_α is given in the notation of I. (29.3) by

$$1 - \int_{-\tau_\alpha}^{\tau_\alpha} h_{n-2}(t)\,dt = \alpha\,. \tag{6.30}$$

The *sample correlation coefficient* is now (for $p=1$) defined by

$$r_{21} = w_{21}/(w_{11}w_{22})^{1/2} \quad \text{and we have}$$

Theorem 6.3. *Under the hypotheses of Theorem 6.2, the r.v.* $r_{12}(n-2)^{1/2}(1-r_{12}^2)^{-1/2}$ *is t-distributed with* $n-2$ *degrees of freedom when* $\rho_{12}=0$.

Combining Theorem 6.1 with Theorem 6.3, we obtain a *test for the independence of two normally distributed* (one-dimensional) *populations*. More precisely, $\{x_{1l}, x_{2l}),\ 1 \leqslant l \leqslant n : |r_{12}(n-2)^{1/2}(1-r_{12}^2)^{-1/2}| \geqslant \tau_\alpha\}$ provides a *similar critical region* with level of significance α for the *null hypothesis*

$$-\infty < a < \infty\,, \qquad 0 < \sigma_i^2 < \infty\,, \qquad i=1,2\,, \qquad \sigma_{12}/(\sigma_1\sigma_2) = \rho_{12} = 0\,.$$

Here, τ_α is determined by (6.30).

We now turn, as announced, to the distribution of the multiple sample correlation coefficient. We first prove a theorem which is also of independent interest.

Theorem 6.4. *Let ξ be a k-dimensional normally distributed r.v. with density in* R_k

$$|D|^{1/2}(2\pi)^{-k/2}e^{-(x-a)'D(x-a)/2} \tag{6.31}$$

where $a \neq 0$ *and D is positive-definite. Then the r.v.* $\xi' D \xi$ *is non-central* χ^2*-distributed with k degrees of freedom and parameter* $a' D a$.

Proof. Combining the transformation III. (11.6) and III. (11.8) with the notation changed one sees that a transformation $x = Bz$ can be so chosen that $x' D x$ is taken into $z' z$. Moreover, B^{-1} exists. Then

$$B' D B = \begin{pmatrix} 1 & 0 \ldots 0 \\ 0 & 1 \ldots 0 \\ \ldots & \ldots \\ 0 & 0 \ldots 1 \end{pmatrix}.$$

We now define a k-dimensional r.v. ζ by $\zeta = B^{-1}\xi$. Then

$$E(\zeta) = B^{-1}a\,. \tag{6.32}$$

For the density of ζ one gets from (6.31) for each $z \in R_k$

$$(2\pi)^{-k/2} \exp\left(-\tfrac{1}{2}(z-E(\zeta))'(z-E(\zeta))\right).$$

Thus, writting $\zeta=(\zeta_1, \ldots, \zeta_k)$, one sees that the r.v.'s ζ_i, $1 \leqslant i \leqslant k$, are independent. But $\zeta'\zeta$ and $\xi' D \xi$ have the same distribution. According to III, p. 163 ff. and (6.32), this is a non-central χ^2-distribution with k degrees of freedom and parameter $(B^{-1}a)'(B^{-1}a)=a'Da$, which was to be proved.

We now arrive at the actual goal of our considerations. We again use the notation introduced on p. 389 and 396 and prove

Theorem 6.5. *Let $\xi_1, \ldots, \xi_n$ be $n \geqslant p+2 \geqslant 3$, $(p+1)$-dimensional, independent r.v.'s, each possessing the same normal distribution with density for $x \in R_{p+1}$ given by*

$$|D|^{1/2}(2\pi)^{-(p+1)/2} e^{-(x-a)'D(x-a)/2}$$

where $D=(d_{ij})_{1\,p+1}^{1\,p+1}$ is positive-definite. Then the density of the distribution of $\hat{K}_{p+1}^2$ is given by

$$\left.\begin{array}{ll} C(n,p)\, u^{\frac{p}{2}-1}(1-u)^{(n-p-3)/2} \displaystyle\sum_{l=0}^{\infty} (u\,c(p))^l \frac{\left[\left((l-1)+\frac{n-1}{2}\right)\cdots\left(\frac{n-1}{2}\right)\right]^2}{l!\left[\left(\frac{p}{2}+(l-1)\right)\cdots\frac{p}{2}\right]} & \\ \qquad\qquad 0<u<1 & \\ 0 \qquad\qquad \textit{otherwise}. & \end{array}\right\} \tag{6.33}$$

Here,

$$C(n,p)=\Gamma((n-1)/2)\left(\Gamma(p/2)\Gamma((n-p-1)/2)\right)^{-1}\left(\frac{\sigma_{p+1.1\ldots p}^2}{\sigma_{p+1\,p+1}}\right)^{(n-1)/2}$$

and

$$c(p)=(\sigma_{p+1\,p+1}-\sigma_{p+1.1\ldots p})/\sigma_{p+1\,p+1},$$

where $(\sigma_{ij})_{1\,p+1}^{1\,p+1}$ denotes the covariance matrix D^{-1}.

Proof. We use the method we used to prove Theorem 3.5. Thus, we view $\hat{K}_{p+1}^2$ initially only as function of the r.v.'s $\xi_{p+1\,1}, \ldots, \xi_{p+1\,n}$, where

$$\xi_l=(\xi_{1\,l}, \ldots, \xi_{p+1\,l}), \quad 1 \leqslant l \leqslant n.$$

The remaining r.v.s' are replaced by the real numbers x_{ij}, $1 \leqslant i \leqslant p$, $1 \leqslant j \leqslant n$. The distribution of $\hat{K}_{p+1}^2$ [29] is then considered as conditional distribution under the hypothesis $\xi_{ij}=x_{ij}$, $1 \leqslant i \leqslant p$, $1 \leqslant j \leqslant n$.

[29] It is not entirely correct to use the symbol $\hat{K}_{p+1}^2$ here, but this should cause no confusion.

We use the notation of **1** and first rewrite the expression

$$(1-|w_{ij}|_{1\,p+1}^{1\,p+1}/(w_{p+1\,p+1}|w_{ij}|_{1\,p}^{1\,p})).$$

From (2.17) and (2.18) we have, in somewhat more detail,

$$w_{p+1\,p+1}=\sum_{l=1}^{n}\left(x_{p+1\,l}-\hat{b}_0-\sum_{j=1}^{p}\hat{b}_j x_{jl}\right)^2+\sum_{i,j=1}^{p} w_{ij}\hat{b}_i\hat{b}_j.$$

In the same notation, (2.20) looks like

$$\sum_{l=1}^{n}\left(x_{p+1\,l}-\hat{b}_0-\sum_{j=1}^{p}\hat{b}_j x_{jl}\right)^2=|w_{ij}|_{1\,p+1}^{1\,p+1}(|w_{ij}|_{1\,p}^{1\,p})^{-1}.$$

Then, with an easy calculation we get

$$\hat{K}_{p+1}^2=\sum_{i,j=1}^{p} w_{ij}\hat{b}_i\hat{b}_j\left(\sum_{l=1}^{n}\left(x_{p+1\,i}-\hat{b}_0-\sum_{j=1}^{p}\hat{b}_j x_{ji}\right)^2+\sum_{i,j=1}^{p} w_{ij}\hat{b}_i\hat{b}_j\right)^{-1}. \tag{6.34}$$

We now determine the joint distribution of $\hat{\boldsymbol{b}}_1,\ldots,\hat{\boldsymbol{b}}_p$. We repeat that the $\hat{\boldsymbol{b}}_i$ are defined as in **1** and are, briefly speaking, functions only of the r.v.'s $\xi_{p+1\,l}$, $1\leqslant l\leqslant n$. Provided $|w_{ij}|_{1\,p}^{1\,p}>0$, the density of the distribution of the p-dimensional r.v. $(\hat{\boldsymbol{b}}_1,\ldots,\hat{\boldsymbol{b}}_p)$ is given for all $(z_1,\ldots,z_p)\in R_p$ by

$$(2\pi)^{-p/2}\frac{(|w_{ij}|_{1\,p}^{1\,p})^{1/2}}{\sigma}\,e^{-\frac{1}{2\sigma^2}\sum_{i,j=1}^{p}(z_i-\beta_i)(z_j-\beta_j)w_{ij}}. \tag{6.35}$$

Here, σ^2 and β_j, $1\leqslant j\leqslant p$, are defined by (1.13). This result follows from the covariance part of Theorem 1.1 in the special form given on p. 357, an application of I, Theorem 36.2, and the remark preceding (1.13).

From Theorem 5.1 and the remark on the regression function of a normal distribution which follows it, we find that the β_j are regression coefficients and are also given by the right side of (5.3). Thus, we also have

$$\sigma^2=|\sigma_{ij}|_{1\,p+1}^{1\,p+1}/|\sigma_{ij}|_{1\,p}^{1\,p}. \tag{6.36}$$

On p. 361, is was shown that $\frac{1}{\sigma^2}\sum_{l=1}^{n}\left(\xi_{p+1\,l}-\hat{\boldsymbol{b}}_0-\sum_{j=1}^{p}\hat{\boldsymbol{b}}_j x_{jl}\right)^2$ is χ^2-distributed with $n-p-1$ degrees of freedom independently of the $(\hat{\boldsymbol{b}}_1,\ldots,\hat{\boldsymbol{b}}_p)$. Since the density of $(\hat{\boldsymbol{b}}_1,\ldots,\hat{\boldsymbol{b}}_p)$ in R_p is given by (6.35), an application of Theorem 6.4 shows that $\sum_{i,j=1}^{p} w_{ij}\hat{\boldsymbol{b}}_i\hat{\boldsymbol{b}}_j$ is non-central χ^2-distributed with p degrees of freedom and parameter $\frac{1}{\sigma^2}\sum_{i,j=1}^{p} w_{ij}\beta_i\beta_j$, and, indeed, independently of $\sum_{l=1}^{n}\left(\xi_{p+1\,l}-\hat{\boldsymbol{b}}_0-\sum_{j=1}^{p}\hat{\boldsymbol{b}}_j x_{jl}\right)^2$. For the density of the joint distribution of these two r.v.'s we thus have

$$K(n,p)e^{-\lambda/2}e^{-\frac{1}{2}y/\sigma^2}(y/\sigma^2)^{\frac{n-p-1}{2}-1}e^{-x/2\sigma^2}(x/\sigma^2)^{(p-2)/2}\int_0^\pi e^{\frac{1}{\sigma}\sqrt{x\lambda}\cos\vartheta}\sin^{p-2}\vartheta\,d\vartheta$$

$$x>0, \quad y>0 \tag{6.37}$$

0 otherwise

with

$$K(n,p)=\sigma^{-4}2^{-(n-1)/2}(\Gamma((n-p-1)/2))^{-1}\pi^{-1/2}(\Gamma((p-1)/2))^{-1}$$

and

$$\lambda=\sum_{i,j=1}^{p}\beta_i\beta_j w_{ij}/\sigma^2. \tag{6.38}$$

We now reformulate (6.37) somewhat. In particular,

$$\int_0^\pi e^{\frac{1}{\sigma}\sqrt{x\lambda}\cos\vartheta}\sin^{p-2}\vartheta\,d\vartheta=\int_0^\pi\sum_{k=0}^\infty\frac{1}{k!}\left(\sqrt{x}\frac{\sqrt{\lambda}}{\sigma}\cos\vartheta\right)^k\sin^{p-2}\vartheta\,d\vartheta,$$

so that

$$\int_0^\pi e^{\frac{1}{\sigma}\sqrt{x\lambda}\cos\vartheta}\sin^{p-2}\vartheta\,d\vartheta=\sum_{l=0}^\infty\left(\frac{x\lambda}{\sigma^2}\right)^l\frac{\Gamma((p-1)/2)\Gamma((2l+1)/2)}{(2l)!\,\Gamma((p+2l)/2)}. \tag{6.39}$$

The term-by-term integration is easily justified by the uniform convergence of the series expansion (for fixed x, σ^2 and λ). Taking (6.34) into consideration, we obtain the conditional distribution of $\hat{K}^2_{p+1}$ from (6.37) by means of the transformation $u=x/(x+y)$, $v=y/\sigma^2$, which takes $x>0, y>0$ into $0<u<1, 0<v<\infty$. The Jacobian is $v\sigma^4(1-u)^{-2}$. Using (6.39) and integrating over v we get with an easy calculation for the density of the sought-for distribution (provided it does not vanish)

$$e^{-\lambda/2}(\Gamma(1/2)\Gamma((n-p-1)/2)^{-1}u^{\frac{p}{2}-1}(1-u)^{(n-p-3)/2}\sum_{l=0}^\infty(2u\lambda)^l$$

$$\times\frac{\Gamma(l+1/2)\Gamma(n+2l-1)/2)}{\Gamma(2l+1)\Gamma(l+p/2)}, \qquad 0<u<1.$$

Now

$$\Gamma\left(l+\frac{n-1}{2}\right)\Gamma(l+1/2)(\Gamma(2l+1)\Gamma(l+p/2))^{-1}$$

$$=\frac{\left(l-1+\frac{n-1}{2}\right)\cdot\ldots\cdot((n-1)/2)\Gamma((n-1)/2)(l-1/2)\cdot\ldots\cdot 1/2\,\Gamma(1/2)}{l!\,2^{2l}(l-1/2)\cdot\ldots\cdot 1/2\left(\frac{p}{2}+l-1\right)\cdot\ldots\cdot\frac{p}{2}\Gamma\left(\frac{p}{2}\right)}$$

so that the density is

$$\left.\begin{array}{ll}\dfrac{\Gamma((n-1)/2)}{\Gamma(p/2)\Gamma((n-p-1)/2)}e^{-\lambda/2}u^{\frac{p}{2}-1}(1-u)^{(n-p-3)/2} & \\ \times\displaystyle\sum_{l=0}^{\infty}\left(\frac{\lambda u}{2}\right)^{l}\frac{\left(l-1+\frac{n-1}{2}\right)\cdot\ldots\cdot\left(\frac{n-1}{2}\right)}{\left(\frac{p}{2}+l-1\right)\cdot\ldots\cdot\frac{p}{2}\,l!}, & 0<u<1 \\ 0 & \text{otherwise.}\end{array}\right\} \quad (6.40)$$

Following the plan for our proof we now view (6.40) as the density of the conditional distribution of $\hat{\boldsymbol{K}}_{p+1}^2$ under the hypothesis $\xi_{il}=x_{il}$, $1\leqslant i\leqslant p$, $1\leqslant l\leqslant n$. Note that we should have written more precisely

$$\lambda(x_{11},\ldots,x_{p1},\ldots,x_{1n},\ldots,x_{pn}) = \sum_{i,j=1}^{p}\beta_i\beta_j w_{ij}(x_{11},\ldots,x_{p1},\ldots,x_{1n},\ldots,x_{pn})/\sigma^2$$

instead of (6.38). In (6.40) the x_{il}, $1\leqslant i\leqslant p$, $1\leqslant l\leqslant n$ appear only in λ. It is thus sufficient to consider the joint distribution of $\hat{\boldsymbol{K}}_{p+1}^2$ and $\sum_{i,j=1}^{p}\beta_i\beta_j\boldsymbol{w}_{ij}$, where $\hat{\boldsymbol{K}}_{p+1}^2$ is now a function of all the r.v.'s ξ_{kl}, $1\leqslant k\leqslant p+1$, $1\leqslant l\leqslant n$, and the $\boldsymbol{w}_{ij}$ for $1\leqslant i,j\leqslant p$ are functions of the r.v.'s ξ_{kl}, $1\leqslant k\leqslant p$, $1\leqslant l\leqslant n$. The density of the sought-for joint distribution is obtained by multiplying (6.40) by the density of the distribution of $\sum_{i,j=1}^{p}\beta_i\beta_j\boldsymbol{w}_{ij}$. This density can be derived as follows: For $1\leqslant l\leqslant n$, the r.v.'s $\sum_{i=1}^{p}\beta_i(\xi_{il}-a_i)$ are independent and normally distributed. Moreover,

$$E\left[\left(\sum_{i=1}^{p}\beta_i(\xi_{il}-a_i)\right)^2\right]=\sum_{i,j=1}^{p}\beta_i\beta_j\sigma_{ij}.$$

Since we have seen that the β_j, $1\leqslant j\leqslant p$, satisfy (5.2), we get $\sum_{i,j=1}^{p}\beta_i\beta_j\sigma_{ij}=\sum_{i=1}^{p}\beta_i\sigma_{ip+1}$ so that finally, by (5.3) and (5.8)

$$\sigma_{p+1\,p+1}-\sigma_{p+1\,.\,1\ldots p}=\sum_{i,j=1}^{p}\beta_i\beta_j\sigma_{ij}. \quad (6.41)$$

Thus, writing $\eta_l=\sum_{i=1}^{p}\beta_i(\xi_{il}-a_i)$, $1\leqslant l\leqslant n$, we have $\sum_{i,j=1}^{p}\beta_i\beta_j\boldsymbol{w}_{ij} =\sum_{l=1}^{n}(\eta_l-\bar{\eta})^2$, and by II, Corollary to Theorem 4.3

$$\sum_{l=1}^{n}(\eta_l-\bar{\eta})^2/(\sigma_{p+1\,p+1}-\sigma_{p+1\,.\,1\ldots p}^2)$$

is χ^2-distributed with $n-1$ degrees of freedom. The density of $\frac{1}{\sigma^2}\sum_{i,j=1}^{p}\beta_i\beta_j w_{ij}$ is thus

$$\sigma^2(\sigma_{p+1p+1}-\sigma^2_{p+1.1\ldots p})^{(n-1)/2}(\Gamma((n-1)/2))^{-1}$$
$$\times\left(\frac{\lambda\sigma^2}{\sigma_{p+1p+1}-\sigma_{p+1.1\ldots p}}\right)^{\frac{n-1}{2}-1}$$
$$\times\exp\left(-\tfrac{1}{2}\lambda\sigma^2/(\sigma_{p+1p+1}-\sigma^2_{p+1.1\ldots p})\right), \qquad \lambda>0$$
$$0 \qquad \lambda<0.$$

Multiply this by (6.40) and integrate w.r.t. λ from 0 to ∞. Term-by-term integration is allowed by absolute convergence of the integrated series and we find (6.33) after simple calculations using (6.36).[30]

We now make some remarks. From (5.10) we easily get

$$\frac{\sigma_{p+1p+1}-\sigma^2_{p+1.1\ldots p}}{\sigma_{p+1p+1}} = K^2_{p+1}.$$

We denote the regression coefficients β_i by $\beta_{p+1i.1\ldots(i-1)(i+1)\ldots p}$, $1\leqslant i\leqslant p$, as before. Because of (6.41), K^2_{p+1} vanishes iff all $\beta_{p+1i.1\ldots(i-1)(i+1)\ldots p}$ vanish. In this case, however, the distribution of $\hat{\boldsymbol{K}}^2_{p+1}$ is $B(p/2,(n-p-1)/2)$, provided the assumptions of Theorem 6.4 remain otherwise unchanged. This follows easily from (6.33). This allows the definition of *a test for the hypothesis* $\beta_{p+1i.1\ldots(i-1)(i+1)\ldots p}=0$, $1\leqslant i\leqslant p$, which does *not assume the knowledge of any other parameters of the distribution* given by (6.31). We thus have a *similar test*.

One easily finds that the *conditional distribution of* $\hat{\boldsymbol{K}}^2_{p+1}$ *under the hypothesis* $\xi_{il}=x_{il}$, $1\leqslant i\leqslant p$, $1\leqslant l\leqslant n$, and assuming that $K_{p+1}=0$, *is likewise given by* $B(p/2,(n-p-1)/2)$ no matter how the real numbers x_{il} are chosen. A similar result has already been encountered on p. 376 ff.[31]

We specialize (6.33) to the case $p=1$. The sample correlation coefficient and the multiple sample correlation coefficients coincide in this case. We then get[32] for the density of $\boldsymbol{r}^2_{21}=\boldsymbol{w}^2_{12}(\boldsymbol{w}_{11}\boldsymbol{w}_{22})^{-1}$, provided

[30] This distribution was first found by R.A. Fisher, Proc. Roy. Soc. London Ser. A, 121, 654—673 (1928).

[31] R.A. Fisher pointed this out in this and other connections. See R.A. Fisher, Metron 3, 90—104 (1925).

[32] R.A. Fisher, Biometrika 10, 507—521 (1915).

it doesn't vanish:

$$\frac{\Gamma((n-1)/2)}{\pi^{1/2}\Gamma((n-2)/2)}(1-\rho_{12}^2)^{(n-1)/2}u^{-1/2}(1-u)^{(n-4)/2}\sum_{l=0}^{\infty}(u\rho_{12}^2)^l \times\frac{\left[\left(l-1+\frac{n-1}{2}\right)\cdots\left(\frac{n-1}{2}\right)\right]^2}{l!(l-1/2)\dots 1/2}, \qquad 0<u<1.$$

We conclude with a general *remark*. For the derivation of most of the distributions connected with the multi-dimensional normal distribution it is essential that the density of $n\geqslant 1$ normal, identically distributed, independent r.v.'s is invariant w.r.t. orthogonal transformations. (See I, Theorem 27.2.) One can thus expect that the properties of the *group of orthogonal transformations* will play an important role in the derivation of the distributions referred to. This idea has been systematically exploited in several papers which we will not go into here[33].

[33] A.T. James, Ann. Math. Statist. 25, 40—75 (1954); A.G. Constantine and A.T. James, Ann. Math. Statist. 29, 1146—1166 (1958); A.T. James, Ann. Math. Statist. 31, 151—158 (1960) and A.T. James, Ann. Math. Statist. 32, 874—882 (1961), and others.

Chapter VII

Introduction to Non-parametric Theories

1. Order statistics. In the previous chapters we have frequently made use of the assumption that the set Γ of parameters is a (open or closed) subset of R_n, $n \geqslant 1$. In addition, in connection with essential results, we have imposed continuity and differentiability requirements on the likelihood function. For some time, so-called non-parametric methods have received considerable attention. Their beginnings, however, lie far in the past. Recent progress is due to Anglo-american, Dutch and Soviet statisticians[1]. The term "non-parametric" is rather unfortunately chosen and it is not easy to give a satisfactory definition of this notion. Roughly speaking, one can call a test or method of estimation non-parametric when no assumptions such as those above are made on Γ[2].

We now introduce the idea of an *order statistic*. Order statistics are important not only for non-parametric methods, but for all of mathematical statistics. Let $(x_1, \ldots, x_n)$ be an arbitrary n-tuple of real numbers, $n \geqslant 1$ (ordered according to their indices). For $j = 1, \ldots, n$, denote by Z_j any choice $(x_{i_1}, \ldots, x_{i_j})$ of j elements from this n-tuple ordered, say, by their indices. It can happen that for certain l with $1 \leqslant l \leqslant j$, x_{i_l} is the same real number. However, the x_{i_l} will be differentiated by means of their indices. In this sense, there are $\binom{n}{j}$ different Z_j which we denote by

---

[1] See the report of D. van Dantzig and J. Hemelrijk, Bull. Inst. Internat. Statist. 34, 239—267 (1954). Another report is in L. Schmetterer, Jber. Deutsch. Math.-Verein 61, 104—126 (1959). A good cross-section is contained in D.A.S. Fraser, Nonparametric Methods in Statistics, John Wiley & Sons-Chapman & Hall, New York-London 1957. A more elementary text is G.E. Noether, Elements of Nonparametric Statistics, John Wiley & Sons, New York-London-Sydney 1967. The most important reference is Hájek J. and Šidák Z., Theory of Rank Tests. Academia Publishing House of the Czechoslovak Academy of Sciences, Prague 1967. See also: Nonparametric Techniques in Statistical Inference. Edited by M.L. Puri. Cambridge at the University Press 1970.

[2] See E. Ruist, Ark. Mat. 3, 133—163 (1955). A detailed analysis of the notion "non-parametric" is in M.G. Kendall and R.M. Sundrum, Rev. Inst. Internat. Statist. 21, 124—134 (1953). Also see III, p. 199.

$Z_j^{(i)}$, $1 \leqslant i \leqslant \binom{n}{j}$. Now define the following functions:

For each $x=(x_1, \ldots, x_n) \in R_n$, let

$$z_1(x) = \min Z_n^{(1)}$$

$$\ldots\ldots\ldots\ldots\ldots\ldots\ldots\ldots\ldots\ldots$$

$$z_j(x) = \max\left(\min Z_{n-j+1}^{(1)}, \ldots, \min Z_{n-j+1}^{\left(\binom{n}{j-1}\right)}\right)$$

$$\ldots\ldots\ldots\ldots\ldots\ldots\ldots\ldots\ldots\ldots$$

$$z_n(x) = \max(\min Z_1^{(1)}, \ldots, \min Z_1^{(n)}).$$

For $2 \leqslant j \leqslant n$, let $Z_j^{(i)}$ be chosen arbitrarily. Denote by $Z_{j-1}^{(i)}$ an arbitrary sub-sequence of $Z_j^{(i)}$ with $j-1$ elements. Then $\min Z_j^{(i)} \leqslant \min Z_{j-1}^{(i)}$, whence for each $x \in R_n$

$$z_1(x) \leqslant z_2(x) \leqslant \cdots \leqslant z_n(x). \tag{1.1}$$

Further one can associate a permutation $(r_1(x), \ldots, r_n(x))$ of $(1, \ldots, n)$ with every $x \in R_n$ which has the following property:

$$z_{r_j}(x) = x_j, \quad 1 \leqslant j \leqslant n. \tag{1.2}$$

If only $<$-signs hold in (1.1), then the permutation $(r_1(x), \ldots, r_n(x))$ is uniquely determined. The $r_i(x)$, $1 \leqslant i \leqslant n$ are called *ranks* of x. The map $x \to (r_1(x), \ldots, r_n(x))$ is not uniquely determined for all $x \in R_n$. However, one can also extend the definition to those x for which at least one equality sign holds in (1.1). To this end, one chooses for such x any one of the permutations of x defined by (1.2). Now let ξ be an n-dimensional r.v. One calls the r.v. $z_i \circ \xi$, $1 \leqslant i \leqslant n$, an *order statistic*. This terminology appears justified by (1.1). Thus, if x is a realization of the r.v. ξ and one orders the sample values according to increasing magnitude, then one gets a realization of $z_1 \circ \xi, \ldots, z_n \circ \xi$. Let S_n be the group of permutations of n elements and Π_n an element of S_n which takes $(1, \ldots, n)$ into, say, $(i_1, \ldots, i_n)$. Then, for each $x \in R_n$ we understand by $\Pi_n x$ the element $(x_{i_1}, \ldots, x_{i_n})$. We have the following simple

Lemma 1.1. *Let F be a d.f. defined over R_n which is invariant w.r.t. the group S_n. Let $M \in \mathfrak{B}_n$ and $\Pi_n M = \{\Pi_n x : x \in M\}$. Then also $\Pi_n M \in \mathfrak{B}_n$ for each $\Pi_n \in S_n$. Moreover, the probability measure P_F corresponding to F is invariant*[3] *w.r.t. the group S_n, i.e., $P_F(M) = P_F(\Pi_n M)$.*

Proof. By assumption, we have for each $x \in R_n$ and each $\Pi_n \in S_n$

$$F(x) = F(\Pi_n x). \tag{1.3}$$

[3] See III, p. 235. Lemma 1.1 can be viewed as an illustration of the arguments carried through there.

First let

$$M=(a_1,b_1]\times\cdots\times(a_n,b_n]\,. \tag{1.4}$$

Denoting the components of $\Pi_n x$ by $(\Pi_n x)_i$, $1\leqslant i\leqslant n$, we get easily

$$\Pi_n M=((\Pi_n a)_1,(\Pi_n b)_1]\times\cdots\times((\Pi_n a)_n,(\Pi_n b)_n]\,.$$

But

$$P_F(M)=\Delta_1\ldots\Delta_n F(a) \tag{1.5}$$

and

$$P_F(\Pi_n M)=\Delta_1\ldots\Delta_n F(\Pi_n a)\,. \tag{1.6}$$

Here, in the notation of I. (9.2), we must choose $h_i=b_i-a_i$ in (1.5), and in (1.6) $h_i=(\Pi_n b)_i-(\Pi_n a)_i$, $1\leqslant i\leqslant n$. From (1.3) the claimed invariance follows for all sets of the form (1.4). Now let

$$\mathfrak{M}=\{M:M\in\mathfrak{B}_n,\,\Pi_n M\in\mathfrak{B}_n \text{ for all } \Pi_n\in S_n\}\,.$$

$\mathfrak{M}$ contains all n-dimensional intervals of the form (1.4) and also R_n. Let $M_i\in\mathfrak{M}$, $i=1,2,\ldots$, and $M_i\cap M_j=\emptyset$, $i\neq j$. Then also $\bigcup_{i=1}^{\infty} M_i\in\mathfrak{M}$. Indeed, from $M_i\cap M_j=\emptyset$ we also have $\Pi_n M_i\cap\Pi_n M_j=\emptyset$, $i\neq j$. Further $\bigcup_{i=1}^{\infty}\Pi_n M_i=\Pi_n\bigcup_{i=1}^{\infty}M_i$, so that $\Pi_n\bigcup_{i=1}^{\infty}M_i\in\mathfrak{B}_n$.

Moreover, if $M,N\in\mathfrak{M}$, then $M-N\in\mathfrak{M}$ which is easy to see. This shows that $\mathfrak{M}$ is a σ-algebra which contains all intervals. Hence, $\mathfrak{M}=\mathfrak{B}_n$. Also, the map $M\to P_F(\Pi_n M)$ is for each $\Pi_n\in S_n$ a probability measure over $(R_n,\mathfrak{B}_n)$ which coincides with P_F on all intervals of the form (1.4). Theorem III then yields the claimed invariance.

For further developments we introduce some useful notation. Let $\mathfrak{F}$ be the set of all, C the set of all continuous, and C_m the set of all strictly monotone increasing d.f.'s. Let $\mathfrak{F}_{(1/2)}$ be the set of all d.f.'s with median 0 and $\mathfrak{F}_s$ the set of all d.f.'s symmetric w.r.t. 0. Further let ${}_*\mathfrak{F}$ be the set of all probability distributions over $(R_1,\mathfrak{B}_1)$ and ${}_*C$ be the subset of all probability distributions with continuous d.f., etc. Let $\mathfrak{F}^{(nn)}$ be the set of all n-dimensional d.f.'s of the form $(x_1,x_2,\ldots,x_n)\to F(x_1)F(x_2)\ldots F(x_n)$ with $F\in\mathfrak{F}$ and ${}_*\mathfrak{F}^{(nn)}$ the set of all n-fold products of a probability measure from ${}_*\mathfrak{F}$ with itself, $n\geqslant 2$, $C^{(nn)}$ will have an analogous meaning, etc. On the other hand, $\prod_{i=1}^{n}{}_*\mathfrak{F}^{(i)}$ stands for the set of all n-dimensional products of probability measures over $(R_1,\mathfrak{B}_1)$ and $\prod_{i=1}^{n}\mathfrak{F}^{(i)}$ for the set of corresponding distribution functions. The meaning of $\prod_{i=1}^{n}C^{(i)}$ and similar symbols is evident.

We prove

Theorem 1.1. *Let* $\xi_1,\ldots,\xi_n, n\geqslant 1$, *be independent, identically distributed r.v.'s with distribution function* $F\in C$. *Then the d.f. of the joint distribution of* $\xi^{(n)}=(\xi_1,\ldots,\xi_n)$ *is invariant w.r.t. the group* S_n. *In addition, the n-dimensional r.v.* $r\circ\xi^{(n)}=(r_1\circ\xi^{(n)},\ldots,r_n\circ\xi^{(n)})$ *possesses a discrete uniform distribution. The definition of the ranks for elements of* R_n *in the case where they are not uniquely determined can be made arbitrarily.*

Proof. The d.f. of $\xi^{(n)}$ is given by $F^{(n)}(x)=\prod_{i=1}^{n} F(x_i)$ for each $x\in R_n$. For each permutation $(i_1,\ldots,i_n)$ of $(1,\ldots,n)$ we have $\prod_{i=1}^{n} F(x_i)=\prod_{j=1}^{n} F(x_{i_j})$, so that $F^{(n)}(x)=F^{(n)}(\Pi_n x)$. For the proof of the second claim we introduce the notation $r\circ\xi^{(n)}=\boldsymbol{r}^{(n)}$ and $r_i\circ\xi^{(n)}=\boldsymbol{r}_i$ which we will also use in the sequel. We must show that $P(\boldsymbol{r}_1=i_1,\ldots,\boldsymbol{r}_n=i_n)=1/n!$ for all permutations $(i_1,\ldots,i_n)$. The ranks are not uniquely determined for exactly those $x\in R_n$ for which there exists a subset $\{j_1,\ldots,j_r\}$ of $\{1,\ldots,n\}$, $2\leqslant r\leqslant n$, with $x_{j_1}=x_{j_2}=\cdots=x_{j_r}$. Now

$$P(\xi_{j_1}=\xi_{j_2}=\cdots=\xi_{j_r})\leqslant P(\xi_{j_1}=\xi_{j_2}). \tag{1.7}$$

But $P(\xi_{j_1}-\xi_{j_2}=0)=0$ for arbitrary $j_1\neq j_2$ since the d.f. of $\xi_{j_1}-\xi_{j_2}$ is continuous, which easily follows from I, Theorem 24.2. Hence, because of (1.7), the union N_n of all sets in R_n for which the ranks are not uniquely determined is a $P_{\xi^{(n)}}$-null set. Denoting by $(j_1,\ldots,j_n)$ the permutation inverse to $(i_1,\ldots,i_n)$, we have

$$P(\boldsymbol{r}_1=i_1,\ldots,\boldsymbol{r}_n=i_n)=\underset{-\infty<x_{j_1}<x_{j_2}<\cdots<x_{j_n}<\infty}{\int\cdots\int} dF^{(n)}(x).$$

Using Lemma 1.1 with $M=\{x\colon -\infty<x_{j_1}<\cdots<x_{j_n}<\infty\}$ we get

$$1=\sum_{\Pi_n\in S_n}\int_{\Pi_n M} dF^{(n)}(x)=n!\,P_{\xi^{(n)}}(M).$$

This completes the proof.

It is advantageous to consider also, in addition to the ranks, the functions which one obtains by going over to the inverse permutations: For each $x\in R_n$ we define the *conjugate ranks* $s_j(x)$, $1\leqslant j\leqslant n$, by $z_j(x)=x_{s_j(x)}$. With respect to the definition of the map $x\to s_j(x)$, there holds a remark similar to that for the ranks. If $\xi^{(n)}$ is an n-dimensional r.v., we again write $\boldsymbol{s}_j$ instead of $s_j\circ\xi^{(n)}$, $1\leqslant j\leqslant n$. For an arbitrary distribution of $\xi^{(n)}$ the n-dimensional r.v.'s $(\boldsymbol{r}_1,\ldots,\boldsymbol{r}_n)$ and $(\boldsymbol{s}_1,\ldots,\boldsymbol{s}_n)$ obviously have the same distribution, provided one defines the ranks and conjugate ranks suitably in N_n. Let n_i, $i=1,2$, be natural numbers with

$$n_1+n_2=n. \tag{1.8}$$

For each $x \in R_n$ and $1 \leqslant i \leqslant n$ we define

$$\varepsilon(s_i(x)) = \begin{cases} 0 & s_i(x) = 1, \ldots, n_1 \\ 1 & s_i(x) = n_1 + 1, \ldots, n, \end{cases} \tag{1.9}$$

and prove

Theorem 1.2. *Let* $\mathfrak{e} = (\varepsilon_1, \ldots, \varepsilon_n)$ *be an n-tuple of real numbers such that* $\varepsilon_i = 0$ *holds exactly* n_1 *times and* $\varepsilon_j = 1$ *exactly* n_2 *times. Then under the assumptions of Theorem* 1.1,

$$P((\varepsilon(s_1), \ldots, \varepsilon(s_n)) = \mathfrak{e}) = 1 \Big/ \binom{n}{n_1}. \tag{1.10}$$

Proof. Let $\varepsilon_{j_1} = \cdots = \varepsilon_{j_{n_1}} = 0$. We must consider the set of all $x \in R_n - N_n$ such that for some permutation $(i_1, \ldots, i_{n_1})$ of $(1, \ldots, n_1)$ and some permutation $(k_1, \ldots, k_{n_2})$ of $(n_1 + 1, \ldots, n)$ each $z_{j_m}(x)$ coincides with x_{i_m}, $1 \leqslant m \leqslant n_1$, and each of the remaining order statistics coincides with exactly one x_{k_l}, $1 \leqslant l \leqslant n_2$. Using the result of Theorem 1.1 we than have

$$P((\varepsilon(s_1), \ldots, \varepsilon(s_n)) = \mathfrak{e}) = n_1!\, n_2! / (n_1 + n_2)!$$

which is obviously the same as (1.10).

The special form of (1.9) obviously plays no rôle in the proof of Theorem 1.2. Replacing 0 by an element a of some set and 1 by a different element b of this set, we again get (1.10) provided the definition of $\mathfrak{e}$ is correspondingly modified.

We will later generalize Theorem 1.2 (see p. 454).

We assumed in Theorem 1.1 that the r.v.'s ξ_i, $1 \leqslant i \leqslant n$, possess the same distribution. For many problems, such as the calculation of the *power function in the two-sample problem* (p. 137), it is important to be able to discard this assumption. In this connection we first want to derive several results on the distribution of order statistics. We prove

Theorem 1.3. *Let* $\xi_1, \ldots, \xi_n$, $n \geqslant 2$ *be independent r.v.'s with the same d.f.* $F \in \mathfrak{F}$. *Then the d.f.* G_i *of the i-th order statistic* $z_i \circ \xi^{(n)}$, $1 \leqslant i \leqslant n$ *is given by*

$$G_i(x) = \frac{n!}{(i-1)!\,(n-i)!} \int_0^{F(x)} t^{i-1}(1-t)^{n-i}\, dt, \qquad x \in R_1. \tag{1.11}$$

Proof. For each $x \in R_1$ we define a r.v. η_x with discrete d.f. Let the latter have mass points l/n, $l = 0, \ldots, n$ and let $P(\eta_x = l/n)$ be the probability that $\xi_j \leqslant x$ for exactly l r.v.'s ξ_j. By p. 89 we then have

$$P(\eta_x = l/n) = \binom{n}{l} (F(x))^l (1 - F(x))^{n-l}, \qquad 0 \leqslant l \leqslant n.$$

Now $G_i(x)=P(z_i\circ\xi^{(n)}\leqslant x)$ for each $x\in R_1$ and this probability coincides with $P(\eta_x\geqslant i/n)$. Indeed, both expressions represent the probability that $\xi_j\leqslant x$ for at least i r.v.'s ξ_j.

Hence, $G_i(x)=\sum_{l=i}^{n}\binom{n}{l}(F(x))^l(1-F(x))^{n-l}$, and an application of I. (32.9) delivers the claim.

We now prove a theorem which provides the basis for the important rôle played by order statistics in non-parametric theory. To this end we first prove

Lemma 1.2. *Let h be a nondecreasing function defined over R_1. Let $\xi_1,\dots,\xi_n$, $n\geqslant 2$, be arbitrary r.v.'s and set $\eta_i=h\circ\xi_i$, $1\leqslant i\leqslant n$. Then, for $i=1,\dots,n$,*

$$h\circ(z_i\circ\xi^{(n)})=z_i(\eta_1,\dots,\eta_n).$$

The *proof* follows at once from the fact that $h(x_i)<h(x_j)$ implies $x_i<x_j$. We now have

Theorem 1.4. *Let $F\in C$ and $F(x)=0$ for $x\leqslant a$, $F(x)=1$ for $x\geqslant b$. Moreover, let F be strictly monotone in $[a,b]$. We allow $a=-\infty$ and $b=\infty$* [4]. *Let $\xi_1,\dots,\xi_n$, $n\geqslant 2$ be independent r.v.'s with the same d.f. F. Then the r.v. $F(z_i\circ\xi^{(n)})$ is distributed according to $B(i,n-i+1)$, $1\leqslant i\leqslant n$. This still holds when we merely assume that $F\in C$.*

Proof. To justify the first claim, we use I, Theorem 11.1. According to it, $\eta_i=F\circ\xi_i$ is uniformly distributed for $1\leqslant i\leqslant n$. By Lemma 1.2, $z_i(\eta_1,\dots,\eta_n)=F(z_i(\xi_1,\dots,\xi_n))$. Then an application of Theorem 1.3 in this case yields the claim. To show the second part, it is sufficient to show, again according to Lemma 1.2, that $\eta_i=F\circ\xi_i$ is uniformly distributed for $1\leqslant i\leqslant n$. If we construct for F the function u_F according to III, Lemma 5.2, then we can easily see that it satisfies $F(u_F(y))=y$ for all $y\in(0,1)$ because of the continuity of F. Hence, we easily get

$$P(F\circ\xi_i<y)=P(\xi_i<u_F(y))=y \quad \text{for } y\in(0,1)$$

which was to be proved.

We next have

Theorem 1.5. *Let $\eta_1,\dots,\eta_n$ be $n\geqslant 2$ independent, (identically) uniformly distributed r.v.'s. Then, with $\eta^{(n)}=(\eta_1,\dots,\eta_n)$ the density of the n-dimensional r.v. $(z_1\circ\eta^{(n)},\dots,z_n\circ\eta^{(n)})$ is given by*

$$h(y_1,\dots,y_n)=\begin{cases} n! & 0<y_1\leqslant y_2\leqslant\cdots\leqslant y_n<1\\ 0 & \text{otherwise}.\end{cases} \tag{1.12}$$

---

[4] For $a=-\infty$ or $b=\infty$, the definition of F is to be trivially modified.

For the marginal distribution of $(z_{l_1} \circ \eta^{(n)}, z_{l_2} \circ \eta^{(n)}, \ldots, z_{l_i} \circ \eta^{(n)})$ *with* $1 \leqslant l_1 < l_2 < \cdots < l_i \leqslant n$ *one gets the density*

$$h^{(i)}(y_{l_1}, y_{l_2}, \ldots, y_{l_i})$$

$$= \begin{cases} \dfrac{n!}{(l_1-1)!(l_2-l_1-1)!\ldots(l_i-l_{i-1}-1)!(n-l_i)!} y_{l_1}^{l_1-1}(y_{l_2}-y_{l_1})^{l_2-l_1-1} \\ \qquad \ldots(y_{l_i}-y_{l_{i-1}})^{l_i-l_{i-1}-1}(1-y_{l_i})^{n-l_i} \\ \qquad \text{for } 0<y_{l_1} \leqslant y_{l_2} \leqslant \cdots \leqslant y_{l_i} < 1 \\ 0 \qquad \textit{otherwise.} \end{cases} \tag{1.13}$$

Proof. The claim concerning the vanishing of h is trivial. Otherwise,

$$z_1 \circ \eta^{(n)} \leqslant y_1, z_2 \circ \eta^{(n)} \leqslant y_2, \ldots, z_n \circ \eta^{(n)} \leqslant y_n$$

holds iff for some permutation $(i_1, \ldots, i_n)$ of $(1, \ldots, n)$,

$$\eta_{i_1} \leqslant y_1, \eta_{i_2} \leqslant y_2, \ldots, \eta_{i_n} \leqslant y_n.$$

Hence, in analogy to the second part of the proof of Theorem 1.1 it follows that

$$P(z_1 \circ \eta^{(n)} \leqslant y_1, z_2 \circ \eta^{(n)} \leqslant y_2, \ldots, z_n \circ \eta^{(n)} \leqslant y_n) = n!\, y_1 y_2 \ldots y_n$$

for $0<y_1 \leqslant y_2 \leqslant \cdots \leqslant y_n < 1$, which implies (1.12). For the density $h^{(i)}$ of the marginal distribution we get from (1.12) for $0<y_{l_1} \leqslant y_{l_2} \leqslant \cdots \leqslant y_{l_i} < 1$

$$h^{(i)}(y_{l_1}, y_{l_2}, \ldots, y_{l_i}) = n! \int_0^{y_{l_1}} dy_{l_1-1} \int_0^{y_{l_1-1}} dy_{l_1-2} \cdots$$

$$\cdots \int_0^{y_2} dy_1 \int_{y_{l_1}}^{y_{l_2}} dy_{l_2-1} \int_{y_{l_1}}^{y_{l_2-1}} dy_{l_2-2} \cdots \int_{y_{l_1}}^{y_{l_1+2}} dy_{l_1+1} \cdots \int_{y_{l_{i-1}}}^{y_{l_i}} dy_{l_i-1}$$

$$\times \int_{y_{l_{i-1}}}^{y_{l_i-1}} dy_{l_i-2} \cdots \int_{y_{l_{i-1}}}^{y_{l_{i-1}+2}} dy_{l_{i-1}+1} \int_{y_{l_i}}^{1} dy_n \int_{y_{l_i}}^{y_n} dy_{n-1} \cdots \int_{y_{l_i}}^{y_{l_i+2}} dy_{l_i+1}$$

whence, after an easy calculation, the nontrivial part of (1.13).

Now consider n independent r.v.'s $\xi_1, \ldots, \xi_n$ with the same density f. Let $f>0$ in (a,b) and $f=0$ otherwise. Denote the corresponding distribution function by F. Then, from (1.13) it follows that the density of the i-dimensional r.v.

$$(z_{l_1} \circ \xi^{(n)}, z_{l_2} \circ \xi^{(n)}, \ldots, z_{l_i} \circ \xi^{(n)}),$$

provided it does not vanish, is given by[5]

$$\left.\begin{array}{l}\dfrac{n!}{(l_1-1)!\,(l_2-l_1-1)!\dots(l_i-l_{i-1})!\,(n-l_i)!}(F(x_{l_1}))^{l_1-1}f(x_{l_1})\\ \times(F(x_{l_2})-F(x_{l_1}))^{l_2-l_1-1}f(x_{l_2})\dots(F(x_{l_i})-F(x_{l_{i-1}}))^{l_i-l_{i-1}}\\ \times f(x_{l_i})(1-F(x_{l_i}))^{n-l_i}.\end{array}\right\} \quad (1.14)$$

Choosing $i=2$, $l_1=1$ and $l_2=n$ in (1.14), we get the distribution of the two-dimensional r.v. $(z_1\circ\xi^{(n)}, z_n\circ\xi^{(n)})$, i.e., the joint distribution of the smallest and largest values of $\xi_1,\dots,\xi_n$. The density of this distribution, provided it does not vanish, is given by

$$n(n-1)f(x_1)f(x_n)(F(x_n)-F(x_1))^{n-2}, \qquad a<x_1\leqslant x_n<b. \quad (1.15)$$

The r.v. $\boldsymbol{R}^{(n)}=z_n\circ\xi^{(n)}-z_1\circ\xi^{(n)}$ is called the n-th *range*. The density of $\boldsymbol{R}^{(n)}$ is obtained from (1.15) by means of the transformation

$$u=x_1,$$
$$R=x_n-x_1.$$

Choosing $a=-\infty$ and $b=\infty$ for the sake of simplicity, we then see that $-\infty<x_1\leqslant x_n<\infty$ is taken by this transformation into $-\infty<u<\infty$, $0\leqslant R<\infty$. We integrate over u and get for the density of $\boldsymbol{R}^{(n)}$:

$$n(n-1)\int_{-\infty}^{+\infty} f(u)f(R+u)(F(R+u)-F(u))^{n-2}\,du, \qquad 0\leqslant R<\infty.$$

Choosing $i=1$ and $l_1=j$ in (1.14), we obtain for the density of $z_j\circ\xi^{(n)}$:

$$\frac{n!}{(j-1)!\,(n-j)!}(F(x))^{j-1}(1-F(x))^{n-j}f(x), \qquad -\infty<x<\infty. \quad (1.16)$$

This naturally follows also from (1.11) even without any special assumptions on f.

Of special practical importance is the example of the normal distribution. For each $z\in R_1$ denote the integral $\int_z^{\infty} e^{-y^2/2}dy$ by $\Psi(z)$ and assume that the r.v.'s $\xi_1,\dots,\xi_n$ are independently $N(a,\sigma^2)$-distributed. Then, using (1.16) for $j=1$ and $j=n$, we get after an easy calculation:

$$E(\boldsymbol{R}^{(n)};(a,\sigma^2))=\sigma n(2\pi)^{-n/2}\int_{-\infty}^{+\infty} z((\Psi(-z))^{n-1}-(\Psi(z))^{n-1})e^{-z^2/2}\,dz.$$

Thus, if $\boldsymbol{R}^{(n)}$ is divided by a suitable constant, we get an *unbiased estimate* for σ, $0<\sigma<\infty$.

Since a realization of $\boldsymbol{R}^{(n)}$ from a sample is extremely easy to calculate, this fact is of practical importance.

[5] More precisely, (1.14) holds, without continuity assumptions on f, only up to a null-set.

We return again to the theory of ranks. In particular, we now generalize the second part of Theorem 1.1. We have

Theorem 1.6[6]. *Let $\xi_1,\dots,\xi_n$, $n\geq 2$, be independent r.v.'s. For $i=1,\dots,n$, let ξ_i possess a distribution with density f_i. Let $\eta_1,\dots,\eta_n$ be independent r.v.'s, each distributed by the same density f. Moreover, let the distribution of ξ_i be absolutely continuous w.r.t. the distribution of η_i. Then, with $\xi^{(n)}=(\xi_1,\dots,\xi_n)$ and $\eta^{(n)}=(\eta_1,\dots,\eta_n)$*

$$P(r_1\circ\xi^{(n)}=i_1,\dots,r_n\circ\xi^{(n)}=i_n)=\frac{1}{n!}E\left[\frac{f_1(z_{i_1}\circ\eta^{(n)})\dots f_n(z_{i_n}\circ\eta^{(n)})}{f(z_{i_1}\circ\eta^{(n)})\dots f(z_{i_n}\circ\eta^{(n)})}\right] \tag{1.17}$$

holds, where $(i_1,\dots,i_n)$ is a permutation of $(1,\dots,n)$.

Proof. If $(j_1,\dots,j_n)$ is the permutation inverse to $(i_1,\dots,i_n)$ then

$$P(r_1=i_1,\dots,r_n=i_n)=\int_{-\infty<x_{j_1}<\dots<x_{j_n}<\infty}\cdots\int f_1(x_1)\dots f_n(x_n)dx.$$

Thus, by assumption, one can also write

$$P(r_1=i_1,\dots,r_n=i_n)=\frac{1}{n!}\int_{-\infty<x_{j_1}<\dots<x_{j_n}<\infty}\cdots\int\frac{\prod_{i=1}^{n}f_i(x_i)}{\prod_{i=1}^{n}f(x_i)}n!\prod_{i=1}^{n}f(x_i)dx$$

$$=\frac{1}{n!}\int_{-\infty<x_1<\dots<x_n<\infty}\cdots\int\frac{\prod_{k=1}^{n}f_k(x_{i_k})}{\prod_{k=1}^{n}f(x_{i_k})}n!\prod_{i=1}^{n}f(x_i)dx.$$

An application of (1.14) for $i=n$, $l_j=j$, $1\leq j\leq n$, to the r.v.'s $\eta_1,\dots,\eta_n$ then yields (1.17).

We then easily have the following

Corollary. *Let ψ be a nondecreasing, differentiable function from $[0,1]$ into $[0,1]$ with $\psi(0)=0$, $\psi(1)=1$. Let n_1,n_2 be natural numbers with $n_1+n_2=n$ and let f be a density with F the corresponding d.f. Let $\xi_1,\dots,\xi_n$ be independent r.v.'s with ξ_i, $1\leq i\leq n_1$, distributed by F and ξ_j, $n_1+1\leq j\leq n$, distributed by $\psi\circ F$. Further let $\zeta^{(n)}$ be an n-tuple of independent r.v.'s,*

[6] We refer especially in this connection to W. Hoeffding, Proceedings of the Second Berkeley Symposium on Mathematical Statistics and Probability 1951, Vol. I, pp. 83—92, University of California Press, Berkeley and Los Angeles.

each uniformly distributed in $[0,1]$. *Then*

$$P(r_1\circ\xi^{(n)}=i_1,\dots,r_n\circ\xi^{(n)}=i_n)=\frac{1}{n!}E\left[\prod_{j=n_1+1}^{n}\psi'(z_{i_j}\circ\zeta^{(n)})\right]. \tag{1.18}$$

The *proof* follows from an application of (1.17). The density of $\psi\circ F$ is $(\psi'\circ F)\cdot f$; Hence, for the left side of (1.18) one gets

$$\frac{1}{n!}\int\cdots\int_{-\infty<x_1<\cdots<x_n<\infty}\prod_{j=n_1+1}^{n}\psi'(F(x_{i_j}))n!\prod_{i=1}^{n}f(x_i)dx.$$

Introducing new variables into this integral by means of the transformation $y_i=F(x_i)$, $1\leqslant i\leqslant n$, one gets (1.18).

The proof of Theorem (1.6) shows immediately that the statement at (1.17) is capable of considerable generalization. In particular, instead of densities, one can start from the R.-N.-densities of measures. We will not go into the details here.

Order statistics suggest the definition of an important *sufficient transformation*. Let $\mathfrak{P}_s^{(n)}$ be the set of all probability measures $P^{(n)}$ over $(R_n,\mathfrak{B}_n)$ with the property:

$$P^{(n)}(\Pi_n A)=P^{(n)}(A), \quad A\in\mathfrak{B}_n, \quad \Pi_n\in S_n. \tag{1.19}$$

Then we have

Theorem 1.7. *The map* $T:x\to(z_1(x),\dots,z_n(x))$ *of* R_n *into* R_n *is sufficient for* $\mathfrak{P}_s^{(n)}$ *and hence also for each subset of* $\mathfrak{P}_s^{(n)}$.

Proof. Let $B\in\mathfrak{B}_n$. Then, $T^{-1}B=\bigcup_{\Pi_n\in S_n}\Pi_n B=B_1$. The set B_1 is naturally symmetric, i.e., $\Pi_n B_1=B_1$ for each $\Pi_n\in S_n$.

Conversely, $T^{-1}B=B$ for each symmetric $B\in\mathfrak{B}_n$. Hence, the σ-algebra $T^{-1}(\mathfrak{B}_n)$ consists of exactly the symmetric Borel sets. Let f be an integrable function over R_n which is $P^{(n)}$-integrable for each $P^{(n)}\in\mathfrak{P}_s^{(n)}$. For each $x\in R_n$ define the function $h(x)=\frac{1}{n!}\sum_{\Pi_n\in S_n}f(\Pi_n x)$. h is $T^{-1}(\mathfrak{B}_n)$-measurable, which is seen immediately. Also, for each $A\in T^{-1}(\mathfrak{B}_n)$ and each $P^{(n)}\in\mathfrak{P}_s^{(n)}$ we obviously have

$$\int_A f\,dP^{(n)}=\int_A h\,dP^{(n)}. \tag{1.20}$$

Hence, $h=E(f\mid T)$ and this holds independently of $P^{(n)}\in\mathfrak{P}_s^{(n)}$.

The construction of this simple sufficient transformation is not limited to R_n. Indeed, let $(R^{(n)},\mathscr{S}^{(n)})$ be a product space of $n\geqslant 1$ copies of the same measurable space $(R,\mathscr{S})$. The map which sends each n-tuple

$(x_1, \ldots, x_n) \in R^{(n)}$ into the set $\{x_1, \ldots, x_n\}$ is sufficient for each set of probability measures over $(R^{(n)}, \mathscr{S}^{(n)})$ which fulfills the condition corresponding to (1.19) for all $A \in \mathscr{S}^{(n)}$.

Theorem 1.7 has many applications in the theory of testing and estimation, those based on V, Theorem 1.6, for example. An important question is whether or not the sufficient transformation T is complete. In this direction we prove

Theorem 1.8[7]. *Let $(R, \mathscr{S})$ be a measurable space and $\mathfrak{H} \subset \mathscr{S}$ a semi-algebra which generates $\mathscr{S}$. Let $\mathfrak{R}$ be the ring of all unions of finitely many pairwise disjoint sets from $\mathfrak{H}$. Let μ be a bounded, atomless measure over $(R, \mathscr{S})$ and $\mathfrak{P}_m$ the set of all probability measures over $(R, \mathscr{S})$ which are dominated by μ and whose R.-N.-densities w.r.t. μ are given by $c_B/\mu(B)$. Here, B runs through all sets in $\mathfrak{R}$ with $\mu(B) \neq 0$. Let $\mathfrak{P}_m^{(n)}$ for $n \geqslant 1$ be the set of all product measures over $(R^{(n)}, \mathscr{S}^{(n)})$ of n copies of the same measure $P \in \mathfrak{P}_m$. Then the map*

$$T: (x_1, \ldots, x_n) \to \{x_1, \ldots, x_n\}$$

defined above is complete (w.r.t. $\mathfrak{P}_m^{(n)}$).

Proof. Let $\mu^{(n)}$ be the n-fold product of μ with itself. Let h be a $\mu^{(n)}$-integrable function defined over $R^{(n)}$ and invariant w.r.t. the group S_n. Thus, in particular, h is $T^{-1}(\mathscr{S}^{(n)})$-measurable. We need to show that

$$E(h; P^{(n)}) = 0 \tag{1.21}$$

for all $P^{(n)} \in \mathfrak{P}_m^n$ implies $h(x) = 0$ $\mu^{(n)}$-a.e. Let $h^+ = \max(h, 0)$ and $h^- = -\min(h, 0)$ thus $h = h^+ - h^-$. We then want to show that (1.21) implies that for arbitrary $A_i \in \mathfrak{H}$, $1 \leqslant i \leqslant n$,

$$\left.\begin{aligned} &\int_{A_1} \cdots \int_{A_n} h^+(x_1, \ldots, x_n)\, d\mu(x_1) \ldots d\mu(x_n) \\ &= \int_{A_1} \cdots \int_{A_n} h^-(x_1, \ldots, x_n)\, d\mu(x_1) \ldots d\mu(x_n), \end{aligned}\right\} . \tag{1.22}$$

In fact, assume that (1.22) holds and for $A^{(n)} \in \mathscr{S}^{(n)}$ denote

$$\int_{A^{(n)}} h^+\, d\mu^{(n)} \text{ by } \mu^+(A^{(n)}) \text{ and } \int_{A^{(n)}} h^-\, d\mu^{(n)} \text{ by } \mu^-(A^{(n)}).$$

Then, (1.22) says that the measures μ^+ and μ^- coincide on $\underbrace{\mathfrak{H} \times \cdots \times \mathfrak{H}}_{n}$.

---

[7] D.A.S. Fraser, Canad. J. Math. 6, 42—45 (1954). Also see P.R. Halmos, Ann. Math. Statist. 17, 34—43 (1946) An interesting general result on sufficient transformations connected with the theory of invariance is in T. S. Pitcher, Trans. Amer. Math. Soc. 85, 166—173 (1957).

Thus, they coincide on $\mathcal{S}^{(n)}$. (See Theorem III and the remark on p. 7.) Hence,

$$\int_{A^{(n)}} (h^+ - h^-)\, d\mu^{(n)} = \int_{A^{(n)}} h\, d\mu^{(n)} = 0 \quad \text{for all } A^{(n)} \in \mathcal{S}^{(n)},$$

whence, the theorem.

We proceed now to the proof of (1.22). Let $A_i \in \mathfrak{H}$, $\mu(A_i) \neq 0$, $1 \leqslant i \leqslant n$, $n \geqslant 1$, and, initially, $A_i \cap A_j = \emptyset$, $i \neq j$. Further let

$$A_{i_1 \dots i_r} = \bigcup_{j=1}^{r} A_{i_j}, \tag{1.23}$$

where $\{i_1, \dots, i_r\}$ is a subset of $\{1, \dots, n\}$. Because of (1.21), we have for $1 \leqslant r \leqslant n$ and each set $A_{i_1 \dots i_r}$

$$\int_{A_{i_1 \dots i_r}} \cdots \int_{A_{i_1 \dots i_r}} h(x_1, \dots, x_n)\, d\mu(x_1) \dots d\mu(x_n) = 0$$

or

$$\sum_{j_1=1}^{r} \cdots \sum_{j_n=1}^{r} \int_{A_{i_{j_n}}} \cdots \int_{A_{i_{j_1}}} h(x_1, \dots, x_n)\, d\mu(x_1) \dots d\mu(x_n) = 0\,. \tag{1.24}$$

We now must show that (1.24) implies

$$\int_{A_1} \int_{A_2} \cdots \int_{A_n} h(x_1, x_2, \dots, x_n)\, d\mu(x_1)\, d\mu(x_2) \dots d\mu(x_n) = 0\,. \tag{1.25}$$

We do this by induction on r. For $r=1$, it follows from (1.24) for $1 \leqslant i \leqslant n$, that

$$\int_{A_i} \int_{A_i} \cdots \int_{A_i} h\, d\mu^{(n)} = 0\,.$$

Now assume that all integrals of the form $\int_{A_{i_1}} \int_{A_{i_2}} \cdots \int_{A_{i_n}} h\, d\mu^{(n)}$ vanish when at most $n-1$ of the indices $i_1, \dots, i_n$ are different from one another. Note that the integrals in (1.24) do not change when the $A_{i_{j_l}}$, $1 \leqslant l \leqslant n$, are permuted in some way. Hence, with the induction hypothesis, we get from (1.24) for $r=n$:

$$n! \int_{A_1} \int_{A_2} \cdots \int_{A_n} h(x_1, x_2, \dots, x_n)\, d\mu(x_1)\, d\mu(x_2) \dots d\mu(x_n) = 0\,,$$

and hence, also (1.25).

We now treat the general case: Let $A_1, \dots, A_n$, $n \geqslant 2$, be arbitrary sets from $\mathfrak{H}$ with $\mu(A_i) \neq 0$.

Let $\varepsilon > 0$ be arbitrary, but so chosen that $\mu(R) > \varepsilon$. The argument of the proof of III, Theorem 9.2 shows the existence of an $M_1 \in \mathcal{S}$ with $\mu(M_1) = \varepsilon$. If $\mu(R - M_1) > \varepsilon$, one can choose in the same way an

$M_2 \in \mathscr{S}$ with $M_2 \subseteq R - M_1$, such that $\mu(M_2) = \varepsilon$. Continuing in this way, one gets a finite sequence of sets $M_1^{(\varepsilon)}, M_2^{(\varepsilon)}, \ldots, M_{n_\varepsilon}^{(\varepsilon)}$ with

$$\bigcup_{i=1}^{n_\varepsilon} M_i^{(\varepsilon)} = R, \tag{1.26}$$

$$M_i^{(\varepsilon)} \cap M_j^{(\varepsilon)} = \emptyset, \quad 1 \leqslant i < j \leqslant n_\varepsilon, \tag{1.27}$$

$$\mu(M_i^{(\varepsilon)}) \leqslant \varepsilon, \quad 1 \leqslant i \leqslant n_\varepsilon. \tag{1.28}$$

Equality holds in (1.28) at least for $1 \leqslant i \leqslant n_\varepsilon - 1$.

First, we assume that $M_i^{(\varepsilon)} \in \mathfrak{R}$, $1 \leqslant i \leqslant n_\varepsilon$. Then, from (1.26)

$$A_1 \times A_2 \times \cdots \times A_n = \bigcup_{i_1=1}^{n_\varepsilon} \bigcup_{i_2=1}^{n_\varepsilon} \cdots \bigcup_{i_n=1}^{n_\varepsilon} (A_1 \cap M_{i_1}^{(\varepsilon)}) \times (A_2 \cap M_{i_2}^{(\varepsilon)}) \times \cdots \times (A_n \cap M_{i_n}^{(\varepsilon)}),$$

so that

$$\left| \int_{A_1} \cdots \int_{A_n} h(x_1, \ldots, x_n) d\mu(x_1) \ldots d\mu(x_n) \right|$$
$$= \left| \sum_{i_1=1}^{n_\varepsilon} \cdots \sum_{i_n=1}^{n_\varepsilon} \int_{A_1 \cap M_{i_1}^{(\varepsilon)}} \cdots \int_{A_n \cap M_{i_n}^{(\varepsilon)}} h(x_1, \ldots, x_n) d\mu(x_1) \ldots d\mu(x_n) \right|.$$

Using (1.25), we find from (1.27) that all integrals in this sum which belong to pairwise different indices $(i_1, \ldots, i_n)$ vanish. Hence, this sum is smaller or equal to

$$\sum_{\substack{1 \leqslant i < j \leqslant n \\ 1 \leqslant k \leqslant n_\varepsilon}} \int_R \cdots \int_{A_i \cap M_k^{(\varepsilon)}} \cdots \int_{A_j \cap M_k^{(\varepsilon)}} \cdots \int_R |h| \, d\mu^{(n)}. \tag{1.29}$$

Now, for all $A^{(n)} \in \mathscr{S}^{(n)}$, the set function ν defined by $\nu(A^{(n)} = \int_{A^{(n)}} |h| \, d\mu^{(n)}$ is absolutely continuous w.r.t. $\mu^{(n)}$. Thus, for given $\delta > 0$, one can choose an $\eta > 0$ such that $\mu(A^{(n)}) < \eta$ implies $\nu(A^{(n)}) < \delta$. But from (1.28) for $1 \leqslant i < j \leqslant n$ and $1 \leqslant k \leqslant n_\varepsilon$

$$\mu^{(n)}((R \times \cdots \times A_i \cap M_k^{(\varepsilon)} \times \cdots \times A_j \cap M_k^{(\varepsilon)} \times \cdots \times R)) \leqslant (\mu(R))^{n-2} \varepsilon^2.$$

From (1.26), (1.28) and the remark following (1.28), we have

$$\sum_{\substack{1 \leqslant i < j \leqslant n \\ 1 \leqslant k \leqslant n_\varepsilon}} (\mu(R))^{n-2} \varepsilon^2 \leqslant \binom{n}{2} (\mu(R))^{n-2} 2\mu(R)\varepsilon.$$

Hence, if ε is chosen sufficiently small, the sum (1.29) becomes arbitrarily small.

If the $M_i^{(\varepsilon)}$ are arbitrary sets from $\mathscr{S}$, one can choose for each i with $1 \leqslant i \leqslant n$ (since $\mathfrak{H}$ generates the σ-algebra $\mathscr{S}$) finitely many pairwise disjoint sets $N_{ij}^{(\varepsilon)} \in \mathfrak{H}$, $1 \leqslant j \leqslant n_i^{(\varepsilon)}$, for which $M_i^{(\varepsilon)} \subseteq \bigcup_{j=1}^{n_i^{(\varepsilon)}} N_{ij}^{(\varepsilon)}$ and $\mu\left(M_i^{(\varepsilon)} - \bigcup_{j=1}^{n_i^{(\varepsilon)}} N_{ij}^{(\varepsilon)}\right)$ is arbitrarily small. Proceed now in the same way with the integrals of the form

$$\int_{A_1 \cap N_{i_1 j_1}^{(\varepsilon)}} \cdots \int_{A_n \cap N_{i_n j_n}^{(\varepsilon)}} h \, d\mu^{(n)}, \quad 1 \leqslant j_1 \leqslant n_{i_1}^{(\varepsilon)}, \ldots, 1 \leqslant j_n \leqslant n_{i_n}^{(\varepsilon)}$$

and take the absolute continuity of ν w.r.t. $\mu^{(n)}$ into account. This completely proves (1.25).

If one is willing to put up with slight complications, σ-finiteness of μ can be assumed in place its boundedness.

As an example we mention the important special case of R_n. The semi-algebra $\mathfrak{H}_1$ (see p. 3) then replaces the semi-algebra $\mathfrak{H}$. The set $\mathfrak{P}_m$ can be replaced for example by the set of all probability distributions $\tilde{\mathfrak{P}}_m$ over $(R_1, \mathfrak{B}_1)$ whose densities are constant but $\neq 0$ on finite unions of pairwise disjoint left-open, right-closed intervals and which vanish on the complement of these unions.

If we go over to the set $\mathfrak{P}$ of all probability distributions for which densities exist, then we have the following assertion:

If $\mathfrak{P}^{(n)}$ is the set of all product measures of n copies of the same probability distribution absolutely continuous w.r.t. Lebesgue measure, then the map

$$T: (x_1, \ldots, x_n) \to (z_1(x), \ldots, z_n(x))$$

is a (w.r.t. $\mathfrak{P}^{(n)}$) sufficient and complete transformation of $(R_n, \mathfrak{B}_n)$ into $(R_n, T^{-1}(\mathfrak{B}_n))$.

One can obtain an analogous result without difficulty if $\tilde{\mathfrak{P}}_m$ is replaced by the *set of all discrete probability distributions over* $(R_1, \mathfrak{B}_1)$.

As an application of V, Theorem 1.7 we note the following result:

Theorem 1.9. *Let $\mathfrak{P}$ be an arbitrary set of probability measures over $(R_1, \mathfrak{B}_1)$ defined by means of densities, which is assumed to contain the set $\tilde{\mathfrak{P}}_m$ mentioned above. For each $P \in \mathfrak{P}$, let the mean $a(P) = \int_{R_1} x \, dP(x)$ and the variance $\int_{R_1} (x - a(P))^2 \, dP(x)$ exist*[8]*. Again, let $\mathfrak{P}^{(n)}$ be the product of n copies of the same distribution from $\mathfrak{P}$. Let $a(P^{(n)}) = a(P)$ for $P^{(n)} \in \mathfrak{P}^{(n)}$. Then the function $x \to \bar{x}_n$ is, for each $n \geqslant 2$, an unbiased and uniformly minimal estimate for the map $P^{(n)} \to a(P^{(n)})$.*

Thus, in short, if the set $\mathfrak{P}$ is large enough, the sample mean is the best unbiased estimate for the population mean in the given sense. This no longer holds if one considers a set of probability distributions which is "too small".

---

[8] This is obviously the case for $P \in \tilde{\mathfrak{P}}_m$.

In this connection we give an example which also shows that one cannot, in general, replace $\mathfrak{R}$ by $\mathfrak{H}$ in Theorem 1.8.

Let $\mathfrak{P}_U$ be the totality of all one-dimensional uniform distributions over all bounded intervals. If $-\infty<a<b<\infty$ and ξ is a r.v. uniformly distributed over (a,b), then

$$E(\xi;(a,b))=(a+b)/2. \tag{1.30}$$

If $\xi_1,\ldots,\xi_n$, $n\geqslant 2$, are independent r.v.'s with the same distribution as ξ we assume that the joint distribution of $\xi^{(n)}=(\xi_1,\ldots,\xi_n)$ runs through the set $\mathfrak{P}_U^{(n)}$ when the distribution of ξ goes through the set $\mathfrak{P}_U$. The density of the distribution of $z_1\circ\xi^{(n)}$ is given by $\frac{n}{(b-a)^n}(b-z_1)^{n-1}$, $a\leqslant z_1\leqslant b$, and that of $z_n\circ\xi^{(n)}$ by $\frac{n}{(b-a)^n}(z_n-a)^{n-1}$, $a\leqslant z_n\leqslant b$, provided these densities do not vanish. An easy calculation then gives

$$E((z_1\circ\xi^{(n)}+z_n\circ\xi^{(n)})/2;(a,b))=(a+b)/2. \tag{1.31}$$

From (1.30) and (1.31) there follows for all $P^{(n)}\in\mathfrak{P}_U^{(n)}$

$$E((z_1\circ\xi^{(n)}+z_n\circ\xi^{(n)})/2-(\xi_1+\cdots+\xi_n)/n;P^{(n)})=0.$$

For $n>2$ it does not, however, hold that

$$((z_1(x_1,\ldots,x_n)+z_n(x_1,\ldots,x_n))/2)-(x_1+\cdots+x_n)/n=0 \quad \text{a.e.}$$

We now calculate the variance of $(z_1\circ\xi^{(n)}+z_n\circ\xi^{(n)})/2$ and note in addition, that the joint distribution of $z_1\circ\xi^{(n)}$ and $z_n\circ\xi^{(n)}$ is given by $n(n-1)(b-a)^{-n}\cdot(z_n-z_1)^{n-2}$ for $a<z_1\leqslant z_n<b$, which follows immediately from (1.15).

A simple calculation leads to

$$E\left[\left(\frac{z_1\circ\xi^{(n)}+z_n\circ\xi^{(n)}}{2}-\frac{a+b}{2}\right)^2;(a,b)\right]=\frac{1}{2(n+1)(n+2)}(b-a)^2. \tag{1.32}$$

Since, on the other hand

$$E[(\bar{\xi}_n-(a+b)/2)^2;(a,b)]=(a-b)^2/12n, \tag{1.33}$$

the variance of $(z_1\circ\xi^{(n)}+z_n\circ\xi^{(n)})/2$ for $n>2$ is always smaller than that of $\bar{\xi}_n$ and even of order $O\left(\frac{1}{n^2}\right)$. Considering the subset of those distributions from $\mathfrak{P}_U$ which, with $b=1$, are characterized by the parameter a, $-\infty<a<1$, one might conclude that this contradicts V, Theorem 1.10. However, it is easy to see that the decisive assumption V, (1.51) is not fulfilled.

2. Tolerance regions. Let $x_1, \dots, x_n$ be a sample of size n from a population with density f. The practically important question arises as to what fraction of the total area 1 under the curve defined by f and the abscissa can be expected to lie between $z_1(x_1, \dots, x_n)$ and $z_n(x_1, \dots, x_n)$. Of special practical interest in this connection are statements which are as independent as possible from the chosen density f. This leads to the following

Definition. Let a set P_Γ of probability measures be given over $(R, \mathscr{S})$. *Let Q be a map of R into $\mathscr{S}$* and the map $\psi_\gamma: x \to P_\gamma(Q(x))$ be $\mathscr{S}$-measurable for each $\gamma \in \Gamma$. Let P_{ψ_γ} be the probability distribution of the r.v. ψ_γ defined for each $B \in \mathscr{S}$ by $P_{\psi_\gamma}(B) = P_\gamma(\psi_\gamma^{-1}(B))$. *If $P_{\psi_{\gamma_1}} = P_{\psi_{\gamma_2}}$ for all $\gamma_1, \gamma_2 \in \Gamma$ holds, then Q is called a tolerance region w.r.t. Γ.*

The connection between this definition and the intuitive description given at the beginning of the section is established by the following considerations: Let Q be a tolerance region w.r.t. Γ and let β be given with $0 < \beta < 1$. Then there exists a δ with $0 \leqslant \delta \leqslant 1$ which by definition can be chosen independently of $\gamma \in \Gamma$ in such a way that $P_{\psi_\gamma}(\psi_\gamma \geqslant \delta) \geqslant \beta$. This means that

$$P_\gamma(\{x : P_\gamma(Q(x)) \geqslant \delta\}) \geqslant \beta. \tag{2.1}$$

β is called the *probability level.*

(2.1) can be interpreted as follows: For all $x \in R$ up to a set of P_γ-measure $\leqslant 1 - \beta$, each $Q(x)$ is covered with at least $100\,\delta\%$ of the total mass 1, independently of the allowed mass distribution P_γ.

Note that the map Q defined by $Q(x) = R$ for all $x \in R$ satisfies trivially the condition (2.1). Naturally one is interested in such a choice of Q for which $Q(x) \neq R$ for every x (up to P_Γ-null sets).

It is obviously of considerable importance to know how to construct the map Q in concrete cases. This construction is simplest in the case of n independent sample variables with the same distribution. If the latter is denoted by $P_\gamma^{(1)}$, $\gamma \in \Gamma$, then one tries to give a map Q of the sample space R_n into $\mathfrak{B}_1$ such that instead of (2.1),

$$P_\gamma^{(n)}(\{x : P_\gamma^{(1)}(Q(x)) \geqslant \delta\}) \geqslant \beta \tag{2.2}$$

holds for each $\gamma \in \Gamma$, where $P_\gamma^{(n)}$ is the n-fold product measure of n copies of $P_\gamma^{(1)}$. The formulation given in (2.2) does not correspond exactly to a specialization of (2.1). However, aside from the fact that this modified formulation is practically important, one can also establish the validity (with appropriate specialization) of (2.1) from that of (2.2) for each δ with $0 < \delta < 1$ and each β with $0 < \beta < 1$.

We now prove

Theorem 2.1. *Let $\xi_1, \dots, \xi_n$ be $n \geqslant 2$ independent, identically distributed r.v.'s with density f. Let f be positive in (a, b), $-\infty \leqslant a < b \leqslant \infty$*

and vanish outside of this interval. Then the map $x \to z_n(x) - z_1(x)$ *from* R_n *into the set of one-dimensional intervals is a tolerance region for all one-dimensional probability distributions whose densities satisfy the given condition. Furthermore, for each* δ *with* $0<\delta<1$ *and each probability level* β *with* $0<\beta<1$, *one can choose an* $n=n(\delta,\beta)$ *so that* (2.2) *holds.*

Proof. We go from (1.15) to a new density by means of the transformation

$$\left.\begin{aligned} u &= \int_{-\infty}^{z_1} f(x)\,dx \\ v &= \int_{z_1}^{z_n} f(x)\,dx \end{aligned}\right\}. \tag{2.3}$$

This sends $a<z_1 \leqslant z_n<b$ into $0<u<1, 0<u+v<1$. Integrating over u, we proceed to the marginal distribution of $\boldsymbol{v} = \int_{z_1 \circ \xi^{(n)}}^{z_n \circ \xi^{(n)}} f(x)\,dx$, where, as always, $\xi^{(n)} = (\xi_1, \dots, \xi_n)$. For the density of $\boldsymbol{v}$, we get

$$\begin{array}{ll} n(n-1)v^{n-2}(1-v) & 0<v<1 \\ 0 & \text{otherwise,} \end{array}$$

so that the distribution of $\boldsymbol{v}$ does not depend on the underlying distribution of the ξ_i. To obtain the last claim of the theorem, one must solve $n(n-1)\int_{\delta}^{1} v^{n-2}(1-v)\,dv = \beta$ for given δ and β. We therefore consider the equation

$$\delta^x(x-1) - x\delta^{x-1} + 1 - \beta = 0 \tag{2.4}$$

for $x>1$. The first derivative of the function $\delta^x(x-1)-x\delta^{x-1}$ is given by

$$\delta^{x-1}\log\delta\,[\delta(x-1)-x] - (1-\delta)\delta^{x-1}$$

for each $x>1$. Since $x(1-\delta)+\delta>1$ for $0<\delta<1$ and $x>1$, we have under the same conditions:

$$\log\delta\,[\delta(x-1)-x] \geqslant -\log\delta > 1-\delta.$$

Hence, the function $x \to \delta^x(x-1) - x\delta^{x-1}$ is strictly monotone increasing for $x \geqslant 1$ has the value -1 at $x=1$ and the limit 0 for $x \to \infty$. For each $\beta, 0<\beta<1$ there thus exists exactly one $x(\beta)>1$ which solves (2.4). If one takes the smallest integer $\geqslant x(\beta)$[9] as $n(\beta,\delta)$ one gets the desired tolerance region.

---

[9] One also writes $n(\beta,\delta) = \{x(\beta)\}$ for this. In general,

$$\{x\} = \begin{cases} [x] & x \text{ an integer} \\ 1+[x] & x \text{ not an integer}. \end{cases}$$

This construction can be extended to multi-dimensional r.v.'s $\xi_1,\dots,\xi_n$, but we will not do this here[10].

3. The asymptotic distribution of order statistics and some theorems of the Kolmogorov-Smirnov type. We turn first to a more detailed study of the structure of order statistics and will obtain therefrom several results on their asymptotic distribution[11]. We have

Theorem 3.1. *Let* $\eta_1,\dots,\eta_n$ *be independent, in* $(0,1)$ *uniformly distributed r.v.'s,* $n \geqslant 2$. *Then the r.v.'s*

$$\boldsymbol{u}_1 = z_1 \circ \eta^{(n)}/z_2 \circ \eta^{(n)},\dots,\boldsymbol{u}_{n-1} = z_{n-1} \circ \eta^{(n)}/z_n \circ \eta^{(n)},\ \boldsymbol{u}_n = z_n \circ \eta^{(n)}$$

are also distributed independently. In addition, $\boldsymbol{u}_i^i$ *is uniformly distributed in* $(0,1)$ *for* $i=1,\dots,n$.

Proof. The density of $(z_1 \circ \eta^{(n)},\dots,z_n \circ \eta^{(n)})$ is given by (1.12). The transformation $u_i = y_i/y_{i+1}$, $1 \leqslant i \leqslant n-1$, $u_n = y_n$, with Jacobian

$$\frac{\partial(y_1,\dots,y_n)}{\partial(u_1,\dots,u_n)} = u_n^{n-1}\, u_{n-1}^{n-2} \dots u_2$$

takes (1.12) into the density[12]

$$n!\, u_n^{n-1}\, u_{n-1}^{n-2} \dots u_2, \qquad 0 < u_i \leqslant 1, \qquad 1 \leqslant i \leqslant n \tag{3.1}$$

of $(\boldsymbol{u}_1,\dots,\boldsymbol{u}_n)$. This demonstrates the claimed independence of the r.v.'s $\boldsymbol{u}_i$, $1 \leqslant i \leqslant n$. The distribution of $\boldsymbol{u}_i$ has the density $i u_i^{i-1}$ for $0 < u_i \leqslant 1$ and is zero elsewhere. Defining the r.v. $\boldsymbol{v}_i = \boldsymbol{u}_i^i$, one easily sees that it is uniformly distributed in $(0,1)$.

The result of Theorem 3.1 makes it possible to represent the r.v. $\log(z_i \circ \eta^{(n)})$, $1 \leqslant i \leqslant n$ as a sum of independent r.v.'s. To this end, we define the r.v.'s $\boldsymbol{w}_i = -\log \boldsymbol{v}_i$, $1 \leqslant i \leqslant n$, which are also independent according to Theorem 3.1 and I, Theorem 13.1. For $1 \leqslant i \leqslant n$, the density of $\boldsymbol{w}_i$ is given by

$$\begin{cases} e^{-x} & 0 < x < \infty \\ 0 & -\infty < x < 0. \end{cases} \tag{3.2}$$

[10] For the literature see foremost the detailed report of S. S. Wilks, Bull. Amer. Math. Soc. 54, 6—50 (1948). See also J.W. Tukey, Ann. Math. Statist. 18, 529—539 (1947) and Ann. Math. Statist. 19, 30—39 (1948); D.A.S. Fraser, Ann. Math. Statist. 24, 44—55 (1953) and D.A.S. Fraser and I. Guttmann, Ann. Math. Statist. 27, 162—179 (1956).

[11] See A. Rényi, Acta Math. Acad. Sci. Hungar. 4, 191—231 (1953) and G. Hajos and A. Rényi, Acta Math. Acad. Sci. Hungar. 5, 1—6 (1954).

[12] Naturally, the density vanishes elsewhere.

One immediately sees from the definition of the r.v.'s $\boldsymbol{u}_i$, $\boldsymbol{v}_i$ and $\boldsymbol{w}_i$ that

$$\log(z_{n-i+1}\circ\eta^{(n)}) = -\sum_{j=n-i+1}^{n} \boldsymbol{w}_j/j \tag{3.3}$$

for $1 \leqslant i \leqslant n$. The proof of the following theorem is based on this fact.

Theorem 3.2. *Under the assumptions of Theorem 3.1, $n(z_i\circ\eta^{(n)})$ is asymptotically gamma-distributed for arbitrary but fixed i and $n\to\infty$. The density of this distribution is given by $e^{-z}z^{i-1}/(i-1)!$ for $z>0$.*

Proof. Denote (for $n>i$) the r.v. $\sum_{j=n-i+1}^{n} \boldsymbol{w}_j/j$ by $\zeta_i^{(n)}$. Using I, Theorem 24.3, one easily gets from (3.2) that the density of $\zeta_i^{(n)}$ is given by

$$\begin{cases} n\binom{n-1}{i-1} e^{-nz}(e^z-1)^{i-1} & z>0 \\ 0 & z\leqslant 0. \end{cases} \tag{3.4}$$

The density of $n\zeta_i^{(n)}$ easily follows from this and we write it for $z>0$ in the form

$$((i-1)!)^{-1} e^{-z}\left[n(e^{z/n}-1)\right]^{i-1} \prod_{j=1}^{i-1} (1-j/n). \tag{3.5}$$

For fixed $z>0$ and fixed i, (3.5) tends to $e^{-z}z^{i-1}/(i-1)!$. Since $|n(e^{z/n}-1)-z| \leqslant e^{z/n}z^2/(2n), 0<z$, the convergence is even uniform in every finite interval so that

$$\lim_{n\to\infty} P(n\zeta_i^{(n)} \leqslant z) = ((i-1)!)^{-1} \int_0^z e^{-y}y^{i-1}dy \tag{3.6}$$

for each nonnegative real z.

However, using (3.3), we get for $z>0$:

$$\begin{aligned} P(n(1-z_{n-i+1}\circ\eta^{(n)}) \leqslant z) &= P(\zeta_i^{(n)} \leqslant -\log(1-z/n)) \\ &= P\left(\zeta_i^{(n)} \leqslant \frac{z}{n} + \frac{z^2}{n^2}\delta(n,z)\right) \end{aligned}$$

with $\lim_{n\to\infty} \delta(n,z) = 1/2$ for each fixed $z>0$. For each $\varepsilon>0$ and each fixed $z>0$ we thus have, for sufficiently large n:

$$P(n\zeta_i^{(n)} \leqslant z-\varepsilon) \leqslant P(n(1-z_{n-i+1}\circ\eta^{(n)}) \leqslant z) \leqslant P(n\zeta_i^{(n)} \leqslant z+\varepsilon).$$

Because of (3.6) we then have, for each nonnegative real z

$$\lim_{n\to\infty} P(n(1-z_{n-i+1}\circ\eta^{(n)}) \leqslant z) = \int_0^z e^{-y}y^{i-1}/((i-1)!)dy. \tag{3.7}$$

An application of (1.13) shows that $z_i \circ \eta^{(n)}$ and $1 - z_{n-i+1} \circ \eta^{(n)}$ have the same distribution. (3.7) then finishes the proof.

Of course the result of Theorem 3.2 can also be derived directly from Theorem 1.3.

We note a consequence of Theorem 3.2:

Theorem 3.3. *Let $F \in C_m$ and $\xi_1, \xi_2, \dots$ a sequence of independent r.v.'s distributed according to F. Then, for each positive real x and fixed natural number i*

$$\lim_{n \to \infty} P(z_i \circ \xi^{(n)} \leqslant F^{-1}(x/n)) = \int_0^x \frac{y^{i-1}}{(i-1)!} e^{-y} dy. \tag{3.8}$$

The *proof* follows from the fact that $F \circ \xi_i$, $1 \leqslant i \leqslant n$, is uniformly distributed (see Theorem 1.4).

Theorem 3.3 is a statement about the asymptotic distribution of order statistics whose index does not depend on n. We now want to prove a theorem on order statistics whose index tends to ∞ [13] along with n itself.

Theorem 3.4. *Let F be a d.f. fulfilling the assumptions of the first part of Theorem 1.4. Moreover, assume the corresponding density f exists. Let μ denote the median of F and let f be continuous at μ. Let $\xi_1, \dots, \xi_{2n+1}$ be independent identically distributed r.v.'s with d.f. F. Then $z_{n+1} \circ \xi^{(2n+1)}$ is asymptotically $N(\mu, (f(\mu) \cdot 2\sqrt{2n+1})^{-2})$-distributed for $n \to \infty$.*

Proof. We start with $2n+1$ independent r.v.'s η_i which are uniformly distributed in $(0,1)$. We will use the notation introduced on p. 423 and p. 424. By means of the density (3.2) of $\boldsymbol{w}_j$, one gets

$$E(\boldsymbol{w}_j) = 1, \quad E[(\boldsymbol{w}_j - E(\boldsymbol{w}_j))^2] = 1, \quad 1 \leqslant j \leqslant 2n+1. \tag{3.9}$$

In addition, for a suitable real number $K > 0$

$$E[|\boldsymbol{w}_j - E(\boldsymbol{w}_j)|^3] \leqslant K, \quad 1 \leqslant j \leqslant 2n+1. \tag{3.10}$$

---

[13] The literature on limit theorems for order statistics is quite large. We mention: N.V. Smirnov, Trudy Mat. Inst. Steklov 1949. Of the many investigations of Gumbel we mention only E.J. Gumbel, Statistics of extremes, Columbia University Press, New York 1958. Investigations have also been made in the case where the r.v.'s $\xi_1, \xi_2, \dots$ mentioned in Theorem 3.3 no longer have the same distribution: D.G. Mejzler, Ukrain. Mat. Ž. 1, 67—84 (1949); M. Loéve, Proceedings of the 3rd Berkeley Symposium on Mathematical Statistics and Probability 1954—1955, Vol. II, pp. 177—194, University of California Press, Berkeley and Los Angeles. Generalizations to the two-dimensional case are in B.V. Finkelštein, Dokl. Akad. Nauk SSSR, n. Ser. 91, 209—211 (1953). A number of individual results, especially helpful in practical work, are brought together in Sarhan and Greenberg, Contributions to Order Statistics, John Wiley & Sons, New York 1962.

It is easy to see that

$$\zeta_{n+1}^{(2n+1)} = \sum_{j=1}^{n+1} w_j/(2n+2-j), \qquad n=1,2,\dots. \tag{3.11}$$

Conditions (3.9)—(3.10) thus suggest an application of I, Theorem 39.3. It is known that $\sum_{k=1}^{\infty} (-1)^{k-1}/k = \log 2$, and since this is an alternating series with monotone decreasing terms,

$$\left| \sum_{k=1}^{2n} \frac{(-1)^{k-1}}{k} - \log 2 \right| \leqslant 1/2n, \qquad n \geqslant 1.$$

Hence, from (3.9) and (3.11) we have because of

$$\frac{1}{m+1} + \dots + \frac{1}{2m} = 1 - \frac{1}{2} + \frac{1}{3} - + \dots + \frac{1}{2m-1} - \frac{1}{2m} \quad \text{for } m \geqslant 1$$

that

$$E(\zeta_{n+1}^{(2n+1)}) = \log 2 + O(n^{-1}). \tag{3.12}$$

From (3.9) and (3.11) we also have

$$\sigma^2(\zeta_{n+1}^{(2n+1)}) = E[(\zeta_{n+1}^{(2n+1)} - E(\zeta_{n+1}^{(2n+1)}))^2] = \sum_{j=1}^{n+1} \frac{1}{[2(n+1)-j]^2}, \qquad n \geqslant 1.$$

Now,

$$\int_n^{2n+1} x^{-2}\,dx > \sum_{k=1}^{n+1} \frac{1}{(n+k)^2} > \int_{n+1}^{2n+1} x^{-2}\,dx, \quad \text{i.e.,}$$

$$\frac{1}{n} - \frac{1}{2n+1} > \frac{1}{(n+1)^2} + \dots + \frac{1}{(2n+1)^2} > \frac{1}{2(n+1)}.$$

Then, we easily get

$$\sigma^2(\zeta_{n+1}^{(2n+1)}) = \frac{1}{2n+1} + O(n^{-2}).$$

With Taylor's theorem this implies that

$$\sigma(\zeta_{n+1}^{(2n+1)}) = (2n+1)^{-1/2} + O(n^{-1}). \tag{3.13}$$

And finally, from (3.10)

$$\sum_{j=1}^{n+1} E[|w_j - E(w_j)|^3/(2n+2-j)^3] \leqslant K/n^2.$$

Combining this with (3.13), we obtain

$$\left(\sum_{j=1}^{n+1} E[|w_j - E(w_j)|^3/(2n+2-j)^3)\right)^{1/3} \Big/ \sigma(\zeta_{n+1}^{(2n+1)}) \to 0 \quad \text{for } n\to\infty\,.$$

An application of I, Theorem 39.3 yields, for each $x \in R_1$,

$$\lim_{n\to\infty} P([(\zeta_{n+1}^{(2n+1)} - E(\zeta_{n+1}^{(2n+1)}))/\sigma(\zeta_{n+1}^{(2n+1)})] \leqslant x) = (2\pi)^{-1/2} \int_{-\infty}^{x} e^{-y^2/2}\,dy\,. \tag{3.14}$$

Now

$$(\zeta_{n+1}^{(2n+1)} - \log 2)[1/(2n+1)]^{-1/2}$$
$$= \frac{\zeta_{n+1}^{(2n+1)} - E(\zeta_{n+1}^{(2n+1)}) - \log 2 + E(\zeta_{n+1}^{(2n+1)})}{\sigma(\zeta_{n+1}^{(2n+1)})}\ \frac{\sigma(\zeta_{n+1}^{(2n+1)})}{(1/(2n+1))^{1/2}}$$

and from (3.12) and (3.13)

$$(\zeta_{n+1}^{(2n+1)} - \log 2)[1/(2n+1)]^{-1/2}$$
$$= \frac{\zeta_{n+1}^{(2n+1)} - E(\zeta_{n+1}^{(2n+1)})}{\sigma(\zeta_{n+1}^{(2n+1)})}(1+O(n^{-1/2})) + O(n^{-1/2})\,.$$

Thus, for each $x \in R_1$

$$P\left(\frac{\zeta_{n+1}^{(2n+1)} - \log 2}{(1/(2n+1))^{1/2}} \leqslant x\right)$$
$$= P([\zeta_{n+1}^{(2n+1)} - E(\zeta_{n+1}^{(2n+1)})](\sigma(\zeta_{n+1}^{(2n+1)}))^{-1} \leqslant (x - \varepsilon_1(n))/(1+\varepsilon_2(n)))$$

with $\varepsilon_i(n) = O(n^{-1/2})$, $i=1,2$. Thus, (3.14) also implies that

$$\lim_{n\to\infty} P\left(\frac{\zeta_{n+1}^{(2n+1)} - \log 2}{(1/(2n+1))^{1/2}} \leqslant x\right) = (2\pi)^{-1/2} \int_{-\infty}^{x} e^{-y^2/2}\,dy\,. \tag{3.15}$$

We turn now to the general case. According to Theorem 1.4, the previous results can be applied to the r.v.'s $\eta_i = F\circ\xi_i$, $1 \leqslant i \leqslant 2n+1$.

(3.3) and (3.11) imply $\zeta_{n+1}^{(2n+1)} = -\log(z_{n+1}\circ\eta^{(2n+1)})$. Considering then the inverse F^{-1} of F, which sends $(0,1)$ into (a,b), we get for $x \in R_1$

$$\left.\begin{aligned} &P(\zeta_{n+1}^{(2n+1)} \leqslant \log 2 + x(2n+1)^{-1/2}) \\ &\quad = P(z_{n+1}\circ\xi^{(2n+1)} \geqslant F^{-1}(\tfrac{1}{2}e^{-x\sqrt{1/(2n+1)}})\,. \end{aligned}\right\} \tag{3.16}$$

By assumption, the derivative of F^{-1} exists and is given by $1/(f\circ F^{-1})$. The Mean Value Theorem of differential calculus yields

$$F^{-1}(\tfrac{1}{2}e^{-x\sqrt{1/(2n+1)}}) - F^{-1}(1/2) = \tfrac{1}{2}(e^{-x\sqrt{1/(2n+1)}} - 1)(f(\mu_n))^{-1} \tag{3.17}$$

with

$$\mu_n = F^{-1}(1/2) + \vartheta(F^{-1}(\tfrac{1}{2}e^{-x\sqrt{1/(2n+1)}}) - F^{-1}(1/2))\,, \qquad |\vartheta| < 1\,.$$

Now, for each $x \in R_1$,

$$\lim_{n\to\infty} (e^{-x\sqrt{1/(2n+1)}} - 1)/(-x(2n+1)^{-1/2}) = 1 . \tag{3.18}$$

Moreover, f is continuous at $\mu = F^{-1}(\frac{1}{2})$ by assumption, so that

$$\lim_{n\to\infty} f(\mu_n) = f(\mu) . \tag{3.19}$$

If (3.15) is combined with (3.16)—(3.19), the claim of the theorem then follows from the continuity of the d.f. of the normal distribution.

In particular, assume that $\xi_1, \ldots, \xi_{2n+1}$ are independent and $N(a, \sigma^2)$-distributed. Then $\mu = a$. Theorem 3.4 then says that $z_{n+1} \circ \xi^{(2n+1)}$ is asymptotically $N(a, \pi\sigma^2/(2(2n+1)))$-distributed. By II, Theorem 3.2 and I, Theorem 39.1,

$$\bar{\xi}_{2n+1} = (\xi_1 + \cdots + \xi_{2n+1})/(2n+1)$$

is asymptotically $N(a, \sigma^2/(2n+1))$-distributed. If one chooses $\sigma^2 = 1$ for the sake of simplicity and if $-\infty < a < \infty$, then $\bar{\xi}_{2n+1}$ is efficient for a by V, p. 290. In addition, in the sense of the definition in V, p. 338, $\bar{\xi}_{2n+1}$ is asymptotically more efficient for a, $-\infty < a < \infty$ than $z_{n+1} \circ \xi^{(2n+1)}$. Indeed, $\lim_{n\to\infty} \frac{1}{2n+1}\left(\frac{\pi}{2(2n+1)}\right)^{-1} = 2/\pi < 1$. In spite of this, the $(n+1)$th order statistic is often used in practical applications as an estimate of the mean of a normal distribution since a realization of it can be determined without calculation from a sample of size n.

We turn now to several theorems of the Kolmogorov-Smirnov type. We will introduce the important notion of an *empirical distribution function*, which is closely connected with order statistics. For $n \geq 1$, let $x = (x_1, \ldots, x_n)$ be real numbers viewed as a sample from some population. For each real y let $A_n(x, y)$ denote the number of x_i, $1 \leq i \leq n$, which are $\leq y$. Then the function

$$F_n(x; y) = A_n(x, y)/n \tag{3.20}$$

defined for each $y \in R_1$ is called the *empirical distribution function* of $(x_1, \ldots, x_n)$. The following definition, which we will give, for the sake of simplicity, only for pairwise different x_i, $1 \leq i \leq n$, is equivalent:

$$\begin{aligned} &F_n(x; y) = 0 , && -\infty < y < z_1(x) , \\ &F_n(x; y) = l/n , && z_l(x) \leq y < z_{l+1}(x) , \quad 1 \leq l \leq n-1 , \\ &F_n(x; y) = 1 , && y \geq z_n(x) . \end{aligned}$$

If $\xi^{(n)}$ is an n-dimensional r.v., then $F_n(\xi^{(n)}; y)$ defines a r.v. for each $y \in R_1$. We will usually denote this r.v. by $\boldsymbol{F}_n(y)$.

From each of these definitions it is easy to see that when $\xi_1,\ldots,\xi_n$ are independent identically distributed with d.f. F, then the r.v. $\boldsymbol{F}_n(y)$ coincides (with $\xi^{(n)}=(\xi_1,\ldots,\xi_n)$) for each $y\in R_1$ with the r.v. η_y (see p. 410).

This immediately yields

Theorem 3.5. *Let $\xi_1, \xi_2,\ldots$ be a sequence of independent identically distributed r.v.'s with d.f. F. Then the sequence $\boldsymbol{F}_n(y)$ converges stochastically to $F(y)$ for each real y.*

This follows from an application of I, Theorem 32.2. An application of I, Theorem 38.5 gives a strong version of Theorem 3.5.

This result lends special support to the notion that F is the distribution function of an infinite population. If the sample is large enough, the empirical distribution function approximates the d.f. of the population arbitrarily closely at every point. The rate of this convergence is given by a theorem of Kolmogorov which we will give below.

First we make a general *remark* connected with the statement of Theorem 3.5. Let Γ be an arbitrary set of indices γ. Let a distribution function F_γ correspond one-to-one to each γ and let $y\in R_1$ be arbitrarily chosen. Let the map d be $\gamma\to F_\gamma(y)$. Then $F_n(y)$ is a sequence of consistent estimates for d. This simple observation forms the basis for the construction of sequences of consistent estimates under quite general conditions[14].

We now give Kolmogorov's theorem[15] without proof and observe the following fact. The supremum of a countable set of r.v.'s is again a r.v. Also $\sup_r \boldsymbol{F}_n(r)=\sup_{y\in R_1}\boldsymbol{F}_n(y)$ when r runs through all rational numbers. Hence, $\sup_{y\in R_1}\boldsymbol{F}_n(y)$ is also a r.v. (see V, Lemma 3.2).

Theorem 3.6. *Let the hypotheses of Theorem 3.5 be fulfilled. In particular, let $F\in C$. Then*

$$\lim_{n\to\infty} P\Big(\sqrt{n}\sup_{-\infty<y<\infty}|F_n(y)-F(y)|\leqslant x; F\Big) = \begin{cases}\sum\limits_{k=-\infty}^{\infty}(-1)^k e^{-2k^2x^2} & x>0\\ 0 & x\leqslant 0.\end{cases}$$

---

[14] See L. Le Cam, Proceedings of the 3rd Berkeley Symposium on Mathematical Statistics and Probability, 1954—1955, Vol. I, pp. 129—156, University of California Press, Berkeley and Los Angeles.

[15] The first proof is due to A. N. Kolmogorov, Giorn. Ist. Ital. Attuari 4, 83—91 (1933). Other proofs are in W. Feller, Ann. Math. Statist. 19, 177—189 (1948) and J. L. Doob, Ann. Math. Statist. 20, 393—403 (1949), augmented by M. D. Donsker, Ann. Math. Statist. 22, 277—281 (1952).

We will prove a weaker one-sided counterpart to Kolmogorov's theorem:

Theorem 3.7[16]. *Let* $\xi_1,\dots,\xi_n$, $n\geqslant 1$ *be independent identically distributed r.v.'s with the d.f.* $F\in C_m$. *Then, for each* $u>0$

$$P\Big(\sup_{y\in R_1}(F(y)-\boldsymbol{F}_n(y))\geqslant u;\,F\Big) \tag{3.21}$$
$$=\sum_{j=0}^{[n(1-u)]} u\binom{n}{j}(u+j/n)^{j-1}(1-u-j/n)^{n-j}.$$

For $u=0$ *the left side of* (3.21) *equals* 1.

Proof. The last statement can immediately be obtained directly or by a limit argument. The rest of the proof is simplified by the fact that the left side of (3.21) (in agreement with the claim) does not depend on the d.f. F under consideration. Indeed, let ψ be a strictly monotone increasing, continuous map from R_1 into R_1. Then, the r.v. $\eta_i=\psi^{-1}\circ\xi_i$, $1\leqslant i\leqslant n$, has the d.f. $G=F\circ\psi$. Lemma 1.2 then implies that the left side of (3.21) is also given by

$$P\Big(\sup_{y\in R_1}(G(y)-\boldsymbol{G}_n(y))\geqslant u;\,G\Big).$$

By I, Theorem 11.1, one can therefore assume that F is the d.f. of a uniform distribution over $(0,1)$. Denote the empirical distribution function of a uniform distribution by $\boldsymbol{H}_n$. For each $u>0$,

$$P\Big(\sup_{0<y<1}(y-\boldsymbol{H}_n(y))\geqslant u\Big) \tag{3.22}$$

is to be calculated. Considering an $x=(x_1,\dots,x_n)\in R_n$ and writing in detail $y\to H_n(x_1,\dots,x_n;\,y)$ for the empirical distribution function, we have that

$$\sup_{0<y<1}(y-H_n(x_1,\dots,x_n;\,y))>0$$

if $0\leqslant x_i<1$, $1\leqslant i\leqslant n$. Since H_n is constant except for finitely many jumps which can only occur among the $z_j(x_1,\dots,x_n)$, $1\leqslant j\leqslant n$, for at least one i with $0\leqslant i\leqslant n-1$

$$\sup_{0<y<1}(y-H_n(x_1,\dots,x_n;\,y))=z_{i+1}(x_1,\dots,x_n)-i/n$$

must hold. Hence, if $\eta_1,\dots,\eta_n$ are independent, uniformly over $(0,1)$ distributed r.v.'s, we then have

$$P\Big(\sup_{0<y<1}(y-\boldsymbol{H}_n(y))\geqslant u\Big)=P\Big(\max_{0\leqslant i\leqslant n-1}((z_{i+1}(\eta_1,\dots,\eta_n)-i/n)\geqslant u)\Big).$$

[16] Z. W. Birnbaum and F. H. Tingey, Ann. Math. Statist. 22, 592—596 (1951). E. Hlawka has applied this result in an investigation of order statistics from a quite different viewpoint: see E. Hlawka, Math. Ann. 150, 259—267 (1963).

But

$$\left.\begin{aligned}&\left\{x: \max_{0 \leqslant i \leqslant n-1} ((z_{i+1}(x_1,\dots,x_n) - i/n) \geqslant u)\right\}\\ &\quad = \bigcup_{j=0}^{n-1} \{x: z_1(x_1,\dots,x_n) < u, \dots, z_j(x_1,\dots,x_n)\\ &\qquad - (j-1)/n < u, z_{j+1}(x_1,\dots,x_n) - j/n \geqslant u\}\end{aligned}\right\}. \tag{3.23}$$

Since the sets on the right are disjoint, the probability we are seeking is given by

$$\left.\begin{aligned}&\sum_{j=0}^{n-1} P(z_1(\eta_1,\dots,\eta_n) < u, \dots, z_j(\eta_1,\dots,\eta_n)\\ &\qquad - (j-1)/n < u, z_{j+1}(\eta_1,\dots,\eta_n) - j/n \geqslant u)\end{aligned}\right\}. \tag{3.24}$$

It thus suffices to show that the j-th summand in (3.24) is given by $u\binom{n}{j}(u+j/n)^{j-1}(1-u-j/n)^{n-j}$, provided $j \leqslant [(1-u)n]$. The density of the joint distribution of the $z_i(\eta_1,\dots,\eta_n)$, $1 \leqslant i \leqslant j+1$ is obtained by applying (1.13). Hence, with $\eta^{(n)} = (\eta_1,\dots,\eta_n)$,

$$\begin{aligned}&P(z_1 \circ \eta^{(n)} < u, \dots, z_j \circ \eta^{(n)} - (j-1)/n < u, z_{j+1} \circ \eta^{(n)} - j/n \geqslant u)\\ &= n!\,[(n-j-1)!]^{-1} \int_0^u dz_1 \int_{z_1}^{u+1/n} dz_2 \dots \int_{z_{j-1}}^{u+(j-1)/n} dz_j \int_{u+j/n}^{1} dz_{j+1} (1-z_{j+1})^{n-j-1}.\end{aligned} \tag{3.25}$$

Now, for each real u:

$$\int_0^u dz_1 \int_{z_1}^{u+1/n} dz_2 = \frac{u}{2!}\left(u + \frac{2}{n}\right).$$

Assuming that

$$\int_0^u dz_1 \int_{z_1}^{u+1/n} dz_2 \dots \int_{z_{j-1}}^{u+(j-2)/n} dz_{j-1} = \frac{u}{(j-1)!}\left(u + \frac{j-1}{n}\right)^{j-2} \tag{3.26}$$

for each real u, it suffices to show that

$$\int_0^u dz_1 \int_{z_1}^{u+1/n} dz_2 \dots \int_{z_{j-1}}^{u+(j-1)/n} dz_j = \frac{u}{j!}(u+j/n)^{j-1}. \tag{3.27}$$

To this end, we make, in order for $i = j, \dots, 2$, the transformations $z_i = z_1 + v_{i-1}$. Then, for the integral on the left in (3.27), we get

$$\int_0^u dz_1 \int_0^{-z_1+1/n} dv_1 \dots \int_{v_{j-2}}^{u-z_1+(j-1)/n} dv_{j-1},$$

which, because of (3.26) is also equal to

$$\int_0^u (u-z_1+1/n)\,(u-z_1+j/n)^{j-2}\,dz_1 .$$

(3.27) then follows easily. Hence, the integral on the right in (3.25) is given by $\binom{n}{j} u(u+j/n)^{j-1}(1-u-j/n)^{n-j}$, provided $u+j/n<1$. For $u+j/n\geqslant 1$, the integral vanishes. The theorem is proved.

Remark. If η is a r.v. uniformly distributed over $(0,1)$ and $F\in C$ but not necessarily $\in C_m$, then III, Lemma 5.2 says that $\xi=u_F(\eta)$ has a distribution with d.f. F. Indeed (see the proof of Theorem 1.4),

$$P(\xi\leqslant x)=P(u_F(\eta)\leqslant x)=P(\eta\leqslant F(x))=F(x)$$

for each real x. This allows the extension of Theorem 3.7 to d.f.'s $F\in C$.

A *one-sided test* for the *null hypothesis* that a sample of size n arises from a population *with given d.f.* $F\in C_m$ (or even $\in C$) can be based on Theorem 3.7.

One can prove in exactly the same way as above, that

$$P\Big(\inf_{0<y<1}(F(y)-F_n(y))\leqslant -u;\,F\Big)$$

is equal to the right side of (3.21) for $n>0$. Since, however,

$$P\Big(\sup_{0<y<1}|F(y)-F_n(y)|\geqslant u\Big)\leqslant P\Big(\sup_{0<y<1}(F(y)-F_n(y))\geqslant u\Big) + P\Big(\inf_{0<y<1}(F(y)-F_n(y))\leqslant -u\Big),$$

one can also easily define a two-sided test for the problem mentioned above. On the other hand, it is rather difficult to calculate

$$P\Big(\sup_{0<y<1}|F(y)-F_n(y)|\geqslant u;\,F\Big)\quad\text{for } n\geqslant 2$$

and each real u with $0<u<1$. The behavior of this probability for $n\to\infty$ is described by Theorem 3.6. An analogous asymptotic result can be derived from (3.21). It is not difficult to show that under the assumptions of Theorem 3.7,

$$\lim_{n\to\infty} P\Big(\sqrt{n}\sup_{y\in R_1}(F(y)-F_n(y))\geqslant u;\,F\Big)=e^{-2u^2}$$

for each $u>0$.

The comparison of two empirical distribution functions is important for the *two-sample problem.* We give in this direction only the simplest one-sided result:

Theorem 3.8. *Let $\xi_1, \ldots, \xi_n; \xi_{n+1}, \ldots, \xi_{2n}$, $n \geqslant 1$, be independent r.v's each distributed by $F \in C$. Let F_n be the empirical distribution function defined by $\xi_1, \ldots, \xi_n$ and G_n that for $\xi_{n+1}, \ldots, \xi_{2n}$. Then*[17]

$$P\left(\sup_{-\infty<y<\infty} (F_n(y) - G_n(y)) \geqslant u; F\right) = \begin{cases} 1 & u \leqslant 0 \\ \binom{2n}{n-\{un\}} \Big/ \binom{2n}{n} & 0 < u \leqslant 1 \\ 0 & u > 1. \end{cases}$$

Proof[18]. For each $u \leqslant 0$ and $u > 1$ the claim is trivial. It is also easy to see that with $\xi^{(2n)} = (\xi_1, \ldots, \xi_{2n})$, the equation

$$\sup_{-\infty<y<\infty} (F_n(y) - G_n(y)) = \max_{1 \leqslant k \leqslant 2n} (F_n(z_k \circ \xi^{(2n)}) - G_n(z_k \circ \xi^{(2n)})) \quad (3.28)$$

holds. We now call on Theorem 1.2 and define, for each $x \in R_{2n}$ and $1 \leqslant i \leqslant 2n$

$$\delta(s_i(x)) = \begin{cases} 1 & s_i(x) = 1, \ldots, n \\ -1 & s_i(x) = n+1, \ldots, 2n. \end{cases}$$

Obviously, for $1 \leqslant i \leqslant 2n$ with $s_i = s_i \circ \xi^{(2n)}$

$$F_n(z_i \circ \xi^{(2n)}) - G_n(z_i \circ \xi^{(2n)}) = \frac{1}{n} \sum_{1 \leqslant j \leqslant i} \delta(s_j). \quad (3.29)$$

The joint distribution of the s_i, $1 \leqslant i \leqslant 2n$, is discrete with probabilities $1 \Big/ \binom{2n}{n}$ at the mass points. Because of (3.28) and (3.29), we must determine among all mass points $e = (e_1, \ldots, e_{2n})$ of this distribution with

$$e_i = \pm 1, \quad 1 \leqslant i \leqslant 2n \quad (3.30)$$

and

$$\sum_{1 \leqslant i \leqslant 2n} e_i = 0, \quad (3.31)$$

those which satisfy the condition

$$\max_{1 \leqslant j \leqslant 2n} \sum_{i=1}^{j} e_i \geqslant un. \quad (3.32)$$

---

[17] For the meaning of $\{un\}$, see[9].

[18] This proof is due to Gnedenko and Koroljuk: B.V. Gnedenko and V.S. Koroljuk, Dokl. Akad. Nauk SSSR, n. Ser. 80, 525—528 (1951). See also V.S. Koroljuk, Izv. Akad. Nauk SSSR, Ser. Mat. 19, 81—96 (1955).

Now, if for a j the inequality $\sum_{i=1}^{j} e_i \geqslant un$ is fulfilled, we have from (3.30) that

$$\sum_{i=1}^{j} e_i \geqslant \{un\} \tag{3.33}$$

and, naturally, vice versa. If (3.33) holds, then there exists a smallest index j_0, $1 \leqslant j_0 \leqslant j \leqslant 2n$, such that

$$\sum_{i=1}^{j_0} e_i = \{un\}. \tag{3.34}$$

Because of (3.31), we then have

$$-\sum_{i=j_0+1}^{2n} e_i = \{un\}. \tag{3.35}$$

Conversely, (3.34) and (3.35) imply the existence of at least one j for which (3.33) holds. Writing

$$e_i^* = e_i, \quad 1 \leqslant i \leqslant j_0, \quad e_i^* = -e_i, \quad j_0 + 1 \leqslant i \leqslant 2n, \tag{3.36}$$

we find from (3.34) and (3.35) that

$$\sum_{i=1}^{2n} e_i^* = 2\{un\}. \tag{3.37}$$

Now, $u > 0$ implies $\{un\} > 0$ so that if $e_i^* = \pm 1$, $1 \leqslant i \leqslant 2n$, and (3.37) holds, there always exists a smallest j_0 with $1 \leqslant j_0 < 2n$ such that (3.34) and (3.35) hold along with (3.36). Obviously, (3.37) holds iff $e_i^* = 1$ $(n + \{un\})$ times and $e_i^* = -1$ $(n - \{un\})$ times. This occurs in $\binom{2n}{n - \{un\}}$ ways and this completes the proof of the theorem.

The following consequence is easy:

Corollary. *Under the assumptions of Theorem* 3.8,

$$\lim_{n \to \infty} P\left(\sqrt{\frac{n}{2}} \sup_{-\infty < y < \infty} (F_n(y) - G_n(y)) \geqslant u; F\right) = \begin{cases} e^{-2u^2} & u > 0 \\ 1 & u \leqslant 0. \end{cases}$$

This has been only a small cross-section of the many results that have been obtained in this area[19]. We present two further results with-

[19] We mention B. V. Gnedenko, Math. Nachr. 12, 29—63 (1954) and Darling's instructive report: D. A. Darling, Ann. Math. Statist. 28, 823—839 (1957). Theorems of a somewhat different type are given by A. Rényi, l. c.[11], as well as I. Vincze, Magyar Tud. Akad. Mat. Kutató Int. Közl. 2, 183—209 (1957) and Magyar Tud. Akad. Mat. Kutató Int. Közl. 4, 29—47 (1959).

out proof. The first can be proved as in Theorem 3.8: *Under the assumptions of Theorem* 3.8:

$$P\Big(\sup_{-\infty<y<\infty}|F_n(y)-G_n(y)|\geqslant u; F\Big)$$

$$=\begin{cases}0 & u\leqslant 1/n\\ \displaystyle\sum_{k=-c(n)}^{c(n)}(-1)^k\binom{2n}{n-k\left[u\sqrt{\frac{n}{2}}\right]}\Big/\binom{2n}{n}, & u>\dfrac{1}{n},\end{cases}$$

where $c(n)=\left[u\sqrt{\frac{n}{2}}\right]+1.$

The second goes as follows[20]: Let $\{\xi_i\}$ and $\{\eta_j\}$ be sequences of independent identically distributed r.v.'s with d.f. $F\in C$ such that ξ_i and η_j for $i,j=1,2,\ldots$ are also independent. Let F_n correspond to the first n ξ_i's and G_m correspond to the first m η_j's. Then

$$\lim_{m,n\to\infty} P\left(\sqrt{\frac{nm}{n+m}}\sup_{-\infty<y<\infty}|F_n(y)-G_m(y)|<u; F\right)$$

$$=\begin{cases}\displaystyle\sum_{j=-\infty}^{\infty}(-1)^j e^{-2j^2u^2} & u>0\\ 0 & u\leqslant 0.\end{cases}$$

In connection with the definition of the empirical distribution function, we want to touch on the problem of *empirical Bayes estimates.* We showed in V, Theorem 3.12 (in the notation used there) that the function h^* defined for almost all $x\in R_n$ by V, (3.132) is a Bayes estimate for d. However, the effective calculation of h^* is only possible when the density p of the a priori distribution of γ is known. If it is not, one is faced with the following problem, which we formulate under the simplest assumptions: In the notation of p. 330, let $n=k=1$. Let $\mathfrak{Q}$ be an arbitrary set of a priori distributions Q of $\boldsymbol{\gamma}$. Then, for each $Q\in\mathfrak{Q}$, the Bayes estimate for d is given by

$$h_Q^*(x)=\int_{R_1} d(\gamma)f(x|\gamma)dQ(\gamma)\Big/\int_{R_1} f(x|\gamma)dQ(\gamma) \tag{3.38}$$

where we assume to avoid inessential complications that (3.38) makes sense for all $x\in R_1$. Now let $\xi_1, \xi_2, \ldots$ be a sequence of independent r.v.'s with the same distribution, whose density is given by $\int_{R_1} f(x|\gamma)dQ(\gamma)$

---

[20] This theorem is due originally to Smirnov: N. V. Smirnov, Mat. Sb., n. Ser. 6, 3—26 (1939), who makes the additional assumption that $\lim_{m,n\to\infty} m/n=c\neq 0$. The version given here was proved by I. I. Gichman, Dokl. Akad. Nauk SSSR, n. Ser. 82, 837—840 (1952).

for all $x \in R_1$. The ξ_n, $n \geq 1$ are to play the rôle of sample variables and the question arises whether one can construct with their help a sequence of estimators $\{h_n(\xi_1, \ldots, \xi_n; x)\}$ for each $x \in R_1$ which is consistent for $h_Q^*(x)$, $Q \in \mathfrak{Q}$. One might try to use the empirical distribution functions $F_n(\xi_1, \ldots, \xi_n; x)$, $n \geq 1$, for this purpose and to construct for each $x \in R_1$ a function ψ_x such that

$$h_n(\xi_1, \ldots, \xi_n; x) = \psi_x(F_n(\xi_1, \ldots, \xi_n; x)) \quad \text{for } n \geq 1 \quad \text{and } x \in R_1.$$

In simple cases, one can actually carry out such constructions[21].

4. Application of invariance theory. The two-sample problem. We have already touched in **1** on connections with the theory developed in III.**11**. and now want to treat them in more detail. In particular, we will consider among other topics the two-sample problem.

As sample space we will almost always consider $(R_n, \mathfrak{B}_n)$ for an $n \geq 1$. Let $\mathfrak{R}^{(n)} \subseteq \prod_{i=1}^{n} {}_*\mathfrak{F}^{(i)}$. An n-dimensional d.f. $F^{(n)} \in \prod_{i=1}^{n} \mathfrak{F}^{(i)}$ corresponds uniquely to each $P^{(n)} \in \mathfrak{R}^{(n)}$. $P^{(n)}$ is thus given by an n-tuple $(F_1, \ldots, F_n)$ of d.f.'s which determines $F^{(n)}$.

It is therefore often convenient to use as parameter set for $\mathfrak{R}^{(n)}$ simply the set of all corresponding n-tuples $(F_1, \ldots, F_n)$. This is especially recommended when $\mathfrak{R}^{(n)} \subseteq {}_*\mathfrak{F}^{(nn)}$. In this case there is thus a one-to-one map from $\mathfrak{R}^{(n)}$ into a subset of ${}_*\mathfrak{J}$. If $\mathfrak{G}$ is a group which acts in the sense of III, p. 235 on $(R_n, \mathfrak{B}_n)$, one obtains from $P^{(n)}_{F^{(n)}} \in \mathfrak{R}^{(n)}$ a probability measure $P^{(n)}_{\bar{\mathfrak{g}}F^{(n)}}$ with $F^{(n)} = (F_1, \ldots, F_n)$ for every $\mathfrak{g}$ when one applies III, (11.26).

We now consider a special group. Let $\mathfrak{G}^*$ be the group of all continuous and strictly increasing transformations of R_1 into itself, where the group operation is composition. Let a transformation $\mathfrak{g}$ of R_n into itself correspond to each $\mathfrak{g}^* \in \mathfrak{G}^*$ for $n \geq 2$: $\mathfrak{g}x = (\mathfrak{g}^* x_1, \ldots, \mathfrak{g}^* x_n)$ for all $x \in R_n$. The set of all these $\mathfrak{g}$ is a group $\mathfrak{G}_n$, where the group operation $\mathfrak{g}_1 \mathfrak{g}$ is naturally defined by $\mathfrak{g}_1 \mathfrak{g} x = (\mathfrak{g}_1^* \mathfrak{g}^* x_1, \ldots, \mathfrak{g}_1^* \mathfrak{g}^* x_n)$, $x \in R_n$. For this group, the distribution function $\bar{\mathfrak{g}} F^{(n)}$ (corresponding to the measure $P^{(n)}_{\bar{\mathfrak{g}}F^{(n)}}$) can be easily determined. Indeed, choosing in III. (11.26) an interval $I \subseteq R_n$ of the form $\prod_{i=1}^{n} (-\infty, y_i)$, one then gets

$$P^{(n)}_{\bar{\mathfrak{g}}F^{(n)}}(I) = P^{(n)}_{F^{(n)}}(\mathfrak{g}^{-1} I) = \prod_{i=1}^{n} F_i((\mathfrak{g}^*)^{-1} y_i) = \prod_{i=1}^{n} \overline{\mathfrak{g}^*} F(y_i).$$

[21] H. Robbins, Proceedings of the 3rd Berkeley Symposium on Mathematical Statistics and Probability, Vol. I, pp. 157—163, University of California Press, Berkeley and Los Angeles. See also M. V. Johns jr. Ann. Math. Statist. 28, 649—669 (1957).

Thus for each $y \in R_n$

$$\bar{\mathfrak{g}} F^{(n)}(y) = \prod_{i=1}^{n} F_i((\mathfrak{g}^*)^{-1} y_i). \tag{4.1}$$

If we want to apply the group $\mathfrak{G}_n$ within the framework of invariance theory, we must first find sets of probability measures which are invariant w.r.t. $\mathfrak{G}_n$.

To this end we prove

Theorem 4.1. *Let $\psi_i, 1 \leqslant i \leqslant n$, be continuous nondecreasing functions over R_1 which satisfy the conditions $\psi_i(0)=0$, $\psi_i(1)=1$. Then the set $\mathfrak{R}_\psi^{(n)}$ of all probability measures over $(R_n, \mathfrak{B}_n)$ whose d.f.'s*[22] *are given by $\prod_{i=1}^{n} \psi_i(F(x_i))$, $F \in C$ for each $(x_1, \dots, x_n) \in R_n$ is invariant w.r.t. $\mathfrak{G}_n$. Further for a $\bar{\mathfrak{g}}^* \in \bar{\mathfrak{G}}^*$,*

$$\tilde{F}_i = \mathfrak{g}^* F_i, \quad \tilde{F}_i, F_i \in C, \quad 1 \leqslant i \leqslant n, \tag{4.2}$$

holds iff there exist functions ψ_i with the given properties and $\tilde{F}, F \in C_m$, such that for $1 \leqslant i \leqslant n$

$$\tilde{F}_i = \psi_i \circ \tilde{F} \tag{4.3}$$

and

$$F_i = \psi_i \circ F. \tag{4.4}$$

Proof. The first claim is almost trivial because of (4.1). In fact, along with $F, y \to F((\mathfrak{g}^*)^{-1} y)$ belongs to C for each $\mathfrak{g}^* \in \mathfrak{G}^*$. The n-dimensional d.f. defined for each $x \in R_n$ by $\prod_{i=1}^{n} \psi_i(F((\mathfrak{g}^*)^{-1} x_i))$ thus defines a measure from $\mathfrak{R}_\psi^{(n)}$.

To show the remaining claims we assume that (4.2) holds. If F is chosen arbitrarily from C_m, then the relation (4.1) holds with $\psi_i = F_i \circ F^{-1}$, $\psi_i(0)=0$, $\psi_i(1)=1$. Defining $\tilde{F}$ by $\tilde{F}(y) = F((\mathfrak{g}^*)^{-1} y)$ for each $y \in R_1$, we see that $\tilde{F} \in C_m$ and (4.2) and (4.4) then imply (4.3). The converse is also quite easy. If (4.3) and (4.4) hold for $1 \leqslant i \leqslant n$, then $\tilde{F}^{-1} \circ F$ is a continuous, strictly increasing map from R_1 onto R_1 and is thus an element of $\mathfrak{G}^*$ which we denote by $(\mathfrak{g}^*)^{-1}$. (4.2) is satisfied by this element.

We have the following

Corollary. *Let ϕ be a test invariant w.r.t. $\mathfrak{G}_n$. Assume ψ_i and $\tilde{\psi}_i$, $1 \leqslant i \leqslant n$, fulfill the conditions in Theorem 4.1. Then ϕ is similar for*

---

[22] We also abbreviate these distribution functions by

$$\prod_{i=1}^{n} \psi_i \circ F.$$

the problem[23]

$$\left(\left\{\prod_{i=1}^{n} \psi_i \circ F : F \in C_m\right\}, \left\{\prod_{i=1}^{n} \tilde{\psi}_i \circ F : F \in C_m\right\}\right).$$

and possesses a constant power function.

This follows at once from the corollary to III, Lemma 11.2. In fact, if F and G are choosen arbitrarily from C_m, then there always exists a $\mathfrak{g}^* \in \mathfrak{G}^*$ such that G coincides with the map $y \to F(\mathfrak{g}^* y)$, $y \in R_1$.

The following argument serves for the construction of invariant tests: For each $x \in R_n$ let $M_x = \{\mathfrak{g}x : \mathfrak{g} \in \mathfrak{G}_n\}$. Any two sets M_x, M_y, $x, y \in R_n$ are either identical or disjoint. $R_n = \bigcup_{x \in R_n} M_x$ and for each $\mathfrak{g} \in \mathfrak{G}_n$ and $x \in R_n$ there holds $\mathfrak{g} M_x = M_x$.

Now let Θ be a map of R_n into some set Q which is constant on each M_x but in such a way that the elements from Q which are assumed on different sets M_x are different. Then, Θ is invariant and we call such maps *maximal invariant*. Indeed, if Θ_1 is an arbitrary invariant map into a set Q_1 such that $\Theta_1(\mathfrak{g}x) = \Theta(x)$ for each $x \in R_n$ and all $\mathfrak{g} \in \mathfrak{G}_n$, then there always exists a map χ from Q into Q_1 with $\Theta_1 = \psi \circ \Theta$. If $\Theta(x) = q$ and $\Theta_1(x) = q_1$ for $x \in R_n$, it is enough to define $\chi(q) = q_1$. χ is well-defined since Θ is maximal invariant so that $\Theta_1(x) \neq \Theta_1(y)$ always implies $\Theta(x) \neq \Theta(y)$. Knowledge of a maximal invariant map thus provides in principle a complete survey of all maps invariant w.r.t. $\mathfrak{G}_n$.

Naturally, one sees at once that the special nature of the group $\mathfrak{G}_n$ and the space R_n upon which it acts, is not essential for these arguments.

We now apply these ideas. Let N_n have the same meaning as on p. 409 and S_n the same as on p. 407. If $(j_1, \ldots, j_n)$ is a permutation of $(1, \ldots, n)$ and

$$M_{(j_1, \ldots, j_n)} = \{x : x \in R_n, x_{j_1} < x_{j_2} < \cdots < x_{j_n}\},$$

then $\bigcup_{(j_1, \ldots, j_n) \in S_n} M_{(j_1, \ldots, j_n)} = R_n - N_n$ and the $M_{(j_1, \ldots, j_n)}$ are pairwise disjoint. Let $x, y \in M_{(j_1, \ldots, j_n)}$ and $x \neq y$. Then there always exists a $\mathfrak{g} \in \mathfrak{G}_n$ with $\mathfrak{g}x = y$. It sufficies to define the corresponding $\mathfrak{g}^*$ as follows:

$$\mathfrak{g}^* z = \begin{cases} y_{j_1} + \dfrac{y_{j_2} - y_{j_1}}{x_{j_2} - x_{j_1}}(z - x_{j_1}), & z \leqslant x_{j_2}, \\[2ex] y_{j_k} + \dfrac{y_{j_{k+1}} - y_{j_k}}{x_{j_{k+1}} - x_{j_k}}(z - x_{j_k}), & x_{j_k} \leqslant z \leqslant x_{j_{k+1}}, \quad 2 \leqslant k \leqslant n-2, \\[2ex] y_{j_{n-1}} + \dfrac{y_{j_n} - y_{j_{n-1}}}{x_{j_n} - x_{j_{n-1}}}(z - x_{j_{n-1}}), & z \geqslant x_{j_{n-1}}. \end{cases}$$

---

[23] We naturally assume that the set of null hypotheses and the set of alternatives are disjoint.

Since there obviously exists no $\mathfrak{g}\in\mathfrak{G}_n$ with $\mathfrak{g}x=y$ for $x,y\in R_n$ which belong to different $M_{(j_1,\ldots,j_n)}$ we have $M_{(j_1,\ldots,j_n)}=M_x$ for each $x\in M_{(j_1,\ldots,j_n)}$.

The map $x\to(r_1(x),\ldots,r_n(x))$ (see p.407) is thus maximal invariant on R_n-N_n. But we have—under the assumptions of Theorem 1.1, say—$P_{\xi^{(n)}}(N_n)=0$ so that in this case, each invariant test is a function of the ranks except on a set of $P_{\xi^{(n)}}$-measure 0. One thus calls tests invariant w.r.t. $\mathfrak{G}_n$ *rank-invariant*.

It is obvious that the map $x\to(s_1(x),\ldots,s_n(x))$ is also maximal invariant on R_n-N_n.

The corollary to Theorem 4.1 says that for each problem of testing referred to there, one can construct a rank-invariant test by means of the Neyman-Pearson lemma which is most powerful for the alternative hypothesis w.r.t. the set of all rank-invariant tests with the same level of significance.

We now turn to rank-invariant tests for the *two-sample problem* which we will again formulate in a quite general way (see II, p. 137): Let $(R_n,\mathfrak{B}_n)$ be the sample space with $n_1+n_2=n$, $n_i\geqslant 1$, assumed to hold. Let the set of admissible hypotheses be of the form $\mathfrak{F}^{(n_1n_1)}\times\mathfrak{F}^{(n_2n_2)}$ and the null hypothesis given by $\mathfrak{F}^{(nn)}$.

We denote the sample variables by $\xi_1,\ldots,\xi_{n_1}$, $\xi_{n_1+1},\ldots,\xi_{n_1+n_2}$. It will usually be expedient in the sense of the remark on p.436 to denote the set $\mathfrak{F}^{(n_1n_1)}\times\mathfrak{F}^{(n_2n_2)}$ (or $C^{(n_1n_1)}\times C^{(n_2n_2)}$) by (F,G) and $\mathfrak{F}^{(nn)}$ (or $C^{(nn)}$) by (F,F) where F, $G\in\mathfrak{F}$ (or F, $G\in C$).

The above formulation of the two-sample problem corresponds to most practical requirements: One wants to test the hypothesis that the sample of size n_1 and that of size n_2 come from the same population. "Complete uncertainty" governs the alternative hypotheses. The theory shows, however, that the handling of alternative hypotheses which are "too comprehensive" is extraordinarily complicated. On the other hand, for alternatives of the form $\{(F,\psi(F)\colon F\in C_m)\}$, where ψ is a continuous, nondecreasing map from $[0,1]$ into $[0,1]$ (which should naturally be different from the identity mapping) the problem is quite simple, as we have just mentioned. In this case—as the corollary to Theorem 1.6 shows—the explicit calculation of the power function can be carried out without difficulty, at least under the conditions given there. For practical purposes, however, such an alternative is usually unrealistic.

Before giving several examples of rank invariant tests for the two-sample problem, we want to simplify the testing problem further by means of a sufficient transformation and the remark of III, p.208.

We prove

Theorem 4.2. *Let $\xi_1,\ldots,\xi_{n_1}$ and $\xi_{n_1+1},\ldots,\xi_{n_1+n_2}$ be independent and identically distributed r.v.'s with d.f.'s F and G, resp. Let F and G run in-*

identically distributed r.v.'s with d.f. F and G, resp. Let F and G run independently of each other through C. Let the set of distributions induced by the (n_1+n_2)-dimensional r.v. $(s_1\circ\xi^{(n)},\ldots,s_{n_1+n_2}\circ\xi^{(n)})$, $n_1+n_2=n$, be denoted by $\mathfrak{M}$. Then the transformation T given by $x\to(\varepsilon(s_1(x)),\ldots,\varepsilon(s_{n_1+n_2}(x)))$ is sufficient for $\mathfrak{M}$.

Proof. Let F,G be arbitrary d.f.'s $\in C$ and $(i_1,\ldots,i_n)$ some permutation of $(1,\ldots,n)$. Now

$$P(s_1=i_1,\ldots,s_n=i_n\,|\,T;(F,G)),$$

provided it does not trivially vanish, is given by

$$\frac{\underset{x_{i_1}<\cdots<x_{i_n}}{\int\cdots\int} dF(x_1)\ldots dF(x_{n_1})dG(x_{n_1+1})\ldots dG(x_n)}{n_1!\,n_2!\underset{x_{i_1}<\cdots<x_{i_n}}{\int\cdots\int} dF(x_1)\ldots dF(x_{n_1})dG(x_{n_1+1})\ldots dG(x_n)}=1/(n_1!\,n_2!)$$

since for all $x^{(1)}=(x_1,\ldots,x_{n_1})\in R_{n_1}$, $x^{(2)}=(x_{n_1+1},\ldots,x_n)\in R_{n_2}$ and each $\Pi_{n_1}\in S_{n_1}$, and $\Pi_{n_2}\in S_{n_2}$ (see p.407)

$$T(x^{(1)},x^{(2)})=T(\Pi_{n_1}x^{(1)},\Pi_{n_2}x^{(2)})\,.$$

This completes the proof.

Hence, for the construction of rank-invariant tests, it is sufficient to consider test statistics of the form $x\to h(\varepsilon(s_1(x)),\ldots,\varepsilon(s_n(x)))$, where h is a map of the set of all vectors $\mathfrak{e}$ (see p.410) into R_1.

Important examples of such test statistics are:

1. $\sum_{i=1}^{n} i\varepsilon(s_i)$. The corresponding test is called the *Wilcoxon*[24] or *Mann-Whitney test.*

2. $\sum_{i=1}^{n}\Phi^{-1}(i/(n+1))\varepsilon(s_i)$, where Φ is the d.f. of an $N(0,1)$. This is *van der Waerden's test*[25].

3. $\sum_{i=1}^{n}E(z_1\circ\eta^{(n)})\varepsilon(s_i)$, where $\eta^{(n)}$ is an n-dimensional r.v. with n independent, $N(0,1)$-distributed random components. This defines the *Terry-Fisher-Yates test*[26].

[24] F. Wilcoxon, Biometrics 1, 80—83 (1945). H. B. Mann and D. R. Whitney, Ann. Math. Statist. 18, 50—60 (1947). A detailed exposition is in J. Hemelrijk and Ph. van Elteren, Cursus Toegepaste Statistiek, Hoofdstuk 8. De Toets van Wilcoxon. Mathematisch Centrum: Amsterdam 1954. See also A. Rényi, Magyar Tud. Akad. Alkalm Mat. Int. Közl. 2, 243—265 (1954). Interesting historical remarks are to be found in W. H. Kruskal, J. Amer. Statist. Assoc. 52, 356—360 (1957).

[25] B. L. van der Waerden, Math. Ann. 126, 93—107 (1953).

[26] M. E. Terry, Ann. Math. Statist. 23, 346—366 (1952).

We now consider the Wilcoxon test in more detail. It is usual to describe this test by means of the test statistic

$$\sum_{i=1}^{n} i\varepsilon(s_i) - \frac{1}{2} n_2(n_2+1). \tag{4.5}$$

The use of this test statistic has certain practical advantages: Indeed, let $x_1,\ldots,x_{n_1}; x_{n_1+1},\ldots,x_{n_1+n_2}$ be a sample and assume that the x_i, $1\leqslant i\leqslant n$, $(n_1+n_2=n)$, are pairwise different and are ordered according to their magnitudes. In this connection, assume that the $x_{n_1+1},\ldots,x_{n_1+n_2}$ after having been ordered are numbered by $i_1,\ldots,i_{n_2}$, where the inequalities $i_j<i_{j+1}$, $1\leqslant j\leqslant n_2-1$ and $1\leqslant i_k\leqslant n$, $1\leqslant k\leqslant n_2$, are to hold. Then, i_1-1 of the remaining x_l are smaller than x_{i_1}, i_2-2 of the x_l smaller than $x_{i_2},\ldots$, and finally, $i_{n_2}-n_2$ of the x_l smaller than x_{n_2} with $1\leqslant l\leqslant n_1$. The number of pairs (x_l,x_k), $1\leqslant l\leqslant n_1$, $n_1+1\leqslant k\leqslant n_1+n_2$ with $x_l<x_k$ is thus given by $\sum_{j=1}^{n_2} i_j-\frac{1}{2}n_2(n_2+1)$ and this is precisely the value assumed by the test statistic (4.5) for this sample. In practice the number of these pairs can be determined rapidly if n is not too large so that the Wilcoxon test has obtained a certain popularity.

The test statistic (4.5) can also be represented in another way. Let $I=[0,\infty)$ and $U(x_1,\ldots,x_n)=\sum_{l=1}^{n_1}\sum_{k=n_1+1}^{n_1+n_2} c_I(x_k-x_l)$ for $x\in R_n$. Then (4.5)—viewed as a map over R_n—and U coincide on R_n-N_n.

The following result is easy to show:

Theorem 4.3. *Let the assumptions of Theorem* 4.2 *hold and set* $U=U(\xi_1,\ldots,\xi_n)$. *Then*

$$E(U;(F,G))=n_1 n_2 \int_{-\infty}^{\infty} F\,dG \tag{4.6}$$

and

$$E[(U-E(U))^2;(F,G)]=n_1 n_2 \int_{-\infty}^{\infty} F\,dG+n_1 n_2(1-n_1-n_2)\left(\int_{-\infty}^{\infty} F\,dG\right)^2$$
$$+n_1(n_1-1)n_2 \int_{-\infty}^{\infty} F^2\,dG+n_2(n_2-1)n_1 \int_{-\infty}^{\infty} (1-G)^2\,dF. \tag{4.7}$$

In particular, for $F=G$,

$$E[U;(F,F)]=n_1 n_2/2 \tag{4.8}$$

and

$$E[(U-E(U))^2;(F,F)]=\tfrac{1}{12} n_1 n_2(n_1+n_2+1). \tag{4.9}$$

Proof. For $1 \leqslant l \leqslant n_1$, $n_1+1 \leqslant k \leqslant n$[27],

$$E(c_I(\xi_k-\xi_l)) = P(\xi_k-\xi_l \geqslant 0) = \int_{-\infty}^{\infty} F\,dG. \tag{4.10}$$

From this follows (4.6). Further

$$E(c_I^2(\xi_k-\xi_l)) = E(c_I(\xi_k-\xi_l)),$$

so that

$$E(c_I^2(\xi_k-\xi_l)) = \int_{-\infty}^{\infty} F\,dG. \tag{4.11}$$

Since the ξ_i are independent,

$$\begin{cases} E(c_I(\xi_k-\xi_l)\,c_I(\xi_{k'}-\xi_{l'})) = \left(\int_{-\infty}^{\infty} F\,dG\right)^2 \\ 1 \leqslant l,\, l' \leqslant n_1,\ n_1+1 \leqslant k,\, k' \leqslant n,\ l \neq l',\ k \neq k'. \end{cases} \tag{4.12}$$

If $l \neq l'$, we have with $M = \{x_1, x_2, x_3 : x_1 - x_2 \geqslant 0, x_1 - x_3 \geqslant 0\}$

$$E(c_I(\xi_k-\xi_l)\,c_I(\xi_k-\xi_{l'})) = \iiint_M dG(x_1)\,dF(x_2)\,dF(x_3)$$

$$= \int_{-\infty}^{+\infty} dG(x_1) \int_{-\infty}^{x_1} dF(x_2) \int_{-\infty}^{x_1} dF(x_3),$$

hence,

$$E(c_I(\xi_k-\xi_l)\,c_I(\xi_k-\xi_{l'})) = \int_{-\infty}^{\infty} F^2\,dG. \tag{4.13}$$

Analogously, for $k \neq k'$

$$E(c_I(\xi_k-\xi_l)\,c_I(\xi_{k'}-\xi_l)) = \int_{-\infty}^{\infty} (1-G)^2\,dF. \tag{4.14}$$

An elementary calculation then leads by means of (4.6) and (4.11)—(4.14) to (4.7). The equations (4.8) and (4.9) are obtained by specializing (4.6) and (4.7) from

$$\int_{-\infty}^{+\infty} F\,dF = 1/2, \qquad \int_{-\infty}^{\infty} F^2\,dF = 1/3 \quad \text{and} \quad \int_{-\infty}^{\infty} (1-F)^2\,dF = 1/3.$$

We now prove a general theorem[28] on the unbiasedness of certain tests. This theorem will then be applied in particular to the Wilcoxon test.

---

[27] Obviously, this and the following expectations are taken w.r.t. (F, G).

[28] E. L. Lehmann, Ann. Math. Statist. 22, 165—179 (1951).

Theorem 4.4. *Let* $\xi_1,\ldots,\xi_{n_1};\ \xi_{n_1+1},\ldots,\xi_{n_1+n_2}$ *be independent r.v.'s which are distributed under the null hypothesis according to* $\{(F,F)\colon F\in C\}$ *and under the alternative according to* $\{(F,G)\colon F,G\in C, G\leqslant F, F\neq G\}$. *Then, each test* ϕ *defined over* R_n, $n_1+n_2=n$, $n_1, n_2\geqslant 1$, *and similar for the null hypothesis (for arbitrary level of significance), which satisfies the condition*

$$\phi(x_1,\ldots,x_{n_1},x_{n_1+1},\ldots,x_n)\geqslant\phi(x_1,\ldots,x_{n_1},y_{n_1+1},\ldots,y_n) \qquad (4.15)$$

for all $(x_1,\ldots,x_{n_1})\in R_{n_1}$ *and all* $(x_{n_1+1},\ldots,x_n)\in R_{n_2}$, $(y_{n_1+1},\ldots,y_n)\in R_{n_2}$ *with* $x_i\geqslant y_i$, $n_1+1\leqslant i\leqslant n$, *is unbiased for the given alternative.*

Proof. Let ϕ be one of the tests mentioned in the assumption for level of significance α. Let $F,G\in C$ be chosen with

$$G\leqslant F. \qquad (4.16)$$

By III, Lemma 5.2, we construct the functions u_G and u_F. According to this construction, we have from (4.16)

$$u_F\leqslant u_G. \qquad (4.17)$$

Let η be a r.v. uniformly distributed in $(0,1)$. It was shown on p. 432 that $u_F\circ\eta$ has the d.f. F and $u_G\circ\eta$ the d.f. G. Thus, if $\eta_1,\ldots,\eta_n$ are independent r.v.'s with the same distribution as η, we find by assumption that

$$E(\phi(u_F\circ\eta_1,\ldots,u_F\circ\eta_n))=\alpha.$$

Then, by (4.17) and (4.15)

$$E(\phi(u_F\circ\eta_1,\ldots,u_F\circ\eta_{n_1},u_G\circ\eta_{n_1+1},\ldots,u_G\circ\eta_{n_1+n_2}))\geqslant\alpha$$

which we wanted to show.

As an application we get almost at once: The *Wilcoxon test is unbiased for the problem*

$$(\{(F,F);F\in C\},\{(F,G);G\leqslant F, G\neq F, F,G\in C\}).$$

Indeed, the test ϕ is given by choosing for given level of significance α suitable real numbers k and c with $0\leqslant c\leqslant 1$ and then defining for $x\in R_n$:

$$\phi(x)=\begin{cases}1 & U(x)>k\\ c & U(x)=k\\ 0 & U(x)<k.\end{cases}$$

If one replaces each x_i by x_i^* with $x_i^*\geqslant x_i$ for $n_1+1\leqslant i\leqslant n$, U can by definition (see p. 441) only increase. ϕ thus satisfies (4.15).

5. The theory of runs and the Wald-Wolfowitz test. We discuss still another test[29] for the two-sample problem and introduce the notion of a *run*. Let n_1 identical objects be given which we label by a and n_2 other identical objects labeled by b. Let $n_1+n_2=n$, $1\leqslant n_1\leqslant n-1$, $1\leqslant n_2\leqslant n-1$. Order all of the objects into a sequence, for example *aaa b a bbb aa bb a* .... Each subsequence of adjacent equal elements is called a *run* and the number of its elements is called its *length*. In the example we have 6 runs: *aaa*; *b*; *a*; *bbb*; *aa*; *bb*. To make the notion of a run useful in a probability framework we proceed as follows. Consider all n-tuples whose components consist of n_1 a's and n_2 b's. Call this set R and let $\mathscr{S}$ be the set of all subsets of R. Let the probability distribution over $(R,\mathscr{S})$ be uniform assigning probability $n_2!\,n_2!/n!$ to each of these n-tuples. Define a r.v. $\boldsymbol{i}$ over $(R,\mathscr{S})$ which gives the number of runs contained in each n-tuple from R. We prove

Theorem 5.1. *Let the above mentioned uniform distribution be given over* $(R,\mathscr{S})$ with $1\leqslant n_1<n$, $1\leqslant n_2<n$, $n_1+n_2=n$. Then[30]

$$\left.\begin{aligned} P(\boldsymbol{i}=2k)&=2\binom{n_1-1}{k-1}\binom{n_2-1}{k-1}\Big/\binom{n_1+n_2}{n_1},\\ P(\boldsymbol{i}=2k+1)&=\left(\binom{n_1-1}{k-1}\binom{n_2-1}{k}+\binom{n_1-1}{k}\binom{n_2-1}{k-1}\right)\Big/\binom{n_1+n_2}{n},\\ k&=1,\ldots,\left[\frac{n}{2}\right].\end{aligned}\right\}\quad(5.1)$$

Proof. It sufficies to show that there exist $2\binom{n_1-1}{k-1}\binom{n_2-1}{k-1}$ elements of R containing exactly $2k$ runs. The first part of (5.1) then follows from I, Theorem 1.1. Split the runs up into a- and b-runs. Since the a- and b-runs in a fixed n-tuple from R must always alternate, such an n-tuple consists of $k\,a$- and $k\,b$-runs. We must now answer the following combinatorial question: In how many ways can one form k runs from n_1 elements a and k runs from n_2 elements b? If we imagine the n_1 a-objects writen in a row, we can then get the k runs by distributing $k-1$ separation bars[31] in some way over the n_1-1 "spaces" between the a's. The first a-run then goes up to the first separation bar and the k-th begins after the last bar (for example, with $n_1=7$, $k=4$: $a|aaa|aa|a$, etc.). We must therefore determine in how many ways $k-1$ objects can be distributed over n_1-1 positions in such a way that each position

[29] A. Wald and J. Wolfowitz, Ann. Math. Statist. 11, 147—162 (1940).

[30] See W. Feller, 61ff. l.c. I[1].

[31] At most one separation bar can be placed in each space.

is occupied by at most one object. One thus gets $\binom{n_1-1}{k-1}$ a-runs and in the same way $\binom{n_2-1}{k-1}$ b-runs which can be combined in a total of $\binom{n_1-1}{k-1}\binom{n_2-1}{k-1}$ ways. However, since two n-tuples with the same a- and b-runs are different when an a-run begins the first and a b-run the second, we get exactly $2\binom{n_1-1}{k-1}\binom{n_2-1}{k-1}$ n-tuples containing $2k$ runs. This proves the first part of (5.1).

For the second part of (5.1) we note that an n-tuple with exactly $2k+1$ runs must contain either k a-runs and $k+1$ b-runs, or vice versa. Now argue as above.

Let us find the expectation of $\boldsymbol{i}$. We have

$$\begin{aligned}
&\binom{n_1+n_2}{n_1}E(\boldsymbol{i})\\
&\quad=\sum_{k=1}^{[\frac{n}{2}]}\left(2\binom{n_1-1}{k-1}\binom{n_2-1}{k-1}2k+\left[\binom{n_1-1}{k}\binom{n_2-1}{k-1}\right.\right.\\
&\qquad\left.\left.+\binom{n_1-1}{k-1}\binom{n_2-1}{k}\right](2k+1)\right)\\
&\quad=\sum_{k=1}^{[\frac{n}{2}]}\binom{n_1-1}{k-1}\binom{n_2-1}{k-1}\left(4k+\left(\frac{n_1+n_2-2k}{k}\right)(2k+1)\right]\\
&\quad=2(n_1+n_2-1)\sum_{k=1}^{[\frac{n}{2}]}\binom{n_1-1}{k-1}\binom{n_2-1}{k-1}+\frac{(n_1+n_2)}{n_2}\sum_{k=1}^{[\frac{n}{2}]}\binom{n_1-1}{k-1}\binom{n_2}{k}\\
&\quad=\left(2n_1+\frac{n_1+n_2}{n_2}\right)\binom{n_1+n_2-1}{n_2-1},
\end{aligned}$$

hence,

$$E(\boldsymbol{i})=1+\frac{2n_1n_2}{n_1+n_2}.$$

Thus, when $n_1\leqslant n_2$, say, we see that

$$1+n_1\leqslant E(\boldsymbol{i})\leqslant 1+n_2.$$

We return to the Wald-Wolfowitz test for the two-sample problem mentioned at the beginning. We start with $n_1+n_2=n$ independent r.v.'s $\xi_1,\dots,\xi_{n_1};\ \xi_{n_1+1},\dots,\xi_n$. Assume that the joint distribution of $\xi_1,\dots,\xi_n$ belongs to $C^{(nn)}$ under the null hypothesis. We now use

Theorem 1.2. When the null hypothesis is true, (1.10) holds. Hence, for each distribution from $C^{(nn)}$, the induced distribution of $(\varepsilon(s_1),\dots,\varepsilon(s_n))$ is exactly the same uniform distribution over the set of all n-tuples $\mathfrak{e}$ (see p. 410). Now choose as test statistic the r.v. $\boldsymbol{i}$ which gives the number of runs in the n-tuple $(\varepsilon(s_1),\dots,\varepsilon(s_n))$. For given level of significance α, one determines for a one-sided test the largest integer, i_α, say, for which

$$P(\boldsymbol{i}\leqslant i_\alpha)\leqslant\alpha$$

holds under the null hypothesis.

The Wald-Wolfowitz test is obviously rank-invariant.

6. Further examples in invariance theory. We consider here several special groups and consider first the so-called *median* or *sign tests*. Let $\mathfrak{G}^{*(m)}$ be the group of all one-to-one measurable transformations $\mathfrak{g}^*$ of R_1 into R_1 satisfying

$$\begin{cases}\mathfrak{g}^*x>0 & x>0\\ \mathfrak{g}^*x=0 & x=0\\ \mathfrak{g}^*x<0 & x<0\,.\end{cases} \tag{6.1}$$

For $n\geqslant 2$ we define the group $\mathfrak{G}_n^{(m)}$ as the direct product of n copies of the group $\mathfrak{G}^{*(m)}$, i.e., each $\mathfrak{g}\in\mathfrak{G}_n^{(m)}$ is of the form $\mathfrak{g}=(\mathfrak{g}_1^*,\dots,\mathfrak{g}_n^*)$ with $\mathfrak{g}_i^*\in\mathfrak{G}^{*(m)}$, $1\leqslant i\leqslant n$, and the group operation in $\mathfrak{G}_n^{(m)}$ is defined by componentwise multiplication. For each $x\in R_n$, let $\mathfrak{g}x=(\mathfrak{g}_1^*x_1,\dots,\mathfrak{g}_n^*x_n)$. For $x\in R_1$, let

$$\operatorname{sign}x=\begin{cases}1 & x>0\\ 0 & x=0\\ -1 & x<0\,.\end{cases}$$

One then sees easily that the map v of R_n into R_n given by $(x_1,\dots,x_n)\to(\operatorname{sign}x_1,\dots,\operatorname{sign}x_n)$ is maximal invariant w.r.t. $\mathfrak{G}_n^{(m)}$. Each test whose test statistic depends only on v is called a sign test. To justify the term "median test", we consider the set of probability measures $\prod_{i=1}^{n}{}_*\mathfrak{F}_{(1/2)}^{(i)}$ over $(R_n,\mathfrak{B}_n)$ and show that it is invariant w.r.t. $\mathfrak{G}_n^{(m)}$. One need only show that if ξ is a r.v. and $P_\xi\in{}_*\mathfrak{F}_{(1/2)}$, then the probability distribution P_η of $\eta=(\mathfrak{g}^*)^{-1}\xi$ also belongs to ${}_*\mathfrak{F}_{(1/2)}$ for each $\mathfrak{g}^*\in\mathfrak{G}^{*(m)}$. Now

$$P(\eta\leqslant 0)=P((\mathfrak{g}^*)^{-1}\xi\leqslant 0)=P(\xi\leqslant\mathfrak{g}^*0)=P(\xi\leqslant 0)=1/2$$

which we wanted to show.

It is easy to see that $\mathfrak{G}^{*(d)} = \mathfrak{G}^* \cap \mathfrak{G}^{*(m)}$ is again a group. $\mathfrak{G}^{*(d)}$ is in fact the set of all continuous, strictly monotone increasing maps of R_1 into itself which satisfy (6.1). The group $\mathfrak{G}_n^{(d)}$, which is the direct product of n copies of the group $\mathfrak{G}^{*(d)}$, then acts as above on R_n, $n \geqslant 2$. It is easy to see that v is also maximal invariant w.r.t. $\mathfrak{G}^{*(d)}$.

Furthermore, the following result is easy to show:

Let $F_1, \tilde{F}_1 \in C$ and assume there exists a $\mathfrak{g} \in \mathfrak{G}^{*(d)}$ with $\tilde{F}_1 = \bar{\mathfrak{g}} F_1$; then there exist. $F, \tilde{F} \in C_m \cap C_{(1/2)}$ and a ψ which, in addition to the conditions given above (p.437 for ψ_i), also satisfies $\psi(1/2) = \vartheta$ for a real ϑ with

$$0 < \vartheta < 1 , \tag{6.2}$$

so that $F_1 = \psi \circ F$ and $\tilde{F}_1 = \psi \circ \tilde{F}$ hold and conversely. This is proved exactly as the corresponding claim of Theorem 4.1.

This suggests considering the set $C_{(\vartheta)}$ of all d.f.'s $\in C$ which satisfy $F(0) = \vartheta$, where ϑ is as in (6.2). We now have

Theorem 6.1. *Let $\xi_1, \ldots, \xi_n$ be independent r.v.'s and let ξ_i be distributed by $F_i \in C_{(\vartheta)}$, $1 \leqslant i \leqslant n$. Then*

$$P(\operatorname{sign} \xi_1 = j_1, \operatorname{sign} \xi_2 = j_2, \ldots, \operatorname{sign} \xi_n = j_n) = \prod_{i=1}^{n} \vartheta^{\frac{1-j_i}{2}} (1-\vartheta)^{\frac{1-j_i}{2}} , \tag{6.3}$$

where $j_i = \pm 1$, $1 \leqslant i \leqslant n$, independently of the choice of the F_i from $C_{(\vartheta)}$.

The *proof* is almost trivial. We have $P(\operatorname{sign} \xi_i = 0) = 0$ since F_i is continuous. Hence,

$$P(\operatorname{sign} \xi_i = j_i) = \begin{cases} P(\xi_i > 0) = 1 - \vartheta , & j_i = 1 , \\ P(\xi_i < 0) = \vartheta , & j_i = -1 , \end{cases}$$

which completes the proof.

It immediately follows that each median test is similar for the null hypothesis $\prod_{i=1}^{n} C_{(\vartheta)}^{(i)}$, where ϑ is a real number satisfying (6.2).

A special, frequently used test is defined by the test statistic $(x_1, \ldots, x_n) \to \sum_{i=1}^{n} \operatorname{sign} x_i$. It is called the *sign test.* III, Theorem 5.1 shows almost immediately that the sign test is most powerful w.r.t. all median tests for the problem

$$\left(\left\{ \prod_{i=1}^{n} C_{(\vartheta)}^{(i)} : \vartheta \geqslant \frac{1}{2} \right\} , \left\{ \prod_{i=1}^{n} C_{(\vartheta)}^{(i)} : \vartheta < \frac{1}{2} \right\} \right) .$$

Indeed, since v is maximal invariant w.r.t. $\mathfrak{G}_n^{(d)}$, it is enough to consider the sample space R whose elements are all 2^n n-tuples with components

± 1. However, since the binomial distributions have monotone likelihood ratios, our claim follows from (6.3)[32].

In practice, the sign test is often used for the following problem. Let $\xi_1, \ldots, \xi_n; \eta_1, \ldots, \eta_n$, $n \geqslant 1$, be independent r.v.'s with continuous d.f. The null hypothesis is the assumption that ξ_i and η_i have the same d.f. $F_i \in C$ for $1 \leqslant i \leqslant n$, where, however, F_i can be $\neq F_j$ for $i \neq j$. When the null hypothesis is true, $\zeta_i = \xi_i - \eta_i$ has a d.f. from $C_{(1/2)}$, which is easy to prove. For the hypothesis $\prod_{i=1}^{n} C_{(1/2)}^{(i)}$, the sign test is similar and is used for the given problem as a two-sided test. One thus chooses as level of significance $\alpha = \frac{k}{2^n}$, $0 \leqslant k \leqslant 2^n$, say, and as critical region in R_n the set

$$\left\{x: \sum_{i=1}^{n} \operatorname{sign} x_i \geqslant \varepsilon_\alpha, \sum_{i=1}^{n} \operatorname{sign} x_i \leqslant -\varepsilon_\alpha\right\},$$

where ε_α is so determined that this critical region has exactly the given level of significance under the null hypothesis.

Remark. If ξ and η are independent, identically distributed r.v.'s with d.f. $\in C$, then the d.f. of $\xi - \eta$ is not only in $C_{(1/2)}$ but even in $C_s = C \cap \mathfrak{F}_s$ and naturally, $C_s \subset C_{(1/2)}$ holds. The problem just treated then leads naturally to the idea of a *test of symmetry*. In the construction of such tests, one can thus limit oneself to a subgroup $\mathfrak{G}^{*(s)}$ of $\mathfrak{G}^{*(m)}$. The elements $\mathfrak{g}^*$ of $\mathfrak{G}^{*(s)}$ should fulfill, in addition to (6.1), also

$$\mathfrak{g}^* x = -\mathfrak{g}^*(-x)$$

for all $x \in R_1$. Hence, the group $\mathfrak{G}^{*(s)}$ consists of all odd, continuous, strictly monotone transformations of R_1 onto itself. A consideration of the group $\mathfrak{G}_n^{(s)}$ in analogy with $\mathfrak{G}_n^{(m)}$ now suggests itself. However, we will consider the more special group $\mathfrak{G}_n^{(ss)}$ whose elements $\mathfrak{g}$ are characterized as follows: For all $(x_1, \ldots, x_n) \in R_n$ let $\mathfrak{g} x = (\mathfrak{g}^* x_1, \ldots, \mathfrak{g}^* x_n)$, where $\mathfrak{g}^*$ runs through all elements of $\mathfrak{G}^{*(s)}$. (Cf. the definition of $\mathfrak{G}_n$, p. 436.) Correspondingly, the null hypothesis for the test of symmetry will be $C_s^{(nn)}$ rather than $\prod_{i=1}^{n} C_s^{(i)}$. It is quite easy to see that $C_s^{(nn)}$ is invariant w.r.t. $\mathfrak{G}_n^{(ss)}$. To construct maximal invariant maps we define the function $p(x) = \sum_{i=1}^{n} c_I(x_i)$ for each $x \in R_n$, where $I = (0, \infty)$. p is thus the number of positive components of x. One sees without difficulty that the map

$$x \to (r_1(|x_1|, \ldots, |x_n|), \ldots, r_n(|x_1|, \ldots, |x_n|), p(x))$$

[32] For the sign test see W. J. Dixon and A. M. Mood, J. Amer. Statist. Assoc. 41, 557—566 (1946) and E. Ruist, l. c.[2].

is maximal invariant w.r.t. $\mathfrak{G}_n^{(ss)}$. We now find, in complete analogy to Theorem 4.2, that it is sufficient for the construction of tests of symmetry to use for $x \in R_n$ only the ranks of the positive components of x. It is thus enough to consider test statistics for the null hypothesis which depend on x only via the map

$$x \to (r_{j_1(x)}(|x_1|, \dots, |x_n|), \dots, r_{j_{p(x)}(x)}(|x_1|, \dots, |x_n|)),$$

where $x_{j_i(x)}$, $1 \leqslant i \leqslant p(x)$ are all positive components of x.

Remark. If $\xi_1, \dots, \xi_n$ are independent r.v.'s each of whose distribution is from C, then with $\xi^{(n)} = (\xi_1, \dots, \xi_n)$, the relation $P\left(p \circ \xi^{(n)} = \sum_{i=1}^{n} \frac{\operatorname{sign} \xi_i + 1}{2}\right) = 1$ holds.

We now turn to a test of symmetry due to Hemelrijk[33] which is concerned with the general null hypothesis $\prod_{i=1}^{n} \mathfrak{F}_s^{(i)}$. We need to define several functions over R_n, $n \geqslant 1$:

For each $x \in R_n$ let $0(x)$ be the number of vanishing, $p(x)$ again the number of positive and $n(x)$ the number of negative components x_i of x. Naturally, for each $x \in R_n$, $0 \leqslant 0(x) \leqslant n$, $0 \leqslant p(x) \leqslant n - 0(x)$ and $0 \leqslant n(x) \leqslant n - 0(x)$. For each $x \in R_n$ let $K(x)$ be the set of all components x_i with $|x_i| > 0$. It contains exactly $p(x) + n(x)$ elements.

Define two disjoint sets $K_1(x)$ and $K_2(x)$, which can be empty, with the following properties: $K(x) = K_1(x) \cup K_2(x)$. If $k_i(x) \geqslant 0$ for $i = 1, 2$ is the number of elements of $K_i(x)$, then $k_2(x) - k_1(x)$ is always $\geqslant 0$. Assume that $|x_i| \in K_1(x)$ and $|x_j| \in K_2(x)$ always implies $|x_i| < |x_j|$. $K_1(x)$ and $K_2(x)$ have the property among all partitions of $K(x)$ satisfying the above conditions, that $k_2(x) - k_1(x)$ assumes a minimum.

It is easy to see that the $K_i(x)$ are uniquely determined. If the elements of $K(x)$ are pairwise different, then

$$k_2(x) = \left[\frac{p(x) + n(x) + 1}{2}\right].^{34} \tag{6.5}$$

(6.5) can be shown as follows: The elements of $K(x)$ can be uniquely ordered by magnitude. If $p(x) + n(x)$ is even, $K_1(x)$ and $K_2(x)$ each contain $(p(x) + n(x))/2$ elements, and since $(p(x) + n(x))/2 = \left[\frac{p(x) + n(x) + 1}{2}\right]$, (6.5) holds. However, if $p(x) + n(x)$ is odd, $K_2(x)$ contains one element more than $K_1(x)$ and hence, (6.5) again holds.

---

[33] J. Hemelrijk, Indag. Math. 12, 340—350 (1950). Also see E. Ruist, l.c.[2]. For another general test of symmetry see C. van Eeden and A. Benard, Indag. Math. 19, 381—408 (1957).

[34] That is: $k_2(x)$ isthe largest integer contained in $\frac{p(x) + n(x) + 1}{2}$.

Let $u(x)$ be the number of positive x_i's in $K_2(x)$. We obviously have

$$0 \leqslant u(x) \leqslant \min(k_2(x), p(x)); \qquad p(x) - u(x) \leqslant n - 0(x) - k_2(x).$$

The announced test of symmetry is based on

Theorem 6.3. *Let* $\xi^{(n)} = (\xi_1, \ldots, \xi_n)$ *be a r.v. whose probability distribution belongs to* $\prod_{i=1}^{n} {}_*\mathfrak{F}_s^{(i)}$, *k, g,* n_1 *and m nonnegative natural numbers with* $\left[\frac{n-m+1}{2}\right] \leqslant k \leqslant n-m, 0 \leqslant g \leqslant \min(k, n_1), n_1 - g \leqslant n-m-k, n_1 + m \leqslant n$. *Then*

$$P(p \circ \xi^{(n)} = n_1, u \circ \xi^{(n)} = g \,|\, 0 \circ \xi^{(n)} = m, k_2 \circ \xi^{(n)} = k) = 2^{-(n-m)} \binom{k}{g} \binom{n-m-k}{n_1 - g}, \tag{6.6}$$

provided the conditional probability is defined.

Proof. We first show that

$$\sum_{i_1, \ldots, i_{n-m}} P(\operatorname{sign} \xi_{i_1} = j_1, \ldots, \operatorname{sign} \xi_{i_{n-m}} = j_{n-m}, \xi_i = 0, i \neq i_1, \ldots, i_{n-m} \,|\, 0 \circ \xi^{(n)} = m) = 2^{-(n-m)}. \tag{6.7}$$

The summation here extends over all $\binom{n}{m}$ combinations of m elements from $(1, \ldots, n)$. Further $j_i = \pm 1$, $1 \leqslant i \leqslant n-m$. In fact, if B_m (with $P_{\xi^{(n)}}(B_m) \neq 0$) is the set $0^{-1}(\{m\})$, then B_m consists, as is easy to see, of $\binom{n}{m}$ disjoint sets C_l. Each of the sets

$$\bigcup_{j_1, \ldots, j_{n-m}} \{x: \operatorname{sign} x_{i_1} = j_1, \ldots, \operatorname{sign} x_{i_{n-m}} = j_{n-m}, \operatorname{sign} x_i = 0, i \neq i_1, \ldots, i_{n-m}\}.$$

where the union extends over all 2^{n-m} sign combinations, coincides with exactly one C_l. However, by assumption,

$$P_{\xi^{(n)}}(\{x: \operatorname{sign} x_{i_1} = j_1, \ldots, \operatorname{sign} x_{i_{n-m}} = j_{n-m}, \operatorname{sign} x_i = 0, i \neq i_1, \ldots, i_{n-m}\})$$

has the same value for all 2^{n-m} combinations of signs $j_1, \ldots, j_{n-m}$. This shows (6.7). In order to prove (6.6), one calls on (6.7) and carries through a quite similar argument of a combinatorial nature: The set $\{0^{-1}(\{m\}) \cap k_2^{-1}(\{k\})\}$ consists of finitely many disjoint sets which are all of the following type:

$$\{x: |x_i| < |x_j|, m+1 \leqslant i < n-k, n-k+1 \leqslant j \leqslant n, x_l = 0, 1 \leqslant l \leqslant m\}.$$

One must take all sets of this type which also fulfill $p(x)=n_1$. Among these sets there exist $\binom{k}{g}\binom{n-m-k}{n_1-g}$ which also satisfy $u(x)=g$.

A critical region for a test similar w.r.t. $\prod_{i=1}^{n}\mathfrak{F}_s^{(i)}$ under the condition $0\circ\xi^{(n)}=m,\ k_2\circ\xi^{(n)}=k$ can be obtained, say, by ordering the values on the right side of (6.6) according to their magnitudes. One then gets for example $p_1\leqslant p_2\leqslant\cdots$. For given level of significance α one then determines the critical region by means of those p_l, $1\leqslant l\leqslant j_\alpha$, which satisfy $\sum_{l=1}^{j_\alpha}p_l\leqslant\alpha<\sum_{l=1}^{j_\alpha+1}p_l$.

Hornich[35] has derived an interesting inequality for the distributions from $\mathfrak{F}_s^{(nn)}$ whose first moments exist:

Theorem 6.4. *Let $\xi_1,\ldots,\xi_n$, $n\geqslant 2$, be independent r.v.'s with the same d.f. $F\in\mathfrak{F}_s$. For the sake of simplicity, assume that F is continuous at zero. Let the first moment exist and assume*

$$E(\xi_i)=0\,,\qquad 1\leqslant i\leqslant n\,. \tag{6.8}$$

Let $E(|\xi_i|)=m$ for all i and $\zeta_n=\sum_{i=1}^{n}\xi_i$. Then

$$E(|\zeta_n|)\geqslant nm2^{-(n-1)}\binom{n-1}{[(n-1)/2]}. \tag{6.9}$$

Proof. Denote by $Z_1,\ldots,Z_{2^n}$, in some order, the sets

$$\{x\in R_n:\operatorname{sign}x_1=\pm 1,\ldots,\operatorname{sign}x_n=\pm 1\}\,,$$

where all sign combinations occur. Let Z_1, say, be the set in whose definition only plus-signs occur. Then, since $F\in\mathfrak{F}_s$,

$$\begin{aligned}
E(|\zeta_n|)&\geqslant\sum_{i=1}^{2^n}\int_{Z_i}|x_1+\cdots+x_n|\,dF(x_1)\ldots dF(x_n)\\
&=\sum_{l=0}^{n}\binom{n}{l}\int_{Z_1}|x_1+\cdots+x_l-x_{l+1}-\cdots-x_n|\,dF(x_1)\ldots dF(x_n)\\
&\geqslant 2\sum_{l=0}^{\left[\frac{n-1}{2}\right]}\int_{Z_1}\binom{n}{l}|x_{l+1}+\cdots+x_n-x_1-\cdots-x_l|\,dF(x_1)\ldots dF(x_n)\\
&\geqslant 2\sum_{l=0}^{\left[\frac{n-1}{2}\right]}\int_{Z_1}\binom{n}{l}\left(\sum_{j=l+1}^{n}x_j-\sum_{i=1}^{l}x_i\right)dF(x_1)\ldots dF(x_n)\,.
\end{aligned}$$

[35] H. Hornich, Mh. Math. Phys. 50, 142—150 (1941). Z. W. Birnbaum and H. S. Zuckerman, Ann. Math. Statist. 15, 328—329 (1944) have couched his formulation (within the framework of risk theory) in the language of probability theory.

From (6.8) and the continuity assumption on F follows $m = 2 \int_{\{x\,:\,x>0\}} x\,dF(x)$.
Thus,

$$E(|\zeta_n|) \geqslant \sum_{l=0}^{\left[\frac{n-1}{2}\right]} \binom{n}{l}(n-2l)\,m\,2^{-(n-1)}.$$

Since $l\binom{n}{l} = n\binom{n-1}{l-1}$, (6.9) follows from an easy manipulation.

The modification of the theorem when the continuity assumption on F is dropped is obvious.

Finally, we touch briefly on the *problem of independence*. One begins with $n \geqslant 1$ two-dimensional r.v.'s (ξ_i, η_i), $1 \leqslant i \leqslant n$. The null hypothesis assumes that all (ξ_i, η_i) have the same distribution for $1 \leqslant i \leqslant n$ and that the distribution of (ξ_i, η_i) belongs to $\prod_{i=1}^{2} C^{(i)}$.

More intuitively, one can say that under the null hypothesis, the n sample variables ξ_i and the n sample variables η_i arise from two independent populations.

Let $\mathfrak{G}^*$ have the same meaning as on p. 436.) We define a group $\mathfrak{G}_{2n}^{(n)}$ acting on R_{2n} and given according to

$$\mathfrak{g}(x_1, y_1, x_2, y_2, \ldots, x_n, y_n) = (\mathfrak{g}^* x_1, \mathfrak{g}_1^* y_1, \mathfrak{g}^* x_2, \mathfrak{g}_1^* y_2, \ldots, \mathfrak{g}^* x_n, \mathfrak{g}_1^* y_n),$$

$\mathfrak{g}^*, \mathfrak{g}_1^* \in \mathfrak{G}^*$. $\mathfrak{G}_{2n}^{(n)}$ leaves the null hypothesis invariant.

It is easy to see that one can base invariant tests on the test statistic

$$(x_1, y_1, \ldots, x_n, y_n) \to (\varepsilon(s_1(x)), \varepsilon(s_1(y)), \ldots, \varepsilon(s_n(x)), \varepsilon(s_n(y))).$$

7. Locally most powerful rank-invariant tests. We consider here the two-sample problem again and will use the notation introduced on p. 439. Assume a null hypothesis of the form $\{(F,F): F \in C_m\}$. We know from the corollary to Theorem 4.1 that each rank-invariant test is similar for this null hypothesis. Imbed this null hypothesis in a one-parameter family of alternatives as follows: Let $\eta > 0$ and $\Gamma = \{\gamma: 0 \leqslant \gamma < \eta\}$. Associate a pair of d.f.'s (F_γ, G_γ) with each $\gamma \in \Gamma$. Let $F_\gamma \in C$ and $G_\gamma \in \mathfrak{F}$. In particular, for $\gamma = 0$ let $F_0 \in C_m$ and

$$F_0 = G_0. \tag{7.1}$$

It will be convenient to denote F_γ also by $x \to F(x, \gamma)$ and likewise for G_γ.

The test problem we are interested in has the form

$$(\{(F_0, G_0): F_0 \in C_m\}, \{(F_\gamma, G_\gamma): F_\gamma \in C, G_\gamma \in \mathfrak{F}, \gamma \in \Gamma, \gamma \neq 0\}). \tag{7.2}$$

Here, (7.1) should hold. We consider the problem of constructing *locally most powerful rank-invariant tests* for this problem[36].

---

[36] See H. Uzawa, Ann. Math. Statist. 31, 685—702 (1960). Uzawa's results include those of E. L. Lehmann, Ann. Math. Statist. 24, 23—43 (1953).

We first rewrite the test problem somewhat. As we mentioned in the proof of Theorem 1.4, for each F_γ there exists a function u_{F_γ}, also denoted by F_γ^{-1},[37] such that for $0<y<1$

$$F(u_{F_\gamma}(y),\gamma)=y. \tag{7.3}$$

In place of F_γ^{-1} we also write $y\to F^{-1}(y,\gamma)$.

Since we consider rank-invariant tests only, we can limit ourselves to the test statistics given on p. 440.

Hence, it is more suitable to start with the sample space $(R,\mathscr{S})$ instead of $(R_{n_1+n_2},\mathfrak{B}_{n_1+n_2})$, where R is the set of all $\binom{n}{n_1}$ vectors e (see p. 410) and $\mathscr{S}$ the set of all subsets of R. Naturally, there always exist measures over $(R,\mathscr{S})$ which dominate the set of all probability measures over $(R,\mathscr{S})$, for example, the uniform distribution over R.

We now call on III, Theorem 6.1 and prove

Theorem 7.1. *Let* $\eta>0$ *and* $\Gamma=\{\gamma:0\leqslant\gamma<\eta\}$. *Assume we have a test problem of the form* (7.2), *where* (7.1) *is fulfilled. Let the level of significance* α *be* $l\Big/\binom{n}{n_1}$, $0\leqslant l\leqslant\binom{n}{n_1}$. *For all* $y\in[0,1]$ *and each* $\gamma\in\Gamma$, *let*

$$H(y,\gamma)=G(F^{-1}(y,\gamma),\gamma). \tag{7.4}$$

Thus, in particular for $y\in[0,1]$,

$$H(y,0)=y. \tag{7.5}$$

The map $\gamma\to H(y,\gamma)$ *is assumed to be right differentiable at* $\gamma=0$ *for each* y.

For each $y\in[0,1]$ *we denote the right derivative* $\dfrac{\partial H(y,0)}{\partial\gamma}$ *by* $Q(y)$ *and assume that* Q *is of bounded variation in* $[0,1]$. *In addition, let an* $M>0$ *exist such that for* $y\in[0,1]$ *and* $\gamma\in\Gamma-\{0\}$,

$$\left|\frac{H(y,\gamma)-H(y,0)}{\gamma}\right|\leqslant M. \tag{7.6}$$

Then, the locally most powerful rank-invariant test for the given test problem is given by the test statistic $(\varepsilon_1,\ldots,\varepsilon_n)\to\sum_{i=1}^{n}a_i\varepsilon_i$ *defined over* R. *The real numbers* a_j *are given for* $1\leqslant j\leqslant n$ *by*

$$a_j=\binom{n-1}{j-1}\int_0^1 t^{j-1}(1-t)^{n-j}\,dQ(t). \tag{7.7}$$

---

[37] For $F_\gamma\in C_m$, F_γ^{-1} is simply the inverse map.

Proof. Let $\varepsilon_i=\varepsilon(s_i)$, $1\leqslant i\leqslant n$ and for each $e=(\varepsilon_1,\ldots,\varepsilon_n)\in R$ and $\gamma\in\Gamma$, set $L(e,\gamma)=P((\varepsilon_1,\ldots,\varepsilon_n)=(\varepsilon_1,\ldots,\varepsilon_n);(F_\gamma,G_\gamma))$.

One then easily finds from a generalization of the proof of Theorem 1.2 that

$$L(e,\gamma)=n_1!n_2!\int\cdots\int_{-\infty<x_1\leqslant\cdots\leqslant x_n<\infty}\prod_{j=1}^{n}d\left[F(x_j,\gamma)^{1-\varepsilon_j}G(x_j,\gamma)^{\varepsilon_j}\right]. \quad (7.8)$$

Because of (7.3) and (7.4), we get from (7.8) that

$$L(e,\gamma)=n_1!n_2!\int\cdots\int_{0\leqslant u_1\leqslant\cdots\leqslant u_n<1}\prod_{j=1}^{n}d\left[u_j^{1-\varepsilon_j}H(u_j,\gamma)^{\varepsilon_j}\right]. \quad (7.9)$$

It follows from (7.6) that $\frac{\partial}{\partial\gamma}\int_a^b f(y)\,dH(y,0)=\int_a^b f(y)\,dQ(y)$, for arbitrary real numbers a,b with $0\leqslant a<b\leqslant 1$ and each continuously differentiable f over $(0,1]$. This can be shown as follows: $y\to\frac{H(y,\gamma)-H(y,0)}{\gamma}$ is of bounded variation for each $\gamma>0$. By partial integration:

$$\int_a^b f(y)\,d\left(\frac{H(y,\gamma)-H(y,0)}{\gamma}\right)$$

$$=\frac{f(b)\,(H(b,\gamma)-H(b,0))-f(a)\,(H(a,\gamma)-H(a,0))}{\gamma}$$

$$-\int_a^b\frac{H(y,\gamma)-H(y,0)}{\gamma}f'(y)\,dy.$$

An application of Theorem VI and another partial integration now deliver the claim.

Hence,

$$\frac{\partial L(e,0)}{\partial\gamma}=n_1!n_2!\int\cdots\int_{0\leqslant u_1\leqslant\cdots\leqslant u_n\leqslant 1}\sum_{j=1}^{n}\varepsilon_j\,du_1\ldots du_{j-1}\,dQ(u_j)\,du_{j+1}\ldots du_n. \quad (7.10)$$

Since for each t in $0\leqslant t\leqslant 1$

$$\int\cdots\int_{0\leqslant u_1\leqslant\cdots\leqslant u_{j-1}\leqslant t}du_1\ldots du_{j-1}=t^{j-1}/(j-1)!$$

and

$$\int\cdots\int_{t\leqslant u_{j+1}\leqslant\cdots\leqslant u_n\leqslant 1}du_{j+1}\ldots du_n=(1-t)^{n-j}/(n-j)!,$$

one gets from (7.10)

$$\frac{\partial L(\mathrm{e},0)}{\partial \gamma} = n_1!n_2! \sum_{j=1}^{n} \varepsilon_j \frac{1}{(j-1)!(n-j)!} \int_0^1 t^{j-1}(1-t)^{n-j} dQ(t). \quad (7.11)$$

Since the $L(\mathrm{e},0)$ for each $\mathrm{e} \in R$ have the same nonzero value by Theorem 1.2, one can also assume that $L(\mathrm{e},\gamma) \neq 0$ for all $\mathrm{e} \in R$ and all $\gamma \in \Gamma$.

We can now apply III, Theorem 6.1 taking the remark on p. 201 into consideration: For the given level of significance α, the locally most powerful test ϕ^* is given by

$$\phi^*(\mathrm{e}) = \begin{cases} 1 & \dfrac{\partial L(\mathrm{e},0)}{\partial \gamma} > k \\ 0 & \dfrac{\partial L(\mathrm{e},0)}{\partial \gamma} < k. \end{cases} \quad (7.12)$$

Here, the real number k is to be suitably chosen as a function of α. Because of (7.11), the claim follows from (7.12).

Remark. If c is an arbitrary positive number and β an arbitrary real number, then the test statistic $c \sum_{i=1}^{n} a_i \varepsilon_i + \beta$ defined for each $(\varepsilon_1, \ldots, \varepsilon_n) \in R$, where the a_i, $1 \leqslant i \leqslant n$ are given by (7.7), defines the same locally most powerful test.

The result of Theorem 7.1 can now be applied to a series of special test problems. We will illustrate this by a two-sample problem with a special alternative:

$$(\{(F,F): F \in C_m\}, \{(F,G): F \in C_m, G \in \mathfrak{F}, F \neq G, F \geqslant G\}).$$

This problem can be called a *general one-sided two-sample problem.*

The condition

$$F \geqslant G, \quad F \neq G \quad (7.13)$$

can also be expressed by means of $H = G \circ F^{-1}$ in such a way that for $y \in [0,1]$,

$$H(y) \leqslant y \quad (7.14)$$

is to hold, where for at least one y, the $<$-sign obtains. In addition, H can be trivially extended to a d.f. over R_1 by choosing

$$H(y) = 0, \ y \leqslant 0; \quad H(y) = 1, \ y \geqslant 1. \quad (7.15)$$

Conversely, if a d.f. H is given satisfying (7.15) and (7.14) so that the $<$-sign holds at least once, then there always exists a $G \in \mathfrak{F}$ and an $F \in C_m$ such that (7.13) is valid and $H = G \circ F^{-1}$. Indeed, it suffices to choose an arbitrary $F \in C_m$ and to define $G = H \circ F$. The test problem is thus described by the set $\mathfrak{H}$ of all d.f.'s H which fulfill (7.14) and (7.15)[38]. Now consider one-parameter families of d.f.'s in $\mathfrak{H}$. As in Theorem 7.1, let $\Gamma=\{\gamma:0\leqslant\gamma<\eta\}$ for some real η with $0<\eta<1$. If $H\in\mathfrak{H}$, then a d.f. is defined for each real $\gamma\in\Gamma$ by

$$H_s(y,\gamma)=\begin{cases}0 & y\leqslant 0\\ (1-\gamma)y+\gamma H(y) & 0\leqslant y\leqslant 1\\ 1 & y\geqslant 1.\end{cases} \tag{7.16}$$

Hence, for $\gamma=0$ one gets from (7.16) the uniform distribution over (0,1). For $\gamma\neq 0$ and $0\leqslant y\leqslant 1$, $H_s(y,\gamma)\leqslant y$ by (7.14), where the $<$-sign holds for at least one y when H is not the uniform distribution over (0,1). If we denote the function $y\to H_s(y,\gamma)$ for each $\gamma\in\Gamma$ by $H_{s\gamma}$, so that H_{s0} is the uniform distribution, then $((H_{s0},H_{s0}),\{(H_{s0},H_{s\gamma}):\gamma>0\})$ defines a one-parameter test problem whose admissible hypothese are a part of the admissible hypotheses for the general one-sided two-sample problem. We have already investigated one-parameter problems of this form in Theorem 7.1. Since $\dfrac{\partial H_s(y,0)}{\partial\gamma}=-y+H(y)$ for $0\leqslant y\leqslant 1$, it follows by the remark at the end of the proof of Theorem 7.1 that the locally most powerful rank-invariant test is given by $\sum_{i=1}^{n} b_i\varepsilon_i$ for each $(\varepsilon_1,\ldots,\varepsilon_n)\in R$, where

$$b_i=b_i(H)=\binom{n-1}{i-1}\int_0^1 t^{i-1}(1-t)^{n-i}\,dH(t),\qquad 1\leqslant i\leqslant n. \tag{7.17}$$

It is easy to show that the subset $\{b_1(H),\ldots,b_n(H): H\in\mathfrak{H}\}$ of R_n is compact.

Indeed, it is bounded, which one sees at once. Let $\{b_i(H_k)\}$, $1\leqslant i\leqslant n$, be a sequence with $H_k\in\mathfrak{H}$ such that $\lim\limits_{k\to\infty} b_i(H_k)=b_i^*$ exists. By I, Lemma 23.2, there exists a sub-sequence $\{H_{k_j}\}$ converging to a nondecreasing (right continuous) function H at all points of continuity of H. H naturally satisfies (7.14), and since all $H_{k_j}, j\geqslant 1$, fulfill (7.15), $H(1)=1$ and H is even a distribution function. But then from I, Lemma 23.1.

$$\lim_{j\to\infty} b_i(H_{k_j})=b_i(H)=\binom{n-1}{i-1}\int_0^1 t^{i-1}(1-t)^{n-i}\,dH(t)$$

so that $b_i^*=b_i(H)$ for $1\leqslant i\leqslant n$. This shows the compactness. Two questions now arise:

[38] The uniform distribution over (0,1) thus also belongs to $\mathfrak{H}$.

1. *Let* $(\{(F_0, G_0): F_0 = G_0\}, \{(F_\gamma, G_\gamma): F_\gamma \neq G_\gamma, \gamma > 0\})$ *be a one-parameter test problem whose admissible hypotheses are a subset of those for the general one-sided two-sample problem mentioned on p.* 455. Hence,

$$F_\gamma \geq G_\gamma \tag{7.18}$$

for all $\gamma \in \Gamma - \{0\}$. Let $H_\gamma = G_\gamma \circ F_\gamma^{-1}$, and assume all the assumptions of Theorem 7.1 are fulfilled. Then for each level of significance given in Theorem 7.1, there exists an n-tuple $(a_1, \ldots, a_n)$ of real numbers which defines a locally most powerful rank-invariant test for the problem mentioned above. *Do there always exist a* $c > 0$, *a real* β *and an* $H \in \mathfrak{H}$ *such that* $c a_i + \beta = b_i(H)$ *holds for* $1 \leq i \leq n$?

2. *If the answer to* 1. *is affirmative, how can one characterize the set* $\{b_1(H), \ldots, b_n(H): H \in \mathfrak{H}\}$?

We turn first to 1. We consider the one-parameter test problem appearing there, so that (7.18) holds. From (7.18), in the notation introduced by (7.4), we have for each y and $\gamma \in \Gamma$

$$H(y, \gamma) \leq H(y, 0) \tag{7.19}$$

and

$$H(0, \gamma) = 0\,, \qquad H(1, \gamma) = 1\,, \tag{7.20}$$

as well as

$$H(y, 0) = y\,, \qquad 0 \leq y \leq 1\,. \tag{7.21}$$

By the definition of Q on p. 453 it follows from (7.19) that

$$Q(y) \leq 0\,, \qquad 0 \leq y \leq 1\,. \tag{7.22}$$

From (7.20)

$$Q(0) = Q(1) = 0\,. \tag{7.23}$$

Assuming initially that Q is differentiable and that Q' is bounded in $(0,1)$, there follows the existence of a positive c for which $H^{(1)}(y) = y + cQ(y)$, $0 \leq y \leq 1$, is a d.f. Because of (7.23) we need only show that one can choose $c > 0$ in such a way that $H^{(1)}$ does not decrease. By assumption, however, there exists a $K > 0$ such that $|Q'| < K$. Now choose $0 < c < 1/K$. Then, for arbitrary $y_1, y_2 \in [0,1]$ with $y_1 < y_2$, the relation $H^{(1)}(y_2) - H^{(1)}(y_1) = y_2 - y_1 + c(Q(y_2) - Q(y_1)) = y_2 - y_1 + c(y_2 - y_1) Q'(y_1 + \vartheta(y_2 - y_1))$, holds, where $0 < \vartheta < 1$. Hence,

$$H^{(1)}(y_2) - H^{(1)}(y_1) \geq (y_2 - y_1)(1 - cK) > 0\,.$$

From (7.22)

$$H^{(1)}(y) \leq y\,, \qquad 0 \leq y \leq 1$$

so that $H^{(1)} \in \mathfrak{H}$.

Defining $H_s^{(1)}(y,\gamma)=(1-\gamma)y+\gamma H^{(1)}(y)$ for $\gamma\in\Gamma$ and $y\in[0,1]$ we find that

$$\frac{\partial H_s^{(1)}(y,0)}{\partial\gamma}=cQ(y) \quad \text{for } y\in[0,1].$$

Hence, at least under the assumptions made on Q, the answer to the first question is always affirmative. We remark that we have not used (7.21) here.

Now drop the assumption that Q is differentiable and assume instead that Q is of bounded variation. Then, for each $\gamma\in\Gamma$ and $k=1,2,\ldots$, d.f.'s $y\to H_k(y,\gamma)$ can be defined which are constant in $(-\infty,0)$ and $(1,\infty)$ and which satisfy (7.6) and (7.19)—(7.21).

Moreover, with

$$Q_k(y)=\frac{\partial H_k(y,0)}{\partial\gamma}, \quad 0\leqslant y\leqslant 1$$

the relation

$$\lim_{k\to\infty} Q_k(y)=Q(y) \tag{7.24}$$

holds for all points of continuity y of Q in $[0,1]$. The Q_k are continuously differentiable in $[0,1]$ for $k\geqslant 1$, uniformly bounded, and fulfill conditions corresponding to (7.22) and (7.23). We sketch the construction of such a sequence $\{H_k\}$ of d.f.'s. First note the boundedness of Q follows from the assumption that it has bounded variation. Hence, from (7.22) for an $M>0$

$$-M\leqslant Q(y)\leqslant 0, \quad 0\leqslant y\leqslant 1. \tag{7.25}$$

We now make the additional assumption that Q is (right) continuous at $y=0$ and (left) continuous at $y=1$. Then it is known that for Q_k, $k\geqslant 1$, one can choose a trigonometric polynomial with real coefficients such that $Q_k(y)=\sum_{j=0}^{k}(\alpha_j\sin 2\pi jy+\beta_j\cos 2\pi jy)$ for $0\leqslant y\leqslant 1$ with real α_j,β_j, $1\leqslant j\leqslant k$.

The condition that the Q_k be uniformly bounded is then fulfilled, because of (7.25), in the form

$$-M\leqslant Q_k(y)\leqslant 0, \quad 0\leqslant y\leqslant 1, \quad k\geqslant 1.$$ [39]

However, if the additional continuity assumptions on Q do not hold, then one can again define trigonometric polynomials Q_k^*, $k\geqslant 1$, which satisfy (7.22), (7.24) at every point of continuity of Q and which are uni-

[39] These statements follow at once from Fejér's theorem on the behavior of the arithmetic mean of a Fourier series. See A. Zygmund, Trigonometric series. 2nd ed., Vol. I, Cambridge University Press, New York 1959, 89.

formly bounded. We illustrate the construction of functions Q_k, $k\geqslant 1$, satisfying all of the required conditions—hence also (7.23)—for the case where e.g. $Q_k^{*\prime}(1/k)>0$ and $Q_k^{*\prime}\left(1-\frac{1}{k}\right)<0$ hold. For other possible cases, the following construction can be easily modified. Define

$$Q_k(y)=\begin{cases}(-k^2 q_k+k q_k')y^2+(2k q_k-q_k')y, & 0\leqslant y\leqslant 1/k,\\ Q_k^*(y), & 1/k\leqslant y\leqslant 1-\dfrac{1}{k},\\ (-k^2\bar{q}_k-k\bar{q}_k')y^2+(2k(k-1)\bar{q}_k+(2k-1)\bar{q}_k')y & \\ \qquad +k(2-k)\bar{q}_k+(1-k)\bar{q}_k', & 1-\dfrac{1}{k}\leqslant y\leqslant 1.\end{cases}$$

Here, $q_k=Q_k^*(1/k)$, $q_k'=Q_k^{*\prime}(1/k)$, $\bar{q}_k=Q_k^*\left(1-\frac{1}{k}\right)$ and $\bar{q}_k'=Q_k^{*\prime}\left(1-\frac{1}{k}\right)$. The uniform boundedness of the Q_k, $k\geqslant 1$, so defined follows from the following known inequality[40]: $\max_{0\leqslant y\leqslant 1}|Q^{*\prime}(y)|\leqslant k\max_{0\leqslant y\leqslant 1}|Q_k^*(y)|$, $k\geqslant 1$. It is clear that the Q_k, $k\geqslant 1$, also satisfy all the other required conditions.

Now simply define

$$H_k(y,\gamma)=y+\gamma Q_k(y),\qquad 0\leqslant y\leqslant 1.$$

Since for each $k\geqslant 1$ there exists an $M_k>0$ with $|Q_k'|<M_k$, $y\to H_k(y,\gamma)$ is a d.f. for sufficiently small positive γ, say $\gamma\leqslant\eta_k$, as we saw on p. 457. This also holds trivially for $\gamma=0$. If $\eta>\eta_k$, define $y\to H_k(y,\gamma)$ for $\eta>\gamma>\eta_k$ arbitrarily but in such a way that (7.19), (7.20) and (7.6) are fulfilled. Thus, for each $k\geqslant 1$ the set of all distribution functions $y\to H_k(y,\gamma)$, $\gamma\in\Gamma-\{0\}$ defines a one-parameter family of hypotheses for the general one-sided two-sample problem, where the corresponding function Q_k is continuously differentiable. Hence, as we saw on p. 456, the corresponding locally most powerful rank-invariant test for $k\geqslant 1$ can be described by an n-tuple of the form

$$(b_1^{(k)}(H^{(k)}),\dots,b_n^{(k)}(H^{(k)}))\quad\text{with } H^{(k)}\in\mathfrak{H},$$

where $H^{(k)}(y)=y+c_k Q_k(y)$, $0\leqslant y\leqslant 1$ holds for suitable $c_k>0$. The same test is, however, also defined by the n-tuple $(a_1^{(k)},\dots,a_n^{(k)}$, where

$$a_j^{(k)}=\binom{n-1}{j-1}\int_0^1 t^{j-1}(1-t)^{n-j}\,dQ_k(t),\qquad 1\leqslant j\leqslant n.$$

[40] This inequality, valid for arbitrary trigonometric polynomials, is due to S. N. Bernstein. See N. K. Bary, A treatise on trigonometric series. Vol. I, Pergamon Press Book. The Macmillan Co., New York 1964, 35.

However, since (7.24) holds at all points of continuity of Q and the Q_k, $k \geqslant 1$, are uniformly bounded we get (by partial integration, for example) that

$$a_j = \lim_{k\to\infty} a_j^{(k)} = \binom{n-1}{j-1} \int_0^1 t^{j-1}(1-t)^{n-j} dQ(t)$$

exists for $1 \leqslant j \leqslant n$. From the compactness proved on p. 456 we thus obtain the existence of an $H \in \mathfrak{H}$ for which $(a_1, \ldots, a_n)$ and $(b_1(H), \ldots, b_n(H))$ define the same test. Hence, the first question is affirmatively answered.

For the second question we call on the arguments of I. **41** and give merely some general indications.

First let b_i be real numbers such that an $H \in \mathfrak{H}$ exists with $b_i = b_i(H)$, $1 \leqslant i \leqslant n$. Since

$$t^i = \sum_{j=i+1}^{n} \binom{n-i-1}{n-j} t^{j-1}(1-t)^{n-j}, \quad 0 \leqslant i \leqslant n-1, \quad 0 \leqslant t \leqslant 1,$$

we get for the moments c_i of H

$$c_i = \sum_{j=i+1}^{n} \binom{n-i-1}{n-j} b_j \Big/ \binom{n-1}{j-1}, \quad 0 \leqslant i \leqslant n-1. \tag{7.26}$$

The identity (7.26) can easily be inverted by means of

$$t^{j-1}(1-t)^{n-j} = \sum_{i=j-1}^{n-1} (-1)^{i-j+1} \binom{n-j}{n-i-1} t^i, \quad 1 \leqslant j \leqslant n, \quad 0 \leqslant t \leqslant 1.$$

The use of I. (41.3) now suggests itself. One then sees from (7.26) that for $k, l \geqslant 0$, $k+l \leqslant n-1$ the following condition must necessarily hold:

$$\sum_{j=0}^{l} (-1)^j \binom{l}{j} \sum_{s=k+j+1}^{n} \binom{n-k-j-1}{n-s} b_s \left[\binom{n-1}{s-1}\right]^{-1} \geqslant 0. \tag{7.27}$$

By interchanging the order of summation one gets for the left side of (7.27)

$$\sum_{s=k+1}^{k+l+1} b_s \left[\binom{n-1}{s-1}\right]^{-1} \sum_{j=0}^{s-k-1} (-1)^j \binom{l}{j} \binom{n-k-j-1}{n-s}$$
$$+ \sum_{s=k+l+2}^{n} b_s \left[\binom{n-1}{s-1}\right]^{-1} \sum_{j=0}^{l} (-1)^j \binom{l}{j} \binom{n-k-j-1}{n-s}.$$

A simple argument shows[41] that for any integers $u, v, w \geqslant 0$

$$\sum_{j=0}^{\infty} (-1)^j \binom{u}{j} \binom{v-j}{w} = \binom{v-u}{v-w}. \tag{7.28}$$

Naturally, only finitely many summands occur on the left side of (7.27).

[41] See, for example, B. E. Netto, Lehrbuch der Combinatorik, Teubner, Leipzig 1901, 252.

An application of (7.28) leads finally to the condition

$$\sum_{s=k+1}^{n} b_s \binom{n-k-l-1}{s-k-1}\left[\binom{n-1}{s-1}\right]^{-1} \geqslant 0,\ l \geqslant 0,\ k \geqslant 0,\ k+l \leqslant n-1. \tag{7.29}$$

However, (7.29) is not sufficient. In order to obtain sufficient conditions that the equations $b_i = b_i(H)$, $1 \leqslant i \leqslant n$, with $H \in \mathfrak{H}$ hold for an n-tuple of real numbers $(b_1, \dots, b_n)$, one can call on the arguments at the end of I. **41**. We will not go into the details here.

8. Some asymptotic results. We turn to some aspects of the asymptotic behavior of the Wilcoxon test. They may serve as a pattern for more general results since the results obtained here can easily be extended to larger classes of rank tests. Literature references in this regard will be given later. We first prove

Theorem 8.1[42]. *Let $\xi_1, \xi_2, \dots$ be a sequence of independent, l-dimensional, $l \geqslant 1$, r.v.'s with the same distribution. Let h be a symmetric function defined over R_{kl}, $k \geqslant 1$ and let $\eta_1, \dots, \eta_k$ be independent, l-dimensional r.v.'s with the same distribution as ξ_i and which are also independent of all the ξ_i, $i \geqslant 1$. Let $E(h^2(\eta_1, \dots, \eta_k))$ exist and let.*

$$E(h(\eta_1, \dots, \eta_k)) = a. \tag{8.1}$$

In addition, let σ_r^2, $1 \leqslant r \leqslant k$ be the variance of $\phi_r(\eta_1, \dots, \eta_r) = E(h(\eta_1, \dots, \eta_r, \eta_{r+1}, \dots, \eta_k) | \eta_1, \dots, \eta_r)$. Let $V(\xi_1, \dots, \xi_n) = \binom{n}{k}^{-1} \sum_{i_1, \dots, i_k} h(\xi_{i_1}, \dots, \xi_{i_k})$ for $n \geqslant k$, where $i_1, \dots, i_k$ runs through all combinations of k elements from the set $\{1, \dots, n\}$. Then $n^{1/2} k^{-1}(V(\xi_1, \dots, \xi_n) - a)$ is asymptotically $N(0, \sigma_1^2)$-distributed if $\sigma_1^2 > 0$.

Proof. $(E(h(\eta_1, \dots, \eta_r, \eta_{r+1}, \dots, \eta_k) | \eta_1, \dots, \eta_r))^2 \leqslant E(h^2(\eta_1, \dots, \eta_r, \eta_{r+1}, \dots, \eta_k) | \eta_1, \dots, \eta_r)$ with probability 1. (see I[26]). Hence,

$$E\big((E(h(\eta_1, \dots, \eta_r, \eta_{r+1}, \dots, \eta_k) | \eta_1, \dots, \eta_r))^2\big) \leqslant E(h^2(\eta_1, \dots, \eta_k))),$$

so that the existence of σ_r^2 for $r \leqslant k$ is assured. Considering $h(\xi_i, \eta_2, \dots, \eta_k)$, $i \geqslant 1$, we denote $E(h(\xi_i, \eta_2, \dots, \eta_k) | \xi_i))$ naturally by $\phi_1(\xi_i)$. Let $\zeta_n = n^{-1/2} \sum_{i=1}^{n} (\phi_1(\xi_i) - a)$ for $n \geqslant 1$. From the assumptions, it follows by an application of I, Theorem 39.1 that ζ_n is asymptotically $N(0, \sigma_1^2)$-distributed. Let $v_n = n^{1/2} k^{-1}(V(\xi_1, \dots, \xi_n) - a)$. If we can prove that

$$E[(\zeta_n - v_n)^2] \to 0, \quad n \to \infty, \tag{8.2}$$

[42] W. Hoeffding, Ann. Math. Statist. 19, 293—325 (1948).

then we are finished. Indeed, by Čebyšev's inequality for each $\varepsilon>0$

$$P(|\zeta_n - v_n| \geqslant \varepsilon) \leqslant \frac{E[(\zeta_n - v_n)^2]}{\varepsilon^2} \to 0$$

for $n\to\infty$ so that $\zeta_n - v_n$ converges in probability to 0. An application of I, Theorem 40.1 then delivers the claim.

We turn to the proof of (8.2) and calculate $E(\zeta_n^2)$, $E(v_n^2)$ and $E(\zeta_n v_n)$ one after the other. First,

$$E(\zeta_n^2) = \sigma_1^2. \tag{8.3}$$

To calculate $E(v_n^2)$ we investigate

$$E[(h(\xi_{i_1},\ldots,\xi_{i_k}) - a)(h(\xi_{j_1},\ldots,\xi_{j_k}) - a)], \tag{8.4}$$

where $I_1 = \{i_1,\ldots,i_k\}$ and $I_2 = \{j_1,\ldots,j_k\}$ are subsets of $\{1,\ldots,n\}$. If $I_1 \cap I_2 = \emptyset$, then (8.4) vanishes because of (8.1) since the ξ_i, $i\geqslant 1$, are independent. If, however, $I_1\cap I_2$ contains exactly $r>0$ elements, then one can assume because of the symmetry of h that $i_l = j_l$, $1\leqslant l\leqslant r$. Hence, for (8.4) one gets

$$E\big[E\big((h(\xi_{i_1},\ldots,\xi_{i_r},\xi_{i_{r+1}},\ldots,\xi_{i_k}) - a) \times \\ \times (h(\xi_{i_1},\ldots,\xi_{i_r},\xi_{j_{r+1}},\ldots,\xi_{j_k}) - a) | \xi_{i_1},\ldots,\xi_{i_r}\big)\big].$$

From this it follows that (8.4) is also equal to $E[(\phi_r(\xi_{i_1},\ldots,\xi_{i_r}) - a)^2]$. Thus, if $I_1 \cap I_2$ contains exactly $r>0$ elements, then

$$E((h(\xi_{i_1},\ldots,\xi_{i_k}) - a)(h(\xi_{j_1},\ldots,\xi_{j_k}) - a)) = \sigma_r^2. \tag{8.5}$$

We have

$$E(v_n^2) = \frac{n}{k^2}\binom{n}{k}^{-2} \sum_{\substack{i_1,\ldots,i_k \\ j_1,\ldots,j_k}} E((h(\xi_{i_1},\ldots,\xi_{i_k}) - a)(h(\xi_{j_1},\ldots,\xi_{j_k}) - a)), \tag{8.6}$$

where $i_1,\ldots,i_k$ and $j_1,\ldots,j_k$ run through all combinations of k elements from the set $\{1,\ldots,n\}$ independently of each other. We must now determine how often $\{i_1,\ldots,i_k\}$ and $\{j_1,\ldots,j_k\}$ have exactly r elements, $0\leqslant r\leqslant k$, in common. Choose any r elements from $\{1,\ldots,n\}$ and hold them fixed. Then there are $\binom{n-r}{k-r}$ combinations $\{i_1,\ldots,i_k\}$ which contain these chosen elements and the r elements can be chosen in $\binom{n}{r}$ ways. Hence, we obtain in all $\binom{n}{r}\binom{n-r}{k-r}$ combinations each with r "marked" elements. For each such combination $\{i_1,\ldots,i_k\}$ there exist $\binom{n-k}{k-r}$ combinations $\{j_1,\ldots,j_k\}$ having exactly the same r marked elements in common with $\{i_1,\dots,i_k\}$, which is easy to check. Defining $\sigma_0^2=0$, one gets from (8.5) and (8.6)

$$E(v_n^2)=\frac{n}{k^2}\binom{n}{k}^{-2}\sum_{r=0}^{k}\binom{n}{r}\binom{n-r}{k-r}\binom{n-k}{k-r}\sigma_r^2 .$$

An easy calculation gives

$$E(v_n^2)=nk^{-2}\binom{n}{k}^{-1}\sum_{r=0}^{k}\binom{k}{r}\binom{n-k}{k-r}\sigma_r^2 .$$

Now for $1\leqslant r\leqslant k$,

$$\binom{n-k}{k-r}\Big/\binom{n}{k}=O(n^{-r}). \tag{8.7}$$

More precisely, for $r=1$ one obtains

$$\binom{n-k}{k-1}\Big/\binom{n}{k}=\frac{k}{n}+O\left(\frac{1}{n^2}\right). \tag{8.8}$$

Hence, because $\sigma_0^2=0$,

$$E(v_n^2)=\sigma_1^2+O\left(\frac{1}{n}\right). \tag{8.9}$$

We now have to calculate

$$E(\zeta_n v_n)=k^{-1}\binom{n}{k}^{-1}\sum_{j=1}^{n}(\phi_1(\xi_j)-a)\sum_{i_1,\dots,i_k}E((h(\xi_{i_1},\dots,\xi_{i_k})-a)).$$

To do this, we argue as before.

For $j\neq i_l$, $1\leqslant l\leqslant k$,

$$E\big((h(\xi_{i_1},\dots,\xi_{i_k})-a)\,(\phi_1(\xi_j)-a)\big)=0.$$

However, if

$$j=i_l \tag{8.10}$$

holds for an l and a j, then

$$E\big((h(\xi_{i_1},\dots,\xi_{i_k})-a)\,(\phi_1(\xi_j)-a)\big)=\sigma_1^2 .$$

Equality (8.10) occurs exactly $n\binom{n-1}{k-1}$ times. Hence,

$$E(\zeta_n v_n)=k^{-1}\binom{n}{k}^{-1}n\binom{n-1}{k-1}\sigma_1^2 ,$$

so that

$$E(\zeta_n v_n)=\sigma_1^2 . \tag{8.11}$$

From (8.3), (8.9) and (8.11) follows $E[(\zeta_n - v_n)^2] = O\left(\frac{1}{n}\right)$ which proves Theorem 8.1.

We now apply Theorem 8.1 to the Wilcoxon test[43]. We will essentially use the notation introduced earlier, in particular that of Theorem 4.3. However, we limit ourselves to the case where $n_1 = n_2 = m$, $n = 2m$, $m \geqslant 1$. It is convenient to write $\xi_{m+i} = \eta_i$ for $1 \leqslant i \leqslant m$. We now denote the r.v. U more exactly by U_m and prove

Theorem 8.2. *Let $F \in C$. For real γ denote the d.f. $x \to F(x-\gamma)$ by F_γ and let $\{\gamma_m\}$ be a null sequence. Let the r.v.'s $\xi_1, \ldots, \xi_m$ for $m = 1, 2, \ldots$ be independently distributed with d.f. F and $\eta_1, \ldots, \eta_m$ independently distributed with d.f. F_{γ_m}. Moreover, let all ξ_i be independent of all η_j. Then*

$$\frac{m^{1/2}\left(m^{-2} U_m - \int\limits_{R_1} F\, dF_{\gamma_m}\right)}{(1/6)^{1/2}} \tag{8.12}$$

is $N(0,1)$-distributed.

Proof. This goes as the proof of Theorem 8.1. Write

$$U_m = \sum_{\substack{i,j=1\\ i \neq j}}^{m} c_I(\eta_i - \xi_j) + \sum_{i=1}^{m} c_I(\eta_i - \xi_i). \tag{8.13}$$

For $1 \leqslant i \leqslant m$ we introduce the notation $(\xi_i, \eta_i) = \xi_i^{(1)}$ and define for $i \neq j$

$$h(\xi_i^{(1)}, \xi_j^{(1)}) = c_I(\eta_i - \xi_j) + c_I(\eta_j - \xi_i). \tag{8.14}$$

Naturally, h is symmetric so that the r.v.

$$V(\xi_1^{(1)}, \ldots, \xi_m^{(1)}) = \binom{m}{2}^{-1} \sum_{i_1, i_2} h(\xi_{i_1}^{(1)}, \xi_{i_2}^{(2)})$$

is of the type considered in Theorem 8.1. Here, $\{i_1, i_2\}$ runs through all combinations of 2 elements from $\{1, \ldots, m\}$. Applying (4.10), we get from (8.14)

$$E[h(\xi_i^{(1)}, \xi_j^{(1)}); (F, F_{\gamma_m})] = 2 \int\limits_{-\infty}^{\infty} F\, dF_{\gamma_m}. \tag{8.15}$$

We now proceed with trivial modifications as in the proof of Theorem 8.1. It is easy to see that $\phi_1(\xi_j^{(1)}) = F(\eta_j) + 1 - F(\xi_j - \gamma_m)$, $1 \leqslant j \leqslant m$,

---

[43] For generalizations see E. L. Lehmann, l. c.[28]. A systematic treatment of the asymptotic theory of rank-tests is given in Hájek and Šidák l. c.[1].

with probability 1. One then finds without difficulty from (8.15) that

$$E[(\phi_1(\xi_j^{(1)})-E(\phi_1(\xi_j^{(1)})))^2]=\int_{-\infty}^{\infty} F_{\gamma_m}^2\,dF-2\int_{\infty}^{\infty} F F_{\gamma_m}\,dF+2\int_{-\infty}^{\infty} F_{\gamma_m}\,dF-2\left(\int_{-\infty}^{\infty} F_{\gamma_m}\,dF\right)^2 \tag{8.16}$$

where the expectation on the left is w.r.t. the pair (F, F_{γ_m}). This will also pertain to the expectations which follow. From $\lim_{m\to\infty}\gamma_m=0$, the continuity of F and the inequality $|F(x-\gamma_m)-F(x)|\leqslant 2$ which holds for $x\in R_1$ and all γ_m, $m\geqslant 1$, one easily finds that the right side of (8.16) converges to 1/6 and is hence positive for sufficiently large m. Since $|\phi_1(\xi_j^{(1)})|\leqslant 2$, $1\leqslant j\leqslant m$, with probability 1, there exists a positive number C independent of m for which

$$E[|\phi_1(\xi_j^{(1)})-E(\phi_1(\xi_j^{(1)}))|^3]\leqslant C, \quad 1\leqslant j\leqslant m.$$

An application of I, Theorem 39.3 shows that

$$m^{-1/2}\sum_{j=1}^{m}(\phi_1(\xi_j^{(1)})-E(\phi_1(\xi_j^{(1)})))/E[(\phi_1(\xi_j^{(1)})-E(\phi_1(\xi_j^{(1)})))^2]$$

is asymptotically $N(0,1)$-distributed. Hence, for each $y\in R_1$,

$$\lim_{m\to\infty} P\left(\frac{m^{-1/2}\sum_{j=1}^{m}(\phi_1(\xi_j^{(1)})-E(\phi_1(\xi_j^{(1)})))}{(1/6)^{1/2}}\leqslant y\right)=(2\pi)^{-1/2}\int_{-\infty}^{y} e^{-x^2/2}\,dx. \tag{8.17}$$

Since one can show exactly as before that

$$\lim_{m\to\infty} E\left[\left(2^{-1/2}m^{1/2}\left(V(\xi_1^{(1)},\ldots,\xi_m^{(1)})-2\int_{-\infty}^{\infty} F\,dF_{\gamma_m}\right)-m^{-1/2}\sum_{j=1}^{m}(\phi_1(\xi_j^{(1)})-E(\phi_1(\xi_j^{(1)})))\right)^2\right]=0$$

we have also shown that

$$\frac{2^{-1/2}m^{1/2}\left(V(\xi_1^{(1)},\ldots,\xi_m^{(1)})-2\int_{-\infty}^{\infty} F\,dF_{\gamma_m}\right)}{(1/6)^{1/2}}$$

is asymptotically $N(0,1)$-distributed.

But from (8.13) it follows that

$$m^{-2}U_m = \frac{m-1}{2m} V(\xi_1^{(1)}, \ldots, \xi_m^{(1)}) + \frac{1}{m^2} \sum_{i=1}^{m} c_I(\eta_i - \xi_i).$$

To prove the theorem, it now suffices to show that

$$\lim_{m\to\infty} P\left(m^{-2} \sum_{i=1}^{m} c_I(\eta_i - \xi_i) \geqslant \varepsilon\right) = 0$$

for each $\varepsilon > 0$.

Since, however, $E\left[\left(\sum_{i=1}^{m} c_I(\eta_i - \xi_i)\right)^2\right] \leqslant m^2$ this is obvious, which proves Theorem 8.2.

Remark. The condition $n_1 = n_2 = m$ in Theorem 8.2 can be replaced by

$$0 < \varliminf_{n\to\infty} \frac{n_1}{n} \leqslant \varlimsup_{n\to\infty} \frac{n_1}{n} < 1 \tag{8.18}$$

(with $n_1 + n_2 = n$). This requires only trivial modifications of the method of proof. With the same method one can show that $\sum_{i=1}^{n_1} \sum_{j=1}^{n_2} c_I(\eta_j - \xi_i)$ is asymptotically normally distributed under the hypothesis (F, G), $F, G \in C$ and the condition (8.18). In particular, this statement holds for the null hypothesis (F, F), $F \in C$.

We now prove a result on the Pitman efficiency of the Wilcoxon test relative to the t-test. We will use III, Lemma 12.3 and show

Theorem 8.3[44]. *Let $F(x) = (2\pi)^{-1/2} \int_{-\infty}^{x} e^{-y^2/2} dy$ for all $x \in R_1$. Again let $\eta > 0$ and $\Gamma = [0, \eta)$. Let $\gamma_m \in \Gamma$ with $\lim_{m\to\infty} \gamma_m = 0$ and $\{\phi_m^{(1)}\}$ the sequence of Wilcoxon tests for the two-sample problem with $n_1 = n_2 = m$ and sequence of hypotheses $\{(F, F_{\gamma_m})\}$. Let $\{\phi_m^{(2)}\}$ be the analogous sequence of (one-sided) t-tests. Assume both sequences of tests have the same asymptotic level of significance, i.e., III. (12.5) is fulfilled. Then*

$$re(\{\phi_m^{(1)}\}, \{\phi_m^{(2)}\}) = \pi/3.$$

Proof. If one defines $T_m^{(2)}$, $m \geqslant 1$, over R_{2m} (in obvious notation) by

$$(\bar{y}_m - \bar{x}_m) \Big/ \left(\frac{s_x^2 + s_y^2}{2}\right)^{1/2},$$

[44] We believe this theorem was first proved by H. R. van der Vaart, Nederl. Akad. Wetensch., Proc. Ser. A, 53, 494—520 (1956).

then by II. (6.3), $\{T_m^{(2)}\}$ is a sequence of test statistics for the t-test. Moreover (with $\bar{\eta}_m=(\eta_1+\cdots+\eta_m)/m$ and $\bar{\xi}_m=(\xi_1+\cdots+\xi_m)/m$),

$$(\bar{\eta}_m-\bar{\xi}_m-\gamma_m)((s_\xi^2+s_\eta^2)/2)^{-1/2}$$

has a t-distribution with $2(m-1)$ degrees of freedom under the hypothesis (F, F_{γ_m}). Since $\sqrt{(s_\xi^2+s_\eta^2)/2}$ converges in probability to 1 for $m\to\infty$ (see Theorem 4.2 of II), we have for each $x\in R_1$

$$\lim_{m\to\infty} P\left(\frac{[(\bar{\eta}_m-\bar{\xi}_m)((s_\xi^2+s_\eta^2)/2)^{-1/2}]-\gamma_m}{(2/m)^{1/2}}\leqslant x\right)=F(x). \tag{8.19}$$

If $c>0$ and $\gamma_m=cm^{-1/2}$, $m\geqslant 1$, then trivially

$$\lim_{m\to\infty} \gamma_m m^{1/2}=c. \tag{8.20}$$

Hence, III. (12.8) and (12.14) are satisfied in this case. We now need to investigate the asymptotic behavior of

$$\int_{-\infty}^{\infty} (F_{-\gamma_m}-F)dF \tag{8.21}$$

for $m\to\infty$. We must thus consider

$$\int_{+\infty}^{+\infty} e^{-x^2/2} \int_{-\infty}^{x} (e^{-(y+\gamma_m)^2/2}-e^{-y^2/2})dy\,dx.$$

For $0<\vartheta_{y,m}<1$,

$$\int_{-\infty}^{x} e^{-y^2/2}(e^{-y\gamma_m-\gamma_m^2/4}-1)dy$$
$$=\int_{-\infty}^{x} e^{-y^2/2}[(-y\gamma_m-\gamma_m^2/4)+\tfrac{1}{2}(-y\gamma_m-\gamma_m^2/4)^2 e^{\vartheta_{y,m}(-y\gamma_m-\gamma_m^2/4)}]dy.$$

Hence,

$$\int_{-\infty}^{x} (e^{-(y+\gamma_m)^2/2}-e^{-y^2/2})dy=-\gamma_m\int_{-\infty}^{x} e^{-y^2/2}y\,dy+O(\gamma_m^2),$$

where "O" does not depend on x. An easy calculation yields

$$-\gamma_m\int_{-\infty}^{x} e^{-y^2/2}y\,dy=\gamma_m e^{-x^2/2}, \qquad x\in R_1.$$

One then gets for (8.21) the value $\gamma_m/(2\sqrt{\pi})+O(\gamma_m^2)$ for each null sequence $\{\gamma_m\}$.

Choosing specially $\gamma_m = c m^{-1/2}$, one gets from the above, along with (8.20), (8.19) and the result of Theorem 8.2 by applying III. (12.17), that

$$\frac{c m^{-1/2}(1/(6 r_m))^{1/2}}{c m^{-1/2}(2\sqrt{\pi})^{-1}(2/m)^{1/2}} \to 1 \quad \text{for } m \to \infty .$$

This implies $\lim_{m \to \infty} m/r_m = 3/\pi$ as was claimed.

One can determine the Pitman efficiency for translations of the mean of a normal distribution for a whole series of rank tests by following the pattern of Theorem 8.3. See the footnote[45] for relevant literature.

We now show that Wilcoxon's test (or better, a suitable sequence of such tests) is consistent for a wide class of hypotheses. To this end, we introduce the following notation: Let $\mathfrak{E}_p$ be the set of all pairs (F,G), $F, G \in C$, such that

$$\int_{-\infty}^{\infty} F \, dG = p . \tag{8.22}$$

We have $0 < p < 1$. When we say in connection with the two-sample problem that the hypothesis $(F,G) \in \mathfrak{E}_p$ holds, then we naturally mean (in the sense of the agreement on p. 439), that $\xi_1, \ldots, \xi_{n_1}$ are independently distributed with d.f. F, $\eta_1, \ldots, \eta_{n_2}$ are independently distributed with d.f. G, and that (8.22) holds. We now prove

Theorem 8.4. *Let (8.18) be fulfilled. Then the sequence of Wilcoxon tests is consistent for the set of hypotheses* $(\{(F,F) : F \in C\}, \bigcup_{p > 1/2} \mathfrak{E}_p)$. *On the other hand, it is not consistent if the alternative hypothesis is replaced by* $\bigcup_{p < 1/2} \mathfrak{E}_p$.

Proof[46]. Write again $U_{n_1,n_2} = \sum_{i=1}^{n_1} \sum_{j=1}^{n_2} c_I(\eta_j - \xi_i)$ and $n = n_1 + n_2$. Let $\{c_{n_1,n_2}\}$ be a with n nondecreasing sequence of real numbers with $\lim_{n \to \infty} c_{n_1,n_2} = \infty$, whose choice we will fix later.

Set

$$\sigma^2_{0,n_1,n_2} = E[(U_{n_1,n_2} - E(U_{n_1,n_2};(F,F)))^2 ; (F,F)] .$$

---

[45] We mention J. L. Hodges and E. L. Lehmann, l. c. III[90], H. Chernoff and I. R. Savage, Ann. Math. Statist. 29, 972—994 (1958), M. Dwass, Ann. Math. Statist. 27, 352—374 (1956). See also F. C. Andrews, Ann. Math. Statist. 25, 724—736 (1954).

[46] D. van Dantzig, Indag. Math. 13, 1—8 (1951).

σ_{0,n_1,n_2}^2 is also given by the right side of (4.9). From the remark following Theorem 8.2, we get by (4.8)

$$\lim_{n\to\infty} P\left(\left(U_{n_1,n_2}-\frac{n_1 n_2}{2}\right)\sigma_{0,n_1,n_2}^{-1} > c_{n_1,n_2};\,(F,F)\right)=0. \tag{8.23}$$

Note that $F^2 \leqslant F$ implies for each $(F,G)\in\mathfrak{C}_p$

$$\int_{-\infty}^{\infty} F^2\,dG \geqslant p. \tag{8.24}$$

Since $(1-G)^2 \leqslant 1-G$ we get likewise

$$\int_{-\infty}^{+\infty} (1-G)^2\,dF \leqslant 1-\int_{-\infty}^{\infty} G\,dF = \int_{-\infty}^{\infty} F\,dG,$$

so that

$$\int_{-\infty}^{\infty} (1-G)^2\,dF \leqslant p. \tag{8.25}$$

Thus, from (4.7) for $(F,G)\in\mathfrak{C}_p$

$$E[(U_{n_1,n_2}-E(U_{n_1,n_2};(F,G)))^2;(F,G)] \leqslant p(1-p)n_1 n_2(n_1+n_2-1).$$

Since $0<p(1-p)<1/4$ we also have

$$E[(U_{n_1,n_2}-E(U_{n_1,n_2};(F,G)))^2;(F,G)] < 3\sigma_{0,n_1,n_2}^2. \tag{8.26}$$

Now

$$\left.\begin{aligned} &P\left(U_{n_1,n_2} \leqslant \frac{n_1 n_2}{2} + \sigma_{0,n_1,n_2} c_{n_1,n_2}\right)\\ &\quad = P\left(U_{n_1,n_2} - p n_1 n_2 \leqslant n_1 n_2\left(\frac{1}{2}-p\right) + \sigma_{0,n_1,n_2} c_{n_1,n_2}\right).\end{aligned}\right\} \tag{8.27}$$

But

$$\left.\begin{aligned} &n_1 n_2\left(\frac{1}{2}-p\right) + \sigma_{0,n_1,n_2} c_{n_1,n_2}\\ &\quad = \left(\frac{1}{2}-p\right) n_1 n_2\left(1+\frac{c_{n_1,n_2}}{2\sqrt{3}(1/2-p)}\left(\frac{n_1+n_2+1}{n_1 n_2}\right)^{1/2}\right).\end{aligned}\right\} \tag{8.28}$$

Choosing

$$c_{n_1,n_2} = o(n^{1/2}) \tag{8.29}$$

we then get from (8.18) that

$$c_{n_1,n_2}((n_1+n_2+1)/(n_1 n_2))^{1/2} = o(n^{1/2})O(n^{-1/2}) = o(1).$$

Hence, $1+\frac{c_{n_1,n_2}}{2\sqrt{3}(1/2-p)}((n_1+n_2+1)/(n_1 n_2))^{1/2}>0$ for sufficiently large n, provided (8.29) holds. Finally, by (8.28) for $p>1/2$,

$$n_1 n_2\left(\frac{1}{2}-p\right)+\sigma_{0,n_1,n_2} c_{n_1,n_2}<0.$$

Čebyšev's inequality for large enough n then yields

$$P\left(U_{n_1,n_2}-p n_1 n_2 \leqslant n_1 n_2\left(\frac{1}{2}-p\right)+\sigma_{0,n_1,n_2} c_{n_1,n_2}\right)$$

$$\leqslant E[(U_{n_1,n_2}-p n_1 n_2)^2]\,\sigma_{0,n_1,n_2}^{-2}\left[\left(\frac{1}{2}-p\right)(12 n_1 n_2/(n_1+n_2+1))^{1/2}+c_{n_1,n_2}\right]^{-2}$$

where the probability (also in the sequel) and the expected value are w.r.t. the pair (F,G).

This implies by (8.26) that

$$P\left(U_{n_1,n_2}-p n_1 n_2 \leqslant n_1 n_2\left(\frac{1}{2}-p\right)+\sigma_{0,n_1,n_2} c_{n_1,n_2}\right)$$

$$\leqslant 3\left[\left(\frac{1}{2}-p\right)(12 n_1 n_2/(n_1+n_2+1))^{1/2}+c_{n_1,n_2}\right]^{-2}.$$

Because of (8.29), we find from (8.27) for $(F,G)\in\bigcup_{p>1/2}\mathfrak{E}_p$:

$$P\left(\left(U_{n_1,n_2}-\frac{n_1 n_2}{2}\right)\sigma_{0,n_1,n_2}^{-1}\leqslant c_{n_1,n_2}\right)\to 0 \quad \text{for } n\to\infty$$

or

$$P\left(\left(U_{n_1,n_2}-\frac{n_1 n_2}{2}\right)\sigma_{0,n_1,n_2}^{-1}> c_{n_1,n_2}\right)\to 1 \quad \text{for } n\to\infty.$$

However, if $(F,G)\in\bigcup_{p>1/2}\mathfrak{E}_p$, then one gets in exactly the same way that

$$P\left(\left(U_{n_1,n_2}-\frac{n_1 n_2}{2}\right)\sigma_{0,n_1,n_2}^{-1}> c_{n_1,n_2}\right)$$

$$=P\left(U_{n_1,n_2}-p n_1 n_2 > n_1 n_2\left(\frac{1}{2}-p\right)+\sigma_{0,n_1,n_2} c_{n_1,n_2}\right).$$

But since $\frac{1}{2}-p>0$,

$$\left(\frac{1}{2}-p\right)n_1 n_2+\sigma_{0,n_1,n_2} c_{n_1,n_2}>0$$

and an application of Čebyšev's inequality gives together with (8.26)

$$P\left(U_{n_1,n_2}-p n_1 n_2>\left(\frac{1}{2}-p\right) n_1 n_2+\sigma_{0,n_1,n_2} c_{n_1,n_2}\right)$$
$$\leqslant 3\left[\left(\frac{1}{2}-p\right)(12 n_1 n_2/(n_1+n_2+1))^{1/2}+c_{n_1,n_2}\right]^{-2}.$$

For $n\to\infty$, however, the right side of this inequality tends to 0, no matter how one chooses the sequence $\{c_{n_1,n_2}\}$ with $c_{n_1,n_2}\to\infty$. The proof of Theorem 8.4 is complete.

9. Stochastic approximation. To close this chapter we turn to a quite different area of non-parametric statistics[47], the construction of consistent sequences of estimates for a special purpose. We will at the same time give an illustration of the ideas of III. **13**. Consider only the very simplest case: Let ξ_x be a r.v. over a probability space $(R,\mathscr{S},P_\gamma)$, $\gamma\in\Gamma$, for each $x\in R_1$. Let $E(\xi_x;\gamma)$ exist for each real x and all $\gamma\in\Gamma$ and denote it by $M_\gamma(x)$. For a given real α, let

$$M_\gamma(x)=\alpha$$

have exactly one real solution $\vartheta(\gamma)$ for all $\gamma\in\Gamma$. Consider the problem of constructing a consistent sequence of estimates for $\vartheta(\gamma)$. We will hold $\gamma\in\Gamma$ fixed in what follows and will suppress reference to this γ. Let $\{a_n\}$ be a sequence of positive real numbers chosen independently of $\gamma\in\Gamma$. Let η_1 be an arbitrary r.v. For $n\geqslant 2$ define the r.v.'s η_n by

$$\eta_{n+1}=\eta_n+a_n(\alpha-\zeta_n)\,. \tag{9.1}$$

Here, ζ_n, $n\geqslant 1$, is assumed to be a r.v. whose conditional distribution under the hypothesis $\eta_1=y_1,\dots,\eta_n=y_n$, $(y_1,\dots,y_n)\in R_n$, coincides with the distribution of ξ_{y_n}. We then also have

$$E(\zeta_n|\eta_1=y_1,\dots,\eta_n=y_n)=E(\zeta_n|\eta_n=y_n)=M(y_n)\,. \tag{9.2}$$

with probability 1. We then prove

Theorem 9.1[48]. *For a* $C>0$ *let*

$$|M(x)|\leqslant C, \qquad -\infty<x<\infty, \tag{9.3}$$

---

[47] The basic reference is H. Robbins and S. Monro, Ann. Math. Statist. 22, 400—407 (1951).

[48] The theorem and its proof in this form are due to J. Wolfowitz, Ann. Math. Statist. 23, 457—461 (1952).

and for a real ϑ:

$$\begin{cases} M(x) < \alpha & x < \vartheta \\ M(x) = \alpha & x = \vartheta \\ M(x) > \alpha & x > \vartheta . \end{cases} \tag{9.4}$$

Assume M *is Borel-measurable and for each* $\varepsilon > 0$ *let there exist a* $\delta(\varepsilon) > 0$, *such that*

$$\inf_{|x-\vartheta|>\varepsilon} |M(x)-\alpha| \geqslant \delta(\varepsilon). \tag{9.5}$$

Let there exist a $C_1 > 0$ *for which*

$$E[(\xi_x - M(x))^2] \leqslant C_1, \qquad -\infty < x < \infty, \tag{9.6}$$

and let the sequence $\{a_n\}$ *of positive numbers satisfy*

$$\sum_{n=1}^{\infty} a_n = \infty \tag{9.7}$$

$$\sum_{n=1}^{\infty} a_n^2 < \infty. \text{ [49]} \tag{9.8}$$

If $E(\eta_1^2)$ *is finite, then* $\{\eta_n\}$ *converges stochastically to* ϑ.

Proof. We first claim that $E[(\eta_n - \vartheta)^2] = b_n$ is finite for $n \geqslant 1$. For $n = 1$, this follows by assumption. Assume it holds for $1 \leqslant i \leqslant n$. We will show it holds for $n+1$. Indeed,

$$b_{n+1} = E[(\eta_{n+1} - \vartheta)^2] = E[((\eta_n - \vartheta) + a_n(\alpha - \zeta_n))^2]$$

because of (9.1). From this, using I. (20.5) along with I, Theorems 20.1 and 20.2, we get

$$b_{n+1} = b_n + a_n^2 E[E((\alpha - \zeta_n)^2 | \eta_n)] - 2a_n E[(\eta_n - \vartheta)(E((\zeta_n - \alpha) | \eta_n))]. \tag{9.9}$$

But from (9.2): $E[(\eta_n - \vartheta)(E((\zeta_n - \alpha)|\eta_n))] = E[(\eta_n - \vartheta)(M(\eta_n) - \alpha)]$.

Furthermore,

$$\begin{aligned} E[(\alpha - \zeta_n)^2 | \eta_n = y_n] &= E[(\zeta_n - M(y_n) + M(y_n) - \alpha)^2 | \eta_n = y_n] \\ &= E[(\zeta_n - M(\eta_n))^2 | \eta_n = y_n] + (M(y_n) - \alpha)^2 \end{aligned}$$

because of (9.2). From (9.3) and (9.6) we then get

$$d_n = E[E((\alpha - \zeta_n)^2 | \eta_n)] \leqslant C_1 + (C + |\alpha|)^2,$$

so that with $C_2 > 0$ for $n \geqslant 1$,

$$0 \leqslant d_n \leqslant C_2. \tag{9.10}$$

---

[49] This naturally means that $\sum_{n=1}^{\infty} a_n^2$ converges.

From (9.4)

$$E[(\eta_n-\vartheta)(M(\eta_n)-\alpha)] = E[|\eta_n-\vartheta|\,|M(\eta_n)-\alpha|] \geqslant 0,$$

so that finally from (9.9)

$$b_{n+1} \leqslant b_n + a_n^2 C_2, \tag{9.11}$$

which implies the finiteness of b_{n+1}.

Writing

$$e_n = E[(\eta_n-\vartheta)(M(\eta_n)-\alpha)]$$

for $n \geqslant 1$, we easily get from (9.9) that

$$b_{n+1} = b_1 + \sum_{i=1}^{n} a_i^2 d_i - 2\sum_{i=1}^{n} a_i e_i. \tag{9.12}$$

Because $b_{n+1} \geqslant 0$,

$$2\sum_{i=1}^{n} a_i e_i \leqslant b_1 + \sum_{i=1}^{n} a_i^2 d_i \leqslant b_1 + C_2 \sum_{i=1}^{n} a_i^2.$$

Since $a_i e_i \geqslant 0$ for $i \geqslant 1$ the convergence of $\sum_{i=1}^{\infty} a_i e_i$ follows. This also implies with (9.12), (9.8) and (9.10), the convergence of the sequence $\{b_n\}$. Now the assumption (9.7) comes into play. It implies that $\varliminf_{i\to\infty} e_i = 0$. If this were not true, then for $i \geqslant i_0$, say, and a real $a > 0$, the inequality $e_i \geqslant a$ would hold, which contradicts the convergence of $\sum_{i=1}^{\infty} a_i e_i$. Hence, there exists a sequence of integers $0 < n_1 < n_2 < \cdots$, for which $\lim_{j\to\infty} e_{n_j} = 0$.

We show that $\{\eta_{n_j}\}$ converges in probability to ϑ. If this were not true, then there would have to exist an $\varepsilon_0 > 0$, a $v_0 > 0$ and a subsequence $\{n_j\}$ of $\{n'_j\}$ such that

$$P(|\eta_{n'_j} - \vartheta| > \varepsilon_0) > v_0.$$

Then

$$e_{n'_j} = E[|\eta_{n'_j} - \vartheta|\,|M(\eta_{n'_j}) - \alpha|] \geqslant \varepsilon_0 \inf_{|x-\vartheta|>\varepsilon_0} |M(x)-\alpha|\, P(|\eta_{n'_j}-\vartheta|>\varepsilon_0),$$

so that by (9.5)

$$e_{n'_j} \geqslant \varepsilon_0 \,\delta(\varepsilon_0)\, v_0,$$

contradicting $\lim_{j\to\infty} e_{n'_j} = 0$.

We now need to show that the entire sequence $\{\eta_n\}$ also converges in probability to ϑ.

From (9.1), for each n_j and $n > n_j$,

$$\eta_n - \eta_{n_j} = \sum_{i=n_j}^{n-1} a_i(\alpha - \zeta_i).$$

This implies, in complete analogy to (9.12), that

$$E[(\eta_n-\vartheta)^2\,|\,\eta_{n_j}=y_{n_j}]=(y_{n_j}-\vartheta)^2+\sum_{i=n_j}^{n-1}a_i^2 d_i-2\sum_{i=n_j}^{n-1}a_i e_i$$

with probability 1. Hence,

$$E[(\eta_n-\vartheta)^2\,|\,\eta_{n_j}=y_{n_j}]\leqslant(y_{n_j}-\vartheta)^2+\sum_{i=n_j}^{n-1}a_i^2 d_i\,.$$

Choosing n_j large enough, and taking (9.8) and (9.10) into consideration, we find

$$E[(\eta_n-\vartheta)^2\,|\,\eta_{n_j}=y_{n_j}]\leqslant(y_{n_j}-\vartheta)^2+\delta \tag{9.13}$$

with probability 1 for each $\delta>0$.

Using (9.13), we get for $\varepsilon>0$ from Čebyšev's inequality

$$P(|\eta_n-\vartheta|>\varepsilon\,|\,|\eta_{n_j}-\vartheta|<\delta^{1/2})\leqslant E[(\eta_n-\vartheta)^2\,|\,|\eta_{n_j}-\vartheta|<\delta^{1/2}]\,\varepsilon^{-2}<\frac{v}{2}$$

with probability 1. Here $v=4\delta\varepsilon^{-2}$ which becomes arbitrarily small when ε is fixed and δ is chosen sufficiently small. Hence,

$$P(|\eta_n-\vartheta|>\varepsilon,\eta_{n_j}-\vartheta|<\delta)$$

$$=P(|\eta_n-\vartheta|>\varepsilon\,|\,|\eta_{n_j}-\vartheta|<\delta)P(|\eta_{n_j}-\vartheta|<\delta)<\frac{v}{2}\cdot 1\,,$$

so that

$$P(|\eta_n-\vartheta|>\varepsilon,|\eta_{n_j}-\vartheta|<\delta)<v/2\,. \tag{9.14}$$

Also,

$$P(|\eta_n-\vartheta|>\varepsilon,|\eta_{n_j}-\vartheta|\geqslant\delta)\leqslant P(|\eta_{n_j}-\vartheta|\geqslant\delta)\,.$$

Since we have shown that η_{n_j} converges in probability to ϑ, we also have

$$P(|\eta_n-\vartheta|>\varepsilon,|\eta_{n_j}-\vartheta|\geqslant\delta)<v/2 \tag{9.15}$$

for large enough n_j.

(9.14) and (9.15) imply

$$P(|\eta_n-\vartheta|>\varepsilon)<v$$

for large enough n, which was to be proved.

Under assumptions considerably weaker as those in Theorem 9.1, one can even show that η_n converges to ϑ with probability 1. We can only give references to numerous other important results such as the order of magnitude of $E(|\eta_n-\vartheta|^p)$ for $p>1$ and the determination of the asymptotic distribution of η_n when it exists[50].

---

[50] See L. Schmetterer, Österreich. Ing.-Arch. 7, 111—117 (1953); K. L. Chung, Ann. Math. Statist. 25, 463—483 (1954); A. Dvoretzky, Proceedings of the 3rd Berkeley Symposium on Mathematical Statistics and Probability 1954—1955, Vol. I, pp. 39—55, University of California Press, Berkeley and Los Angeles (1956). A rather complete summary is in L. Schmetterer, Proc. 4th Berkeley Sympos. Math. Statist. and Prob. Vol. I, 587—609 (1960) Univ. California Press, Berkeley, Calif. and Multivariate Analysis, Vol. 2, Academic Press Inc., New York 1969, 443—460. Related problems are also treated in A. Špaček, Czechosl. Math. J. 5, 462—466 (1955).

We do want to treat briefly an important application of stochastic approximation: the determination of the so-called LD 50 in biology. Let x be the dosage (or the log of the dosage) of a substance whose effectiveness (for example, on a mouse) is to be checked. Let $M(x)$ be that part of the population (e.g. of mice) which reacts to the dosage x. One can assume that M is nondecreasing, that $M(0)=0$, and that $M(x)\to 1$ for $x\to\infty$. M is thus a d.f. The LD 50 is then the (assumed unique) median ϑ, i.e., a solution of the equation $M(x)=\frac{1}{2}$. M is in general unknown but it is assumed that for each dosage x, an experiment can be carried out whose outcome is a r.v. ξ_x which, with nonvanishing probability, can only assume the values 1 or 0 and which has the following meaning: $\xi_x=1$ if after prescription of the dosage x, a reaction can be observed, and $\xi_x=0$ when this is not the case. We now have $P(\xi_x=1)=M(x)$, $P(\xi_x=0)=1-M(x)$, so that $E(\xi_x)=M(x)$. Hence, beginning with a completely arbitrary dosage, one can determine the LD 50 by means of the method of stochastic approximation. It turns out in practice that it yields an extraordinarily good approximation to the LD 50 after only a few steps.

Appendix

A thoughtful comparison of the material developed in Chapters III, IV and V uncovers many similarities. This coincidence is not accidently. The theory of testing (and its dual, the theory of confidence regions), as well as the theory of estimation, can be understood within a unified framework. The recognition of this fact is due to A. Wald[1] who developed what has come to be called decision theory in a series of fundamental papers.

In statistical applications one tries to make decisions based on the evidence of samples. These decisions can be of a quite varied nature. For instance, in the theory of tests one want to decide from which population a sample has been taken. In the theory of estimation one wants to find the best way to determine the true value of a parameter by means of a sample. In so far as problems of testing with fixed sample size are considered, the set of decisions consists of only two elements, which can be interpreted as "rejection" or "acceptance" of a null hypothesis. In the theory of estimation, in which one wants to estimate a parameter γ from a set Γ, the set of all decisions consists of the set Γ.

A certain risk is connected with each decision, and it is the main problem of decision theory to minimize this risk in some sense.

We proceed to a precise formulation of these ideas and an introduction to decision theory. But it should be understood that this introduction only serves the purpose explained above. Today decision theory is a highly developed branch of mathematical statistics. For a good summary one may consult for example the book by Ferguson[2]. Let $(R, \mathscr{S})$ be a sample space and P_Γ a set of probability measures defined over it. Let K be a nonempty set, which we will call the set of decisions k. Each map κ of R into K is called a *decision function*. The choice of such a

---

[1] Abraham Wald, Statistical Decision Functions, John Wiley, New York, N.Y., Chapman & Hall, Ltd., London 1950.

[2] T. S. Ferguson, Mathematical Statistics: A decision theoretic approach, Probability and Mathematical Statistics, Vol. 1, Academic Press, New-York-London 1967.

decision function determines the strategy of the statistician by associating some decision with each sample.

Let w be a function defined over $K \times \Gamma$, the so-called *loss function*[3]. For each $x \in R$, each κ and each $\gamma \in \Gamma$, this function gives the loss $w(\kappa(x), \gamma)$ suffered when one considers the sample x and takes the decision κ, when γ is the true parameter. Under the assumption that the function $x \to w(\kappa(x), \gamma)$ is integrable w.r.t. P_γ for each $\gamma \in \Gamma$, one can define the so-called *risk function* by means of

$$L(\kappa, \gamma) = \int_R w(\kappa(x), \gamma) \, dP_\gamma(x), \tag{1}$$

The left side of (1) is an expected value and can accordingly be viewed as the mean loss suffered when one chooses the decision function κ and γ is the true parameter.

In the sequel, we consider only the case where $w \geqslant 0$. Further we allow only decision functions κ for which the right side of (1) exists for all $\gamma \in \Gamma$. Denote the set of these decision functions by $\mathfrak{K}$. The risk function is thus defined over $\mathfrak{K} \times \Gamma$, and the problem is then to minimize it in some sense by means of a suitable choice of $\kappa \in \mathfrak{K}$.

We now give a number of definitions.

A $\kappa \in \mathfrak{K}$ is said to be *at least as good* in Γ *as* $\kappa_1 \in \mathfrak{K}$ if

$$L(\kappa, \gamma) \leqslant L(\kappa_1, \gamma) \tag{2}$$

for all $\gamma \in \Gamma$. We then write $\kappa_1 \preccurlyeq \kappa$. κ is said to be *better than* κ_1 when the $<$*-sign holds at least once in* (2). In this case we write $\kappa_1 \prec \kappa$. Obviously, $\kappa_1 \prec \kappa_2$ and $\kappa_2 \prec \kappa_3$ imply $\kappa_1 \prec \kappa_3$.

κ is called *optimal* in Γ if $\kappa_1 \preccurlyeq \kappa$ *for all* $\kappa_1 \in \mathfrak{K}$.

The decision function $\kappa \in \mathfrak{K}$ is called *admissible* in Γ if there exists no $\kappa_1 \in \mathfrak{K}_1$ with $\kappa \prec \kappa_1$.

A subset $\mathfrak{K}_1$ of $\mathfrak{K}$ *is complete* if for each $\kappa \in \mathfrak{K} - \mathfrak{K}_1$ there exists a $\kappa_1 \in \mathfrak{K}_1$ with $\kappa \prec \kappa_1$.

Under certain circumstances there exists a *minimal complete set* $\mathfrak{C}$ which contains no complete subset $\mathfrak{K}_1$ with $\mathfrak{C} \neq \mathfrak{K}_1$. We have

Theorem 1. *If a minimal complete set* $\mathfrak{C}$ *of decision functions exists, then it consists of exactly the set of admissible decision functions.*

For the simple *proof* note that each admissible decision function must naturally belong to $\mathfrak{C}$. On the other hand, if there existed a $\kappa \in \mathfrak{C}$ which were not admissible, then there would exist a $\kappa_1 \in \mathfrak{K}$ with $\kappa \prec \kappa_1$. Either $\kappa_1 \in \mathfrak{C}$ or there exists a $\kappa_2 \in \mathfrak{C}$ with $\kappa_1 \prec \kappa_2$. In both cases, $\mathfrak{C} - \{\kappa\}$ is likewise a complete set which contradicts the definition of $\mathfrak{C}$.

---

[3] See V, p. 330.

We give finally a

Definition. *A decision function* $\kappa \in \mathfrak{K}$ is called *minimax* if

$$\sup_{\gamma \in \Gamma} L(\kappa, \gamma) \leqslant \sup_{\gamma \in \Gamma} L(\kappa_1, \gamma) \quad \text{for all } \kappa_1 \in \mathfrak{K}.$$

This definition contains the concept introduced in III as a special case. (See below.)

Before we specialize these ideas, obtaining a whole series of optimality principles for the theory of testing and estimation, we will extend the theory thus far developed somewhat. This will be of special importance for applications to theory of testing.

In short, we will no longer associate a definite decision with each sample $x \in R$, but rather a random mechanism with whose help the associated decision will be determined.

Let $(K, \mathfrak{S})$ be a measurable space. Let a set $\mathfrak{P}$ of probability measures be defined over this space. As random decision function η, we define any map from R into $\mathfrak{P}$. Denote by η_x the probability measure from $\mathfrak{P}$ associated with the element $x \in R$ by means of η. Then, a (generalized) risk function is defined by

$$R(\eta, \gamma) = \int_R \int_K w(k, \gamma)\, d\eta_x(k)\, dP_\gamma(x)\,, \tag{3}$$

provided the integral on the right exists for all $\gamma \in \Gamma$. We again consider only the set E of all η for which this is the case.

If one makes the assumption (which we will call (A)) that the set $\{k\}$ belongs to $\mathfrak{S}$ for each $k \in K$ and in addition, that for each $k \in K$, $\mathfrak{P}$ contains the probability measure P_k degenerate at k, then each decision function $\kappa \in \mathfrak{K}$ belongs to E.

More precisely: If $\kappa \in \mathfrak{K}$ denotes the map $x \to k_x$, then the map $x \to P_{k_x}$ belongs to E and can be identified with κ. Both maps possess the same risk function.

The extension of the definitions on p. 477 and on this page to random decision functions is obvious.

Consider an example. Let $K = \{k_1, k_2\}$ and $\mathfrak{S}$ the set of all subsets of K. Each probability distribution P over $(K, \mathfrak{S})$ is uniquely determined by $P(\{k_2\})$. Each map ϕ of R into $[0, 1]$ can thus be viewed as a generalized decision function. Consider a nonempty subset Γ_0 of Γ and define a loss function w over $K \times \Gamma$ as follows:

$$w(k_1, \gamma) = \begin{cases} 0 & \gamma \in \Gamma_0 \\ d_1 & \gamma \in \Gamma - \Gamma_0\,; \end{cases} \qquad w(k_2, \gamma) = \begin{cases} d_2 & \gamma \in \Gamma_0 \\ 0 & \gamma \in \Gamma - \Gamma_0\,. \end{cases}$$

Here, d_1, d_2 are positive real numbers. The risk function is given in this special case by

$$R(\phi,\gamma)=\int_R w(k_1,\gamma)(1-\phi(x))\,dP_\gamma(x)+\int_R w(k_2,\gamma)\phi(x)\,dP_\gamma(x).$$

Hence, for $\gamma\in\Gamma_0$, one gets $R(\phi,\gamma)=d_2 E(\phi;\gamma)$ and for $\gamma\in\Gamma-\Gamma_0$, $R(\phi,\gamma)=d_1 E(1-\phi;\gamma)$. Obviously, ϕ must be assumed $\mathscr{S}$-measurable. Let α be a real number with $0\leqslant\alpha\leqslant 1$. If one chooses $d_2=1$ and allows only decision functions fulfilling $R(\phi,\gamma)\leqslant\alpha$ for all $\gamma\in\Gamma_0$, then III. (1.1) says that each such decision function ϕ is a test for the problem $(\alpha,\Gamma_0,\Gamma-\Gamma_0)$. The set of all these tests is denoted by $\mathfrak{K}_\alpha$. The function $\gamma\to 1-R(\phi,\gamma)/d_1$ is identical with the power function of ϕ on $\Gamma-\Gamma_0$.

If ϕ_1 is a most powerful test for the problem $(\alpha,\Gamma_0,\Gamma-\Gamma_0)$ and one allows only the $\phi\in\mathfrak{K}_\alpha$, then ϕ_1 is optimal in $\Gamma-\Gamma_0$ w.r.t. $\mathfrak{K}_\alpha$. The optimality principles of the theory of testing are thus subordinate to those of decision theory.

We indicate briefly how the theory of estimation developed in V can be incorporated into decision theory. For the sake of simplicity, assume that maps from Γ into R_1 are to be estimated. It is convenient to assume at the outset that $\Gamma\subseteq R_1$ and to limit oneself to the estimation of $\gamma\in\Gamma$. As set of decisions K we choose Γ itself. If one defines the loss function w over $K\times\Gamma$ as $(k-\gamma)^2$, then one gets for the risk function

$$L(\kappa,\gamma)=\int_R (\kappa(x)-\gamma)^2\,dP_\gamma(x).$$

Almost all of the investigations of V were based on this risk function. We denoted the decision functions κ there as estimates. A somewhat more general risk function is obtained when one defines the loss function w over $K\times\Gamma$ by $\psi(k-\gamma)$. Here, ψ is a convex function defined over R_1. Such a loss function was introduced in V, Theorem 3.13.

It is obvious that the notions in V such as uniformly minimal, admissible and complete are specializations of the corresponding notions in general decision theory.

In III. **11** we introduced the invariance principle in the theory of testing. This can be generalized to an invariance principle for general decision theory[4]. Let $\mathfrak{G}$ be an arbitrary group which acts on a measurable space $(R,\mathscr{S})$ in the sense of III, p. 235. Let $\overline{\mathfrak{G}}$ be a homomorphic image[5] of $\mathfrak{G}$, such that $\bar{\mathfrak{g}}\Gamma=\Gamma$ for each $\bar{\mathfrak{g}}\in\overline{\mathfrak{G}}$. Moreover, let $\mathfrak{G}_1$ likewise be a homomorphic image of $\mathfrak{G}$ such that $\mathfrak{G}_1$ acts on the measurable space $(K,\mathfrak{S})$. Denote the image of $\mathfrak{g}\in\mathfrak{G}$ in $\overline{\mathfrak{G}}$ (in $\mathfrak{G}_1$) by $\bar{\mathfrak{g}}$ (by $\mathfrak{g}_1$). $\mathfrak{G}$ is called *admissible for a decision problem* with loss function w if

$$P_{\bar{\mathfrak{g}}\gamma}(\mathfrak{g}A)=P_\gamma(A) \tag{4}$$

[4] See J. Kiefer, Ann. Math. Statist. 28, 573—601 (1957) and the literature cited there.

[5] See III[80].

for all $A \in \mathscr{S}$, all $\gamma \in \Gamma$ and all $\mathfrak{g} \in \mathfrak{G}$ and if

$$w(\mathfrak{g}_1 k, \bar{\mathfrak{g}} \gamma) = w(k, \gamma) \tag{5}$$

for all $k \in K$, all $\gamma \in \Gamma$ and all $\mathfrak{g} \in \mathfrak{G}$.

Each $\mathfrak{g} \in \mathfrak{G}$ defines a map $\mathfrak{g}_{\mathfrak{P}}$ from $\mathfrak{P}$ into itself according to the following correspondence: If η is a map $x \to \eta_x$ from R into $\mathfrak{P}$, then for $x \in R$ and $C \in \mathfrak{S}$ set

$$\mathfrak{g}_{\mathfrak{P}} \eta_x(C) = \eta_{\mathfrak{g}x}(\mathfrak{g}_1 C). \tag{6}$$

From (3), (4) and (5) we then have (under the corresponding measurability assumptions)

$$\begin{aligned} R(\mathfrak{g}_{\mathfrak{P}} \eta, \gamma) &= \int_K \int_R w(k, \gamma) \, d\mathfrak{g}_{\mathfrak{P}} \eta_x(k) \, dP_\gamma(x) \\ &= \int_R \int_K w(\mathfrak{g}_1^{-1} k, \gamma) \, d\eta_{\mathfrak{g}x}(k) \, dP_\gamma(x) \\ &= \int_R \int_K w(\mathfrak{g}_1^{-1} k, \gamma) \, d\eta_x(k) \, dP_\gamma(\mathfrak{g}^{-1} x) \\ &= \int_R \int_K w(k, \bar{\mathfrak{g}} \gamma) \, d\eta_x(k) \, dP_{\bar{\mathfrak{g}}\gamma}(x) = R(\eta, \bar{\mathfrak{g}} \gamma). \end{aligned}$$

A decision function η is called *invariant* (w.r.t. $\mathfrak{G}$) if $\mathfrak{g}_{\mathfrak{P}} \eta_x = \eta_x$ for all $x \in R$.

By specializing this, one gets the invariance principle for the theory of testing. If we have the test problem $(\Gamma_0, \Gamma - \Gamma_0)$, then one must require in addition that the elements of $\mathfrak{G}$ fulfill III.(11.28) and (11.29). $\mathfrak{G}_1$ consists only of the unit element. One can then write in place of (6): $\mathfrak{g}_{\mathfrak{P}} \phi(x) = \phi(\mathfrak{g} x)$. The invariance of the decision function ϕ is then expressed by $\phi(\mathfrak{g} x) = \phi(x)$ for $\mathfrak{g} \in \mathfrak{G}$ and $x \in R$. (See p. 236.)

In IV, p. 265 and V, p. 322 ff. we treated the Bayesian concept. This can also be brought into the general decision theoretic framework. To do so, one must naturally start with the assumption that, in addition to Γ, a σ-algebra $\mathfrak{C}$ of subsets of Γ is also given along with a measure ν over $(\Gamma, \mathfrak{C})$ which need not be a probability measure. Beginning with the risk function L defined over $\mathfrak{K} \times \Gamma$, one arrives at a modified risk function of the form

$$l(\kappa, \nu) = \int_\Gamma L(\kappa, \gamma) \, d\nu(\gamma),$$

provided it is well-defined for at least certain $\kappa \in \mathfrak{K}$. We assume for the sake of simplicity that this is the case for all $\kappa \in \mathfrak{K}$. Each $\kappa_0 \in \mathfrak{K}$ satisfying

$$\inf_{\kappa \in \mathfrak{K}} l(\kappa, \nu) = l(\kappa_0, \nu) \tag{7}$$

is called a *Bayes solution w.r.t. ν*.

Under special assumptions, we showed in V, Theorem 3.12 how one can determine a Bayes solution. Bayes solution do not always exist. However, they exist under quite general conditions[6].

It is clear that this definition can also be extended to generalized decision functions. One then gets another risk function according to

$$r(\eta, v) = \int_{\Gamma} R(\eta, \gamma)\, dv(\gamma).$$

A Bayes solution η_0 then satisfies a condition analogous to (7).

The practical value of the Bayes solution for a given measure v is in general slight since in concrete problems one will only know that v is an element of a set V of measures over $(\Gamma, \mathfrak{C})$.

On the other hand, there exist interesting connections between a number of previously defined notions and Bayes solutions which allow the meaning of the Bayesian concept to appear in a new light. As an illustration, we prove, maintaining the notation above

Theorem 2. *Let V be the set of all probability measures over* $(\Gamma, \mathfrak{C})$. *Let $\mathfrak{C}$ fulfill assumption* (A) (p. 478). *Assume there exists a $v_0 \in V$ (a so-called least favorable a priori distribution*[7]*) satisfying*

$$\inf_{\eta \in E} r(\eta, v_0) = \sup_{v \in V} \inf_{\eta \in E} r(\eta, v). \tag{8}$$

Moreover, let η_0 be minimax, i.e.,

$$\sup_{\gamma \in \Gamma} R(\eta_0, \gamma) = \inf_{\eta \in E} \sup_{\gamma \in \Gamma} R(\eta, v). \tag{9}$$

If

$$\sup_{v \in V} \inf_{\eta \in E} r(\eta, v) = \inf_{\eta \in E} \sup_{v \in V} r(\eta, v), \tag{10}$$

then η_0 is a Bayes solution w.r.t. v_0.

Proof. For each $\eta \in E$ and $v \in V$

$$r(\eta, v) = \int_{\Gamma} R(\eta, \gamma)\, dv(\gamma) \leqslant \sup_{\gamma \in \Gamma} R(\eta, \gamma),$$

whence

$$\sup_{v \in V} r(\eta, v) \leqslant \sup_{\gamma \in \Gamma} R(\eta, \gamma). \tag{11}$$

Further for each $\eta \in E$ and each $\varepsilon > 0$, there exists a $\gamma_{\eta,\varepsilon} \in \Gamma$ with

$$\sup_{\gamma \in \Gamma} R(\eta, \gamma) - \varepsilon \leqslant R(\eta, \gamma_{\eta,\varepsilon}). \tag{12}$$

[6] See A. Wald, l.c.[1], 89.

[7] Cf. III. 182.

If one considers the probability measure over $(\Gamma, \mathfrak{C})$ degenerate at $\gamma_{\eta,\varepsilon}$, then it easily follows from (12) and assumption (A) that $\sup_{\gamma\in\Gamma} R(\eta,\gamma) - \varepsilon \leqslant \sup_{v\in V} r(\eta, v)$ and hence also $\sup_{\gamma\in\Gamma} R(\eta,\gamma) \leqslant \sup_{v\in V} r(\eta, v)$ for each $\eta\in E$. This implies with (11) that

$$\sup_{v\in V} r(\eta, v) = \sup_{\gamma\in\Gamma} R(\eta,\gamma). \tag{13}$$

Using (9) one obtains

$$\inf_{\eta\in E} \sup_{\gamma\in\Gamma} R(\eta,\gamma) = \sup_{v\in V} r(\eta_0, v),$$

and again because of (13), also

$$\inf_{\eta\in E} \sup_{v\in V} r(\eta, v) = \sup_{v\in V} r(\eta_0, v).$$

Together with (10), this yields

$$r(\eta_0, v_0) \leqslant \sup_{v\in V} r(\eta_0, v) = \sup_{v\in V} \inf_{\eta\in E} r(\eta, v),$$

so that from (8), $r(\eta_0, v_0) \leqslant \inf_{\eta\in E} r(\eta, v_0)$. That is, η_0 is a Bayes solution w.r.t. v_0.

Relation (10) played a decisive role in this proof. The assumptions guaranteeing its validity are the subject of game-theoretic investigations[8]. We will not go into this here.

We have previously always assumed that a fixed sample space $(\mathrm{R}, \mathscr{S})$ has been given. Wald[9], however, developed his theory from the outset so generally that the case of sequences of sample spaces is also included. This brings one to the interesting asymptotic questions in decision theory. An illustration is offered by V, Theorem 3.13.

[8] The basic reference in game-theory is J. v. Neumann and O. Morgenstern, Theory of Games and Economic Behavior 3. ed. Princeton University Press, Princeton, N.J., 1953. We also mention S. Karlin l.c. III[11], Vol. I and II. For deeper mathematical considerations see also H. W. Kuhn and A. W. Tucker, Contributions to the Theory of Games I (Annals of Mathematics Studies no. 24), Princeton University Press, Princeton, N.J., 1950; II (Annals of Mathematics Studies no. 28), Princeton University Press, Princeton, N.J., 1953; M. Dresher, A. W. Tucker and P. Wolfe, III Ann. Math. Studies 38, Princeton University Press, Princeton, N.J., 1957; A. W. Tucker and R. D. Luce, IV Ann. Math. Studies 40, Princeton University Press, Princeton, N.J., 1959. Applications of game theory to statistics are contained in Blackwell D. and M. A. Girshick, Theory of games and statistical decisions, John Wiley and Sons, New York. Chapman and Hall, London, 1954, as well as A. Wald, l.c.[1] and Advances in Game Theory, Edited by M. Dresher, L. S. Shapley, A. W. Tucker, Ann. Math. Studies 52, Princeton University Press, Princeton, N. J., 1964.

[9] A. Wald, l.c.[1]. Of his many relevant papers, we mention only Ann. Math. Statist. 20, 165—205 (1949).

Bibliography

(The page numbers at the end of each quotation refer to the positions in the text where that paper is cited. The journals are cited here according to the abbreviation code of the Mathematical Reviews).

A.C. Aitken: On least squares and linear combination of observations. Proc. Roy. Soc. Edinburgh Sect. A, 55, 42—48 (1935); p. 357

T.W. Anderson: An introduction to multivariate statistical analysis. John Wiley & Sons, Inc., New York; Chapman & Hall, Ltd., London, 1958; p. 370

F.C. Andrews: Asymptotic behavior of some rank tests for analysis of variance. Ann. Math. Statist. 25, 724—736 (1954); p. 468

P. Armitage: A comparison of stratified with unrestricted random sampling from a finite population. Biometrika 34, 273—280 (1947); p. 152

R.R. Bahadur: Sufficiency and statistical decision functions. Ann. Math. Statist. 25, 423—462 (1954); p. 211

R.R. Bahadur: On unbiased estimates of uniformly minimum variance. Sankhyā 18, 211—224 (1957); p. 276, 277, 278, 289

R.R. Bahadur: Examples of inconsistency of maximum likelihood estimates. Sankhyā 20, 207—210 (1958); p. 307

R.R. Bahadur: Stochastic comparison of tests. Ann. Math. Statist. 31, 276—295 (1960); p. 247

R.R. Bahadur: On the asymptotic efficiency of tests and estimates. Sankhyā 22, 229—253 (1960); p. 342

R.R. Bahadur: On Fisher's bound for asymptotic variances. Ann. Math. Statist. 35, 1545—1552 (1964); p. 352

R.R. Bahadur: Rates of convergence of estimates and test statistics. Ann. Math. Statist. 38, 303—324 (1967); p. 352

R.R. Bahadur and E.L. Lehmann: Two comments on sufficiency and statistical decision functions. Ann. Math. Statist. 26, 139—142 (1955); p. 60

R.R. Bahadur and R. Ranga Rao: On deviation of the sample mean. Ann. Math. Statist. 31, 1015—1027 (1960); p. 109

E.W. Barankin: Locally best unbiased estimates. Ann. Math. Statist. 20, 477—501 (1949); p. 275, 284

E.W. Barankin: On systems of linear equations, with applications to linear programming and the theory of tests of statistical hypotheses. Univ. California Publ. Statist. 1, 161—214 (1951); p. 176

E.W. Barankin and J. Gurland: On asymptotically normal, efficient estimators, I. Univ. California Publ. Statist. 1, 89—129 (1951); p. 342

E.W. Barankin and M. Katz: Sufficient statistics of minimal dimension. Sankhyā 21, 217—246 (1959); p. 218

E. W. Barankin and A. P. Maitra: Generalization of the Fisher-Darmois-Koopman-Pitman theorem on sufficient statistics, Sankhyā 25, 217—244 (1963); p. 218

G. A. Barnard: Sequential tests in industrial statistics. Suppl. J. Roy. Statist. Soc. 8, 1—21 (1946); p. 254

O. Barndorff-Nielsen and K. Pedersen: Sufficient data reduction and exponential families. Math. Scand, 22, 197—202 (1968); p. 218

M. S. Bartlett: On the theory of statistical regression. Proc. Roy. Soc. Edinburgh, Sect. A, 53, 260—283 (1932—1933); p. 391

N. K. Bary: A treatise on trigonometric series. Vol. I. A. Pergamon Press Book. The Macmillan Co., New York, 1964; p. 459

D. Basu: The concept of asymptotic efficiency. Sankhyā 17, 193—196 (1956); p. 342

H. Bauer: Wahrscheinlichkeitstheorie und Grundzüge der Maßtheorie. Walter de Gruyter & Co., Berlin, 1968; p. 25

V. Baumann: Eine parameterfreie Theorie der ungünstigsten Verteilungen für das Testen von Hypothesen. Z. Wahrscheinlichkeitstheorie Verw. Gebiete 11, 41—60 (1968); p. 178

Th. Bayes: An essay toward solving a problem in the doctrine of chance. Philos. Trans. Roy. Soc. 53, 376—398 (1763) and 54, 248—310 (1764); p. 28

W. U. Behrens: Ein Beitrag zur Fehlerberechnung bei wenigen Beobachtungen. Landwirtschaftliche Jahrbücher 48, 807—837 (1929); p. 224

A. Berger: On uniformly consistent tests. Ann. Math. Statist. 22, 289—293 (1951); p. 241

P. J. Bickel and J. A. Yahav: Some contributions to the asymptotic theory of Bayes solutions. Z. Wahrscheinlichkeitstheorie und Verw. Gebiete 11, 257—276 (1969); p. 320

Z. W. Birnbaum and F. H. Tingey: One-sided confidence contours for probability distribution functions. Ann. Math. Statist. 22, 592—596 (1951); p. 430

Z. W. Birnbaum and H. S. Zuckermann: An inequality due to H. Hornich. Ann. Math. Statist. 15, 328—329 (1944); p. 451

D. H. Blackwell: Conditional expectation and unbiased sequential estimation. Ann. Math. Statist. 18, 105—110 (1947); p. 277

D. H. Blackwell: On a class of probability spaces. Proceedings of the Third Berkeley Symposium on Mathematical Statistics and Probability 1954—1955, Vol. II, pp. 1—6, University of California Press, Berkeley and Los Angeles 1956; p. 57

D. H. Blackwell and M. A. Girshick: Theory of games and statistical decisions. John Wiley & Sons, Inc., New York; Chapman & Hall, Ltd., London 1954; p. 482

D. H. Blackwell and C. Ryll-Nardzewski: Non existence of everywhere proper conditional distributions. Ann. Math. Statist. 34, 223—225 (1963); p. 57

A. Blanc-Lapierre et R. Fortet: Theorie des Fonctions aléatoires. Masson & Cie, Editeurs, Paris 1953; p. 25

L. N. Bol'šev: The refinement of the Cramér-Rao inequality. Teor. Verojatnost. i. Primenen. 6, 319—326 (1961) (Russian); p. 284

R. Borges: Subjektiv trennscharfe Konfidenzbereiche. Z. Wahrscheinlichkeitstheorie und Verw. Gebiete 1, 47—69 (1962); p. 266

L. Breiman: Probability. Addison-Wesley Publ. Comp., Reading London 1968; p. 25

O. Bunke: Neue Konfidenzintervalle für den Parameter der Binomialverteilung. Wiss. Z. Humboldt-Univ. Berlin, Math.-Natur. Reihe 9, 335—363 (1959/60); p. 260

D. L. Burkholder: Sufficiency in the undominated case. Ann. Math. Statist. 32, 1191—1200 (1961); p. 218

D. L. Burkholder: On the order structure of the set of sufficient subfields. Ann. Math. Statist. 33, 596—599 (1962); p. 218

P. Cantelli, W. Feller, M. Fréchet, R. v. Mises, J. F. Steffensen, and A. Wald: Les Fondements du Calcul des Probabilités (Actualités scientifiques et industrielles 735), Hermann & Cie, Paris 1938; p. 22

K.C. Chanda: A Note on the consistency and maxima of the roots of likelihood equation. Biometrika 41, 56--61 (1954); p. 298

D.G. Chapman and H. Robbins: Minimum variance estimation without regularity assumptions. Ann. Math. Statist. 22, 581—586 (1951); p. 284

H. Chernoff and I. R. Savage: Asymptotic normality and efficiency of certain nonparametric test statistics. Ann. Math. Statist. 29, 972—994 (1958); p. 468

H. Chernoff and H. Scheffé: A generalisation of the Neyman-Pearson fundamental lemma. Ann. Math. Statist. 23, 213--225 (1952); p. 166

K.L. Chung: On stochastic approximation method. Ann. Math. Statist. 25, 463--483 (1954); p. 474

K. L. Chung: Markov chains with stationary transition probabilities. Die Grundlehren der mathematischen Wissenschaften Bd. 104. Springer Verlag, Berlin-Göttingen-Heidelberg, 1960; p. 25

K.L. Chung: A course in probability theory. Harcourt, Brace & World, Inc., New York, 1968; p. 25

J. Clopper and E.S. Pearson: The use of confidence or fiducial limits illustrated in the case of binomial. Biometrika 26, 404--413 (1934); p. 260

W.G. Cochran: The distribution of quadratic forms in a normal system, with applications to the analysis of covariance. Proc. Cambridge Philos. Soc. 30, 178--191 (1933--1934); p. 128

W.G. Cochran: The chi-square test of goodness of fit. Ann. Math. Statist. 23, 315--345 (1952); p. 230

A.G. Constantine and A.T. James: On the general canonical correlation distribution. Ann. Math. Statist. 29, 1146--1166 (1958); p. 405

H. Cramér: Mathematical Methods of Statistics. Princeton Mathematical Series, Vol. 9. Princeton University Press, Princeton, N.J., 1946; p. 116, 122, 230, 298

H. Cramér: A contribution to the theory of statistical estimation. Skand. Aktuarietidskr. 29, 85--94 (1946); p. 284

G.B. Dantzig: On the non-existence of tests of "Student's" hypothesis having power functions independent of σ, Ann. Math. Statist. 11, 186—192 (1940); p. 224

G.B. Dantzig and A. Wald: On the fundamental lemma of Neyman and Pearson. Ann. Math. Statist. 22, 87—93 (1951); p. 166, 170

D. van Dantzig: On the consistency and the power of Wilcoxon's two sample test. Indagationes Math. 13, 1—8 (1951); p. 468

D. van Dantzig and J. Hemelrijk: Statistical methods based on a few assumptions. Bull. Inst. Internat. Statist. 24, 2ième livraison, 239—267 (1954); p. 406

D.A. Darling: The Kolmogorov-Smirnov, Cramér-von Mises tests. Ann. Math. Statist. 28, 823—839 (1957); p. 434

F.N. David and J. Neyman: Extension of the Markoff theorem on least squares. Statist. Res. Mem., Univ. London 2, 105—116 (1938); p. 268, 359

J. L. Denny: A continuous real-valued function on E^n almost everywhere 1—1. Fundamenta Math. 55, 95—99 (1964); p. 208

J. L. Denny: Sufficient conditions for a family of probabilities to be exponential. Proc. Nat. Acad. Sci., USA. 57, 1184—1187 (1967); p. 218

J.L. Denny: Chauchy's equation and sufficient statistics on arcwise connected spaces. Ann. Math. Statist. 41, 401—411 (1970); p. 218

W.J. Dixon and A.M. Mood: The statistical sign test. J. Amer. Statist. Assoc. 41, 557--566 (1946); p. 448

H.F. Dodge and H.G. Romig: Sampling inspection tables: Single and double sampling. 2nd ed., John Wiley & Sons, Inc., New York, Chapman & Hall, Ltd., London, 1959; p. 145

M. D. Donsker: Justification and extension of Doob's heuristic approach to the Kolmogorov-Smirnov theorems. Ann. Math. Statist. 23, 277—281 (1952); p. 429

J. L. Doob: Probability and statistics. Trans. Amer. Math. Soc. 36, 759—775 (1934); p. 310

J. L. Doob: Heuristic approach to the Kolmogorov-Smirnov theorems. Ann. Math. Statist. 20, 393—403 (1949); p. 429

J. L. Doob: Application of the theory of martingales. Le Calcul des Probabilités et ses Applications. Colloques Internationaux du Centre National de la Recherche Scientifique, no 13, pp. 23—27. Centre National de la Recherche Scientifique, Paris 1949; p. 295

J. L. Doob: Stochastic processes. John Wiley & Sons, Inc., New York; Chapman & Hall, Ltd., London, 1953; p. 25

M. Dresher, A. W. Tucker and P. Wolfe: Contributions to the theory of games III. Ann. Math. Studies 38, Princeton University Press, Princeton, N.J., 1959; p. 482

M. Dresher: Advances in Game theory. Edited by M. Dresher, L. S. Shapley, A. W. Tucker. Ann. Math. Studies 52, Princeton, N.J., 1964; p. 482

H. Drygas: The coordinate-free approach to Gauss-Markov estimation. Lecture Notes in Operations Research and Mathematical Systems, VOL. 40. Springer-Verlag, Berlin-New York, 1970; p. 359

D. Dugué: Traité de statistique théorique et appliquée: analyse aléatoire, algèbre aléatoire. Masson et Cie, Paris, 1958; p. 122

A. Dvoretzky: On stochastic approximation. Proceedings of the Third Bekely Symposium on Mathematical Statistics and Probability 1954—1955 Vol. I, pp. 39—55. University of California Press, Berkeley and Los Angeles, 1956; p. 474

M. Dwass: The large sample power of rank order tests in the two-sample problem. Ann. Math. Statist. 27, 352—374 (1956); p. 468

E. B. Dynkin: Necessary and sufficient statistics for a family of probability distributions. Uspehi Mat. Nauk (N.S.) 6, no. 1 (41), 68—90 (1951); p. 216

E. B. Dynkin: Die Grundlagen der Theorie der Markoffschen Prozesse. Die Grundlehren der mathematischen Wissenschaften. Bd. 108, Springer-Verlag, Berlin-Göttingen-Heidelberg, 1961; p. 25

E. B. Dynkin: Markov processes. Vol. I. Die Grundlehren der mathematischen Wissenschaften, Bd. 121. Springer-Verlag, Berlin-Göttingen-Heidelberg, 1965; p. 25

C. van Eeden and A. Benard: A general class of distributionfree tests for symmetry containing the tests of Wilcoxon and Fisher. I, II, III. Indag. Math. 19, 381—408 (1957); p. 449

C. G. Esseen: Fourier analysis of distribution functions-A mathematical study of the Laplace-Gaussian law. Acta Math. 77, 1—125 (1944); p. 109

W. Feller: Note on regions similar to the sample space. Statist. Res. Mem. Univ. London 2, 117—125 (1938); p. 222

W. Feller: On the normal approximation to the binominal distribution. Ann. Math. Statist. 16, 319—329 (1945); p. 115

W. Feller: On the Kolmogorov-Smirnov limit theorems for empirical distributions. Ann. Math. Statist. 19, 177—189 (1948); p. 429

W. Feller: Errata for "On the Kolmogorov-Smirnov limit theorem for empirical distributions". Ann. Math. Statist. 21, 301 (1950); p. 115

W. Feller: An introduction to probability theory and its applications. Vol. I, 3rd ed. and Vol. II. John Wiley & Sons, Inc., New York-London-Sidney 1966; p. 25, 444

T. S. Ferguson: A method of generating best asymptotically normal estimates with applications to the estimation of bacterial densities. Ann. Math. Statist. 29, 1046—1062 (1958); p. 342

T. S. Ferguson: Mathematical Statistics: A decision theoretic approach, Probability and Mathematical Statistics, Vol. 1. Academic Press. New York-London 1967; p. 476

B. V. Finkelstein: On the limiting distributions of the extreme terms of a variational series of a two-dimensional random quantity. Doklady Akad. Nauk SSSR, (N.S.) 91, 209—211 (1953); p. 425

R. A. Fisher: On an absolute criterion for fitting frequency curves. Messenger of Math. 41, 150—160 (1912); p. 295

R. A. Fisher: Frequency distribution of the values of the correlation coefficient in samples from an indefinitely large population. Biometrika 10, 507—521 (1915); p. 83

R. A. Fisher: On the "probable error" of a coefficient of correlation deduced from a small sample. Metron 1, 1—32 (1921); p. 85

R. A. Fisher: On the interpretation of χ^2 from contingency tables and the calculation of P. J. Roy. Statist. Soc. 85, 87—94 (1922); p. 230

R. A. Fisher: On the mathematical foundations of theoretical statistics. Philos. Trans. Roy. Soc. London, Ser. A 222, 309—368 (1922); p. 338

R. A. Fisher: Applications of Student's distribution. Metron 5, 90—104 (1925); p. 136

R. A. Fisher: Theory of statistical estimation. Proc. Cambridge Philos. Soc. 22, 700—725 (1925); p. 338

R. A. Fisher: The general sampling distribution of the multiple correlation coefficient. Proc. Roy. Soc. London. Ser. A. 121, 654—673 (1928); p. 404

R. A. Fisher: The use of multiple measurements in taxonomic problems. Ann. Eugenics 7, 179—188 (1936); p. 379

M. Fisz: The limiting distributions of the multinomial distribution. Studia Math. 14, 272—275 (1954); p. 113

M. Fisz: Probability Theory and Mathematical Statistics, 3rd. ed., John Wiley & Sons. New York; Chapman & Hall, Ltd., London 1963; p. 25

D. A. S. Fraser: Nonparametric tolerance regions. Ann. Math. Statist. 24, 44—55 (1953); p. 423

D. A. S. Fraser: Completeness of order statistics. Canad. J. Math. 6, 42—45 (1954); p. 416

D. A. S. Fraser: Nonparametric methods in statistics. John Wiley & Sons-Chapman & Hall, New York-London 1957; p. 406

D. A. S. Fraser and L. Guttman: Bhattacharyya bounds without regularity assumptions. Ann. Math. Statist. 23, 629—632 (1952); p. 284

D. A. S. Fraser and I. Guttman: Tolerance regions. Ann. Math. Statist. 27, 162—179 (1956); p. 423

M. Fréchet: Sur l'extension de certaines evaluation statistiques au cas de petits échantillons. Rev. Inst. Internat. Statist. 11, 182—205 (1943); p. 284

R. C. Geary: Distribution of Student's ratio for non-normal samples. J. Roy. Statist. Soc., Supp. 3, 178 (1936); p. 134

I. I. Gihmann: On the empirical distribution function in the case of grouping of the data. Dokl. Akad. Nauk SSSR, (N.S.) 82, 837—840 (1952); p. 435

B. V. Gnedenko: Kriterien für die Unveränderlichkeit der Wahrscheinlichkeitsverteilung von zwei unabhängigen Stichprobenreihen. Math. Nachr. 12, 29—63 (1954) (Russian); p. 434

B. V. Gnedenko: The theory of probability. Chelsea Publishing Co., New York, 1967; p. 25

B. V. Gnedenko and A. V. Kolmogorov: Limit Distributions for Sums of independent Random Variables, Cambridge, Mass. 1954; p. 91, 106

B. V. Gnedenko and V. S. Koroljuk: On the maximum discrepancy between two empirical distributions. Doklady Akad. Nauk SSSR, (N. S.) 80, 525—528 (1951); p. 433

M. H. de Groot and M. M. Rao: Bayes estimation with convex loss. Ann. Math. Statist. 34, 839—846 (1963); p. 331

E. J. Gumbel: Statistics of extremes. Columbia University Press, New York, 1958; p. 425

J. Hájek: Asymptotically most powerful rank order tests. Ann. Math. Statist. 33, 1124—1147 (1962); p. 342

J. Hájek and Z. Šidák: Theory of Rank Test. Academia Publishing House of the Czechoslovak Academy of Sciences, Prague 1967; p. 406, 464

G. Hajos and A. Rényi: Some Fundamental connections between elementary proofs in the theory of ordered samples. Acta Math. Acad. Sci. Hungar. 5, 1—6 (1954); p. 423

P. R. Halmos: The theory of unbiased estimation. Ann. Math. Statist. 17, 34—43 (1946); p. 416

P. R. Halmos: Measure Theory. D. van Nostrand Company, Inc., New York, N.Y., 1950; p. 5

P. R. Halmos and L. J. Savage: Application of the Radon-Nikodym theorem to the theory of sufficient statistics. Ann. Math. Statist. 20, 225—241 (1949); p. 186, 211

H. Hamburger: Beiträge zur Konvergenztheorie der Stieltjesschen Kettenbrüche. Math. Z. 4, 186—222 (1919); p. 119

H. Hamburger: Über die Konvergenz eines mit einer Potenzreihe assoziierten Kettenbruchs. Math. Ann. 81, 31—45 (1920); p. 119

H. Hamburger: Über eine Erweiterung des Stieltjesschen Momentenproblems I. Math. Ann. 81, 235—319 (1920); p. 119

H. Hamburger: Über eine Erweiterung des Stieltjesschen Momentenproblems II. Math. Ann. 82, 120—164 (1921); p. 119

H. Hamburger: Über eine Erweiterung des Stieltjesschen Momentenproblems III. Math. Ann. 82, 168—187 (1921); p. 119

M. H. Hansen and W. N. Hurwitz: On the theory of sampling from finite populations. Ann. Math. Statist. 14, 333—362 (1943); p. 156

F. Hausdorff: Summationsmethoden und Momentfolgen I. Math. Zeitschr. 9, 74—109 (1921); p. 119

E. R. Helmert: Über die Wahrscheinlichkeit der Potenzsummen der Beobachtungsfehler und über einige damit im Zusammenhang stehende Fragen. Zeitschr. für Math. und Physik 21, 192—219 (1876); p. 80

J. Hemelrijk: A family of parameterfree tests for symmetry with respect to a given point I. Indagationes Math. 12, 340—350 (1950); p. 449

J. Hemelrijk and Ph. van Elteren: A course in applied statistics. Math. Centrum Amsterdam. Statist. Afdeling. Rep. S 120 (1954); p. 440

P. L. Hennequin and A. Tortrat: Théorie des probabilités et quelques applications. Masson et Cie, Editeurs, Paris, 1965; p. 25

A. J. Hinčin: Sur la loi des grands nombres. C. R. Acad. Sci., Paris 189, 477—479 (1929); p. 103

E. Hlawka: Integrale auf konvexen Körpern. Monatsh. Math. 55, 105—137 (1951); p. 109

E. Hlawka: Geordnete Schätzfunktionen und Diskrepanz. Math. Ann. 150, 259—267 (1963); p. 430

J. L. Hodges, jr. and E. L. Lehmann: The efficency of some nonparametric competitors of the t-tests. Ann. Math. Statist. 27, 324—335 (1956); p. 247

J. L. Hodges, jr. and E. L. Lehmann: Comparison of the normal scores and Wilcoxon tests. Proc. Fourth Berkeley Sympos. Math. Statist. and Prob. Vol. I, pp 307—317, Univ. California Press, Berkeley Calif. 1961; p. 244, 468

W. Hoeffding: A class of statistics with asymptotically normal distribution. Ann. Math. Statist. 19, 293—325 (1948); p. 461

W. Hoeffding: "Optimum" nonparametric test. Proceedings of the Second Berkeley Symposium on Mathematical Statistics and Probability, 1950, pp. 83—92. University of California Press, Berkeley and Los Angeles, 1951; p. 414

P. Hoel: Testing the homogeneity of Poisson-frequencies. Ann. Math. Statist. 16, 362—368 (1945); p. 226

N. Hofreiter und W. Gröbner: Intergraltafel. Zweiter Teil: Bestimmte Intergrale. 2. Aufl. Springer-Verlag, Wien, 1961; p. 164

H. Hornich: Zur Theorie des Risikos. Monatsh. Math. Phys. 50, 142—150 (1941); p. 451

H. Hornich: Zur Auflösung von Gleichungssystemen. Monatsh. Math. 54, 130—134 (1950); p. 300

H. Hotelling: The generalization of Student's ratio. Ann. Math. Statist. 2, 360—378 (1931); p. 378

H. Hotelling: A generalized T test and measure of multivariate dispersion. Proceedings of the Second Berkeley Symposium on Mathematical Statistics and Probability, 1950, pp. 23—41. University of California Press, Berkeley and Los Angeles, 1951; p. 378

P.L. Hsu: A new proof of the joint product moment distribution. Proc. Cambridge Philos. Sec. 35, 336—338 (1939); p. 370

P.J. Huber: The behavior of maximum likelihood estimates under nonstandard conditions. Proc. Fifth Berkeley Sympos. Math. Statist. and Probability (Berkeley, Calif. 1965/66), Vol. I: Statistics, pp. 221—233. Univ. California Press, Berkeley, Calif., 1967; p. 316

P.J. Huber: Théorie de l'inférence statistique robuste. Séminaire de mathématiques supérieures 31; Montréal, Canada. Les Presses de l'Université de Montréal, 1969; p. 140

V.S. Huzurbazar: The likelihood function. Ann. Eugenics 14, 185—200 (1948); p. 305

I.A. Ibragimov and J.V. Linnik: Independent and stationary sequences of random variables. Ed. by F.C. Kingman, Groningen. Wolters-Noordhoff 1971; p. 109, 115

I. van Ijzeren: The theoretical aspect of least squares. Statistica, Rijswijk 8, 21—45 (1954); p. 357

St.L. Isaacson: On the theory of unbiased tests of simple statistical hypotheses specifying the values of two or more parameters. Ann. Math. Statist. 22, 217—234 (1951); p. 205

G.S. James: Notes on a theorem of Cochran. Proc. Cambridge Philos. Soc. 48, 443—446 (1952); p. 129

A.T. James: Normal multivariate analysis and the orthogonal group. Ann. Math. Statist. 25, 40—75 (1954); p. 405

A.T. James: The distribution of the latent roots of the covariance matrix. Ann. Math. Statist. 31, 151—158 (1960); p. 405

A.T. James: The distribution of noncentral means with known covariance. Ann. Math. Statist. 32, 874—882 (1961); p. 405

M. Jiřina: Conditional probabilities on strictly separable σ-algebras. Czechoslovak. Math. J. 4 (79), 372—380 (1954); p. 58

M. Jiřina: On regular conditional probabilities. Czechoslovak. Math. J. 9 (84), 445—451 (1959); p. 58

M.V. Johns, jr.: Nonparametric empirical Bayes procedures. Ann. Math. Statist. 28, 649—669 (1957); p. 436

S. Kakutani: On equivalence of infinite product measures. Ann. of Math. (2) 49, 214—224 (1948); p. 241

D.A. Kappos: Strukturtheorie der Wahrscheinlichkeitsfelder und Räume. Ergebnisse der Mathematik und ihre Grenzgebiete; Neue Folge; Heft 24. Springer-Verlag Berlin-Göttingen-Heidelberg, 1960; p. 25

S. Karlin: Mathematical methods and theory of games. Programming and Economics I, II, Pergamon Press-Addison Wesley Publishing Company, Oxford-London-New York-Paris, 1959; p. 166, 482

S. Karlin and L.S. Shaple: Geometry of moment Spaces. Mem. Amer. Math. Soc. No. 12, Providence 1953; p. 119

T. Kawata and H. Sakamoto: On the characterisation of the normal population by the independence of the sample mean and the sample variance. J. Math. Soc. Japan 1, 111—115 (1949); p. 134

H. Kellerer: Zur Existenz analoger Bereiche. Z. Wahrscheinlichkeitstheorie und Verw. Gebiete 1, 240--246 (1963); p. 222

O. Kempthorne: The design and analysis of experiments. John Wiley & Sons, Inc. New York; Chapman & Hall, Ltd., London, 1952; p. 357

M. Kendall and A. Stuart: The advanced theory of statistics I, II, III. Griffin, London, 1969, 1967, 1968; p. 122

M.G. Kendall and R.M. Sundrum: Distribution-free methods and order properties. Rev. Inst. Internat. Statist. 21, 124—134 (1953); p. 406

J. Kiefer: On minimum variance estimators. Ann. Math. Statist. 23, 627--629 (1952); p. 284

J. Kiefer: Invariance, minimax sequential estimation, and continuous time processes. Ann. Math. Statist. 28, 573--601 (1957); p. 479

A.N. Kolmogorov: Grundbegriffe der Wahrscheinlichkeitsrechnung. Ergebnisse der Mathematik und ihrer Grenzgebiete, Springer-Verlag, Berlin, 1933; p. 23

A.N. Kolmogorov: Sulla determinazione empirica di una legge di distribuzione. Giorn. Ist. Ital. Attuari 4, 83—91 (1933); p. 429

A.N. Kolmogorov: On the proof of the method of least squares. Uspehi Mathem. Nauk (N.S.) 1 (11), no. 1, 57—70 (1946); p. 357

A.N. Kolmogorov: The real meaning of the results of the analysis of variance. Proceedings of the Second All-Union Congress on Mathematical Statistics, Sept. 27-Oct. 2 1948, pp. 240—268. Acad. Sci. Uzbekistan Soviet Socialist Republic, Tashkent 1949; p. 229

A.N. Kolmogorov: Unbiased estimates. Izvestiya Akad. Nauk SSSR, Ser. Mat. 14, 303--326 (1950); p. 271, 277

B.O. Koopman: On distributions admitting a sufficient statistic. Trans. Amer. Math. Soc. 39, 399—409 (1936); p. 216

V.S. Koroljuk: On the discrepancy of empiric distributions for the case of two independent samples. Izv. Akad. Nauk SSSR. Ser. Math. 19, 81—96 (1955); p. 433

Ch. Kraft: Some conditions for consistency and uniform consistency of statistical procedures. Univ. California Publ. Statist. 2, 125—141 (1955); p. 241

Ch. Kraft and L. Le Cam: A remark on the roots of the maximum likelihood equation. Ann. Math. Statist. 27, 1174—1177 (1956); p. 307

O. Krafft and H. Witting: Optimale Tests und ungünstige Verteilungen. Z. Wahrscheinlichkeitstheorie und Verw. Gebiete 7, 289—302 (1967); p. 178

A. Krámli: A remark to a paper of L. Schmetterer. Studia Sci. Math. Hung. 2, 159--161 (1967); p. 274

K. Krickeberg: Probability Theory. Addison-Wesley, Reading-London, 1965; p. 6, 105

K. Krickeberg: On Cramér's theorems concerning weak convergence of distributions. Metrika 10, 179—181 (1966); p. 118

W.H. Kruskal: Historical notes on the Wilcoxon unpaired two sample test. J. Amer. Statist. Assoc. 52, 356—360 (1957); p. 440

A.M. Kshirsagar: Bartlett decomposition and Wishart distribution. Ann. Math. Statist. 30, 239—241 (1959); p. 391

H.W. Kuhn and A.W. Tucker: Contributions to the theory of games I. Annals of Mathematical Studies no. 24, Princeton University Press, Princeton N.J., 1950; p. 482

H.W. Kuhn and A.W. Tucker: Contributions to the theory of games II. Annals of Mathematics Studies no. 28, Princeton University Press, Princeton, N.J., 1953; p. 482

L. Le Cam: On some asymptotic properties of maximum likelihood estimates and related Bayes estimates. Univ. California Publ. Statist. 1, 277—329 (1953); p. 318, 320, 322, 333, 338

L. Le Cam: On the asymptotic theory of estimation and testing hypotheses. Proceedings of the Third Berkeley Symposium on Mathematical Statistics and Probability.

1954—1955, Vol. I, pp. 129—156. University of California Press, Berkeley and Los Angeles, 1956; p. 342, 429
L. Le Cam: Locally asymptotically normal families of distributions. Univ. California Publ. Statist. 3, 37—98 (1960); p. 342
L. Le Cam: On the assumptions used to prove asymptotic normality of maximum likelihood estimates. Ann. Math. Statist. 41, 802—828 (1970); p. 316, 318
L. Le Cam and L. Schwartz: A necessary and sufficient condition for the existence of consistent estimates. Ann. Math. Statist. 31, 140—150 (1960); p. 295
E. L. Lehmann: Some comments on large sample tests. Proceedings of the Berkeley Symposium on Mathematical Statistics and Probability 1949, pp. 451—457, University of California Press, Berkeley and Los Angeles; p. 342
E. L. Lehmann: Consistency and unbiasedness of certain nonparametric tests. Ann. Math. Statist. 22, 165—179 (1951); p. 442, 464
E. L. Lehmann: The power of rank tests. Ann. Math. Statist. 24, 23—43 (1953); p. 452
E. L. Lehmann: Testing statistical hypotheses. John Wiley & Sons, Inc., New York; Chapman & Hall, Ltd., London, 1959; p. 14, 160, 206, 223, 239
E. L. Lehmann and H. Scheffé: Completeness, similar regions and unbiased estimation, part I. Sankhyā 10, 305—340 (1950); p. 218, 222, 278
P. Lévy: Calcul des probabilités. Gauthier-Villars et Cie, Paris, 1925; p. 67, 69, 106
P. Lévy: Théorié de l'addition des variables aléatoires. Gauthier-Villars et Cie, 2nd Ed., Paris, 1954; p. 106
P. Lévy: Processus stochastiques et mouvement brownien. Gauthier-Villars et Cie, Paris 1965; p. 25
J. Lindenstrauss: A short proof of Liapunoff's convexity theorem. J. Math. Mech. 15, 971—972 (1966); p. 6
A. Linder: Statistische Methoden für Naturwissenschaftler, Mediziner und Ingenieure, 3rd ed., Birkhäuser Verlag, Basel und Stuttgart, 1960; p. 122, 355
J. V. Linnik: On the probability of large deviations for the sums of independent variables. Proc. Fourth Berkeley Sympos. Math. Statist. and Prob. Vol. II, pp. 289—306. Univ. California Press, Berkeley, Calif. (1960); p. 109
J. V. Linnik: Die Methode der kleinsten Quadrate in moderner Darstellung. VEB Deutscher Verlag der Wissenschaften, Berlin 1961; p. 357
J. V. Linnik: Statistical problems with nuisance parameters. Translations of Mathematical Monographs Vol. 20, Amer. Math. Soc., Providence, R. I., 1968; p. 225
A. Ljapunov: Sur les fonctions-vecteurs complètement additives. Bull. Acad. Sci. URSS, Ser. Math. 4, 465—478 (1940); p. 6
M. Loève: Ranking limit problem. Proceedings of the 3rd Berkeley Symposium on Mathematical Statistics and Probability 1954—1955, Vol. II, pp. 177—194. University of California Press, Berkeley and Los Angeles 1956; p. 425
M. Loève: Probability theory. 3rd ed., D. van Nostrand, Princeton-Toronto-New York-London 1963; p. 25
F. Lösch and F. Schoblik: Die Fakultät (Gammafunktion) und verwandte Funktionen. Teubner, Leipzig, 1951; p. 115
E. Lukacs: A characterization of the normal distribution. Ann. Math. Statist. 13, 91—93 (1942); p. 134
E. Lukacs: Stochastic convergence. Math. Monographs, D.C. Heath, Lexington, Mass., 1968; p. 104, 116
E. Lukacs: Characteristic functions. Second Edition, Griffin, London, 1970; p. 67
P.C. Mahalanobis: On the generalized distance in statistics. Proc. Nat. Inst. Sci. India 2, 49—55 (1936); p. 379
P.C. Mahalanobis: Historical note on the D^2 statistic. Sankhyā 9, 237—239 (1949); p. 379
H. B. Mann: The algebra of a linear hypothesis. Ann. Math. Statist. 31, 1—15 (1960); p. 228

H. B. Mann and A. Wald: On the choice of the number of class-intervals in the application of the χ^2 test. Ann. Math. Statist. 13, 306—317 (1942); p. 230
H. B. Mann and D. R. Whitney: On a test of whether one of two random variables is stochastically larger than the other. Ann. Math. Statist. 18, 50—60 (1947); p. 440
A. Markov: Wahrscheinlichkeitsrechnung, 2nd ed. Leipzig-Berlin, 1912; p. 359
D. G. Meĭzler: On a problem of B. V. Gnedenko. Ukrain. Mat. Žurnal 1, no. 2, 67—84 (1949); p. 425
M. Métivier: Notions fondamentales de la théorie des probabilités. Maitrises de mathématiques, Dunod, Paris, 1968; p. 25
A. Meyer: Vorlesungen über Wahrscheinlichkeitsrechnung. B. G. Teubner. Leipzig, 1879; p. 91
P. A. Meyer: Probabilités et potentiel. Actualités Scientifiques et Industrielles 1318, Hermann Paris, 1966; p. 25
H. Midzuno: On the sampling system with probability proportionate to sum of sizes. Ann. Inst. Statist. Math. 3, 99—107 (1951/52); p. 156
R. v. Mises: Fundamentalsätze der Wahrscheinlichkeitsrechnung. Math. Z. 4, 1—97 (1919); p. 22
R. v. Mises: Wahrscheinlichkeitsrechnung und ihre Anwendung in der Statistik und theoretischen Physik. (Vorlesungen aus dem Gebiet der angewandten Mathematik, Vol. I); Deuticke, Leipzig-Wien 1931; p. 22
A. M. Mood: Introduction to the theory of statistics. McGraw-Hill, New York 1950; p. 122
D. Morgenstern: Einführung in die Wahrscheinlichkeitsrechnung und Mathematische Statistik. Springer-Verlag, Berlin-Göttingen-Heidelberg 1964; p. 122
H. K. Nandi: On type B_1 and type B regions. Sankhyā 11, 13—22 (1951); p. 205
B. E. Netto: Lehrbuch der Combinatorik. Teubner Leipzig, 1901; p. 460
J. v. Neumann and O. Morgenstern: Theory of games and economic behavior. 3 ed. Princeton University Press, Princeton, New York, 1953; p. 482
J. Neveu: Mathematical foundations of the calculus of Probability. Holden Day, San Franzisco-London-Amsterdam 1965; p. 25
J. Neyman: On two different aspects of the representative method: the method of stratified sampling and the method of purpositive selection. J. Roy. Statist. Soc. 109, 558—606 (1934); p. 149
J. Neyman: On the problem of confidence intervals. Ann. Math. Statist. 6, 111—116 (1935); p. 255
J. Neyman: Sur la vérification des hypothèses statistiques composées. Bull. Soc. Math. France 63, 246—266 (1935); p. 205
J. Neyman: Su un teorema concernente le cosidette statistiche sufficienti. Giorn. Ist. Ital. Attuari 6, 320—334 (1935); p. 211
J. Neyman: Outline of a theory of statistical estimation based on the classical theory of probability. Philos. Trans. Roy. Soc. London, Ser. A 236, 333—380 (1937); p. 255
J. Neyman: L'estimation statistique traitée comme un problème classique de probabilité. Actualités scientifiques et industrielles 739, 25—57; Herman & Cie, Paris, 1938; p. 255
J. Neyman: Fiducial argument and the theory of confidence intervals. Biometrika 32, 128—150 (1941); p. 255
J. Neyman: Contribution to the theory of χ^2 test. Proceedings of the Berkeley Symposium on Mathematical Statistics and Probability pp. 239—273. University of California Press, Berkeley and Los Angeles (1949); p. 342
J. Neyman: Sur une famille de tests asymptotiques des hypothèses statistiques composées. Trabajos Estadist. 5, 161—168 (1954); p. 342
J. Neyman: First course in probability and statistics. John Wiley & Sons, Inc., New York; Chapman & Hall Ltd., London, 1962; p. 122

J. Neyman and E.S. Pearson: On the use and interpretation of certain test criteria for purposes of statistical inference. Biometrika 20 A, 175—240 (1928); p. 160

J. Neyman and E.S. Pearson: On the use and interpretation of certain test criteria for the purposes of statistical inference. Biometrika 20 A, 263—294 (1928); p. 160, 225

J. Neyman and E.S. Pearson: On the problem of the most efficient tests of statistical hypotheses. Philos. Trans. Roy. Soc. London, Ser. A 231, 289—337 (1933); p. 160, 221

J. Neyman and E.S. Pearson: Contributions to the theory of testing statistical hypotheses. Statistical Res. Mem. Univ. London 1, 1—37 (1936); p. 166

J. Neyman and E.S. Pearson: Contributions to the theory of testing statistical hypotheses. Statist. Res. Mem. Univ. London 2, 25—57 (1938); p. 204

J. Neyman and E. Scott: Consistent estimates based on partially consistent observations. Econometrica 16, 1—32 (1948); p. 240

G. Noelle: Zur Theorie der bedingten Tests. Z. Wahrscheinlichkeitstheorie und Verw. Gebiete 11, 208—229 (1969); p. 223

G. Noelle and D. Plachky: Zur schwachen Folgenkompaktheit von Testfunktionen. Z. Wahrscheinlichkeitstheorie und Verw. Gebiete 8, 182—184 (1967); p. 14

G.E. Noether: On a theorem of Pitman. Ann. Math. Statist. 26, 64—68 (1955); p. 247

G.E. Noether: Elements of nonparametric statistics. John Wiley & Sons, Inc., New York-London-Sydney, 1967; p. 406

E. Parzen: Modern probability theory and its applications. John Wiley & Sons, Inc., New York; Chapman & Hall Ltd., London, 1960; p. 25

K. Pearson: Contribution to the mathematical theory of evolution. Philos. Trans. Roy. Soc. London, Ser. A 185, 71—110 (1894); p. 87

K. Pearson: On the criterion that a given system of deviations from the probable in the case of a correlated systems of variables is such that it can be reasonable supposed to have arisen from random sampling. Philos. Mag. 50, Ser. 5, 157—175 (1900); p. 80, 230

K. Pearson: Researches on the mode of distribution of the constants of samples taken at random from a bivariate normal population. Proc. Soc. London, Ser. A 112, 1—14 (1926); p. 398

K. Pearson: Tables of the incomplete B-function. Cambridge University Press, London, 1934; p. 91

J. Pfanzagl: Überall trennscharfe Tests und monotone Dichtequotienten. Z. Wahrscheinlichkeitstheorie und Verw. Gebiete 1, 109—115 (1963); p. 191

J. Pfanzagl: A characterisation of the one parameter exponential family by existence of uniformly most powerful tests. Sankhyā, Ser. A 30, 147—156 (1968); p. 197

J. Pfanzagl: Allgemeine Methodenlehre der Statistik I, 5th ed; II, 3rd ed. Sammlung Göschen, Walter de Gruyter & Co, Berlin 1972, 1968; p. 122

J. Pfanzagl: On measurability and consistency of minimum contrast estimates. Metrika 14, 249—272 (1969); p. 316

J. Pfanzagl: On the existence of product measurable densities. Sankhyā, Ser. A 31, 13—18 (1969); p. 179

T.S. Pitcher: Positivity in H-systems and sufficient statistics. Trans. Amer. Math. Soc. 85, 166—173 (1957); p. 218, 416

J.G. Pitman: Lecture notes on nonparametric inference. Columbia University New York, 1949; p. 247

M.L. Puri: Nonparametric techniques in statistical inference (M.L. Puri editor). Cambridge University Press, 1970; p. 406

J. Radon: Theorie und Anwendungen der absolut additiven Mengenfunktionen. Österr. Akad. Wiss. Math.-Naturw. Kl. S-Ber. 122 Abt. IIa 1295—1438 (1913); p. 11

C.R. Rao: Information and the accuracy attainable in the estimation of statistical parameters. Bull. Calcutta Math. Soc. 37, 81—91 (1945); p. 284

C.R. Rao: Tests of significance in multivariate analysis. Biometrika 35, 58—79 (1948); p. 379

C.R. Rao: On some problems arising out of discrimination with multiple characters. Sankhyā 9, 343—366 (1949); p. 375, 379, 383

C.R. Rao: A note on the distribution of $D^2_{p+q} - D^2_p$ and some computational aspects of D^2 statistic and discriminant function. Sankhyā 10, 257—268 (1950); p. 379

C.R. Rao: Some theorems of minimum variance estimation. Sankhyā 12, 27—42 (1952); p. 273

C.R. Rao: Asymptotic efficiency and limiting information. Proc. 4th Berkeley Sympos. Math. Statist. and Prob., Vol. I, pp. 531—545 Univ. California Press, Berkeley Calif., (1960); p. 341

C.R. Rao: Efficient estimates and optimum inference procedures in large samples. J. Roy. Statist. Soc., Ser. B. 24, 46—72 (1962); p. 341

C.R. Rao: Apparent anomalies and irregularities in maximum likelihood estimation. Sankhyā Ser. A 24, 73—101 (1962); p. 341

C.R. Rao: Criteria of estimation in large samples. Sankhyā Ser. A 25, 189—206 (1963); p. 341

A. Rényi: On the theory of order statistics. Acta. Math. Sci. Hungar. 4, 191—231 (1953); p. 423, 434

A. Rényi: Neue Kriterien zum Vergleich zweier Stichproben. Magyar Tud. Akad. Alkam. Math. Int. Közl. 2, 243—265 (1954); p. 440

A. Rényi: Wahrscheinlichkeitsrechnung. Mit einem Anhang über Informationstheorie. (Hochschulbücher für Mathematik, Vol. 54) VEB Deutscher Verlag der Wissenschaften, Berlin 1962; p. 23

H. Richter: Verallgemeinerung eines in der Statistik benötigten Satzes der Maßtheorie. Math. Ann. 150, 85—90 (1963); p. 307

H. Richter: Beweisergänzung zu der Arbeit: „Verallgemeinerung eines in der Statistik benötigten Satzes der Maßtheorie". Math. Ann. 150, 440—441 (1963); p. 307

H. Richter: Wahrscheinlichkeitstheorie, Second Edition, Die Grundlehren der mathematischen Wissenschaften, Bd. 66, Springer-Verlag, Berlin-Heidelberg-New York 1966; p. 5

W. Richter: Über Wahrscheinlichkeiten großer Abweichungen für Summen unabhängiger Zufallsgrößen. Wiss. Z. Techn. Hochsch. Dresden 10, 7—14 (1961); p. 109

E. Ricker: The concept of confidence or fiducial limits applied to the Poisson distribution. J. Amer. Statist. Assoc. 32, 349—356 (1937); p. 260

H. Robbins: An empirical Bayes approach. Proceedings of the 3rd Berkeley Symposium on Mathematical Statistics and Probability 1954—1955, Vol. I. pp. 157—163. University of California Press, Berkeley and Los Angeles 1956; p. 436

H. Robbins and S. Monro: A stochastic approximation method. Ann. Math. Statist. 22, 400—407 (1951); p. 471

V. Romanovskij: On the distribution of the regression coefficient in samples from normal population. Bull. Acad. Sci. Leningrad (6) 20, 643—648 (1926); p. 398

J. Roy and I.M. Chakravarti: Estimating the mean of a finite population. Ann. Math. Statist. 31, 392—398 (1960); p. 280

E. Ruist: Comparison of tests for non-parametric hypotheses. Ark. Mat. 3, 133—163 (1955); p. 406, 448, 449

A.E. Sarkan and B.G. Greenberg: Contributions to order statistics. John Wiley & Sons, New York 1962; p. 425

H. Scheffé: A useful convergence theorem for probability distribution. Ann. Math. Statist. 18, 434—438 (1947); p. 199

H. Scheffé: The Analysis of Variance. John Wiley & Sons, Inc., New York; Chapman & Hall, Ltd., London 1959; p. 226, 359, 360

L. Schmetterer: Bemerkungen zum Verfahren der stochastischen Iteration. Österreich. Ing.-Arch. 7, 111—117 (1953); p. 474

L. Schmetterer: Bemerkungen zur Theorie der erwartungstreuen Schätzfunktionen. Mitteilungsbl. Math. Statist. 9, 147—152 (1957); p. 275

L. Schmetterer: Über nichtparametrische Methoden in der mathematischen Statistik. Jber. Deutsch. Math. Verein. 61 (Abt. 1), 104—126 (1958); p. 406

L. Schmetterer: On unbiased estimation. Ann. Math. Statist. 31, 1154—1163 (1960); p. 274, 275, 277, 289

L. Schmetterer: Stochastic approximation, Proc. 4th Berkeley Sympos. Math. Statist. and Prob., Vol. I. pp. 587—609. Univ. California Press, Berkeley, Calif., 1961; p. 474

L. Schmetterer: Über eine allgemeine Theorie der erwartungstreuen Schätzungen. Publ. Math. Inst. Hungar, Acad. Sci. Ser. A. 6, 295—300 (1961); p. 274, 359

L. Schmetterer: Some remarks on the power of a most powerful test. Sankhyā 25, 207—210 (1963); p. 171

L. Schmetterer: Zur Theorie der Stichproben aus endlichen Gesamtheiten. Abh. Deutsch. Akad. Wiss. Berlin Kl. Math. Phys. Techn. 1964, nr. 4, 117—120; p. 280

L. Schmetterer: On the asymptotic efficiency of estimates. Research Papers in Statistics (Festschrift J. Neyman) pp. 301—317, John Wiley, London 1966; p. 352

L. Schmetterer: On superefficiency. Symposium on Probability Methods in Analysis (Loutraki, 1966) pp. 291—295. Springer, Berlin, 1967; p. 352

L. Schmetterer: Multidimensional stochastic approximation. Multivariate Analysis II (Proc. Second Internat. Sympos., Dayton, Ohio, 1968) pp. 443—460. Academic Press, New York, 1969; p. 474

L. Schmetterer and R. Stender: H. Behnke, K. Fladt und W. Süss (Herausgeber). Grundzüge der Mathematik für Lehrer an Gymnasien sowie für Mathematiker in Industrie und Wirtschaft. Band III. Analysis 96—132; p. 25

C.P. Schnorr: Zufälligkeit und Wahrscheinlichkeit. Eine algorithmische Begründung der Wahrscheinlichkeitstheorie (Lecture Notes in Mathematics) Springer-Verlag, Berlin-Heidelberg-New York 1971; p. 22

L. Schwartz: On Bayes procedures. Z. Wahrscheinlichkeitstheorie und Verw. Gebiete 4, 10—26 (1965); p. 331

W. Sendler: Einige maßtheoretische Sätze bei der Behandlung trennscharfer Tests. Z. Wahrscheinlichkeitstheorie und Verw. Gebiete 18, 183—196 (1971); p. 171

C.R. Seth: On the variance of estimates. Ann. Math. Statist. 20, 1—27 (1949); p. 284

J.A. Shohat and J.D. Tamarkin: The Problem of Moments. American Mathematical Society Mathematical Surveys, Vol. I. American Mathematical Society, New York, 1943; p. 119

M. Sion: On uniformisation of sets in topological spaces. Trans. Amer. Math. Soc. 96, 237—246 (1960); p. 307

N.V. Smirnov: Sur les écarts de la courbe de distribution empirique. Rec. Math. N.S. (Mat. Sbornik) 6 (48) 3—26 (1939); p. 435

N.V. Smirnov: Limit distributions for the terms of a variational series. Trudy Mat. Inst. Steklov. 25 (1949); p. 425

A. Spaček: Zufällige Gleichungen. Czechoslovak, Math. J. 5 (80), 462—466 (1955); p. 474

H. Steinhaus and L. Kaczmarz: Theorie der Orthogonalreihen. Monografie Matematyczne VI, Warschau, 1935; p. 219

T.J. Stieltjes: Note sur l'intégrale $\int_0^\infty e^{-u^2} du$. Nouv. Ann. Math., ser. 3, 9, 479—480 (1890); p. 74

Student: The probable error of a mean. Biometrika 6, 1—25 (1908); p. 83

H. Teicher: Identifiability of finite mixtures. Ann. Math. Statist. 34, 1265—1269 (1963); p. 271

M. E. Terry: Some rank order tests which are most powerful against specific parametric alternatives. Ann. Math. Statist. 23, 346—366 (1952); p. 440

A. W. Tucker and R. D. Luce: Contributions to the theory of games IV. Ann. Math. Studies 40, Princeton University Press, Princeton, New York 1959; p. 482

J. W. Tukey: Nonparametric estimation II. Statistically equivalent blocks and tolerance regions-the continuous case. Ann. Math. Statist. 18, 529—539 (1947); p. 423

J. W. Tukey: Nonparametric estimation III. Statistically equivalent blocks and multivariate tolerance regions-the discontinuous case. Ann. Math. Statist. 19, 30—39 (1948); p. 423

H. R. van der Vaart: Some remarks on the power function of Wilcoxon's test for the problem of two samples. I, II. Nederl. Akad. Wetensch., Proc. Ser. A. 53, 494—520 (1950); p. 466

S. Vajda: Theory of games and linear programming. John Wiley & Sons, Inc., New York; Chapman & Hall Ltd., London 1956; p. 176

I. Vincze: Einige zweidimensionale Verteilungs- und Grenzverteilungssätze in der Theorie der zugeordneten Stichproben. Magyar Tud. Akad. Mat. Kutato Int. Közl 2, 183—209 (1957); p. 434

I. Vincze: On some joint distributions and joint limiting distributions in the theory of order statistics. II. Magyar Tud. Akad. Mat. Kutato Int. Közl. 4, 29—47 (1959); p. 434

D. Voelker und G. Doetsch: Die zweidimensionale Laplace-Transformation. Verlag Birkhäuser, Basel 1950; p. 220

B. L. van der Waerden: Empirische Bestimmung von Wahrscheinlichkeiten und physiologische Konzentrationsauswertung. Ber. Verh. sächs. Akad. Leipzig, Math.-Phys. Kl. 87, 353—364 (1935); p. 330

B. L. van der Waerden: Eine einfache Herleitung der Stirlingschen Formel $n! \sim n^n \cdot e^{-n}\sqrt{2\pi n}$. Nieuw. Arch. Wiskde 18, 40—45 (1936); p. 112

B. L. van der Waerden: Ein neuer Test für das Problem der zwei Stichproben. Math. Ann. 126, 93—107 (1953); p. 440

B. L. van der Waerden: Mathematical statistics. Die Grundlehren der mathematischen Wissenschaften, Band 156. Springer-Verlag, New York-Heidelberg, 1969; p. 122

A. Wald: Tests of statistical hypotheses concerning several parameters when the number of observations is large. Trans. Amer. Math. Soc. 54, 462—482 (1943); p. 230, 243

A. Wald: Sequential tests of statistical hypotheses. Ann. Math. Statist. 16, 117—186 (1945); p. 254

A. Wald: Sequential analysis. John Wiley & Sons, Inc., New York; Chapman & Hall, Ltd., London 1947; p. 254

A. Wald: Statistical decision functions. Ann. Math. Statist. 20, 165—205 (1949); p. 482

A. Wald: Note on the consistency of the maximum likelihood estimate. Ann. Math. Statist. 20, 595—601 (1949); p. 310

A. Wald: Statistical decision functions. John Wiley & Sons, Inc., New York, Chapman & Hall Ltd., London, 1950; p. 476, 482

A. Wald and J. Wolfowitz: On a test whether two samples are from the same population. Ann. Math. Statist. 11, 147—162 (1940); p. 444

A. Wald and J. Wolfowitz: Optimum character of the sequential probability ratio test. Ann. Math. Statist. 19, 326—339 (1948); p. 247

A. Weil: L'intégration dans le groupes topologiques et ses applications. Actualités scientifiques et industrielles 869—1145, Hermann & Cie, 2nd ed., Paris 1953; p. 239

O. Wesler: Invariance theory and modified minimax principle. Ann. Math. Statist. 30, 1—20 (1959); p. 239
R.A. Wijsman: Random orthogonal transformations and their use in some classical distribution problems in multivariate analysis. Ann. Math. Statist. 28, 415—422 (1957); p. 391
F. Wilcoxon: Individual comparisons by ranking methods. Biometrics 1, 80—83 (1945); p. 440
S.S. Wilks: Order statistics. Bull. Amer. Math. Soc. 54, 6—50 (1948); p. 423
S.S. Wilks: Mathematical statistics. John Wiley & Sons, Inc., New York; Chapman & Hall. Ltd., London 1962; p. 122
W. Winkler: Grundriß der Statistik I, Second ed. Manz'sche Verlagsbuchhandlung. Vienna, 1947; p. 18, 124
J. Wishart: The generalized product moment distribution in samples from a normal multivariate population. Biometrika 20 A, 32—52 (1928); p. 370
H. Witting: Über einen χ^2-Test, dessen Klassen durch geordnete Stichprobenfunktionen festgelegt werden. Arch. Math. 10, 468—479 (1959); p. 230
H. Witting: Mathematische Statistik. B.G. Teubner Stuttgart 1966; p. 122
H. Witting und G. Nölle: Angewandte mathematische Statistik. B.G. Teubner, Stuttgart 1970; p. 122
J. Wolfowitz: On Wald's proof of the consistency of the maximum likelihood estimate. Ann. Math. Statist. 20, 601—602 (1949); p. 310
J. Wolfowitz: On the stochastic approximation method of Robbins and Monro, Ann. Math. Statist. 23, 457—461 (1952); p. 471
J. Wolfowitz: Asymptotic efficiency of the maximum likelihood estimator. Theor. Probab. Appl. 10, 247—260 (1965); p. 352
H. Uzawa: Locally most powerful rank tests for two sample problems. Ann. Math. Statist. 31, 685—702 (1960); p. 452
G.U. Yule: On the theory of correlation for any number of variables treated by a new system of notation. Proc. Roy. Soc. London Ser. A. 79, 182—193 (1907); p. 389
A. Zygmund: Trigonometric series. 2nd ed., Vol. I. Cambridge University Press, New York 1959; p. 458

Name and Subject Index

Die Grundlehren der mathematischen Wissenschaften in Einzeldarstellungen mit besonderer Berücksichtigung der Anwendungsgebiete

Eine Auswahl

---

23. Pasch: Vorlesungen über neuere Geometrie. In Vorbereitung
41. Steinitz: Vorlesungen über die Theorie der Polyeder. In Vorbereitung
45. Alexandroff/Hopf: Topologie. Band 1
46. Nevanlinna: Eindeutige analytische Funktionen
63. Eichler: Quadratische Formen und orthogonale Gruppen
102. Nevanlinna/Nevanlinna: Absolute Analysis
114. MacLane: Homology
123. Yosida: Functional Analysis
127. Hermes: Enumerability, Decidability, Computability
131. Hirzebruch: Topological Methods in Algebraic Geometry
135. Handbook for Automatic Computation. Vol. 1/Part a: Rutishauser: Description of ALGOL 60
136. Greub: Multilinear Algebra
137. Handbook for Automatic Computation. Vol. 1/Part b: Grau/Hill/Langmaack: Translation of ALGOL 60
138. Hahn: Stability of Motion
139. Mathematische Hilfsmittel des Ingenieurs. 1. Teil
140. Mathematische Hilfsmittel des Ingenieurs. 2. Teil
141. Mathematische Hilfsmittel des Ingenieurs. 3. Teil
142. Mathematische Hilfsmittel des Ingenieurs. 4. Teil
143. Schur/Grunsky: Vorlesungen über Invariantentheorie
144. Weil: Basic Number Theory
145. Butzer/Berens: Semi-Groups of Operators and Approximation
146. Treves: Locally Convex Spaces and Linear Partial Differential Equations
147. Lamotke: Semisimpliziale algebraische Topologie
148. Chandrasekharan: Introduction to Analytic Number Theory
149. Sario/Oikawa: Capacity Functions
150. Iosifescu/Theodorescu: Random Processes and Learning
151. Mandl: Analytical Treatment of One-dimensional Markov Processes
152. Hewitt/Ross: Abstract Harmonic Analysis. Vol. 2: Structure and Analysis for Compact Groups. Analysis on Locally Compact Abelian Groups
153. Federer: Geometric Measure Theory
154. Singer: Bases in Banach Spaces I
155. Müller: Foundations of the Mathematical Theory of Electromagnetic Waves
156. van der Waerden: Mathematical Statistics
157. Prohorov/Rozanov: Probability Theory. Basic Concepts. Limit Theorems. Random Processes
158. Constantinescu/Cornea: Potential Theory on Harmonic Spaces
159. Köthe: Topological Vector Spaces I
160. Agrest/Maksimov: Theory of Incomplete Cylindrical Functions and their Applications
161. Bhatia/Szegö: Stability Theory of Dynamical Systems
162. Nevanlinna: Analytic Functions
163. Stoer/Witzgall: Convexity and Optimization in Finite Dimensions I
164. Sario/Nakai: Classification Theory of Riemann Surfaces

165. Mitrinović/Vasić: Analytic Inequalities
166. Grothendieck/Dieudonné: Eléments de Géométrie Algébrique I
167. Chandrasekharan: Arithmetical Functions
168. Palamodov: Linear Differential Operators with Constant Coefficients
169. Rademacher: Topics in Analytic Number Theory
170. Lions: Optimal Control of Systems Governed by Partial Differential Equations
171. Singer: Best Approximation in Normed Linear Spaces by Elements of Linear Subspaces
172. Bühlmann: Mathematical Methods in Risk Theory
173. Maeda/Maeda: Theory of Symmetric Lattices
174. Stiefel/Scheifele: Linear and Regular Celestial Mechanics. Perturbed Two-body Motion—Numerical Methods—Canonical Theory
175. Larsen: An Introduction to the Theory of Multipliers
176. Grauert/Remmert: Analytische Stellenalgebren
177. Flügge: Practical Quantum Mechanics I
178. Flügge: Practical Quantum Mechanics II
179. Giraud: Cohomologie non abélienne
180. Landkof: Foundations of Modern Potential Theory
181. Lions/Magenes: Non-Homogeneous Boundary Value Problems and Applications I
182. Lions/Magenes: Non-Homogeneous Boundary Value Problems and Applications II
183. Lions/Magenes: Non-Homogeneous Boundary Value Problems and Applications III
184. Rosenblatt: Markov Processes. Structure and Asymptotic Behavior
185. Rubinowicz: Sommerfeldsche Polynommethode
186. Handbook for Automatic Computation. Vol. 2. Wilkinson/Reinsch: Linear Algebra
187. Siegel/Moser: Lectures on Celestial Mechanics
188. Warner: Harmonic Analysis on Semi-Simple Lie Groups I
189. Warner: Harmonic Analysis on Semi-Simple Lie Groups II
190. Faith: Algebra: Rings, Modules, and Categories I
192. Mal'cev: Algebraic Systems
193. Pólya/Szegö: Problems and Theorems in Analysis I
194. Igusa: Theta Functions
195. Berberian: Baer *-Rings
196. Athreya/Ney: Branching Processes
197. Benz: Vorlesungen über Geometrie der Algebren
198. Gaal: Linear Analysis and Representation Theory
199. Nitsche: Vorlesungen über Minimalflächen
200. Dold: Lectures on Algebraic Topology
201. Beck: Continuous Flows in the Plane
203. Schoeneberg: Elliptic Modular Functions
204. Popov: Hyperstability of Control Systems
205. Nikol'skii: Approximation of Functions of Several Variables and Embedding Theorems
206. André: Homologie des Algèbres Commutatives
207. Donoghue: Monotone Matrix Functions and Analytic Continuation
208. Lacey: The Isometric Theory of Classical Banach Spaces
209. Ringel: Map Color Theorem
210. Gihman/Skorohod: The Theory of Stochastic Processes I
211. Comfort/Negrepontis: The Theory of Ultrafilters
212. Switzer: Algebraic Topology—Homotopy and Homology
213. Shafrevich: Basic Algebraic Geometry
214. van der Waerden: Group Theory and Quatum Mechanics
215. Schaefer: Banach Lattices and Positive Operators.
216. Pólya/Szegő: Problems and Theorems in Analysis II. In preparation